Applications of Power Electronics

Applications of Power Electronics

Volume 1

Special Issue Editors

Frede Blaabjerg
Tomislav Dragičević
Pooya Davari

MDPI • Basel • Beijing • Wuhan • Barcelona • Belgrade

MDPI

Special Issue Editors

Frede Blaabjerg
Aalborg University
Denmark

Tomislav Dragičević
Aalborg University
Denmark

Pooya Davari
Aalborg University
Denmark

Editorial Office
MDPI
St. Alban-Anlage 66
4052 Basel, Switzerland

This is a reprint of articles from the Special Issue published online in the open access journal *Electronics* (ISSN 2079-9292) from 2018 to 2019 (available at: https://www.mdpi.com/journal/electronics/special_issues/appli_power_elec)

For citation purposes, cite each article independently as indicated on the article page online and as indicated below:

LastName, A.A.; LastName, B.B.; LastName, C.C. Article Title. *Journal Name* **Year**, *Article Number*, Page Range.

Volume 1
ISBN 978-3-03897-974-6 (Pbk)
ISBN 978-3-03897-975-3 (PDF)

Volume 1-2
ISBN 978-3-03921-148-7 (Pbk)
ISBN 978-3-03921-149-4 (PDF)

Cover image courtesy of Frede Blaabjerg.

Contents

About the Special Issue Editors

Frede Blaabjerg (professor) worked with ABB-Scandia, Randers, Denmark, from 1987 to 1988. He obtained his Ph.D. in electrical engineering at Aalborg University in 1995. He became an assistant professor in 1992, an associate professor in 1996, and a full professor of power electronics and drives in 1998. In 2017, he became a Villum Investigator. He is honoris causa at University Politehnica Timisoara (UPT), Romania and Tallinn Technical University (TTU) in Estonia. His current research interests include power electronics and its applications, such as in wind turbines, PV systems, reliability, harmonics, and adjustable-speed drives. He has published more than 600 journal papers in the field of power electronics and its applications. He is the co-author of four monographs and editor of ten books in power electronics and its applications. He has received 29 IEEE Prize Paper Awards, the IEEE PELS Distinguished Service Award in 2009, the EPE-PEMC Council Award in 2010, the IEEE William E. Newell Power Electronics Award in 2014, and the Villum Kann Rasmussen Research Award in 2014. He was the Editor-in-Chief of *IEEE Transactions on Power Electronics* from 2006 to 2012. He was a distinguished lecturer for the IEEE Power Electronics Society from 2005 to 2007 and for the IEEE Industry Applications Society from 2010 to 2011, as well as from 2017 to 2018. In 2018, he became the president elect of the IEEE Power Electronics Society. He serves as vice-president of the Danish Academy of Technical Sciences. He was nominated in 2014, 2015, 2016, and 2017 by Thomson Reuters as one of the world's 250 most cited researchers in engineering.

Tomislav Dragičević (associate professor) received his M.Sc. degree in 2009 and his industrial Ph.D. degree in electrical engineering in 2013 from the Faculty of Electrical Engineering and Computing, University of Zagreb, Zagreb, Croatia. From 2013 to 2016, he was a postdoctoral research associate at Aalborg University, Aalborg, Denmark. Since March 2016, he has been an associate professor at Aalborg University. He was a visiting guest professor at Nottingham University, Nottingham, U.K., during the spring/summer of 2018. His principal fields of interest include overall system design of autonomous and grid-connected dc and ac microgrids and application of advanced modeling and control concepts to power electronic systems. Dr. Tomislav has authored and coauthored more than 140 technical papers (more than 55 of them are published in international journals, mostly *IEEE Transactions*) in his domain of interest, eight book chapters, and a book. His research interests include overall system design of autonomous and grid-connected dc and ac microgrids and application of advanced modeling and control concepts to power electronic systems. Dr. Dragičević is the recipient of the Končar Prize for the best industrial Ph.D. thesis in Croatia and the Robert Mayer Energy Conservation Award. He serves as an associate editor for the *IEEE Transactions on Industrial Electronics* and for the *Journal of Power Electronics*.

Pooya Davari (associate professor) received B.Sc. and M.Sc. degrees in electronic engineering from the University of Mazandaran, Babolsar, Iran, in 2004 and 2008, respectively, and a Ph.D. degree in power electronics from Queensland University of Technology (QUT), Brisbane, Australia, in 2013. From 2005 to 2010, he was involved in several electronics and power electronics projects as a development engineer. From 2010 to 2014, he investigated and developed high-power high-voltage power electronic systems for multidisciplinary projects, such as ultrasound application, exhaust gas emission reduction, and tissue-materials sterilization. From 2013 to 2014, he was a lecturer with QUT. He joined the Department of Energy Technology, Aalborg University, Aalborg, Denmark, in 2014, as a postdoctoral researcher and he is currently an associate professor there. His current research interests include EMI/EMC in power electronics, WBG-based power converters, active frontend rectifiers, harmonic mitigation in adjustable-speed drives, and pulsed power applications. Dr. Davari received a research grant from the Danish Council of Independent Research in 2016.

electronics

MDPI

Editorial

Applications of Power Electronics

Frede Blaabjerg, Tomislav Dragicevic and Pooya Davari *

Department of Energy Technology, Aalborg University, 9220 Aalborg, Denmark; fbl@et.aau.dk (F.B.); tdr@et.aau.dk (T.D.)
* Correspondence: pda@et.aau.dk

Received: 19 April 2019; Accepted: 22 April 2019; Published: 25 April 2019

1. Introduction

Power electronics technology is still an emerging technology, and it has found its way into many applications, from renewable energy generation (i.e., wind power and solar power) to electrical vehicles (EVs), biomedical devices, and small appliances such as laptop chargers. In the near future, electrical energy will be provided and handled by power electronics and consumed through power electronics; this not only will intensify the role of power electronic technology in power conversion processes, but also implies that power systems are undergoing a paradigm shift, from centralized distribution to distributed generation. Today, more than 1000 gigawatts (GW) renewables (photovoltaic (PV) and wind) have been installed, all of which are handled by power electronics technology. However, areas such as energy saving and electrification transportation are booming, creating a huge market not only for power devices but also for packaging technology and power converter design. Some of the driving forces of the technology are their cost, volume, weight, functionality as well as reliability. At the moment, the technology is seeing a change from being purely silicon-based to being built upon wide bandgap (WBG) technology, such as silicon carbide (SiC) and gallium nitride (GaN), which demands a completely new paradigm in power converter design and layout, as those devices can operate at least an order of magnitude faster.

The main aim of this Special Issue was to seek high-quality submissions that highlight and address recent breakthroughs over the whole range of emerging applications of power electronics, the harmonic and electromagnetic interference (EMI) issues of the devices and system levels, as also discussed in [1–4], robust and reliable power electronics technologies, including fault prognosis and diagnosis techniques [5–7], the stability of grid-connected converters [8,9], and the smart control of power electronics for devices, microgrids and at system levels [10–13].

2. The Present Special Issue

This special issue with 49 published articles has gained a great deal of attention from both academia and industry, clearly showing the growth in significance of "Applications of Power Electronics" in the current research and development arena. The accepted articles cover broad topics in the field of power electronics, and they are categorized into seven different focus areas:

T1: Fault Diagnosis, Reliability and Condition Monitoring [14–17];
T2: Modeling, Control and Design of Power Electronic Converters [18–25];
T3: Electrical Machines, Drives and Traction Systems [26–34];
T4: Distributed Power Generation and e-Grid [35–46];
T5: Emerging Power Electronic Technologies (Pulsed Power, Energy Storage, Others) [47–51];
T6: Energy Access and Micro-Grids [52–56];
T7: Wireless Power Transfer Systems [57–62].

2.1. Fault Diagnosis, Reliability and Condition Monitoring (T1)

In order to extend overall system lifetimes, fault diagnosis, fault-tolerant control and health management systems are of significant importance, and these have been one of the major focus areas in the power electronics field in the last decades. Open circuit fault diagnosis and the fault tolerance control of three-phase active rectifiers as an inherent stage of many power electronics applications have been addressed in [14]. For induction motors, an automatic fault diagnosis system under a transient situation is developed in [15] and a fault-tolerant control strategy for five-phase induction motors under four and three-phase operation is addressed in [16]. Lastly, in [17], a review is provided on a health management system for lithium-ion batteries with a specific focus on electric vehicle applications.

2.2. Modeling, Control and Design of Power Electronic Converters (T2)

In [18], a synchronous reference frame control design methodology is provided for shunt power filters, while the sliding mode control and one-cycle controller design and stability performance of a class-D amplifier and boost power factor correction are discussed in [19,20], respectively.

Performance evaluation and the improvement of a dual active bridge converter as one of the suitable topologies for isolated power converters are discussed in [21,22]. Lastly, digital control techniques for voltage source inverters in renewable energy applications are summarized in [23].

Hardware-in-the-loop (HIL) techniques are identified as effective methods for validation of power converter and/or its controller prior to full system implementation. In [24], two different HIL implementation methods suitable for nonlinear control methods are addressed, while the application of FPGA for HIL implementation and its limitations are discussed in [25].

2.3. Electrical Machines, Drives and Traction Systems (T3)

With the continuous cost reduction of power semiconductor devices, and due to be controllability of power electronics-based systems, more and more motor-driven applications are being equipped with power electronics. Thereby, there is a focus on improving the performance and stability of motor drive systems through control, utilizing multi-phase motors and the proper modeling of motors over a wide range of loading conditions. A robust control with auto-tuned closed loop control is discussed in [26]. In [27], a comparative analysis of different control structures in improving the performance of dual three-phase permanent magnet synchronous motors (PMSM) is addressed. Extending the Kalman filter-based sliding mode control of a parallel-connected two five-phase PMSM drive system is explained in [28]. Since a slim DC-link drive provides a compact drive system, improving the motor drive performance and control stability through modulation and active damping is proposed in [29]. Utilizing composite active vector modulation in improving a direct torque control scheme for PMSM is introduced in [30]. Lastly, the suitability of utilizing a line starter PMSM for industrial applications from the efficiency point of view is discussed in [31].

The knowledge of motor behavior through proper modeling plays an important role for motor drive system design and control. The frequency-dependent behavior of an induction motor's equivalent inductance and its importance on the output current ripple and total harmonic distortion (THD) is analyzed in [32]. In [33], 3D finite element analysis (FEA) is used to analyze the profile effect of various magnet shapes in axial flux PM motors to obtain higher efficiency. Finally, a series active filter design based on a hybrid modular multi-level converter (MMC) suitable for traction systems is introduced and analyzed in [34].

2.4. Distributed Power Generation and e-Grid (T4)

Photovoltaic (PV) applications, as they utilize renewable energy resources to reduce the carbon footprint, are being employed more and more for distributed power generation. Applying the power electronics technique for PV application is the main focus in [35–37], covering different design aspects. While [35] addresses the practical implementation of a three-level boost converter using FPGA, in [36], the importance of using wide band-gap devices such as silicon carbide (SiC) to achieve better performance and power density is addressed. Lastly, the application of a modular multi-level converter (MMC) based on a cascaded connection is described in [37] for PV applications. The application and optimal design of MMC is further extended in [38] for high voltage direct current (HVDC) systems. Another design aspect which has attracted attention is efficiency. In [39], the possibility of improving voltage source converter efficiency through optimal switching frequency selection is discussed. In [40], a new technique to improve power converter efficiency by reducing the switching count for a distribution static compensator (DSTATCOM) and induction motor drive applications is addressed.

With the high penetration of power electronic systems, another aspect that again is attracting increased attention is the mitigation and control of the harmonics and EMI noise emissions of power converters. In [41], the utilization of an online selective harmonic elimination (SHE) method and particle swarm optimization to reduce harmonics is addressed, while in [42], a comprehensive review on control strategies for mitigating the dead-time effect on power converters to improve the total harmonic distortion of output waveforms is presented. With respect to EMI, modeling and proper EMI filter design in order to comply with international standards, which is of high importance, is addressed in [43,44]. Furthermore, applying active spectral shaping can maintain the generated EMI noise while reducing the size of an EMI filter in an effective method presented in [45]. Finally, applying optimization techniques for EMI filter design that not only increase the converter power density but also can make the design process automatic (reducing the time-to-market, to name one benefit) is addressed in [46].

2.5. Emerging Power Electronic Technologies (Pulsed Power, Energy Storage, Others) (T5)

In this sub-topic, the first article addresses the 10 kV high-frequency switching power supply known as a pulsed power supply for plasma generation [47]. In this article, a pulsed power supply is developed for water purification.

As the second focus of this sub-topic in the energy storage area, four articles were accepted and published. The first one provides a review of the electrical circuit modeling of double layer capacitors for energy storage [48]. The reduction of battery cell inconsistency using a composite equalizer to improve overall system performance is addressed in [49]. Improvements of state-of-charge (SoC) estimation using optimization and proper filtering methods are introduced in [50]. Lastly, a review and future challenges of SoC estimation for lithium–ion batteries are provided in [51].

2.6. Energy Access and Micro-Grids (T6)

Five articles have been accepted in the area of microgrids. In all of these articles, innovative control strategies have been proposed for AC, DC and hybrid AC–DC microgrids. In [52], a harmonic linearization technique has been deployed to analyze the stability of the AC microgrid in the sequence domain. Focusing more on the higher-level control, an innovative switching control strategy has been developed for EV charging stations to minimize their effect on the performance of a hybrid microgrid system in [53]. In [54], a power electronic converter interface has been used to allow for the variable-speed operation of hydro-pumped energy storage. On the grid side, this achieved better frequency and voltage regulation compared to scenarios without power electronic interfaces. An islanding mechanism has been proposed for renewable-based microgrids in [55], while an accurate load-sharing even in the presence of faulty communication links was developed for DC microgrids in [56].

2.7. Wireless Power Transfer Systems (T7)

Wireless power transfer (WPT) has emerged as an innovative technology to simplify the charging process, and this is the focus of the last four articles. The importance of synchronization in mitigating power oscillations and ensuring system stability is introduced in [57]. In [58], WPT system efficiency improvement and size reduction is considered by adding a single-switch boost stage at the secondary side, while in [59], the possibility of efficiency improvement using a current-fed inverter is discussed. The simulation modeling of an induction power transfer (IPT) as a replica of a 2-kW IPT charger for an electric vehicle battery charger is addressed in [60,61]. Lastly, in [62], a comprehensive review of WPT system topologies, structures and EMI diagnostics is presented.

3. Concluding Remarks

Although the 21st century can be identified as the golden age of power electronics applications, more in-depth research and development still need to be carried out in this area in order to accelerate the deployment of power electronics applications. This requires further improvements in the areas of power converter reliability, control stability and efficiency, and also the proper modeling of the system itself as well as the system around the application. Furthermore, providing electromagnetic compatibility at both the device level and system level is necessary to ensure interoperability and compatibility, which can be a challenging issue with WBG-based power electronic systems, as mentioned in the Introduction. In addition, the interactions among multitude power converters and the presence of non-ideal conditions, which may lead to instability issues, especially in distributed generation systems, call for further investigation. Combined multi-disciplinary efforts from both academia and industry are essential to provide a brighter future for power electronics applications and enable smarter and carbon-free future power grids.

Author Contributions: The authors worked together and contributed equally during the editorial process of this special issue.

Funding: This research received no external funding.

Acknowledgments: The guest editors would like to thank all authors for their excellent contribution to this Special Issue on Applications of Power Electronics. They are also thanful for the dedicated effort of all the reviewers who contributed to the reviewing process. Finally, our gratitude goes to the editorial board of MDPI *Electronics* journal for giving us the opportunity to host this special issue, and to the *Electronics* editorial office staff for their hard and precise work.

Conflicts of Interest: The authors declare no conflict of interest.

References

1. Davari, P.; Yang, Y.; Zare, F.; Blaabjerg, F. A Multipulse Pattern Modulation Scheme for Harmonic Mitigation in Three-Phase Multimotor Drives. *IEEE J. Emerg. Sel. Top. Power Electron.* **2016**, *4*, 174–185. [CrossRef]
2. Zare, F.; Soltani, H.; Kumar, D.; Davari, P.; Delpino, H.A.M.; Blaabjerg, F. Harmonic Emissions of Three-Phase Diode Rectifiers in Distribution Networks. *IEEE Access* **2017**, *5*, 2819–2833.
3. Davari, P.; Yang, Y.; Zare, F.; Blaabjerg, F. Predictive Pulse-Pattern Current Modulation Scheme for Harmonic Reduction in Three-Phase Multidrive Systems. *IEEE Trans. Ind. Electron.* **2016**, *63*, 5932–5942. [CrossRef]
4. Davari, P.; Blaabjerg, F.; Hoene, E.; Zare, F. Improving 9–150 kHz EMI Performance of Single-Phase PFC Rectifier. In Proceedings of the CIPS 2018, 10th International Conference on Integrated Power Electronics Systems, Stuttgart, Germany, 20–22 March 2018; pp. 1–6.
5. Wang, H.; Davari, P.; Wang, H.; Kumar, D.; Zare, F.; Blaabjerg, F. Lifetime Estimation of DC-Link Capacitors in Adjustable Speed Drives Under Grid Voltage Unbalances. *IEEE Trans. Power Electron.* **2019**, *34*, 4064–4078. [CrossRef]
6. Dragicevic, T.; Wheeler, P.; Blaabjerg, F. Artificial Intelligence Aided Automated Design for Reliability of Power Electronic Systems. *IEEE Trans. Power Electron.* **2019**. [CrossRef]
7. Davari, P.; Kristensen, O.; Iannuzzo, F. Investigation of acoustic emission as a non-invasive method for detection of power semiconductor aging. *Microelectron. Reliab.* **2018**, *88–90*, 545–549. [CrossRef]
8. Taul, M.G.; Wang, X.; Davari, P.; Blaabjerg, F. An overview of assessment methods for synchronization stability of grid-connected converters under severe symmetrical grid faults. *IEEE Trans. Power Electron.* **2019**. [CrossRef]
9. Dragicevic, T.; Novak, M. Weighting Factor Design in Model Predictive Control of Power Electronic Converters: An Artificial Neural Network Approach. *IEEE Trans. Ind. Electron.* **2019**. [CrossRef]

10. Peyghami, S.; Davari, P.; Mokhtari, H.; Blaabjerg, F. Decentralized Droop Control in DC Microgrids Based on a Frequency Injection Approach. *IEEE Trans. Smart Grid* **2019**. [CrossRef]

11. Heydari, R.; Gheisarnejad, M.; Khooban, M.H.; Dragicevic, T.; Blaabjerg, F. Robust and Fast Voltage-Source-Converter (VSC) Control for Naval Shipboard Microgrids. *IEEE Trans. Power Electron.* **2019**. [CrossRef]

12. Dragičević, T. Dynamic Stabilization of DC Microgrids with Predictive Control of Point-of-Load Converters. *IEEE Trans. Power Electron.* **2018**, *33*, 10872–10884. [CrossRef]

13. Dragičević, T. Model Predictive Control of Power Converters for Robust and Fast Operation of AC Microgrids. *IEEE Trans. Power Electron.* **2018**, *33*, 6304–6317. [CrossRef]

14. Cheng, H.; Chen, W.; Wang, C.; Deng, J. Open Circuit Fault Diagnosis and Fault Tolerance of Three-Phase Bridgeless Rectifier. *Electronics* **2018**, *7*, 291. [CrossRef]

15. Burriel-Valencia, J.; Puche-Panadero, R.; Martinez-Roman, J.; Sapena-Bano, A.; Pineda-Sanchez, M.; Perez-Cruz, J.; Riera-Guasp, M. Automatic Fault Diagnostic System for Induction Motors under Transient Regime Optimized with Expert Systems. *Electronics* **2019**, *8*, 6. [CrossRef]

16. Rangari, S.; Suryawanshi, H.; Renge, M. New fault-tolerant control strategy of five-phase induction motor with four-phase and three-phase modes of operation. *Electronics* **2018**, *7*, 159. [CrossRef]

17. Omariba, Z.; Zhang, L.; Sun, D. Review on health management system for lithium-ion batteries of electric vehicles. *Electronics* **2018**, *7*, 72. [CrossRef]

18. Balasubramanian, R.; Parkavikathirvelu, K.; Sankaran, R.; Amirtharajan, R. Design, Simulation and Hardware Implementation of Shunt Hybrid Compensator Using Synchronous Rotating Reference Frame (SRRF)-Based Control Technique. *Electronics* **2019**, *8*, 42. [CrossRef]

19. Zaman, H.; Zheng, X.; Wu, X.; Khan, S.; Ali, H. A fixed-frequency sliding-mode controller for fourth-order class-D amplifier. *Electronics* **2018**, *7*, 261. [CrossRef]

20. Zhang, R.; Ma, W.; Wang, L.; Hu, M.; Cao, L.; Zhou, H.; Zhang, Y. Line Frequency Instability of One-Cycle-Controlled Boost Power Factor Correction Converter. *Electronics* **2018**, *7*, 203. [CrossRef]

21. Zhang, Y.; Li, X.; Sun, C.; He, Z. Improved Step Load Response of a Dual-Active-Bridge DC–DC Converter. *Electronics* **2018**, *7*, 185. [CrossRef]

22. Lu, M.; Li, X. Performance Evaluation of a Semi-Dual-Active-Bridge with PPWM Plus SPS Control. *Electronics* **2018**, *7*, 184. [CrossRef]

23. Tahir, S.; Wang, J.; Baloch, M.; Kaloi, G. Digital control techniques based on voltage source inverters in renewable energy applications: A review. *Electronics* **2018**, *7*, 18. [CrossRef]

24. Rosa, A.; Silva, M.; Campos, M.; Santana, R.; Rodrigues, W.; Morais, L. Shil and dhil simulations of nonlinear control methods applied for power converters using embedded systems. *Electronics* **2018**, *7*, 241. [CrossRef]

25. Sanchez, A.; Todorovich, E.; de Castro, A. Exploring the limits of floating-point resolution for hardware-in-the-loop implemented with fpgas. *Electronics* **2018**, *7*, 219. [CrossRef]

26. Kim, S.K.; Lee, K.B. Robust DC-Link Voltage Tracking Controller with Variable Control Gain for Permanent Magnet Synchronous Generators. *Electronics* **2018**, *7*, 339. [CrossRef]

27. Ahmad, M.; Wang, Z.; Yan, S.; Wang, C.; Wang, Z.; Zhu, C.; Qin, H. Comparative Analysis of Two and Four Current Loops for Vector Controlled Dual-Three Phase Permanent Magnet Synchronous Motor. *Electronics* **2018**, *7*, 269. [CrossRef]

28. Kamel, T.; Abdelkader, D.; Said, B.; Padmanaban, S.; Iqbal, A. Extended Kalman filter based sliding mode control of parallel-connected two five-phase PMSM drive system. *Electronics* **2018**, *7*, 14. [CrossRef]

29. Aksoz, A.; Song, Y.; Saygin, A.; Blaabjerg, F.; Davari, P. Improving Performance of Three-Phase Slim DC-Link Drives Utilizing Virtual Positive Impedance-Based Active Damping Control. *Electronics* **2018**, *7*, 234. [CrossRef]

30. Yuan, T.; Wang, D. Performance Improvement for PMSM DTC System through Composite Active Vectors Modulation. *Electronics* **2018**, *7*, 263. [CrossRef]

31. Zöhra, B.; Akar, M.; Eker, M. Design of A Novel Line Start Synchronous Motor Rotor. *Electronics* **2019**, *8*, 25. [CrossRef]

32. Srndovic, M.; Fišer, R.; Grandi, G. Analysis of Equivalent Inductance of Three-phase Induction Motors in the Switching Frequency Range. *Electronics* **2019**, *8*, 120. [CrossRef]

33. Cetin, E.; Daldaban, F. Analyzing the profile effects of the various magnet shapes in axial flux PM motors by means of 3D-FEA. *Electronics* **2018**, *7*, 13. [CrossRef]

34. Ali, M.; Khan, M.; Xu, J.; Faiz, M.; Ali, Y.; Hashmi, K.; Tang, H. Series Active Filter Design Based on Asymmetric Hybrid Modular Multilevel Converter for Traction System. *Electronics* **2018**, *7*, 134. [CrossRef]
35. Sulake, N.; Devarasetty Venkata, A.; Choppavarapu, S. FPGA Implementation of a Three-Level Boost Converter-fed Seven-Level DC-Link Cascade H-Bridge inverter for Photovoltaic Applications. *Electronics* **2018**, *7*, 282. [CrossRef]
36. Öztürk, S.; Canver, M.; Çadırcı, I.; Ermiş, M. All SiC grid-connected PV supply with HF link MPPT converter: System design methodology and development of a 20 kHz, 25 kVA prototype. *Electronics* **2018**, *7*, 85. [CrossRef]
37. Karthikeyan, D.; Vijayakumar, K. Generalized Cascaded Symmetric and Level Doubling Multilevel Converter Topology with Reduced THD for Photovoltaic Applications. *Electronics* **2019**, *8*, 161.
38. Lu, J.; Huang, Q.; Mao, X.; Tan, Y.; Zhu, S.; Zhu, Y. Optimized Design of Modular Multilevel DC De-Icer for High Voltage Transmission Lines. *Electronics* **2018**, *7*, 204. [CrossRef]
39. Albatran, S.; Smadi, I.; Ahmad, H.; Koran, A. Online optimal switching frequency selection for grid-connected voltage source inverters. *Electronics* **2017**, *6*, 110. [CrossRef]
40. Jibhakate, C.; Chaudhari, M.; Renge, M. A Reduced Switch AC-AC Converter with the Application of D-STATCOM and Induction Motor Drive. *Electronics* **2018**, *7*, 110. [CrossRef]
41. Güvengir, U.; Çadırcı, I.; Ermiş, M. On-Line Application of SHEM by Particle Swarm Optimization to Grid-Connected, Three-Phase, Two-Level VSCs with Variable DC Link Voltage. *Electronics* **2018**, *7*, 151. [CrossRef]
42. Ji, Y.; Yang, Y.; Zhou, J.; Ding, H.; Guo, X.; Padmanaban, S. Control Strategies of Mitigating Dead-time Effect on Power Converters: An Overview. *Electronics* **2019**, *8*, 196. [CrossRef]
43. Varajão, D.; Esteves Araújo, R.; Miranda, L.; Peças Lopes, J. EMI Filter Design for a Single-stage Bidirectional and Isolated AC–DC Matrix Converter. *Electronics* **2018**, *7*, 318. [CrossRef]
44. Zhu, H.; Liu, D.; Zhang, X.; Qu, F. Reliability of Boost PFC Converters with Improved EMI Filters. *Electronics* **2018**, *7*, 413. [CrossRef]
45. Nguyen, V.; Huynh, H.; Kim, S.; Song, H. Active EMI Reduction Using Chaotic Modulation in a Buck Converter with Relaxed Output LC Filter. *Electronics* **2018**, *7*, 254. [CrossRef]
46. Giglia, G.; Ala, G.; Di Piazza, M.; Giaconia, G.; Luna, M.; Vitale, G.; Zanchetta, P. Automatic EMI filter design for power electronic converters oriented to high power density. *Electronics* **2018**, *7*, 9. [CrossRef]
47. Krishna, T.; Sathishkumar, P.; Himasree, P.; Punnoose, D.; Raghavendra, K.; Naresh, B.; Rana, R.; Kim, H.J. 4T Analog MOS Control-High Voltage High Frequency (HVHF) Plasma Switching Power Supply for Water Purification in Industrial Applications. *Electronics* **2018**, *7*, 245. [CrossRef]
48. Jiya, I.; Gurusinghe, N.; Gouws, R. Electrical circuit modelling of double layer capacitors for power electronics and energy storage applications: A review. *Electronics* **2018**, *7*, 268. [CrossRef]
49. Lai, X.; Jiang, C.; Zheng, Y.; Gao, H.; Huang, P.; Zhou, L. A novel composite equalizer based on an additional cell for series-connected lithium-ion cells. *Electronics* **2018**, *7*, 366. [CrossRef]
50. Lai, X.; Yi, W.; Zheng, Y.; Zhou, L. An all-region state-of-charge estimator based on global particle swarm optimization and improved extended kalman filter for lithium-ion batteries. *Electronics* **2018**, *7*, 321. [CrossRef]
51. Rivera-Barrera, J.; Muñoz-Galeano, N.; Sarmiento-Maldonado, H. SoC estimation for lithium-ion batteries: Review and future challenges. *Electronics* **2017**, *6*, 102. [CrossRef]
52. Rahman, A.; Syed, I.; Ullah, M. Small signal stability of a balanced three-phase ac microgrid using harmonic linearization: Parametric-based analysis. *Electronics* **2019**, *8*, 12. [CrossRef]
53. Kamal, T.; Karabacak, M.; Hassan, S.; Fernández-Ramírez, L.; Riaz, M.; Khan, M.; Khan, L. Energy management and switching control of PHEV charging stations in a hybrid smart micro-grid system. *Electronics* **2018**, *7*, 156. [CrossRef]
54. Bitew, G.; Han, M.; Mekonnen, S.; Simiyu, P. A Variable Speed Pumped Storage System Based on Droop-Fed Vector Control Strategy for Grid Frequency and AC-Bus Voltage Stability. *Electronics* **2018**, *7*, 108. [CrossRef]
55. Hashmi, K.; Mansoor Khan, M.; Jiang, H.; Umair Shahid, M.; Habib, S.; Talib Faiz, M.; Tang, H. A virtual micro-islanding-based control paradigm for renewable microgrids. *Electronics* **2018**, *7*, 105. [CrossRef]
56. Shahid, M.; Khan, M.; Hashmi, K.; Habib, S.; Jiang, H.; Tang, H. A control methodology for load sharing system restoration in islanded DC micro grid with faulty communication links. *Electronics* **2018**, *7*, 90. [CrossRef]

Electronics **2019**, *8*, 465

57. Liu, X.; Jin, N.; Yang, X.; Wang, T.; Hashmi, K.; Tang, H. A Novel Synchronization Technique for Wireless Power Transfer Systems. *Electronics* **2018**, *7*, 319. [CrossRef]
58. Liu, X.; Jin, N.; Yang, X.; Hashmi, K.; Ma, D.; Tang, H. A Novel Single-switch Phase Controlled Wireless Power Transfer System. *Electronics* **2018**, *7*, 281. [CrossRef]
59. Wang, T.; Liu, X.; Jin, N.; Tang, H.; Yang, X.; Ali, M. Wireless Power Transfer for Battery Powering System. *Electronics* **2018**, *7*, 178. [CrossRef]
60. Vázquez, J.; Roncero-Sánchez, P.; Parreño Torres, A. Simulation Model of a 2-kW IPT Charger with Phase-Shift Control: Validation through the Tuning of the Coupling Factor. *Electronics* **2018**, *7*, 255. [CrossRef]
61. Vázquez, J.; Roncero-Sánchez, P.; Parreño Torres, A. Correction: Vázquez, J. et al. Simulation Model of a 2-kW IPT Charger with Phase-Shift Control: Validation through the Tuning of the Coupling Factor. *Electronics* 2018, 7, 255. *Electronics* **2018**, *7*, 385. [CrossRef]
62. Abou Houran, M.; Yang, X.; Chen, W. Magnetically coupled resonance WPT: Review of compensation topologies, resonator structures with misalignment, and EMI diagnostics. *Electronics* **2018**, *7*, 296. [CrossRef]

electronics

MDPI

Article

Open Circuit Fault Diagnosis and Fault Tolerance of Three-Phase Bridgeless Rectifier

Hong Cheng, Wenbo Chen *, Cong Wang and Jiaqing Deng

School of Mechanical Electronic & Information Engineering, China University of Mining and Technology Beijing, Ding No. 11 Xueyuan Road, Haidian District, Beijing 100083, China; chengh@cumtb.edu.cn (H.C.); wangc@cumtb.edu.cn (C.W.); dengjiaqingrae@163.com (J.D.)
* Correspondence: sglm.chen@gmail.com; Tel.: +86-176-0013-1007

Received: 5 October 2018; Accepted: 30 October 2018; Published: 1 November 2018

Abstract: Bridgeless rectifiers are widely used in many applications due to a unity power factor, lower conduction loss and high efficiency, which does not need bidirectional energy transmission. In this case, the potential failures are threatening the reliability of these converters in critical applications such as power supply and electric motor driver. In this paper, open circuit fault is analyzed, taking a three-phase bridgeless as an example. Interference on both the input and output side are considered. Then, the fault diagnosis method including detection and location, and fault tolerance through additional switches are proposed. At last, simulation and experiments based on the hardware in loop technology are used to validate the feasibility of fault diagnosis and fault tolerance methodology.

Keywords: three-phase bridgeless rectifier; fault diagnosis; fault tolerant control; hardware in loop

1. Introduction

Multilevel converters have been widely used in middle- and high-voltage application fields in the past decades, such as renewable energy, adjustable speed drive, power transmission network, electric vehicle [1] etc. Topologies of these converters including H Bridge-based, neutral point clamping-based and bridgeless-based are most popular in literatures. Recently, bridgeless-based topologies have drawn increasing attention from industry and academia due to its high efficiency, low loss and simplification control strategy [2–6]. Compared with H Bridge-based converters, these bridgeless-based converters cannot work as inverters. However, considering that the applications are mostly pumps, fans and compressors which only need the power flowing unidirectionally [7], H-bridge-based converter has gradually been substituted by bridgeless-based converters as a pulse width modulation (PWM) rectifier in these fields. The bridgeless-based converter as a rectifier provides a sinusoidal input current at unity power factor and a controllable dc output voltage.

With the growing power switch numbers and power density, reliability of power electronic converters is increasingly important because the malfunctions are unacceptable and cause serious losses (e.g., nonscheduled downtime) in the critical applications. As a kind of electric energy conversion device, three-phase bridgeless converter (3-BLC) also endures high frequency voltage shock, over temperature impact, overload and improper driving signal. Semi-conductor devices, especially power switches, will fail more easily than other components. As discussed in Reference [8], the power switches contribute to 31% of failures, which are the most fragile components among capacitors, gate drivers, resistors and inductors. Power switches faults are usually caused by bond-wire lift-off or solder cracking, which will lead to an open circuit or short circuit of converters. The faults are named open circuit faults (OCF) and short circuit faults (SCF) respectively. An SCF will cause a large current and result in system shutdown, so hardware-based approaches such as fast fuses or breakers to transfer an SCF to an OCF are generally used. An OCF will not shutdown a system immediately but it degrades the performance inconspicuously. These may, in turn, cause secondary faults. Therefore, it is necessary to study a fault diagnosis and tolerance method for power switches OCF of the rectifier in this article.

In the past decade, numerous fault diagnosis and tolerance methods have been proposed for power electronic converters in the literature [9–16]. However, there are a few research works for ac-dc rectifiers, especially bridgeless rectifiers. For example, a diagnosis method based on a mixed logical dynamic model and residual generation was applied to a single-phase rectifier in a railway electrical traction drive system [17]. This method was fast, simple and stable, but it does not suit three-phase systems. For three-phase conditions, a current waveforms-based similarity analysis method for a three-phase PWM rectifier was proposed in Reference [18]; current waveforms were analyzed pairwise to diagnose an open circuit fault. There was a critical drawback of this method that it ignored the three-phase voltage imbalance, which is often seen for grid. In Reference [19], a fault tolerance control with additional devices for three-phase soft-switching mode rectifier is proposed. The circuit configuration had two extra center-tapped autotransformers and three more toggle switches compared with the traditional system. This method was more suitable for new design, but more retrofit cost was demanded for the existing systems. Considering the dc voltage decrease by OCF, Reference [20] proposed a fault tolerant method for the three-level rectifier in a wind turbine system, which was implemented by adding a compensation value to the reference voltages. The proposed method preserved the power factor under faulty conditions utilizing the redundancy of the switching devices, where the 3-BLC does not have such ability.

This study aims for an OCF diagnosis and tolerance method for bridgeless-based rectifiers [21–23], especially the 3-BLC. The contribution of this paper is to propose a fault feature extraction method for 3-BLC, and a fault tolerant method based on an extra two switches. The fault features were extracted from the three-phase currents. Load sudden change, source voltage imbalance or fluctuation and harmonic interference have been considered to prevent the impact on the proposed method. After that, the OCF is identified by a mixed logical model-based algorithm. When a fault was diagnosed, the drive signals of the faulty phase were redistributed artificially by the additional switches. Thus, it will maintain the current path in failure condition and make the 3-BLC still work as normal.

The rest of this paper is organized as follows. In Section 2, the mathematical model of single and three-phase bridgeless converters are analyzed. In Section 3, the fault diagnosis with an improvable feature extraction method is proposed. Section 4 details the fault tolerant implementation through additional devices. System validation using simulation and experiment data is provided in Section 5. Finally, conclusions are drawn in Section 6.

2. Basic Principles of Three-Phase Bridgeless Converters

2.1. Structure and Operation of 3-BLC

A three-phase bridgeless converter (3-BLC) is shown in Figure 1a. As can be seen, this three-phase converter is expanded from a single-phase dual-boost bridgeless structure, which has additional slow-recovery diodes D5 & D6 and two boost inductors L1 & L2 to reduce common mode noise [24,25]. Compared to the H bridge structure, bridgeless structure reduces 50% of fully controlled switches. Therefore, the control circuits, gate drivers, as well as protection units are greatly reduced, thus decreasing the system complexity and switching losses drastically [26]. The equivalent ac side circuit is depicted in Figure 1b and the mathematical model of 3-BLC can be expressed as

$$\begin{cases} L\frac{di_{sA}}{dt} = u_{sA} - u_{acA} - u_{NO} \\ L\frac{di_{sB}}{dt} = u_{sB} - u_{acB} - u_{NO} \\ L\frac{di_{sC}}{dt} = u_{sC} - u_{acC} - u_{NO} \end{cases} \tag{1}$$

where $u_{acA}, u_{acB}, u_{acC}$ are the ac voltages of phase A, B and C; u_{NO} is the neutral point voltage; L is the inductance of L1 and L2.

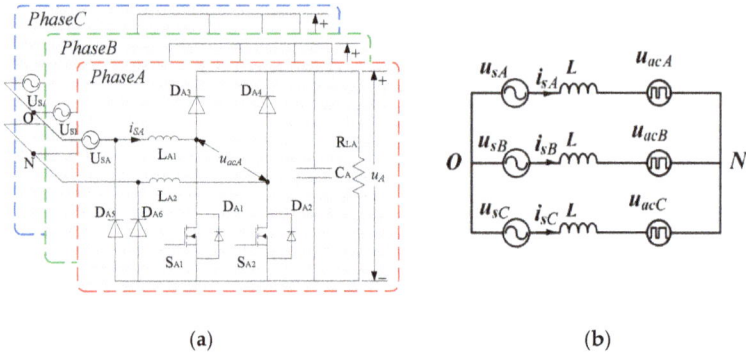

Figure 1. Three-phase bridgeless converter topology and equivalent ac side circuit. (**a**) Circuit topology; (**b**) the equivalent ac side circuit.

In order to clarify the analysis process, in this study, the slow-recovery diodes are temporarily substituted by the body diodes and the boost inductors are equivalent to an inductor L. Therefore, taking a single-phase bridgeless rectifier as an example, Figure 2 shows its principles on four operating modes. In this study, the power switches S1 and S2 are synchronously turned ON and OFF, which is named synchronous control scheme. Define S as a switch function of this converter. When $S = 1$, S1 and S2 are both turned ON, the power is transferred to power storage inductor L as shown in Figure 2a,c. When $S = 0$, S1 and S2 are both turned OFF, the power stored in inductor L is transferred to the load side as shown in Figure 2b,d. Then, a bulky electrolytic capacitor C is employed to buffer the power and, hence, smooth the output voltage. Thus, the steady-state mathematic model can be yielded by applying KVL and KCL as following.

$$\begin{cases} L\frac{di_L}{dt} = u_s - (1-S)u_C \\ C\frac{du_C}{dt} = (1-S)i_L - \frac{u_C}{R_L} \end{cases} \tag{2}$$

where L is the inductance of the power storage inductor, C is the capacitance of the output capacitor, R_L is the resistance of load, i_L is the input current, u_C is the capacitor voltage, u_s is the ac voltage, and $S^* = 1 - S$.

Applying the volt-second balance and ampere-second balance principles to Equation (2) derives

$$\begin{cases} i_L = \frac{u_s}{R_L(1-d)^2} \\ u_C = \frac{u_s}{1-d} \end{cases} \tag{3}$$

where d is the duty cycle.

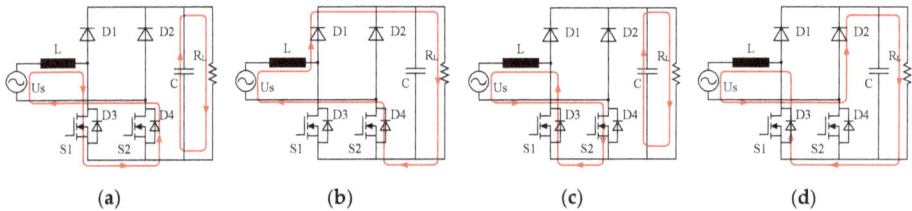

Figure 2. Basic bridgeless PFC rectifier topology and operating mode. (**a**) Mode 1 in positive ac cycle; (**b**) Mode 2 in positive ac cycle l; (**c**) Mode 3 in negative ac cycle; (**d**) Mode 4 in negative ac cycle.

2.2. Open Circuit Fault Analysis

As described above, there are four operating modes which generate through the ac voltage polarity and the switching states. When a switch fails due to physical damage or improperly driving signal, the corresponding operating mode no longer exists. This will cause changes on the related signals, which is called fault features, and is the theoretical foundation of fault detection and location. In the rest of this section, fault features of S1 OCF were analyzed as an example. Because power switches are the most fragile component and an SCF can be converted to an OCF by fast fuses immediately, the current and voltage waveforms shown in this section were acquired by simulations. The fault times were set in the positive and negative ac cycle of the input voltage respectively.

When an OCF occurs on switch S1, the power storage path as shown in Figure 2a is disconnected. Therefore, the converter works only in power discharging mode as shown in Figure 2b. However, in the negative ac cycle, the converter works properly because S1 is not in both a power charging and discharging path. The waveform of the input current and capacitor voltage during S1 open circuit fault at different half cycles are shown in Figure 3. Because it is a non-resonant circuit, the large impedance makes the input current fell into nearly zero after S1 failed in negative ac cycle, but it seems normal in positive ac cycles. The capacitor voltage resembles the input current which double frequency ripple disappears obviously after 0.5446 s, but it seems no change after 0.5346s until a positive ac cycle. The features of input current and capacitor voltage before and after switch S2 fails are just like S1.

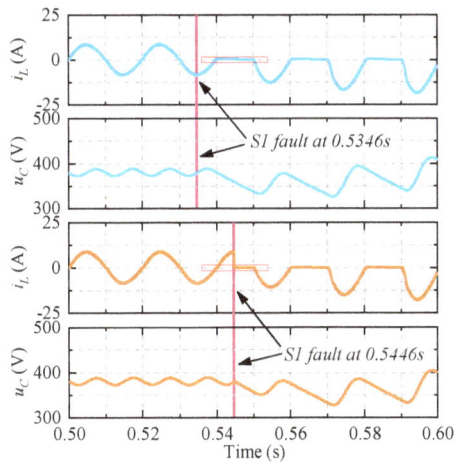

Figure 3. Input current and capacitor voltage waveforms at the time of S1 fault in positive and negative cycles.

The dynamic details of the input current inside the red frame of Figure 3 are also shown in Figure 4. According to Equation (3), the duty cycle of one switch will mutate to zero when it occurred an OCF. Since the switching period no longer existed, the assumption that the input voltage is constant during the switching cycle was no longer valid. Due to the impedance of the inductor to the low frequency signal, the input current became zero after the energy in the inductor was released. The load voltage was maintained by the DC capacitor until another ac cycle which was not affected by the fault.

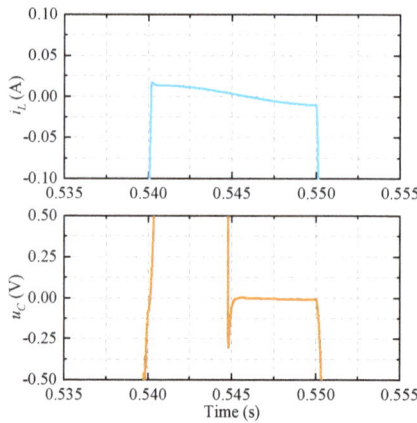

Figure 4. Dynamic details of input current and capacitor voltage waveforms at the time of S2 fault.

2.3. Input Side Interferences

Voltage fluctuations and harmonic pollution are unavoidable on the input side. They will affect the characteristics of current and voltage signals which have an impact on the diagnosis of the fault. Thus, this study discusses the features of input current and capacitor voltage under input voltage fluctuation and harmonic pollution, which are shown in Figure 5. About 7% harmonic was injected to input voltage at 0.3176 s, consequently, the THD of the input current increased from 4% to 8% but the impact on the capacitor voltage was limited. Still, from this figure, the input voltage fluctuated about 10% higher at 0.7273 s. The input current decreased a little and the capacitor voltage jittered rapidly at the same time. Different from the failure conditions, these varieties recovered in a short time. The features are significantly different from those failures discussed in the previous paragraphs.

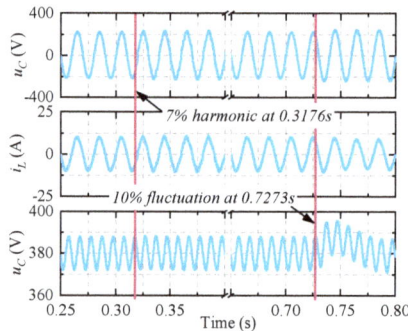

Figure 5. Input current and capacitor voltage waveforms on input voltage fluctuation and harmonic injection.

2.4. Load Side Interferences7

During the operation of the converters, the power on the load side was not always constant. The load sudden changed sometimes and the reference dc voltage also changed when transmitting different powers on the three-phase imbalance condition. This will lead to failure of fault diagnosis. In this study, output power variations due to reference voltage changed and load sudden change were also involved.

As shown in Figure 6, the reference output voltage increased from 300 V to 380 V at 0.5346 s, then the magnitude of the input current and capacitor voltage rose rapidly and became stable in two ac cycles. The output power saw a 60% growth from 400 W to 640 W. Still, in this figure, the load became heavier abruptly at the same time, which resulted in output power rising up to 1000 W with a 56% increment. The input current increased smoothly and rapidly, meanwhile the capacitor voltage decreased a little but restored fast with a higher ripple.

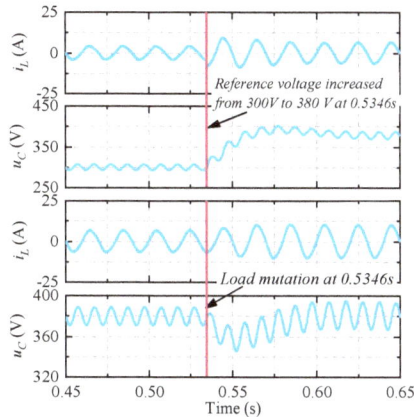

Figure 6. Input current and capacitor voltage during the time of load sudden change and reference voltage change.

In summary, an OCF will result in input current falling to about zero and increasing ripple of the capacitor voltage. The intensity of current oscillation depended on the values of *L* and *C*. The features of input or load side interferences were essentially similar, and did not affect the sinusoidal characteristic of the input current and capacitor voltage signals. It has to be noticed that in these analyses, when switch S1 failed in a negative ac cycle, the converter maintained regular operation until a positive ac cycle, and vice versa for switch S2. The duration of this condition lasted up to a half ac cycle, which is related to the time of failures.

3. Fault Diagnosis Technique

Fault diagnosis is the basis of fault tolerant control and consisted of fault detection and fault location [27]. It is important to detect and locate a malfunction switch rapidly, taking into account the fault features. In addition, the performance within fault tolerant behavior, such as redundancy control, is not only affected by fault diagnosing time directly, but also fault diagnosing accuracy including misdiagnosis and missed diagnosis [28]. Therefore, the fault diagnosis algorithm needs to be simple and effective.

3.1. Fault Features Extraction

As aforementioned, the sinusoidal characteristic of input current is damaged when an OCF occurs, but is reserved under other interferences. Therefore, it is feasible to select the input current as the characteristic signal of fault diagnosis. Generally, it is straightforward to utilize frequency domain characteristics as fault features [29]. However, most of the frequency domain methods require Fast Fourier Transform (FFT), which costs large amounts of computation. In order to improve the diagnostic efficiency and reduce the computational cost, this study used a direct time domain analysis method. Considering a three-phase converter, an $abc - \alpha\beta$ transformation, named Concordia transformation, is applied to analyze conveniently in a two-phase stationary coordinate system. This transformation can be expressed as

$$
\begin{bmatrix} i_\alpha \\ i_\beta \end{bmatrix} = \begin{bmatrix} \sqrt{\frac{2}{3}} & -\frac{1}{\sqrt{6}} & -\frac{1}{\sqrt{6}} \\ 0 & \frac{1}{\sqrt{2}} & -\frac{1}{\sqrt{2}} \end{bmatrix} \begin{bmatrix} i_A \\ i_B \\ i_C \end{bmatrix} \tag{4}
$$

Transformed input current signals are shown in Figure 7, which took i_α as the abscissa and i_β as the ordinate. The gradient of color represented the increase of time; therefore, the characteristics of i_α and i_β before and after failure were revealed in these figures. Single switch failures of each switch were shown in Figure 7, as well as normal and load side interference conditions. Input side interference was similar with the load side.

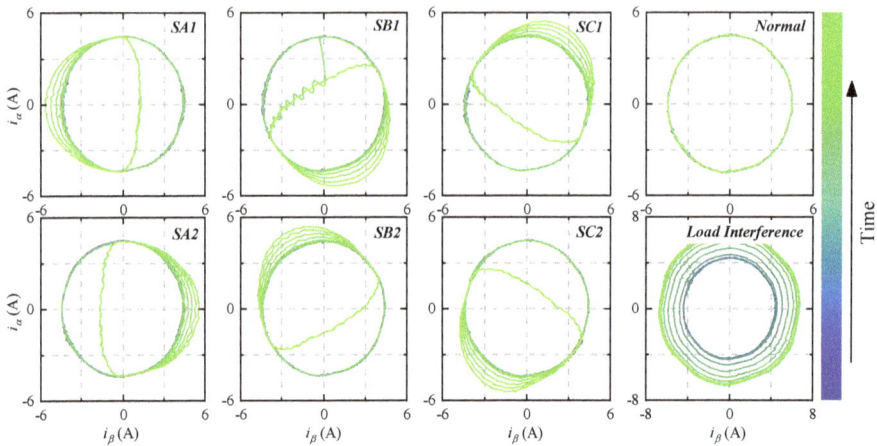

Figure 7. Input current features in a two-phase coordinate system under eight conditions.

3.2. Fault Detection and Location Method

According to the features shown in the above figure, i_α and i_β constitute a circular trajectory under normal and interference conditions. However, an OCF will change the trajectory and different fault location will result in different trajectories. In other words, vectors from $(0,0)$ to (i_α, i_β) contain distinguished features of OCFs. Define R as the length of this vector, an interval increment-based fault detection method was proposed and can be expressed as

$$
S = \int_{t-t_s}^{t} |R_0 - R(t)| dt
$$
$$
if \ S - S_{th} > res, \ then \ an \ OCF \ occurred \tag{5}
$$

where S is the accumulative bias of R, S_{th} is a reference value for detecting an OCF, res is the threshold of residual for decision, and t_s is the length of the interval. $R_0 = \frac{1}{T}\int_{t-2T}^{t-T} R(t)dt$ is the reference of R which lags one ac cycle to increase the sensitivity, and T is the power frequency cycle.

Review the angle between this vector and the positive direction of the abscissa axis. It can be found that the abnormal R occurred at a specific angular interval corresponding to different fault locations. The central values of these intervals, defined as θ_{th}, are ideally taken as $\left(0, \pi, \frac{2\pi}{3}, \frac{5\pi}{3}, \frac{4\pi}{3}, \frac{\pi}{3}\right)$ corresponding to SA1, SA2, SB1, SB2, SC1, SC2 OCFs one by one. Then, a fault location method after fault detection was proposed based on this vector. It selected the minus central value between the angle when a fault was detected and θ_{th} as the location judgment, which can be expressed as

$$
\min_{i} |\theta_{th}(i) - \theta_d| \quad \theta \in [0, 2\pi] \tag{6}
$$

where $i = 1, 2, \cdots, 6$, which represents SA1, SA2, SB1, SB2, SC1, and SC2 respectively. θ_d is the angle when an OCF was detected.

The flow chart including fault detection and fault location algorithm are shown in Figure 8.

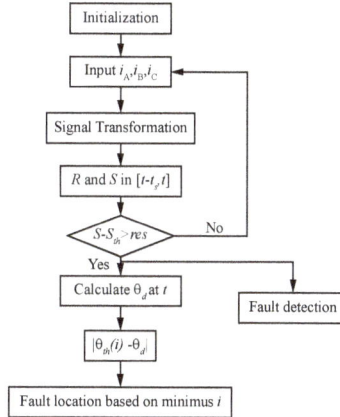

Figure 8. Fault detection and location algorithm flow chart based on current residual.

4. Fault Tolerance Method

An OCF will not cause the converter to shutdown immediately, but it slowly degrades the performance and reliability. Actually, a converter is desired to have the capability to operate in a quasi-normal condition in the post-failure period. Therefore, a fault tolerance method includes two aspects: fault tolerant topology with an extra two switches and a corresponding fault tolerant control method were proposed.

Also in the case of a single phase, the converter consists of two fast recovery diodes and two power switches with body diodes [25]. A current loop is broken due to an open circuit fault. Therefore, in order to obtain fault tolerant capability, additional devices must be added to restore the original current loop in the fault state. A fault tolerant topology with an additional two power switch which connected in parallel across the two fast recovery diodes is shown in Figure 9a. As aforementioned, more than two-thirds of ac-dc rectifiers in motor drives of industry only require a single direction for energy transmission. Therefore, considering the efficiency, loss, and control algorithms, this topology still operates as a bridgeless converter in the normal state, although it is similar to the H-bridge structure. The fault tolerant control diagram with drive signal distribution is shown in Figure 9b.

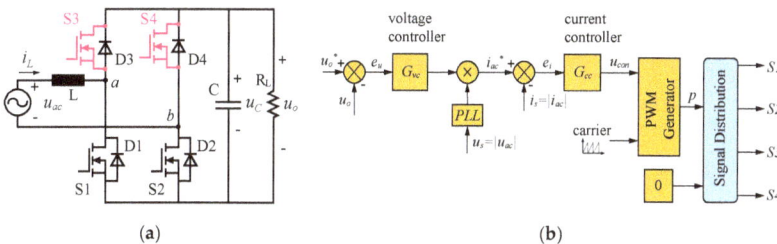

(a)   (b)

Figure 9. Fault tolerant control topology and control method. (**a**) Proposed fault tolerant topology with two additional switches; (**b**) block diagram of the fault tolerant control method with signal distribution.

The control method diagram includes two control loops: an inner current loop and an outer voltage loop. The role of the inner loop is to realize a unity power factor, and the role of the outer loop is to provide a controllable output DC voltage. The voltage controller G_{vc} and current controller G_{cc} are obtained as

$$G_{vc} = \frac{K_{Pv}s + K_{Iv}}{s}$$
$$G_{cc} = \frac{K_{Pi}s + K_{Ii}}{s}$$

(7)

The output of the double PI loop was compared with the carrier. Hence, in turn, the PWM signal was generated for signal distribution. At the same time, constant zero which means low level driving signal, was also generated.

The converter achieved fault tolerance through dynamic structure reconfiguration [30,31]. According to the location of the faulty device, there are three substructure types after an OCF. Each of them has a corresponding drive signal distribution rule to maintain the normal operation of the converter. The circuit reconfigurations are shown in Figure 10.

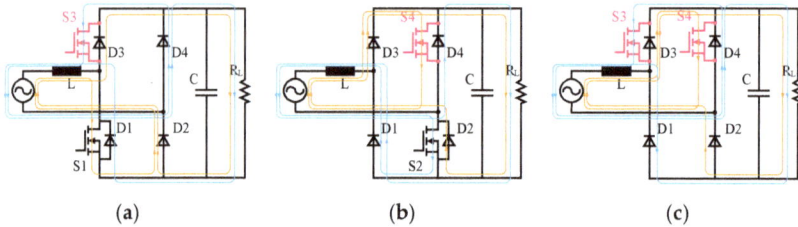

(a)   (b)   (c)

Figure 10. Circuit type reconfigurations. (**a**) Totem pole bridgeless; (**b**) Symmetry totem pole bridgeless; (**c**) Symmetry boost bridgeless.

In normal operation, both S3 and S4 are driven off by the low level, and S1 and S2 are driven by the PWM signal p in a synchronous drive mode. When it is detected that an open circuit fault occurs in S2, S3 will be driven by $-p$ which means S1 and S3 are driven by complementary signals. At this point, the circuit topology is converted into a totem pole bridgeless structure, as shown in Figure 10a. When an open circuit fault occurs in S1, the drive signal of S4 will also be replaced by $-p$, and S2 and S4 will continue to operate in a complementary signal drive mode. At this point, the circuit topology is converted to a symmetry totem pole bridgeless structure, as shown in Figure 10b. If an open circuit fault occurs in both S1 and S2, the circuit will be converted into a symmetry boost bridgeless structure consisting of S3 and S4, which will be driven by p synchronously. This situation is shown in Figure 10c. The current path for the different topologies is also shown in Figure 10, with orange representing the path for the positive ac cycle and blue for the negative. The drive signal distribution table corresponding to each fault state is shown in Table 1.

Table 1. Drive signal distribution of different fault switch.

Faulty Switch	Drive Signal			
	S1	S2	S3	S4
None	p	p	0	0
S1	/	p	0	$-p$
S2	p	/	$-p$	0
S1&S2	/	/	p	p

"0" respects low level, "p" respects the PWM signal, "/" respects an OCF.

5. Simulation and Experiment Results

It is essential to demonstrate the proposed fault diagnosis and tolerant control method function as expected in a real converter. However, failures of a real device will lead to uncontrollable consequences such as burning or explosion. Therefore, hardware in loop (HIL) simulation technology is suitable for device failure experiments. In this study, simulation and experiment-based MATLAB/Simulink and NI platform were realized, as can be seen in Figure 11a; and the details concerning the equipment used in this setup are shown in Figure 11b.

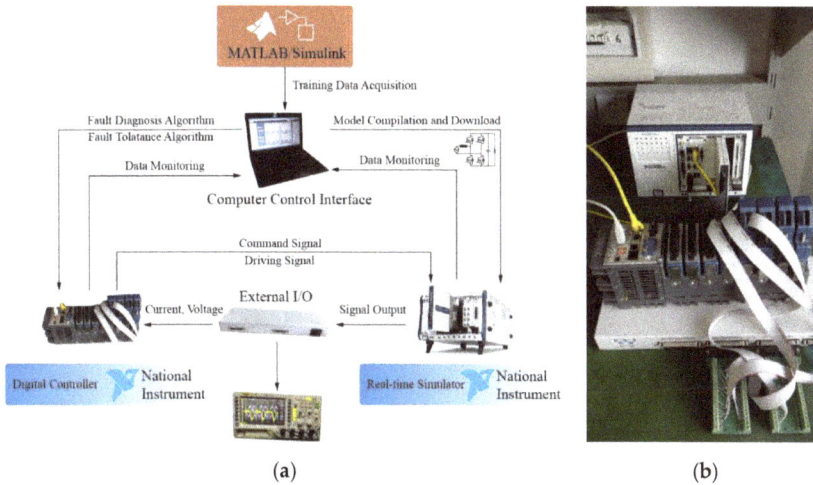

Figure 11. Diagram of simulation and experiment setup.

The simulation model of the three-phase bridgeless converter was built in MATLAB/Simulink and used to verify the proposed fault diagnosis and tolerant control method. Then HIL simulation based on NI-PXI platform was employed to emulate physical experiment, which is widely recognized and adopted in the field of power electronic device failure researches [32]. The power system model was set up as shown in Figure 1, and the parameters are presented in Table 2. The simulation and experiment results are revealed in two aspects: fault diagnosis results and fault tolerant control results. It is noticed that there was no real fault and all the OCFs were emulated by focusing the low level drive signal of specific switches.

Table 2. Specification of Simulation and Experiment.

Parameter	Value
Input ac voltage	220 V 50 Hz
Reference dc voltage	380 V
Boost inductor	5 mH
dc capacitor	330 μF
Normal power	2000 W
Switching frequency	10 kHz
Sampling frequency	10 kHz
K_{Pv}, K_{Iv}	0.021, 0.55
K_{Pi}, K_{Ii}	10.1, 200

For the diagnostic algorithm proposed in this paper, the selection of t_s has a direct impact on diagnostic resolution and diagnostic performance. Figure 12 shows the value of S when t_s is equal to 3 ms and 6 ms, plotted in red and green, respectively. Obviously, the diagnostic frequency (or resolution) of the proposed online real-time fault diagnosis is positively correlated with t_s. In the normal state, the value of S was about zero. When a fault occurred, S began to fluctuate greatly. The amplitude was also positively correlated with t_s. That is to say, a larger t_s meaning larger diagnostic interval or lower diagnostic frequency can make it easier to identify a fault trigger, reduce misdiagnosis or missed diagnosis.

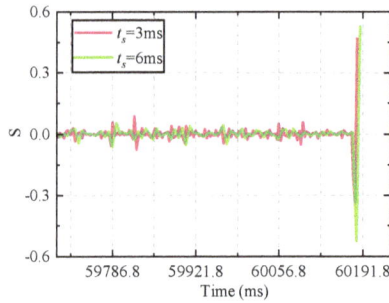

Figure 12. Plot of S with two different t_s until fault detection.

In order to make the method proposed in this study applicable with different system parameters, both i_α and i_β were normalized in the actual process. Furthermore, Figure 13 shows the results of 200 samples with different t_s from 2 to 6 ms, where the average detection time is connected in green and the average accuracy rates in red. In this figure, the influence of different S_{th} are also shown, in which the solid line represents $S_{th} = 0.2$ and the dashed dotted line represents $S_{th} = 0.4$. When $S_{th} = 0.2$, as can be seen, the fault detection time increased as t_s grew because it slowed down the diagnostic frequency, and the accuracy rates were maintained at around 98%. When $S_{th} = 0.4$, the detection time generally increased by about 20 ms, but there was greater decline in accuracy rates with t_s growing. The reason is that longer integration time makes the value of S closer to zero, and the larger threshold is gradually more unsuitable, resulting in a drop in the accuracy rate.

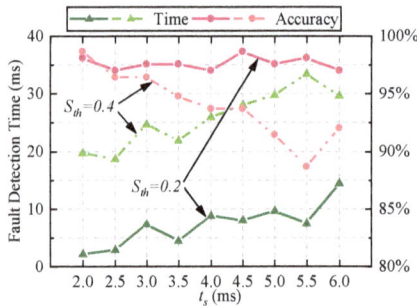

Figure 13. Detection time and accuracy with two S_{th}.

After selecting the appropriate parameters, the current signal i_A and corresponding trigger signals collected by the oscilloscope were as shown in Figure 14.

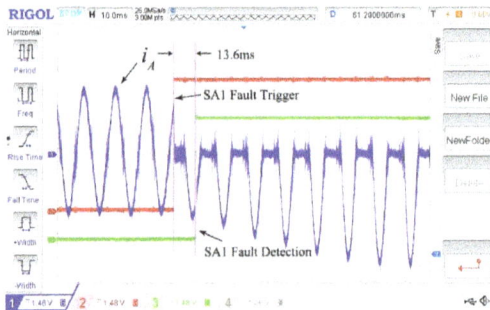

Figure 14. Input current and SA1 fault trigger and detection signals of phase A.

18

After a fault was detected, the fault location algorithm needed to be employed to locate the fault. Since the detected time t_d was not the actual time t_f at which a failure occurred. The θ_d calculated using $i_\alpha(t_d)$ and $i_\beta(t_d)$ was disorderly, and therefore, it was impossible to locate the failure in practice. In this study, this problem was solved by sacrificing a certain fault tolerant time that the value of θ_d was taking as the time minimizing R within a power cycle after t_d. It can be expressed as

$$\theta_d = \theta(t_{\min})$$
$$t_{\min} = \min_t R(t), t \in [t_d, t_d + T] \tag{8}$$

Figure 15 shows the scatter plots of $i_\alpha(t_d)$ and $i_\beta(t_d)$ for each of the 30 samples. The pink auxiliary line identifies the angles corresponding to each fault under ideal conditions. The data points generated by original θ_d were scattered and could not be used for fault location. The optimized data points were concentrated in the vicinity of the ideal values which could be used for accurate fault location.

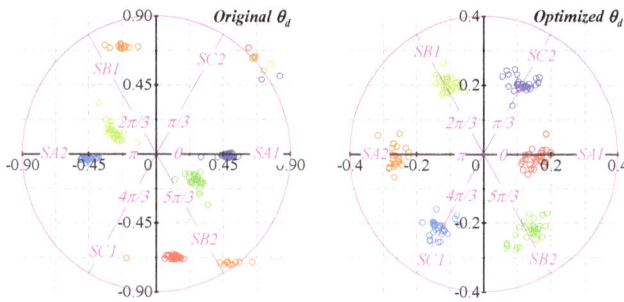

Figure 15. Fault location data distribution with original and optimized θ_d.

Under the condition of $t_s = 5$ ms, $S_{th} = 0.5$, 30 fault tolerance experiments were implemented where the faulty switch was randomly selected. The times from fault occurrence to fault tolerant control execution are shown in Figure 16. It can be found that the fault tolerance time increased by 20 ms relative to the detection time from Figure 13, where the delay of one ac cycle is caused by Equation (8).

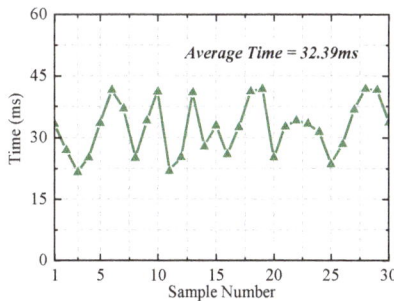

Figure 16. Fault tolerance times of thirty samples.

An example of fault diagnosis and tolerant control is shown in Figure 17. The signals were monitored by the host computer of the system built in this study.

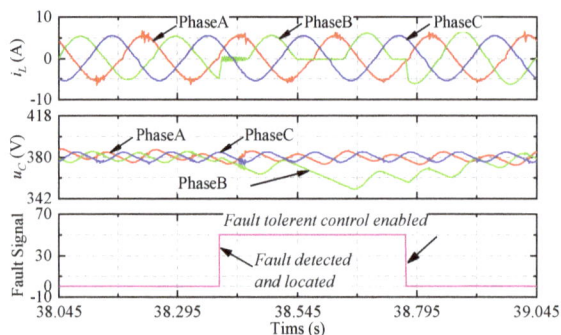

Figure 17. Signals collected by the host computer.

In summary, the proposed fault diagnosis and fault tolerant control method were verified in simulation and experimentation. An OCF will be detected within at least 20 ms in general conditions, and tolerated with a 20 ms delay. Various interferences have limited impact on this algorithm. The accuracy of fault diagnosis was over 98%, and the error rate of the fault location and fault tolerance was zero.

6. Conclusions

Reliability is one of the primary concerns for the three-phase bridgeless converter. This paper presented an open circuit fault diagnosis and tolerant control method to maintain the converter running. As the basis for fault tolerance, the open circuit fault is detected and located accurately within a few milliseconds, thus, the abnormal operation time is reduced. Only two additional switches are needed to maintain the normal operation by structure reconfiguration. A lookup table is built for switch drive signals reconfiguration. Finally, the feasibility and effect of the proposed method on reliability promotion was verified by simulations and experiments. Furthermore, the interference analysis in this paper is still insufficient that the proposed method is not robust enough in practice. This requires further work.

Author Contributions: W.C., H.C. and C.W. conceptualized the main idea of this project; W.C. proposed the methods and designed the work; W.C. conducted the experiments and analyzed the data; J.D. checked the results; W.C. wrote the whole paper; and H.C., C.W., and J.D. reviewed and edited the paper.

Funding: This research was funded by National Natural Science Foundation of China under Grant 51577187.

Conflicts of Interest: The authors declare no conflict of interest.

References

1. Blahnik, V.; Kosan, T.; Peroutka, Z.; Talla, J. Control of a Single-Phase Cascaded H-Bridge Active Rectifier under Unbalanced Load. *IEEE Trans. Power Electron.* **2018**, *33*, 5519–5527.
2. Wang, C.; Zhuang, Y.; Jiao, J.; Zhang, H.; Wang, C.; Cheng, H. Topologies and Control Strategies of Cascaded Bridgeless Multilevel Rectifiers. *IEEE J. Emerg. Sel. Top. Power Electron.* **2017**, *5*, 432–444. [CrossRef]
3. Kremes, W.D.J.; Font, C.H.I. Proposal of a three-phase bridgeless PFC SEPIC rectifier with MPPT for small wind energy systems. In Proceedings of the IEEE International Conference on Industry Applications, Curitiba, Brazil, 20–23 November 2016; pp. 1–8.
4. Silva, C.E.A.; Oliveira, D.S.; Barreto, L.H.S.C.; Bascopé, R.P.T. A novel three-phase rectifier with high power factor for wind energy conversion systems. In Proceedings of the Power Electronics Conference (COBEP '09), Bonito-Mato Grosso do Sul, Brazil, 27 September–1 October 2009; pp. 985–992.
5. Park, S.M.; Park, S. Versatile Control of Unidirectional AC–DC Boost Converters for Power Quality Mitigation. *IEEE Trans. Power Electron.* **2015**, *30*, 4738–4749. [CrossRef]

6.   Whitaker, B.; Barkley, A.; Cole, Z.; Passmore, B.; Martin, D.; McNutt, T.R.; Lostetter, A.B.; Lee, J.S.; Shiozaki, K. A High-Density, High-Efficiency, Isolated On-Board Vehicle Battery Charger Utilizing Silicon Carbide Power Devices. *IEEE Trans. Power Electron.* **2014**, *29*, 2606–2617. [CrossRef]

7.   Malinowski, M.; Gopakumar, K.; Rodriguez, J.; Pérez, M.A. A Survey on Cascaded Multilevel Inverters. *IEEE Trans. Ind. Electron.* **2010**, *57*, 2197–2206. [CrossRef]

8.   Yang, S.; Bryant, A.; Mawby, P.; Xiang, D. An Industry-Based Survey of Reliability in Power Electronic Converters. *IEEE Trans. Ind. Appl.* **2011**, *47*, 1441–1451. [CrossRef]

9.   Yu, Y.; Zhao, Y.; Wang, B.; Huang, X.; Xu, D. Current sensor fault diagnosis and tolerant control for VSI-based induction motor drives. *IEEE Trans. Power Electron.* **2018**, *33*, 4238–4248. [CrossRef]

10.   Choi, U.; Lee, J.S.; Blaabjerg, F.; Lee, K.B. Open-Circuit Fault Diagnosis and Fault-Tolerant Control for a Grid-Connected NPC Inverter. *IEEE Trans. Power Electron.* **2016**, *31*, 7234–7247. [CrossRef]

11.   Wang, T.; Xu, H.; Han, J.; Bouchikhi, E. Cascaded H-Bridge Multilevel Inverter System Fault Diagnosis Using a PCA and Multi-class Relevance Vector Machine Approach. *IEEE Trans. Power Electron.* **2015**, *30*, 7006–7018. [CrossRef]

12.   Freire, N.; Estima, J.O.; Cardoso, A. A Voltage-Based Approach without Extra Hardware for Open-Circuit Fault Diagnosis in Closed-Loop PWM AC Regenerative Drives. *IEEE Trans. Ind. Electron.* **2014**, *61*, 4960–4970. [CrossRef]

13.   Freire, N.; Estima, J.O.; Cardoso, A. Open-Circuit Fault Diagnosis in PMSG Drives for Wind Turbine Applications. *IEEE Trans. Ind. Electron.* **2013**, *60*, 3957–3967. [CrossRef]

14.   Estima, J.O.; Cardoso, A.J.M. A New Approach for Real-Time Multiple Open-Circuit Fault Diagnosis in Voltage-Source Inverters. *IEEE Trans. Ind. Appl.* **2011**, *47*, 2487–2494. [CrossRef]

15.   Zidani, F.; Diallo, D.; Benbouzid, M.E.H.; Nait-Said, R. A fuzzy-based approach for the diagnosis of fault modes in a voltage-fed PWM inverter induction motor drive. *IEEE Trans. Ind. Electron.* **2008**, *55*, 586–593. [CrossRef]

16.   Aly, M.; Ahmed, E.; Shoyama, M. A New Single Phase Five-Level Inverter Topology for Single and Multiple Switches Fault Tolerance. *IEEE Trans. Power Electron.* **2018**, *33*, 9198–9208. [CrossRef]

17.   Gou, B.; Ge, X.; Wang, S.; Feng, X. An Open-Switch Fault Diagnosis Method for Single-Phase PWM Rectifier using a Model-based Approach in High-Speed Railway Electrical Traction Drive System. *IEEE Trans. Power Electron.* **2016**, *31*, 3816–3826. [CrossRef]

18.   Feng, W.; Jin, Z. Current Similarity Analysis Based Open-Circuit Fault Diagnosis for Two-Level Three-Phase PWM Rectifier. *IEEE Trans. Power Electron.* **2016**, *32*, 3935–3945.

19.   Chao, K.; Hsieh, C. Mathematical Modeling and Fault Tolerance Control for a Three-Phase Soft-Switching Mode Rectifier. *Math. Probl. Eng.* **2013**, *2013*, 598130. [CrossRef]

20.   Lee, J.S.; Lee, K.B. Open-Circuit Fault-Tolerant Control for Outer Switches of Three-Level Rectifiers in Wind Turbine Systems. *IEEE Trans. Power Electron.* **2016**, *31*, 3806–3815. [CrossRef]

21.   Wang, C.; Zhuang, Y.; Kong, J.; Tian, C.; Cheng, H. Revised topology and control strategy of three-phase cascaded bridgeless rectifier. In Proceedings of the IEEE Industrial Electronics Society, Beijing, China, 29 October–1 November 2017; pp. 1499–1504.

22.   Zhang, Y.; Kotecha, R.; Mantooth, H.A.; Balda, J.C.; Zhao, Y.; Farnell, C. Cascaded bridgeless totem-pole multilevel converter with model predictive control for 400 V dc-powered data centers. In Proceedings of the Applied Power Electronics Conference and Exposition, Tampa, FL, USA, 26–30 March 2017; pp. 2745–2750.

23.   Jiao, J.; Wang, C. The research of new cascade bridgeless multi-level rectifier. In Proceedings of the Power Electronics and Application Conference and Exposition, Shanghai, China, 5–8 November 2014; pp. 1513–1518.

24.   Gopinath, M.; Prabakaran; Ramareddy, S. A brief analysis on bridgeless boost PFC converter. In Proceedings of the International Conference on Sustainable Energy and Intelligent Systems, Chennai, India, 20–22 July 2011; pp. 242–246.

25.   Kolar, J.W.; Friedli, T. The Essence of Three-Phase PFC Rectifier Systems. In Proceedings of the IEEE Conference of Telecommunications Energy, Amsterdam, The Netherlands, 9–13 October 2011; pp. 179–198.

26.   Wang, C.; Jiao, J.; Wang, C.; Jiang, X.; Cheng, H. Research on New Multi-level Cascaded Bridgeless Rectifier. *J. Power Supply* **2015**, *13*, 10–16.

27.   Gao, Z.; Cecati, C.; Ding, S.X. A Survey of Fault Diagnosis and Fault-Tolerant Techniques-Part I: Fault Diagnosis with Model-Based and Signal-Based Approaches. *IEEE Trans. Ind. Electron.* **2015**, *62*, 3757–3767. [CrossRef]

28. Lezana, P.; Pou, J.; Meynard, T.A.; Rodriguez, J.; Ceballos, S.; Richardeau, F. Survey on Fault Operation on Multilevel Inverters. *IEEE Trans. Ind. Electron.* **2010**, *57*, 2207–2218. [CrossRef]

29. Qiao, W.; Lu, D. A Survey on Wind Turbine Condition Monitoring and Fault Diagnosis—Part II: Signals and Signal Processing Methods. *IEEE Trans. Ind. Electron.* **2015**, *62*, 6546–6557. [CrossRef]

30. Rahmani-Andebili, M.; Fotuhi-Firuzabad, M. An Adaptive Approach for PEVs Charging Management and Reconfiguration of Electrical Distribution System Penetrated by Renewables. *IEEE Trans. Ind. Inf.* **2018**, *14*, 2001–2010. [CrossRef]

31. Rahmani-Andebili, M. Dynamic and adaptive reconfiguration of electrical distribution system including renewables applying stochastic model predictive control. *IET Gener. Transm. Distrib.* **2017**, *11*, 3912–3921. [CrossRef]

32. Jamshidpour, E.; Poure, P.; Gholipour, E.; Saadate, S. Single-switch DC-DC converter with fault-tolerant capability under open-and short-circuit switch failures. *IEEE Trans. Power Electron.* **2015**, *30*, 2703–2712. [CrossRef]

electronics

MDPI

Article

Automatic Fault Diagnostic System for Induction Motors under Transient Regime Optimized with Expert Systems

Jordi Burriel-Valencia, Ruben Puche-Panadero, Javier Martinez-Roman, Angel Sapena-Bano *, Manuel Pineda-Sanchez, Juan Perez-Cruz and Martin Riera-Guasp

Institute for Energy Engineering, Universitat Politècnica de València, 46022 Valencia, Spain;
joburva@die.upv.es (J.B.-V.); rupucpa@die.upv.es (R.P.-P.); jmroman@die.upv.es (J.M.-R.);
mpineda@die.upv.es (M.P.-S.); juperez@die.upv.es (J.P.-C.); mriera@die.upv.es (M.R.-G.)
* Correspondence: asapena@die.upv.es; Tel.: +34-96-387-7597

Received: 31 October 2018; Accepted: 19 December 2018; Published: 21 December 2018

Abstract: Induction machines (IMs) power most modern industrial processes (induction motors) and generate an increasing portion of our electricity (doubly fed induction generators). A continuous monitoring of the machine's condition can identify faults at an early stage, and it can avoid costly, unexpected shutdowns of production processes, with economic losses well beyond the cost of the machine itself. Machine current signature analysis (MCSA), has become a prominent technique for condition-based maintenance, because, in its basic approach, it is non-invasive, requires just a current sensor, and can process the current signal using a standard fast Fourier transform (FFT). Nevertheless, the industrial application of MCSA requires well-trained maintenance personnel, able to interpret the current spectra and to avoid false diagnostics that can appear due to electrical noise in harsh industrial environments. This task faces increasing difficulties, especially when dealing with machines that work under non-stationary conditions, such as wind generators under variable wind regime, or motors fed from variable speed drives. In these cases, the resulting spectra are no longer simple one-dimensional plots in the time domain; instead, they become two-dimensional images in the joint time-frequency domain, requiring highly specialized personnel to evaluate the machine condition. To alleviate these problems, supporting the maintenance staff in their decision process, and simplifying the correct use of fault diagnosis systems, expert systems based on neural networks have been proposed for automatic fault diagnosis. However, all these systems, up to the best knowledge of the authors, operate under steady-state conditions, and are not applicable in a transient regime. To solve this problem, this paper presents an automatic system for generating optimized expert diagnostic systems for fault detection when the machine works under transient conditions. The proposed method is first theoretically introduced, and then it is applied to the experimental diagnosis of broken bars in a commercial cage induction motor.

Keywords: fault diagnosis; condition monitoring; induction machines; support vector machines; expert systems; neural networks

1. Introduction

Induction machines (IMs) power most modern industrial processes (induction motors) and generate an increasing portion of our electricity (doubly fed induction generators). Therefore, fault diagnosis of IMs has become an important area of condition-based maintenance (CBM) programs, to avoid the high economic losses generated by unexpected breakdowns of IMs and sudden stoppages of the production lines that they drive. Specifically, fault diagnosis techniques based on the analysis of the MCSA [1–6] have gained a wide industrial deployment, due to their simplicity, low requirements of hardware and software, and capability for on-line simultaneous detection of a wide range of machine faults. Despite its advantages, industrial application of MCSA in harsh industrial environments, under real working conditions, is challenging. The spectral lines, whose amplitude signals the presence of a fault, can be difficult to evaluate under the myriad of spectral lines in the spectrum of the machine current, especially in case of incipient faults, where the fault harmonics have small amplitudes, or in case of low slip working conditions, where the leakage of the fundamental can bury the fault harmonics appearing at very close frequencies. To deal with these difficulties, several ongoing research works [7,8] propose the development of expert systems that can improve the diagnostic hit ratio, mostly where the fault features information obtained to detect these faults is scarce or unrepresentative. Nevertheless, developing and combining expert systems with fault diagnostic methods to improve hit ratio it is not trivial at all. An optimum combination of both elements can lead to a significant hit ratio improvement in fault detection, but an inadequate combination can even result in a misdiagnose. In the scientific literature, some works such as [9–13] are focused on the analysis, explanation, and development of recommendations, techniques, and methodologies to achieve a correct expert system implementation with optimal problem resolution. Following these recommendations, two main aspects are relevant to build an accurate expert system for the diagnosis of a faulty IM: on the one hand it is necessary to use a method able to detect and obtain features of the motor that can characterize a given type of fault; on the other hand, some algorithm or methodology able to interpret these features to discern about fault existence must be developed.

The design of an expert system for fault detection of IMs, taking into account both aspects, is a complex task, with many design variables that can influence the performance of such a diagnostic system. In this context, this paper proposes the automation of this design stage, through the development an automated system (the supra-system) which automatically generates custom fault diagnostic systems with high precision rate for fault detection. The proposed, so called supra-system is based on the exhaustive comparison of different combinations of fault diagnostic methods and optimized expert systems. It has been applied with success to the generation of an expert system for the detection of broken bars in a squirrel cage IM, both in steady-state regime and in transient state. The application to the detection of other types of fault is straightforward.

The paper is structured as follows: Section 2 introduces methods to detect faults in the induction machine under transient conditions. Section 3 describes the expert systems more commonly used in this field. Section 4 describes the development of the supra-system to generate optimized diagnostic systems. Section 5 is devoted to the experimental results and validation, and finally, in Section 6 the conclusions are presented.

2. Components of the Generator of Expert Systems for Fault Diagnosis of IMs

The three main components of an expert system for fault diagnosis of IMs, to be generated by the proposed supra- system, are the following ones:

- The quantity measured in the IM.
- The method used for extracting fault representative features from the measured quantities.
- The type of expert system used to perform the fault diagnosis from the selected IM features.

These three characteristics are analyzed in the following sections.

2.1. Quantities Measured in the IM

Fault diagnostic methods for IMs can obtain representative fault features from different motor quantities, like phase currents or voltages, acoustic, temperature, vibrations, etc. Using these features, these methods must able to detect the presence or absence of failures like broken rotor bars, winding short circuits, bearing damages, eccentricities, etc. In addition, these methods must work under any machine operation regimes (standstill, start-up, steady-state or transient regime). In this paper, the selected quantity is the machine stator current, because it is non-invasive (a Hall sensor or a current transformer placed in the line feeding the machine can acquire the current), can be acquired on-line, without disturbing the machine work, and can identify a wide range of simultaneous machine faults. Each type of fault generates a characteristic signature in the stator current, and, thus, it is possible to detect different type of faults by the on-line analysis of the machine current. Furthermore, it is easy to adapt this approach to other IM quantities.

2.2. Methods to Obtain Fault Representative Features

Under steady-state regime, the stator current analysis is performed using the frequency spectrum of the stator current, via the FFT [14]. Nevertheless, in many modern industrial processes, IMs operate under transient conditions (due to varying loads, action of controllers, etc.). The methods used for obtaining fault representative features and analyzing them in steady-state regime are usually not valid in transient state (for example, the FFT cannot be used under varying speed conditions). On the contrary, most methods designed for transient state diagnosis are also valid for the detection of failures in steady-state condition, but at the cost of a higher computational complexity both of the set of diagnostic features and of the analysis algorithms. For these reasons, this research focuses on the fault diagnosis both in steady-state regime and in transient state.

Recently, the development of diagnostic techniques focused on machines working in transient regimes have attracted the attention of many researches, giving raise to works as [15–17] dealing with this subject. In [15] the empirical mode decomposition (EMD) is used to obtain the fault features. The proposed approach, to obtain the fault features not only needs to compute the consecutive intrinsic mode functions (IMFs) containing the fault related components, but also the average value and the zero crossing of the sum of the IMFs. This process adds disturbances losing information about the fault evolution. In [16] the discrete wavelet transform (DWT) is used to decompose the Park vector of the stator currents in 12 levels; this approach implies a dyadic decomposition of the frequency bands and a computation of the energy contained in each range. Therefore, the information used (the energy of frequency band) is not only related to the fault features but also to other components or effects such as the spectral leakage that can significantly influence on the results. In a second step, the standard deviation of this data is used as fault features being less precise. Therefore, the use of a self-organizing map (SOM) network is proposed to detect the presence or absence of a given fault.

In [17] the time-frequency plane is used to detect the fault. Nevertheless, the proposed method is only valid for the detection of the rotor broken bar fault in induction motors and only for the start-up transient. The fault feature is a complex representation of the characteristic V-shape pattern (with specific width and angle) for this fault (rotor broken bar) and this regime (start-up transient). It requires complex image treatment (dilatations, erosions, thresholds, subtraction of images, etc.) to obtain the fault features and implies a loss of fault information and a very limited field of application. Therefore, although these techniques can be also applied to obtain the fault features in steady-state regimes theses fault features would be more complex and less precise than the specific methods for steady state regimes.

A recent development in fault diagnosis of IMs allows to apply the same tool for analyzing the spectral content of the motor current in both regimes. This technique, known as harmonic order tracking analysis (HOTA) [4,5,18] allows to design a unique expert system for both regimes. Therefore, it is chosen as the basis of the diagnostic system generator proposed in this paper.

2.3. The Harmonic Order Tracking Analysis (HOTA) Method

In this section, the HOTA method, introduced in [4,5,18], is briefly described. Unlike other methods where the frequency position of fault features (obtained from the current or its envelope) depend on other variables, in the HOTA method the fault frequencies are normalized to an integer, harmonic k-order scale, which is independent of the motor supply frequency and of the motor slip. This further simplifies and accelerates the procedure of conditioning the current signal prior to its processing by the expert system.

HOTA method for transient state is a method based on reducing the 2D time-frequency content of the fault harmonics of the motor current to a much simpler 1D harmonic *k*-order domain. As explained in [5], this simplification process is implemented using a Gaussian window and a short time Fourier transform (STFT), by iteratively moving the Gaussian window along the time domain and performing the frequency axis re-scaling at each step (see an example in Figure 1).

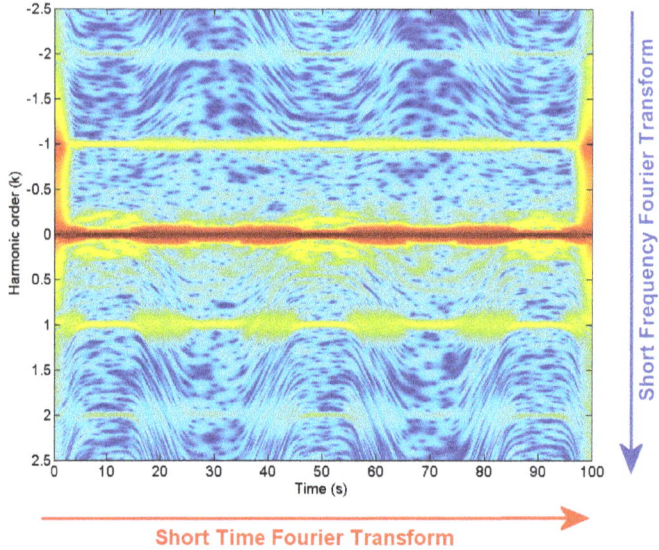

Figure 1. Example of time/harmonic *k*-order space generated with HOTA method for induction motors fault diagnosis in transient state.

An alternative option is to replace the STFT transform in HOTA by the short-frequency Fourier transform (SFFT), as in [6]. SFFT generates a time-frequency Gaussian window which is displaced along the frequency axis, instead of the time axis (Figure 1). Both the STFT and the SFFT transform generate the same time/frequency representations, although the SFFT has speed computing advantages when applied to fault diagnosis of IMs, as will be shown in Section 4.1.

In the last step of HOTA algorithm, once the time/frequency space is generated, a conversion into the harmonic *k*-order domain is made, obtaining as a result a single vector. Inside this vector each fault component is clearly shown through its *k* component number, as shown in Figure 2. This figure shows an example of the final result generated from the time-harmonic *k*-order space example shown in Figure 1. This graphic shows a fault *k* component within the harmonic *k*-order domain corresponding to a rotor broken bar fault.

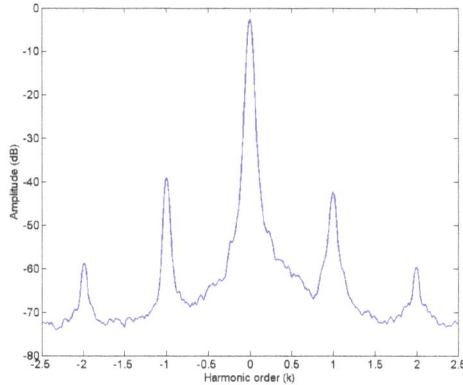

Figure 2. Final result vector in harmonic k-order domain obtained with HOTA method.

3. Expert Systems for Features Classification

Expert systems implementation to automate decision-making process are applied to interpret complex data or correlation features with some degree of uncertainty. Although diverse types of expert systems are used to resolve problems with different origins, in the field of IMs fault diagnosis the classification expert systems prevail. Classification expert systems generate discriminant results (decisions made regarding features), precluding the generation of ambiguous results.

As shown in [7,8,19], the support vector machine (SVM) expert system and artificial neural network (ANN) expert system are the most commonly used expert systems to solve features classification problems, due to their high learning coefficient of the problem. With a failure representative features database, obtained via fault diagnostic methods, these expert systems analyze and interpret the data to assess the presence or absence of an IM fault.

SVM and ANN expert systems have an input interface to the system where fault features are inserted. With these features, the expert system deduces a solution for the fault diagnosis problem. The internal part of the system is then adjusted to perform the interpretation of input data, and an output of one or more results is finally generated. Every solution is related to its respective input data.

However, in a different way about procedural programs, an expert system is not an execution of a sequence of commands that finally generate a result. Both SVM and ANN systems need to develop a previous training of the expert system to "learn" the mechanics of the problem to find the optimal solution. Adjusting the properties and parameters of the expert system to obtain a good training and good failure rates is a highly complex task, which in many cases is carried on by a manual trial and error procedure.

3.1. Support Vector Machine

SVM is an expert system based on n-dimensional spaces where each n parameter of the feature corresponds to the n dimension of the space (n feature parameters $= n$ dimensions of the space). The learning stage is based on the generation of a $(n-1)$ hyperplane that divides the n-dimensional space into two subspaces where each one represents one solution of the problem, which is sufficient to discern the occurrence or absence of fault.

In Figure 3 it is shown an example of a 2-dimensional space divided by a hyperplane generated with previous training. This SVM space has been generated with the supra-system during the experimental test described in Section 5.1.

The SVM optimization for obtaining the best system for analysis and diagnosis of failures depends on configuring the best kernel method, on the proper parameter fitting, and on the development of an optimal learning of the problem.

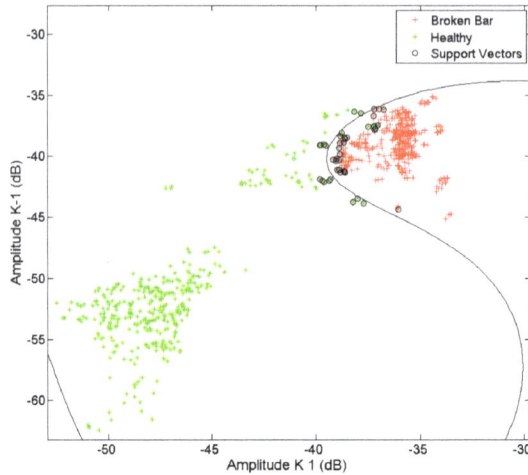

Figure 3. Space of a 2-dimensional SVM (2 parameters on each failure representative feature) where a hyperplane is splitting the space in subspaces of healthy features and failure features.

3.2. Artificial Neural Network

ANN expert systems are based on the emulation of biological neuronal networks. An ANN is composed by a set of "neurons" distributed in several interconnected layers. Although there are several types of neural networks structures and configurations, the most used for features classification is the Multilayer Perceptron. Figure 4 shows the layer structure of a Multilayer Perceptron, with an input data terminal, no or some hidden layers, and one output layer.

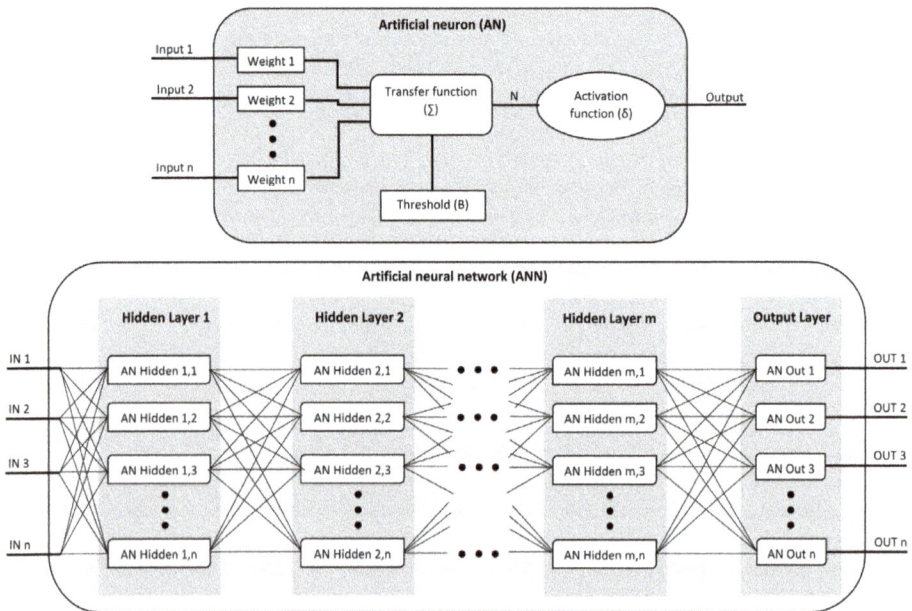

Figure 4. Scheme of Multilayer Perceptron neural network.

ANNs also must be trained before being able to correctly solve the problem. On ANN optimization, this learning is based on finding the best structure for a given problem (number of layers and number of neurons per layer) as well as the correct weight for each neuron within this network.

4. Development of the Supra-System, a System Able to Generate and Optimize Failure Diagnostic Systems Improved with Expert Systems

As presented in the previous sections, fault representative features finding methods and expert systems need to be optimized to obtain an acceptable diagnostic system. Finding the best fitting is not a trivial task since the accuracy of the expert system depends critically on the selection of its parameters. Therefore, to perform these tasks, in this paper, an autonomous system (the supra-system) is proposed. It is able not only of generating fault diagnostic systems, but also to optimize them through a set of fault representative features finding methods (Figure 5).

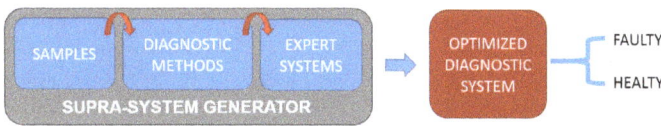

Figure 5. General scheme of the supra-system, the system generator of optimized fault diagnostic system.

4.1. Optimization of HOTA Method for IMs Fault Diagnosis

In HOTA method, as in the other methods that develop time/frequency spaces, it is essential to adjust the proportion of window filter to maximize the resolution of the space and the desired frequencies. Depending on the proportion applied to the Gaussian window filter, the space resolution can be improved (Figure 6a,c) or be worsened (Figure 6b,d), hindering the fault detection.

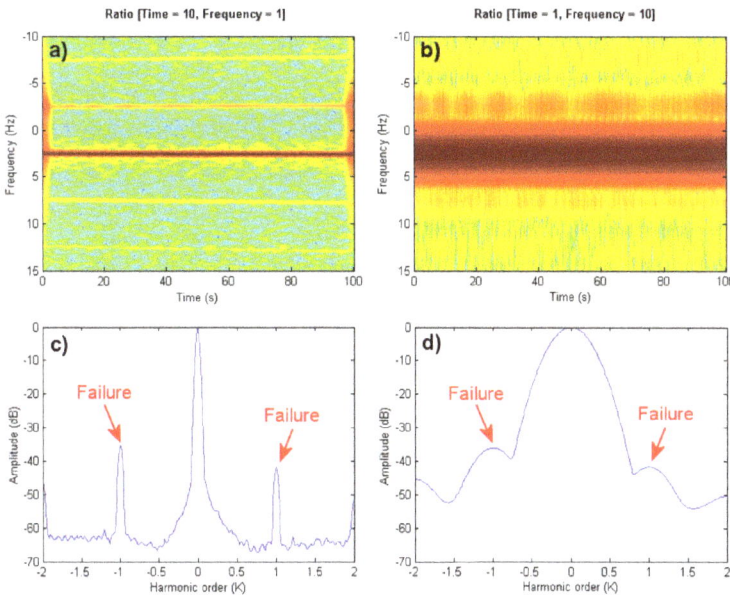

Figure 6. Examples of the time/frequency space generated with a phase current using a Gaussian window of ratio 10/1 (**a**) and its HOTA solution (**c**). The time/frequency space obtained with the same current but using a Gaussian window of ratio 1/10 is shown in (**b**), and its HOTA solution is shown in (**d**).

Figure 6 shows an example of a motor with a broken bar fault, with the fault components at $k = 1$ and $k = -1$; To detect properly these components in the joint time-frequency domain (t-f domain) is a challenging task, because their amplitude is much smaller than the main component amplitude. The performance of the fault diagnostic system depends critically on the analyzing window used for obtaining the t-f spectrum of the stator current. In this work, the Gaussian window has been selected, because it achieves the highest power density in the t-f domain. Nevertheless, different width to height ratios of a Gaussian window with the same minimum area in the t-f domain can alter substantially the shape of the fault harmonics. For each fault component, it can be observed that exists a specific window whose width to height ratio gives the sharpest fault components in the t-f spectrum [20], and consequently the fault components on HOTA result are also maximal (most representative fault features). Moving away from this ideal window size ratio, either in time or frequency, the resulting fault components on HOTA are less representatives for the fault. One of the goals of the supra-system presented in this paper is precisely to obtain this optimal window ratio, for generating the best t-f spectrum and the best fault representatives features in HOTA result.

The sharpness of the fault components is calculated with the mean value of the amplitude in the frequency range between the fault components ($k = 1$ and $k = -1$) and the main component. In the proposed system, the search for the best window size ratio has been optimized using a binary search algorithm [21], which is more specialized than a linear search algorithm.

As discussed above, the t-f spectrum is generated with a STFT transform. Nevertheless, as explained in [6], a SFFT transform can be used instead STFT, obtaining the same t-f spectrum. It is remarkable that the use of the SFFT in diagnostic applications reduces greatly the required computational power, regarding the STFT. This is due to the fact that the fault frequencies to be analyzed are known and the SFFT is able to generate the t-f spectrum by moving the Gaussian window in the frequency domain within a limited range close to the desired frequencies. This can be seen in Figure 7, where the results of the SFFT are shown in color, whereas the gray zone shows the results of the t-f spectrogram as obtained with the traditional STFT.

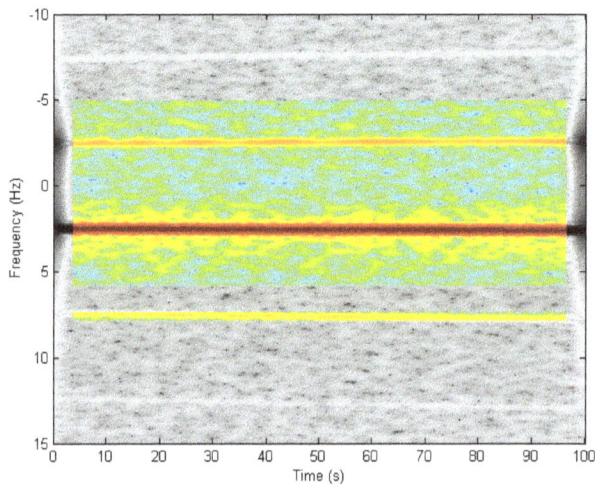

Figure 7. Example of t-f spectrum where the range of frequencies that would be necessary to obtain broken bar fault features has been marked.

4.2. Optimization of the Learning Process of the SVM Expert System for IMs Fault Diagnosis

The SVM expert system is quite optimal in solving problems that show a poorly linear distribution in its solutions. Nevertheless, its flexibility implies that it is necessary to adjust several coefficients during the training process to obtain a satisfactory learning of the expert system. The goal of the supra-system presented in this paper is also to optimize the value of such coefficients in an automatic way.

SVM allows different kernels to transform the space form to improve the discrimination of different features with the use of a hyperplane. In [10] several common kernels are discussed. In particular, with the Polynomial Kernel (1) and Radial Basis Function (RBF) kernel (2), it is possible to solve almost all cases where the solution distribution of the problem goes from totally linear to scarcely linear.

$$K(x_i, x_j) = (1 + x_i^T x_j)^d \tag{1}$$

$$K(x_i, x_j) = exp\left[-\left(\frac{\|x_i - x_j\|}{2\sigma} \right)^2 \right] \tag{2}$$

To use one of these two kernels one must set two parameters in order to obtain an optimum learning SVM. If a Polynomial Kernel is used, it is necessary to set the pair of parameters $\delta(C, d)$, where C is the box limit parameter for the search box during training and d is the polynomial order. If a RBF kernel is used, the pair of parameters that must be set are $\delta(C, \sigma)$, where C is the box limit parameter and σ is the scale factor of the Gaussian function.

In [10] it is explained how to fit these parameters by a mesh search method. Although this fitting method is less optimal than others with lower execution time, it guarantees a good approximation to the global solution avoiding local solutions that can be found by other heuristic fitting methods.

Like the scheme in Figure 8, the mesh search method divides each parameter reachable range into subsets (similar to a mesh). On each subset an SVM is trained with the parameters of this subset, obtaining an estimation of its hit rate. Finally, the subset with the trained SVM with highest hit rate is selected and the described process is iterated in a depth search until the subset with the highest hit rate (without over-training) is obtained.

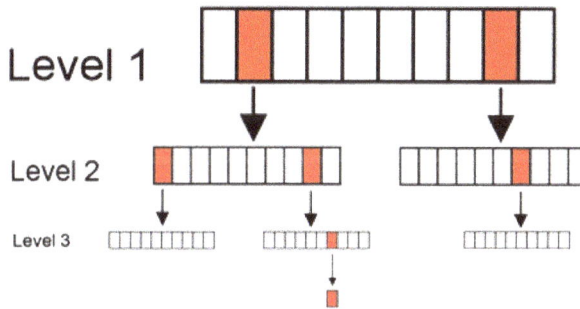

Figure 8. Schematic example of setting a parameter with the mesh search method.

It is relevant that all these trained SVM, generated during the mesh search, have had an optimal training to obtain a good approximation to the real hit rate. Achieving an optimal training for each SVM is another issue that is addressed in the supra-system presented in this paper.

An expert system cannot be trained directly with the whole training set of fault features to obtain its hit ratio. Otherwise, the hit ratio would be erroneous. There are several methods to train and validate an SVM, but the most reliable for obtaining the best hit ratio approximation for SVM is the *'leave one out cross validation'* method. With this method, the number of SVMs trained are equal to the number of fault features. In the particular field of IMs fault diagnosis, the number of fault features are not very high and therefore the computational complexity is acceptable.

As described the scheme for the *'leave one out cross validation'* (Figure 9), for each feature k, a trained SVM is generated with the remaining features, that is, excluding k feature. Once the SVM has been trained, then the *k* feature is validated. Repeating this process for the whole set of fault features, it yields finally an approximate hit ratio quite close to the actual hit ratio.

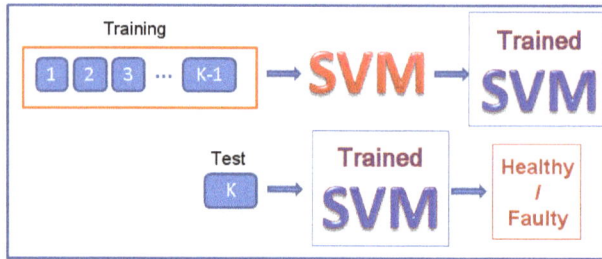

Figure 9. Scheme about training and classification for the iteration of *k* feature in the *'leave one out cross validation'* training algorithm.

4.3. Optimization of the Learning Process of ANN Expert System for IMs Fault Diagnosis

ANN expert systems are quite optimal in problem solving and may even be better than SVM when the distribution of problem solutions is highly non-linear. In contrast, the complexity to train and optimize an ANN expert system that solves the problem is greater than in the case of an SVM system.

The optimization of the structure of an ANN expert system implies the selection of the number of neurons per layer and the number of layers. These choices are relevant, since they influence the quality of learning and the accuracy of the results. For each concrete problem to solve, there is an optimal structural configuration. Training an unoptimized structure may lead to under-training or over-training the problem.

As of today, there is not a well defined method to find an optimal structural network in a direct way. However, in the case of neural networks for classification, several methodological rules have been published that discuss the structural limits of the network [10,11]. For classification problems these researches recommend a hidden layer range in the network between one and two layers, a heuristic search algorithm for fitting optimal network structure, and the best internal configuration for each neuron [9].

A stochastic search with heuristic optimization using the pyramid rule has been chosen to search the optimal structural network for IMs fault diagnosis. In the pyramid rule it is assumed that the best initial structure for searching optimal network has a trapezoidal pyramid shape where the network base are the inputs and the top are the outputs.

In the case of classification problems for ANN developed in the artificial intelligence field, as shown in Table 1, for each hidden layer's neuron is assigned a "hyperbolic tangent Sigmoid" transfer function. On the other hand, on the output layer, for each neuron a 'Competitive *SoftMax*' transfer function is used, whose output acquires the maximum value while the other outputs are cancelled.

Table 1. Transfer functions used to solve classification problems with ANN expert systems for IMs fault diagnosis.

Layer Type	Transfer Function	In(i)/Out(o) Tie
Hidden Layer	Hyperbolic Tangent Sigmoid	$o = \frac{i^n - i^{-n}}{i^n + i^{-n}}$
Output Layer	Competitive Softmax	$o = 1$, n max $o = 0$, the others

It must be emphasized that both optimizing methods for ANN network structure and configuration used in this work are only valid and optimal when the ANN is used to solve classification problems. Therefore, these methods are valid for IMs fault diagnosis since it is a classification problem (healthy/faulty).

In each step of this ANN structural optimization algorithm, a specific network structure is generated (specific number of layers and number of neurons per layer). For each generated structure an optimal training and validation must be carried out to obtain a valid approximation hit rate for this ANN structure. The highest hit rate when the ANN structural optimization algorithm has finished is declared as the best approximation hit rate and its structure is the optimum one to solve the IMs classification problem.

For ANN, the *'leave one out cross validation'* method is not optimal to obtain a good approximation of the hit ratio, due to the operative of the algorithms used to train ANN expert systems. Accordingly, in this research, it has been implemented for training the ANN a *'training/validation/test cross validation'* method. This method, shown in the scheme of Figure 10, iterates in a loop of 100 iterations (100 basic trainings) where the fault characteristics are distributed in each iteration between the training, validation, and test sets. Between 60 and 80 percent of features are assigned to the training set, between 10 and 20 percent features for the test set, and the rest is assigned to the validation set.

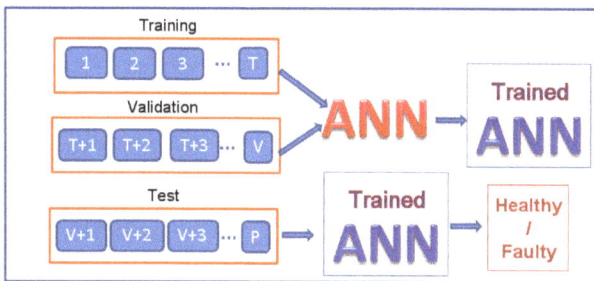

Figure 10. Scheme about training and classification for *X* iteration in the *'training/validation/test cross validation'* training algorithm.

In the *'training/validation/test cross validation'* method, it is important that the features selected on each set are representative of the whole range of values that the fault features can reach. In this research, this issue has been solved with the generation of a features space and its segmentation in a random sequence.

As analyzed in [12], there are several learning algorithms based on a gradient back-propagation, where each one has its advantages and disadvantages. The scaled conjugate gradient back-propagation algorithm has been chosen in this work as it guarantees a global optimal learning for classification problems with a moderate computational complexity, and without over-training locally (for the specific ANN structure trained).

The cost function used to minimize the gradient error is the *'cross entropy'* formulated as:

$$FC = -\frac{1}{n} \cdot \sum_{i=1}^{n} [y_i \cdot ln(a_i) + (1 - y_i) \cdot ln(1 - y_i)] \tag{3}$$

where the variable *n* denotes the total number of features for training, variable *a* stands for actual outputs and variable *y* stands for desired outputs.

5. Experimental Validation

The experimental validation of the supra-system has been carried out to detect rotor broken bars as in [1,3,22]. Nonetheless, this supra-system can be easily adapted to detect other types of faults such as eccentricity, stator inter-turn short circuits and bearing faults.

5.1. Test Bed

Figure 11 shows the test bed used in this paper. Two squirrel cage IMs whose main characteristics are given in Appendix A were tested, one in healthy conditions the other with a rotor broken bar fault. The broken bar fault was created by drilling a hole in one bar, in the junction of the bar with the end ring, as can be seen in Figure 12. This way to produce the bar breakage avoids damaging the magnetic circuit of the rotor and enables an easy verification of the complete disconnection between the bar and the end ring. To cover a wide scenario of industrial situation, the motor under test has been feed, alternatively, through variable speed drives (VSDs) of two different brands (ABB model ACS800 and Siemens model M440) with up to four different control strategies (scalar, scalar with slip compensation, field oriented control (FOC) and direct torque control (DTC)) and direct on-line (DOL) through an autotransformer. A permanent magnet synchronous machine (PMSM) is used as a mechanical load controlled by a drive ABB model ACSM1-04AS-024A-4. The test bed is controlled using a programmable logic controller (PLC) and a system control and data acquisition (SCADA) system which allows to perform the test in an automatic way and to repeat accurately the same conditions with different motors under test. A digital oscilloscope model Yokogawa DL750 has been used to acquire the currents during the different performed tests.

In this case, the test bed has been used for detecting broken bars fault, as in [1,3,22], but it can be easily adapted to detect other faults such as inter-turn short circuits, bearing damages or eccentricities. Indeed, other IMs with different faults can be coupled in the test bed and the tests set can be performed automatically.

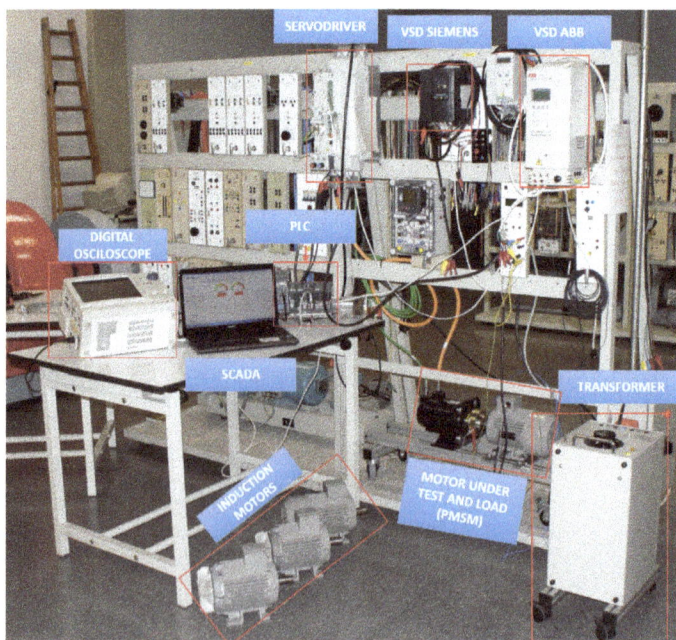

Figure 11. Test Bed used to obtain the experimental signals.

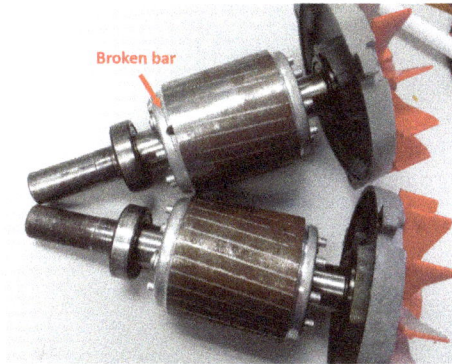

Figure 12. Top rotor with an artificially rotor broken bar fault and bottom rotor in healthy conditions.

5.2. Experimental Results

The supra-system generator has been implemented in MATLAB R2014 platform for evaluating its feasibility, the supra-system generator is applied to a large set of experimental current samples, obtained through the test bed described in the previous section, testing healthy and faulty IMs under different conditions (supply conditions, load conditions).

The samples database generated in this research comprises 726 phase current samples, with 369 healthy rotor samples and 357 faulty rotor samples. During the sampling process the frequency and load torque has been changed to obtain also transient regime samples. Table 2 summarizes the conditions under which the tests were conducted.

Table 2. Summary of the tests performed for the experimental validation.

Regime	Supply		Load	Tests
Steady	DOL	50 Hz	Constant	12
	VSD	25 Hz	Constant	24
		50 Hz	Constant	24
Transient	DOL	50 Hz	Pulse	8
			Ramps	11
	VSD	25 Hz	Pulse	15
			Ramps	19
		50 Hz	Pulse	15
			Ramps	19
		Ramps	Constant	60
			Pulse	33
			Ramps	12

The supra-system developed in this work has explored and optimized a search space given by all the possible combinations of the following diagnostic systems options:

- two HOTA implementations (with STFT or SFFT).
- two SVM expert system variants (with Polynomial or RBF Kernel).
- two ANN expert system variants (one or two hidden layers)

By an optimized search in this search space, applying the techniques introduced in Section 4.1, the global optimum result (the best fault diagnostic system) is obtained, that is, the optimized expert system for IMs fault diagnosis which has the highest diagnosis hit rate.

5.3. Time Required by the Supra-System to Generate Locals and Global Optimized IMs Fault Diagnostic Systems

In this experimental test, 286044 SVM expert systems and 252000 ANN expert systems have been analyzed in the process of finding the final optimum fault diagnostic system by the supra-system. The total time used for training (Table 3) and for classification of the IM condition (Table 4) shows that the fastest approach is to use the SFFT technique for implementing the HOTA method.

Table 3. Time needed by the supra-system to generate each local optimal diagnostic system (hours).

Optimization Time (hours)	HOTA + STFT	HOTA + SFFT
SVM with Polynomial Kernel	13.81	2.70
SVM with RBF Kernel	12.57	1.83
ANN with 1 hidden layer	13.58	3.03
ANN with 2 hidden layers	34.02	20.16

Table 4. Time used by the diagnostic system generated by the supra-system to classify the IM condition (seconds).

Classification Time (seconds)	HOTA + STFT	HOTA + SFFT
SVM with Polynomial Kernel	36.64	5.60
SVM with RBF Kernel	36.64	5.60
ANN with 1 hidden layer	36.65	5.60
ANN with 2 hidden layers	36.65	5.60

5.4. Hit Rates Obtained by the Expert Systems for IMs Fault Diagnosis Generated by the Supra-System

The hit rates obtained by the expert systems generated by the supra-system Table 5, are very high, giving a very efficient IM fault diagnostic system. It is relevant that, although SVM and ANN expert systems have different implementations, in both cases the hit rates are very similar. This similarity means that:

- In both expert systems the maximum hits ratio has been reached (the optimal diagnostic system).
- None of these local diagnostic system shows over-training. Otherwise it would show a higher hit ratio much closer to 100% and hit ratios values would be more different between local diagnosis systems.

Table 5. Hit rates obtained in each of the local optimal diagnostic systems generated by the supra-system.

Hit Rate (%)	HOTA + STFT	HOTA + SFFT
SVM with Polynomial Kernel	98.62	98.89
SVM with RBF Kernel	97.24	97.38
ANN with 1 hidden layer	97.10	97.38
ANN with 2 hidden layers	98.89	98.89

HOTA implementing SFFT with SVM with Polynomial Kernel and HOTA implementing SFFT with ANN with 2 hidden layers lead to the same result. In both cases a global optimum diagnostic system is obtained, with the best hit ratio and with the lowest diagnosis time. Therefore the use of one or another for diagnosis would be only a user decision.

Even so, the others diagnostic systems (optimal local but not optimal global) also show an excellent hit ratio with a low diagnostic time. All of them are very well optimized and could be used alike.

6. Conclusions

In this research a supra-system implementation has been proposed to generate an optimized fault diagnostic system. This supra-system has shown the following advantages regarding to traditional expert systems:

- All the diagnostic systems generated are suitable to be used in transient regime operation.
- The generation process is totally automated. That is, starting on the samples input until finishing the generation of optimum diagnostic system no user intervention is required at all.
- The generation process is totally autonomous. That is, it is not necessary to carry out any control or adjustment task on the supra-system for a successful development of the optimum diagnostic system.

This supra-system has been experimentally tested and validated, confirming that it achieves the proposed goals. The supra-system approach solves a practical industrial problem in the field of IMs fault diagnosis area, especially in transient regime, where the generation of a conventional expert diagnostic system must be manually customized for each specific motor structure.

The application of the supra-system approach to the design of IMs fault diagnostic systems addressing other types of faults (bearings faults, eccentricity, etc.) is straightforward by using the fault features of this types of fault which can be extracted with the same methods as those proposed in this paper.

Author Contributions: Conceptualization, J.M.-R. and M.R.-G.; Data curation, J.B.-V. and J.M.-R.; Formal analysis, J.B.-V. and R.P.-P.; Investigation, A.S.-B. and J.P.-C.; Methodology, J.B.-V., R.P.-P. and J.P.-C.; Project administration, M.P.-S.; Resources, M.P.-S.; Supervision, R.P.-P., M.P.-S. and M.R.-G.; Validation, J.B.-V., J.M.-R., A.S.-B., J.P.-C. and M.R.-G.; Writing—original draft, J.B.-V. and M.P.-S.; Writing—review and editing, A.S.-B., M.P.-S. and M.R.-G. These authors contributed equally to this work.

Funding: This research received no external funding.

Conflicts of Interest: The authors declare no conflict of interest.

Appendix A

Squirrel Cage Induction Motor Three-phase squirrel cage induction motor, star connection. Rated characteristics: P = 1.5 kW, f = 50 Hz, U = 400 V, I = 3.25 A, n = 2860 rpm and cos φ = 0.85.

References

1. Puche-Panadero, R.; Pineda-Sanchez, M.; Riera-Guasp, M.; Roger-Folch, J.; Hurtado-Perez, E.; Perez-Cruz, J. Improved resolution of the MCSA method via Hilbert transform, enabling the diagnosis of rotor asymmetries at very low slip. *IEEE Trans. Energy Convers.* **2009**, *24*, 52–59. [CrossRef]
2. Abd-el Malek, M.; Abdelsalam, A.K.; Hassan, O.E. Induction motor broken rotor bar fault location detection through envelope analysis of start-up current using Hilbert transform. *Mech. Syst. Signal Process.* **2017**, *93*, 332–350. [CrossRef]
3. Martinez, J.; Belahcen, A.; Muetze, A. Analysis of the vibration magnitude of an induction motor with different numbers of broken bars. *IEEE Trans. Ind. Appl.* **2017**, *53*, 2711–2720. [CrossRef]
4. Sapena-Bano, A.; Pineda-Sanchez, M.; Puche-Panadero, R.; Perez-Cruz, J.; Roger-Folch, J.; Riera-Guasp, M.; Martinez-Roman, J. Harmonic order tracking analysis: A novel method for fault diagnosis in induction machines. *IEEE Trans. Energy Convers.* **2015**, *30*, 833–841. [CrossRef]
5. Sapena-Bano, A.; Burriel-Valencia, J.; Pineda-Sanchez, M.; Puche-Panadero, R.; Riera-Guasp, M. The Harmonic Order Tracking Analysis Method for the Fault Diagnosis in Induction Motors Under Time-Varying Conditions. *IEEE Trans. Energy Convers.* **2017**, *32*, 244–256. [CrossRef]
6. Burriel-Valencia, J.; Puche-Panadero, R.; Martinez-Roman, J.; Sapena-Bano, A.; Pineda-Sanchez, M. Short-Frequency Fourier Transform for Fault Diagnosis of Induction Machines Working in Transient Regime. *IEEE Trans. Instrum. Meas.* **2017**, *66*, 432–440. [CrossRef]

7. Yin, Z.; Hou, J. Recent advances on SVM-based fault diagnosis and process monitoring in complicated industrial processes. *Neurocomputing* **2016**, *174*, 643–650. [CrossRef]

8. Bazan, G.H.; Scalassara, P.R.; Endo, W.; Goedtel, A.; Godoy, W.F.; Palácios, R.H.C. Stator fault analysis of three-phase induction motors using information measures and artificial neural networks. *Electr. Power Syst. Res.* **2017**, *143*, 347–356. [CrossRef]

9. Beale, M.H.; Hagan, M.T.; Demuth, H.B. Neural network toolbox 7. In *User's Guide*; MathWorks: Natick, MA, USA, 2010.

10. Hsu, C.W.; Chang, C.-C.; Lin, C.-J. *A Practical Guide to Support Vector Classification*; Technical Report; Department of Computer Science, National Taiwan University: Taipei City, Taiwan, 2013.

11. Bishop, C.M. *Neural Networks for Pattern Recognition*; Oxford University Press: Oxford, UK, 1995.

12. Mustafidah, H.; Hartati, S.; Wardoyo, R.; Harjoko, A. Selection of Most Appropriate Backpropagation Training Algorithm in Data Pattern Recognition. *Int. J. Comput. Trends Technol.* **2014**, *2*, 92–95. [CrossRef]

13. Godoy, W.F.; da Silva, I.N.; Goedtel, A.; Palácios, R.H.C.; Lopes, T.D. Application of intelligent tools to detect and classify broken rotor bars in three-phase induction motors fed by an inverter. *IET Electr. Power Appl.* **2016**, *10*, 430–439. [CrossRef]

14. Ghorbanian, V.; Faiz, J. A survey on time and frequency characteristics of induction motors with broken rotor bars in line-start and inverter-fed modes. *Mech. Syst. Signal Process.* **2015**, *54*, 427–456. [CrossRef]

15. Valles-Novo, R.; de Jesus Rangel-Magdaleno, J.; Ramirez-Cortes, J.M.; Peregrina-Barreto, H.; Morales-Caporal, R. Empirical mode decomposition analysis for broken-bar detection on squirrel cage induction motors. *IEEE Trans. Instrum. Meas.* **2015**, *64*, 1118–1128. [CrossRef]

16. Vitor, A.L.; Scalassara, P.R.; Endo, W.; Goedtel, A. Induction motor fault diagnosis using wavelets and coordinate transformations. In Proceedings of the 2016 12th IEEE International Conference on Industry Applications (INDUSCON), Curitiba, Brazil, 20–23 November 2016; pp. 1–8.

17. De Santiago-Perez, J.J.; Rivera-Guillen, J.R.; Amezquita-Sanchez, J.P.; Valtierra-Rodriguez, M.; Romero-Troncoso, R.J.; Dominguez-Gonzalez, A. Fourier transform and image processing for automatic detection of broken rotor bars in induction motors. *Meas. Sci. Technol.* **2018**, *29*, 095008. [CrossRef]

18. Perez-Cruz, J.; Perez-Vazquez, M.; Pineda-Sanchez, M.; Puche-Panadero, R.; Sapena-Bano, A. The Harmonic Order Tracking Analysis (HOTA) for the Diagnosis of Induction Generators Working Under Steady State Regime. In Proceedings of the 2017 Asia-Pacific Engineering and Technology Conference (APETC 2017), Kuala Lumpur, Malaysia, 25–27 May 2017; pp. 1864–1869.

19. Merabet, H.; Bahi, T.; Drici, D.; Halam, N.; Bedoud, K. Diagnosis of rotor fault using neuro-fuzzy inference system. *J. Fundam. Appl. Sci.* **2017**, *9*, 170–182. [CrossRef]

20. Riera-Guasp, M.; Pineda-Sanchez, M.; Pérez-Cruz, J.; Puche-Panadero, R.; Roger-Folch, J.; Antonino-Daviu, J.A. Diagnosis of induction motor faults via Gabor analysis of the current in transient regime. *IEEE Trans. Instrum. Measur.* **2012**, *61*, 1583–1596. [CrossRef]

21. Gambhir, A.; Vijarania, M.; Gupta, S. Implementation and Application of Binary Search in 2-D Array. *Int. J. Inst. Ind. Res.* **2016**, *1*, 30–31.

22. Gyftakis, K.N.; Cardoso, A.J.M.; Antonino-Daviu, J.A. Introducing the Filtered Park's and Filtered Extended Park's Vector Approach to detect broken rotor bars in induction motors independently from the rotor slots number. *Mech. Syst. Signal Process.* **2017**, *93*, 30–50. [CrossRef]

electronics

MDPI

Article

New Fault-Tolerant Control Strategy of Five-Phase Induction Motor with Four-Phase and Three-Phase Modes of Operation

Sonali Chetan Rangari [1,*], Hiralal Murlidhar Suryawanshi [2] and Mohan Renge [1]

[1] Department of Electrical Engineering, Shri Ramdeobaba College of Engineering and Management, Nagpur 440013, India; rangaris@rknec.edu
[2] Department of Electrical Engineering, Visvesvaraya National Institute of Technology, Nagpur 44013, India; hms_1963@rediffmail.com
* Correspondence: renkey10@yahoo.co.in; Tel.: +91-982-247-0026

Received: 20 June 2018; Accepted: 7 August 2018; Published: 23 August 2018

Abstract: The developed torque with minimum oscillations is one of the difficulties faced when designing drive systems. High ripple torque contents result in fluctuations and acoustic noise that impact the life of a drive system. A multiphase machine can offer a better alternative to a conventional three-phase machine in faulty situations by reducing the number of interruptions in industrial operation. This paper proposes a unique fault-tolerant control strategy for a five-phase induction motor. The paper considers a variable-voltage, variable-frequency control five-phase induction motor in one- and two-phase open circuit faults. The four-phase and three-phase operation modes for these faults are utilized with a modified voltage reference signal. The suggested remedial strategy is the method for compensating a faulty open phase of the machine through a modified reference signal. A modified voltage reference signal can be efficiently executed by a carrier-based pulse width modulation (PWM) system. A test bench for the execution of the fault-tolerant control strategy of the motor drive system is presented in detail along with the experimental results.

Keywords: five-phase machine; fault-tolerant control; induction motor; one phase open circuit fault (1-Ph); adjacent two-phase open circuit fault (A2-Ph); volt-per-hertz control (scalar control)

---

1. Introduction

In electric drives and machines, a three-phase machine is the default implementation in industrial applications. Emphasis should be placed on possibilities with more than a three-phase machine which is difficult to achieve with conventional three-phase machines. The simple expansion of three-phase drives to multiphase drives is not sufficient. It is highly important to investigate inventive employment of the extra degrees of flexibility. Incorporation of more than three phases is advised to improve performance. The advantages that can be achieved with the utilization of multiphase systems are investigated in [1]. Numerous endeavors concluded that multiphase machines have some inherent advantages such as higher reliability, higher frequency of torque pulsation with lower amplitude, lower rotor harmonic current, reduction in current per phase without expanding the voltage per phase, and less current ripple in the DC link [1–5].

Multiphase system reliability is most important in safety-critical applications, such as, electric ships, compressors, pumps, electric aircraft, hybrid vehicles and marine applications. In recent high power industrial applications, a multi-leg voltage source inverter (VSI) was used for multi-phase induction motors for variable-voltage, variable-frequency control.

In many industrial applications, if open-circuit fault exists in any phase of three-phase machine, it leads to considerably large torque oscillations. These oscillations are double the electrical line frequency, which may affect the shaft of the machine. For fault-tolerant control of three-phase machine requires separate current control for remaining healthy phases by enabling the connection of motor star point to the DC link midpoint [6]. Broad investigations have been accounted for open-circuit fault-tolerant mode of operation for three-phase AC machines [7–14]. Increasing the number of phases provides better sinusoidal Magneto Motive Force (MMF) distribution, which decreases torque ripples and harmonic currents compared to three-phase machines [15,16]. A five-phase machine is superior to a three-phase machine for fault-tolerant operation modes. When single-phase (1-Ph) or adjacent double-phase (A2-Ph) open circuit faults occur, the machines can remain in operation using other healthy phases without additional hardware and control [17–20].

A five-phase machine with the star-connected stator winding with no neutral connection can work as a four-phase machine when a single-phase open circuit fault (1-Ph) occurs. Similarly, it works as a three-phase machine when adjacent double phase open circuit faults (A2-Ph) occur. These faulty conditions generate torque oscillations due to unbalanced rotating MMF present in the air gap [21]. Connecting a load neutral point to the DC link midpoint reduces the negative sequence MMF component in the air gap and the oscillation without any additional control strategy.

Phase sequences are highly important when considering AC motors, as the production of the torque via the sequential "rotation" of the applied five-phase power is responsible for the mechanical rotation of the rotor. The frequency of positive-sequence is used to drive the rotor in the required direction, whereas the frequency of negative-sequence operates motor in the opposite direction of the rotation of the rotor. However, the frequency of the zero-sequence neither adds to nor detracts from the torque of the rotor. Because of the distortion in the current, an excessive number of harmonics of negative-sequence (5th, 11th, 17th and/or 23rd) is observed in the power, and if this power is applied to a five-phase AC machine, it will result in deterioration of the performance as well as possible overheating.

Many investigations have been accounted for the open-phase fault-tolerant operation of multiphase induction machines [6,20,21], developed fault-tolerant control algorithm including non-linearities of machine and converter in the modeling of open-phase fault drive system. The speed control of five-phase induction motor by using finite-control set model-based predictive control for fault-tolerant condition is introduced in [22]. The fundamental and third-harmonic component of current is used as a fault-tolerant control technique for the excitation of healthy stator phases has been proposed in [23]. The aim of this work is to represent reconfiguration of motor phase currents under one-phase and two-phase open fault condition. This paper presents the implementation of a remedial strategy to neutralize ripple in the torque and analyzed the motor-performance in four-phase and three-phase modes of operation.

The contribution of this work is

i.   Insight into the asymmetrical post-fault mode of operation and the remedial strategy compensates the unbalanced rotating MMF present in the air gap of the machine by a modified reference signal.

ii.   The control strategy is emphasized on the reduction in torque oscillations and verified with a reduction in unbalanced line current.

iii. By using volt-per-hertz (V/f) scalar control, a voltage compensation control algorithm is developed in the dsPIC33EP256MU810 Digital Signal Controller.

iv.   Pre-fault and post-fault mode of operation with fault remedial technique is experimentally verified and discussed.

v.   The method presented here enhances the continuity of the star-connected five-phase induction motor in case of one-phase and two-phase open faults.

It is assumed that the stator winding is opened in a five-phase induction motor because of gate failure of the inverter, i.e., an open switch condition.

2. Fault-Tolerant Remedial Strategy of a Five-Phase System

The schematic arrangement for a five-leg inverter with an induction motor is represented in Figure 1. The arrangement is composed of a five-phase voltage source inverter (VSI) with dc-link. Based on the industrial application dc-link voltage (VDC) can be supplied through a DC source. Five-phase motor drive system consists of phase shift of 72o symmetrical connection of the stator windings and separate neutral connection (n) [24]. The switch S1 denotes a gate drive open fault, and the switch S2 denotes the short-circuit device fault. It is not recommended to run the drive under foresaid faulty conditions even though another device in the same leg of the inverter is in healthy condition. The switch S3 is included in phase "a" to isolate the faulty leg in order to analyze the continuous operation of drive in four-phase mode of operation under a healthy and open-phase fault condition. Similarly, a gate drive open fault or switch short circuit fault in two adjacent or alternate legs of the inverter, may cause a two-phase open fault. The two faulty legs should be isolated and the drive runs in three-phase mode operation [22]. The modeling and performance of the five-phase voltage source inverter with the five-phase induction machine in pre-fault and post-fault conditions is briefly explained in section A, B, and C, respectively.

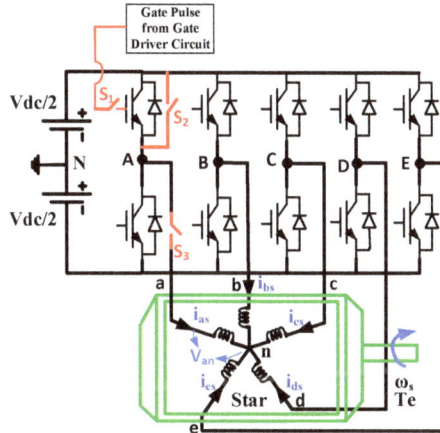

Figure 1. Five-phase system with 1-Ph fault.

2.1. Voltage Source Inverter

A five-phase drive for a machine can be obtained by developing a five-leg voltage source inverter (VSI). The phase voltages of the motor with this inverter are signified in (1).

Lowercase alphabetical letters (a–e) represent the phase voltages and the inverter leg voltages are represented by capital letters (A, B, C, D, E). Each switch conducts for 180°, giving a ten-step mode of operation.

The phase difference between two conducting switches in any sequential two phases is 720 [3,20]. For star-connected load phase-to-neutral voltages are obtained by determining the difference between the voltage of the neutral point 'n' of the load and the negative point of the dc-bus 'N'.

$$V_i = V_j + V_{nN} \tag{1}$$

where, i is {A, B, C, D, E} and j is {a, b, c, d, e}.

Since in a star-connected load the aggregate of phase voltages equals to zero and the sum of the equations yields.

$$V_{nN} = \frac{1}{5} \times (V_A + V_B + V_C + V_D + V_E) \tag{2}$$

Replacing (2) into (1), the loads with phase-to-neutral voltages are as follows:

$$\left.\begin{aligned}
V_{nN} &= \tfrac{1}{5} \times (V_A + V_B + V_C + V_D + V_E) \\
V_b &= \tfrac{4}{5}V_B - \tfrac{1}{5}(V_A + V_C + V_D + V_E) \\
V_c &= \tfrac{4}{5}V_C - \tfrac{1}{5}(V_A + V_B + V_D + V_E) \\
V_d &= \tfrac{4}{5}V_D - \tfrac{1}{5}(V_A + V_B + V_C + V_E) \\
V_e &= \tfrac{4}{5}V_E - \tfrac{1}{5}(V_E + V_B + V_C + V_D)
\end{aligned}\right\} \tag{3}$$

The phase values of inverter-leg voltages are $\pm 0.5 V_{DC}$. For a fixed modulation index M and dc-link voltage VDC, the fundamental inverter leg voltages analogous to a star-connected winding can be given as [3,20].

$$V_{Ph\ Star} = M \frac{V_{DC}}{2} \times \sin(\omega_s t) \tag{4}$$

2.2. Modeling of a Five-Phase Induction Motor

The five-phase induction motor is provided with an IGBT-based five-phase voltage source converter (VSCs) drive system. A DC-link voltage is provided from a diode bridge rectifier, which exclusively permits unidirectional power flow. The Clarke matrix for this particular case is:

$$T(\theta) = \frac{2}{5}\begin{bmatrix}
Cos\,\theta & Cos\left(\theta - \tfrac{2\pi}{5}\right) & Cos\left(\theta - \tfrac{4\pi}{5}\right) & Cos\left(\theta + \tfrac{4\pi}{5}\right) & Cos\left(\theta + \tfrac{2\pi}{5}\right) \\
Sin\,\theta & Sin\left(\theta - \tfrac{2\pi}{5}\right) & Sin\left(\theta - \tfrac{4\pi}{5}\right) & Sin\left(\theta + \tfrac{4\pi}{5}\right) & Sin\left(\theta + \tfrac{2\pi}{5}\right) \\
Cos\,\theta & Cos\left(\theta + \tfrac{4\pi}{5}\right) & Cos\left(\theta - \tfrac{2\pi}{5}\right) & Cos\left(\theta + \tfrac{2\pi}{5}\right) & Cos\left(\theta - \tfrac{4\pi}{5}\right) \\
Sin\,\theta & Sin\left(\theta + \tfrac{4\pi}{5}\right) & Sin\left(\theta - \tfrac{2\pi}{5}\right) & Sin\left(\theta + \tfrac{2\pi}{5}\right) & Cos\left(\theta - \tfrac{4\pi}{5}\right) \\
\tfrac{1}{2} & \tfrac{1}{2} & \tfrac{1}{2} & \tfrac{1}{2} & \tfrac{1}{2}
\end{bmatrix} \tag{5}$$

$$\begin{bmatrix} i_{ds} & i_{qs} & i_{xs} & i_{ys} & i_{os} \end{bmatrix}^t = T(\theta)\begin{bmatrix} i_{as} i_{bs} & i_{cs} & i_{ds} & i_{es} \end{bmatrix}^t \tag{6}$$

Clark transformation is used to disintegrate the phase a, b, c, d, e variable into two subspaces, the d-q, x-y and the zero variable components. The d-q subspaces are orthogonal to each other and provide basic torque and flux production. In healthy operation, they can be independently controlled; the x-y subspace is not coupled with the d-q subspace. In case of 1-Ph and A2-Ph open circuit faults, the d-q and x-y components are coupled with each other. The mapping of the various harmonics with the subspaces are as follows: order of the harmonics $10n \pm 1$ (where n = 1, 2, 3, 4 ...) are mapped with the q-d subspace including the fundamental component while the order of the harmonics $5n \pm 1$ (n = 1, 3, 5, 7 ...) are mapped with the x-y subspace.

The mathematical modelling equations of the machine assuming sinusoidal distributed symmetrical windings and linear flux path are represented below. Using vector space decomposition phase voltage equations of the stator winding in a stationary reference frame [1] are,

$$\left.\begin{aligned}
V_{qs} &= r_s i_{qs} + \frac{d\{L_{ls}i_{qs} + L_m(i_{qs} + i_{qr})\}}{dt} + \omega\{L_{ls}i_{ds} + L_m(i_{ds} + i_{dr})\} \\
V_{ds} &= r_s i_{ds} + \frac{d\{L_{ls}i_{ds} + L_m(i_{ds} + i_{dr})\}}{dt} - \omega\{L_{ls}i_{qs} + L_m(i_{qs} + i_{qr})\} \\
V_{xs} &= r_s i_{xs} + \frac{d(L_{ls}i_{xs})}{dt} \\
V_{ys} &= r_s i_{ys} + \frac{d(L_{ls}i_{ys})}{dt} \\
V_{os} &= r_s i_{os} + \frac{d(L_{ls}i_{os})}{dt}
\end{aligned}\right\} \tag{7}$$

Since the neutral of the five-phase winding disconnected, zero-sequence currents i_0 cannot flow and are precluded from the investigation. The x-y currents are not connected with the rotor side, leaving only circulating currents which flow in the stator winding and create stator copper loss. The production of torque is associated with the d-q subspace as in the case of three-phase. By using Equation (7), the electromagnetic torque and machine's rotor speed can be determined as follows.

$$T_e = \frac{5}{2}\frac{p}{2}\frac{L_m}{(L_{lr} + L_m)}(\lambda_{dr}i_{qs} - \lambda_{qr}i_{ds})$$

$$\omega_r = \int \frac{p}{2J}(T_e - T_L) \tag{8}$$

where,

$$\lambda_{qr} = L_{ls}i_{qr} + L_m(i_{qs} + i_{qr}) \text{ and } \lambda_{dr} = L_{ls}i_{dr} + L_m(i_{ds} + i_{dr})$$

The five-phase machine drive works on normal/healthy operation with zero x-y currents. The conventional variable-voltage, variable-frequency (V/f) controller is used to control the d-q currents rotating in the positive reference frame apart from the faulty five-phase drive system required to incorporate a controller to control the circulating x-y currents. The control scheme of V/f and the equations of the d-q plane remain same as in three-phase machines.

2.3. Four-Phase and Three-Phase Modes of Operation

In the incident of an open-gate drive circuit and switch short circuit it is compulsory to diagnose and isolate the faulty leg before the control strategy is reconstructed [24]. For the concept of fault-tolerant control technique, suppose the induction motor carries regulated balanced five-phase sinusoidal currents, which gives positive sequence rotating MMF.

$$i_a = I_m \cos(\omega_e t)$$

$$i_b = I_m \cos\left(\omega_e t - \frac{2\pi}{5}\right)$$

$$i_c = I_m \cos\left(\omega_e t - \frac{4\pi}{5}\right) \tag{9}$$

$$i_d = I_m \cos\left(\omega_e t + \frac{4\pi}{5}\right)$$

$$i_e = I_m \cos\left(\omega_e t + \frac{2\pi}{5}\right)$$

By considering stator winding sinusoidal distribution, the stator current generates rotating MMF, hence effective resultant rotating MMF is the summation of the MMFs generated by each of the five phases. Under normal healthy operation, five-phase stator currents give balanced healthy positively rotating MMF. The resultant MMF is specified by,

$$F_s = \frac{5}{2}NI_m \cos(\omega_e t - \phi) = i_a + ai_b + a^2i_c + a^3i_d + a^4i_e \tag{10}$$

where, $a = e^{j2\pi/5}$ and N is the active stator turns per phase with spatial angle denoted by Ø. For "disturbance-free" operation during 1-Ph, A2-Ph, or A3-Ph open circuit faults, the winding of themachine carries harmonic distributed currents. This current produces MMF that should be the same as that in the healthy condition. For example, if phase "a" is isolated due to an open gate drive fault or device fault or machine windings fault, a rotating positive forward field is feasible by setting ia, equal to zero, Equation (10) becomes,

$$\frac{5}{2}NI_m \cos(\omega_e t - \varnothing) = ai_b' + a^2i_c' + a^3i_d' + a^4i_e' \tag{11}$$

By separating real and imaginary terms of the Equation

$$5\frac{I_m}{2}\cos(\omega t) = \cos\left(\frac{2\pi}{5}\right)(i_b' + i_e') + \cos\left(\frac{4\pi}{5}\right)(i_c' + i_d')$$

$$5\frac{I_m}{2}\sin(\omega t) = \sin\left(\frac{2\pi}{5}\right)(i_b' - i_e') + \sin\left(\frac{4\pi}{5}\right)(i_c' - i_d') \tag{11a}$$

To find a solution by assuming that each winding has the same current magnitude, so that

$$i_b' = -i_d' \text{ and } i_c' = -i_e' \tag{11b}$$

which gives the currents in the remaining phases are

$$\left.\begin{array}{l}
i_b' = \dfrac{5i_m}{4(\sin\frac{2\pi}{5})^2}\cos\left(\omega t - \frac{\pi}{5}\right) = 1.382 I_m \cos\left(\omega t - \frac{\pi}{5}\right) \\[2mm]
i_c' = \dfrac{5i_m}{4(\sin\frac{2\pi}{5})^2}\cos\left(\omega t - \frac{4\pi}{5}\right) = 1.382 I_m \cos\left(\omega t - \frac{4\pi}{5}\right) \\[2mm]
i_d' = \dfrac{5i_m}{4(\sin\frac{2\pi}{5})^2}\cos\left(\omega t + \frac{4\pi}{5}\right) = 1.382 I_m \cos\left(\omega t + \frac{4\pi}{5}\right) \\[2mm]
i_e' = \dfrac{5i_m}{4(\sin\frac{2\pi}{5})^2}\cos\left(\omega t + \frac{\pi}{5}\right) = 1.382 I_m \cos\left(\omega t + \frac{\pi}{5}\right)
\end{array}\right\} \tag{12}$$

With the modified stator currents, d-q currents get modified which gives the electromagnetic torque,

$$T_e' = \frac{5}{2}\frac{p}{2}\frac{L_m}{(L_{lr} + L_m)}\left(\lambda_{dr}' i_{qs}' - \lambda_{qr}' i_{ds}'\right) \tag{12a}$$

To clarify this, by assuming phase "a" is isolated, the resultant rotating MMF produced by the stator winding currents will be composed of a negative-sequence component and a positive-sequence component. The remaining phase currents are expressed in such a way that there is only forward rotating MMF [13]. Hence, if any phase is open-circuited, "disturbance-free" control is possible with the modification of adjacent phases. If phase "a" is an open phase "b" advanced by 360 and phase "c" is retarded by 360. Figure 2b shows the phasor relationships before and after phase "a" is suddenly open-circuited.

However, due to the open circuit fault coupled d-q and x-y current components, the x-y currents cannot be zero. Because the d-q currents remain unchanged, it is necessary to maintain a forward rotating MMF and smooth post-fault operation (Figure 2b) [3]. By using the Clark transformation, "x" current is equal to the "d" current with the negative value: $i_x = -i_d$ and it is possible to remove the "x" component by using (12). If two adjacent phases open i.e., "a" and "b" as shown in Figure 2c,d, then the equation with real and imaginary terms of,

$$\frac{5}{2}NI_m \cos(\omega_e t - \varnothing) = a^2 i_c'' + a^3 i_d'' + a^4 i_e'' \tag{13}$$

With the assumption is that no neutral connection is required

$$i_c'' + i_d'' + i_e'' = 0 \tag{14}$$

Solving Equations (13) and (14)

$$i_e' = \frac{5i_m}{4(\sin\frac{2\pi}{5})^2}\cos\left(\omega t + \frac{\pi}{5}\right) = 1.382 I_m \cos\left(\omega t + \frac{\pi}{5}\right)$$

$$i_d'' = \frac{5I_m \cos\left(\frac{\pi}{5}\right)^2}{\left(\sin\frac{2\pi}{5}\right)^2} \cos\left(\omega t + \frac{4\pi}{5}\right) = 3.618 I_m \cos\left(\omega t + \frac{4\pi}{5}\right)$$

$$i_e'' = \frac{5I_m \cos\left(\frac{\pi}{5}\right)}{2\left(\sin\frac{2\pi}{5}\right)^2} \cos\left(\omega t\right) = 2.236 I_m \cos\left(\omega t\right) \tag{15}$$

Similarly with modified stator currents under two adjacent open-phase fault, d-q currents get modified which gives the electromagnetic torque,

$$T_e'' = \frac{5}{2}\frac{p}{2}\frac{L_m}{(L_{lr} + L_m)}\left(\lambda_{dr}'' i_{qs}'' - \lambda_{qr}'' i_{ds}''\right) \tag{15a}$$

If three phases are open circuited i.e., "a", "b" and "c", for remedial strategy and disturbance-free operation the motor neutral must be connected to the dc mid-point so that remaining two phase currents can be individually controlled.

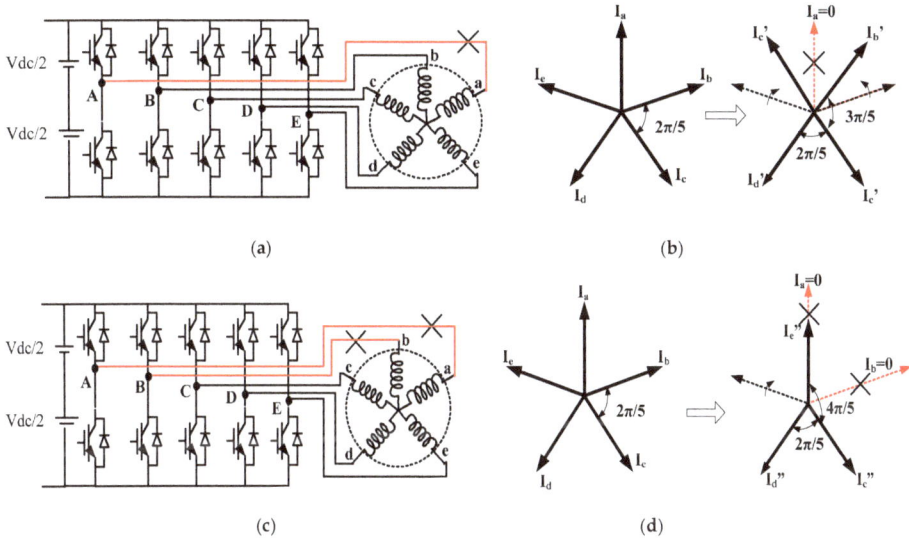

(a)

(b)

(c)

(d)

Figure 2. (**a**) five-phase system with 1-Ph circuit fault; (**b**) vector reconfiguration during 1-Ph open fault; (**c**) five-phase system with A2-Ph open circuit fault; (**d**) vector reconfiguration during A2-Ph open circuit fault.

3. Description of Fault-Tolerant Control Strategy

The general block diagram of a volt/Hz controlled power circuit of a five-phase induction motor drive is shown in Figure 3, which is in the fault mode condition. In most drives, conventional scalar control is used. Hence, the conventional scalar control method for fault-tolerant control strategy is represented. In conventional scalar control, reference signals are given to the pulse-width modulator which operates based on the common speed reference. Generation of these reference signals is as follows. V/f block is multiplied by the reference frequency (ω).

Generally, this is a fixed value depending on the rating of frequency and voltage of the machine. Five sinusoidal reference signals, vref a, vref b, vref c, vref d and vref e are the outputs of this block. These reference signals have an appropriate magnitude and operating frequency with a phase shift of 72° (Refer Figure 2b). These reference signals are fed to the modulator, which depends on the operating speed of the machine. In a modulator using a comparator, the reference signal and a saw-tooth signal are compared. The frequency of the saw-tooth signal is equal to the required switching frequency. The reference signal changes either at the healthy condition or at a faulty four-phase or three-phase modes of operation with the reference operating speed.

The cause of a negative-sequence component in the distorted stator current is due to the disconnection of a faulty phase in the motor winding. Accordingly, the negative-sequence current appears in the x-y component, which gives MMF in the negative sequence reference frame [13]. The proposed control strategy will try to reduce this x-y component to zero or a minimum. The output signal of the control strategy is the modified reference signal (refer Equations (12) and (13)), which is generated by using a standard constant V/f (scalar) control system. The resultant modified signal is then given to the modulator to get the proper switching pattern. In the healthy condition the modified reference signal making the negative-sequence component zero results in balanced five-phase line currents which are equal in peak values with a phase shift of 72°. If a 1-Ph open fault occurs, the remaining four active phase currents are rearranged by the controller so that they are equal in peak and have a phase shift of 72°, 108°, 72° and 108°, which cause only clockwise (i.e., positive-sequence) rotating MMF (refer to Equation (12)).

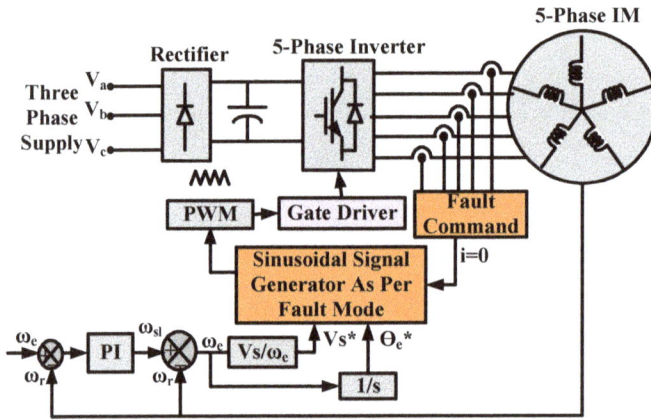

Figure 3. Block diagram of fault-tolerant control strategy.

This can be described as a virtual four-phase connected winding, with phase "a" isolated (refer to Figure 2a). The equivalent active four-phase currents of phase-b, c, d, e in the stator winding are equal in peak value (magnitude) with each other and a phase shift equal to 72°, 108°, 72° and 108° between them (refer to Figure 2b). Hence by keeping the position of vector "c" and "d" as it is and moving phase "b" vector in advance by 360 and phase "c" is retarded by 360 modified the switching pattern. In a similar manner with two-phase "a" and "b" isolated, the other two active currents of phases "c" and "e" in the stator winding are the same in magnitude. The magnitude of phase "d" current is 1.62 time of magnitude of other two currents. By keeping the position of vector "c" and "d" as it is and moving phase "e" vector at the position of "a" modified the switching pattern for two-phase open. (Refer to Figure 2c,d).

In a practical case, if a gate driver open circuit fault or switched short-circuit fault is occurring then it is recommended to disconnect the power lines of the inverter. Isolation switches are inserted in between inverter power lines and motor for laboratory experimentation. The current sensor is used to measure the line current. If the summation of all measured current is zero, then the induction motor drive is working in healthy condition. The opening of particular phase can be done using isolation switch for single phase or two phase open fault. Hence, phase current in specific phase becomes zero. Now summation of other phase currents are no longer being zero identifies the faulty condition. Also, it can be seen that due to an open-circuit fault, torque control is lost, i.e., torque is oscillating which oscillates the speed. Hence in V/f control, the speed control loop gets weak and oscillation in torque.

By checking each current with zero detects particular phase open fault. Proposed control logic provides the switching pattern to the adjacent phases of the specific open phase. Similarly, by checking adjacent current with zero gives an idea about adjacent phase open fault. For this, the proposed control logic is used to get the switching pattern for remaining three phases of the inverter. Accordingly, the proposed new fault-tolerant control strategy decide required switching pattern for four-phase and three-phase modes of operation in V/f control technique. The complete control logic is shown in the flowchart of Figure 4.

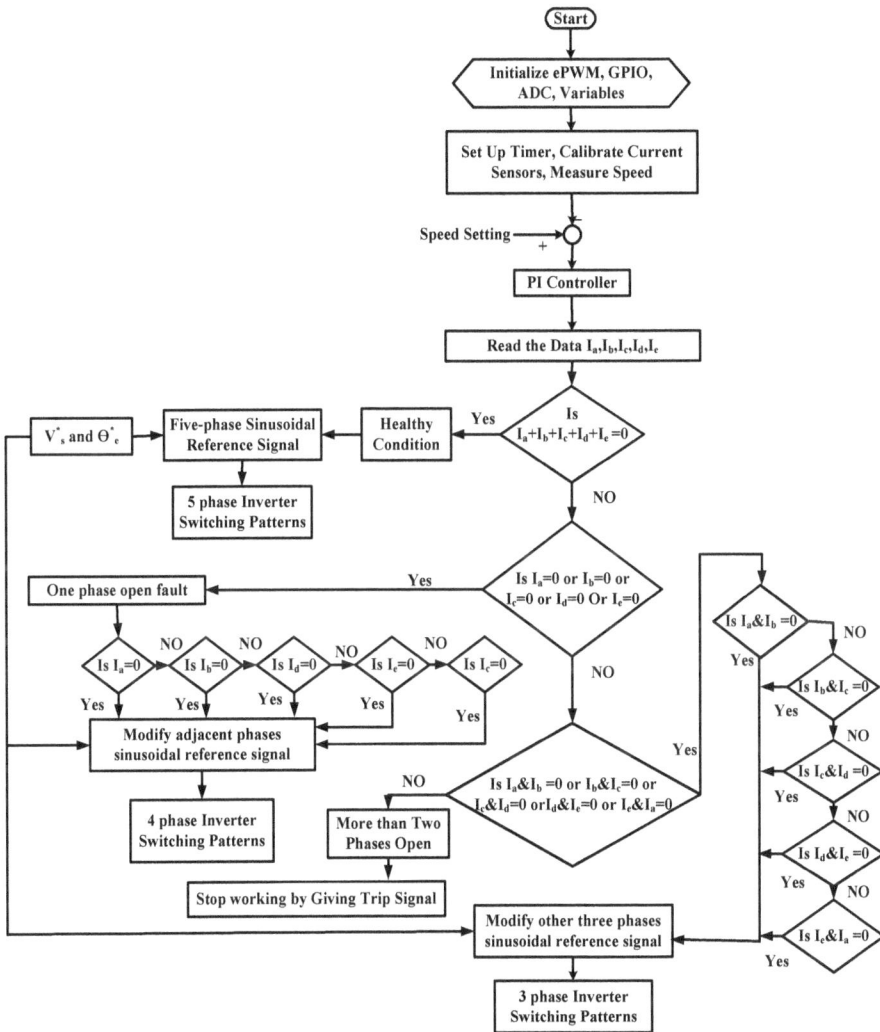

Figure 4. Flow-chart of fault-tolerant control strategy.

4. Experimental Results

4.1. Test Experimental Setup

To investigate the performance of the proposed fault-tolerant control technique, the drive comprises a 1 Hp five-phaseinduction squirrel cage symmetrically distributed induction motor. It has rs = 0.499 Ω, Lls = 2.7 mH, rr = 0.926 Ω, Llr =2.7 mH, Lm = 223mH, P = 2 and rotor inertia (J) = 0.047 kg-m^2.

The motor was designed so that it can be configured either as a star, pentagon or pentacle-connected stator [19]. The motor is composed of 40 stator slots with closed rotor bars. The five-phase induction machine is provided by an IGBT-based two-level five-phase inverter (Fairchild, Sunnyvale, CA, USA). The machine is driven in normal/healthy operation with modified reference signal controlled by a variable-voltage, variable-frequency (V/f) controller Vdc = 300 V (refer to Figure 5).

Figure 5. Experimental setup.

The control circuit of the drive is performed by dsPIC33EP256MU810 Digital Signal Controller, of MICROCHIP (MICROCHIP, Chandler, AZ, USA). This controller has 83 I/O pins, 12 PWM outputs, 2 ADC modules with 32 channels which are useful for motor-control applications. Code Composer Studio software (Version 7, Texas Instruments, Dallas, TX, USA) is used for programming of control unit. It is capable of simultaneously controlling two two-level three-phase inverters. A two-level, five-phase inverter requires only 10 gate signals; hence, PWM output signals can be directly given to the gate driver circuit of the five-phase inverter (Micrel, San Jose, CA, USA). Five hall-effect current sensors (LEM, Geneva, Switzerland) are used to measure the line current. The main processor has the fault-tolerant control technique. The PWM switching frequency was set to 10 kHz.

4.2. Experimental Results

The fault-tolerant strategy described in Section 3 is experimentally performed in the laboratory and results are shown in Figures 5–7. The load condition was at one fourth, i.e., 2 Nm, 50 Hz. This is because while doing experimentation the motor will be under loaded during the fault conditions, i.e., four-phase and three-phase modes of operation.

The motor-drive performance with line current, torque and circulating d-q current is indicated in these figures. For the healthy and faulty conditions, Figures 6–8 with a,b and d show the results of 1-Ph and A2-Ph open circuit faults without a control strategy. Figures 6–8 with c, e show when a control strategy was introduced. The experimental test results of line current for the 1-Ph open circuit fault are illustrated in Figure 6b,c) without and with a control strategy, respectively. Similarly, the line current waveform for the A2-Ph open circuit fault is represented in Figure 6d,e without and with a control strategy, respectively.

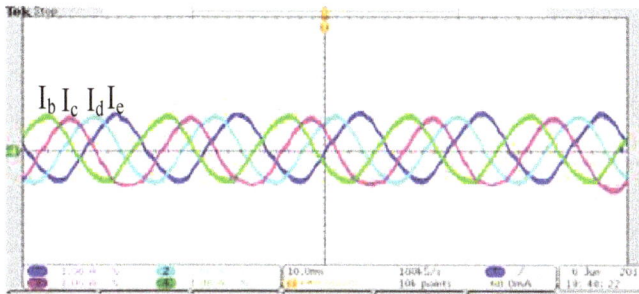

(**a**) Line current in amp (Scale: 1 A/div, 10 ms/div)

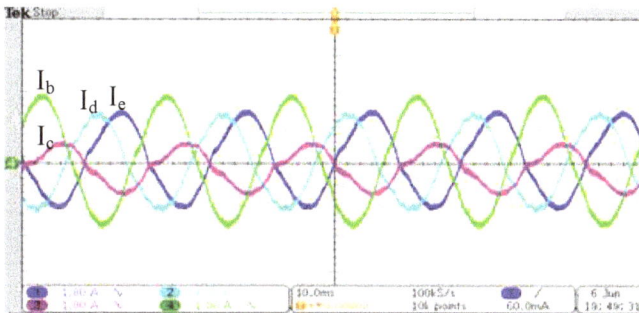

(**b**) Line current in amp (Scale: 1 A/div, 10 ms/div)

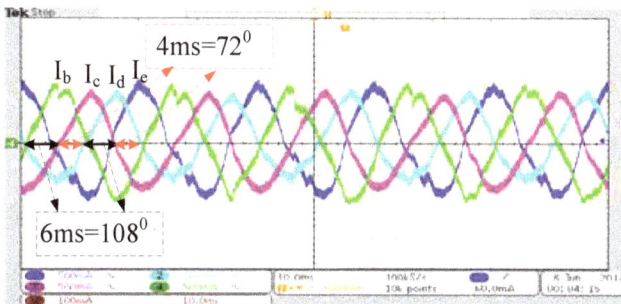

(**c**) Line current in amp (Scale: 500 mA/div, 10 ms/div)

Figure 6. *Cont.*

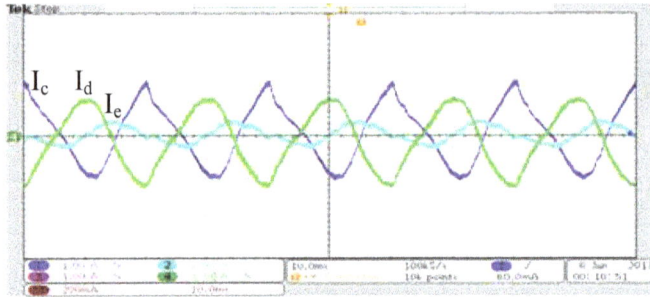

(**d**) Line current in amp (Scale: 1 A/div, 10 ms/div)

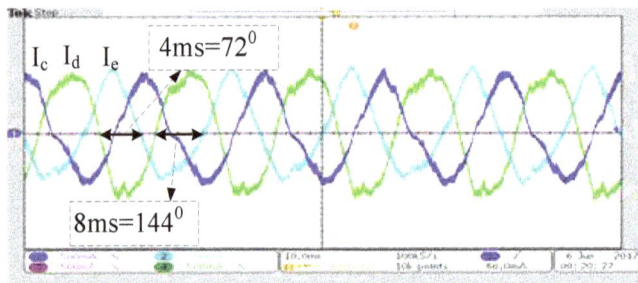

(**e**) Line current in amp (Scale: 500 mA/div, 10 ms/div)

Figure 6. Line current at (**a**) Healthy operation; (**b**) 1-Ph open circuit fault without control strategy; (**c**) 1-Ph open circuit fault with control strategy; (**d**) A2-Ph open circuit fault without control strategy; (**e**) A2-Ph open circuit fault with control strategy.

Unequal increase in the peak value of the line currents at any instant is due to one of the phases being disconnected. This increases the negative sequence component of the current. This control strategy maintained the equal magnitude of the active current and phase displacement which ensured that the torque pulsation was reduced (As shown in Figure 7c,e). Additionally, the current reduces the negative sequence component. The capability of this control strategy is verified by analyzing the unbalanced line current with balanced line current for both the cases of open phase fault (as shown in Figure 6). The line current obtained from the experiment for the four-phase and three-phase modes of operation with a deactivated control strategy is represented in Figure 6b,d. For the case of the activated control strategy, the line current waveforms are represented in Figure 6c,e. The effectiveness of this controller shows in the reduced magnitude of the line current with properly balanced current, which remarkably enhances the quality of the output torque of the five-phase machine.

Induction motor torque profile during the transition of a five-phase healthy mode to the four-phase and three-phase faulty modes of operation is represented in Figure 7. These figures show the effectiveness of this fault-tolerant method and corresponding quality of the fault control strategy, which maintains the quality of the motor's torque under faulty conditions. The motor output torque waveforms with control strategy deactivated for 1-Ph and A2-Ph faults is shown in Figure 7b,d. The motor output torque waveforms with control strategy activated is shown in Figure 7c,e. The torque pulsation is of approximately 3 N-m, while the developed torque is 2 N-m when this control strategy was not used at the steady-state condition. The torque pulsation decreased to less than 2.5 N-m when the control strategy was introduced, as shown in Figure 7c,e. Since for smooth post-fault operation, the MMF remains unchanged, the d-q currents describe nearly a circle as in healthy operation has shown Figure 8a. In contrast, currents cannot be circular, as shown in Figure 8b,c. The use of the control strategy current makes a near circular current as in the case of Figure 8d.

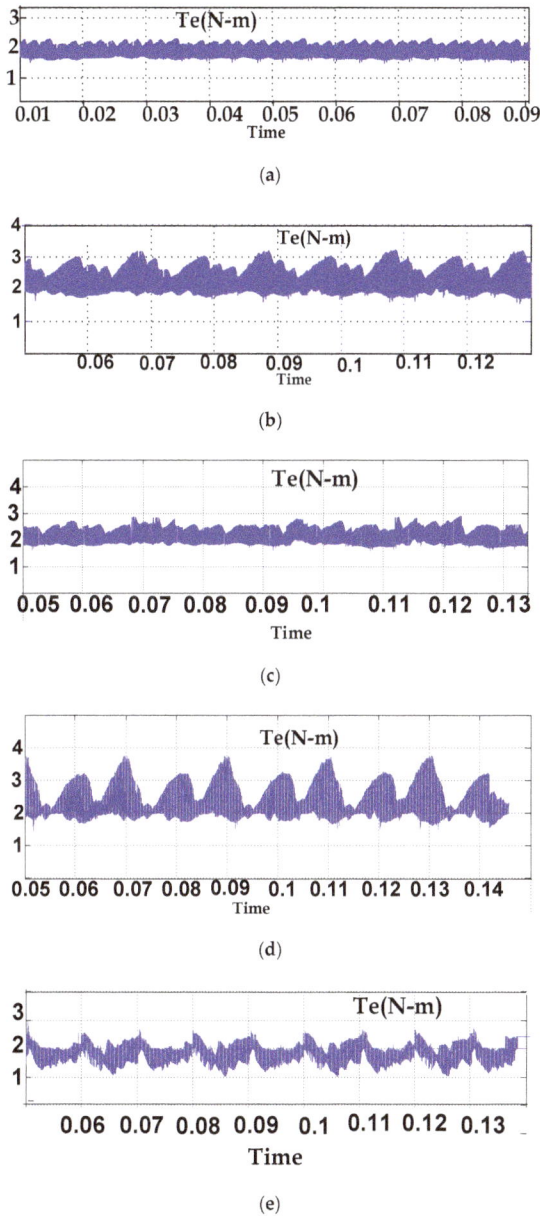

Figure 7. Torque (N-m) under (**a**) Healthy operation; (**b**) 1-Ph open circuit fault without control strategy; (**c**) 1-Ph open circuit fault with control strategy; (**d**) A2-Ph open circuit fault without control strategy; (**e**) A2-Ph open circuit fault with control strategy.

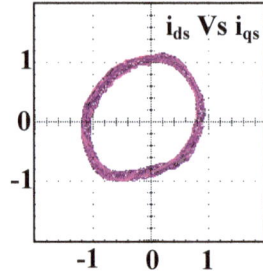

(a)

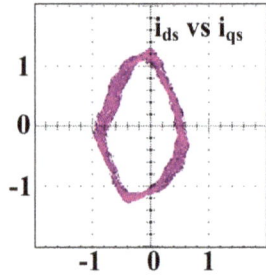

(b)

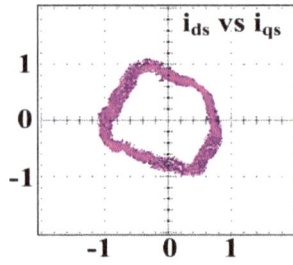

(c)

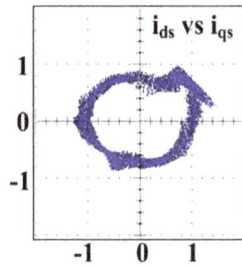

(d)

Figure 8. *Cont.*

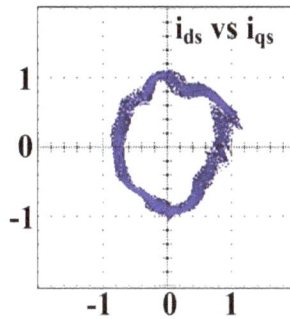

(e)

Figure 8. d-q current in amp (both the scale is in amperes) under (**a**) Healthy operation; (**b**) 1-Ph open circuit fault without control strategy; (**c**) A2-Ph open circuit fault without control strategy; (**d**) 1-Ph open circuit fault with control strategy; (**e**) A2-Ph open circuit fault with control strategy.

5. Conclusions

The theoretical and conceptual background of a new control technique validated with experimental results is presented here. The control technique enables the four-phase and three-phase operation modes of a star-connected induction motor. The experimental results show that a five-phase induction motor drive supplied by a faulty five-phase voltage source inverter can be successfully operated in the four-phase operation mode when 1-Ph open circuit fault occurs, and can be operated in the three-phase operation mode when an A2-Ph open circuit fault occurs. Smoothly controlling the speed of the machine using V/f control, improves the reliability control. The machine performance in the four-phase and three-phase operations was thoroughly analyzed. It shows that the current in the remaining active phases is independently controlled and can produce a positive-sequence rotating MMF component. The proposed controller is able to remarkably reduce the torque pulsations. The available torque in the four-phase and three-phase operation modes is considerably smaller than that of the five-phase drive in the healthy condition. This control strategy is suitable for a drive in steady state operation (or slow acceleration/deceleration, such as transportation drives) with minimum losses in the stator winding. Furthermore, the torque capacity can be enhanced by over-designing the power circuit and control strategy.

Author Contributions: S.C.R. and H.M.S. developed the concept; H.M.S. and S.C.R. performed the experiment; S.C.R. and H.M.S. wrote the paper; M.R. analyzed the data. These authors contributedequally to this work.

Funding: This research received no external funding.

Acknowledgments: Authors acknowledges the VNIT, Nagpur for providing infrastructure support and RCOEM, Nagpur for all institutional facilities.

Conflicts of Interest: The authors declare no conflict of interest.

References

1. Levi, E.; Bojoi, R.; Profumo, F.; Toliyat, H.A.; Williamson, S. Multiphase induction motor drives—A technology status review. *IET Electr. Power Appl.* **2007**, *1*, 489–516. [CrossRef]
2. Levi, E.; Barrero, F.; Duran, M.J. Multiphase machines and drives-revisited. *IEEE Trans. Ind. Electron.* **2016**, *63*, 429–432. [CrossRef]
3. Barrero, F.; Duran, M.J. Recent advances in the design, modeling, and control of multiphase machines—Part I. *IEEE Trans. Ind. Electron.* **2016**, *63*, 449–458. [CrossRef]

4.   Duran, M.J.; Barrero, F. Recent advances in the design, modeling, and control of multiphase machines—Part II. *IEEE Trans. Ind. Electron.* **2016**, *63*, 459–468. [CrossRef]

5.   Levi, E. Advances in converter control and innovative exploitation of additional degrees of freedom for multiphase machines. *IEEE Trans. Ind. Electron.* **2016**, *63*, 433–448. [CrossRef]

6.   Liu, T.H.; Fu, J.R.; Lipo, T.A. A strategy for improving reliability of field-oriented controlled induction motor drives. *IEEE Trans. Ind. Appl.* **1993**, *29*, 910–918.

7.   Wallmark, O.; Harnefors, L.; Carlson, O. Control algorithms for a fault-tolerant PMSM drive. *IEEE Trans. Ind. Electron.* **2007**, *54*, 1973–1980. [CrossRef]

8.   Zhao, W.; Cheng, M.; Hua, W.; Jia, H.; Cao, R. Back-EMF harmonic analysis and fault-tolerant control of flux-switching permanent-magnet machine with redundancy. *IEEE Trans. Ind. Electron.* **2011**, *58*, 1926–1935. [CrossRef]

9.   De Lillo, L.; Empringham, L.; Wheeler, P.W.; Khwan-On, S.; Gerada, C.; Othman, M.N.; Huang, X. Multiphase power converter drive for fault-tolerant machine development in aerospace applications. *IEEE Trans. Ind. Electron.* **2010**, *57*, 575–583. [CrossRef]

10. Bianchi, N.; Bolognani, S.; Zigliotto, M.; Zordan, M.A.Z.M. Innovative remedial strategies for inverter faults in IPM synchronous motor drives. *IEEE Trans. Energy Convers.* **2003**, *18*, 306–314. [CrossRef]

11. Errabelli, R.R.; Mutschler, P. Fault-tolerant voltage source inverter for permanent magnet drives. *IEEE Trans. Power Electron.* **2012**, *27*, 500–508. [CrossRef]

12. Aghili, F. Fault-tolerant torque control of BLDC motors. *IEEE Trans. Power Electron.* **2011**, *26*, 355–363. [CrossRef]

13. Sayed-Ahmed, A.; Mirafzal, B.; Demerdash, N.A. Fault-tolerant technique for Δ-connected AC-motor drives. *IEEE Trans. Energy Convers.* **2011**, *26*, 646–653. [CrossRef]

14. Mendes, A.M.; Cardoso, A.M. Fault-tolerant operating strategies applied to three-phase induction-motor drives. *IEEE Trans. Ind. Electron.* **2006**, *53*, 1807–1817. [CrossRef]

15. Abdel-Khalik, A.S.; Ahmed, S.; Elserougi, A.A.; Massoud, A.M. Effect of stator winding connection of five-phase induction machines on torque ripples under open line condition. *IEEE/ASME Trans. Mechatron.* **2015**, *20*, 580–593. [CrossRef]

16. Yepes, A.G.; Riveros, J.A.; Doval-Gandoy, J.; Barrero, F.; López, O.; Bogado, B.; Jones, M.; Levi, E. Parameter identification of multiphase induction machines with distributed windings—Part 1: Sinusoidal excitation methods. *IEEE Trans. Energy Convers.* **2012**, *27*, 1056–1066. [CrossRef]

17. Mecrow, B.C.; Jack, A.G.; Haylock, J.A.; Coles, J. Fault-tolerant permanent magnet machine drives. *IEE Proc.-Electr. Power. Appl.* **1996**, *143*, 437–442. [CrossRef]

18. Parsa, L. On advantages of multi-phase machines. In Proceedings of the 31st Annual Conference of IEEE Industrial Electronics Society, Raleigh, NC, USA, 6–10 November 2005; pp. 1574–1579.

19. Parsa, L.; Toliyat, H.A. Fault-tolerant interior-permanent-magnet machines for hybrid electric vehicle applications. *IEEE Trans. Veh. Technol.* **2007**, *56*, 1546–1552. [CrossRef]

20. Mohammadpour, A.; Sadeghi, S.; Parsa, L. A generalized fault-tolerant control strategy for five-phase PM motor drives considering star, pentagon, and pentacle connections of stator windings. *IEEE Trans. Ind. Electron.* **2014**, *61*, 63–75. [CrossRef]

21. Jasim, O.; Sumner, M.; Gerada, C.; Arellano-Padilla, J. Development of a new fault-tolerant induction motor control strategy using an enhanced equivalent circuit model. *IET Electr. Power Appl.* **2011**, *5*, 618–627. [CrossRef]

22. Guzman, H.; Duran, M.J.; Barrero, F.; Bogado, B.; Sergio, L.; Marín, T. Speed control of five-phase induction motors with integrated open-phase fault operation using model-based predictive current control techniques. *IEEE Trans. Ind. Electron.* **2014**, *61*, 4474–4484. [CrossRef]

23. Dwari, S.; Parsa, L. Fault-tolerant control of five-phase permanent-magnet motors with trapezoidal back EMF. *IEEE Trans. Ind. Electron.* **2011**, *58*, 476–485. [CrossRef]

24. Kastha, D.; Bose, B.K. Fault mode single-phase operation of a variable frequency induction motor drive and improvement of pulsating torque characteristics. *IEEE Trans. Ind. Electron.* **1994**, *41*, 426–433. [CrossRef]

electronics

MDPI

Review

Review on Health Management System for Lithium-Ion Batteries of Electric Vehicles

Zachary Bosire Omariba [1,2], Lijun Zhang [1,*] and Dongbai Sun [1]

[1]   National Center for Materials Service Safety, University of Science and Technology Beijing,
    Beijing 100083, China; zomariba@egerton.ac.ke (Z.B.O.); dbsun@ustb.edu.cn (D.S.)
[2]   Computer Science Department, Egerton University, Egerton 20115, Kenya
*   Correspondence: ljzhang@ustb.edu.cn; Tel.: +86-10-6232-1017

Received: 15 April 2018; Accepted: 10 May 2018; Published: 15 May 2018

Abstract: The battery is the most ideal power source of the twenty-first century, and has a bright future in many applications, such as portable consumer electronics, electric vehicles (EVs), military and aerospace systems, and power storage for renewable energy sources, because of its many advantages that make it the most promising technology. EVs are viewed as one of the novel solutions to land transport systems, as they reduce overdependence on fossil energy. With the current growth of EVs, it calls for innovative ways of supplementing EVs power, as overdependence on electric power may add to expensive loads on the power grid. However lithium-ion batteries (LIBs) for EVs have high capacity, and large serial/parallel numbers, when coupled with problems like safety, durability, uniformity, and cost imposes limitations on the wide application of lithium-ion batteries in EVs. These LIBs face a major challenge of battery life, which research has shown can be extended by cell balancing. The common areas under which these batteries operate with safety and reliability require the effective control and management of battery health systems. A great deal of research is being carried out to see that this technology does not lead to failure in the applications, as its failure may lead to catastrophes or lessen performance. This paper, through an analytical review of the literature, gives a brief introduction to battery management system (BMS), opportunities, and challenges, and provides a future research agenda on battery health management. With issues raised in this review paper, further exploration is essential.

Keywords: lithium-ion batteries; electric vehicles; battery management system; electric power

1. Introduction

Lithium-ion batteries (LIBs) are one of the most promising technologies due to advantages like high efficiency, lower volume, small weight, temperature sensitivity, and maintenance [1–3]. They are the most ideal power source of the twenty-first century and have a bright future in many applications, such as portable electronic devices, electric vehicles (EVs) [4], aerospace systems, and power storage for renewable energy sources, like solar and wind turbines. However, there are many shortfalls, such as lack of safety, fragility, and aging, which may restrict the extensive use of LIBs. The consequences of battery failure can lead to catastrophes and inconveniences, which have turned to be popular and challenging issues [5], as reliability of LIBs is yet to be improved. However, determining the remaining useful life (RUL) of LIBs can aid to some level in curbing this problem [6,7].

There are many different techniques proposed in the literature that capture these crucial parameters to determine the battery state, to ensure that the battery delivers its specified output while optimizing the charge/discharge processes, and must be communicated to on-board systems. The battery management system (BMS) plays a significant role in the prediction of RUL for LIBs, as it acts as a connector between the battery and the EVs. The main goal of the BMS is three-fold: to protect the battery system from damage by detecting malfunctions, such as overcharge, excessive rise in temperature, and electric leak; to predict and increase the battery life; and to maintain the battery system in an accurate and reliable operational condition [8,9]. The BMS is a combination of sensors, controllers, communication, and computation hardware, with software algorithms designed to decide the maximum charge/discharge current and duration from the estimation of state-of-charge (SOC) and state-of-health (SOH) of the battery pack [10]. From this definition BMS performs the two main roles of monitoring the battery to determine information, such as SOH, SOC, and RUL, as well as to operate the battery in a safe, efficient, and non-damaging way [11–13].

In many industrial applications that make use of LIBs as one of the main power sources, the BMS has proven useful. The BMS contains a set of activities that monitor and perform SOH, SOC, state-of-life (SOL), end-of-life (EOL), and state-of-power (SOP) estimation throughout the battery's entire life, and make a suitable decision to predict RUL. Thus, the BMS for LIBs is a decision process to intelligently perform maintenance, logistics, and system configuration activities on the basis of diagnostic and/or prognostics with the aim of producing actionable information to enable timely decisions [14] on maintenance optimization support, and reduce the costs of maintenance [7]. The BMS, therefore, implements state monitoring and evaluation, charge control, and cell balancing functionalities in order to maintain the safety and reliability of batteries [15]. Failure to perform these functionalities can result in battery failure, which can lead to reduced performance, operational impairment, and even catastrophic failure [16], making the performance of accurate prediction of RUL essential. RUL has attracted a great deal of interest from researchers and funding agencies around the world to mitigate the challenges associated with LIB use, in many high-impact applications, while protecting the environment.

There is a growing increase of EVs according to Bruen et al. [17], dependent on LIBs due to their numerous advantages, as compared to other batteries. This is further accelerated with the climate change concerns having a focus on a spotlight to EVs, and LIBs are believed to be the future to widespread EV adoption. However EVs are also faced with a number of drawbacks, as illustrated in Table 1, although technology is advancing fast to curb these challenges.

Table 1. Advantages and disadvantages of EVs from selected review papers.

Advantages		Disadvantages	
-	Highly efficient	-	Electricity storage is still expensive
-	Reduced emissions	-	Battery charging is time consuming
-	High performance and low maintenance	-	Primary resource depletion for some elements of the LIB
-	Very responsive and have very good torque	-	Range anxiety
-	EV motors are quiet and smooth	-	Battery degradation costs
-	Are more digitally connected than conventional vehicles	-	Sufficient public charging infrastructure is still lacking
-	Simplified powertrain	-	Causes indirect pollution
-	Low electricity consumption	-	Lacks the power to accelerate and climb quickly
-	Good acceleration	-	Are heavy due to overloaded batteries
-	Can be charged overnight on low cost electricity produces by any type of power station, including renewables			

(Source: [11,18–28]).

The BMS contains a portion responsible to monitor and control the SOH of a battery pack, and it is also referred to as the battery health management system (BHMS). However, according to Saha et al. [29], the BMS is a hardware designed to be a low-cost analog-to-digital data acquisition system. This hardware has three components: the signal conditioning board, the data acquisition board, and the embedded processor board. However, the BMS' main function is to monitor, control, and report the SOH of a battery. This review work will be a comprehensive collation of existing prognostic methods, and will provide convenience and inspiration for scholars to study and conduct further research. This review paper is organized into five sections. Section 2 talks about the BMS. Section 3 is about opportunities and challenges with respect to battery health and prognostics. Finally, Section 4 is about the critical future battery prognostic research work, and the conclusion are provided in Section 5.

2. Health Management Systems for Batteries

There is continuous increase in EV stock, but annual growth rates have been reducing consistently since 2011. In 2016 the EV stock growth was 59%, down from 76% in 2015, and 84% in 2014, but statistics shows that battery electric vehicles (BEVs) still account for the majority of the electric car stock, at 60%, as per the "Global EV Outlook 2017: Two Million and Counting" report [30]. According to this report the number of EVs increased from the previous report of a projection of one million EVs in 2016, to two million, projected in 2017. This trend shows that the number of EVs has been doubling over the years and this will put more pressure in the demand for LIBs, which has proved to be the main source of efficient power.

The current trend clearly demonstrates that with proper a BMS the number of BEVs will continue to rise, and LIBs are the main source of energy. This is because there are many aspects of the reliability process, such as requirement analysis, modelling and simulation, control strategy research, and online hardware testing of developing a BMS which requires a model to identify the characteristics of LIBs [31]. In recent years, a tremendous growth in sales of battery electric cars has been experienced and this puts more pressure on the battery technology. Table 2 shows the battery electric car stock by country, 2005–2016.

Table 2. Battery electric car stock by country, 2005–2016 (thousands).

	2005	2006	2007	2008	2009	2010	2011	2012	2013	2014	2015	2016
Canada							0.22	0.84	2.48	5.31	9.69	14.91
China					0.48	1.57	6.32	15.96	30.57	79.48	226.19	483.19
France	0.01	0.01	0.01	0.01	0.12	0.30	2.93	8.60	17.38	27.94	45.21	66.97
Germany	0.02	0.02	0.02	0.09	0.10	0.25	1.65	3.86	9.18	17.52	29.60	40.92
India				0.37	0.53	0.88	1.33	2.76	2.95	3.35	4.35	4.80
Japan					1.08	3.52	16.13	29.60	44.35	60.46	70.93	86.39
Korea						0.06	0.34	0.85	1.45	2.76	5.67	10.77
Netherlands				0.01	0.15	0.27	1.12	1.91	4.16	6.83	9.37	13.11
Norway			0.01	0.26	0.40	3.35	5.38	9.55	19.68	41.80	72.04	98.88
Sweden						0.18	0.45	0.88	2.12	5.08	8.03
United Kingdom	0.22	0.55	1.00	1.22	1.40	1.65	2.87	4.57	7.25	14.06	20.95	31.46
United States	1.12	1.12	1.12	2.58	2.58	3.77	13.52	28.17	75.86	139.28	210.33	297.06
Others					0.64	0.80	3.17	5.83	10.60	19.43	36.20	52.41
Total	**1.37**	**1.70**	**2.16**	**4.54**	**7.48**	**16.42**	**55.16**	**112.95**	**226.79**	**420.34**	**745.61**	**1208.90**

(Source: Global EV Outlook 2017 report: two million and counting [30]).

Due to the promising growth in sales of battery-powered cars, there is an increased research interest towards the BMS of LIBs. This is attributed to the need of models and technologies for accurate estimation of a battery's RUL for different high-impact applications, including mobility applications in EVs [32]. Additionally, LIBs are the most promising power source for EVs due to their numerous benefits, like being lightweight, their high energy density, and relatively low self-discharge compared to nickel-cadmium (NI-cad) and NiMH batteries [33,34]. Battery health management and RUL estimation are performed in order to ascertain the current and previous battery states and to predict the future state and RUL.

The BMS is a hardware and software system that is in charge of battery protection and SOH, SOC, and state-of-function (SOF) estimation [6,35]. The key performance parameters tradeoffs, like safety, life span, performance, charging time, and cost, are managed by the BMS. Among those, an accurate quantification of the battery state is one of the most critical tasks for the BMS, along with the task of supervising lithium-ion cells when they are used in large battery packs [10]. In LIB systems, the BMS is used to maintain safety specifications of the battery system by ensuring that each cell is equally charged and voltage balance exists in the battery pack [36]. This means that a reliable BMS is crucial; otherwise the LIBs can move into the danger zone below the threshold area, which can be catastrophic, or lead to reduced performance.

2.1. Battery Terminologies

RUL: RUL is the remaining time or number of load cycles that the battery has during which it will be able to meet its operating requirements [6,7,37], or it is simply the length of time from the present time to the end of life [38]. RUL has attracted major emphasis in research and manufacturing vehicles' BMS so as to meet the requirement of reduced costs, increased accuracy and reliability, and avoidance of catastrophic failure. RUL can be computed as:

$$RUL = T_f - T_c \tag{1}$$

where T_f is a random variable of time of failure when degradation is detected, and T_C is the current time when the predicted signal passes the failure time with some confidence to show uncertainty of the prediction [14]. The different sources of uncertainty, however, must be propagated together with the confidence of prediction and RUL estimation, since inherent uncertainties of the degradation process, measurements, environmental/operational conditions, and modeling errors exist.

SOC: SOC represents the available capacity and is one of the most important states that needs to be monitored to optimize the performance and extend the lifetime of the batteries [6]. The battery SOC is an expression of the present battery capacity as a percentage of the maximum capacity. SOC is estimated according to such conditions as working current, temperature, and voltage [39]. The SOC is generally calculated using the current integration to determine the change in battery capacity over time. If we consider a completely discharged battery, with $I_b(\tau)$ as the charging current, the charge delivered to the battery is $\int_{t_0}^{t} I_b(\tau) d\tau$. The SOC of the battery is simply expressed as:

$$SOC(t) = \frac{\int_{t_0}^{t} I_b(\tau) d\tau}{Q_0} \times 100\% \tag{2}$$

as the charging current, the charge where Q_0 is the battery capacity at time t. According to Saxena et al. [40] estimation of SOC is, by far, the most popular approach where charge counting or current integration is used in different ways to estimate battery capacity. This makes SOC estimation the most important approach in battery management since it represents the available battery capacity, which enables performance optimization and extension of battery lifetime [6]. BMS prevents the battery from discharging below a certain SOC and charging when it is full [41]. From specifications of EV batteries, as shown in Table 3, the safety range for charging and discharging is about $-20\,°C$ to $60\,°C$.

DOD: The depth-of-discharge (DOD) is the percentage (%) of the battery capacity that has been discharged, expressed as a percentage of maximum capacity. A discharge of at least 80% DOD [42], is referred to as deep discharge, and many studies assume a fixed cycle lifetime. This is a strong simplification of reality as a traction battery will not be fully discharged every single time until the allowed minimum SOC of 20%. The battery DOD is given by the equation:

$$\text{DOD}(t) = 1 - \text{SOC}(t) = \frac{Q_0 - \int_0^t I_b(\tau)d\tau}{Q_0} \times 100\% \tag{3}$$

It is common that when the DOD is higher the shorter the cycle life. To achieve this higher cycle life, a larger battery can be used for a lower DOD during normal operations [10].

SOH: The SOH is a function of depth-of-charge, and is defined as the ratio of the maximum charge capacity of an aged battery to the maximum charge capacity when the battery is new [10]. The indicator that the battery capability to store energy is deteriorating, and decreases over the battery lifetime, is the measured SOH [6]. The BMS of EVs ensures that the battery cells charge within the safety ranges. SOH is tracked by measuring the internal resistance, since the internal resistance increases as the capacity increases. The SOH is computed as:

$$\text{SOH} = \frac{Q_{act}}{Q_R} \times 100\% \tag{4}$$

where Q_R is the rated capacity and Q_{act} is the actual battery capacity.

EOL: Prognostics are focused on predicting when a fault, damage, or wear of a component, subsystem, or system will progress to a point that is deemed unsafe, or which a system will not function as specified [43,44]. This time point is called EOL, and the time remaining until that point is reached is known as RUL [45]. At the EOL the system value of performance is deemed to be unreliable, or can lead to failure of the EVs. When the battery usage cycle reaches the EOL, the prediction accuracy increases, and prediction variance gradually decreases [32].

Table 3. Specifications of EVs LIB.

Energy density (W/Kg)	72 to 200 (chargeable electric energy per weight of battery pack)
Nominal voltage	3.7 V
Power density	1800 (proportion of dischargeable electric energy to charged energy)
Overcharge tolerance	Very low
Cycle life	500 to 1000 (Number of charge/discharge cycles in battery's entire life)
Operating rate of temperature	−20 °C to 60 °C
Energy efficiency	85 to 98%
Energy cost	500–2500 $/kWh
Lifetime	5 to 15 years
Limitation	High energy cost/safety

(Source: [46,47]).

EOD: End-of-discharge (EOD) is the reading of the lower battery capacity that is occasioned by the energy loss that occurs inside the battery, and a drop in the voltage that causes the battery to reach the low-end voltage cut-off sooner. EOD means the battery is empty under discharge, and that the SOC should be 0%. It can be an absolute voltage level or a variable which is compensated by loading. The prediction of EOD times for a battery has been investigated recently, to predict the time when a predefined cut-off threshold voltage is reached and the power source is no longer available [40], thus indicating that the battery pack has run out of charge [29].

2.2. Architecture of the BMS

The architecture of the BMS comprises both the hardware and software parts. This system is built to control the operational conditions of the battery to prolong its life, guarantee its safety, and provide an accurate estimation of different states of the battery for the energy management modules [6] at all battery states, whether it is in use or not, during charging/discharging. Consequently, the BMS improves battery security, reliability, and prevents overshooting and overcharge, while optimizing the performance of EVs. To meet this, the hardware architecture of the BMS for LIBs comprises six components, namely: cell monitoring (e.g., temperature, charge/discharge monitoring); passive/active cell balancing; current measurement; contactor and interlock control and monitoring; isolation monitoring; and communication interfaces to peripherals and the environment [13]. The software structure of the BMS consists of four functionalities to aid in state determination, power capability prediction, load balancing, and safety monitoring [2]. All of these tasks are run by the BMS controller, which also extracts high-level battery pack information from the individual cell within the pack, and serves as an interface between the battery pack and the vehicle system controller [48].

2.3. Stages of Performing BMS

In relation to battery management, the functional tree of a state-of-the-art BMS for large lithium-ion battery packs, as shown in Figure 1, consists of five main stages. Each phase performs a unique task from each other, but in a coordinated way to guarantee the overall objective of performing an efficient management of battery health. There are five main stages of performing BMS: condition monitoring, hazard protection, charge/discharge management, diagnosis, and data management and assessment, as explained below. Freischer et al. [49] captured all of these main stages, as shown in Figure 1, in their functional tree of a state-of-the-art BMS for large LIB packs. As a tree with one stem, but many branches and leaves, battery management contains five components, as shown in the functional state-of-the-art BMS for large LIB packs, but these components are broken down into smaller units. It starts with monitoring the battery conditions and extends to data management, which aids the decision-making process about the batteries. The batteries' overall state can reflect the kind of action to be taken, which includes regular maintenance checks. Any BMS product chosen must provide most of the functions identified in the functional tree in Figure 1.

2.3.1. Condition Monitoring

The BMS's first main function is to monitor measurable states of the battery, like pack voltage and current, cell voltage, temperature, isolation, and interlocks [50]. Data from battery states are collected at regular intervals through a procedure of monitoring carefully-selected physical parameters, which indicate the health condition or state of the equipment under given profiles [14]. The battery conditions can be influenced by both internal and external parameters, like temperature and vibrations. The literature shows that vibration load and temperature influences performance of LIBs, leading to a significant decrease in cell capacity, and deterioration inconsistencies [51]. Whenever any abnormal conditions, such as over-voltage or overheating, are detected, the BMS should notify the user and execute the preset correction procedure. In addition to these functions, the BMS also monitors the system temperature to provide a better power consumption scheme, and communicates with individual components and operators [15].

2.3.2. Hazard Protection

The battery is protected against hazards by first defining the system settings which can predict events that are hazardous. The hazards can be as a result of undercharge, overcharge [9], rise in temperature, and other unforeseen factors. The BMS monitors the battery state and the obtainable measurement data can be used to detect or predict these events, which are either internal or external. Timely response to hazardous events can be achieved through running of fault tolerance routines, cooling/heating management, interlock loop management, and external communication. The BMS should contain accurate algorithms to measure and estimate the functional status of the battery and, at the same time, be equipped with state-of-the-art mechanisms to protect the battery from hazardous and inefficient operating conditions [50]. Therefore, it is of importance to identify and quantify substances being released from the battery during tests representing misuse and abuse events, and to ensure that the amounts released are not hazardous to vehicle occupants and first aid responders [52].

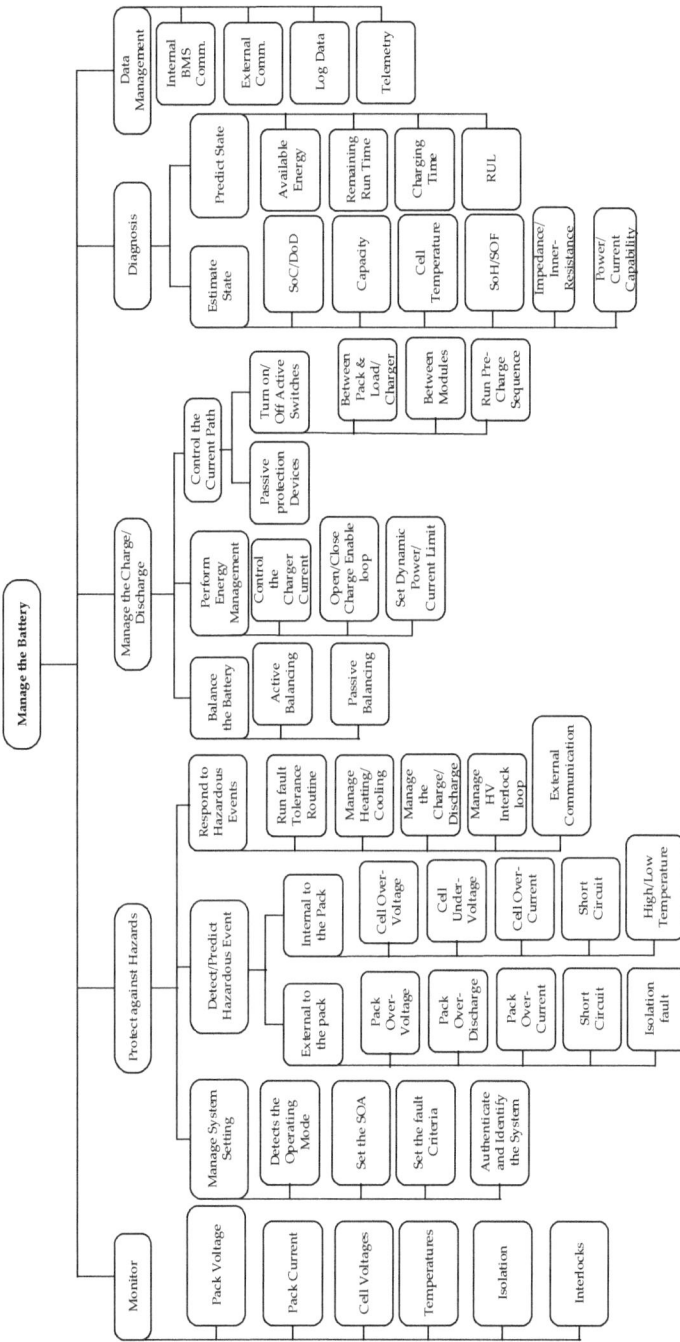

Figure 1. Functional tree of a state-of-the-art BMS for large LIB packs (Reproduced with permission from [49], Copyright Publisher, 2016).

2.3.3. Charge/Discharge Management

The requirements of electrified vehicles brought up the challenge in the charge/discharge rate, making battery degradation during charge/discharge optimization extremely important. The charge/discharge starts when the demanded power crosses the threshold, and is evaluated per unit time, and according to [53] cell balancing equalizes the voltage on each cell of the battery pack. This constitutes three major sub-functions for current control, energy management, and battery balancing. The current path is controlled by employing passive protection devices, or by turning on/off active switches between the pack and load/charger, between modules, and running of a pre-charge sequence. On performing energy management the charger current is controlled, as well as opening/closing the charge-enable loop, and sets the dynamic power of current limits. At the same time battery balancing is achieved through active/passive balancing to equalize the states of cell charge. Since the main energy source of EVs is the battery, it is very crucial to have proper battery protection during charge and discharge. More so, when the EVs are travelling long distances, it is proper to predict the remaining driving distance, as this usually involves discharge of up to 80% or more. Consequently, if batteries are discharged in brine, their initial voltage will be above the electrolysis voltage of water, and hydrogen gas will be produced, thus requiring ventilation to avoid an explosion. If discharged by a resistor the current must be kept low enough so that the batteries do not overheat [54].

2.3.4. Diagnosis

Battery health diagnosis is the process of monitoring the underlying degradation to be able to track the actual performance and take countermeasures if developing faults occur [55]. In this phase the battery states are mainly estimated or predicted based on capacity, cell temperature, charge/discharge time, impedance, and power. This diagnostic mode helps to determine or predict the RUL or to observe safe and reliable battery operation while aging. The diagnosis contains functions to estimate and predict battery states. Therefore, information is used, on the one hand, to observe the safe operation of the battery while aging and, on the other hand, to perform complex algorithms, e.g., for a range estimation in EVs [2]. When a component suddenly fails and the system cannot perform its functions, maintenance actions are automatically carried out to restore the system to working order [56]. Various techniques, such as electrochemical impedance spectroscopy, slow rate cyclic voltammetry, differential thermal voltammetry, incremental capacity (IC) and differential voltage (DV) could identify and quantify degradation modes in real-time applications and could be suitable for implementation within the BMS [57].

2.3.5. Data Management and Assessment

Data are important sources of information to build prognostics models, but accuracy of prognostics suffers from inherent data uncertainties. This is attributed to factors, like lack of sufficient data, sensor noise, and unknown environmental and operating conditions, together with engineering variations [14]. Thus, the battery system data are managed in order to make decisions or take actions which can deter system failures or which can lead to catastrophes. This is achieved by internal or external communication that involves human and machine. The algorithm that performs all the tasks is shown in Figure 2.

There are various indirect methods that propose connecting the measured battery parameters (voltage, current, and temperature) with the battery SOC employing a battery model. A high-fidelity battery model is required to capture the characteristics of the real-life battery and predict its behavior under a wide variety of conditions. In the BMS state estimation algorithm, using the parameters as model inputs, the model can be used to calculate the SOC, and other states of the battery [6] and, therefore, determines the battery's more critical states [58]. A great deal of research is being done to improve the performance of estimation algorithms, as shown in Figure 2.

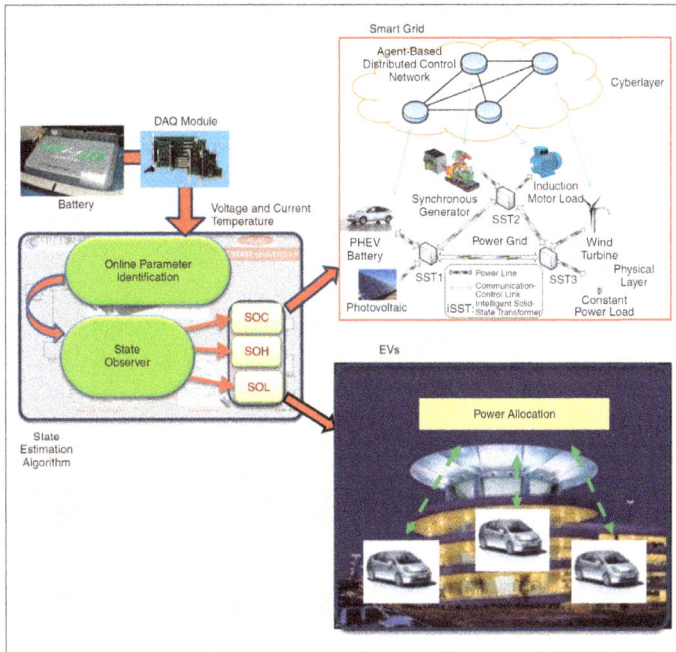

Figure 2. Battery management algorithms function in both the smart grid and EVs (Reproduced with permission from [50], Copyright Publisher, 2013).

The battery parameters are captured through online parameter identification. The BMS calculates the percentage of the cell health, by monitoring the cell internal resistance, together with tracking down the weakest cell in the battery pack. The user is, therefore, provided with the information on the overall SOH of the battery pack. This demonstrates the internality of the BMS in the EVs. The EVs receive their power from the centralized power allocation system, which ensures that every vehicle's battery pack is charged to full capacity before it starts its operations. The smart grid (SG) contains various power sources, like the plug-in hybrid electric vehicle (PHEV) battery, synchronous generator, photovoltaic cell, constant power load, and wind turbine.

The emergence of the SG presents the next generation of electrical power systems, and will enable residents to have the opportunities to manage their home energy usage and reduce expenditures on energy [59]. This is due to considering sustainable development and the crisis of energy, and renewable energy production becomes an important factor in the electricity generation system. The power load data is communicated through the cyber layer, which is an agent-based distributed control network. However, if kinetic energy generated when the vehicle is in motion is transformed into charging the battery, then a great deal of savings can be made on the side of power usage, and efficiency can be improved.

2.4. Issues of the BMS

The hardware and software of the BMS have various issues that have to be addressed in order to meet the demand for secure and safe usage of LIBs. For an ideal BMS and its development process, several issues arise that must be taken into consideration seriously during the design of the BMS. These issues are illuminated below.

2.4.1. Diversity of Battery Management Applications

There are a diverse number of battery management applications. These applications include applications for monitoring of battery tests, evaluation of various state diagnosis and state predictions, rapid assembly of battery system demonstrators, tests for new sensor technologies, like sensor-less temperature measurement, etc. [60]. This variant in battery management applications calls for the development of a modular and flexible BMS. The modularized balancing system should have different equalization systems that operate inside and outside of the modules [8].

2.4.2. Handling of Potential, but Unprecedented, Hazards

LIB cells need to be monitored continuously. To maintain the safety and reliability of battery cells and the safety of people, the use of battery systems is of great use. However, some unprecedented hazards may occur which may turn out to be catastrophic. There is a missing literature on the safety analysis of the hazards of opening lithium batteries, and how discharging them mitigates these hazards. The main safety hazard of opening LIB in air is the exothermic reaction of lithium ions or lithium metal (if present) with oxygen. Opening in water results in an exothermic reaction and the generation of hydrogen gas, which is explosive [54]. Since damaged cells cannot be safely opened, this calls for recycling. Another safety hazard is discharging of the batteries themselves. If the batteries are discharged in brine, their initial voltage will be above the electrolysis voltage of water and hydrogen and oxygen gases will be produced. These gases must be ventilated to avoid an explosion. If the batteries are discharged by a resistor the current must be kept low enough so that the batteries do not overheat.

Some spent batteries are classified as hazardous wastes, increasing transportation, treatment, and disposal costs, as well as the effort needed to achieve regulatory compliance [61]. These hazards form the major function of the BMS to protect the battery against hazardous situations, while maintaining each cell of the battery within its safe and reliable operating range. However, despite the hazardous nature of spent LIBs, they also contain valuable metals, such as copper, aluminum, and cobalt with commercial potential, and the increased mining of natural ores for these metals is leading to shortages, creating a market for recycled LIBs [62,63].

2.4.3. Lack of Safe Operating Areas for Specific Battery Cells

Due to the continuous change of both the internal and external environment for the batteries, there is no single existence of a safe operating area for specific battery cells. When cells are connected in series, some discrepancies in cell internal resistance and differences in cell capacity may occur, and might lead to cell overcharging or undercharging. This inconsistency leads to unreliability and unstable efficiency of the cells, and this poses a serious problem. To solve this problem cell balancing can be used to achieve long battery life and to ensure reliability and safety [60,64,65]. Two types of algorithms are used in cell balancing: voltage-based algorithm and state of charge-based balancing algorithms [66]. The BMS is imperative for active or passive balance circuits to overcome any inconsistency problems among the serially-connected cells [67]. Therefore, it is required that the BMS is designed and developed to take control of the unending control of the change of environment that is sometimes unpredictable.

2.4.4. Ensuring an Efficient Operational State of the Peripheral Control Units and the Power Converters

Since the meaningful hardware and electrochemical properties of the battery cells are impacted by many diverse factors, it is difficult to ensure an efficient operational state of the peripheral control units [60,68]. This will, consequently, impact the design and development of the BMS for EVs. Prognostics, themselves, are useful because they supply the decision-maker with an early warning about the expected time to system/subsystem/component failure and let them decide the appropriate actions to deal with the failure. The benefit from prognostics can flourish if its information is used as the main source of system health management. The BMS not only controls the operational conditions of the battery to prolong its life and guarantee its safety, but also provides accurate estimation of the SOC and SOH for the energy management modules in the smart grid and EVs. To fulfill these tasks, a BMS has several features to control and monitor the operational state of the battery at different battery cell, battery module, and battery pack levels [50].

2.5. Prognostic Methods

Prognostics and health management (PHM) is a set of activities that monitor and estimate the system's SOH throughout its entire life and take suitable decisions at favorable times to extend the system's RUL [69]. Prognostics, itself, are useful because they supply the decision-maker with an early warning about the expected time to system/subsystem/component failure and let them decide the appropriate actions to deal with this failure [70]. The benefits from prognostics can flourish if their information are used as the main source for system health management. The least mature element, and chief component of PHM, are prognostics, which attempt to estimate the RUL of a component when a given abnormal condition has been detected [71]. The key factor is to estimate the RUL, as well as assess the confidence estimate. This makes a prognostic failure a relatively recent area of research to which the scientific community is beginning to give increasing importance, contrary to diagnostics [56,72].

Various factors, like storage voltage, internal and external battery temperatures, rate of discharge, depth-of-discharge, vibrations, etc. [73], must be taken into account when performing battery capacity degradation monitoring. According to Wu et al. [74], this LIB degradation is a nonlinear and time-varying dynamic electrochemical process, and in-depth mechanism analysis is clear in physical significance and concepts. It involves a large number of parameters and complex calculations for accurate modelling making it unsuitable for real-time monitoring and accurate modeling. The capacity degradation of LIBs is often used as a health indicator to establish degradation models. LIB failure, however, occurs when the capacity drops below a normal capacity value or failure threshold value [75]. To perform this task for safe and reliable use of LIBs [76], there are basically three methods classified as physical methods, data driven methods, and hybrid methods that are used to realize an accurate BMS. The summary of these methods is illustrated in Table 4.

Table 4. Summary of prognostics methods.

Prognostic Approaches	Categories of Approaches	Pros	Cons
Physical Approaches ([5,6,37])	Electrical Circuit Model-Based Estimation (ECM) and Electro-chemical Model-Based Estimation (EChM)	- Gives accurate predictions of the temperature distribution - Shows better performance - Simplicity	- The test has to be conducted under exact conditions - Some measurements must be conducted via invasive operation - Some instruments cannot be utilized into real application - Hard to identify the parameters in the model - Parameters may change along with the working condition.
Data-based Approaches ([6,72,74])	Machine Learning Approaches	- Does not need a data model - The algorithms are simple and feasible - The algorithms are the best solution for non-linear systems	- The point estimated value of RUL - Does not describe the uncertainty of measurement results
	Filtering Approaches	- Can be used in any form of state-space model - Best solution for non-linear, Gaussian, and non-Gaussian systems	- Needs data mode (state-space model) - The point estimated value of RUL

Table 4. *Cont.*

Prognostic Approaches	Categories of Approaches	Pros	Cons
	Stochastic Approaches	- Considers the time-dependence of the degradation process - Describes the uncertainty of predictable results	- Higher calculation complexity - Considers uncertain factors
Hybrid Approaches [26,77–82]	Series/Parallel Approach	- Achieves higher accuracy than conventional methods - Increases process reliability and robustness	- Reliability is valid only for given conditions and a period of time.

2.5.1. Physical Methods

The physical method is also called the model-based method. It comprises of an electrical circuit model-based estimation (ECM) method, and an electrochemical circuit model-based estimation (EChM) method. In the ECM discrete-time identification methods are less robust due to undesired sensitivity issues in the transformation of discrete domain parameters. This method promises simplicity by way of enabling easy implementation on a low-cost target microcontroller. It shows better performance in a low SOC range compared with one that uses average SOC in the ECM. The battery's nonlinear dynamic behavior identification could increase significantly as this method is quite accurate. When it comes to temperature distribution through the cell surface, and the behavior under various operating conditions, the ECM gives accurate predictions as it could be used in enhanced SOC estimation procedures.

The EChM includes dependence of the battery behavior on SOC and temperature. However excess temperature can greatly accelerate the battery aging process and even cause fire or explosion in the battery pack under severe cases. The tests are to be conducted under exact conditions despite the same measurements being conducted under invasive operations. The parameters are difficult to be identified in this model, as they change along with working conditions. In general the battery degradation increases if it is kept at a high SOC. Currently, countries and vehicle manufacturers are announcing aggressive targets for completely phasing out internal combustion engines, and EVs will get a great boost [83].

2.5.2. Data-Based Methods

The data-driven prognostics typically require sufficient offline training datasets for accurate remaining useful life for engineering products [84]. Data-based approaches of battery modeling use the battery's SOH data, which can be measured through advanced sensor technology to extract effective feature information, and construct the degradation model to predict RUL [74]. The data-based models are based on three methods, namely, machine learning or artificial intelligence (AI), filtering, and stochastic approaches. The machine learning method is a probabilistic method meant to improve the performance of estimation algorithms. There are four approaches under this algorithm, namely, particle swarm optimization (PSO), genetic algorithm-based estimation (GA), fuzzy-based neural networks (ANFIS), and fuzzy logic-based estimation (FL). The AI method does not need a data model, is simple, feasible, and is the best solution for non-linear systems [37], but this method does not describe the uncertainty of measurement results.

The filtering technique is used in any form of space model and is the best solution for non-linear, Gaussian, and non-Gaussian systems. However, this model needs a data model (state-space model). On the other hand, the stochastic technique is desired because it describes the uncertainty of the predictable results and considers the time dependence of the degradation process. However, this approach involves calculation complexity, and considers uncertain factors. The main advantage of this approach is its precision, since the predictions are achieved based on a mathematical model of the degradation. However, the derived degradation model is specific to a particular kind of component or material and, thus, cannot be generalized to all the system's components. In addition to this, obtaining a mathematical model of degradation is not an easy task and needs well-instrumented test benches, which can be expensive [85]. The advantage of using a data-driven prognostic approach is its applicability, cost, and implementation.

2.5.3. Hybrid Methods

This method constitutes the series and parallel approaches. This method is usually based on combining various physical and data-based approaches to leverage the strengths from both categories. This method proves to be exhibiting more strength than its predecessors since it narrows down their weaknesses. Thus, this method increases process reliability and robustness by combining the complementary information from different prognostic methods in intelligent ways compared to a single model-based method [77,78,81,82]. However, this method has a limitation of offering reliable validity only for certain given conditions and periods of time. These benefits of hybrid methods explain why it is gaining popularity compared to its counterparts.

2.6. Battery Management System Framework

Figure 3 is an elaborate basic framework with descriptions of the software and hardware of the BMS of EVs. The hardware component is embedded into the EV equipment, and is coupled with instructions on how to perform certain basic operations to ensure smooth running of EVs. This system will send signal warnings to the driver and risk responders whenever they sense some element of danger so that the necessary action can be taken. Close monitoring and frequent maintenance operations for this kind of system is crucial, so that any eventuality can be countered well in advance. The BMS achieves this by rigorously opening the contactors in the case of harsh limits violations to prevent the battery operating beyond its limits [12].

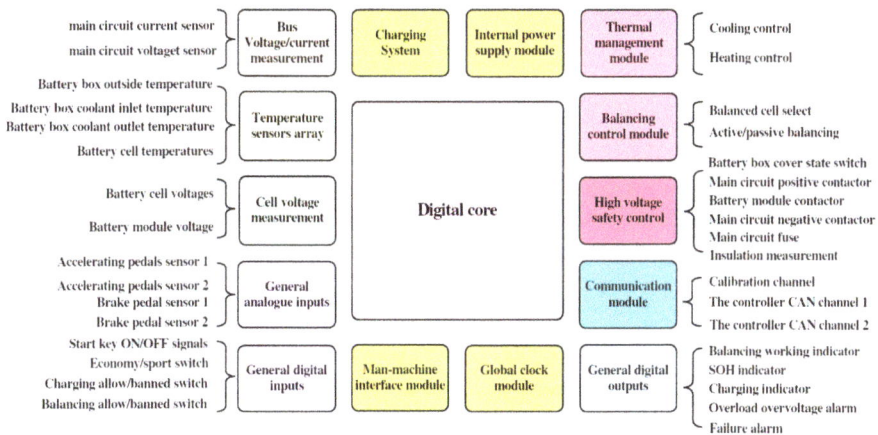

Figure 3. Basic framework of software and hardware of the BMS in vehicles (Reproduced with permission from [39], Copyright Publisher, 2013).

The inputs that the BMS should have are a main circuit current sensor and voltage sensor to measure the main current and voltage; temperature sensors to measure the cell's temperature, the temperature outside the battery box, and maybe also the temperature at the battery coolant inlet and outlet; general analog inputs, like an accelerator pedal sensor and brake pedal sensor; and general inputs, like start key (ON/OFF) signals, charging allow/banned switch, etc. Consequently, the BMS outputs are to the thermal management modules, like fans and electric heaters, to provide cooling and heating control; balancing modules, like capacitors plus switch arrays and dissipation resistance to provide battery equalization; voltage safety management, like a main circuit contactor and battery module contactor; general digital outputs, like a charging indicator and failure alarm;

and a communication module [39]. The global clock module and the internal power supply module, together with the charging system and man-machine interface module also exist in the BMS.

In Figure 3, the software of the BMS covers various functions, as shown. First the battery parameters are detected, which includes total and individual cell current and voltage, temperature, smoke, insulation, collision, impedance, and so on. This to prevent overcharge or over-discharge. Once all parameters have been detected, the internal battery states (SOH, DOD, SOC, SOF) are estimated and the temperature is controlled according to such conditions as working current, voltage, and temperature. This is done to prevent the battery from operating in hazardous conditions and maintains its performance for a long time [86]. However, the variations of internal parameters can be very high and, thus, affect the battery performance. This can be due to production processes as each cell in a battery pack differs and does not reach the same full charge at the same time during the charge period, and also reaches different SOC during the discharge time. This calls upon the BMS to perform cell balancing.

The BMS also performs an onboard diagnosis to identify any fault that may arise from sensors, actuation, battery, network, overcharge, over-discharge, overload, insulation, extreme temperature rise or fall, loose connections, and so on. After the faults have been diagnosed the EV's system control unit is informed through the network. If a certain threshold value is exceeded, or is likely to be, the BMS can also cut off the power supply to prevent any possible damage from taking place.

The BMS also controls the charger when charging the batteries and adopts reliable battery equalization methods based on the information of each cell that is available. This is done so that each cell's SOC is made as consistent as possible during the charging and discharging period. This ensures that the circuit should deliver a current high enough to perform the required charge redistribution during battery runtime. In order to evaluate the functionality of the overcharge/over-discharge protection system, charging or discharging of the battery is performed beyond the recommended limits by the manufacturer [52].

The heating and cooling process is determined by the BMS based on the overall temperature distribution within the battery pack and the requirements of charge/discharge. However, due to variability in manufacturing, cooling, heating, and other operating conditions, some of the crucial thermal and electrical parameters can vary from cell to cell [87]. Excessively high, low, or uneven temperature will do harm to battery performance, thus, a reasonable battery thermal management system with local thermal control management must be designed, with which it cools down under high temperature conditions, or heats up under low temperature conditions [88]. A safe battery operation is ensured for both surface temperature and internal temperature, which proves to be crucial since the battery's internal temperature can reach a critical condition much quicker than the surface temperature [6]. The EV's LIB system is monitored online, and it is usually real-time. The overall data from the EVs, such as SOC, SOH, charge/discharge, faults, and so on, is stored by the BMS. In the battery arrangement for EVs, one battery within the pack must be assigned as the master controller, while the rest are assigned as the slaves. In summary, the real-time battery terminal voltage, cell working temperatures, and load current are measured by the BMS, and the online parameters of the battery pack model can be identified by the reduced labeled samples (RLS) based on the real-time data provided by the BMS [89].

3. Opportunities and Challenges on Prognosis of LIB Health

Since prognostics are still considered relatively immature (as compared to diagnostics), more focus, so far, has been on developing prognostic methods rather than evaluating and comparing their performances. Consequently, there is a need for dedicated attention towards developing standard methods to evaluate prognostic performance from a viewpoint of how post-prognostic reasoning will be integrated into the health management decision-making process [90]. There are many opportunities for prognostics and health management of LIBs, but they have been met with various challenges in equal measure. In this research we will look at these opportunities and challenges from four perspectives, namely, technological, financial/cost, security, and environmental. This paper evaluates each aspect from the opportunities and challenges perspective that LIB research and development is facing. Table 5 shows a summary of opportunities and challenges of LIBs.

Table 5. Summary of opportunities and challenges of LIBs.

Aspect	Opportunities	Challenges	References
Technological	- Existence of recycling technologies - Growth in demand for LIB	• Performance of higher power, larger energy capacity • RUL Prediction, SOH diagnosis, Aging • Enduring adverse circumstances • Development of low weight battery • Reliability • Degradation uncertainties	[1,9,23,60–70]
Financial/Cost	- Intelligent cost management - Cost savings	• Lower cost • Battery degradation costs	[14,24,38,61,62,71]
Security	- Battery state monitoring system	• More safety assurance	[48,52,74,88,91]
Environmental	- 3R-principle: Recycle, reuse, reduce - Mitigated emissions - Integration of renewable energies	• Disposing waste batteries • How to reduce production of new batteries	[64,68,72–85]

3.1. Technological Aspects

Currently, the vehicle manufacturers use the high-power LIB technology to supply electric and hybrid vehicles [92]. These EVs are supplied with power from a battery pack made up of modules connected in series and/or parallel, depending on the desired voltage supply and storage capacity. Out of this, there are many technological opportunities in the use of LIBs. This includes the existence of recycling technologies, and extensive demand of LIBs. The LIB technology is considered as one of the most promising for the near future by a majority of literary sources [11]. However, a great deal has to be done to ensure that batteries of higher power performance and larger energy capacity are realized, as well as improving the batteries to endure adverse circumstances. Consequently, the prediction of RUL, monitoring of battery aging, and SOH diagnosis requires technologies that are accurate in order to avoid catastrophic failures. LIBs provide a lower range of kilometers for EVs as compared to gasoline vehicles, thus, technologies require improvements towards ensuring that higher power energy density is achieved [93].

LIB technologies are very promising for the development of future-generation EVs. These sets of batteries exhibit several advantages as demonstrated, together with various appealing features making them a darling use in many applications. This is explained by the way it has caught the market share in commercialization in consumer electronics, such as cell phones, laptops, video cameras, digital cameras, power tools, and other portable electronic devices [46]. LIBs have many advantages which include [1,34,46,91,94,95]:

- Light weight: applications that make use of LIB go farther and faster due to their lightweight.
- High energy density: EV operates longer between charges while still consuming the same amount of power. LIBs are highly efficient and can be charged with electricity or renewable energies [11].
- Low self-density: the rate of self-discharge is far lower than that of lead acid batteries [96].
- No maintenance: LIBs require little to no maintenance to maintain high-performing products.
- Faster recharge: LIBs have little to no resistance, which allows you to charge at a much higher rate.
- Customizable: not only are LIBs more powerful, lighter, and hold charge longer than lead acid batteries, but they are customizable to fit your needs.

The distance travelled by EVs after full charging is determined by the size of the battery pack (Wh) against the energy consumption (Wh/km):

$$\text{Distance travelled (km)} = \frac{\text{Battery pack size(Wh)}}{\text{Energy consumption (Wh/km)}} \tag{5}$$

According to the study conducted by Martinez et al. it is revealed through the literature that the total distance an EV can cover after a full recharge process is 200 km due to their weight and number of batteries [97]. This signifies that more research is required on how the power efficiency can be increased to offer higher distances travelled by EVs.

3.2. Cost Aspects

The cost and performance of the LIBs are the most expensive component in a vehicle, and is directly linked with the adoption of EVs [18,98], however, in the recent past this cost have been reduced significantly by almost 65% from 2010 [65]. Several countries are working around the clock to see how best they can substitute fossil-based powered energies for technological options of greener/renewable energy to further cut the cost, as well as to conserve the environment. Manufacturers are working on how to employ intelligent cost management methods in order to produce low-cost LIBs. This is partly achieved by recycling used LIBs, which could have gone into waste dumped in a landfill. However, the development of LIBs, and its certification of safety-critical applications, are very expensive.

These costs can be reduced by encapsulating safety-critical components, and safety measures can be restricted to the respective parts [35]. Accounting for battery aging is crucial as the cost of LIBs has a crucial significant impact on overall system cost. The modeled battery degradation cost includes the impacts of the battery temperature, the average SOC, and the DOD on the fading LIB capacity [99]. However, the general cost of batteries reduced tremendously from 1000 $/kWh in 2008, to 268 $/kWh in 2015, which is a 73% reduction in seven years [100]. If this trend continues, it looks promising to consumers of EVs in the future, but if alternative measures to sources of energy are not sought, then the meager resources for the manufacture of LIBs will be depleted.

3.3. Security Aspects

Safety of LIBs is paramount to ensure confidence and widespread adoption of electro-mobility in our society, as they are a proven technology for automotive applications and their continuous use in the future is undeniable. For enhanced security and more accurate SOC estimation, the parameter values of the equivalent circuit models (ECMs) should be continually updated since surface temperature measurements alone might not be sufficient to ensure safe battery operation [58]. During cycling, cells within a pack exhibit non-uniform properties, which may lead to some imbalances (e.g., voltage variations between cells) that may trigger a safety hazard [52]. When EVs are used outdoors, poor pavement conditions, changes in temperature and load can cause performance degradation in LIBs. Battery degradation may lead to leakage, insulation damage, and partial short-circuit. If there is no online detection of degradation, further battery usage will cause serious situations, such as spontaneous combustion and explosions, especially if the current state of health has not been assessed in a timely fashion, or the future battery health state has not been estimated [74]. The main thermal safety issues of LIBs to be addressed are overheating, combustion, explosion, and cycle life. To avoid any catastrophic incidences caused by degradation of LIBs, and to predictively maintain the safety of vehicles, carrying out research on RUL prognostics of LIBs is of great importance [48,52,74,88].

3.4. Environmental Aspects

The use of LIBs will be the next big thing as many governments are fighting against production and sale of vehicles powered only by fossil fuels in favor of cleaner vehicles. This is in a bid to clean up the country's air, or in fighting against global warming, thus ensuring zero-emission vehicles (ZEV).

EVs are seen as an alternative to the conventional transportation based on combustion engines, looking to contribute to solving environmental issues related with zero emissions policies. For this case the EVs are projected as the most sustainable solutions for future transportation [101]. EVs have many advantages over conventional hydrocarbon internal combustion engines, including energy efficiency, environmental friendliness, noiselessness, and less dependence on fossil fuels [102].

According to the International Energy Agency (IEA) report, there is an alarming statistic that shows just how far many other countries have to go in expunging the use of fossil-powered vehicles from their roads. Globally, 95% of EVs are sold in only 10 countries: China, the U.S., Japan, Canada, Norway, the U.K., France, Germany, the Netherlands, and Sweden [103]. This global statistic still poses a challenge to researchers and environmental enthusiasts in regard to the uptake of zero-emission vehicles or green vehicles globally. There are several opportunities in relation to environmental aspects in the use, production, and sale of green vehicles, or at least hybrid ones, which is a boost to the fight against global warming and puts pressure on the extensive use, production, and sale of EVs. Several environmental opportunities that exist include, but are not limited to, the 3R principle: recycle, reuse, reduce, mitigated emissions, and integration of renewable energies.

3.4.1. 3R Principle: Recycle, Reuse, and Reduce

The sharp growing volume of spent LIBs [63,104,105], requires a well-functioning collection and recycling infrastructure to minimize associated environmental impacts and maximize the batteries' reuse potential [62]. The recycling of LIBs reduces energy consumption, reduces greenhouse gas emissions, and results in considerable natural resource savings when compared to landfill [106]. However, it is unclear which recycling processes have the least impact on the environment. There is need for incentives from government and non-governmental agencies to LIB recyclers as a motivating factor to improve recycling, thus mitigating the pressure on the scarce raw material. If the raw materials come from ores, significant negative environmental factors can occur from ore mining and processing, and these can be avoided if the material can be recycled.

Spent $LiFePO_4$ battery packs will retain approximately 80% of their performance, allowing the pack to be applied in a second application, such as a stationary energy storage system [98]. In these spent $LiFePO_4$ batteries, whenever they are recycled at the end of their useful life, a great deal of valuable metals can be recovered, such as copper, aluminum, magnesium, nickel, cobalt, and lithium, thus reducing the pressure on mining the ore, environmental contamination problems [107], and the costs of production. Consequently the LIB consumers require awareness on taking part on the 3R-principle. Since many consumers prefer new batteries, resulting in spent batteries having little potential for reuse, end up being dumped along with other urban solid waste.

Taking into account the importance of key parameters for the environmental performance of LIBs, research efforts should not only focus on energy density, but also on maximizing cycle life and charge/discharge efficiency [42]. The application of the 3R principle to LIBs will bring savings quantified in terms of energy and cumulative energy extracted from the natural environment [108]. It will be seen how material or cell recovery from existing cells will be another source of future materials for LIBs [83]. Therefore, the 3R rule for reuse, recycle, and reduce should be employed purposely to reduce the extinction of the rare iron ores, and this will go along with the conservation of the environment. Current research is aimed towards these principles to improve on the technologies around it.

3.4.2. Mitigated Emissions

The benefits of mitigated emissions include reduced air pollution and climate change, and increased integration and penetration of renewable sources of energy [109]. There are many government policies which are being set out to mitigate emissions from vehicles, as shown in Table 5. One of the policies is the encouragement of EVs, green energy-powered vehicles, and/or hybrid vehicle production, sale, and usage. Some states worldwide have set timelines and others are planning on the shift to EVs. Table 6 shows a summary of the progress so far.

Table 6. Countries that want to ban gas and diesel.

Country	Year	Expectations
Norway	2025	• All new passenger cars and vans sold by 2025 should be zero emission vehicles. As per 2016, 40% of cars sold were electric and hybrid. it is leading the way
India	2030	• Projection of having every vehicle sold to be powered by electricity
France	2040 After-2040	• To end the sale of gas and diesel powered vehicles as it fights global warming • Automakers will only be allowed to sell cars that run on electricity/other cleaner power. Hybrid cars will also be permitted
Britain	2040 2050	• To ban sales of new gasoline and diesel cars in a bid to clean up the country's air. • All cars on the road will need to be have zero emissions
China	-	• Working on a plan to ban production and sale of vehicles powered only by fossil fuels.

(Source: [103,110]).

Other countries in the league are Austria, Denmark, Ireland, Netherlands, Japan, Portugal, Korea, and Spain, who have set official targets for EV sales. The USA does not have a federal policy, but at least eight states have set out goals. According to the IEA, India will join China and the USA to account for 2/3 of the world's expansion in renewable power sources from solar and wind [110]. This is a clear indicator that the demand for LIBs will soar high when new players join the league, as well as the current ones up their game.

3.4.3. Integration of Renewable Energies

The demand for electricity storage capability for EVs is on the rise, and it will increase even more in the future. To support EVs' electric storage there have recently been increases in the contribution of renewable energies to the electrical supply mainly from the installation of photovoltaic modules and wind turbines [111]. Therefore, the internal combustion engine can be replaced with small-sized photovoltaic (PV) modules located on the roof of the EVs, and a micro-wind turbine located in front of the EVS, behind the condenser of the air conditioning system [112]. This technique will improve the power efficiency, regulate the DC-link accurately, and produce suitable stator currents for the traction motor. This is followed by the fact that, by 2030, the demand for energy consumption for EVs will increase to 50% and 40% in USA and Europe, and double for India and China, respectively. Therefore, renewable energy remains the only important resource to consider [59].

4. Future Research Agenda

Currently, there is a scarcity of real prognostics to meet industrial challenges. This may be due to inherent uncertainties associated with deterioration processes, lack of sufficient quantities of data, sensor noise, unknown environmental and operating conditions, and engineering variations, etc., which prevents building prognostic models that can accurately capture the evolution of degradation [14]. This makes research in battery health management a worthy area of future research. This is attributed to the over-emphasis on electrification of our vehicles on our roads today in many countries so as to reduce emissions, and the environmental impact on the depletion of fossil fuels [113]. However, this vehicle electrification has to be regulated to ensure that desirable benefits are achieved in the long run. The solution is, perhaps, the investment on EV powered by alternative solutions.

Cost Effective Production: this is mainly an industrial topic, but a very important future research direction to reduce the price of EVs on the market and thereby widen the customer base [113]. Investment on better technologies which can cause less degradation, but lead to higher energy efficiency on a large scale, and lower long-term costs, is required to be researched. The cost and performance of the battery, the most expensive component in a vehicle, is directly linked with the adaption of electric vehicles. The adoption towards battery electric vehicles mainly depends on the willingness to pay for the extra cost of the traction battery [93]. However the cost can significantly be reduced by economies of scale, and implementing an accurate SOC estimation strategy [114]. The technological breakthroughs in battery life, abuse tolerance and drive range will eventually result in the development of cost effective, long-lasting LIBs.

Disposal of replaced battery: Since spent LiBs retain some of their electrical power, their improper disposal can cause explosions, posing massive environmental, human health or safety hazards and necessitating expensive clean-up and mitigation measures [62]. A lot has to be done to see how this challenge is over done in future. How will the replaced batteries be disposed without posing an environmental impact is another research area in future? Research has to be conducted on how to further extract their useful energy capacity to make an extra profit, while saving out the rare battery resource. Also since pent batteries are defined as hazardous waste, further improvements on basic requirements for packaging, collection, storage, transport, and disposal should be addressed adequately, mainly from scientific research [115] and industrial practices.

Life cycle assessment: research on the LIBs used in EVs must be conducted so that the life cycle study on system boundary [115] can be assessed. Materials used in LIBs production that are rare, toxic, and difficult to recycle should be avoided for improvement of the environment. According to [43], if the model simulations show that cell change-out extends pack life indefinitely while maintaining pack performance at a steady-state, the concept would be of interest to EV and battery manufacturers for its economic benefits, and will hopefully lead to a reduced load of batteries on the recycling and disposal infrastructure.

Identifiability: identifiability in battery SOH estimation is required to establish how to estimate the progress of different battery aging effects. This is perhaps one of the largest challenges in battery SOH estimation [116]. Specifically, the fact that different battery aging dynamics are complex, intertwined, and similar in their time constants means that it is fundamentally very difficult to estimate the progress of different battery aging effects online using voltage, current, and temperature measurements. This, in turn, makes it very challenging to estimate the health of LIBs online, predict their death, and control them in a manner that postpones such death. Battery identifiability remains a very open research area whose exploration can shed light on the extremely important question of what additional sensors, beyond terminal current, temperature, and voltage, can provide with respect to the best means for onboard battery health prognostics, diagnostics, and control [117], using various model estimation algorithms. There are a number of SOH estimation algorithms in battery management systems, but one of the important classes of estimation algorithms is the equivalent circuit model [51].

Developing second-use technology of retired EVs' LIBs: the already retired EVs' LIBs require some research to establish how they can be best put into meaningful second use. According to Wang et al. [94] recycling of spent LiFePO$_4$ batteries is important not only for the treatment of waste, but also for the recovery of useful resources. However they further observed that the treatment of spent LiFePO$_4$ batteries is challenging because LiFePO$_4$ batteries do not contain any precious metals, treatment is complex using traditional recycling processes, and the number of spent batteries recovered from the public has been very small recently, and is faced with challenges. The major challenge that must to be addressed by the recycling industry is developing economical ways to extract and process these metals [118]. Furthermore, there is still a lack of adequate policy and feasible technology for addressing retired LIBs [104], thus, there needs to be recycling processes developed that have economic advantages in terms of chemical costs and added value [34].

Lowering capacity degradation: the LIB capacity is influenced by many factors; among them are temperature, vibrations, and other unforeseen environmental factors. Thus, the efficiency of energy required keeps varying over time [119]. This energy efficiency in LIBs reflects charging and discharging energy powers of the same cycle, and so it is closely related to the battery capacity. Energy efficiency is defined as the percentage of energy use which actually achieves the energy service required [16]. It is given by:

$$\eta = \frac{W_D}{W_C} \times 100\% \qquad (6)$$

where W_D is the energy efficiency during discharge, and W_C is the energy efficiency during charging. Due to high energy density per weight, LIBs are a better option than lead acid batteries in EV application [34,36,120,121]. However, much degradation cannot be prevented, and the real lifetime of the LIBs can be extended by using various other types of approaches.

5. Summary and Closing Remarks

EVs are considered one of the novel solutions to the transport system since they reduce over-dependence on fossil energy. This, in return, will reduce carbon emissions, as EVs act as greener solutions in the transport industry. However, research has shown that the major challenge of the LIBs is battery life. This, however, can be extended by cell balancing to ensure safety of the systems, as well as reliability. The manner in which BMS offers this safety is two-fold: safety of persons and safety of cells. The current growth of EVs is anticipated to lead to enormous penetration into the electric power grid, thus calling for innovative ways of supplementing the EV's power. This is feared because the over-dependence on electric power may add to extensive loads on the power grid, which will have extensive effects on existing distribution networks.

This paper presented a comprehensive review in terms of battery health management for EVs. First the health management systems for batteries are introduced by battery terminologies, BMS architecture, stages of performing BMS, and the prognostic methods used in performing battery health. Furthermore, the opportunities and challenges of BMS from three perspectives, namely technological, cost, security, and environmental aspects was reviewed. Finally, the future research agendas are discussed. In the future the production of cost-effective LIBs, the disposal of replaced LIBs, life-cycle assessment, development of second-use technologies of retired EV batteries, and how battery degradation can be reduced should be the focus of research. Therefore, in regard to the issues raised for research in this paper, further exploration is essential.

Author Contributions: All authors contributed to the paper. Z.B.O. wrote the manuscript with the supervision from L.Z. and D.S., L.Z. acted as a corresponding author.

Funding: This work was financially supported by the China Scholarship Council (CSC). This work was also financially supported by the National Key Research and Development Program of China (No. 2016YFF0203804) and the National Natural Science Foundation of China (No. 51775037).

Conflicts of Interest: The authors declare no conflict of interest. The funding sponsors had no role in the design of the study; in the collection, analyses, or interpretation of data; in the writing of the manuscript; or in the decision to publish the results.

References

1. Raszmann, E.; Baker, K.; Shi, Y.; Christensen, D. Modeling stationary lithium-ion batteries for optimization and predictive control. In Proceedings of the 2017 IEEE Power and Energy Conference at Illinois (PECI), Champaign, IL, USA, 23–24 February 2017. [CrossRef]
2. Liu-Henke, X.; Scherler, S.; Jacobitz, S. Verification oriented development of a scalable battery management system for lithium-ion batteries. In Proceedings of the 2017 Twelfth International Conference on Ecological Vehicles and Renewable Energies (EVER), Monte Carlo, Monaco, 11–13 April 2017. [CrossRef]
3. Ordoñez, J.; Gago, E.J.; Girard, A. Processes and technologies for the recycling and recovery of spent lithium-ion batteries. *Renew. Sustain. Energy Rev.* **2016**, *60*, 195–205. [CrossRef]
4. Dubarry, M.; Devie, A.; Liaw, B.Y. The value of battery diagnostics and prognostics. *J. Energy Power Sources* **2014**, *1*, 242–249.
5. Song, Y.; Liu, D.; Yang, C.; Peng, Y. Data-driven hybrid remaining useful life estimation approach for spacecraft lithium-ion battery. *Microelectron. Reliab.* **2017**, *75*, 142–153. [CrossRef]

6. Muñoz-Galeano, R.-B.N.; Pablo, J.; Sarmiento-Maldonado, H.O. SoC estimation for lithium-ion batteries: review and future challenges. *Electronics* **2017**, *6*, 102. [CrossRef]

7. Zhang, L.; Mu, Z.; Sun, C. Remaining useful life prediction for lithium-ion batteries based on exponential model and particle filter. *IEEE Access* **2018**, *6*, 17729–17740. [CrossRef]

8. Daowd, M.; Antoine, M.; Omar, N.; Lataire, P.; van den Bossche, P.; van Mierlo, J. Battery management system-balancing modularization based on a single switched capacitor and bi-directional DC/DC converter with the auxiliary battery. *Energies* **2014**, *7*, 2897–2937. [CrossRef]

9. Fujita, Y.; Hirose, Y.; Kato, Y.; Watanabe, T. Development of battery management system. *FUJITSU TEN Tech. J.* **2016**, *42*, 68–80.

10. Allam, A.; Onori, S.; Marelli, S.; Taborelli, C. Battery health management system for automotive applications: A retroactivity-based aging propagation study. In Proceedings of the American Control Conference (ACC), Chicago, IL, USA, 1–3 July 2015; pp. 703–716. [CrossRef]

11. Andwari, A.M.; Pesiridis, A.; Rajoo, S.; Martinez-Botas, R.; Esfahanian, V. A review of battery electric vehicle technology and readiness levels. *Renew. Sustain. Energy Rev.* **2017**, *78*, 414–430. [CrossRef]

12. Akdere, M.; Giegerich, M.; Wenger, M.; Schwarz, R.; Koffel, S.; Fühner, T.; Waldhör, S.; Wachtler, J.; Lorentz, V.R.H.; März, M. Hardware and software framework for an open battery management system in safety-critical applications. In Proceedings of the IECON 2016—42nd Annual Conference of the IEEE Industrial Electronics Society, Florence, Italy, 23–26 October 2016; pp. 5507–5512. [CrossRef]

13. Brandl, M.; Gall, H.; Wenger, M.; Lorentz, V.; Giegerich, M.; Baronti, F.; Fantechi, G.; Fanucci, L.; Roncella, R.; Saletti, R.; et al. Batteries and battery management systems for electric vehicles. In Proceedings of the 2012 Design, Automation & Test in Europe Conference & Exhibition (DATE), Dresden, Germany, 12–16 March 2012; pp. 971–976. [CrossRef]

14. Javed, K.; Gouriveau, R.; Zerhouni, N. State of the art and taxonomy of prognostics approaches, trends of prognostics applications and open issues towards maturity at different technology readiness levels. *Mech. Syst. Signal Process.* **2017**, *94*, 214–236. [CrossRef]

15. Xing, Y.; Ma, E.W.M.; Tsui, K.L.; Pecht, M. Battery management systems in electric and hybrid vehicles. *Energies* **2011**, *4*, 1840–1857. [CrossRef]

16. Wang, S.; Zhao, L.; Su, X.; Ma, P. Prognostics of lithium-ion batteries based on battery performance analysis and flexible support vector regression. *Energies* **2014**, *7*, 6492–6508. [CrossRef]

17. Bruen, T.; Hooper, J.M.; Marco, J.; Gama, M.; Chouchelamane, G.H. Analysis of a battery management system (BMS) control strategy for vibration aged Nickel Manganese Cobalt Oxide (NMC) Lithium-Ion 18650 battery cells. *Energies* **2016**, *9*, 255. [CrossRef]

18. Shareef, H.; Islam, M.M.; Mohamed, A. A review of the stage-of-the-art charging technologies, placement methodologies, and impacts of electric vehicles. *Renew. Sustain. Energy Rev.* **2016**, *64*, 403–420. [CrossRef]

19. Castro, T.S.; de Souza, T.M.; Silveira, J.L. Feasibility of electric vehicle: electricity by grid × photovoltaic energy. *Renew. Sustain. Energy Rev.* **2017**, *69*, 1077–1084. [CrossRef]

20. Coffman, M.; Bernstein, P.; Wee, S. Integrating electric vehicles and residential solar PV. *Transp. Policy* **2017**, *53*, 30–38. [CrossRef]

21. Gough, R.; Dickerson, C.; Rowley, P.; Walsh, C. Vehicle-to-grid feasibility: A techno-economic analysis of EV-based energy storage. *Appl. Energy* **2017**, *192*, 12–23. [CrossRef]

22. Hannan, M.A.; Azidin, F.A.; Mohamed, A. Hybrid electric vehicles and their challenges: A review. *Renew. Sustain. Energy Rev.* **2014**, *29*, 135–150. [CrossRef]

23. Palinski, M. A Comparison of electric vehicles and conventional automobiles: Costs and quality perspective. Bachelor's Thesis, Novia University of Applied Sciences, Vaasa, Finland, 2017.

24. Salisbury, M. Economic and Air Quality Benefits of Electric Vehicles in Nevada Executive Summary. Southwestern Energy Efficiency Project (SWEEP). 2014. Available online: http://www.swenergy.org/data/sites/1/media/documents/publications/documents/Economic_and_AQ_Benefits_of_EVs_in_NV-Sept_2014.pdf (accessed on 10 April 2018).

25. Helmers, E.; Patrick, M. Electric cars: Technical characteristics and environmental impacts. *Environ. Sci. Eur.* **2012**, *24*, 14. [CrossRef]

26. Pappas, J.C.K. A New Prescription for electric cars. *Energy Law J.* **2014**, *35*, 151–198.

27. Hall, D.; Moultak, M.; Lutsey, N. *Electric Vehicle Capitals of the World: Demonstrating the Path to Electric Drive*; International Council on Clean Transportatio: Washington, DC, USA, 2017.

28.	Varun, M.; Kumar, C. Problems in electric vehicles. *Int. J. Appl. Res. Mech. Eng.* **2012**, *2*, 63–73.

29.	Saha, B.; Quach, C.C.; Hogge, E.F.; Strom, T.H.; Hill, B.L.; Goebel, K. Battery health management system for electric UAVs. In Proceedings of the 2011 Aerospace Conference, Big Sky, MT, USA, 5–12 March 2011. [CrossRef]

30.	Cazzola, P.; Gorner, M.; Munuera, L. Global EV Outlook 2017: Two Million and Counting. International Energy Agency (IEA). 2017. Available online: https://webstore.iea.org/global-ev-outlook-2017 (accessed on 22 February 2018).

31.	Zhang, L.; Peng, H.; Ning, Z.; Mu, Z.; Sun, C. Comparative research on RC equivalent circuit models for lithium-ion batteries of electric vehicles. *Appl. Sci.* **2017**, *7*, 1002. [CrossRef]

32.	Rezvani, M.; Abuali, M.; Lee, S.; Lee, J.; Ni, J. A comparative analysis of techniques for electric vehicle battery Prognostics and Health Management (PHM). *SAE Int.* **2011**, *11*. [CrossRef]

33.	Xi, Z.; Jing, R.; Yang, X.; Decker, E. State of charge estimation of lithium-ion batteries considering model Bias and parameter uncertainties. In Proceedings of the ASME 2014 International Design Engineering Technical Conferences & Computers and Information in Engineering Conference, Buffalo, NY, USA, 17–20 August 2014; pp. 1–7. [CrossRef]

34.	Li, L.; Bian, Y.; Zhang, X.; Xue, Q.; Fan, E.; Wu, F.; Renjie Chen, R. Economical recycling process for spent lithium-ion batteries and macro- and micro-scale mechanistic study. *J. Power Sources* **2018**, *377*, 70–79. [CrossRef]

35.	Li, J.; Lai, Q.; Wang, L.; Lyu, C.; Wang, H. A method for SOC estimation based on simplified mechanistic model for LiFePO$_4$ battery. *Energy* **2016**, *114*, 1266–1276. [CrossRef]

36.	Podder, S.; Khan, M.Z.R. Comparison of lead acid and Li-ion battery in solar home system of Bangladesh. In Proceedings of the 2016 5th International Conference on Informatics, Electronics and Vision, ICIEV, Dhaka, Bangladesh, 13–14 May 2016; pp. 434–438. [CrossRef]

37.	Leone, G.; Cristaldi, L.; Turrin, S. A data-driven prognostic approach based on statistical similarity: An application to industrial circuit breakers. *Measurement* **2017**, *108*, 163–170. [CrossRef]

38.	Wei, J.; Dong, G.; Chen, Z. Remaining useful life prediction and state of health diagnosis for lithium-ion batteries using particle filter and support vector regression. *IEEE Trans. Ind. Electron.* **2017**, *65*, 5634–5643. [CrossRef]

39.	Lu, L.; Han, X.; Li, J.; Hua, J.; Ouyang, M. A review on the key issues for lithium-ion battery management in electric vehicles. *J. Power Sources* **2013**, *226*, 272–288. [CrossRef]

40.	Saxena, A.; Celaya, J.R. Roychoudh loading profiles: Some lessons learned. In Proceedings of the European Conference on Prognostics and Health Management Society, Dresden, Germany, 3–5 July 2012.

41.	Chen, A.; Sen, P.K. Advancement in battery technology: a state-of-the-art review. In Proceedings of the 2016 IEEE Industry Applications Society Annual Meeting, Portland, OR, USA, 2–6 October 2016. [CrossRef]

42.	Peters, J.F.; Baumann, M.; Zimmermann, B.; Braun, J.; Weil, M. The environmental impact of Li-Ion batteries and the role of key parameters—A review. *Renew. Sustain. Energy Rev.* **2017**, *67*, 491–506. [CrossRef]

43.	Mathew, M.; Kong, Q.H.; McGrory, J.; Fowler, M. Simulation of lithium ion battery replacement in a battery pack for application in electric vehicles. *J. Power Sources* **2017**, *349*, 94–104. [CrossRef]

44.	Su, X.; Wang, S.; Pecht, M.; Zhao, L.; Ye, Z. Interacting multiple model particle filter for prognostics of lithium-ion batteries. *Microelectron. Reliab.* **2017**, *70*, 59–69. [CrossRef]

45.	Khorasgani, H.; Biswas, G.; Sankararaman, S. Methodologies for system-level remaining useful life prediction. *Reliab. Eng. Syst. Saf.* **2016**, *154*, 8–18. [CrossRef]

46.	Chen, X.; Shen, W.; Vo, T.T.; Cao, Z.; Kapoor, A. An overview of lithium-ion batteries for electric vehicles. In Proceedings of the 2012 10th International Power & Energy Conference, Ho Chi Minh City, Vietnam, 12–14 December 2012; pp. 230–235. [CrossRef]

47.	Piromjit, P.; Tayjasanant, T. Peak-demand management for improving undervoltages in distribution systems with electric vehicle connection by stationary battery. In Proceedings of the 2017 IEEE Transportation Electrification Conference and Expo, Asia-Pacific (ITEC Asia-Pacific), Harbin, China, 7–10 August 2017. [CrossRef]

48.	Zhang, F.; Rehman, M.M.U.; Zane, R.; Maksimovic, D. Hybrid balancing in a modular battery management system for electric-drive vehicles. In Proceedings of the 2017 IEEE Energy Conversion Congress and Exposition (ECCE), Cincinnati, OH, USA, 1–5 October 2017; pp. 578–583. [CrossRef]

49. Fleischer, C.; Sauer, D.U.; Barreras, J.V.; Schaltz, E.; Christensen, A.E. Development of software and strategies for battery management system testing on HIL simulator. In Proceedings of the 2016 Eleventh International Conference on Ecological Vehicles and Renewable Energies (EVER), Monte Carlo, Monaco, 6–8 April 2016. [CrossRef]

50. Rahimi-Eichi, H.; Ojha, U.; Baronti, F.; Chow, M.-Y. Battery management system: An overview of its application in the smart grid and electric vehicles. *IEEE Ind. Electron. Mag.* **2013**, 4–16. [CrossRef]

51. Zhang, L.; Ning, Z.; Peng, H.; Mu, Z.; Sun, C. Effects of vibration on the electrical performance of lithium-ion cells based on mathematical statistics. *Appl. Sci.* **2017**, *7*, 802. [CrossRef]

52. Ruiz, V.; Pfrang, A.; Kriston, A.; Omar, N.; van den Bossche, P.; Boon-Brett, L. A review of international abuse testing standards and regulations for lithium ion batteries in electric and hybrid electric vehicles. *Renew. Sustain. Energy Rev.* **2018**, *81*, 1427–1452. [CrossRef]

53. Artakusuma, D.D.; Afrisal, H.; Cahyadi, A.I.; Wahyunggoro, O. Battery management system via bus network for multi battery electric vehicle. In Proceedings of the 2014 IEEE International Conference on Electrical Engineering and Computer Science, Kuta, Indonesia, 24–25 November 2014; pp. 179–181. [CrossRef]

54. Sonoc, A.; Jeswiet, J.; Soo, V.K. Opportunities to improve recycling of automotive lithium ion batteries. *Procedia CIRP* **2015**, *29*, 752–757. [CrossRef]

55. Nuhic, A.; Terzimehic, T.; Soczka-guth, T.; Buchholz, M.; Dietmayer, K. Health diagnosis and remaining useful life prognostics of lithium-ion batteries using data-driven methods. *J. Power Sources* **2013**, *239*, 680–688. [CrossRef]

56. Belkacem, L.; Simeu-abazi, Z.; Dhouibi, H.; Gascard, E.; Messaoud, H. Diagnostic and prognostic of hybrid dynamic systems: Modeling and RUL evaluation for two maintenance policies. *Reliab. Eng. Syst. Saf.* **2017**, *164*, 98–109. [CrossRef]

57. Pastor-Fernández, C.; Widanage, W.D.; Chouchelamane, G.H.; Marco, J. A SoH diagnosis and prognosis method to identify and quantify degradation modes in li-ion batteries using the IC/DV technique. In Proceedings of the 6th Hybrid and Electric Vehicles Conference (HEVC 2016), London, UK, 2–3 November 2016. [CrossRef]

58. Hannan, M.A.; Lipu, M.S.H.; Hussain, A.; Mohamed, A. A review of lithium-ion battery state of charge estimation and management system in electric vehicle applications: Challenges and recommendations. *Renew. Sustain. Energy Rev.* **2017**, *78*, 834–854. [CrossRef]

59. Melhem, F.Y.; Grunder, O.; Hammoudan, Z.; Moubayed, N. Optimization and energy management in smart battery storage system with integration of electric vehicles. *Can. J. Electr. Comput. Eng.* **2017**, *40*, 128–138. [CrossRef]

60. Giegerich, M.; Akdere, M.; Freund, C.; Fuhner, T.; Grosch, J.L.; Koffel, S.; Schwarz, R.; Waldhor, S.; Wenger, M.; Lorentz, V.R.H.; et al. Open, flexible and extensible battery management system for lithium-ion batteries in mobile and stationary applications. In Proceedings of the 2016 IEEE 25th International Symposium on Industrial Electronics (ISIE), Santa Clara, CA, USA, 8–10 June 2016; pp. 991–996. [CrossRef]

61. Gaines, L. The future of automotive lithium-ion battery recycling: Charting a sustainable course. *Sustain. Mater. Technol.* **2014**, *1–2*, 2–7. [CrossRef]

62. Zeng, X.; Li, J.; Liu, L. Solving spent lithium-ion battery problems in China: Opportunities and challenges. *Renew. Sustain. Energy Rev.* **2015**, *52*, 1759–1767. [CrossRef]

63. Wei, J.; Zhao, S.; Ji, L.; Zhou, T.; Miao, Y.; Scott, K.; Li, D.; Yang, J.; Wu, X. Reuse of Ni-Co-Mn oxides from spent Li-ion batteries to prepare bifunctional air electrodes. *Resour. Conserv. Recycl.* **2018**, *129*, 135–142. [CrossRef]

64. Yusof, M.S.; Toha, S.F.; Kamisan, N.; Hashim, N.N.W.N.; Abdullah, M. Battery cell balancing pptimisation for battery management system. *IOP Conf. Ser. Mater. Sci. Eng.* **2017**, *184*, 012021. [CrossRef]

65. Rahman, A.; Rahman, M.; Rashid, M. Wireless battery management system of electric transport. *IOP Conf. Ser. Mater. Sci. Eng.* **2017**, *260*, 012029. [CrossRef]

66. Piao, C.; Wang, Z.; Cao, J.; Zhang, W.; Lu, S. Lithium-ion battery cell-balancing algorithm for battery management system based on real-time outlier detection. *Math. Probl. Eng.* **2015**, *2015*. [CrossRef]

67. Lin, J.-C.M. Development of a new battery management system with an independent balance module for electrical motorcycles. *Energies* **2017**, *10*, 1289. [CrossRef]

68. Liang, Y.; Wang, Y.; Han, D. Design of energy storage management system based on FPGA in design of energy storage management system based on FPGA in Micro-Grid. *IOP Conf. Ser. Earth Environ. Sci.* **2017**, *108*, 052040. [CrossRef]

69. Sutharssan, T.; Montalvao, D.; Chen, Y.K.; Wang, W.-C.; Pisac, C.; Elemara, H. A review on prognostics and health monitoring of proton exchange membrane fuel cell. *Renew. Sustain. Energy Rev.* **2017**, *75*, 440–450. [CrossRef]

70. Elattar, H.M.; Elminir, H.K.; Riad, A.M. Prognostics: A literature review. *Complex Intell. Syst.* **2016**, *2*, 125–154. [CrossRef]

71. Goebel, K.; Saha, B.; Saxena, A.; Celaya, J.R.; Christophersen, J.P. Prognostics in battery health management. *IEEE Instrum. Meas. Mag.* **2008**, *11*, 33–40. [CrossRef]

72. Tsui, K.L.; Chen, N.; Zhou, Q.; Hai, Y.; Wang, W. Prognostics and health management: A review on data driven approaches. *Math. Probl. Eng.* **2014**, *2015*. [CrossRef]

73. Wang, D.; Yang, F.; Zhao, Y.; Tsui, K.L. Battery remaining useful life prediction at different discharge rates. *Microelectron. Reliab.* **2017**, *78*, 212–219. [CrossRef]

74. Wu, L.; Fu, X.; Guan, Y. Review of the remaining useful life prognostics of vehicle lithium-ion batteries using data-driven methodologies. *Appl. Sci.* **2016**, *6*, 166. [CrossRef]

75. Zhou, D.; Xue, L.; Song, Y.; Chen, J. On-Line Remaining useful life prediction of lithium-ion batteries based on the optimized Gray Model GM(1,1). *Batteries* **2017**, *3*, 21. [CrossRef]

76. Arachchige, B.; Perinpanayagam, S.; Jaras, R. Enhanced prognostic model for lithium ion batteries based on particle filter state transition model modification. *Appl. Sci.* **2017**, *7*, 1172. [CrossRef]

77. Zhou, D.; Gao, F.; Breaz, E.; Ravey, A.; Miraoui, A. Degradation prediction of PEM fuel cell using a moving window based hybrid prognostic approach. *Energy* **2017**, *138*, 1175–1186. [CrossRef]

78. Chang, Y.; Fang, H.; Zhang, Y. A new hybrid method for the prediction of the remaining useful life of a lithium-ion battery. *Appl. Energy* **2017**, *206*, 1564–1578. [CrossRef]

79. Skima, H.; Medjaher, K.; Varnier, C.; Dedu, E.; Bourgeois, J. Microelectronics Reliability A hybrid prognostics approach for MEMS: From real measurements to remaining useful life estimation. *Microelectron. Reliab.* **2016**, *65*, 79–88. [CrossRef]

80. Wu, X.; Ye, Q.; Wang, J. A hybrid prognostic model applied to SOFC prognostics. *Int. J. Hydrog. Energy* **2017**, *42*, 25008–25020. [CrossRef]

81. Li, Z.; Wu, D.; Hu, C.; Terpenny, J. An ensemble learning-based prognostic approach with degradation-dependent weights for remaining useful life prediction. *Reliab. Eng. Syst. Saf.* **2018**. [CrossRef]

82. Khalastchi, E.; Kalech, M.; Rokach, L. A hybrid approach for improving unsupervised fault detection for robotic systems. *Expert Syst. Appl.* **2017**, *81*, 372–383. [CrossRef]

83. Olivetti, E.A.; Ceder, G.; Gaustad, G.G.; Fu, X. Lithium-ion battery supply chain considerations: Analysis of potential bottlenecks in critical metals. *Joule* **2017**, *1*, 229–243. [CrossRef]

84. Xi, Z.; Zhao, X. Data driven prognostics with lack of training data sets. In Proceedings of the ASME 2015 International Design Engineering Technical Conferences & Computers and Information in Engineering Conference, Boston, MA, USA, 2–5 August 2015. [CrossRef]

85. Medjaher, K.; Zerhouni, N.; Baklouti, J. Data-driven prognostics based on health indicator construction: Application to PRONOSTIA's data. In Proceedings of the 2013 European Control Conference (ECC), Zurich, Switzerland, 17–19 July 2013; pp. 1451–1456.

86. Orcioni, S.; Ricci, A.; Buccolini, L.; Scavongelli, C.; Conti, M. Effects of variability of the characteristics of single cell on the performance of a lithium-ion battery pack. In Proceedings of the 2017 13th Workshop on Intelligent Solutions in Embedded Sysems (WISES), Hamburg, Germany, 12–13 June 2017; pp. 15–21. [CrossRef]

87. Dey, S.; Perez, H.E.; Moura, S.J. Model-based battery thermal fault diagnostics: Algorithms, analysis, and experiments. *IEEE Trans. Control Syst. Technol.* **2017**, 1–12. [CrossRef]

88. Ye, X.; Zhao, Y.; Quan, Z. Thermal management system of lithium-ion battery module based on micro heat pipe array. *Int. J. Energy Res.* **2017**, *42*, 648–655. [CrossRef]

89. Zhang, X.; Wang, Y.; Liu, C.; Chen, Z. A novel approach of remaining discharge energy prediction for large format lithium-ion battery pack. *J. Power Sources* **2017**, *343*, 216–225. [CrossRef]

90. Saxena, A.; Celaya, J.; Saha, B.; Saha, S.; Goebel, K. Metrics for offline evaluation of prognostic performance. *Int. J. Progn. Health Manag.* **2010**, *1*, 1–20.

91. Sreejith, R.; Rajagopal, K.R. An insight into motor and battery selections for three-wheeler electric vehicle. In Proceedings of the 2016 IEEE 1st International Conference on Power Electronics, Intelligent Control and Energy Systems (ICPEICES), Delhi, India, 4–6 July 2016; pp. 9–14. [CrossRef]

92. Rizoug, N.; Sadoun, R.; Mesbahi, T.; Bartholumeus, P.; LeMoigne, P. Aging of high power Li-ion cells during real use of electric vehicles. *IET Electr. Syst. Transp.* **2017**, *7*, 14–22. [CrossRef]

93. Fotouhi, A.; Auger, D.J.; Propp, K.; Longo, S.; Purkayastha, R.; O'Neill, L.; Walus, S. Lithium-sulfur cell equivalent circuit network model parameterization and sensitivity Analysis. *IEEE Trans. Veh. Technol.* **2017**, *66*, 7711–7721. [CrossRef]

94. Wang, W.; Wu, Y. An overview of recycling and treatment of spent LiFePO₄ batteries in China. *Resour. Conserv. Recycl.* **2017**, *127*, 233–243. [CrossRef]

95. Gong, Y.; Yu, Y.; Huang, K.; Hu, J.; Li, C. Evaluation of lithium-ion batteries through the simultaneous consideration of environmental, economic and electrochemical performance indicators. *J. Clean. Prod.* **2018**, *170*, 915–923. [CrossRef]

96. Amjad, S.; Neelakrishnan, S.; Rudramoorthy, R. Review of design considerations and technological challenges for successful development and deployment of plug-in hybrid electric vehicles. *Renew. Sustain. Energy Rev.* **2010**, *14*, 1104–1110. [CrossRef]

97. Martínez, D.A.; Poveda, J.D.; Montenegro, D. Li-Ion battery management system based in fuzzy logic for improving electric vehicle autonomy. In Proceedings of the 2017 IEEE Workshop on Power Electronics and Power Quality Applications (PEPQA), Bogota, Colombia, 31 May–2 June 2017. [CrossRef]

98. Berckmans, G.; Messagie, M.; Smekens, J.; Omar, N.; Vanhaverbeke, L.; van Mierlo, J. Cost projection of state of the art lithium-ion batteries for electric vehicles up to 2030. *Energies* **2017**, *10*, 1314. [CrossRef]

99. Badawy, M.O.; Sozer, Y. Power flow management of a grid tied PV-battery system for electric vehicles charging. *IEEE Trans. Ind. Appl.* **2017**, *53*, 1347–1357. [CrossRef]

100. Cazzola, P.; Gorner, M.; Yi, J.T.; Yi, W. Global EV Outlook 2016 Electric Vehicles Initiative. International Energy Agency. 2016. Available online: https://www.iea.org/publications/freepublications/.../Global_EV_Outlook_2016.pdf (accessed on 10 February 2018).

101. Purwadi, A.; Shani, N.; Heryana, N.; Hardimasyar, T.; Firmansyah, M.; Sr, A. Modelling and analysis of electric vehicle DC fast charging infrastructure based on PSIM. In Proceedings of the 2013 1st International Conference on Artificial Intelligence, Modelling and Simulation, Kota Kinabalu, Malaysia, 3–5 December 2013; pp. 359–364. [CrossRef]

102. Saw, L.H.; Tay, A.A.O.; Zhang, L.W. Thermal management of lithium-ion battery pack with liquid cooling. In Proceedings of the 2015 31st Thermal Measurement, Modeling & Management Symposium (SEMI-THERM), San Jose, CA, USA, 15–19 March 2015; pp. 298–302. [CrossRef]

103. Petroff, A. These Countries Want to Ditch Gas and Diesel Cars–26 July 2017. *CNNMoney*, 11 September 2017. Available online: http://money.cnn.com/2017/07/26/autos/countries-that-are-banning-gas-cars-for-electric/index.html (accessed on 4 February 2018).

104. Wang, M.M.; Zhang, C.C.; Zhang, F.S. Recycling of spent lithium-ion battery with polyvinyl chloride by mechanochemical process. *Waste Manag.* **2017**, *67*, 232–239. [CrossRef] [PubMed]

105. Chen, X.; Ma, H.; Luo, C.; Zhou, T. Recovery of valuable metals from waste cathode materials of spent lithium-ion batteries using mild phosphoric acid. *J. Hazard. Mater.* **2017**, *326*, 77–86. [CrossRef] [PubMed]

106. Boyden, A.; Soo, V.K.; Doolan, M. The environmental impacts of recycling portable lithium-ion batteries. *Procedia CIRP* **2016**, *48*, 188–193. [CrossRef]

107. Fan, B.; Chen, X.; Zhou, T.; Zhang, J.; Xu, B. A sustainable process for the recovery of valuable metals from spent lithium-ion batteries. *Waste Manag. Res.* **2016**, *34*, 474–481. [CrossRef] [PubMed]

108. Dewulf, J.; Van der Vorst, G.; Denturck, K.; Van Langenhove, H.; Ghyoot, W.; Tytgat, J.; Vandeputte, K. Recycling rechargeable lithium ion batteries: Critical analysis of natural resource savings. *Resour. Conserv. Recycl.* **2010**, *54*, 229–234. [CrossRef]

109. Sovacool, B.K.; Axsen, J.; Kempton, W. The future promise of Vehicle-to-Grid (V2G) integration: A sociotechnical review and research agenda. *Annu. Rev. Environ. Resour.* **2017**, *42*, 377–406. [CrossRef]

110. OECD/IEA. World Energy Outlook 2017 Executive Summary. International Energy Agency, 2017. Available online: http://www.iea.org/publications/freepublications/publication/world-energy-outlook-2017---executive-summary---english-version.html (accessed on 7 February 2018).

111. Helbig, C.; Bradshaw, A.M.; Wietschel, L.; Thorenz, A.; Tuma, A. Supply risks associated with lithium-ion battery materials. *J. Clean. Prod.* **2018**, *172*, 274–286. [CrossRef]
112. Fathabadi, H. Plug-in Hybrid Electric Vehicles (PHEVs): Replacing Internal Combustion Engine with Clean and Renewable Energy Based Auxiliary Power Sources. *IEEE Trans. Power Electron.* **2018**. [CrossRef]
113. Grunditz, E.A.; Thiringer, T. Performance analysis of current BEVs based on a comprehensive review of specifications. *IEEE Trans. Transp. Electrification* **2016**, *2*, 270–289. [CrossRef]
114. Bashash, S.; Moura, S.J.; Fathy, H.K. Charge trajectory optimization of plug-in hybrid electric vehicles for energy cost reduction and battery health enhancement. In Proceedings of the 2010 American Control Conference, Baltimore, MD, USA, 30 June–2 July 2010. [CrossRef]
115. Wang, Q.; Liu, W.; Yuan, X.; Tang, H.; Tang, Y.; Wang, M.; Zuo, J.; Song, Z.; Sun, J. Environmental impact analysis and process optimization of batteries based on life cycle assessment. *J. Clean. Prod.* **2018**, *174*, 1262–1273. [CrossRef]
116. Hatzell, K.B.; Sharma, A.; Fathy, H.K. A survey of long-term health modeling, estimation, and control of Lithium-ion batteries: Challenges and opportunities. In Proceedings of the 2012 American Control Conference (ACC), Montreal, QC, Canada, 27–29 June 2012; pp. 584–591.
117. Berecibar, M.; Gandiaga, I.; Villarreal, I.; Omar, N.; van Mierlo, J.; van den Bossche, P. Critical review of state of health estimation methods of Li-ion batteries for real applications. *Renew. Sustain. Energy Rev.* **2016**, *56*, 572–587. [CrossRef]
118. Winslow, K.M.; Laux, S.J.; Townsend, T.G. A review on the growing concern and potential management strategies of waste lithium-ion batteries. *Resour. Conserv. Recycl.* **2018**, *129*, 263–277. [CrossRef]
119. Kong, Q.; Ruan, M.; Zi, Y. A health management system for marine cell group. *IOP Conf. Ser. Earth Environ. Sci.* **2017**, *69*, 012081. [CrossRef]
120. Hu, J.; Zhang, J.; Li, H.; Chen, Y.; Wang, C. A promising approach for the recovery of high value-added metals from spent lithium-ion batteries. *J. Power Sources* **2017**, *351*, 192–199. [CrossRef]
121. Peng, X.; Garg, A.; Zhang, J.; Shui, L. Thermal management system design for batteries packs of electric vehicles: A Survey. In Proceedings of the 2017 Asian Conference on Energy, Power and Transportation Electrification (ACEPT), Singapore, 24–26 October 2017; pp. 1–5. [CrossRef]

electronics

MDPI

Article

Design, Simulation and Hardware Implementation of Shunt Hybrid Compensator Using Synchronous Rotating Reference Frame (SRRF)-Based Control Technique

R. Balasubramanian, K. Parkavikathirvelu, R. Sankaran and Rengarajan Amirtharajan *

School of Electrical & Electronics Engineering (SEEE)S, SASTRA Deemed University, Thirumalaisamudram, Thanjavur 613401, India; rbalu@eee.sastra.edu (R.B.); to_parkavi@eee.sastra.edu (K.P.); rs@eee.sastra.edu (R.S.)
* Correspondence: amir@ece.sastra.edu; Tel.: +91-4362-264101

Received: 26 September 2018; Accepted: 21 December 2018; Published: 1 January 2019

Abstract: This paper deals with the design, simulation, and implementation of shunt hybrid compensator to maintain the power quality in three-phase distribution networks feeding different types balanced and unbalanced nonlinear loads. The configuration of the compensator consists of a selective harmonic elimination passive filter, a series-connected conventional six-pulse IGBT inverter, acting as the active filter terminated with a DC link capacitor. The theory and modelling of the compensator based on current harmonic components at the load end and their decomposition in d-q axis frame of reference are utilized in the reference current generation algorithm. Accordingly, the source current waveform is made to follow the reference current waveform using a high-frequency, carrier-based controller. Further, this inner current control loop is supported by a slower outer voltage control loop for sustaining desirable DC link voltage. Performance of the compensator is evaluated through MATLAB simulation covering different types of loads and reduction of harmonic currents and THD at the supply side along with excellent regulation of DC link voltage are confirmed. The performance of a hybrid compensator designed and fabricated using the above principles is evaluated and corroborated with the simulation results.

Keywords: harmonics; hybrid power filter; active power filter; power quality; total harmonic distortion

---

1. Introduction

One major area of research that has gained attention in recent times is maintaining power quality of distribution systems. The power quality issues arise due to the widespread usage of processed power in industrial applications and commercial/domestic applications [1–3]. For example, variable speed drives are implemented through power modulators which consist of high-power controlled/uncontrolled rectifiers feeding variable voltage and variable frequency multiphase inverters. Similarly, commercial power consumption is characterised by appliances like computers, photocopiers, and fax machines, along with fluorescent and CFL lamps. All the above represent nonlinear loads, resulting in lower supply-side power factor and waveform distortion, indicated by harmonic components in voltage and current. The adverse effect of harmonics includes heating and extra losses, saturation and malfunctioning of distribution transformers, interference with communication signals, damages to consumer utilities, and in extreme cases, the failure of supply-side equipment [4].

The initial steps towards mitigation of the above problems were focused on the low power factor at the supply side only, whereby a passive power filter (PPF) connected in shunt compensates for the lagging reactive current, which was extended with selective harmonic elimination. The disadvantage of this approach is insensitivity with load current changes and fluctuations in supply-side voltage.

The filter performance also depends on the load power factor, which may be variable. For example, in [5], the load is a motor that can work at different conditions. As a means of overcoming these problems, various active compensator topologies comprising of series and shunt elements have been proposed which have gained wide acceptance. However, some of the disadvantages, like high initial and operating cost due to the use of high-rating semiconductors and also the need for maintaining a high DC link voltage, have limited the application of pure shunt active power filter in medium- and high-power installations [6,7]. As a result, the stage was set for development of shunt hybrid power filters (HPFs), which represent a judicious combination of both passive and active power filters (APFs). The various hybrid compensator configurations are: (i) series active power filter and shunt passive power filter, (ii) shunt active power filter and shunt passive power filter and (iii) a series combination of passive power filter and active power compensator connected in shunt with the system. The series element in configuration (i) has to be rated for maximum load current and is not flexible for many applications. The configuration (ii) with two independent compensators in parallel requires both blocks to be rated at the supply voltage leading to high DC link capacitor voltage rating [8–10].

In comparison, the third topology where the passive power filter elements appear in series with the standard active power filter circuitry poses important advantages in terms of reflecting nearly zero impedance of the passive power filter for load current harmonics and at the same time high impedance for system side voltage harmonics. Further, it leads to absorbing the fundamental voltage component across the passive power filter, thereby reducing the voltage rating of the DC link capacitor to only the harmonic components which are to be suppressed. Accordingly, the rating and cost of the capacitor and the power semiconductor switches in the compensator are considerably reduced [8]. Further, this topology is effective in preventing system resonance, reducing switching noise and avoidance of any circulating current in the compensator.

The control requirements for all configurations involving active power filter boil down to the generation of reference current waveform, switching and triggering timings for semiconductor switches to match the source current and also for maintaining the desired voltage across the DC link capacitor. Accordingly, a variety of control schemes supported by related algorithms have been reported in the literature [11–20]. Time domain methods provide fast response, compared with frequency domain methods. Accordingly, many authors have proposed control techniques such as instantaneous reactive power theory, synchronous rotating reference frame (SRRF) theory, sliding mode controller, neural network techniques and feedforward control to improve the performance of both active as well as hybrid filters.

In this paper, the third topology has been utilised, where the passive power filter is designed with the aim of 5th and 7th selective harmonic elimination along with an active power compensator connected in series. The active component of the compensator is modelled in the stationary abc frame of mains and further transformed to the rotating dq frame to avoid time dependence of parameters to reduce the control complexity. A control technique using PI controller, based on decoupled currents, is used to inject currents from the compensator to ensure tracking of the reference waveform [13,14,21]. An independent outer control loop using another PI controller regulates the DC link voltage for sustained operation of the compensator. The parallel combination of 5th and 7th tuned passive harmonic filters connected in series with active filter configuration, the fundamental voltage of the system mainly drops on the PPF capacitor, not in the APF. Hence, the APF DC link voltage has been reduced with an objective of APF voltage rating reduction. The passive filter parameters present in this topology not only function as harmonic filter but also act as a filter for the switching ripples present in the system. Finally, the entire power and control circuits are fabricated, where the FPGA development kit SPARTAN-6 has been employed as the controller. Further, the performance improvement of the supply system along with the fabricated compensator has been evaluated by carrying out by a series of experiments on a prototype system and the results are presented covering different loading conditions.

2. System Configuration of Hybrid Compensator

The system shown in Figure 1 consists of a standard three-phase 400 V, 50 Hz mains connected to a pair of three-phase uncontrolled bridge rectifiers feeding individual R-L, R-C loads, which introduce harmonic currents in the supply system due to the nonlinearity. The shunt compensator contains 5th and 7th tuned passive filters, which are in parallel and the combination is connected in series with low voltage rated active power filter. The parallel connected passive filters having selective harmonic elimination are meant for effective reactive power compensation. This system eliminates the inherent disadvantages of both active and passive filters. The hybrid compensator is operated such that the distortion currents generated due to nonlinear loads are confined within the PPF and do not flow in the AC supply mains. The fundamental voltage drop across passive power filter components permits shunt APF to operate at low DC link voltage. The three-phase source voltage, the voltage at the point of common coupling, load and compensator currents are denoted as V_{abc}, V_{s123}, i_{L123} and i_{c123}, respectively, and shown individually in Figure 1.

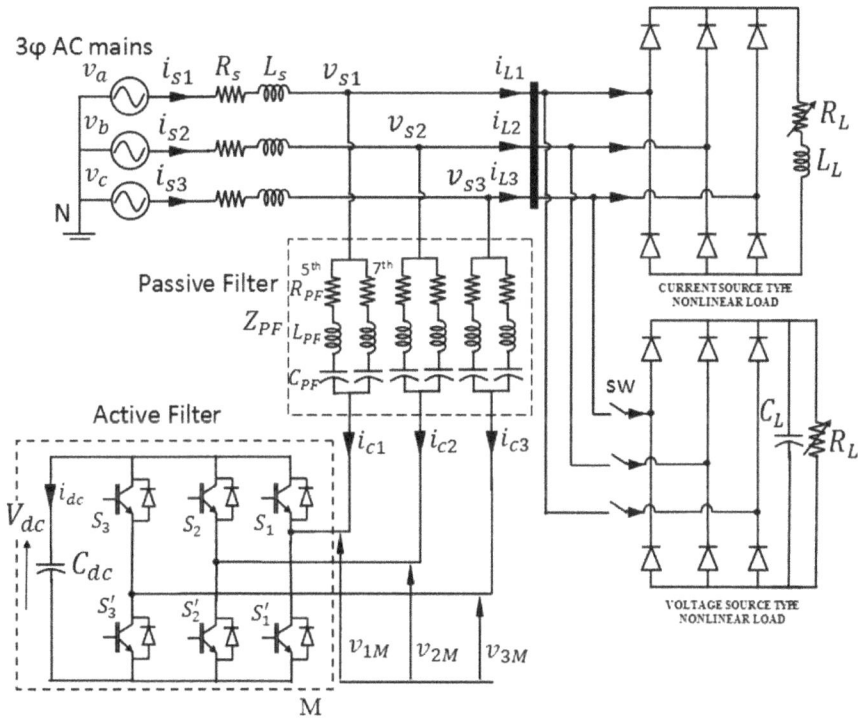

Figure 1. Configuration of the shunt hybrid compensator.

2.1. Modelling of Shunt Hybrid Power Compensator

By applying Kirchhoff's laws in Figure 1, the following equations in differential form in stationary three-phase reference frame are obtained, where the R_{PFe}, L_{PFeq} and C_{PFeq} are equivalent parameter values of the 5th and 7th selective harmonic filters.

$$v_{sk} = L_{PFeq}\frac{di_{ck}}{dt} + R_{PFeq}i_{ck} + (1/C_{PFeq})\int i_{ck}dt + v_{kM} + v_{MN} \tag{1}$$

where k = 1, 2, 3 represent the three phases.

Differentiation of Equation (1) to eliminate the integral term yields

$$\frac{dv_{sk}}{dt} = L_{PFeq}\frac{d^2 i_{ck}}{dt^2} + R_{PFeq}\frac{di_{ck}}{dt} + \frac{1}{C_{PFeq}}i_{ck} + \frac{dv_{kM}}{dt} + \frac{dv_{MN}}{dt} \tag{2}$$

Assuming balanced three-phase supply voltage yields

$$v_{s1} + v_{s2} + v_{s3} = 0$$

Summing the three equations included in (1) for $k = 1, 2, 3$ and assuming nonexistence of the zero-sequence current into three-wire system [16] results in

$$v_{MN} = -\frac{1}{3}\sum_{k=1}^{3} v_{kM} \tag{3}$$

The switching function C_k [12] of the kth leg of the inverter is the state of the power semiconductor devices S_k and S'_k and is defined as

$$C_k = \begin{cases} 1, \; if \; S_k \; is \; On \; and \; S'_k \; is \; Off \\ 0, \; if \; S_k \; is \; Off \; and \; S'_k \; is \; On \end{cases} \tag{4}$$

Thus, with $v_{kM} = C_k V_{dc}$ and differentiation of the same, this leads to

$$\frac{dv_{kM}}{dt} = C_k\frac{dV_{dc}}{dt} \tag{5}$$

By differentiating Equation (3), we get

$$\frac{dv_{MN}}{dt} = -\frac{1}{3}\sum_{k=1}^{3}\frac{d}{dt}(C_k V_{dc}) \tag{6}$$

Substitution of Equations (5) and (6) into (2) and rearranging the same yields

$$\frac{d^2 i_{ck}}{dt^2} = \frac{R_{PFeq}}{L_{PFeq}}\frac{di_{ck}}{dt} - \frac{1}{C_{PFeq}L_{PFeq}}i_{ck} - \frac{1}{L_{PFeq}}\left(C_k - \frac{1}{3}\sum_{m=1}^{3} C_m\right)\frac{dV_{dc}}{dt} + \frac{1}{L_{PFeq}}\frac{dv_{sk}}{dt} \tag{7}$$

defining the switching state function as $q_{nk} = \left(C_k - \frac{1}{3}\sum_{m=1}^{3} C_m\right)_n$, where, $n = 0$ or 1.

In other words, the vector q_n epends on the parameters C_1, C_2, C_3 through a matrix transformation given below, indicating the interaction among the three phases [15].

$$\begin{bmatrix} q_{n1} \\ q_{n2} \\ q_{n3} \end{bmatrix} = \frac{1}{3}\begin{bmatrix} 2 & -1 & -1 \\ -1 & 2 & -1 \\ -1 & -1 & 2 \end{bmatrix}\begin{bmatrix} C_1 \\ C_2 \\ C_3 \end{bmatrix} \tag{8}$$

$$\frac{d^2 i_{ck}}{dt^2} = -\frac{R_{PFeq}}{L_{PFeq}}\frac{di_{ck}}{dt} - \frac{1}{C_{PFeq}L_{PFeq}}i_{ck} - \frac{1}{L_{PFeq}}q_{nk}\frac{dV_{dc}}{dt} + \frac{1}{L_{PFeq}}\frac{dv_{sk}}{dt} \tag{9}$$

The capacitor current $i_{dc} = i_{c1} + i_{c2} + i_{c3}$ and is related to V_{dc} by

$$\frac{dV_{dc}}{dt} = \frac{1}{C_{dc}}i_{dc} \tag{10}$$

Expressing the DClink capacitor current i_{dc} in terms of the switching and compensator currents, the following equation is obtained

$$\frac{dV_{dc}}{dt} = \frac{1}{C_{dc}} \sum_{k=1}^{3} q_{nk} i_{ck} = \frac{1}{C_{dc}} [q_{n123}]^T [i_{c123}] \tag{11}$$

In the nonexistence of zero-sequence currents, the variables i_{c3} and q_{n3} can be eliminated by the substitution of $i_{c3} = -(i_{c1} + i_{c2})$ and $q_{n3} = -(q_{n1} + q_{n2})$ so that Equation (11) for the modelling of the capacitor is modified as follows.

$$\frac{dV_{dc}}{dt} = \frac{1}{C_{dc}} [2q_{n1} + q_{n2}] i_{c1} + \frac{1}{C_{dc}} [q_{n1} + 2q_{n2}] i_{c2} \tag{12}$$

The complete model of the shunt hybrid power compensator in *abc* reference frame is indicated by Equations (9) and (12).

2.2. Equations in dq Frame

The model given by Equations (9) and (12) is transformed into synchronous orthogonal frame using the transformation matrix [13]

$$T_{dq}^{123} = \sqrt{\frac{2}{3}} \begin{bmatrix} \cos\theta & \cos\left(\theta - \frac{2\pi}{3}\right) & \cos\left(\theta - \frac{4\pi}{3}\right) \\ -\sin\theta & -\sin\left(\theta - \frac{2\pi}{3}\right) & -\sin\left(\theta - \frac{4\pi}{3}\right) \end{bmatrix} \tag{13}$$

where $\theta = \omega t$ and ω represents the mains frequency.

Since T_{dq}^{123} is orthogonal, $\left(T_{dq}^{123}\right)^{-1} = \left(T_{dq}^{123}\right)^T$ and Equation (12) can be written as

$$\frac{dV_{dc}}{dt} = \frac{1}{C_{dc}} \left(T_{dq}^{123} [q_{ndq}]\right)^T \left(T_{dq}^{123} [i_{dq}]\right) = \frac{1}{C_{dc}} [q_{ndq}][i_{dq}] \tag{14}$$

On the other hand, Equation (9) can be written as

$$\frac{d^2}{dt^2} [i_{c12}] = -\frac{R_{PFeq}}{L_{PFeq}} \frac{d}{dt} [i_{c12}] - \frac{1}{C_{PFeq} L_{PFeq}} [i_{c12}] - \frac{1}{L_{PFeq}} [q_{n12}] \frac{dV_{dc}}{dt} + \frac{1}{L_{PFeq}} \frac{d}{dt} [v_{s12}] \tag{15}$$

The three-phase current with the absence of zero-sequence components can be converted into *d-q* frame using reduced transformation matrix. Applying the transformations in Equation (15), the complete *d-q* frame dynamic model of the system is obtained as follows;

$$L_{PFeq} \frac{d^2 i_d}{dt^2} = -R_{PFeq} \frac{di_d}{dt} + 2\omega L_{PFeq} \frac{di_q}{dt} - \left(-\omega^2 L_{PFeq} + \frac{1}{C_{PFeq}}\right) i_d + \omega R_{PFeq} i_q - q_{nd} \frac{dV_{dc}}{dt} + \frac{dv_d}{dt} - \omega v_q \tag{16}$$

$$L_{PFeq} \frac{d^2 i_q}{dt^2} = -R_{PFeq} \frac{di_q}{dt} - 2\omega L_{PFeq} \frac{di_d}{dt} - \left(-\omega^2 L_{PFeq} + \frac{1}{C_{PFeq}}\right) i_q - \omega R_{PFeq} i_d - q_{nq} \frac{dV_{dc}}{dt} + \frac{dv_q}{dt} + \omega v_d \tag{17}$$

$$C_{dc} \frac{dV_{dc}}{dt} = q_{nd} i_d + q_{nq} i_q \tag{18}$$

The role of i_d in Equation (16) is interpreted as the component for meeting the switching losses in the compensator, whereas the component i_q is utilised to supply reactive power and maintain the DClink voltage across the capacitor for sustaining the compensator action.

It is specifically noted that this set of Equations (16)–(18) contain nonlinear terms involving the control variables q_{nd} and q_{nq}. Accordingly, the implementation of this control strategy is termed as nonlinear control technique by many authors in the literature [12,15,16].

2.3. Control of Harmonic Currents

Based on the load and compensator models presented in Section 2.2, the control problem is formulated with the objective of minimizing supply-side current harmonics and improving the power factor. Also, for maintaining the performance during load fluctuations, it is necessary to maintain a desired DC link capacitor voltage. The control law is derived using the following approach.

Rewriting Equations (16) and (17) in a more convenient form, we get

$$L_{PFeq}\frac{d^2i_d}{dt^2} + R_{PFeq}\,\frac{di_d}{dt} + \left(-\omega^2 L_{PFeq} + \frac{1}{C_{PFeq}}\right)i_d$$
$$= 2\omega L_{PFeq}\frac{di_q}{dt} + \omega R_{PFeq}i_q - q_{nd}\frac{dV_{dc}}{dt} + \frac{dv_d}{dt} - \omega v_q \tag{19}$$

$$L_{PFeq}\frac{d^2i_q}{dt^2} + R_{PFeq}\,\frac{di_q}{dt} + \left(-\omega^2 L_{PFeq} + \frac{1}{C_{PFeq}}\right)i_q$$
$$= -2\omega L_{PFeq}\frac{di_q}{dt} - \omega R_{PFeq}i_d - q_{nq}\frac{dV_{dc}}{dt} + \frac{dv_q}{dt} + \omega v_d \tag{20}$$

The control variables u_d and u_q are defined as

$$u_d = 2\omega L_{PFeq}\frac{di_q}{dt} + \omega R_{PFeq}i_q - q_{nd}\frac{dV_{dc}}{dt} + \frac{dv_d}{dt} - \omega v_q \tag{21}$$

$$u_q = -2\omega L_{PFeq}\frac{di_d}{dt} - \omega R_{PFeq}i_d - q_{nq}\frac{dV_{dc}}{dt} + \frac{dV_q}{dt} + \omega v_d \tag{22}$$

Using the idea of decoupling the current harmonic components for the purpose of tracking the reference current, the error signals $\bar{i}_d = i_d^* - i_d$ and $\bar{i}_q = i_q^* - i_q$ are generated and processed through a pair of PI controllers [12,14] to obtain u_d and u_q signals which are given below.

i_d^* and i_q^* are the reference currents deduced from the load current, i_d and i_q are the actual compensator currents. Load current in d-q coordinate is processed using a pair of fourth-order Butterworth low-pass filters with cut-off frequency set at 60 Hz to extract the harmonic current references alone.

$$u_d = K_P\bar{i}_d + K_I\int \bar{i}_d dt$$

$$u_q = K_P\bar{i}_q + K_I\int \bar{i}_q dt$$

From Equations (19) and (20), we obtain the transfer function as follows

$$\frac{I_d(s)}{U_d(s)} = \frac{I_q(s)}{U_q(s)} = \frac{1}{L_{PFeq}\left(s^2 + \frac{R_{PFeq}}{L_{PFeq}}s + \frac{1}{C_{PFeq}L_{PFeq}} - \omega^2\right)} \tag{23}$$

The current control of the closed-loop system is shown in Figure 2 and represents the signal flow of the variables $i_q(s)$; the other component $i_d(s)$ is obtained concurrently in a similar manner. The transfer function of the full closed-loop control module is derived as

$$\frac{I_q(s)}{I_q^*(s)} = \frac{K_P}{L_{PFeq}}\left[\frac{s + \frac{K_I}{K_P}}{s^3 + \frac{R_{PFeq}}{L_{PFeq}}s^2 + \left(\frac{1}{C_{PFeq}L_{PFeq}} - \omega^2 + \frac{K_P}{L_{PFeq}}\right)s + \frac{K_I}{L_{PFeq}}}\right] \tag{24}$$

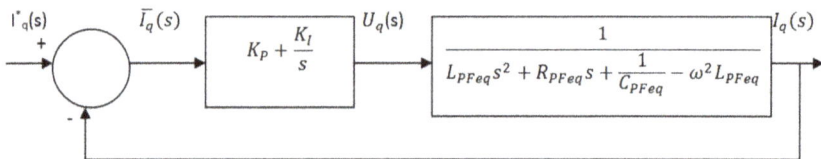

Figure 2. Structure of current control loop for i_q and i_d.

From Equations (21) and (22), the control variables of the proposed system are defined by the following equations.

$$q_{nd} = \frac{2\omega L_{PFeq}\frac{di_q}{dt} + \omega R_{PFeq}i_q + \frac{dV_d}{dt} - \omega v_q - u_d}{\frac{dV_{dc}}{dt}} \tag{25}$$

$$q_{nq} = \frac{-2\omega L_{PFeq}\frac{di_d}{dt} + \omega R_{PFeq}i_d + \frac{dV_q}{dt} - \omega v_d - u_q}{\frac{dV_{dc}}{dt}} \tag{26}$$

The above equations represent both the decoupled linear compensation part and cancellation of nonlinearity.

2.4. DC Link Voltage Control

The capacitor in the APF does not need any external DC source but gets charged through the rectifier action of the built-in reverse diodes across the six Insulated-gate bipolar transistors (IGBTs). The power loss in the capacitor and switching losses in the inverter have to be met by the active component id of the compensator current from the mains, while the component i_q supplies the reactive power stored in the capacitor. The power losses in this circuit can reduce the DC link capacitor voltage, thereby weakening the function of the active filter. Hence, it is necessary to maintain the voltage across the DC link capacitor at a designed reference value by an additional voltage regulator, which modifies the PWM signals appropriately. This regulator is implemented by using a PI controller [12], which processes the error between the reference voltage V_{dc}^* and the actual capacitor voltage V_{dc}. The parameters of the PI regulator are chosen in such a way that the DC voltage is maintained around its desired value. The design values for the PI controller parameters have been obtained following the approach suggested by Salem Rahmani et al. [13,14]. The overall transfer function of this controller is incorporated as a subsystem in the simulation schematic.

The control scheme of the proposed hybrid power compensator is shown in Figure 3, where the various signals such as V_{dc}, V_{dc}^*, i_d, i_q, i_d^* and i_q^* are processed using Equations (21) to (26) to obtain the gate trigger signals.

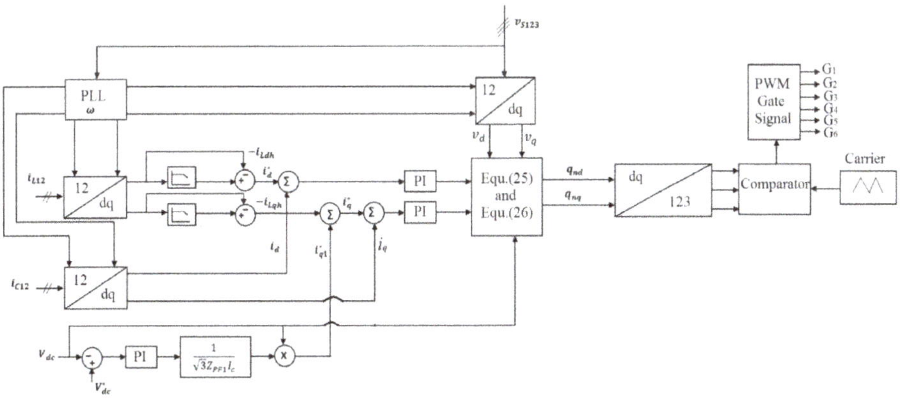

Figure 3. Control structure of the shunt hybrid compensator.

2.5. Simulink Model of the Shunt Hybrid Power Compensator

The Simulink schematic of the power distribution system along with the proposed shunt hybrid power compensator, which translates the entire system equations presented in Section 2 into functional blocks in the Simulink model, has been developed using MATLAB software. The nonlinear load has been modelled using a three-phase full bridge diode rectifier feeding RL and RC loads separately as subsystems and connected to the three phase mains. At the source side, a resistor–inductor combination is used to represent the line impedance of each phase before PCC. The unbalance case in the nonlinear load has been formed by connecting a single-phase diode rectifier feeding RL load between phase1 and ground. The synchronization of the subsystem with main supply frequency is accomplished by using a three-phase discrete PLL block [22–25]. The PLL block detects the supply frequency; the detected supply frequency is used to synchronize the compensator *d-q* axis current and the distorted load *d-q* axis current to extract harmonic component needs to be compensated. During load disturbance and supply voltage distortions, the variation in the supply frequency is detected by the PLL block and synchronizes the subsystems accordingly.

The hybrid filter consists of a parallel connection of selective 5th and 7th harmonic elimination passive filters as depicted in Figure 1 along with APF. The entire system simulation is carried out in a discrete mode, variable step size with ode45 (Dormand–Prince) solver.

3. Simulation Results

To evaluate the performance of the shunt HPF controlled by the proposed control algorithm, a Simulink-based schematic was created so as to operate from a 400 V, 50 Hz supply. The load consists of a set of balanced and unbalanced nonlinear loads, which are selectively connected for successive simulation runs. In this work, the performances of the compensator corresponding to the following loads are analysed: (i) three-phase rectifier feeding RL load, (ii) three-phase rectifier feeding RC load, (iii) dynamic load variation and (iv) unbalanced load. Table 1 indicates the specification of the system parameters used in the simulation.

Table 1. Parameters of the system.

Phase Voltage and Frequency	Vsrms = 230 V and fs = 50 Hz
Impedance of the line	Rs = 0.1 Ω, Ls = 4 mH
Nonlinear load of current source type	RL = 50 Ω, LL = 10 mH
Nonlinear load of voltage source type	RL = 32 Ω, CL = 1000 μF
5th tuned PPF parameters	R = 0.1 Ω, L = 10 mH, C = 40 μF
7th tuned PPF parameters	R = 0.1 Ω, L = 7 mH, C = 30 μF
DClink voltage and capacitance	V_{dc} = 25 V, C_{dc} = 6600 μF
Parameters of the outer loop PI controller	k1 = 0.22 and k2 = 15.85
Inner loop PI controller parameters	K_P = 0.6 and K_I = 1.2

3.1. Performance of Shunt HPF to the Nonlinear Load of Current Source Type

The three-phase mains supply a diode bridge rectifier, whose output is wired to a balanced R-L load, which imposes a typical nonlinearity, resulting in harmonic currents at the supply end. The proposed hybrid shunt compensator is wired at PCC as shown in Figure 1. The results are shown in Figure 4a–e covering the waveforms of the supply voltage, currents of the load, source current after compensation, currents of the compensator and DC link voltage, respectively. The waveform depicts that the supply current waveform is almost sinusoidal after compensation and the DC link capacitor voltage is maintained constant. Figure 5 depicts the simulation results covering the harmonic spectrum of the supply current before and after compensation. The results indicate that the THD of the supply current is reduced from 25.74% without compensator to 4.68% with compensator. It is seen that the proposed control strategy with shunt HPF can mitigate the current harmonics present in the supply system within the limit specified by IEEE 519-1992 standards.

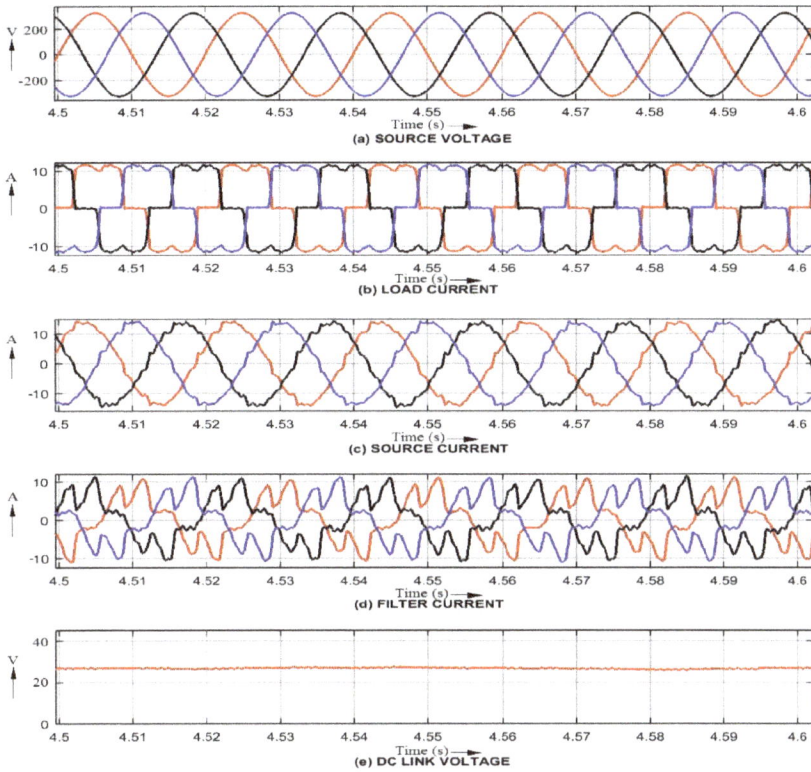

Figure 4. Performance of the compensator to the nonlinear load of current source type.

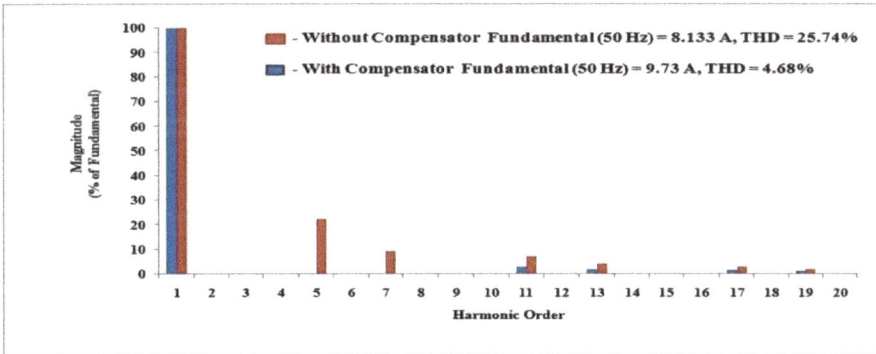

Figure 5. FFT analysis of the supply current with and without compensator.

3.2. Varying Three-Phase Rectifier-Fed RL Load

Since the load in a distribution system can vary concerning time, it is essential to verify the suitability of the designed filter under varying load conditions. The simulation results for sudden change of load current from 9.74 A (rms) to 17.17 A at $t = 4$ s and restoration of the same to the previous value at $t = 4.1$ s are presented in Figure 6a–e covering source voltage, load current, source current, compensator current and DC link voltage waveforms, respectively. The THD of the supply current before step change in the load current has been improved from 25.74% without compensator to 4.55%

with compensator. Similarly, after a step change in the load current, the supply current THD has been reduced from 23.06% to 4.14% with compensator. In both the cases, the compensator is capable of maintaining the THD of the source-side current within the limit specified by the IEEE 519-1992 standards. It is observed from Figure 6e,c that the compensator is able to maintain the DC link voltage as constant and supply current waveforms to be sinusoidal.

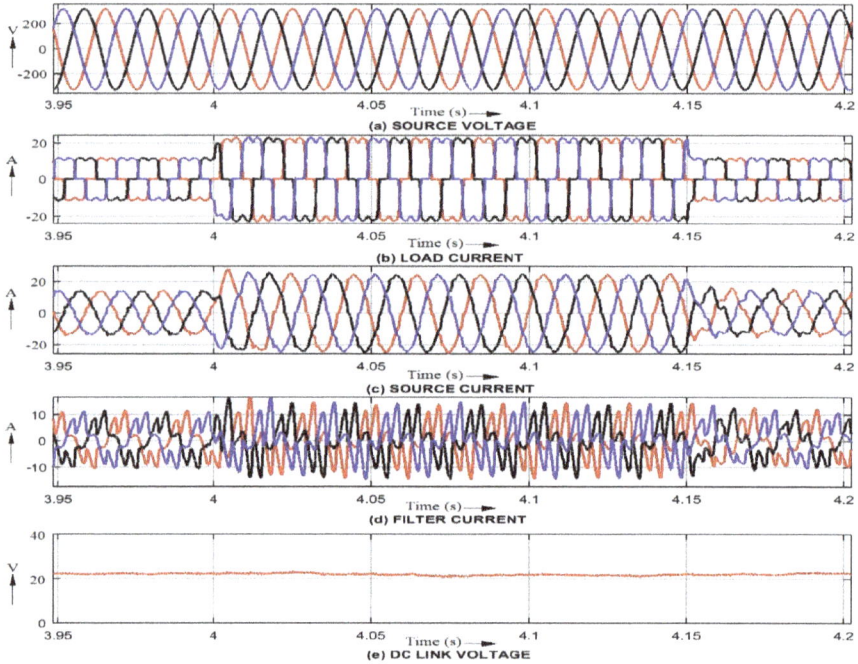

Figure 6. Response of the compensator to sudden change of load current.

3.3. Three-Phase Rectifier-Fed RC Load

The load in the simulation schematic is replaced by an R-C circuit with parameter values as shown in Table 1 and the simulation is executed to obtain the performance of the compensator. Figure 7a–e shows the source voltage, load current, supply current after compensation, compensator current and DC link voltage waveforms, respectively. It is observed from the figures that the DC link voltage of the compensator is maintained constant during compensation, and the supply current variation approaches a sinusoidal waveform. Figure 8 shows the harmonic spectrum of line currents from mains before and after compensation. It shows that the THD of the source current is reduced considerably from 27.49% to 4.08%. The system is highly nonlinear and subjected to disturbances due to loading. Hence, in practice, the DC link voltage varies slightly due to ripples and disturbances inherent in the operation of the compensator.

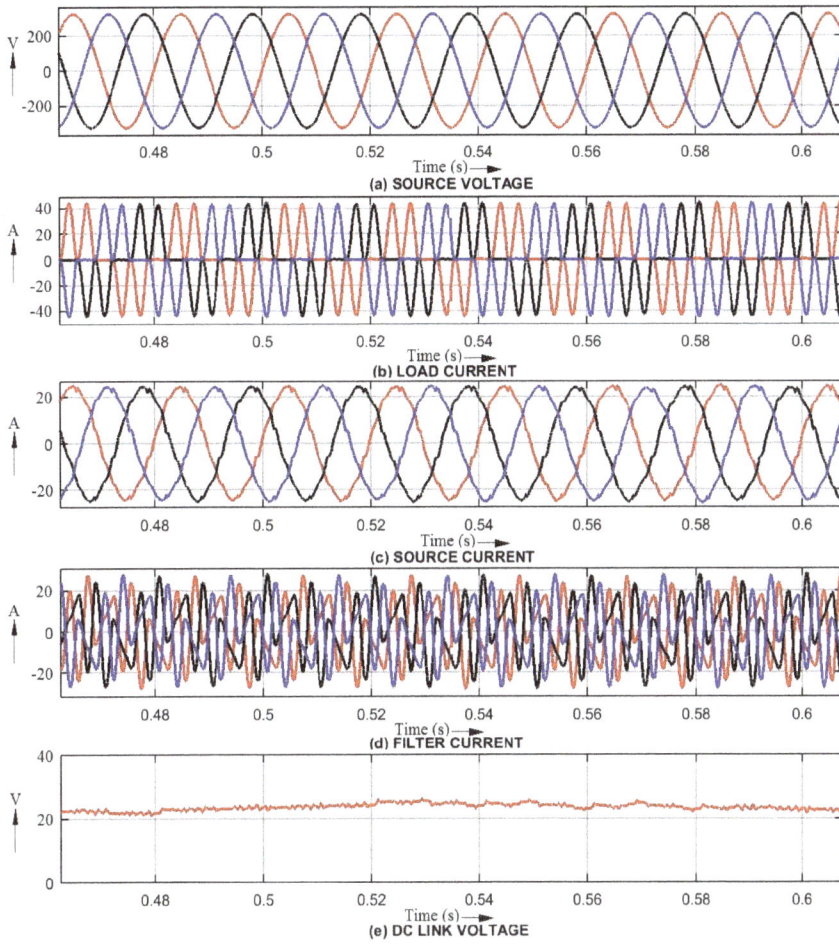

Figure 7. Performance of compensator for voltage source type nonlinear load.

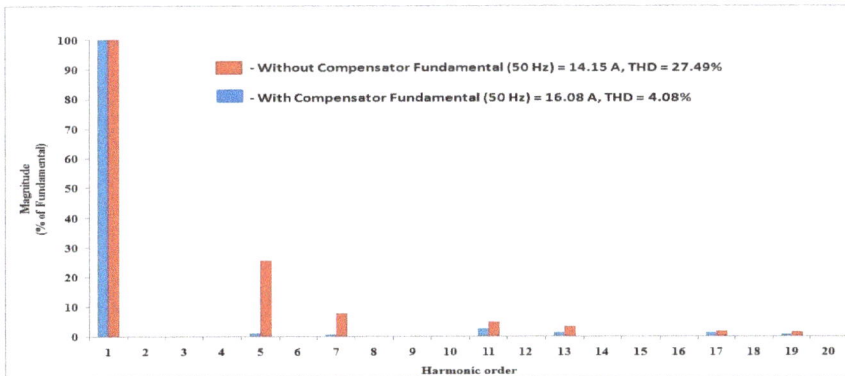

Figure 8. Harmonic spectrum of supply current before and after compensation.

3.4. Varying Three-Phase Rectifier-Fed RC Load

By introducing a 100% step increase followed by a similar decrease in the load current of the R-C circuit, the ability of the hybrid compensator to mitigate the harmonic current under varying load conditions is examined in the simulation setup. The results of this simulation, where the load current changes from 16.05 A (rms) to 30.32 A at t = 2 s and restoration of the same to 16.05 A are shown in Figure 9a–e. These depict the supply voltage, load current, supply current after compensation, compensator current and the DC link voltage, respectively. The obtained results indicate that the desirable features in the performance of the hybrid filter are maintained even after the step changes in the load within a very short time. The THD of the source current has been reduced from 27.49% without compensator to 3.46% with compensator before the step change is made in the load current. Similarly, the THD has been reduced from 20.91% to 4.77% after the step change. It is seen that the compensator operation reduces the harmonic content within a THD of 5%, in addition to maintaining a steady DC link voltage.

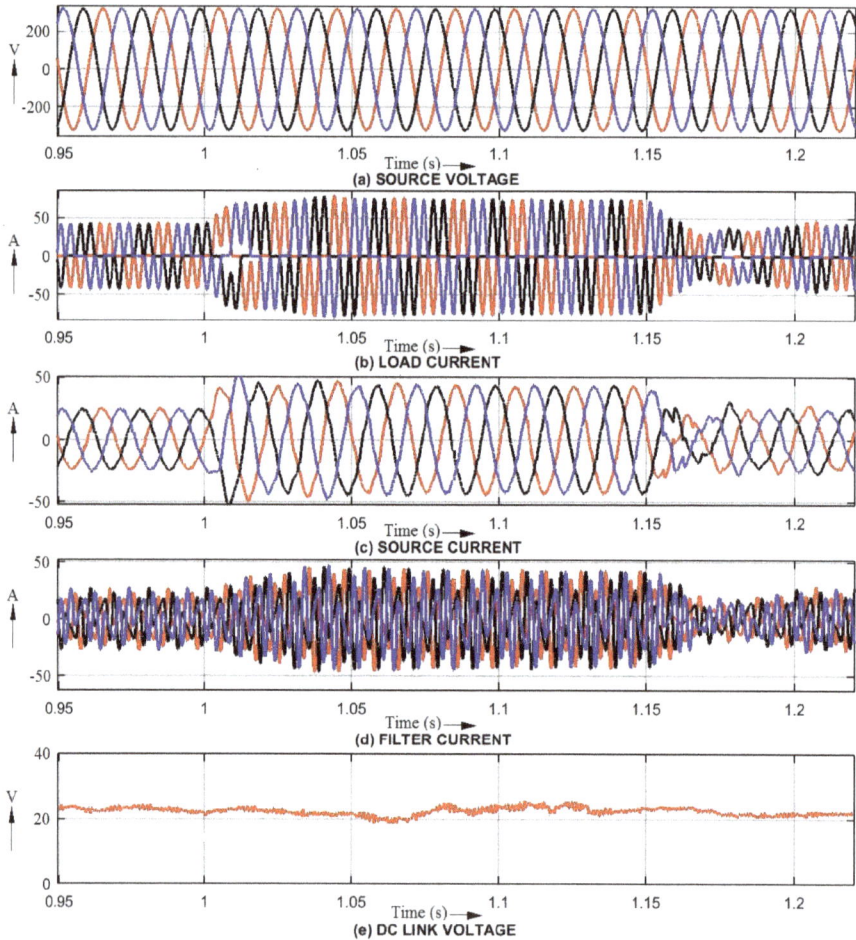

Figure 9. Performance of the compensator for variation in three-phase rectifier-fed RC load.

3.5. Unbalanced Loading Condition

To study the performance of a shunt hybrid power compensator employing synchronous rotating reference-based nonlinear control method, an unbalance in load is created by connecting a single-phase rectifier feeding R-L load across the phase 'a' and phase 'b' of the supply system. The compensation performance at steady state of the proposed compensator, such as the supply voltage, load current, supply current, compensator current and DC link voltage waveforms, is depicted in Figure 10a–e. The THD of supply phase 'a' current is reduced from 9.46% to 2.54% after compensation. Similarly, supply phase 'b' current has been reduced from 7.04% to 1.91%, while phase 'c' current has been reduced from 25.47% to 5.37%.

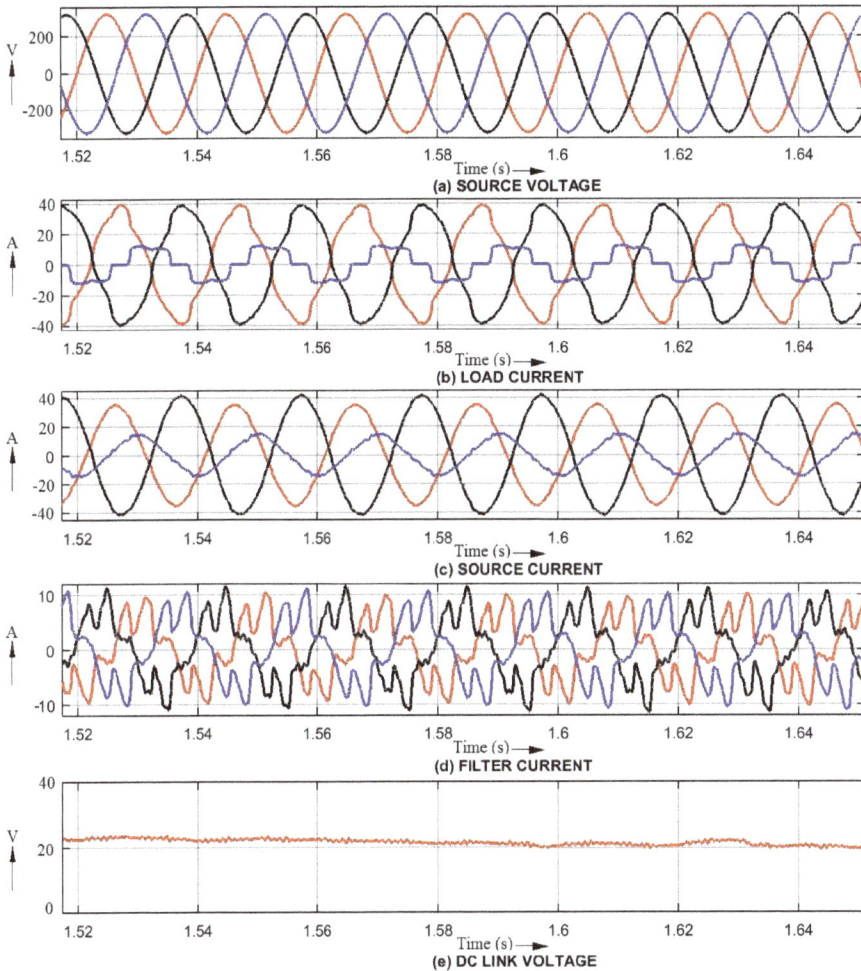

Figure 10. Performance of the compensator to the unbalanced nonlinear load.

It is seen from the obtained results that the proposed control method with shunt hybrid compensator effectively compensates the harmonic distortion present in the supply current under unbalanced loading conditions.

Since the load in a distribution system can vary with respect to time, it is essential to verify the suitability of the designed filter under varying load conditions. During dynamic variations in the load current and unbalance, as shown in Figures 6, 9 and 10, the compensator is able to maintain constant DC link voltage as well as reduce the amount of harmonics present in the system.

4. Hardware Fabrication

A prototype of the distribution system along with the passive and active parts of the hybrid compensator has been fabricated and experimental work on the same was carried out. Figure 11 shows the overall setup of the experimental work. The active filter part of the prototype has been developed using intelligent power module PEC16DSMO1. The IPM consists of a three-phase six-pulse inverter with six IGBT semiconductor power switches and is controlled in real time based on the software control program from the SPARTAN-6 FPGA development board. These IGBT switches are capable of operating at high frequencies of the carrier waveform up to 20 kHz.

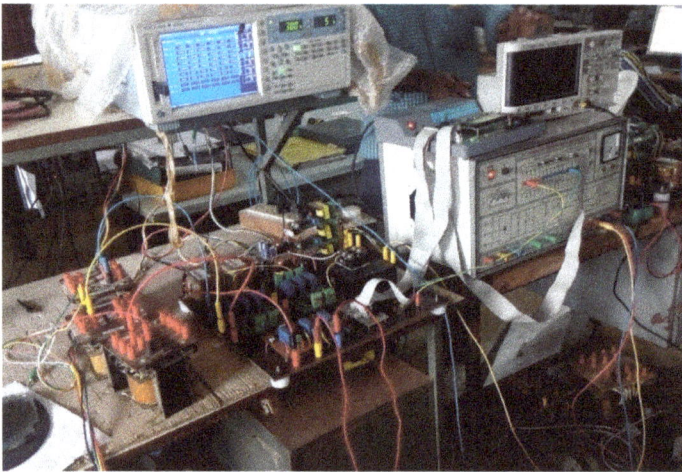

Figure 11. Experimental setup.

4.1. Hardware Resultsand Discussion

A series of experiments involving different types of loads and their dynamic variations have been conducted, and the performance of the compensator has been evaluated. A six-channel YOKOGAWA Power Quality Analyser has been used to capture the signals such as supply voltage, source current, load current, compensator current and DC link voltage. The hardware system parameters are shown in Table 2

Table 2. Experimental circuit parameters.

Active Power Filter	Intelligent Power Module (IPM) PEC16DSMO1 with 6 IGBT Switches
IGBT rating	25 A, 1200 V
Switching frequency of APF switches	2 kHz
Current sensors	LTS 25-NP
Voltage sensors	LV 25-P
Filter inductors	7 mH and 10 mH
Filter capacitors	30 μF and 40 μF
DC link capacitor	6400 μF

4.1.1. Performance of the Compensator for Current Source Type Nonlinear Load

Figure 12 shows the experimental response of the compensator for the current source type nonlinear load. The compensation performance of the hybrid filter has been verified by turn on the compensator at the point 'S' as shown in Figure 12. It is seen from the figure that the source current becomes sinusoidal after the operation of compensator at point 'S'. The harmonic spectrum of the source current without and with compensator is shown in Figure 13a,b and it is observed that the THD of the source current has been reduced from 26.225% to 2.212%.

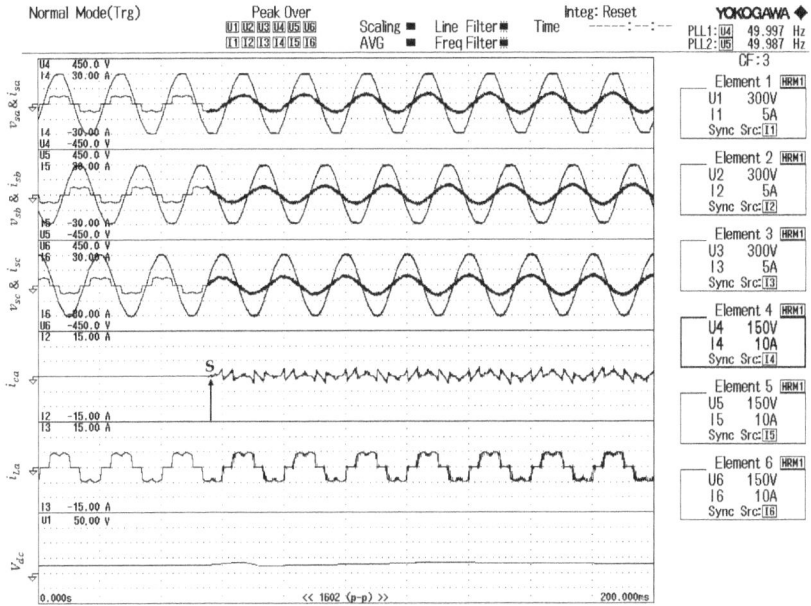

Figure 12. Hardware response of the compensator for current source type load.

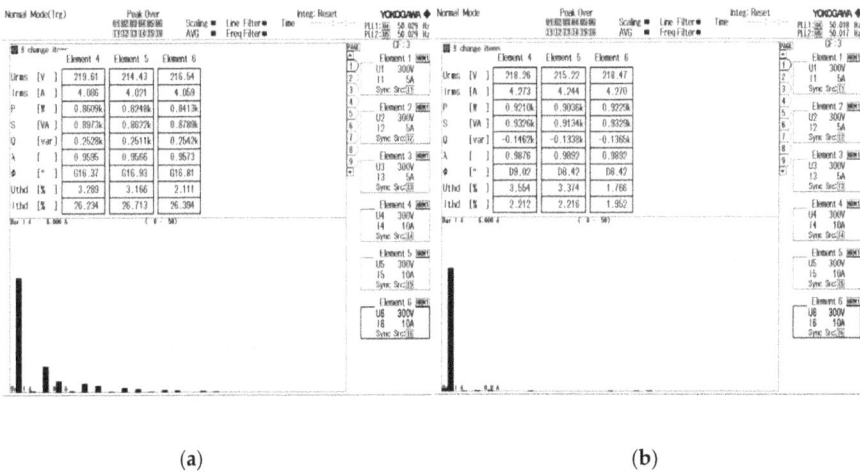

Figure 13. FFT spectrum of source current (**a**) before compensation and (**b**) after compensation.

4.1.2. Performance of the Compensator for Varying Current Source Type Nonlinear Load

The performance of the hybrid compensator has been verified experimentally by creating a step change in the load current at 'A' as shown in Figure 14. It is observed that the source current in all the three phases are sinusoidal and the compensator is able compensate the harmonic distortion even after a step change in the load current. The THD of the source current after compensation has been observed as 3.26%, which is less than 5% as specified by IEEE 519-1992 standards.

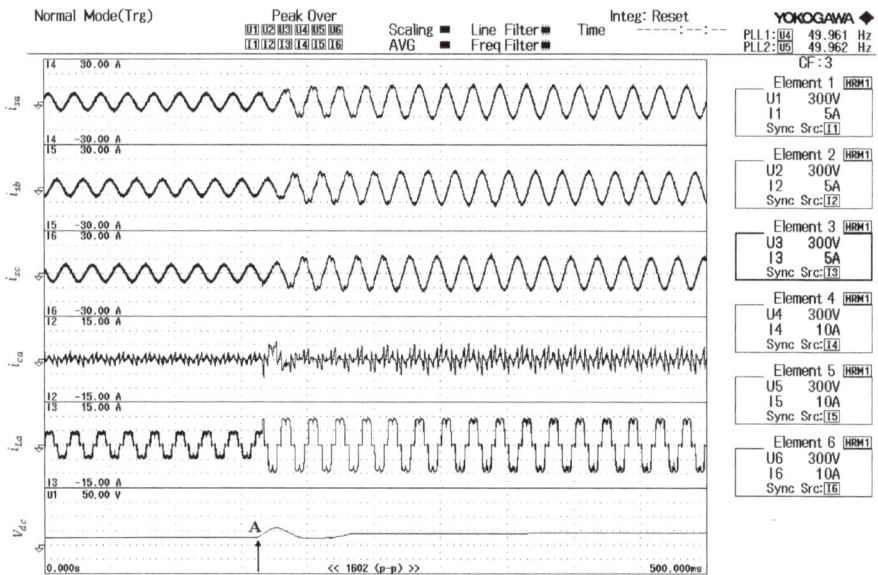

Figure 14. Performance of the compensator for time-varying current source type nonlinear load.

4.1.3. Performance of the Compensator for Voltage Source Type Nonlinear Load

The performance of the compensator for voltage source type nonlinear load has been experimentally verified by connecting the nonlinear load in the three-phase supply system. Figure 15 shows the three-phase source currents of all phases individually, compensator current of one phase, load current of one phase and DC link voltage, respectively. It is seen that the supply current after compensation is sinusoidal without distortion and the DC link voltage of the compensator is also maintained constant.

The FFT spectrum of the source current before and after compensation is shown in Figure 16a,b, respectively. It is observed from the FFT spectrum that the supply current harmonics have been reduced from 67.795% without compensator to 4.136% with compensator. It has been confirmed from the above results that the designed control strategy with proposed shunt hybrid power compensator is able to effectively compensate both the types of nonlinear load distortions present in the supply current. It is observed that the experimental work results are closely matching with simulation work results in all cases and the obtained performance waveforms are also matching with each other.

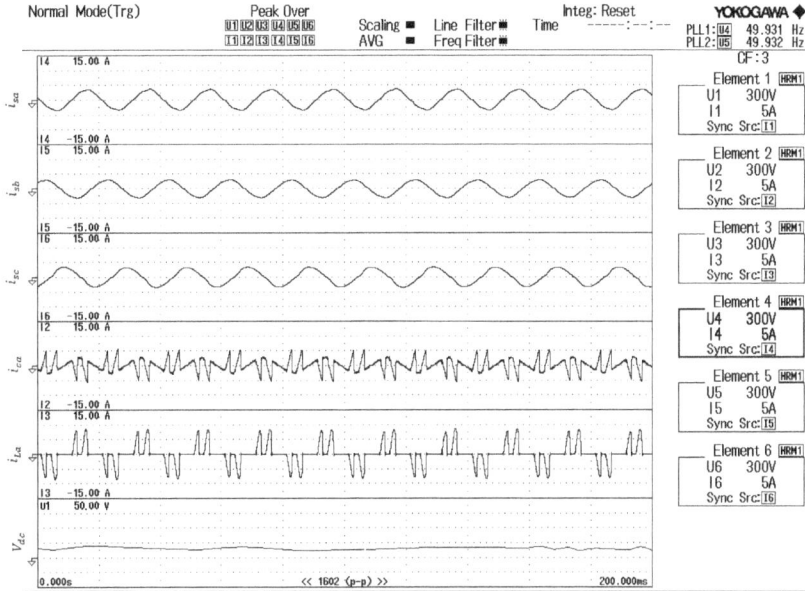

Figure 15. Performance of the compensator for voltage source type nonlinear load.

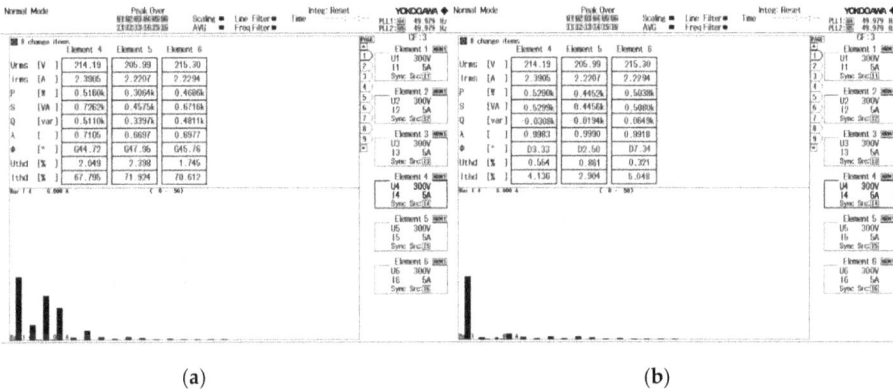

(a)   (b)

Figure 16. FFT Spectrum of source current (**a**) before compensation and (**b**) after compensation.

4.1.4. Performance Comparison and Discussion

Performance of the proposed hybrid filter with the adopted control strategy has been compared with some other control methods proposed in the literature. Table 3 shows the comparison of various control techniques proposed in the literature for harmonic reduction. It is clearly understood from Table 3 that the synchronous rotating reference theory based nonlinear control with the adopted hybrid filter topology proposed in this work gives better THD in the source current during supply feeds of different types of nonlinear loads. Most of the techniques proposed in the literature cover only current source type nonlinear load. In this work, different types of nonlinear loads and various loading conditio..ns are considered during simulation and experimental work. The experimental work has been carried out for low current rating nonlinear loads compared to the simulation work. Hence, it has been observed that the experimental THD values are better than the simulation results.

Furthermore, the hardware setup has been developed for low rating loads only due to the cost. The results of simulation indicate effectiveness of the compensator topology and versatility of the control scheme for meeting the requirements of a set of nonlinear loads indicated above. In addition, the *kVA* ratings of the designed active filter for both types of load are only a small percentage of the respective load ratings, which is a major advantage of this scheme. Although the overall performance of the hybrid compensator has been highly satisfactory, while covering alternate cases of load, their fluctuations and unbalance, it appears that the PI controller-based algorithm shows a minor deficiency, when the passive filter parameter values undergo even minor changes. This can be traced to the presence of fixed values of the PI controller gain and time constant parameters. A possible and feasible solution to this issue of filter parameter variations is to adaptively adjust the PI parameters' value. Although a three-phase distribution system for power quality enhancement has been considered in this paper, it is limited to a three-wire system only with isolated neutral. Hence, issues arising out of zero-sequence currents due to load unbalance have not been considered. This research work can be extended for a three-phase, four-wire system. Further, incorporation of controllers utilizing various other soft computing techniques like neural, fuzzy-neural and genetic algorithms can be carried out as further work.

Table 3. Comparison of the performance of various control techniques.

Control Methods	SRF Theory-Based Nonlinear Control for SHAPF [13]	p-q Theory-Based Control for SHAPF [26]	SRF Theory-Based Control for SHAPF [27]	Parallel Connected SHAPF [28]
THD% Three-phase rectifier-fed RL load	4.6	4.32	-	4.3
THD% Three-phase rectifier-fed RC load	4.08	4.15	-	-
Unbalanced load % THD	2.29 to 4.80	-	1.18 to 2	4.5 to 4.7
DC link voltage	25 V APF rating is less	50 V APF rating is moderate	220 V APF rating is more	26 V APF rating is less

5. Conclusions

A three-phase hybrid compensator scheme discussed in this paper is configured with 5th and 7th selective harmonic elimination passive filter in series with an active filter in the form of an IGBT-based PWM inverter, triggered using a control algorithm, based on a synchronously rotating reference frame (SRRF) in the *d-q* axis. The overall objective is to improve the steady state power factor and to ensure dynamic harmonic compensation at the supply side of three-phase mains. Accordingly, the entire system is simulated using Simulink and the results are presented, covering a set of nonlinear loads. The simulation results indicate the effectiveness of the compensator topology and versatility of the control scheme for meeting the requirements of a set of nonlinear loads. It is seen that significant reduction of source current THD is obtained within limits specified by IEEE 519-1992 standard along with excellent regulation of the DC link voltage.

A series of experiments were carried out using the above prototype along with alternate nonlinear loads and the overall performance of the hybrid filter was recorded using six channels DSO. The THD values as measured by the DSO over a series of experiments were found to be closely matching with corresponding THD values obtained during simulation for all loading conditions. It is further verified that the experimental results compared favourably with the simulation waveforms, thereby validating the simulation model. The experimental results confirm the effectiveness of the SRRF-based control strategy for the hybrid compensator scheme to achieve source side current harmonics reduction and power factor improvement to near unity.

Author Contributions: Conceptualization, R.B. and K.P.; methodology, R.B. and K.P.; software, R.B. and K.P.; validation, R.B., K.P.; and R.S.; formal analysis, R.B.; investigation, R.B. and K.P.; data curation, R.B., and K.P.; writing—original draft preparation, R.B., K.P., R.S. and R.A.; writing—review and editing, R.S. and R.A.; supervision, R.S. and R.A.

Funding: This research received no external funding.

Acknowledgments: Authors wish to express their sincere thanks to acknowledge SASTRA Deemed University, Thanjavur, India 613401 for extending infrastructural support to carry out this work.

Conflicts of Interest: The authors declare no conflict of interest.

References

1. Peng, F.Z. Harmonic sources and filtering approaches. *IEEE Ind. Appl. Mag.* **2001**, *7*, 18–25. [CrossRef]
2. Dugan, R.C.; Mark, F.M.; Surya, S.; Wayne, H.B. *Electrical Power Systems Quality*, 3rd ed.; Tata McGraw-Hill Education: Noida, India, 2012; ISBN 9780071761550.
3. Singh, B.; Singh, B.N.; Chandra, A.; Al-Haddad, K.; Pandey, A.; Kothari, D.P. A review of three-phase improved power quality AC-DC converters. *IEEE Trans. Ind. Electron.* **2004**, *51*, 641–660. [CrossRef]
4. Wagner, V.E.; Balda, J.C.; Griffith, D.C.; McEachern, A.; Barnes, T.M.; Hartmann, D.P.; Phileggi, D.J.; Emannuel, A.E.; Horton, W.F.; Reid, W.E.; et al. Effects of harmonics on equipment. *IEEE Trans. Power Deliv.* **1993**, *8*, 672–680. [CrossRef]
5. Das, J.C. Passive filters-potentialities and limitations. *IEEE Trans. Ind. Appl.* **2004**, *40*, 232–241. [CrossRef]
6. Singh, B.; Al-Haddad, K.; Chandra, A. A review of active filters for power quality improvement. *IEEE Trans. Ind. Electron.* **1999**, *46*, 960–971. [CrossRef]
7. Akagi, H. New trends in active filters for power conditioning. *IEEE Trans. Ind. Appl.* **1996**, *32*, 1312–1322. [CrossRef]
8. Lam, C.-S.; Wong, M.-C. *Design and Control of Hybrid Active Power Filters*; Springer: Berlin, Germany, 2014.
9. Singh, B.; Verma, V.; Chandra, A.; Al-Haddad, K. Hybrid filters for power quality improvement. *IEE Proc.-Gener. Transm. Distrib.* **2005**, *152*, 365–378. [CrossRef]
10. Rivas, D.; Morán, L.; Dixon, J.W.; Espinoza, J.R. Improving passive filter compensation performance with active techniques. *IEEE Trans. Ind. Electron.* **2003**, *50*, 161–170. [CrossRef]
11. Samadaei, E.; Lesan, S.; Cherati, S.M. A new schematic for hybrid active power filter controller. In Proceedings of the 2011 IEEE Applied Power Electronics Colloquium (IAPEC), Johor Bahru, Malaysia, 18–19 April 2011; pp. 143–148.
12. Mendalek, N.; Al-Haddad, K.; Fnaiech, F.; Dessaint, L.A. Nonlinear control technique to enhance dynamic performance of a shunt active power filter. *IEE Proc.-Electr. Power Appl.* **2003**, *150*, 373–379. [CrossRef]
13. Rahmani, S.; Hamadi, A.; Mendalek, N.; Al-Haddad, K. A new control technique for three-phase shunt hybrid power filter. *IEEE Trans. Ind. Electron.* **2009**, *56*, 2904–2915. [CrossRef]
14. Rahmani, S.; Mendalek, N.; Al-Haddad, K. Experimental design of a nonlinear control technique for three-phase shunt active power filter. *IEEE Trans. Ind. Electron.* **2010**, *57*, 3364–3375. [CrossRef]
15. Zouidi, A.; Fnaiech, F.; Al-Haddad, K. Voltage source inverter based three-phase shunt active power filter: Topology, Modeling and control strategies. In Proceedings of the 2006 IEEE International Symposium on Industrial Electronics, Montreal, QC, Canada, 9–13 July 2006; Volume 2, pp. 785–790.
16. Mendalek, N.; Al-Haddad, K. Modeling and nonlinear control of shunt active power filter in the synchronous reference frame. In Proceedings of the Ninth International Conference on Harmonics and Quality of Power. Proceedings (Cat. No.00EX441), Orlando, FL, USA, 1–4 October 2000; Volume 1, pp. 30–35.
17. Balasubramanian, R.; Sankaran, R.; Palani, S. Simulation and performance evaluation of shunt hybrid power filter using fuzzy logic based non-linear control for power quality improvement. *Sadhana* **2017**, *42*, 1443–1452. [CrossRef]
18. Palandöken, M.; Aksoy, M.; Tümay, M. Application of fuzzy logic controller to active power filters. *Electr. Eng.* **2004**, *86*, 191–198. [CrossRef]
19. Dey, P.; Mekhilef, S. Current controllers of active power filter for power quality improvement: A technical analysis. *Automatika* **2015**, *56*, 42–54. [CrossRef]
20. Unnikrishnan, A.K.; Subhash Joshi, T.G.; Manju, A.S.; Joseph, A. Shunt hybrid active power filter for harmonic mitigation: A practical design approach. *Sadhana* **2015**, *40*, 1257–1272.

21. Chauhan, S.K.; Shah, M.C.; Tiwari, R.R.; Tekwani, P.N. Analysis, design and digital implementation of a shunt active power filter with different schemes of reference current generation. *IET Power Electron.* **2013**, *7*, 627–639. [CrossRef]

22. Cataliotti, A.; Cosentino, V. A Time-domain strategy for the measurement of IEEE standard 1459–2000 Power quantities in Non sinusoidal Three–phase and Single–phase systems. *IEEE Trans. Power Deliv.* **2008**, *23*, 2113–2123. [CrossRef]

23. Cataliotti, A.; Cosentino, V. Disturbing Load Identification in Power Systems: A Single-Point Time–Domain Method Based on IEEE 1459–2000. *IEEE Trans. Instrum. Meas.* **2009**, *58*, 1436–1445. [CrossRef]

24. Golestan, S.; Gurrero, J.M.; Vasquez, J.C. Three–Phase PLLs: A Review of Recent Advances. *IEEE Trans. Power Electron.* **2017**, *32*, 1894–1907. [CrossRef]

25. Golestan, S.; Ge, J.M.; Vasquez, J.; Abusorrah, A.M.; Al-Turki, Y.A. Modeling, Tuning, and Performance Comparison of Second–Order-Generalized-Integrator–Based FLLs. *IEEE Trans. Power Electron.* **2018**, *33*, 10229–10239. [CrossRef]

26. Balasubramanian, R.; Palani, S. Simulation and performance evaluation of shunt hybrid power filter for power quality improvement using PQ theory. *Int. J. Electr. Comput. Eng.* **2016**, *6*, 2603–2609. [CrossRef]

27. Dey, P.; Mekhilef, S. Synchronous reference frame based control technique for shunt hybrid active power filter under non-ideal voltage. In Proceedings of the 2014 IEEE Innovative Smart Grid Technologies-Asia (ISGT ASIA), Kuala Lumpur, Malaysia, 20–23 May 2014; pp. 481–486.

28. Bhattacharya, A.; Chakraborty, C.; Bhattacharya, S. Parallel-connected shunt hybrid active power filters operating at different switching frequencies for improved performance. *IEEE Trans. Ind. Electron.* **2012**, *59*, 4007–4019. [CrossRef]

electronics

MDPI

Article

A Fixed-Frequency Sliding-Mode Controller for Fourth-Order Class-D Amplifier

Haider Zaman *, Xiancheng Zheng, Xiaohua Wu, Shahbaz Khan and Husan Ali

School of Automation, Northwestern Polytechnical University, Xi'an 710000, China; zxcer@nwpu.edu.cn (X.Z.); wxh@nwpu.edu.cn (X.W.); muhd_shahbaz@yahoo.com (S.K.); engr.husan@gmail.com (H.A.)
* Correspondence: hdrzaman@hotmail.com; Tel.: +86-1302-2995-582

Received: 20 September 2018; Accepted: 17 October 2018; Published: 19 October 2018

Abstract: Since the parasitic voltage ringing and switching power losses limit the operation of active devices at elevated frequencies; therefore, a higher-order inductor-capacitor (LC) filter is commonly used, which offers extended attenuation above the cutoff frequency and thus, improves the total harmonic distortion (THD) of the amplifier. This paper applies the concept of integral sliding-mode control to a fourth-order class-D amplifier. Two fixed-frequency double integral sliding-mode (FFDISM) controllers are proposed, where one uses the inductor current while the other involves the capacitor current feedback. Their equivalent control equations are derived, but from the realization viewpoint, the controller using the capacitor current feedback is advantageous and, therefore, is selected for final implementation. The performance of the proposed FFDISM controller for fourth-order GaN class-D amplifier is confirmed using simulation and experimental results.

Keywords: fixed-frequency double integral sliding-mode (FFDISM); class-D amplifier; Q-factor; GaN cascode

1. Introduction

For decades, silicon transistors have dominated the power amplifiers industry due to the low-cost and well-established fabrication technology. Since the transistors in a linear power amplifier operate in the active region where power dissipation is significant, thereby they experience poor efficiency. In addition to advanced fabrication techniques like laterally diffused metal oxide semiconductor (LDMOS) [1,2], different control strategies such as Doherty's architecture and load-modulation were adopted to improve the efficiency of a linear amplifier [3]. Due to a narrow margin for improvement left in Si, the demand for high operating voltage, temperature and efficiency has enabled the trend towards wide band-gap (WBG) materials. The attractive features such as a high electric breakdown field, low thermal impedance, and saturated electron drift velocity, motivated their rapid substitution for Si counterparts [4]. Particularly, GaN high electron mobility transistor (HEMT) has become a potential candidate for large bandwidth and low-noise power amplifiers [5].

The earlier release of high-power GaN HEMT was a depletion-mode device also referred to as normally-on FET [6]. Since it requires additional control and protection circuitry for a safe power-up of power converters built with normally-on FETs, therefore, enhancement-mode FETs are preferred over depletion-type devices. There were several attempts made to fabricate a normally-off GaN HEMT, including a recessed gate structure [7], Si substrate with p-type GaN [8], and fluorine plasma treatment [9]. However, due to a low threshold and gate breakdown voltages, the proposed HEMTs are vulnerable to spurious turn-on and gate failure. Alternatively, to achieve a normally-off GaN HEMT, a cascode configuration has been proposed by combining the GaN HEMT with a low-voltage Si metal oxide semiconductor field effect transistor (MOSFET) [10]. Due to the low-cost leaded packages and superior characteristics, GaN cascode is a dominating power device and is preferred over the enhancement-mode HEMTs [11]. They are commonly available in TO-220 package, which enables easy assembling of the heat-sink and does not need special equipment for soldering to printed circuit board (PCB).

The class-D amplifier motivated by its high-efficiency (ideally 100%) encodes the reference signal into a pulse-width modulation output using a switching power circuit, with pulse-width proportional to the amplitude of the reference signal [12]. Since the operation of active devices in either the cutoff or saturation region significantly reduces the power dissipation, therefore, the heat sink requirement relaxes. In the case of battery-powered devices, high efficiency means longer battery life. Thus, class-D amplifier is the ideal choice for miniaturized high-power amplification as compared to the class-A, class-B and class-AB. Today, in addition to the stereo system, class-D amplifiers are also used in a high-precision control application, including the wafer positioning system, magnetic resonance imaging (MRI), and power hardware-in-loop simulation (PHIL) [13,14].

An LC filter demodulates the output pulse-train by attenuating high-frequency content and produces the amplified output with minimum distortion. More LC stages are commonly added at the output of a class-D amplifier to meet the desired level of THD at a given switching frequency [15]. Low THD amplifier with high-power capability has applications in AC power sources and is used for the emulation of certain characteristics of an electrical system. However, the higher-order class-D amplifier causes an irregular-shaped frequency response due to multiple resonant frequencies in the uncompensated architecture [16]. Furthermore, the peaking at resonant frequency increases with an increase in load resistance and approaches zero-damping under the no-load condition. Therefore, it is required to have well-damped characteristics of a higher-order class-D amplifier, almost independent of load variations.

The feedback compensation of a class-D amplifier with a single LC stage is extensively investigated in the literature [17,18]. However, for the fourth-order system, the reported passive damping uses a low-valued resistor in series with filter capacitors to flatten the frequency response, at the cost of reduced efficiency [19]. Feedback controller supplemented with passive damping was adopted in Reference [20], which results in relatively lower power losses. In addition to an RL-branch between the capacitive filter of the first and inductive filter of the second LC stage, authors in Reference [21] proposed a multi-loop controller. Employment of such networks degrades the efficiency, which is the sole advantage of the class-D amplifier. Since passive damping negatively affects the efficiency of the amplifier, the application of high-cost GaN cascode becomes vestigial.

A purely feedback-controlled fourth-order class-D amplifier presented in Reference [22], achieved a peak efficiency of 87%. However, the controller was extremely complex as it requires feedback from all four state-variables. Feedback compensation using an integral sliding-mode (ISM) controller of the fourth-order class-D amplifier has been recently reported in Reference [23]. Nevertheless, it is inhibited for use due to the variable switching frequency nature of hysteretic modulation (HM) [24,25]. Therefore, a promising feedback controller with pulse-width modulation (PWM) is required, which ensures fixed switching frequency, flatter frequency response, reduced tracking error and a high-efficiency.

Since the equivalent control of a fixed-frequency sliding-mode controller is extracted from the sliding-surface by a differentiation operation [26]. Therefore, it is obvious to add a double integral term in the sliding-surface essential for ensuring reduced steady-state error [27,28]. However, the design of the sliding-surface becomes more challenging and would require tedious manipulations while deriving the equivalent control [29,30]. Moreover, due to a number of feedback signals, the resulting controller may not be practical. Thus, the FFDISM is proposed here, with two different control structures based on sliding-surfaces: One uses the inductor current while the other utilizes the capacitor current feedback to flatten the frequency response. The controller that offers high efficiency and realization using reduced opamp count is implemented for experimental verification.

The rest of this paper is organized as follows: Section 2 presents the mathematical model of the fourth-order class-D amplifier, which is an essential step for filter and controller design. Section 3 focuses on the derivation of the equivalent control for the two controllers where one uses the inductor current while the other involves the capacitor current feedback. In Section 4, the circuit realization of the proposed controllers is discussed. Simulation and experimental results are given in Section 5, followed by conclusion in Section 6.

2. Modeling of Fourth-Order Class-D Amplifier

The role of the second LC stage is to improve the THD by reducing the residuals of switching harmonics [31]. It is important to investigate the effect of the additional LC stage on frequency response using average modeling, in order to ease the controller design process. Modeling of the power stage presented here neglects the non-idealities such as the forward voltage drop in the diode, conduction resistance of MOSFET and dead-time delay. Figure 1a,b shows the equivalent circuits during the two class-D operation modes determined by the control signals, i.e., u_H and u_L. These switching signals can attain the binary values; 1 and 0 for the on and off state of MOSFET respectively. For $u_H = 1$, the low-side GaN cascode S_L is in the off-state while the high-side S_H is in the conduction state, and the voltage v_s clamps to $V_{IN}/2$ as shown in Figure 1a. Similarly, for $u_H = 0$, voltage v_s clamps to $-V_{IN}/2$ as shown in Figure 1b. The dynamic equations of the converter for each switching state are expressed as

$$\begin{cases} L_A d\frac{di_{LA}}{dt} = d\frac{V_{IN}}{2} - v_A & \text{when} \quad u_H = 1 \\ L_A d\frac{di_{LA}}{dt} = -d\frac{V_{IN}}{2} - v_A & \text{when} \quad u_H = 0 \end{cases} \tag{1}$$

The subscript "A" and "B" depict parameters of the first and second LC stage. Thus, L_A, i_{LA}, and v_A are the inductance of the first LC stage, the current through L_A, and the voltage across C_A respectively. V_{IN} is the source, and v_o is the output voltage. The average-model of the class-D amplifier, by combining (1) using duty cycle $d_H = \text{avg}(u_H)$ is expressed as

$$L_A \frac{di_{LA}}{dt} = \frac{V_{IN}}{2}(2d_H - 1) - v_A$$
$$C_A \frac{dv_A}{dt} = i_{LA} - i_{LB}$$
$$L_B \frac{di_{LB}}{dt} = v_A - v_B \tag{2}$$
$$C_B \frac{dv_B}{dt} = i_{LB} - \frac{v_B}{R}$$

where L_B, i_{LB}, and v_B are the inductance of the second LC stage, the current through L_B, and the voltage across C_B, respectively, and the load resistance is denoted by R. The dynamics of the class-D amplifier may also be written in state-space form as

$$\dot{x} = f(x,t) + g(x,u,t) \tag{3}$$

where x is the state vector, and f and g are functions of the state vector explicitly given in (4).

$$\begin{cases} x = \begin{bmatrix} i_{LA} & v_A & i_{LB} & v_B \end{bmatrix}^T \\ f(.) = \begin{bmatrix} 0 & \frac{-1}{L_A} & 0 & 0 \\ \frac{1}{C_A} & 0 & \frac{-1}{C_A} & 0 \\ 0 & \frac{1}{L_B} & 0 & \frac{-1}{L_B} \\ 0 & 0 & \frac{1}{C_B} & \frac{1}{RC_B} \end{bmatrix}, g(.) = \begin{bmatrix} \frac{V_{IN}}{2L}(2d_H - 1) & 0 & 0 & 0 \end{bmatrix}^T \end{cases} \tag{4}$$

The open-loop transfer function from v_S to output voltage v_o can be deduced from (4) as:

$$\frac{v_o(s)}{v_S(s)} = \frac{R}{s^4 L_A L_B C_A C_B R + s^3 L_A L_B C_A + s^2(L_A C_A + L_B C_B + L_A C_B)R + s(L_A + L_B) + R} \tag{5}$$

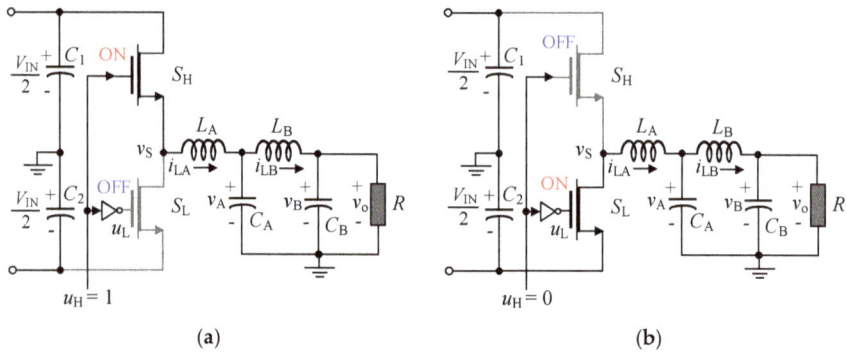

Figure 1. Equivalent circuits of fourth-order GaN class-D amplifier when (a) $u_H = 1$ (b) $u_H = 0$.

Figure 2 shows the frequency response of the fourth-order class-D amplifier in an open-loop configuration. Using the filter values set (i) in Table 1 results in resonant frequencies at 14 kHz and 51 kHz, as shown in Figure 2a. The two frequencies are somewhat close to each other as the relations between the inductances and capacitances of the two stages are $L_A = 2L_B$ and $C_B = 2C_A$ respectively. The frequency response is repeated in Figure 2b using the filter values from the set (ii) in Table 1, resulting in resonant frequencies at 14 kHz and 114 kHz. It indicates that the first resonant frequency is independent of L_B and merely depends on L_A, C_A, and C_B. Moreover, the separation between the two frequencies determines the damping of the second resonance. The second frequency can be displaced adequately by using an appropriate integer multiplier $n \geq 2$ such that $L_A = nL_B$ [9]. It is also noted that the resonant peaking increases with the load resistance and approaches zero-damping under a no-load condition.

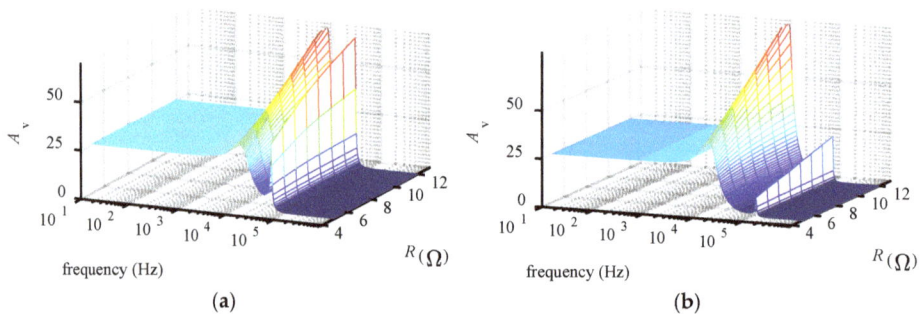

Figure 2. Open-loop frequency response of GaN class-D amplifier with fourth-order filter using (a) value set-i in Table 1 and (b) value set-ii in Table 1.

Table 1. Parameters of GaN class-D amplifier with fourth-order filter.

Parameter	Symbol	Value Set	
		i	ii
source voltage	V_{IN}	100 V	100 V
first stage filter	L_A, C_A	36 μH, 1 μF	36 μH, 1μF
second stage filter	L_B, C_B	18 μH, 2 μF	3 μH, 2 μF

To improve the frequency response of fourth-order GaN class-D amplifier, the FFDISM controller is proposed. Two different sliding-surfaces are proposed; each includes an extra current term (either inductor or capacitor current of the first LC stage) in addition to voltage error in the state variables.

The coefficients of the sliding-surfaces are determined using the stability condition. The controller using capacitor current is compared to the controller with an inductor current in terms of high-efficiency and ease of implementation. The simulation and experimental results are presented to verify the performance of the FFDISM controller using capacitor current feedback.

3. FFDISM Controller of Fourth-Order GaN Amplifier

The design starts by proposing a sliding-surface as a function of state-variables followed by the derivation of the equivalent control, and finally choosing the coefficients of the sliding-surfaces using stability condition [32,33]. Two different sliding-surfaces are proposed here; the first includes the inductor current feedback while the later employs capacitor current feedback.

3.1. Inductor Current Feedback

The state-variables for FFDISM controller are denoted by δ_1 and δ_2 and expressed as

$$\begin{cases} \delta_1 = \int e\, dt - \rho L_A i_{LA} \\ \delta_2 = \int \{ (\int e\, dt) - \rho L_A i_{LA} \} dt \end{cases} \tag{6}$$

where e is the voltage error between reference signal v_{ref} and scaled down output voltage βv_o, i_{LA} is inductor current feedback, ρ and β are scaling factors. Differentiating the set of equations in (6), substituting from (4) gives:

$$\begin{cases} \dot{\delta}_1 = (v_{ref} - \beta v_o) - \rho \left(\frac{V_{IN}}{2} m - v_A \right) \\ \dot{\delta}_2 = \int \left(v_{ref} - \beta v_o \right) dt - \rho L_A i_{LA} \end{cases} \tag{7}$$

where $m = 2d_H - 1$ is the modulation signal.

The sliding-surface is proposed as the weighted sum of δ_1 and δ_2 and expressed as

$$S_1 = \gamma_1 \delta_1 + \gamma_2 \delta_2 \tag{8}$$

where the weights γ_1 and γ_2 are sliding coefficients. By differentiating the sliding-surface in (8), using (4), and finally solving $\dot{S}_1 = 0$ for m_{equ} gives the equivalent control [34] as

$$m_{equ} \left(\frac{\beta V_{IN}}{2} \right) = - \left(L_A \frac{\gamma_2 \beta}{\gamma_1} \right) i_{LA} + \beta v_A + \left(\frac{\beta}{\rho} \right) (v_{ref} - \beta v_o) + \left(\frac{\gamma_2 \beta}{\gamma_1 \rho} \right) \int \left(v_{ref} - \beta v_o \right) dt \tag{9}$$

where the equivalent control is bounded as $|m_{equ}| \leq 1$. By further manipulation, the equivalent control signal as a function of controller gains can be expressed as

$$m_{equ} \left(\frac{\beta V_{IN}}{2} \right) = v_m = -\alpha_1 i_{LA} + \beta v_A + \alpha_2 (v_{ref} - \beta v_o) + \alpha_3 \int \left(v_{ref} - \beta v_o \right) dt \tag{10}$$

where

$$\alpha_1 = \left(L_A \frac{\gamma_2 \beta}{\gamma_1} \right), \alpha_2 = \left(\frac{\beta}{\rho} \right), \alpha_3 = \left(\frac{\gamma_2 \beta}{\gamma_1 \rho} \right) \tag{11}$$

The gains α_1, α_2, and α_3 derived using the necessary existence and stability condition determines the controller performance. The selected range of gains using existence condition determines the region where state trajectory will always be directed to the sliding-surface. By satisfying the Lyapunov condition $\dot{S}_1 \times S_1 < 0$, gives the existence condition:

$$\begin{cases} \beta \left(\frac{V_{IN} - 2v_A}{2} \right) > -\alpha_1 i_{LA} + \alpha_2 (v_{ref} - \beta v_o) + \alpha_3 \int \left(v_{ref} - \beta v_o \right) dt \\ \beta \left(\frac{V_{IN} + 2v_A}{2} \right) > \alpha_1 i_{LA} - \alpha_2 (v_{ref} - \beta v_o) - \alpha_3 \int \left(v_{ref} - \beta v_o \right) dt \end{cases} \tag{12}$$

The stability condition can be derived by substituting $\dot{S}_1 = 0$, which results in the following characteristic equation

$$\gamma_1 \dot{e}_1 + \gamma_2 \dot{e}_2 = 0$$
$$\Rightarrow \gamma_1 \dot{\delta}_1 + \gamma_2 \dot{\delta}_1 = 0 \tag{13}$$

where the straightforward criterion $\gamma_2/\gamma_1 > 0$ ensures the asymptotic stability. Therefore, the ratio of sliding coefficients is chosen as $\gamma_2/\gamma_1 = 1.5\pi f_b$ where f_b is the natural frequency of the closed-loop system. Using $L_A = 33~\mu H$, $f_b = 10$ kHz and $\beta = \rho$ in (11) gives the controller gains as:

$$\alpha_1 = \left(L_A \frac{\gamma_2 \beta}{\gamma_1} \right) = 0.067, \alpha_2 = \left(\frac{\beta}{\rho} \right) = 1, \alpha_3 = \left(\frac{\gamma_2 \beta}{\gamma_1 \rho} \right) = 47124 \tag{14}$$

3.2. Capacitor Current Feedback

The controller design procedure is repeated using a different sliding-surface which involves capacitor current i_{CA}. The state-variables for FFDISM controller are redefined as Δ_1 and Δ_2

$$\begin{cases} \Delta_1 = \int e\,dt - \sigma L_A i_{CA} \\ \Delta_2 = \int \{ (\int e\,dt) - \sigma L_A i_{CA} \} dt \end{cases} \tag{15}$$

where e is the voltage error between reference signal v_{ref} and scaled down output voltage βv_o, β and σ are scaling factors. Differentiating Equation (15) and substituting (4) gives

$$\begin{cases} \frac{d\Delta_1}{dt} = (v_{ref} - \beta v_o) - \sigma L_A \frac{d}{dt}(i_{LA} - i_{LB}) \\ \frac{d\Delta_2}{dt} = \int \left(v_{ref} - \beta v_o \right) dt - \sigma L_A i_{CA} \end{cases} \tag{16}$$

The sliding-surface for the FFDISM controller is defined as

$$S_2 = \varsigma_1 \Delta_1 + \varsigma_2 \Delta_2 \tag{17}$$

where ς_1 and ς_2 are sliding coefficients. The equivalent control [34] is obtained by differentiating the sliding-surface S_2, using (4) and finally solving $\dot{S}_2 = 0$ for equivalent control m_{equ}

$$m_{equ}\left(\frac{\beta V_{IN}}{2} \right) = -\left(L_A \frac{\varsigma_2 \beta}{\varsigma_1} \right) i_{CA} + \beta v_o + \frac{\beta L_A}{L_B}(\beta v_A - \beta v_o) + \left(\frac{\beta}{\sigma} \right)(v_{ref} - \beta v_o) + \left(\frac{\varsigma_2 \beta}{\varsigma_1 \sigma} \right) \int \left(v_{ref} - \beta v_o \right) dt \tag{18}$$

By further manipulation, the equivalent control signal as a function of controller gains can be expressed as

$$m_{equ}\left(\frac{\beta V_{IN}}{2} \right) = v_m = -\lambda_1 i_{CA} + \beta v_o + \lambda_2(v_{ref} - \beta v_o) + \lambda_4(\beta v_A - \beta v_o) + \lambda_3 \int \left(v_{ref} - \beta v_o \right) dt \tag{19}$$

where

$$\lambda_1 = \left(L_A \frac{\varsigma_2 \beta}{\varsigma_1} \right), \lambda_2 = \left(\frac{\beta}{\sigma} \right), \lambda_3 = \left(\frac{\varsigma_2 \beta}{\varsigma_1 \sigma} \right), \lambda_4 = \left(\frac{\beta L_A}{L_B} \right) \tag{20}$$

The controller gains λ_1, λ_2, λ_3, and λ_4 are derived using existence and stability conditions. Satisfying the Lyapunov condition $\dot{S}_2 \times S_2 < 0$, ensures the existence condition:

$$\begin{cases} \beta\left(\frac{V_{IN} - 2v_o}{2} \right) > -\lambda_1 i_{CA} + \lambda_2(v_{ref} - \beta v_o) + \lambda_4(\beta v_A - \beta v_o) + \lambda_3 \int \left(v_{ref} - \beta v_o \right) dt \\ \beta\left(\frac{V_{IN} + 2v_o}{2} \right) > \lambda_1 i_{LA} - \lambda_2(v_{ref} - \beta v_o) - \lambda_4(\beta v_A - \beta v_o) - \lambda_3 \int \left(v_{ref} - \beta v_o \right) dt \end{cases} \tag{21}$$

The stability condition can be derived using $\dot{S} = 0$, which gives the characteristic equation as

$$\varsigma_1 \frac{d\Delta_1}{dt} + \varsigma_2 \frac{d\Delta_2}{dt} = 0$$
$$\Rightarrow \varsigma_1 \frac{d\Delta_1}{dt} + \varsigma_2 \Delta_1 = 0 \tag{22}$$

The criterion $\varsigma_2/\varsigma_1 > 0$ ensures stability. Therefore, the ratio of sliding coefficients is chosen as $\varsigma_2/\varsigma_1 = \pi f_b$ where f_b is the natural frequency of the closed-loop system. The controller parameters, using $f_b = 10$ kHz and $\beta = \sigma$ in (20), are given as:

$$\lambda_1 = \left(L_A \frac{\varsigma_2 \beta}{\varsigma_1}\right) = 0.045, \lambda_2 = \left(\frac{\beta}{\sigma}\right) = 1, \lambda_3 = \left(\frac{\varsigma_2 \beta}{\varsigma_1 \sigma}\right) = 31416, \lambda_4 = \left(\frac{\beta L_A}{L_B}\right) = 0.08 \tag{23}$$

The two FFDISM configurations corresponding to sliding-surfaces S_1 and S_2 are implemented by translating their respective equivalent control equations to analog systems. The FFDISM controller implemented using (10) is shown in Figure 3a, where the difference between the scaled output voltage βv_o and reference signal v_{ref}, is applied to the proportional integral (PI-type) controller. In addition to v_o, it also requires feedback voltage v_A and current i_{LA} from the first LC stage. The closed-loop response is determined by the gains $\alpha_2 - \alpha_3$ and plays the important role of the error signal e processing. Similarly, Figure 3b shows the FFDISM controller involving the capacitor current i_{CA} feedback implemented using (19) where the gains $\lambda_1 - \lambda_4$ determine the response of the closed-loop system. It is observed that the FFDISM controller for fourth-order class-D amplifier requires more feedback variables, i.e., beside i_{LA} and i_{CA} the controller in Figure 3a,b respectively requires βv_A and $\beta(v_A - v_B)$. These additional variables make the circuit implementation more challenging. Therefore, it is important to select the controller that in addition to shaping the frequency response also offers ease of circuit implementation.

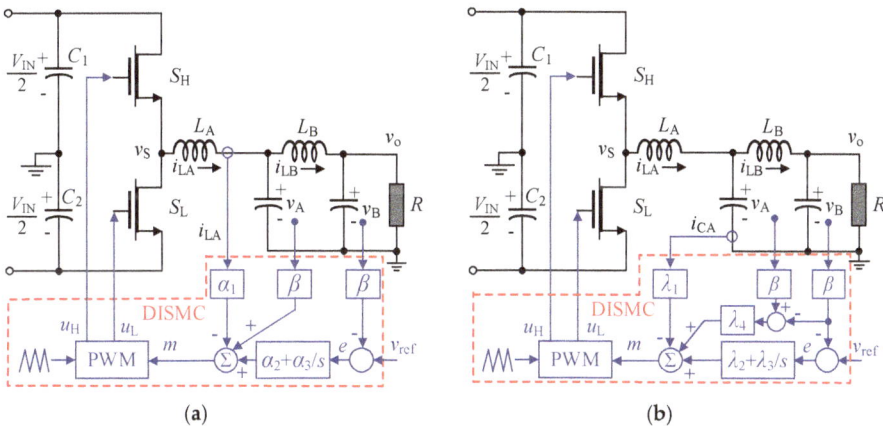

Figure 3. FFDISM controller of fourth-order class-D amplifier using (a) i_{LA} feedback (CNRL1) (b) i_{CA} feedback (CNRL2).

4. Realization of FFDISM Controller

High switching frequency necessitates the analog implementation of the controller using low-cost, single-supply opamps. Figures 4 and 5 show the circuit realization of FFDISM controllers for fourth-order GaN class-D amplifier, with additional i_{LA} and i_{CA} feedback respectively. For ease of reference, the controller involving inductor current and capacitor current is pointed as CNRL1 and CNRL2 respectively. Single-supply opamps are used due to its rail-to-rail input and low power consumption as compared to the dual-supply counterparts. The V_{cc} and V_b are opamps dc supply and mid-point bias voltage respectively. Here, both FFDISM designs are evaluated to find the one that offers ease of implementation, determined by the required number of opamps. The additional voltage feedback signals, i.e., v_A in CNRL1 and $\beta(v_A - v_B)$ in CNRL2 unanimously increases the opamp count.

However, dc-biasing of i_{LA} is required, which increases the opamp count in CNRL1 as compared to CNRL2.

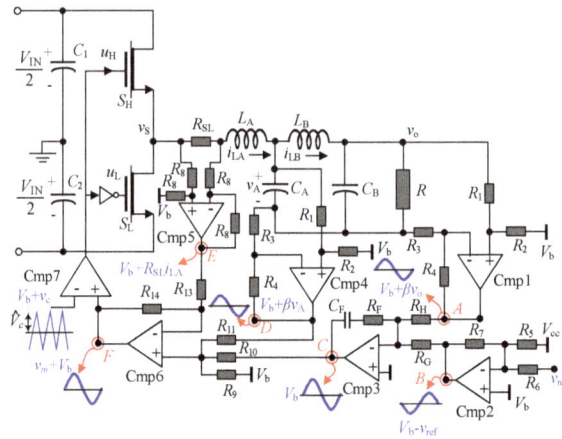

Figure 4. Circuit realization of FFDISM controller with i_{LA} feedback (CNRL1).

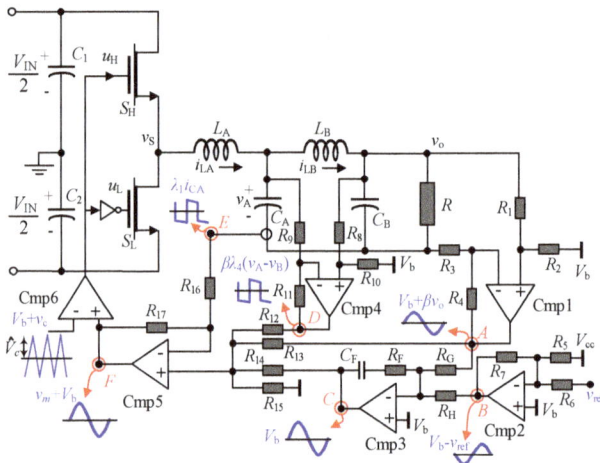

Figure 5. Circuit realization of FFDISM controller with i_{CA} feedback (CNRL2).

Furthermore, the capacitor current i_{CA} in CNRL2 is bidirectional and therefore, can be sensed using a low-cost current transformer as there is no saturation problem. A sense resistor R_{SC} across the secondary of the current transformer converts the current signal into the voltage signal and adequately scales, i.e., $\lambda_1 = 0.01R_{SC}$. This R_{SC} carries low-current and thus, does not affect the efficiency. On the other hand, the sense resistor for the inductor current i_{LA} feedback in CNRL1 can cause a significant reduction in overall efficiency. It is concluded that CNRL2 offers improvement in efficiency and requires fewer numbers of opamps as compared to CNRL1. Therefore, FFDISM controller using the capacitor current i_{CA} is finally implemented for experimental testing.

In Figure 5, the voltage at node A is the scaled and biased output voltage, with the scaling factor determined by the ratio $\beta = R_2/R_1 = R_4/R_3 = 4\,k\Omega/100\,k\Omega$. Meanwhile, Cmp2 inverts and adds bias V_b to v_{ref} as the waveform shows at node B, with the resistors $R_5 = R_6 = R_7 = 200\,k\Omega$. The opamp Cmp3 serves as a proportional integral (PI-type) controller, where capacitor C_F and resistors R_F, R_G, and R_H together determine the proportional and integral gains as $\lambda_2 = R_F R_G$ and $\lambda_3 = 1/C_F R_G$. Since modulating signal is referenced to $V_b = +2.5$ V, the carrier signal is also biased to the same dc-level, where its peak, denoted as $\hat{V_c}$ is given by $\beta V_{IN}/2 = 2$ V.

A prototype of the fourth-order GaN class-D amplifier has been built using the GaN cascode transistor, with the top and bottom sides shown in Figure 6a,b respectively. GaN cascode is realized by combining a low-voltage Si MOSFET with a high-voltage GaN HEMT to exhibit low-conduction and switching power losses. The TPH3006PD (Transphorm Inc., San Jose, CA, USA) available in TO-220 package is used to realize the power circuit. A conventional totem-pole gate driver with a low-valued gate resistance of 3 Ω is used to enable fast switching. The parasitic inductances of the package are high, which is responsible for drain voltage overshoot followed during turn-off transient. Therefore, an resistor-capacitor (RC) snubber across the low-side GaN cascode is used to suppress undesired oscillations. Three 150 μF electrolytic capacitors are connected in parallel to achieve high bus capacitance and reduce the effect of interconnection inductances.

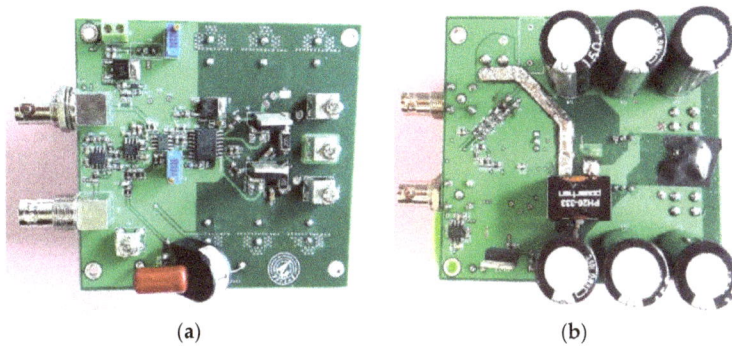

Figure 6. Prototype of GaN class-D amplifier (**a**) top-side (**b**) bottom side.

5. Results and Discussion

In this section, the performance of the proposed FFDISM controller is evaluated using simulation and experimental results. First, the frequency response and step response of the closed-loop system based on control diagram reduction are presented. This fundamental technique is useful to investigate the effect of the inner-loop on the compensation of resonance. Furthermore, computer–based circuit simulator i.e., Plexim Plecs is used to analyze the transient behavior of class-D amplifier using different resistive loads. Finally, experimental results are presented for validation of FFDISM controller.

5.1. Simulation Results

Figure 7a shows the multi-loop control diagram of the class-D amplifier, while the equivalent reduced form is shown in Figure 7b with the open-loop transfer functions H_o and G_o given in (24) and (25) respectively:

$$H_o = \frac{v_o(s)}{v_A(s)} = \frac{R}{s^2 L_B C_B R + s L_B + R} \tag{24}$$

$$\frac{v_A(s)}{v_S(s)} = \frac{v_A(s)}{v_o(s)} \times \frac{v_o(s)}{v_S(s)} = \frac{s^2 L_B C_B R + s L_B + R}{s^4 L_A L_B C_A C_B R + s^3 L_A L_B C_A + s^2 (L_A C_A + L_B C_B + L_A C_B) + s(L_A + L_B) + R} \tag{25}$$

where the parameters of the amplifier are given in Table 1 (value set i), selected using second-order Butterworth approximation. The feedback gains employed are $\lambda_4 = 0.08$ and $\beta = 0.04$. Figure 8a shows the frequency response of the closed-loop class-D amplifier using frequency sweep ranging from 10 Hz to 300 kHz. The frequency response of the FFDISM-controlled amplifier is presented to investigate the significance of the inner-loop for different values of the capacitor current gain λ_1. It is observed that the controller effectively mitigates the resonant peaks and promises a flatter frequency response, almost independent of the load. Further, the proposed control strategy is verified by analyzing the step response in Figure 8b. The voltage overshoot and oscillations decrease with an increase in the capacitor current gain λ_1.

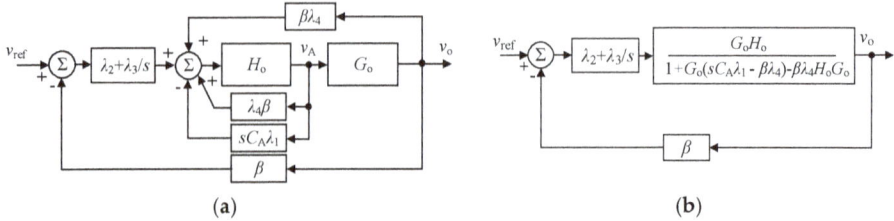

(a)   (b)

Figure 7. Closed-loop schematic of the class-D amplifier: (**a**) multi-loop form (**b**) reduced form.

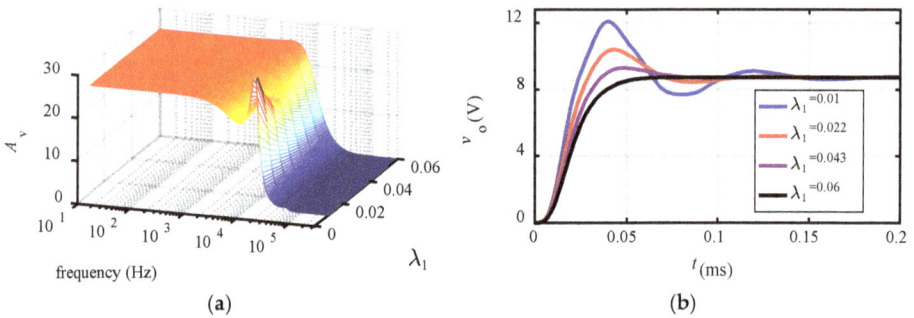

(a)   (b)

Figure 8. Effect of capacitor current i_C feedback on damping of fourth-order class-D amplifier (**a**) closed-loop frequency response for different λ_1 and (**b**) step response for different λ_1.

The transient response of fourth-order GaN class-D amplifier is analyzed using a circuit simulator, which offers a chance to evaluate the correctness of the proposed design. A close match to the experimental results can be obtained by adding the nonidealities to the simulation. Therefore, parasitic resistances of 500 mΩ and 200 mΩ are added to the filter inductors and capacitors, respectively. A sampling time of 1ns is used to capture the simulation results with adequate accuracy at a switching frequency of 100 kHz. Furthermore, a dead-time of 50 ns is introduced before every switching transition. Other factors such as delay in gate driver stage, jitters in PWM, finite rise and fall time of the gate signal are ignored; otherwise, the model takes a long time for solving.

The reference signal is a square wave of 1 kHz frequency, 50% duty cycle and 2 V peak-to-peak amplitude. Such a reference acts as a series of periodic step changes where the slew rate of the rising edge indicates the amplifier's bandwidth. Simulations are performed using the resistive load of 7 Ω and 14 Ω to observe the change in the voltage overshoot and the settling time with the load. Figure 9a indicates that the response of the open-loop GaN class-D amplifier is strongly dependent on the load, and the observed voltage overshoot and settling time are 9.34 V and 0.15 ms, respectively, for $R = 7 \Omega$. By increasing the load resistance to 14 Ω, voltage overshoot and settling time rises to 13.25 V and 0.21 ms respectively. The FFDISM controller using inductor current is also simulated, and results are shown in Figure 9b. The improvement in response is observed using the controller gains derived in (11). Finally, the simulation results for GaN class-D amplifier with FFDISM using the capacitor current obtained under different loads are presented in Figure 9c. The proposed FFDISM controller reduces the voltage overshoot and steady-state error to 0.41 V and 0.3 V, respectively, thereby proving its superiority.

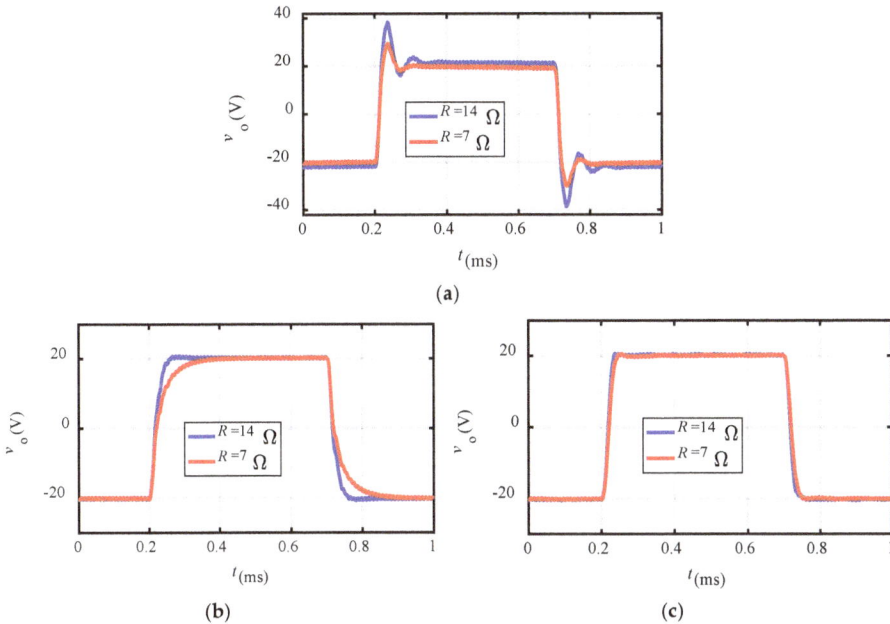

Figure 9. Simulation-based transient response of (**a**) open-loop GaN class-D amplifier; (**b**) FFDISM controller using inductor current; (**c**) FFDISM controller using capacitor current.

5.2. Experimental Results

For experimental validation, a prototype of fourth-order class-D amplifier was implemented on PCB using GaN, available in TO-220 package. Single-supply opamps were used to translate the equivalent control equation in (19) into the analog controller.

For measurement of the frequency response, sinusoidal reference signals (v_{ref}) of different frequencies ranging from 10 Hz to 50 kHz were applied using a signal generator, and the corresponding outputs were listed. The voltage gain (A_v) was computed for open and closed-loop configurations, as the ratio of the reference v_{ref} to output voltage v_o. Figure 10a shows the voltage gain plotted against the frequency of the reference signal when there is a 10 Ω resistive load. It is noted in open-loop (shown in blue), that the voltage gain at the 33 kHz resonant frequency is 29.3 which is successfully compensated by FFDISM controller as depicted in red.

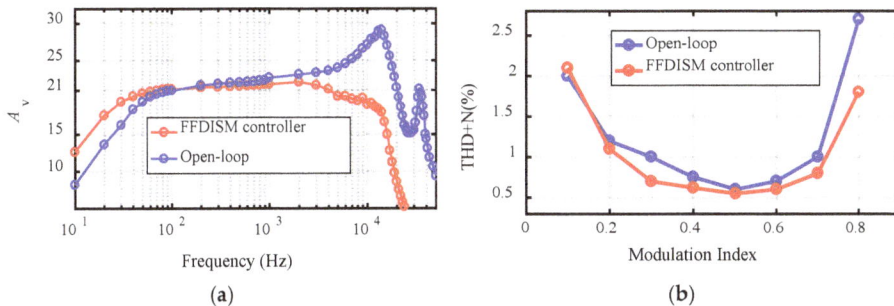

Figure 10. (**a**) Frequency response of the open-loop GaN class-D amplifier and with FFDISMC (**b**) THD+N the of open-loop GaN class-D amplifier and with FFDISMC.

Figure 10b shows the THD+N of GaN amplifier against the modulation index, defined as the ratio of the carrier to reference voltage and can range from 0 to 1. The output voltage v_o was recorded using the data-acquisition card at 5 MSa/s sampling frequency, and the fast Fourier transform (FFT) was used to extract THD+N. It was observed that FFDISM controlled fourth-order class-D amplifier achieved improved THD+N as compared to open-loop architecture.

Figure 11a shows the experimental transient response of open-loop GaN class-D amplifier to a reference square wave of 1 kHz frequency. It is found that with a load resistance of 10 Ω, the resonant frequencies result in voltage overshoot of 5.2 V. Similarly, the transient response with FFDISM controller to 1 kHz reference square wave is shown in Figure 11b, with a recorded overshoot of 1.3 V. Thus it is validated, that the proposed controller improves the transient response of the fourth-order class-D amplifier.

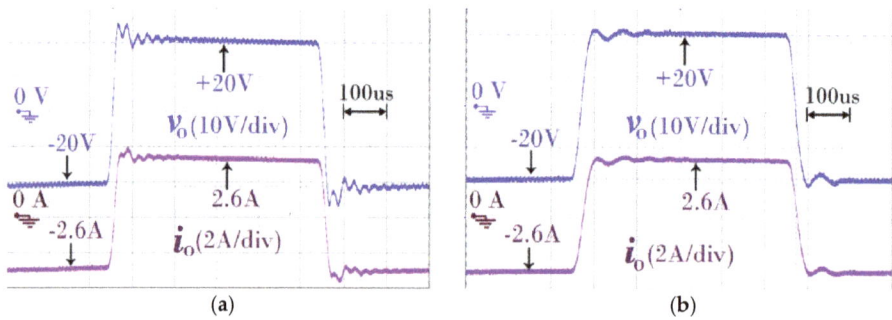

Figure 11. Experimental results transient response of: (**a**) open-loop GaN class-D amplifier; (**b**) GaN class-D amplifier with FFDISMC.

Furthermore, the scalogram analysis was performed, which illustrated a combined time and frequency-domain response of the amplifier. The wavelet transform was applied to represent the output voltage v_o as a weighted sum of the limited duration wavelet functions. Figure 12a shows the scalogram of the transient response in Figure 11a. A horizontal line in the figure represents a particular harmonic in the output waveform with its frequency on the y-axis and magnitude given on the color-bar. Since the square wave is a weighted sum of odd harmonics, the parallel lines below 16 kHz can synthesize it with adequate accuracy. Similarly, the line at 100 kHz is due to residual switching noise. There is a noticeable activity in the time interval 2–4 ms and 7–9 ms due to voltage ringing, corresponding to the resonant frequencies at 12.7 kHz and 33 kHz. Figure 12b shows the scalogram of the transient response in Figure 11b. The shrunken oval-shaped region and the reduction in magnitude noted on the color-bar verify the effectiveness of the FFDISM controller.

Finally, sinusoidal and triangular signals were generated using the FFDISM controlled GaN class-D amplifier. Figure 13a shows the 1 kHz sinusoidal output voltage v_o and the corresponding current i_o of the fourth-order amplifier. Similarly, for a 1 kHz triangular reference, output voltage v_o and current i_o are shown in Figure 13b. Thus, the effectiveness of the FFDISM controller, proposed for fourth-order class-D amplifier has been validated. Figure 14 illustrates the efficiency of GaN class-D amplifier at different operating powers. It is observed that at 150 W power, the amplifier achieves 93% efficiency. Moreover, the efficiency and bandwidth of the proposed class-D amplifier are compared in Table 2, against the hysteretic modulation-based implementation. It indicates that the class-D amplifier with PWM not only gets rid of the variable switching frequency but also improves the efficiency by reducing switching losses.

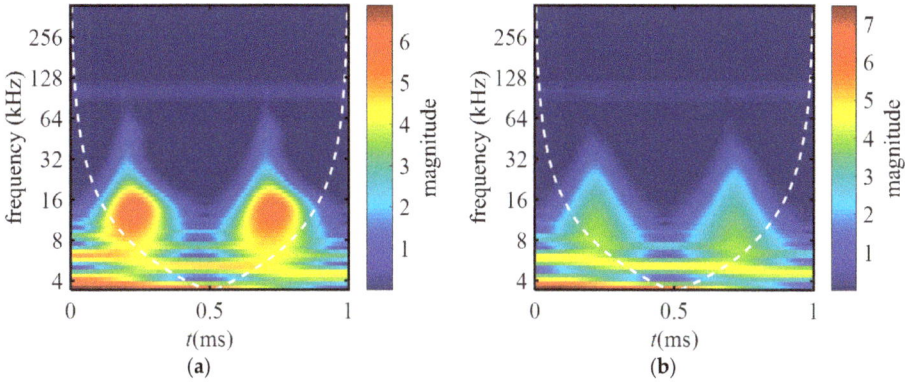

Figure 12. Experimental magnitude scalogram of GaN class-D amplifier: (**a**) open-loop; (**b**) with FFDISM controller.

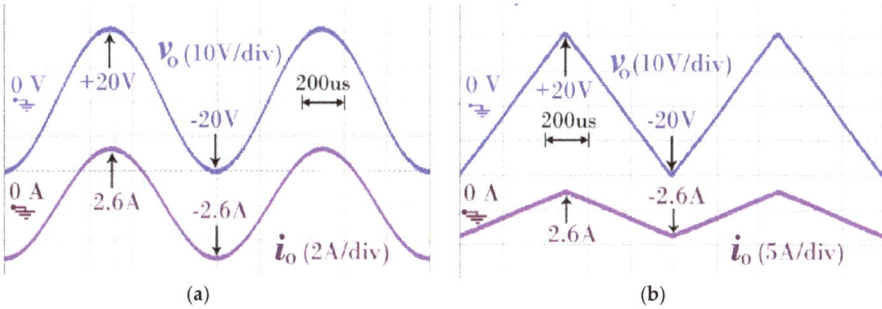

Figure 13. (**a**) Experimental result of GaN class-D amplifier with FFDISM controller output voltage v_o and load current i_o and (**b**) output voltage v_o for reference sine wave v_{ref}.

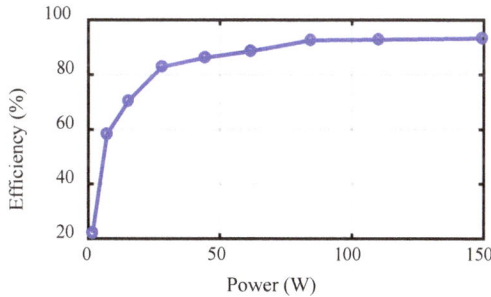

Figure 14. The efficiency of GaN class-D amplifier at different power.

Table 2. Comparative analysis of PWM and Hysteretic modulation based class-D amplifier.

Design	Modulation	Efficiency (%)	Bandwidth (kHz)
[20]	Hysteretic	81	4.6
[22]	Hysteretic	87	3.5
[35]	Hysteretic	84	5
This work	PWM	93.2	10

6. Conclusions

The FFDISM controller has been successfully applied to the GaN-based fourth-order class-D amplifier. The fourth-order filter has been used to improve the THD of class-D amplifiers, operating at a relatively lower switching frequency. Two different FFDISM structures were proposed using the inductor and capacitor current. The advantages offered by the FFDISM controller using the capacitor current over the counterpart using inductor current feedback are listed as: The bidirectional nature of the capacitor current demands a low-cost current-transformer for feedback; the current sense circuit does not affect the overall efficiency; and reduced opamp are required for circuit realization. A prototype of the amplifier was implemented on a PCB. The experimental results revealed that the proposed FFDISM controller effectively flattens the frequency response of the fourth-order amplifier, and results in THD and voltage overshoot of 0.6% and 1 V respectively.

Author Contributions: H.Z. and X.Z. contributed equally to the research presented here. H.Z. and X.Z. conceived and designed the experiments. H.Z. performed the experiments, analyzed the data and wrote the paper. X.W. and H.A. provided significant comments and technical feedback throughout the research. S.K. reviewed and improved the paper.

Conflicts of Interest: The authors declare no conflict of interest.

References

1. Bagger, R.; Sjoland, H. Broadband LDMOS 40 W and 55 W integrated power amplifiers. In Proceedings of the 2017 IEEE MTT-S International Microwave Symposium, Honolulu, HI, USA, 4–9 June 2017; pp. 1950–1952.
2. Bosi, G.; Raffo, A.; Trevisan, F.; Vadalà, V.; Crupi, G.; Vannini, G. Nonlinear-embedding design methodology oriented to LDMOS power amplifiers. *IEEE Trans. Power Electron.* **2018**, *33*, 8764–8774. [CrossRef]
3. Ramella, C.; Piacibello, A.; Quaglia, R.; Camarchia, V.; Pirola, M. High Efficiency Power Amplifiers for Modern Mobile Communications: The Load-Modulation Approach. *Electronics* **2017**, *6*, 96. [CrossRef]
4. Thorsell, M.; Fagerlind, M.; Andersson, K.; Billström, N.; Rorsman, N. An X-band AlGaN/GaN MMIC receiver front-end. *IEEE Microw. Wirel. Compon. Lett.* **2010**, *20*, 55–57. [CrossRef]
5. Nalli, A.; Raffo, A.; Crupi, G.; D'Angelo, S.; Resca, D.; Scappaviva, F.; Salvo, G.; Caddemi, A.; Vannini, G. Gan hemt noise model based on electromagnetic simulations. *IEEE Trans. Microw. Theory Tech.* **2015**, *63*, 2498–2508. [CrossRef]
6. Recht, F.; Huang, Z.; Wu, Y. Characteristics of Transphorm GaN Power Switches. Application Note AN-0002. Available online: https://www.transphormchina.com/en/ (accessed on 19 October 2018).
7. Saito, W.; Takada, Y.; Kuraguchi, M.; Tsuda, K.; Omura, I. Recessed-gate structure approach toward normally off high-voltage AlGaN/GaN HEMT for power electronics applications. *IEEE Trans. Electron Devices* **2006**, *53*, 356–362. [CrossRef]
8. Su, L.Y.; Lee, F.; Huang, J.J. Enhancement-mode GaN-based high-electron mobility transistors on the Si substrate with a p-type GaN cap layer. *IEEE Trans. Electron Devices* **2014**, *61*, 460–465. [CrossRef]
9. Quan, S.; Hao, Y.; Ma, X.; Xie, Y.; Ma, J. Enhancement-mode AlGaN/GaN HEMTs fabricated by fluorine plasma treatment. *J. Semicond.* **2009**, *30*, 124002.
10. Jung, D.Y.; Park, Y.; Lee, H.S.; Jun, C.H.; Jang, H.G.; Park, J.; Kim, M.; Ko, S.C.; Nam, E.S. Design and evaluation of cascode GaN FET for switching power conversion systems. *ETRI J.* **2017**, *39*, 62–68. [CrossRef]
11. Persson, E. How 600 V GaN Transistors Improve Power Supply Efficiency and Density. *Power Electron. Eur.* **2015**, *2*, 22–24.
12. Jiang, X. Fundamentals of Audio Class D Amplifier Design: A Review of Schemes and Architectures. *IEEE Solid-State Circuits Mag.* **2017**, *9*, 14–25. [CrossRef]
13. Cai, W.; Huang, L.; Wang, S. Class D Power Amplifier for Medical Application. *Inform. Eng. Int. J.* **2016**, *4*, 9–15. [CrossRef]
14. Strasser, T. Real-Time Simulation Technologies for Power Systems Design, Testing, and Analysis. *IEEE Power Energy Technol. Syst. J.* **2015**, *2*, 63–73.
15. Boillat, D.O.; Friedli, T.; Mühlethaler, J.; Kolar, J.W.; Hribernik, W. Analysis of the design space of single-stage and two-stage LC output filters of switched-mode AC power sources. In Proceedings of the 2012 IEEE Power and Energy Conference at Illinois, Champaign, IL, USA, 24–25 February 2012; pp. 1–8.

16. Bloechl, M.; Bataineh, M.; Harrell, D. Class D Switching Power Amplifiers: Theory, Design, and Performance. In Proceedings of the IEEE SoutheastCon, Greensboro, NC, USA, 26–29 March 2004; pp. 123–146.

17. Yu, Y.; Tan, M.T.; Goh, W.L.; Cox, S.M. A dual-feedforward carrier-modulated second-order class-D amplifier with improved THD. *IEEE Trans. Circuits Syst. II Express Briefs* **2012**, *59*, 35–39. [CrossRef]

18. Chun, K.L.; Meng, T.T. A class D amplifier output stage with low THD and high PSRR. In Proceedings of the 2009 IEEE International Symposium on Circuits and Systems, Taipei, Taiwan, 24–27 May 2009; pp. 1945–1948.

19. Künzi, R. Passive Power Filters. In Proceedings of the CAS-CERN Accelerator School: Power Converters, Baden, Switzerland, 7–14 May 2014; pp. 265–289.

20. Cortes, P.; Boillat, D.O.; Ertl, H.; Kolar, J.W. Comparative evaluation of multi-loop control schemes for a high-bandwidth AC power source with a two-stage LC output filter. In Proceedings of the 2012 International Conference on Renewable Energy Research and Applications, Nagasaki, Japan, 11–14 November 2012; pp. 1–10.

21. Boillat, D.; Krismer, F.; Kolar, J. Design Space Analysis and ρ-η Pareto Optimization of LC Output Filters for Switch-Mode AC Power Sources. *Power Electron. IEEE Trans.* **2015**, *30*, 6906–6923. [CrossRef]

22. Nielsen, D.; Knott, A.; Andersen, M.A.E. Class D audio amplifier with 4th order output filter and self-oscillating full-state hysteresis based feedback driving capacitive transducers. In Proceedings of the 2014 16th European Conference on Power Electronics and Applications, Lappeenranta, Finland, 26–28 August 2014; pp. 1–7.

23. Ablay, G. Robust integral controllers for high-order class-D power amplifiers. *Power Electron. IET* **2017**, *11*, 1–7. [CrossRef]

24. Ge, T.; Chang, J.S.; Shu, W.; Tan, M.T. Modeling and Analysis of PSRR in Analog PWM Class D Amplifiers. In Proceedings of the 2006 IEEE International Symposium on Circuits and Systems, Island of Kos, Greece, 21–24 May 2006; pp. 1386–1389.

25. Shu, W.; Chang, J.S. THD of closed-loop analog PWM Class-D amplifiers. *IEEE Trans. Circuits Syst. I Regul. Pap.* **2008**, *55*, 1769–1777. [CrossRef]

26. Tan, S.C.; Lai, Y.M.; Tse, C.K. *Sliding Mode Control of Switching Power Converters-Techniques and Implementation*; CRC Press: Boca Raton, FL, USA, 2012.

27. Tan, S.-C.; Lai, Y.M.; Tse, C.K.; Cheung, M.K.H. A fixed-frequency pulsewidth modulation based quasi-sliding-mode controller for buck converters. *IEEE Trans. Power Electron.* **2005**, *20*, 1379–1392. [CrossRef]

28. Pradhan, R.; Subudhi, B. Double integral sliding mode MPPT control of a photovoltaic system. *IEEE Trans. Control Syst. Technol.* **2016**, *24*, 285–292. [CrossRef]

29. Jiao, Y.; Luo, F.L.; Zhu, M. Generalised modelling and sliding mode control for n-cell cascade super-lift DC–DC converters. *IET Power Electron.* **2011**, *4*, 532–540. [CrossRef]

30. Chincholkar, S.H.; Chan, C.Y. Design of fixed-frequency pulsewidth-modulation-based sliding-mode controllers for the quadratic boost converter. *IEEE Trans. Circuits Syst. II Express Briefs* **2017**, *64*, 51–55. [CrossRef]

31. Maislinger, F.; Ertl, H.; Stojcic, G.; Holzner, F. Control Loop Design for Closed-Loop Class-D Amplifiers with 4th Order Output Filter. In Proceedings of the PCIM Europe 2017; International Exhibition and Conference for Power Electronics, Intelligent Motion, Renewable Energy and Energy Management, Nuremberg, Germany, 16–18 May 2017; pp. 1193–1200.

32. Bacha, S.; Munteanu, I.; Bratcu, A. Power Electronic Converters Modeling and Control. *Adv. Textb. Control Singal Process.* **2014**, *454*, 454.

33. Labbe, B.; Allard, B.; Shi, X.-L. Design and stability analysis of a frequency controlled sliding-mode buck converter. *IEEE Trans. Circuits Syst. I Regul. Pap.* **2014**, *61*, 2761–2770. [CrossRef]

34. Rojas-Gonzalez, M.A.; Sanchez-Sinencio, E. Design of a Class D Audio Amplifier IC Using Sliding Mode Control and Negative Feedback. *IEEE Trans. Consum. Electron.* **2007**, *53*, 609–617. [CrossRef]

35. Lu, J.; Gharpurey, R. Design and analysis of a self-oscillating class D audio amplifier employing a hysteretic comparator. *IEEE J. Solid-State Circuits* **2011**, *46*, 2336–2349. [CrossRef]

electronics

MDPI

Article

Line Frequency Instability of One-Cycle-Controlled Boost Power Factor Correction Converter

Rui Zhang [1,*], Wei Ma [2,*], Lei Wang [3], Min Hu [2], Longhan Cao [4], Hongjun Zhou [2] and Yihui Zhang [2]

[1] College of Safety Engineering, Chongqing University of Science and Technology, Chongqing 401331, China
[2] College of Electrical Engineering, Chongqing University of Science and Technology, Chongqing 401331, China; minhuxzr@outlook.com (M.H.); zhjzsj007@163.com (H.Z.); zhangyihui@163.com (Y.Z.)
[3] The School of Electric Power Engineering, South China University of Technology, Guangzhou 510641, China; w.lei19@mail.scut.edu.cn
[4] Department of Electrical Egineering, Chongqing Institute of Communication, Chongqing 400035, China; lhcao1@hotmail.com
* Correspondence: zhangrui@cqust.edu.cn (R.Z.); me16888@163.com (W.M.); Tel.: +86-23-6502-3730 (R.Z.)

Received: 26 July 2018; Accepted: 4 September 2018; Published: 17 September 2018

Abstract: Power Factor Correction (PFC) converters are widely used in engineering. A classical PFC control circuit employs two complicated feedback control loops and a multiplier, while the One-Cycle-Controlled (OCC) PFC converter has a simple control circuit. In OCC PFC converters, the voltage loop is implemented with a PID control and the multiplier is not needed. Although linear theory is used in designing the OCC PFC converter control circuit, it cannot be used in predicting non-linear phenomena in the converter. In this paper, a non-linear model of the OCC PFC Boost converter is proposed based on the double averaging method. The line frequency instability of the converter is predicted by studying the DC component, the first harmonic component and the second harmonic component of the main circuit and the control circuit. The effect of the input voltage and the output capacitance on the stability of the converter is studied. The correctness of the proposed model is verified with numerical simulations and experimental measurements.

Keywords: power factor correction; line frequency instability; one cycle control; non-linear phenomena; bifurcation; boost converter

1. Introduction

Power Factor Correction (PFC) plays an important role in electrical engineering [1]. A PFC converter takes AC voltage as its input and outputs DC voltage. Different from traditional diode rectifiers, a PFC converter in average current mode has a high power factor. In electrical engineering, the average current mode Boost PFC converter is widely used. Although the topology of the Boost PFC is simple, the control circuit is complicated [2–4]. The control circuit consists of two loops. The first is the current control loop, with the aim of forcing the inductor current to be in the same phase as the reference. The second loop is the voltage control loop. The design of the voltage control loop is of great importance because its main objective is achieving a stable system and a near unity power factor [4]. The dynamics of the PFC converter depends on these two control loops. The traditional implementation for the PFC converter requires a multiplier, whose output is the reference current added to the current control loop. The existence of the multiplier increases the control complexity. The dynamics of the PFC converter has interested many researchers and some non-linear phenomena have been observed in the last few years [5–13]. In general, there are two kinds of non-linear dynamics in the PFC converter. The first is the so-called switching frequency instability, which is mainly the result of bifurcation and chaos caused by the current control loop [5]. The second is the line frequency instability, which is the result of bifurcation and chaos caused by the voltage control loop [7–11,14]. Among them, the line frequency instabilities are more detrimental to the normal operation of the PFC converter, as it changes the power factor to an unacceptable value. The power factor of the converter is much less than one due to the line frequency instabilities and thus, it is of great importance to select the appropriate parameters in the design process. In the traditional design, the researchers adopted the linear system theory and it has been shown that this design cannot predict the line frequency instabilities in the PFC converter [7]. Therefore, the researchers developed some powerful methods to compute the boundaries of line frequency stabilities. Among them, the method of harmonic balance needs an exact computation of the unstable periodic orbit of the control voltage [9]. Harmonic balance is applied to the model of the converter incorporating the multiplier and Floquet theory is adopted to decide the stability of the converter. According to Floquet theory, the stability of the converter is identified by calculating the eigenvalues of the transition matrix of the system. Another important method is the method of double averaging, which is based on the first harmonic component in the PFC converter line frequency model [14]. This method is more familiar to many researchers and engineers. In this paper, the later method is adopted to study the non-linear dynamics of the continuous conduction mode One-Cycle-Controlled (OCC) Boost PFC converter.

Different from the traditional average current mode PFC converter, the OCC PFC converter simplifies the control circuit [3]. The one-cycle control belongs to non-linear controls. When this control method is applied to the PFC converter, the current control loop is replaced by a resettable integrator. Therefore, only one voltage control loop is required and the multiplier is not needed. It has been shown that the control circuit of the OCC Boost PFC converter saves space and cost compared to the traditional PFC Boost converter. In most applications, the voltage control loop is designed based on the linear system theory and the prediction of dynamics of the converter is also based on the linear system theory. Therefore, non-linear dynamics of the converter are uncovered. In many applications, bifurcation and chaos are observed but are not addressed. The reason is that the non-linear systems theory is not applied to the OCC PFC converter. In this paper, the method of double averaging is adopted to predict the non-linear dynamics of the OCC Boost PFC converter. Although this method has been applied to the traditional PFC converter, there is still a problem when applying it to the OCC PFC converter, because the control circuits in the two converters are totally different and as a result, some new consequences will occur in the OCC PFC converter. It is important to note that in a previous study [15], the non-linear dynamics of the OCC PFC converter were observed by experiments and no effective computation was provided. In the present paper, the computation is based on the exact non-linear model of the OCC PFC converter and therefore, the conclusions are meaningful in the design process of the converter.

2. The OCC Boost PFC Converter

2.1. The OCC Boost PFC Converter and Its Control Circuit

The OCC Boost PFC converter consists of a diode rectifier and a boost converter, as shown in Figure 1. In some applications, the load of the PFC converter is another DC-DC converter. In this paper, the load of the PFC converter is a resistor, because the emphasis of this paper is on the non-linear dynamics of the PFC converter. The control circuit in Figure 1 is equivalent to a commercial control IC IR1150, which is used to verify the theoretical results in this paper. Apparently, this control circuit has fewer resistors and capacitors than the traditional average current mode PFC converter, where both the current loop and the voltage loop have at least one resistor and one capacitor. The output of the converter is divided by R_{f1} and R_{f2}. The divided voltage is connected to the input of an Operational Amplifier (OA), whose other input is the reference V_{ref}. In IR1150, the OA is a trans-conductance type amplifier. The output of the OA is v_m, which is the input to a resettable integrator. The output of the integrator is compared with another voltage composed of v_m and the voltage across the current sense resistor R_{s1}. The integrator is reset by the output $\overline{Q}$ of the flip-flop. The sensed voltage is amplified by a DC gain $G_{DC} = 2.5$. The operation of the control circuit is described here. In the beginning of every switching period T_s, the clock sets the flip-flop and the output Q of the flip-flop turns on the switch S. At the same time, the integrator outputs the value of the integral of its input signal. When the output of the integrator exceeds the sum of v_m and the sensed voltage, the comparator resets the flip-flop and the output Q turns off the switch S. Therefore, the diode D turns on. Furthermore, the output $\overline{Q}$ resets the integrator until the next clock signal.

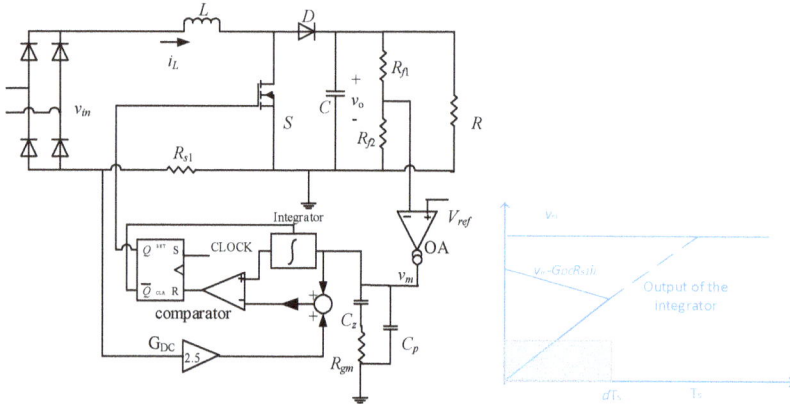

Figure 1. Circuit diagram of the OCC Boost PFC converter and its operating principle.

2.2. The OCC Boost PFC Converter Model

In this paper, the line frequency dynamics are studied. Therefore, the method of double averaging is adopted. The method is composed of two averaging processes. The first averaging is applied to the switching period [16–22]. The converter has two topology structures during one switching period, and is described by:

$$\begin{cases} \frac{di_L}{dt} = \frac{1}{L}v_{in} \\ \frac{dv_o}{dt} = \frac{-1}{RC}v_o \end{cases} \text{ (the switch S is on) or } \begin{cases} \frac{di_L}{dt} = \frac{1}{L}(v_{in} - v_o) \\ \frac{dv_o}{dt} = \frac{1}{C}(i_L - \frac{v_o}{R}) \end{cases} \text{ (the switch S is off).} \tag{1}$$

By averaging over one switching period, one obtains the following:

$$\begin{cases} (1-d)v_o = v_{in} - L\frac{di_L}{dt} \\ (1-d)i_L = C\frac{dv_o}{dt} + \frac{v_o}{R} \end{cases}. \tag{2}$$

From Equation (2), we obtain:

$$\frac{C}{2}\frac{dv_o^2}{dt} = -\frac{v_o^2}{R} + i_L v_{in} - \frac{L}{2}\frac{di_L^2}{dt}. \tag{3}$$

It is important to note that the dynamics of the inductor during one switching period can be omitted when the converter operates stably. Therefore, one has

$$\frac{C}{2}\frac{dv_o^2}{dt} = -\frac{v_o^2}{R} + i_L v_{in}. \tag{4}$$

Based on the operating principle of the converter [3–8], one has:

$$R_s i_L(t) = v_m/T(d), \tag{5}$$

where:

$$T(d) = v_0/v_{in}, \quad R_s = R_{s1} \times 2.5 \tag{6}$$

From Equations (5) and (6), we can obtain:

$$i_L(t) = v_{in}v_m/(R_s v_0). \tag{7}$$

It is important to note that $v_{in} = V_m|\sin \omega_m t|$. Substituting Equation (7) into (4), we obtain:

$$\frac{C}{2}\frac{dv_0^2}{dt} = -\frac{v_0^2}{R} + \frac{v_m}{R_s v_0}V_m^2(1 - \cos 2\omega_m t). \tag{8}$$

On the other hand, the control loop includes the OA. Figure 1 provides the following transfer function of the OA:

$$H(s) = \frac{g_m(1 + sR_{gm}C_z)}{s(C_z + C_p + sR_{gm}C_zC_p)}. \tag{9}$$

As $C_z \gg C_p$, Equation (9) can be written as:

$$H(s) = \frac{g_m(1 + sR_{gm}C_z)}{sC_z}. \tag{10}$$

Therefore, the voltage control loop in Figure 1 is described by:

$$C_z\frac{dv_m}{dt} = g_m\left(V_{ref} - \frac{R_{f2}}{R_{f1} + R_{f2}}v_0\right) - g_m R_{gm}C_z\frac{R_{f2}}{R_{f1} + R_{f2}}\frac{dv_0}{dt}, \tag{11}$$

where g_m is the trans-conductance of the amplifier.

From Figure 1, one has:

$$i_L(t) = \frac{v_{in}v_m R_{f2}}{R_s\left(R_{f1} + R_{f2}\right)V_{ref}}. \tag{12}$$

Therefore, the OCC Boost PFC converter is described by:

$$\begin{cases} \frac{C}{2}\frac{dx^2}{dt} = -\frac{x^2}{R} + \frac{y}{R_s(1+\beta)V_{ref}}V_m^2(1 - \cos 2\omega_m t) \\ C_z\frac{dy}{dt} = g_m\left(V_{ref} - \frac{1}{1+\beta}x\right) - g_m R_{gm}C_z\frac{1}{1+\beta}\frac{dx}{dt} \end{cases}, \tag{13}$$

where $v_m = y$, $v_0 = x$, $\beta = R_{f1}/R_{f2}$.

The next step is applying the second averaging for Equation (13). The second averaging involves taking the moving average over the main period. For any variable $u(t)$ in Equation (13), we have the following expression based on Fourier analysis [23]:

$$u(t) \approx u_0 + u_1 e^{j\omega_m t} + u_{-1}e^{-j\omega_m t} + u_2 e^{j2\omega_m t} + u_{-2}e^{-j2\omega_m t}. \tag{14}$$

where

$$u_k \approx \frac{\omega_m}{2\pi}\int_{t-\frac{2\pi}{\omega_m}}^{t} u(\tau)\exp(-jk\omega_m\tau)d\tau (k = 0, \pm1, \pm2). \tag{15}$$

It is important to note that $u_{-1} = u_1^*$, $u_{-2} = u_2^*$, where $*$ stands for complex conjugate. Taking the second averaging on Equation (13) based on Equation (14) and Equations (A1)–(A5) in Appendix A, one has:

$$\begin{aligned} &\frac{C}{2}\frac{d}{dt}\left(x_0^2 + 2x_{1r}^2 + 2x_{1i}^2 + 2x_{2r}^2 + 2x_{2i}^2\right) + \frac{1}{R}\left(x_0^2 + 2x_{1r}^2 + 2x_{1i}^2 + 2x_{2r}^2 + 2x_{2i}^2\right) \\ &= \frac{V_{in}^2}{R_s(1+\beta)V_{ref}}(y_0 - y_{2r}) \end{aligned}. \tag{16}$$

$$\frac{C}{2}\frac{d}{dt}(x_0x_1 + x_{1r}x_{2r} + x_{1i}x_{2i} + j(x_{1r}x_{2i} - x_{1i}x_{2r})) \\ + \left(j\frac{\omega_m C}{2} + \frac{1}{R}\right)(x_0x_1 + x_{1r}x_{2r} + x_{1i}x_{2i} + j(x_{1r}x_{2i} - x_{1i}x_{2r})) \\ = \frac{V_{in}^2}{R_s(1+\beta)V_{ref}}\left(\frac{y_1}{2} - \frac{y_1^*}{4}\right)$$

(17)

$$\frac{C}{2}\frac{d}{dt}\left((x_1^*)^2 + 2x_0x_2\right) + \left(j\omega_m C + \frac{1}{R}\right)\left((x_1^*)^2 + 2x_0x_2\right) \\ = \frac{V_{in}^2}{R_s(1+\beta)V_{ref}}\left(y_2 - \frac{1}{2}y_0\right)$$

(18)

$$C_z\frac{d}{dt}y_0 = g_m\left(V_{ref} - \frac{1}{1+\beta}x_0\right) - g_m R_{gm}C_z\frac{1}{1+\beta}\frac{d}{dt}x_0.$$

(19)

$$C_z\left(\frac{d}{dt}y_1 + j\omega_m y_1\right) = -g_m\frac{1}{1+\beta}x_1 - g_m R_{gm}C_z\frac{1}{1+\beta}\left(\frac{d}{dt}x_1 + j\omega_m x_1\right).$$

(20)

$$C_z\left(\frac{d}{dt}y_2 + j2\omega_m y_2\right) = -g_m\frac{1}{1+\beta}x_2 - g_m R_{gm}C_z\frac{1}{1+\beta}\left(\frac{d}{dt}x_2 + j2\omega_m x_2\right).$$

(21)

Equations (16)–(21) describe the DC component, the first harmonic component and the second harmonic component of the main circuit and the control circuit, respectively.

3. Stability of the OCC Boost PFC Converter

The stability of the OCC Boost PFC converter was studied based on Equations (16)–(21). To do this, the DC component, the first harmonic component and the second harmonic component are studied.

3.1. The First Harmonic Component

We obtain the steady-state solution by making all time-derivatives in Equations (16)–(21) equal to zero. Therefore, Equation (17) becomes:

$$\left(j\frac{\omega_m C}{2} + \frac{1}{R}\right)(x_0x_{1r} + x_{1r}x_{2r} + x_{1i}x_{2i} + j(x_0x_{1i} + x_{1r}x_{2i} - x_{1i}x_{2r})) \\ = \frac{V_m^2}{R_s(1+\beta)V_{ref}}\left(\frac{y_{1r}}{4} + j\frac{3y_{1i}}{4}\right)$$

(22)

Considering the real and imaginary part of Equation (22), one has:

$$\begin{cases} \frac{1}{R}(x_0x_{1r} + x_{1r}x_{2r} + x_{1i}x_{2i}) - \frac{\omega_m C}{2}(x_0x_{1i} + x_{1r}x_{2i} - x_{1i}x_{2r}) = \frac{V_m^2}{R_s(1+\beta)V_{ref}}\frac{y_{1r}}{4} \\ \frac{\omega_m C}{2}(x_0x_{1r} + x_{1r}x_{2r} + x_{1i}x_{2i}) + \frac{1}{R}(x_0x_{1i} + x_{1r}x_{2i} - x_{1i}x_{2r}) = \frac{V_m^2}{R_s(1+\beta)V_{ref}}\frac{3y_{1i}}{4} \end{cases}$$

(23)

Equation (23) has another form, which is the following:

$$\begin{pmatrix} x_{1r} \\ x_{1i} \end{pmatrix} = \frac{\frac{V_m^2}{4R_s(1+\beta)V_{ref}}}{\left(\frac{1}{R^2} + \frac{\omega_m^2 C^2}{4}\right)(x_0^2 - (x_{2r}^2 + x_{2i}^2))} \\ \times \begin{pmatrix} \frac{x_0 - x_{2r}}{R} + \frac{\omega_m C x_{2i}}{2} & -3\left(\frac{x_{2i}}{R} - \frac{\omega_m C(x_0 - x_{2r})}{2}\right) \\ -\frac{x_{2i}}{R} - \frac{\omega_m C(x_0 + x_{2r})}{2} & 3\left(\frac{x_0 + x_{2r}}{R} - \frac{\omega_m C x_{2i}}{2}\right) \end{pmatrix}\begin{pmatrix} y_{1r} \\ y_{1i} \end{pmatrix}.$$

(24)

By making all time derivatives in Equation (20) equal to zero, one has:

$$C_z j\omega_m(y_{1r} + jy_{1i}) = -g_m\frac{1}{1+\beta}(x_{1r} + jx_{1i}) - g_m R_{gm}C_z\frac{1}{1+\beta}j\omega_m(x_{1r} + jx_{1i}).$$

(25)

Considering the real and imaginary part of Equation (25), one obtains:

$$\begin{cases} -C_z\omega_m y_{1i} = -g_m \frac{1}{1+\beta} x_{1r} + g_m R_{gm} C_z \frac{1}{1+\beta} \omega_m x_{1i} \\ C_z\omega_m y_{1r} = -g_m \frac{1}{1+\beta} x_{1i} - g_m R_{gm} C_z \frac{1}{1+\beta} \omega_m x_{1r} \end{cases}. \tag{26}$$

Equation (26) can be written as:

$$\begin{pmatrix} y_{1r} \\ y_{1i} \end{pmatrix} = \frac{1}{C_z\omega_m} \begin{pmatrix} -g_m R_{gm} C_z \frac{1}{1+\beta} \omega_m & -g_m \frac{1}{1+\beta} \\ g_m \frac{1}{1+\beta} & -g_m R_{gm} C_z \frac{1}{1+\beta} \omega_m \end{pmatrix} \begin{pmatrix} x_{1r} \\ x_{1i} \end{pmatrix}. \tag{27}$$

Equations (24) and (27) describe the transfer function of the first harmonic component in the main circuit and the control circuit of the converter, respectively. By integrating them, one has the total transfer function as follows:

$$M = \frac{1}{C_z\omega_m} \begin{pmatrix} -g_m R_{gm} C_z \frac{1}{1+\beta} \omega_m & -g_m \frac{1}{1+\beta} \\ g_m \frac{1}{1+\beta} & -g_m R_{gm} C_z \frac{1}{1+\beta} \omega_m \end{pmatrix}$$
$$\times \frac{\frac{V_m^2}{4R_s(1+\beta)V_{ref}}}{\left(\frac{1}{R^2} + \frac{\omega_m^2 C^2}{4}\right)\left(x_0^2 - \left(x_{2r}^2 + x_{2i}^2\right)\right)} \tag{28}$$
$$\times \begin{pmatrix} \frac{x_0 - x_{2r}}{R} + \frac{\omega_m C x_{2i}}{2} & -3\left(\frac{x_{2i}}{R} - \frac{\omega_m C(x_0 - x_{2r})}{2}\right) \\ -\frac{x_{2i}}{R} - \frac{\omega_m C(x_0 + x_{2r})}{2} & 3\left(\frac{x_0 + x_{2r}}{R} - \frac{\omega_m C x_{2i}}{2}\right) \end{pmatrix}.$$

One needs the DC component and the second harmonic component before studying Equation (28).

3.2. The DC Component and the Second Harmonic Component

The DC component and the second harmonic component are computed from Equations (16) and (18). It is important to note that the first harmonic component is smaller than the DC component and the second harmonic component in Equations (16) and (18). Therefore, one has:

$$\frac{C}{2}\frac{d}{dt}\left(x_0^2 + 2x_{2r}^2 + 2x_{2i}^2\right) + \frac{1}{R}\left(x_0^2 + 2x_{2r}^2 + 2x_{2i}^2\right) = \frac{V_m^2}{R_s(1+\beta)V_{ref}}(y_0 - y_{2r}). \tag{29}$$

$$\frac{C}{2}\frac{d}{dt}(2x_0 x_2) + \left(j\omega_m C + \frac{1}{R}\right)(2x_0 x_2) = \frac{V_m^2}{R_s(1+\beta)V_{ref}}\left(y_2 - \frac{1}{2}y_0\right). \tag{30}$$

Equations (19), (21), (29) and (30) form the model describing the DC component and the second harmonic component. By making all time-derivatives in those four equations equal to zero, one obtains:

$$\frac{1}{R}\left(x_0^2 + 2x_{2r}^2 + 2x_{2i}^2\right) = \frac{V_m^2}{R_s(1+\beta)V_{ref}}(y_0 - y_{2r}). \tag{31}$$

$$\left(j\omega_m C + \frac{1}{R}\right)(2x_0 x_2) = \frac{V_m^2}{R_s(1+\beta)V_{ref}}\left(y_2 - \frac{1}{2}y_0\right). \tag{32}$$

$$g_m\left(V_{ref} - \frac{1}{1+\beta}x_0\right) = 0. \tag{33}$$

$$j2C_z\omega_m y_2 = -g_m\frac{1}{1+\beta}x_2 - j2\omega_m g_m R_{gm} C_z \frac{1}{1+\beta}x_2. \tag{34}$$

From Equations (31)–(34), one obtains the DC component and the second harmonic component. The steady-state value of the DC component is:

$$x_0 = (1+\beta)V_{ref}. \tag{35}$$

When Equation (35) is satisfied, the second harmonic component is zero. Now we can study Equation (28).

3.3. Stability of the OCC Boost PFC Converter

Based on the DC component and the second harmonic component computed in Section 3.2, one can simplify Equation (28) into:

$$M = \frac{g_m}{C_z \omega_m} \frac{\frac{V_m^2}{4R_s(1+\beta)^2 V_{ref}}}{\left(\frac{1}{R^2} + \frac{\omega_m^2 C^2}{4}\right) x_0} \left(\begin{array}{cc} -\frac{R_{gm} C_z \omega_m}{R} + \frac{\omega_m C}{2} & -\frac{3R_{gm} C_z \omega_m^2 C}{2} - \frac{3}{R} \\ \frac{1}{R} + \frac{R_{gm} C_z \omega_m^2 C}{2} & \frac{3\omega_m C}{2} - \frac{3R_{gm} C_z \omega_m}{R} \end{array} \right). \tag{36}$$

It is important to note that M in (36) is the round-trip signal transfer function of the first harmonic component. When all eigenvalues of M are less than 1, the first harmonic component converges to zero. At the same time, the DC component and the second harmonic component are almost constant, as shown in Section 3.2 and Figure 2. Therefore the converter operates in a stable manner. When the absolute values of one eigenvalue of M is more than 1, the first harmonic component does not converge to zero. The converter begins to exhibit period-doubling bifurcation at the line frequency [24]. Therefore, the criterion of the stability of the converter is the eigenvalues of matrix M.

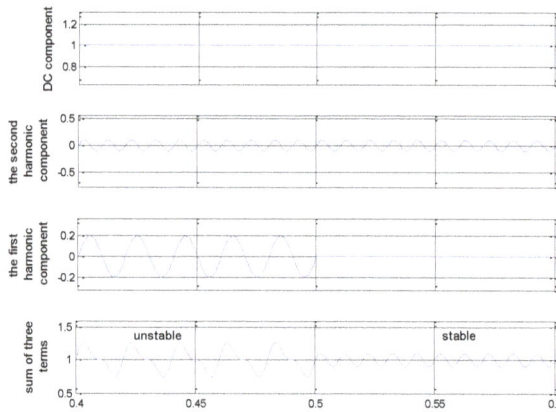

Figure 2. Illustration of stable and unstable operation of the converter.

4. Non-Linear Phenomena of the OCC Boost PFC Converter

To verify the above-mentioned theory, simulations and experiments are conducted. The same circuit topology is adopted (Figure 1). The parameters in the converter are shown in Table 1, unless otherwise specified.

Table 1. Parameters in the OCC Boost PFC converter.

Symbol	Quantity	Unit
T_s	15	μs
ω_m	100π	rad/s
L	2	mH
C	100	μF
R_{f1}	849	$k\Omega$
R_{f2}	37.3	$k\Omega$
R_{gm}	10.25	$k\Omega$
C_z	32	nF

Table 1. *Cont.*

Symbol	Quantity	Unit
C_p	32	pF
V_{ref}	7	V
R_s	0.645	Ω
$R_{(load)}$	1600	Ω
g_m	40	μS

In the converter, the input voltage and the output capacitance are two important parameters, which are selected in the design process. In this paper, the effect of these parameters on the non-linear phenomena of the converter is studied. Figure 3 shows the stability boundaries obtained from theoretical calculation based on Equation (36) and simulation experiment. From Figure 3, we have the following conclusions.

1. The effect of the input voltage on the stability of the converter. Figure 3 shows that when the capacitance is fixed and the input voltage increases, the converter may lose stability.
2. The effect of the output capacitance on the stability of the converter. Figure 3 shows that when the input voltage is increased, a larger output capacitance is needed in order to assure stable operation of the converter. This result is important because a larger output capacitance affects the dynamic performance of the converter.
3. The difference between the two boundaries lies in the fact that some approximations are taken in the analysis, and only the first and the second harmonic components are taken into consideration.

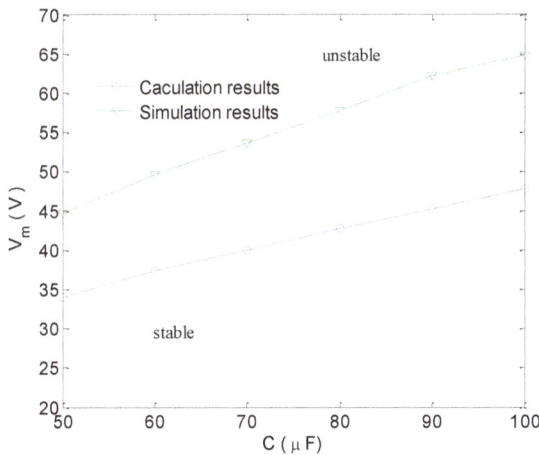

Figure 3. Stability boundaries of the input voltage obtained from the theoretical calculation and simulation experiment.

The simulation waveforms of the output voltage and the inductor current are shown in Figures 4 and 5 when the input voltage $V_m = 40$ V and $V_m = 66.5$ V, respectively. (For the MATLAB model file, please contact the corresponding author by e-mail: zhangrui@cqust.eud.cn.) In Figure 4, the converter operate stably. In Figure 5, the converter exhibits line frequency instability as a result of the period-doubling bifurcation at the line frequency. The instability reduces the power factor of the converter to be considerably lower than 1. If the input voltage increases further, the converter may exhibit chaotic phenomena.

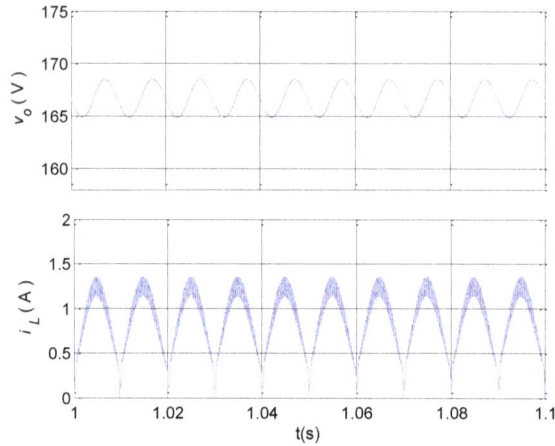

Figure 4. Simulation waveforms of the output voltage and the inductor current when the input voltage $V_m = 40$ V, the load $R = 1600$ Ω and the output capacitance $C = 100$ μF.

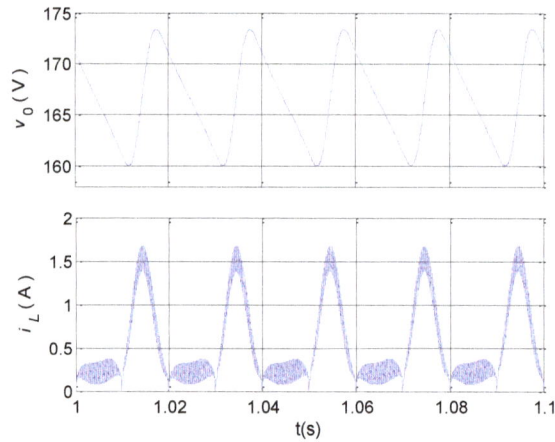

Figure 5. Simulation waveforms of the output voltage and the inductor current when the input voltage $V_m = 66.5$ V, the load $R = 1600$ Ω and the output capacitance $C = 100$ μF.

5. Experimental Verifications

To verify the line frequency instability from the theoretical analysis, an experimental circuit prototype was implemented using IR1150 (Infineon Technologies AG, Neubiberg, Germany), which is a classical OCC Boost PFC IC. In our experiment, the circuit parameters are identical to the above theoretical analysis. The current probe (FLUKE i5s, Everett, WA, USA) AC (400 mV/A) current clamp is used to detect the line current. The input voltage, the inductor current and the output voltage (AC coupling) are shown in Figures 6 and 7 when the input voltage $V_m = 40$ V and $V_m = 68$ V, respectively. In Figure 6, the converter operates in a stable manner and the frequencies of all waveforms are 100 Hz. In Figure 7, the converter exhibits period-doubling phenomena. Furthermore, the first harmonic frequency of the inductor current and the output voltage ripple are 50 Hz, which is half of the rectified AC voltage. As shown in Figures 6 and 7, the experimental results are consistent with the analytical results. It is important to note that when the PFC converter exhibits period-doubling bifurcation, the voltage ripple of the output voltage is larger compared with the normal operation. This value is important for the performance and lifetime of a electrolytic capacitor.

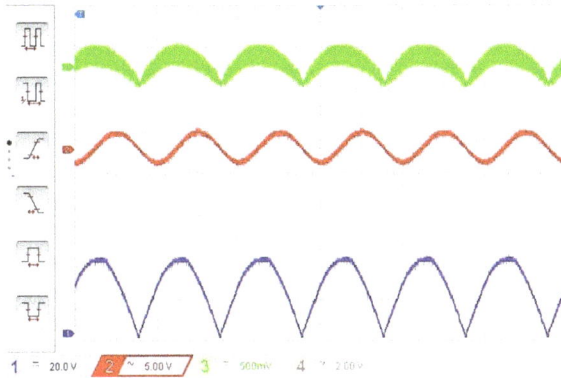

Figure 6. Experimental waveforms of the converter when the input voltage $V_m = 40$ V, the load $R = 1600\ \Omega$ and the output capacitance $C = 100\ \mu F$. CH1: the input voltage (20 V/div), CH3: the inductor current (500 mV/div) and CH2: the output voltage (5 V/div) (AC coupling).

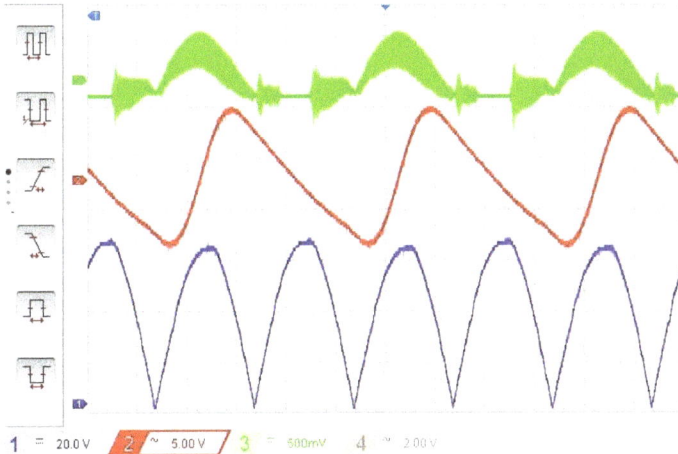

Figure 7. Experimental waveforms of the converter when the input voltage $V_m = 68$ V, the load $R = 1600\ \Omega$ and the output capacitance $C = 100\ \mu F$. CH1: the input voltage (20 V/div), CH3: the inductor current (500 mV/div) and CH2: the output voltage (5 V/div) (AC coupling).

6. Conclusions

The OCC PFC converters have a simpler control circuit compared to the traditional averaged current mode PFC converters. In this paper, the method of double averaging was adopted to study the dynamics of an OCC Boost PFC converter. The first averaging is applied to the switching period, and the second averaging is applied to the line period. We derived the round-trip signal transfer function of the first harmonic component in the converter, and the stability of the converter is decided by the eigenvalues of the round-trip signal transfer function. By calculating the eigenvalues, we gave theoretical prediction of the stability of the converter under different output capacitors. Simulation and experimental results verified theoretical prediction. The method of double averaging can predict nonlinear phenomena which traditional method cannot predict. It is important to note that when the OCC PFC converter exhibits line frequency instabilities, the power factor decreases dramatically. Therefore, theoretical analysis in this paper is of great importance in designing the OCC PFC converter.

Author Contributions: R.Z. and W.M. designed the experiment, drafted the manuscript; L.W. and M.H. conducted the experiment; L.C., H.Z. and Y.Z. reviewed and refined the paper.

Acknowledgments: This research is supported by the National Natural Science Foundation of China under Grant 51377188.

Appendix A

Some important properties:

$$\left(\frac{du(t)}{dt}\right)_k = \frac{du_k(t)}{dt} + jk\omega_m u_k. \tag{A1}$$

$$\left(u(t)\cos 2\omega_m t\right)_k = \frac{1}{2}\left(u_{k-2} + u_{k+2}\right). \tag{A2}$$

$$\left(u^2(t)\right)_0 = u_0^2 + 2|u_1|^2 + 2|u_2|^2. \tag{A3}$$

$$\left(u^2(t)\right)_1 = 2u_0 u_1 + 2(u_{1r}u_{2r} + u_{1i}u_{2i} + j(u_{1r}u_{2i} - u_{1i}u_{2r})). \tag{A4}$$

$$\left(u^2(t)\right)_2 = (u_1^*)^2 + 2u_0 u_2. \tag{A5}$$

where $u_1 = u_{1r} + ju_{1i}$, $u_2 = u_{2r} + ju_{2i}$.

References

1. García, O.; Cobos, J.A.; Prieto, R.; Alou, P.; Uceda, J. Single phase power factor correction: A survey. *IEEE Trans. Power Electron.* **2003**, *18*, 749–755. [CrossRef]
2. Giaouris, D.S.; Banerjee, B.; Zahawi, V. Pickert, Control of fast scale bifurcations in power-factor correction converters. *IEEE Trans. Circuits Syst. II* **2007**, *54*, 805–809. [CrossRef]
3. Lai, Z.; Smedley, K.M. A family of continuous-conduction-mode power-factor-correction controllers based on the general pulse-width modulator. *IEEE Trans. Power Electr.* **1998**, *13*, 501–510.
4. Orabi, M.; Nimoniya, T. Non-linear dynamic of power factor correction converter. *IEEE Trans. Ind. Electron.* **2003**, *50*, 1116. [CrossRef]
5. Iu, H.H.C.; Zhou, Y.F.; Tse, C.K. Fast-scale instability in a PFC boost converter under average current-mode control. *Int. J. Circ. Theor. App.* **2003**, *31*, 611–624. [CrossRef]
6. Orabi, M.; Ninomiya, T. Stability investigation of the cascade twostage PFC converter. *IEICE Trans. Commun.* **2004**, *E87-B*, 3506–3514.
7. Chu, G.; Tse, C.K.; Wong, S.C. Line-frequency instability of PFC power supplies. *IEEE Trans. Power Electron.* **2009**, *24*, 469–482. [CrossRef]
8. El Aroudi, A.; Orabi, M.; Haroun, R.; Martínez-Salamero, L. Asymptotic slow-scale stability boundary of PFC AC-DC power converters: Theoretical prediction and experimental validation. *IEEE Trans. Ind. Electron.* **2011**, *58*, 3448–3460. [CrossRef]
9. Wang, F.; Zhang, H.; Ma, X. Analysis of slow-scale instability in boost PFC converter using the method of harmonic balance and floquet theory. *IEEE Trans. Circuits Syst. Regul. Pap.* **2010**, *57*, 405–414. [CrossRef]
10. El Aroudi, A.; Orabi, M. Stabilizing technique for AC-DC boost PFC converter based on time delay feedback. *IEEE Trans. Circuits Syst. Express Briefs* **2010**, *57*, 56–60. [CrossRef]
11. Ma, W.; Wang, M.; Liu, S.; Li, S.; Yu, P. Stabilizing the average current-mode-controlled boost PFC converter via washout-filter-aided method. *IEEE Trans. Circuits Syst. Express Briefs* **2011**, *58*, 595–599. [CrossRef]
12. Zou, J.; Ma, X.; Tse, C.K.; Dai, D. Fast-scale bifurcation in power-factor-correction buck-boost converters and effects of incompatible periodicities. *Int. J. Circuit Theory Appl.* **2006**, *34*, 251–264. [CrossRef]
13. Wu, X.; Tse, C.K.; Dranga, O.; Lu, J. Fast-scale instability of single-stage power-factor correction power supplies. *IEEE Trans. Circuits Syst. Regul. Pap.* **2006**, *53*, 204–213.

14. Wong, S.-C.; Tse, C.K.; Orabi, M.; Ninomiya, T. The Method of Double Averaging: An Approach for Modeling Power-Factor-Correction Switching Converters. *IEEE Trans. Circuits Syst. Regul. Pap.* **2006**, *53*, 454–464. [CrossRef]

15. Orabi, M.; Haron, R.; Youssef, M.Z. Stability analysis of PFC converters with one-cycle control. In Proceedings of the 31st International Telecommunications Energy Conference, Incheon, Korea, 18–22 October 2009.

16. Smedley, K.M.; Cuk, S. One-Cycle Control of Switching Converters. *IEEE Trans. Power Electr.* **1995**, *10*, 625–633. [CrossRef]

17. Smedley, K.M.; Cuk, S. Dynamics of One-Cycle controlled cuk converters. *IEEE Trans. Power Electr.* **1995**, *10*, 634–639. [CrossRef]

18. Fang, C.C.; Abed, E.H. Robust feedback stabilization of limit cycles in PWM DC-DC converters. *Nonlinear Dyn.* **2002**, *27*, 295–309. [CrossRef]

19. Fang, C.-C. Sampled-Data modeling and analysis of One-Cycle control and charge control. *IEEE Trans. Power Electr.* **2001**, *16*, 345–350. [CrossRef]

20. Lee, F.; Iwens, R.; Yu, Y.; Triner, J. Generalized computer-aided discrete time domain modeling and analysis of dc-dc converters. In Proceedings of the 1977 IEEE Power Electronics Specialists Conference, Palo Alto, CA, USA, 14–16 June 1977.

21. Sanders, S.R.; Verghese, G.C. Synthesis of averaged circuit models for switched power converters. *IEEE Trans. Circuits Syst.* **1991**, *8*, 905–915. [CrossRef]

22. Maksimovic, D.; Zane, R.; Erickson, R. Impact of digital control in power electronics. In Proceedings of the 16th International Symposium on Power Semiconductor Devices & ICs, Kitakyushu, Japan, 24–27 May 2004.

23. Caliskan, V.A.; Verghese, O.C.; Stankovic, A.M. Multifrequency averaging of DC/DC converters. *IEEE Trans. Power Electron.* **1999**, *1*, 124–133. [CrossRef]

24. Dorf, R.C.; Bishop, R.H. *Modern Control Systems*, 11th ed.; Pearson Education, Inc.: Upper Saddle River, NJ, USA, 2008.

electronics

MDPI

Article

Improved Step Load Response of a Dual-Active-Bridge DC–DC Converter

Yifan Zhang, Xiaodong Li *, Chuan Sun and Zhanhong He

Faculty of Information Technology, Macau University of Science and Technology, Macau, China;
zhangyifan0329@outlook.com (Y.Z.); sunchuanmust@163.com (C.S.); calvinhzh@163.com (Z.H.)
* Correspondence: xdli@must.edu.mo; Tel.: +0853-8897-2195

Received: 10 August 2018; Accepted: 6 September 2018; Published: 9 September 2018

Abstract: This paper proposes a fast load transient control for a bidirectional dual-active-bridge (DAB) DC/DC converter. It is capable of maintaining voltage–time balance during a step load change process so that no overshoot current and DC offset current exist. The transient control has been applied for all possible transition cases and the calculation of intermediate switching angles referring to the fixed reference points is independent from the converter parameters and the instantaneous current. The results have been validated by extended experimental tests.

Keywords: transient control; DC–DC conversion; bidirectional converter

1. Introduction

In recent decades, with the increasing concern in environment issue and energy crisis, the power conversion systems (PCSs) have been using widely in renewable generation facilities. The dual-active-bridge (DAB) converter seems to be a preferred choice of PCSs in various bidirectional DC/DC applications, such as energy storage systems, electric vehicles (EV) and solid state transformers, because of its high power density, low cost and zero-voltage-switching (ZVS) features [1–14].

Interfacing two different DC sources, the DAB converter is a kind of bidirectional DC–DC converter, which consists of two full bridges linked by a high-frequency (HF) transformer with the turn ratio of $n_t : 1$, whose circuit layout is shown in Figure 1. Including the leakage inductance of the transformer, the inductor L_s is connected on the primary side as the main energy transfer device. The resistor r_s, which normally is small enough to be neglected in steady-state analysis, is an equivalent resistance of L_s and the total winding resistance of HF transformer. The eight active switches can be controlled by their gating signals with 50% duty cycle and fixed switching frequency. The voltage gain of DAB converter is defined as $M = n_t V_2 : V_1$, where V_1 is the input voltage and V_2 is the output voltage. v_p and v_s are two HF voltages generated on the primary side and the secondary side, respectively. The power is manipulated by controlling the phase-shift angles among each switch arms.

Depending on the number of varying phase-shifts, there are several control schemes for a DAB converter, which are single-phase-shift (SPS) control, extended-phase-shift control (EPS), dual-phase-shift (DPS) control and triple-phase-shift (TPS) control. The SPS control is the simplest control strategy that is easy to implement. However, it has many disadvantages such as high circulating power and loss of ZVS if the converter gain is away from unity [1,2]. In addition, the TPS control has three independent phase-shifts to be controlled, which makes it costly and more complex in real implementation [3,4]. In contrast, EPS control and DPS control are compromised ones with both enough flexibility and easy implementation [5–14]. To make a step load change in a DAB converter, one or more phase-shift angles should be adjusted accordingly. The detailed procedure to adjust those phase-shift angles have direct influence on the transient responses. Improper transient control may cause temporal overcurrent and DC offset in inductor current which can arise extra losses and saturation in magnetic components. Therefore, it is necessary to propose some methods to eliminate DC offset and minimize the load transient period.

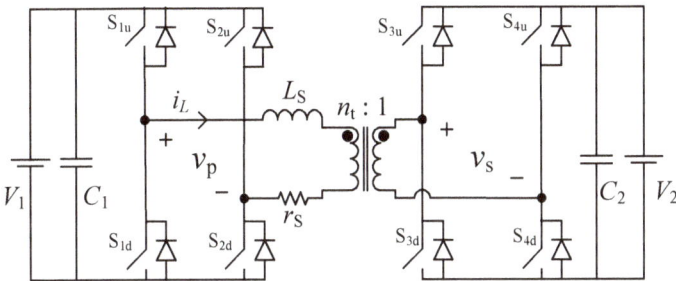

Figure 1. A dual-active-bridge DC/DC converter.

Inserting a capacitor before the transformer is an easy way to block DC bias in the HF transformer current with the increase in the cost and the size of the circuit. Current oscillation might be induced due to change of stored energy in the capacitor at the step-change of load. An effective solution by means of a special magnetic sensor and active compensator is presented in [15], which is quite costly and complicated regardless of the excellent performance. Based on SPS control, a transient control named asymmetric-double-side modulation proposed in [16] distributes the required phase-shift adjustment to both two bridges according to an optimized ratio to depress the DC bias current in the transition process. The same technique is then extended to EPS control in [17,18]. However, the obtained results in [17] can not deal with the operation when the converter gain is close to unity. Another solution under SPS control in [19] manipulates the gating signals of two switch arms in the same bridge with different phase-shifts during the transient process. The method was applied in a three-phase DAB converter too [20]. Although the calculation of this method is easy and is independent on the converter gain, it can be proved that a single current pulse can be induced for specified load transient conditions. In [21], a novel approach to keep transient voltage–time balance is to introduce a small zero-voltage duration in one of the two HF voltages. It is capable of eliminating DC bias in both inductor current and magnetizing current of the transformer. Generally, it is seen that most of the reported solutions are limited in the scenario of SPS control in a steady state. However, two or more phase-shifts are needed for power manipulation in applications with wide variation in converter gain. Thus, to explore new transient control with multiple phase-shifts will be meaningful. In this work, a load transient modulation for EPS control will be proposed for depressing DC bias current in the load-changing process, which stems from the approach in [21]. The proposed transient control will be applied to different transition cases between the two steady-state EPS modes. It will be shown that it is able to not only reduce transient period, but also depress the DC bias effectively.

The paper is organized as follows. In Section 2, the proposed transient control method will be analyzed in detail for each transition case. The values of all gating signal angles would be given before, during and after the transition process. The theoretical analysis is then verified by experimental tests on a lab prototype converter in Section 3. The final conclusion is presented in Section 4.

2. The Proposed Transient Control Method

Under the EPS control scheme, there are two phase shift angles to be used: φ_1—the phase delay between the turn-on moment of S_{1d} and that of S_{2u}; φ_2—the phase delay between the turn-on moment of S_{1d} and that of S_{3d}, S_{4u}. While working under EPS, the converter may have two different steady-state modes according to the different relationship of φ_1, φ_2 shown in Figure 2. Mode A is defined with $0 \leq \varphi_1 \leq \varphi_2 \leq \pi$ and mode B is defined with $0 \leq \varphi_2 \leq \varphi_1 \leq \pi$.

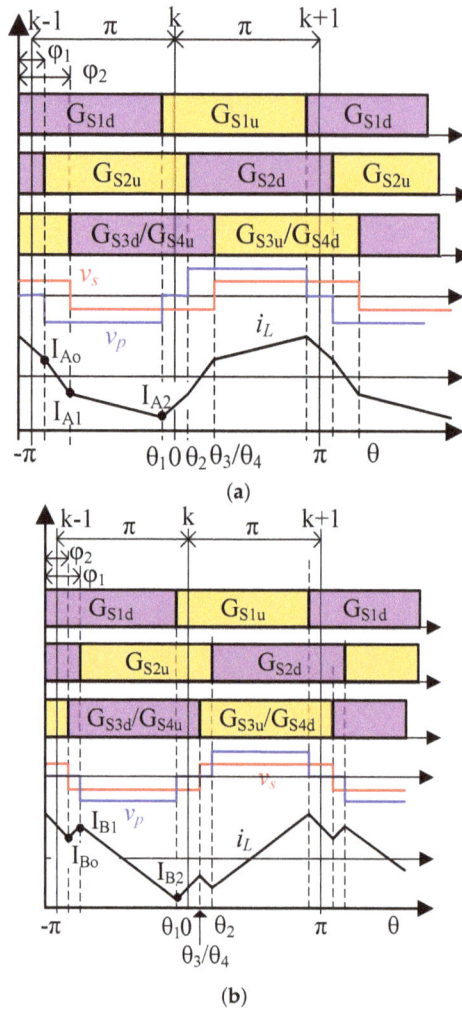

Figure 2. Defined control timing in (**a**) mode A and (**b**) mode B.

2.1. Reference Points and Definition of Switching Angles

By referring to the fixed reference point, $\theta_1, \theta_2, \theta_3, \theta_4$ are defined as the switching angles where S_{1u}/S_{1d}, S_{2u}/S_{2d}, S_{3u}/S_{3d}, S_{4u}/S_{4d} are turned on, respectively.

If the original mode is mode A, the reference points are defined in the mid of φ_1 and the interval between each of them is π. Thus, the switching angles in mode A are expressed as follows:

$$\theta_1 = -\frac{\varphi_1}{2}, \theta_2 = \frac{\varphi_1}{2}; \ \theta_3 = \theta_4 = \varphi_2 - \frac{\varphi_1}{2}. \tag{1}$$

If the original mode is mode B, the reference points are defined in the mid of φ_2. Thus, the switching angles in mode B can be expressed as follows:

$$\theta_1 = -\frac{\varphi_2}{2}, \theta_2 = \varphi_1 - \frac{\varphi_2}{2}; \ \theta_3 = \theta_4 = \frac{\varphi_2}{2}. \tag{2}$$

It is known that in each time interval the change of inductor current is proportional to the voltage difference across it: $\Delta i_L = \frac{v_p - v_s}{\omega_s L_s} \Delta\theta$. Therefore, the instantaneous inductor currents at switching moments in mode A and B can be found as:

$$I_{Ao}(\varphi_1, \varphi_2) = \pi - \varphi_1 + M(2\varphi_2 - 2\varphi_1 - \pi), \tag{3}$$

$$I_{A1}(\varphi_1, \varphi_2) = \pi + \varphi_1 - 2\varphi_2 - M\pi, \tag{4}$$

$$I_{A2}(\varphi_1, \varphi_2) = \varphi_1 - \pi + M(\pi - 2\varphi_2), \tag{5}$$

$$I_{Bo}(\varphi_1, \varphi_2) = \pi - \varphi_1 - M\pi, \tag{6}$$

$$I_{B1}(\varphi_1, \varphi_2) = \pi - \varphi_1 + M(2\varphi_1 - 2\varphi_2 - \pi), \tag{7}$$

$$I_{B2}(\varphi_1, \varphi_2) = \varphi_1 - \pi + M(\pi - 2\varphi_2). \tag{8}$$

When the load level is changed abruptly, there are four condition of DAB converter in EPS control to be dealt with as shown in Table 1.

Table 1. Different load transition conditions of EPS control.

	Initial Mode	Final Mode
Condition 1	Mode A	Mode A
Condition 2	Mode A	Mode B
Condition 3	Mode B	Mode B
Condition 4	Mode B	Mode A

2.2. Load Transient Control within Mode A

As shown in Figure 3, a step load transition happened around the kth reference point. In other words, the switching behaviour referring to $(k-1)^{th}$ reference point is the original steady state with phase angles φ_1, φ_2, while the switching behaviour referring to $(k+1)^{th}$ reference point is the destination steady state with phase angles φ'_1, φ'_2. The phase angles referring to the kth reference point should be selected properly to complete the transition process as fast as possible.

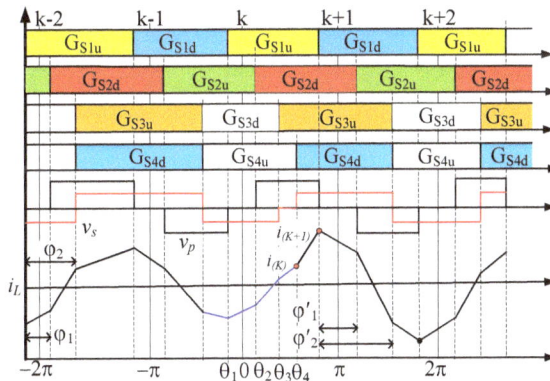

Figure 3. Gate signals, voltage and current waveforms of transient control within mode A.

With the unknown $\theta_1 \sim \theta_4$, two important instant currents are calculated then:

$$i_{(k)} = (-\pi - 2\theta_1 - 2\theta_2 + 2\theta_4) + M(\pi + \varphi_1 - 2\varphi_2 + 2\theta_3), \tag{9}$$

$$i_{(k+1)} = (\pi - 2\theta_1 - 2\theta_2 - \varphi'_1) + M(-\pi + \varphi_1 - 2\varphi_2 + 2\theta_3 + 2\theta_4 + \varphi'_1), \tag{10}$$

where $i_{(k)}$ is the instant current at the last switching point of kth reference point, and $i_{(k+1)}$ is the first current at the first switching point of $(k+1)^{th}$ reference point.

To keep voltage–time balance, the average inductor current should be zero while the inductor current enter into the new steady state at once, which indicates:

$$i_{(k)} = -I_{A1}(\varphi_1', \varphi_2'), \ i_{(k+1)} = -I_{A2}(\varphi_1', \varphi_2'). \tag{11}$$

Therefore, the following conditions can be obtained by substituting Equations (4), (5), (9) and (10) into (11):

$$\theta_1 = -\theta_2, \ \theta_3 = \varphi_2 - \frac{\varphi_1}{2}, \ \theta_4 = \varphi_2' - \frac{\varphi_1'}{2}. \tag{12}$$

It is seen that v_s is set to zero during the interval $[\theta_3 \ \theta_4]$. As shown in Equation (12), the phase shift angles θ_1 and θ_2 during transitions are free to be chosen. However, they should also satisfy the requirement that $\mid \theta_1 \mid = \mid \theta_2 \mid < \min\{\pi - \varphi_2 + \varphi_1/2, \varphi_2' - \varphi_1'/2\}$ lets the converter work in mode A.

2.3. Load Transient Control within Mode B

In this condition, the DAB converter is working in mode B from beginning to end, and the reference points are defined in the mid of φ_2. As is shown in Figure 4, the transition is done at the kth reference point and the converter is expected in the destination steady state at the $(k+1)^{th}$ reference point. The instant currents after the intermediate adjustment are calculated as:

$$\begin{align} i_{(k)} &= (-\pi - 2\theta_1 - \varphi_2 + \varphi_1) + M(\pi - \varphi_2 + 2\theta_3 + 2\theta_4 - 2\theta_2), \tag{13} \\ i_{(k+1)} &= (\pi - 2\theta_1 - 2\theta_2 - \varphi_2' + \varphi_1 - \varphi_2) + M(-\pi + \varphi_2' - \varphi_2 + 2\theta_3 + 2\theta_4). \tag{14} \end{align}$$

To meet such an expectation, the followed equation should be satisfied:

$$i_{(k)} = -I_{B1}(\varphi_1', \varphi_2'), \ i_{(k+1)} = -I_{B2}(\varphi_1', \varphi_2'). \tag{15}$$

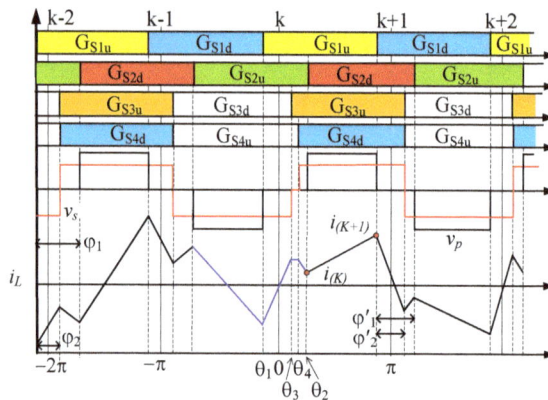

Figure 4. Gate signals, voltage and current waveforms of the proposed control method in within mode B.

Combining Equations (7), (8), (13), (14) and (15), the following switching angles referring to the kth reference point can be calculated to achieve a fast transition within mode B:

$$\theta_1 = \frac{\varphi_1 - \varphi_2 - \varphi_1'}{2}, \theta_2 = \varphi_1' - \frac{\varphi_2'}{2}, \ \theta_3 + \theta_4 = \frac{\varphi_2 + \varphi_2'}{2}. \tag{16}$$

It is seen that v_s is set to zero during the interval $[\theta_3\ \theta_4]$. Equation (16) reveals that θ_3 and θ_4 can be selected flexibly. However, they should also satisfy $\max\{\theta_3, \theta_4\} < \pi - \varphi'_2/2$ to let the converter work in mode B.

2.4. Load Transient Control from Mode A to Mode B

In this condition, as shown in Figure 5, the phase shift angles are changed from mode A to mode B by means of the intermediate adjustment around the kth reference point. Different from the previous cases, the definition of switching angles are changed since that the original mode is not the same as the destination one. The instant currents after the intermediate adjustment are calculated as:

$$i_{(k)} = (-\pi - 2\theta_1) + M(\pi + \varphi_1 - 2\varphi_2 - 2\theta_2 + 2\theta_3 + 2\theta_4), \tag{17}$$

$$i_{(k+1)} = (\pi - 2\theta_1 - 2\theta_2 - \varphi'_2) + M(-\pi + \varphi_1 - 2\varphi_2 + 2\theta_3 + 2\theta_4 + \varphi'_2). \tag{18}$$

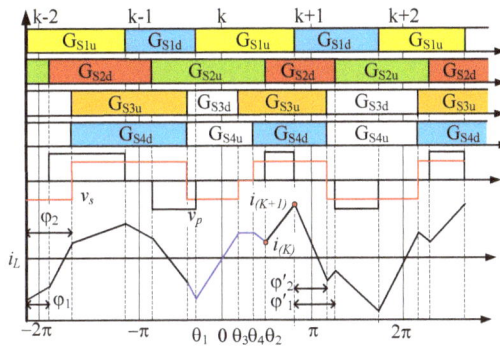

Figure 5. Gate signal, voltage and current waveforms of the proposed control method from mode A to mode B.

To minimize the transient period, the same requirement shown in Equation (15) should be met. Combining Equations (7), (8), (17), (18) and (15), the intermediate switching angles for transition from mode A to mode B are given as:

$$\theta_1 = -\frac{\varphi'_1}{2}, \ \theta_2 = \varphi'_1 - \frac{\varphi'_2}{2}, \ \theta_3 + \theta_4 = \frac{\varphi'_2}{2} + \varphi_2 - \frac{\varphi_1}{2}. \tag{19}$$

It is seen that v_s is set to zero during the interval $[\theta_3\ \theta_4]$. From the results, it can be found that switching angels θ_3 and θ_4 can be determined flexibly. Meanwhile, the condition of mode boundary should be met too: $\max\{\theta_3, \theta_4\} < \pi - \varphi'_1/2$.

2.5. Load Transient Control from Mode B to Mode A

In this condition shown in Figure 6, the original state is mode B while the final state is mode A. Therefore, the instant currents after the kth reference point can be calculated starting from mode B:

$$i_{(k)} = (-\pi - 2\theta_1 + \varphi_1 - \varphi_2) + M(\pi - \varphi_2 - 2\theta_2 + 2\theta_3 + 2\theta_4), \tag{20}$$

$$i_{(k+1)} = (\pi - 2\theta_1 - 2\theta_2 - \varphi'_1 + \varphi_1 - \varphi_2) + M(-\pi + \varphi'_1 - \varphi_2 + 2\theta_3 + 2\theta_4). \tag{21}$$

As the switching angles are redefined in mode A at the $(k+1)^{th}$ reference point, the requirement for the expected fast transition is the same as (11). Substituting Equations (4), (5), (20), (21) into Equation (11), switching angles during transient process referring to the kth reference point for the transition from mode B to mode A are:

$$\theta_1 = \frac{\varphi_1 + \varphi_1' - \varphi_2 - 2\varphi_2'}{2}, \ \theta_2 = \varphi_2' - \frac{\varphi_1'}{2}, \ \theta_3 + \theta_4 = \frac{\varphi_2}{2} + \varphi_2' - \frac{\varphi_1'}{2}. \tag{22}$$

In this case, θ_1 and θ_2 can be chosen flexibly, but is subject to $| \ \theta_1 \ | < \pi - (\varphi_1 - \varphi_2/2)$ and $\theta_2 < \pi - \varphi_2'/2$.

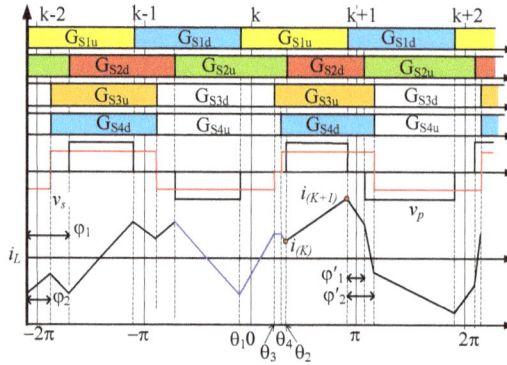

Figure 6. Gate signal, voltage and current waveforms of the proposed control method from mode B to mode A.

3. Validation by Experimental Results

In order to validate the theoretical results, load transition cases using the proposed control method were tested on a lab prototype DAB converter. Table 2 shows the specifications of the converter used in the experiments. The circuit adopts four power MOSFETs (STP40NF20, 200 V, 40 A, 0.038 Ω) on the primary side and the other four power MOSFETs (IPP200N15N3G, 150 V, 50 A, 0.020 Ω) on the secondary side as the switches. The input terminals were connected to a DC power supply, while the output terminals were connected to a DC electronic load. The inductor is made of a toroidal CM400125 MPP core with litz wire winding. The proposed transient control is implemented in a TI-F2812 DSP development board (Texas Instruments, Dallas, USA) and the flowchart is shown in Figure 7. The converter power level is monitored continuously. If no change is to be made to the power, the current φ_1, φ_2 are used to generate $\theta_1 \sim \theta_4$. If a new power command is received and confirmed, the destination will be calculated based on some preset algorithm optimized for better efficiency, which is out of scope of the current work. Then, $\theta_1 \sim \theta_4$ for the next reference point will be updated by $\varphi_1, \varphi_2, \varphi_1', \varphi_2'$ by using one of Equations (12), (16), (19), (22).

Table 2. Specifications of the prototype converter.

Parameters	Value
DC input voltage V_1	120 V
DC output voltage V_2	72 V
Transformer turns ratio n_t : 1	1:1
Transformer ferrite core	PC40ETD49
Series inductor L_s	121.875 μH
HF filter capacitance C_1, C_2	330 μF
Primary-side MOSFETs	STP40NF20
Secondary-side MOSFETs	IPP200N15N3G
Switching frequency f_{sw}	100 kHz

MOSFET: metal-oxide-semiconductor field-effect transistor

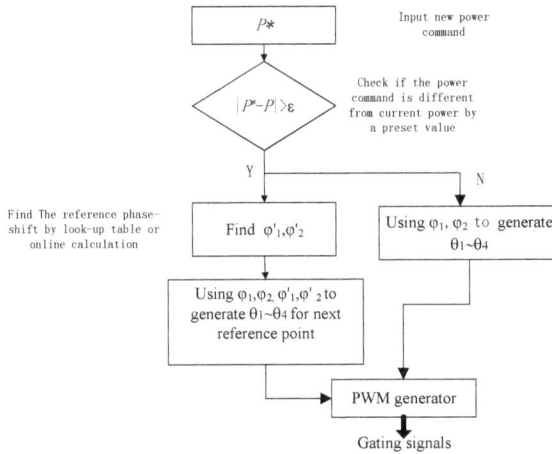

Figure 7. Flowchart for implementation of the proposed transient control method.

The detailed parameters of phase-shift angles of each mode in experiment are shown in Tables 3–6. For each tested load-change condition, a condensed view of the transition is given firstly with a time scale of 300 μs/div while the details are shown later with a time scale of 5 μs/div. In each plot, captured waveforms of v_p, v_s (100 V/div) and i_L (2 A/div) are shown from top to bottom.

Table 3. Transition within mode A.

Phase-Shift-Angles	Initial	Transient	Final
ϕ_1	30°		47.28°
ϕ_2	60°		112.8°
θ_1	−15°	−15°	−23.64°
θ_2	15°	15°	23.64°
θ_3	45°	45°	89.16°
θ_4	45°	89.16°	89.16°

Table 4. Transition within mode B.

Phase-Shift-Angles	Initial	Transient	Final
ϕ_1	60°		88.8°
ϕ_2	42°		82.32°
θ_1	−21°	−35.4°	−41.16°
θ_2	39°	47.64°	47.64°
θ_3	21°	21°	41.16°
θ_4	21°	41.16°	41.16°

Table 5. Transition from mode A to mode B.

Phase-Shift-Angles	Initial	Transient	Final
ϕ_1	30°		90.48°
ϕ_2	60°		81.6°
θ_1	−15°	−45.24°	−40.8°
θ_2	15°	49.68°	49.68°
θ_3	45°	49.68°	40.8°
θ_4	45°	45°	40.8°

Table 6. Transition from mode B to mode A.

Phase-Shift-Angles	Initial	Transient	Final
ϕ_1	114°		30°
ϕ_2	79.2°		60°
θ_1	−39.6°	−27.6°	−15°
θ_2	74.4°	45°	15°
θ_3	39.6°	39.6°	45°
θ_4	39.6°	45°	45°

As an example for comparison, the first case tested (Figure 8) is a transition within mode A by directly changing the phase-shifts. The phase-shift angles φ_1, φ_2 are changed from 30°, 60° to 47.28°, 112.8°. The inductor peak current is expected to rise from 1.56 A to 2.13 A. However, an abnormal peak current 2.95 A results in a transient process and it takes about 20 HF cycles to be absorbed. In the duration of transient process, a temporal DC bias current decays from about 0.69 A until zero.

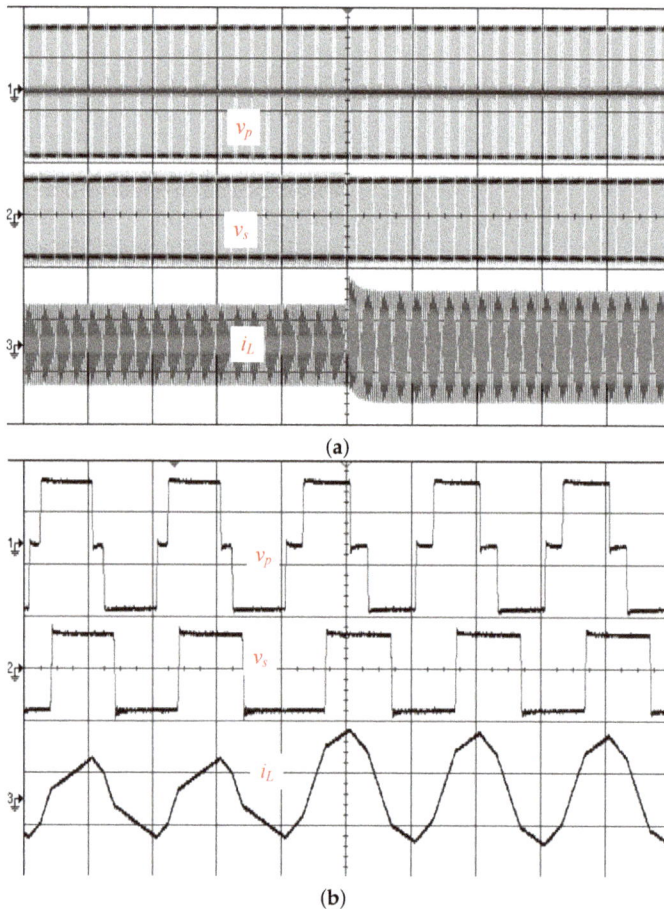

(a)

(b)

Figure 8. Experimental plots of transition within mode A without proposed control method. The signals shown from top to bottom are: v_p (100 V/div), v_s (100 V/div) and i_L (2 A/div). (a) condensed view (300 μs/div); (b) expanded view (5 μs/div).

The same transition is then repeated with proposed transient control in Figure 9. According to (12), θ_3, θ_4 are calculated as 45° and 89.16° during the transient process. While satisfying (12), the switch angles θ_1 and θ_2 are selected as $-\phi_1/2$ and $\phi_1/2$ during transient for the purpose of convenience. In addition, the final values of $\theta_1 \sim \theta_4$ are −23.64°, 23.64°, 89.16° and 89.16°. The transient process now can be completed almost instantly as shown in Figure 9. It is seen that there is no noticeable overshoot current.

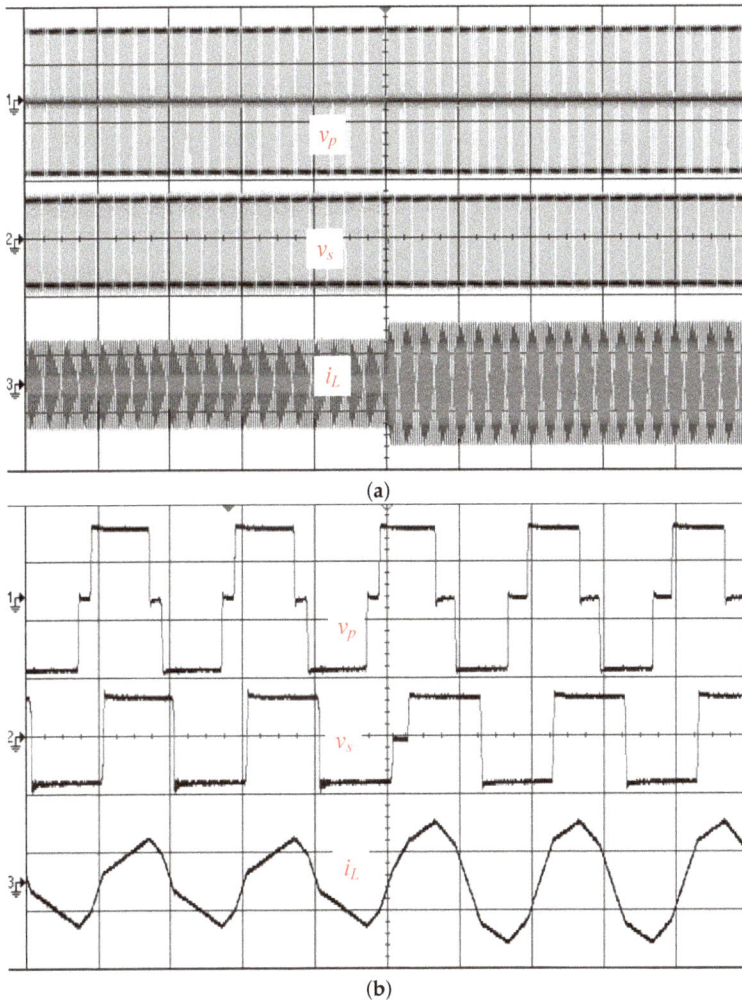

Figure 9. Experimental plots of transition within mode A with proposed control method. The signals shown from top to bottom are: v_p (100 V/div), v_s (100 V/div) and i_L (2A/div). (**a**) condensed view (300 μs/div); (**b**) expanded view (5 μs/div).

For the load transition from mode B to mode B in Figure 10, the phase-shift ϕ_1, ϕ_2 are changed from 60°, 42° to 88.8°, 82.32°. After the transient modulation, the switching angles θ_1, θ_2, θ_3 and θ_4 are changed from −21°, 39°, 21°, 21° to −41.16°, 47.64°, 41.161° and 41.16°, respectively. During the transient process, θ_1, θ_2 are calculated −35.4°, 47.64° directly according to (16). In addition, the transient θ_3 and θ_4 are selected as $\phi_2/2 = 21°$ and $\phi_2'/2 = 41.16°$, respectively.

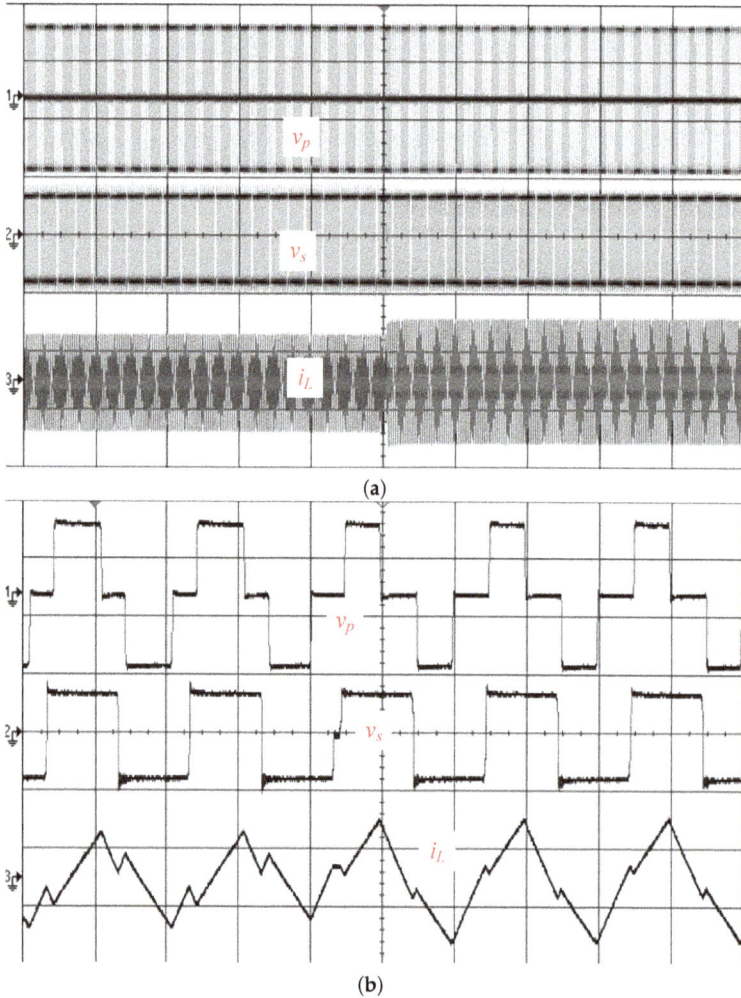

Figure 10. Experimental plots of transition within mode B with proposed control method. The signals shown from top to bottom are: v_p (100 V/div), v_s (100 V/div) and i_L (2 A/div). (**a**) condensed view (300 μs/div); (**b**) expanded view (5 μs/div).

In the condition of transition from mode A to mode B in Figure 11, the phase-shift ϕ_1, ϕ_2 are changed from 30°, 60° to 90.48°, 81.6°. The initial angles θ_1, θ_2, θ_3 and θ_4 are $-15°$, 15°, 45°, 45° and the final angles should be $-40.8°$, 49.68°, 40.8° and 40.8°, respectively. At the kth reference point, θ_1, θ_2 are determined to be $-45.24°$ and 49.68° based on (19). Under the constraint given in (19), the switching angles θ_3 and θ_4 are selected as $\phi_2'/2 = 49.68°$ and $\phi_2 - \phi_1/2 = 45°$, respectively, for the purpose of convenience.

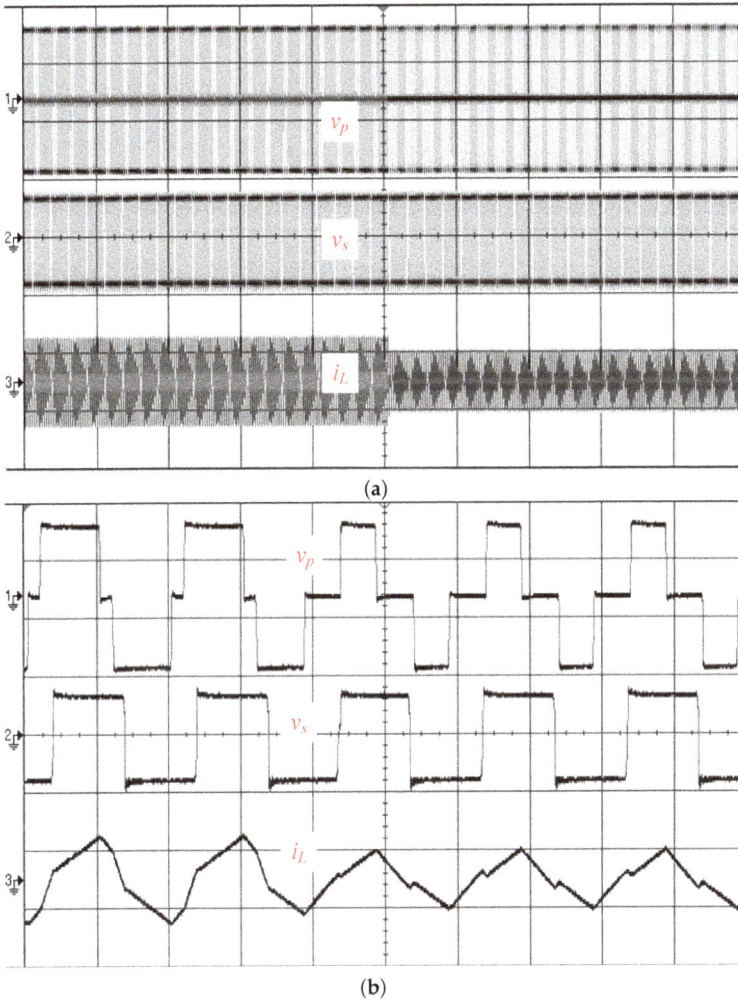

Figure 11. Experimental plots of transition from mode A to mode B with proposed control method. The signals shown from top to bottom are: v_p (100 V/div), v_s (100 V/div) and i_L (2 A/div). (**a**) condensed view (300 µs/div); (**b**) expanded view (5 µs/div).

In Figure 12, the condition of transition from mode B to mode A is presented, in which the phase-shift ϕ_1, ϕ_2 are changed from 114°, 79.2° to 30°, 60°. The initial angles θ_1, θ_2, θ_3 and θ_4 are −39.6°, 74.4°, 39.6°, 39.6° and the final angles should be −15°, 15°, 45° and 45°, respectively. At the kth reference point, θ_1, θ_2 are determined to be −27.6° and 45° based on (22). Under the constrain given in (22), the switch angles θ_3 and θ_4 are selected as $\phi_2'/2 = 39.6°$ and $\phi_2 - \phi_1/2 = 45°$ for the purpose of convenience.

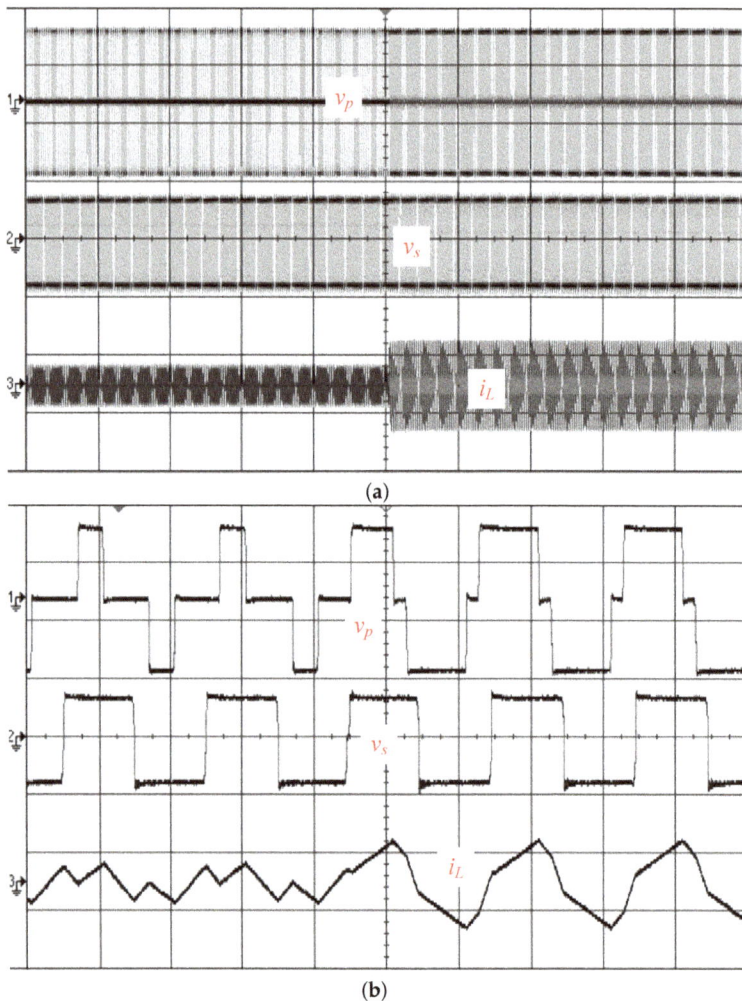

Figure 12. Experimental plots of transition from mode B to mode A with proposed control method. The signals shown from top to bottom are: v_p (100 V/div), v_s (100 V/div) and i_L (2 A/div). (a) condensed view (300 μs/div); (b) expanded view (5 μs/div).

4. Conclusions

In this work, a fast transient control is proposed for a DAB converter that is able to improve the step-load transient response in terms of response time and overshoot current. This transient control is implemented based on the definition of switching angles for each switch arm, which makes it easy for implementation in pulse-width-modulation (PWM) units of common micro-controller platforms. A small duration of zero-voltage is introduced in the transformer voltage during the transition process to keep the voltage-second balance of the inductor. All the transient switching angles can be calculated from the original and final phase-shift angles directly and are not affected by the converter parameters. Though the proposed control method aims to modulate the transient inductor current, no information about the instantaneous current is needed. With this proposed transient control method, the DAB converter can transfer from one steady state to another quickly and smoothly and causes no DC offset in inductor current, which has been validated successfully by a series of experimental tests.

Author Contributions: Y.Z. did most of the theoretical analysis, derivation and paper writing. C.S. contributed to circuit implementation and experimental test. X.L. is responsible for planning, coordination and proofreading. Z.H. contributed to drawing figures and formatting paper.

Funding: This work was supported by Fundo para o Desenvolvimento das Ciências e da Tecnologia under Grant No. 060/2017/A.

Conflicts of Interest: The authors declare no conflict of interest.

References

1. Kheraluwala, M.H.; Gascoigne, R.; Divan, D.M.; Baumann, E. Performance characterization of a high power dual active bridge DC-to-DC converter. *IEEE Trans. Ind. Appl.* **1992**, *28*, 1294–1301. [CrossRef]
2. De Doncker, R.W.; Divan, D.M.; Kheraluwala, M.H. A three-phase soft-switched high-power-density DC/DC converter for high-power applications. *IEEE Trans. Ind. Appl.* **1991**, *27*, 63–73. [CrossRef]
3. Huang, J.; Wang, Y.; Li, Z.; Lei, W. Unified triple-phase-shift control to minimize current stress and achieve full soft-switching of isolated bidirectional DC–DC converter. *IEEE Trans. Ind. Electron.* **2016**, *63*, 4169–4179. [CrossRef]
4. Wu, K.; de Silva, C.W.; Dunford, W.G. Stability analysis of isolated bidirectional dual active full-bridge DC–DC converter with triple phase-shift control. *IEEE Trans. Power Electron.* **2012**, *27*, 2007–2017. [CrossRef]
5. Moonem, M.A.; Pechacek, C.L.; Hernandez, R.; Krishnaswami, H. Analysis of a multilevel dual active bridge (ML-DAB) DC–DC converter using symmetric modulation. *Electronics* **2015**, *4*, 239–260. [CrossRef]
6. Khan, M.A.; Zeb, K.; Sathishkumar, P.; Ali, M.U.; Uddin, W.; Hussian, S.; Ishfaq, M.; Khan, I.; Cho, H.-G.; Kim, H.-J. A Novel Supercapacitor/Lithium-Ion Hybrid Energy System with a Fuzzy Logic-Controlled Fast Charging and Intelligent Energy Management System. *Electronics* **2018**, *7*, 63. [CrossRef]
7. Wang, Y.-C.; Ni, F.-M.; Lee, T.-L. Hybrid Modulation of Bidirectional Three-Phase Dual-Active-Bridge DC Converters for Electric Vehicles. *Energies* **2016**, *9*, 492. [CrossRef]
8. Zhao, B.; Song, Q.; Liu, W.; Sun, Y. Overview of dual-active-bridge isolated bidirectional DC–DC converter for high-frequency-link power-conversion system. *IEEE Trans. Power Electron.* **2014**, *29*, 4091–4106. [CrossRef]
9. Krismer, F.; Kolar, J. Accurate power loss model derivation of a highcurrent dual active bridge converter for an automotive application. *IEEE Trans. Ind. Electron.* **2010**, *57*, 881–891. [CrossRef]
10. Oggier, G.G.; Garcia, G.O.; Oliva, A.R. Modulation strategy to operate the dual active bridge DC–DC donverter under soft switching in the whole operating range. *IEEE Trans. Power Electron.* **2011**, *26*, 1228–1236. [CrossRef]
11. Zhao, B.;Yu, Q.; Sun, W. Extended-phase-shift control of isolated bidi-rectional DC–DC converter for power distribution in microgrid. *IEEE Trans. Power Electron.* **2012**, *27*, 4667–4680. [CrossRef]
12. Jain, A.K.; Ayyanar, R. PWM control of dual active bridge: Comprehensive analysis and experimental verification. *IEEE Trans. Power Electron.* **2011**, *26*, 1215–1227. [CrossRef]
13. Bai, H.; Mi, C. Eliminate reactive power and increase system efficiency of isolated bidirectional dual-active-bridge DC–DC converters using novel dual- phase-shift control. *IEEE Trans. Power Electron.* **2008**, *23*, 2905–2914. [CrossRef]
14. Zhao, B.; Song, Q.; Liu, W. Efficiency characterization and optimization of isolated bidirectional DC–DC converter based on dual-phase-shift control for DC distribution application. *IEEE Trans. Power Electron.* **2013**, *28*, 1711–1727. [CrossRef]
15. Ortiz, G.; Fassler, L.; Kolar, J.W.; Apeldoorn, O. Flux Balancing of Isolation Transformers and Application of "The Magnetic Ear" for Closed-Loop Volt-Second Compensation. *IEEE Trans. Power Electron.* **2014**, *29*, 4078–4090. [CrossRef]
16. Li, X.; Li, Y.-F. An optimized phase-shift modulation for fast transient response in a dual-active-bridge converter. *IEEE Trans. Power Electron.* **2014**, *29*, 2661–2665. [CrossRef]
17. Lin, S.-T.; Li, X.; Sun, C.; Tang, Y. Fast transient control for power adjustment in a dual-active-bridge converter. *Electron. Lett.* **2017**, *53*, 1130–1132. [CrossRef]

18. Sun, C.; Li, X. Fast Transient Modulation for a Step Load Change in a Dual-Active-Bridge Converter with Extended-Phase-Shift Control. *Energies* **2018**, *11*, 1569. [CrossRef]
19. Zhao, B.; Song, Q.; Liu, W.; Zhao, Y. Transient DC Bias and Current Impact Effects of High-Frequency-Isolated Bidirectional DC–DC Converter in Practice. *IEEE Trans. Power Electron.* **2016**, *31*, 3203–3216. [CrossRef]
20. Engel, S.P.; Soltau, N.; Stagge, H.; De Doncker, R.W. Dynamic and Balanced Control of Three-Phase High-Power Dual-Active Bridge DC–DC Converters in DC-Grid Applications. *IEEE Trans. Power Electron.* **2013**, *28*, 1880–1889. [CrossRef]
21. Takagi, K.; Fujita, H. Dynamic Control and Performance of a Dual-Active-Bridge DC–DC Converter. *IEEE Trans. Power Electr.* **2017**, *33*, 7858–7866. [CrossRef]

electronics

MDPI

Article

Performance Evaluation of a Semi-Dual-Active-Bridge with PPWM Plus SPS Control

Ming Lu and Xiaodong Li *

Faculty of Information Technology, Macau University of Science and Technology, Taipa, Macau 999078, China; 1409853pii30001@student.must.edu.mo
* Correspondence: xdli@must.edu.mo; Tel.: +853-8897-2195

Received: 24 August 2018; Accepted: 7 September 2018; Published: 9 September 2018

Abstract: In this paper, a semi-dual-active-bridge (S-DAB) DC/DC converter with primary pulse-width modulation plus secondary phase-shifted (PPWM + SPS) control for boost conversion is analyzed in detail. Under the new control scheme, all effective operation modes are identified at first. Then, the working principle, switching behaviour, and operation range in each mode are discussed. Compared with conventional secondary phase-shifted control, PPWM + SPS control with two controllable phase-shift angles can extend the zero-voltage switching (ZVS) range and enhance control flexibility. In addition, an effective control route is also given that can make the converter achieve at the global minimum root-mean-square (RMS) current across the whole power range and avoid the voltage ringing on the transformer secondary-side at a light load. Finally, a 200 W prototype circuit is built and tested to verify correctness and effectiveness of theoretical results.

Keywords: DC–DC conversion; zero-voltage switching (ZVS)

1. Introduction

With the development of modern technology, there is a constant rise in energy use. Therefore, the concerns regarding the availability of fossil energy and the associated pollution in the mining and consumption process is continuously growing too. In order to alleviate the energy crisis and environmental pollution, the use of renewable energy (solar energy, wind energy, etc.) has developed rapidly around the world. As an important component for the application of renewable energy, the DC/DC converter with higher performance has been one of the most popular research fields [1–5]. So far, a number of DC/DC converter topologies have been proposed according to the various application requirements. In these converters, the phase-shift full-bridge converter is more attractive due to high power density, electrical isolation, easy to realize soft-switching commutation, high efficiency and low electromagnetic interference (EMI) [6–13]. However, it still suffers from high voltage ringing, reverse recovery on the secondary-side rectifier diodes, limited zero-voltage switching (ZVS) range and duty cycle loss.

To extend the ZVS range, a series of the full-bridge converters with various resonant tanks are presented. Among them, the converters with LC or LLC resonant tank are more attractive [14–19]. Nevertheless, the parameters of resonant tank should be selected carefully to achieve higher performance. Meanwhile, the design of magnetic components becomes complicated. On the other hand, when the phase-shift full-bridge converter works at high-output voltage and high-power case, the reverse-recovery problem of the rectifier diodes becomes more serious. In order to solve this problem, two active switches are introduced into the secondary-side rectifier of the converter, which is named the semi-dual-active-bridge (S-DAB) converter [20–22]. On this basis, two modified S-DAB topologies in [23,24] are proposed to only reduce the voltage stress on primary-side and secondary-side semiconductor devices, respectively. Furthermore, the S-DAB converter with an LC resonant tank is also presented in [25,26]. However, when the S-DAB converters work in a discontinuous-current mode (DCM) for boost operation, the voltage ringing phenomenon is generated on the transformer secondary-side. In particular, at the high switching frequency, high power and voltage levels, the excessive ringing might result in strong EMI, distorted gating signals and abnormal high peak voltages across the switches. In addition, the amount of power loss from snubber/parasitic capacitor will also increase during the ringing process [27,28]. Although a customized RC snubber is helpful to alleviate this problem, extra loss will be introduced in continuous-current mode (CCM) operation and the overall efficiency would be lower.

To the authors' best knowledge, the voltage ringing problem in S-DAB has yet to be resolved. In this paper, PPWM + SPS control is applied on an S-DAB converter for boost operation to further improve performance, which also avoids the voltage ringing problem. The rest of this paper is organized as follows: in Section 2, each steady-sate mode of an S-DAB converter with PPWM + SPS control is analyzed comprehensively, including working principle, switching behaviour, and operation range. In Section 3, an effective control route across the whole power range is presented in order to achieve minimum root-mean-square (RMS) current and no voltage ringing. Experimental results are provided in Section 4. Conclusions are drawn in Section 5.

2. Operation Principle of an S-DAB Converter with PPWM + SPS Control

2.1. Basic Operation Principle

The schematic of an S-DAB converter is shown in Figure 1. The primary H-bridge consists of four switches ($M_1 - M_4$), while the secondary H-bridge is realized by a diode leg (D_{S1} and D_{S2}) and a switch leg (M_5 and M_6). The high frequency transformer T with a turns ratio of $n_t : 1$ not only provides galvanic isolation, but also matches voltage level. The voltage gain is defined as $M = n_t V_o / V_{in}$, and $M > 1$ refers to boost operation. The inductor L_s includes the leakage inductance of the transformer and an external inductance. The filter capacitor C_o is connected in parallel with the load R_{load} to depress the output voltage ripple. In this paper, PPWM + SPS control with two phase shift angles is employed on an S-DAB converter. All switches operate at the same frequency f_s with 50% duty cycle, and switches in each switch leg are turned on/off complementarily. α is defined as the inner phase-shift by which the gating signal of M_4 lags that of M_1. Similarly, ϕ is defined as the outer-phase-shift by which the gating signal of M_6 lags that of M_1. Two pulse-width-modulated voltages v_{AB} and v_{CD} are generated by the two bridges, respectively. The pulse-width of v_{AB} is determined by α solely. However, the waveform of v_{CD} is associated with not only phase-shift angles but also the load level.

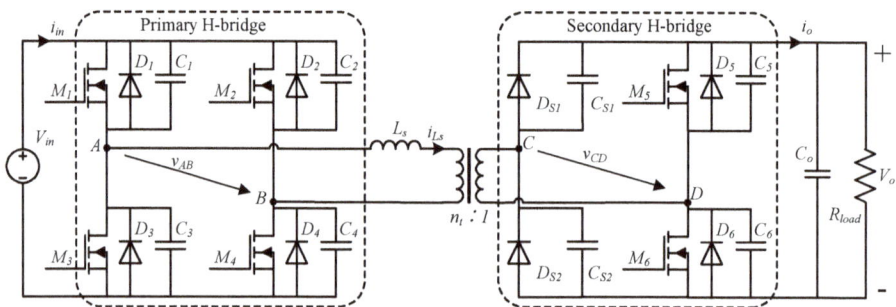

Figure 1. The circuit configuration of a S-DAB converter.

Depending on the relationship between two phase-shifts, an S-DAB converter with PPWM + SPS control can operate in three steady-state working modes, including one CCM (Mode A) and two DCMs, (Mode B and C). In the following part, each mode will be analyzed in detail one by one. In order to simplify the analysis process, four assumptions are made as follows:

1. All components, such as switches, diodes are ideal and lossless.
2. The magnetizing inductance of the transformer is infinity.
3. The snubber/parasitic capacitors and dead-times influence are neglected.
4. The filter capacitor is large enough to maintain constant voltage on the load.

2.2. Steady-State Analysis of Continuous-Current Mode

The ideal steady-state waveforms in Mode A are shown in Figure 2, where β denotes the first zero-crossing points referred to the turn-on moment of M_1. It can be seen that Mode A is featured with $\alpha < \beta < \phi < \pi$; and there are eight different intervals in one switching period. The corresponding equivalent circuits of the first four intervals are presented in Figure 3, respectively. The other four intervals are almost the same except for the directions of voltage/current and involved conducting devices.

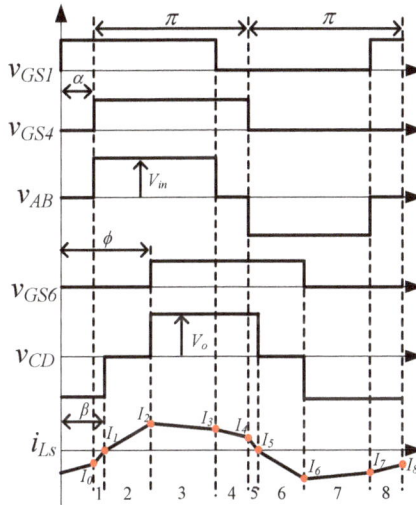

Figure 2. Steady-state waveforms in Mode A.

Interval 1 [Figure 3a]: At the beginning, M_2 is turned off and M_4 is turned on with ZVS. In this interval, the conducting devices are M_1 and M_4, M_5 and D_{s2}. Thus, the voltage across the inductor is clamped at $(V_{in} + n_t V_o)$, the value of i_{Ls} decreases linearly from the negative value I_0 to I_1. The power stored in the inductor is delivered to input DC power and load during this interval:

$$I_1 = I_0 + \frac{V_{in} + n_t V_o}{2\pi f_s Ls}(\beta - \alpha) = 0. \tag{1}$$

Interval 2 [Figure 3b]: At $\beta - \alpha$, the polarity of i_{Ls} is changed, the current flowing diode-leg is shifted naturally from D_{s2} to D_{s1}, i.e., D_{s2} is turned off with zero current. The secondary-side of transformer is shorted now by M_5 and D_{s1}. Meanwhile, the primary current flows from V_{in} to L_s through primary switches M_1 and M_4. Thus, the voltage across the inductor is clamped at V_{in}, the value of i_{Ls} increases linearly from I_1 to the positive maximum I_2. The power is being stored in the inductor L_s during this interval:

$$I_2 = I_1 + \frac{V_{in}}{2\pi f_s L_s}(\phi - \beta). \tag{2}$$

Interval 3 [Figure 3c]: At $(\phi - \alpha)$, M_5 is turned off and M_6 is turned on with ZVS. In this interval, the situation on the primary side does not change; the secondary current is shifted to D_{s2} and M_6, flowing to the load. Thus, during this interval, the voltage across inductor is clamped at $(V_{in} - n_t V_o)$, and the power is transmitted to the load. Due to $(V_{in} < n_t V_o)$, the value of i_{Ls} starts to decrease linearly from I_2 to the positive value I_3:

$$I_3 = I_2 + \frac{V_{in} - n_t V_o}{2\pi f_s L_s}(\pi - \phi). \tag{3}$$

Interval 4 [Figure 3d]: At $(\pi - \alpha)$, M_1 is turned off and M_3 is turned on with ZVS. In this interval, the primary side is shorted by M_3 and M_4; and no change happens on the secondary side. Thus, the voltage across inductor is clamped at $-n_t V_o$. During this interval, the value of i_{Ls} starts to decrease linearly until it reaches $-I_0$. The power stored in the inductor is delivered to load:

$$I_4 = I_3 - \frac{n_t V_o}{2\pi f_s L_s}\alpha = -I_0. \tag{4}$$

Based on Equations (1)–(4), the instantaneous current values at the moments of transition can be calculated as functions of α and ϕ. Furthermore, the output power P_o and the inductor RMS current $I_{Ls,rms}$ can be obtained too. These results are listed in Table 1, where the current and power values are normalized by the following base values:

$$I_b = \frac{V_{in}}{2\pi f_s L_s}, P_b = \frac{V_{in}^2}{2\pi f_s L_s}. \tag{5}$$

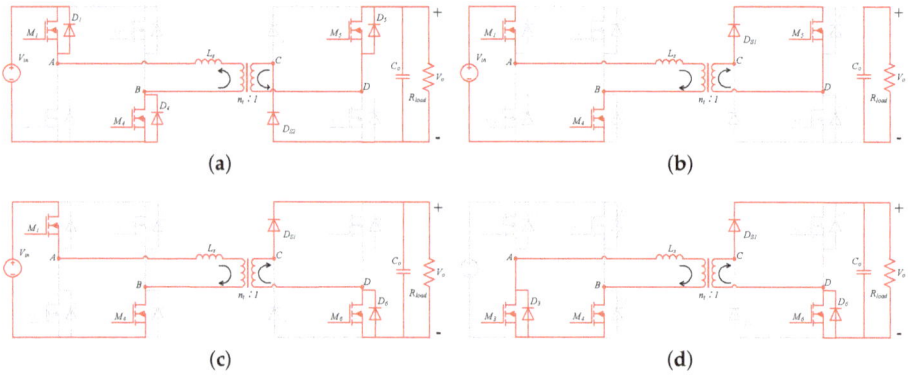

Figure 3. Equivalent circuits corresponding to the first four intervals in Mode A: (a) Interval 1 $[0, \beta - \alpha]$; (b) Interval 2 $[\beta - \alpha, \phi - \alpha]$; (c) Interval 3 $[\phi - \alpha, \pi - \alpha]$; (d) Interval 4 $[\pi - \alpha, \pi]$.

Table 1. Theoretical values of inductor current and output power in Mode A.

Value	Expression
$I_{0,pu}$	$\frac{(1+M)(\alpha-\pi+\alpha M-\phi M+\pi M)}{2+M}$
$I_{1,pu}$	0
$I_{2,pu}$	$\pi - \frac{\alpha-2\phi+3\pi}{2+M}$
$I_{3,pu}$	$\pi + (\pi - \phi)\,(1-M) - \frac{\alpha-2\phi+3\pi}{2+M}$
β	$\frac{\pi+\alpha+\phi M-\pi M}{2+M}$
$P_{o,pu}$	$\frac{M}{2\pi(2+M)^2}\left(\begin{array}{l} 2\alpha\phi M^2 + 4\pi\phi M^2 + 4\alpha\phi M - \alpha^2 M^2 - 2\pi\alpha M^2 - 2\phi^2 M^2 - 2\pi^2 M^2 \\ +4\pi\phi M + \pi^2 M - 4\phi^2 M - 3\alpha^2 M - 2\pi\alpha M \\ +\pi^2 + 4\alpha\phi + 4\pi\phi - 3\alpha^2 - 2\pi\alpha - 4\phi^2 \end{array} \right)$
$I_{Ls,rms,pu}$	$\frac{1}{\sqrt{3\pi(M+2)^2}}\sqrt{ \begin{array}{l} M^3\alpha^3 - 3M^3\alpha^2\phi + 3M^3\alpha^2\pi + 3M^3\alpha\phi^2 - 6M^3\alpha\phi\pi \\ +3M^3\alpha\pi^2 - 2M^3\phi^3 + 6M^3\phi^2\pi - 6M^3\phi\pi^2 + 2M^3\pi^3 \\ +3M^2\alpha^3 - 6M^2\alpha^2\phi + 3M^2\alpha^2\pi + 3M^2\alpha\phi^2 - 3M^2\alpha\pi^2 \\ -2M^2\phi^3 + 6M^2\phi\pi^2 - 3M^2\pi^3 + 4M\alpha^3 - 6M\alpha^2\phi+ \\ 6M\alpha\phi^2 - 6M\alpha\phi\pi - 4M\phi^3 + 6M\phi^2\pi + 2\alpha^3 - 3\alpha^2\pi + \pi^3 \end{array}}$

2.3. Steady-State Analysis of Discontinuous-Current Mode

Different from CCM, the inductor current i_{Ls} in DCM remains at zero for a small duration in each switching period. Steady-state waveforms of two DCMs are shown in Figure 4, where γ denotes the second zero-crossing points referring to the turn-on moment of M_1. It can be found that the difference between those two DCMs can be concluded as: $\alpha < \phi < \pi < \gamma$ for Mode B and $\alpha < \phi < \gamma < \pi$ for Mode C.

Mode B [Figure 4a]

According to the steady-state waveforms in Mode B, the equivalent circuits in the first four intervals are shown in Figure 5. It can be seen that the first three intervals in Mode B are almost the same as Intervals 2–4 in Mode A, except that the inductor current at the end of Interval 3 (Mode B) can arrive again at zero.

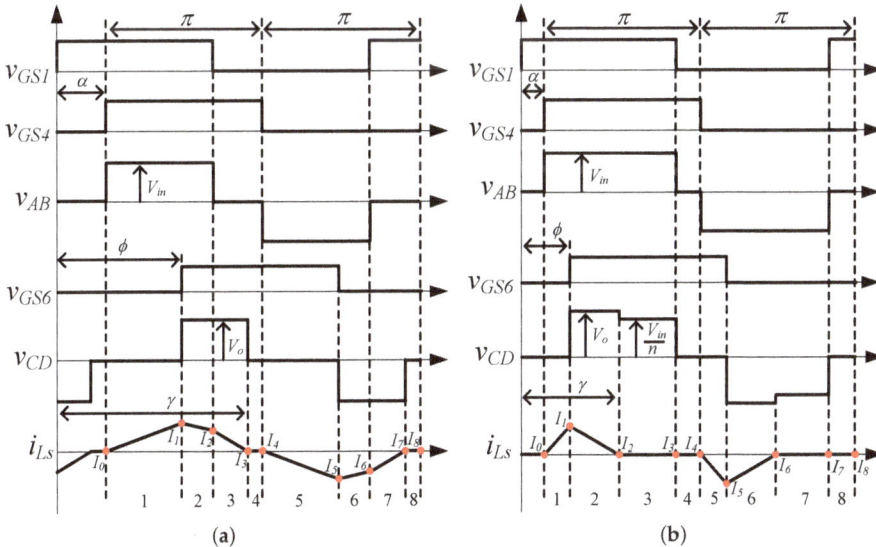

Figure 4. Steady-state waveforms in: (**a**) Mode B; (**b**) Mode C.

Interval 4 [Figure 5d]: After $\gamma - \alpha$, all secondary diodes are reversed biased, which will result in the secondary-side of transformer being open-circuited. Meanwhile, the primary switches M_3 and M_4 are still conducting. Thus, the transformer secondary voltage is clamped at 0, and i_{Ls} is kept at zero. This interval ends up with M_2 turned on at zero current. In this zero-current interval, there is no power transferring in the converter.

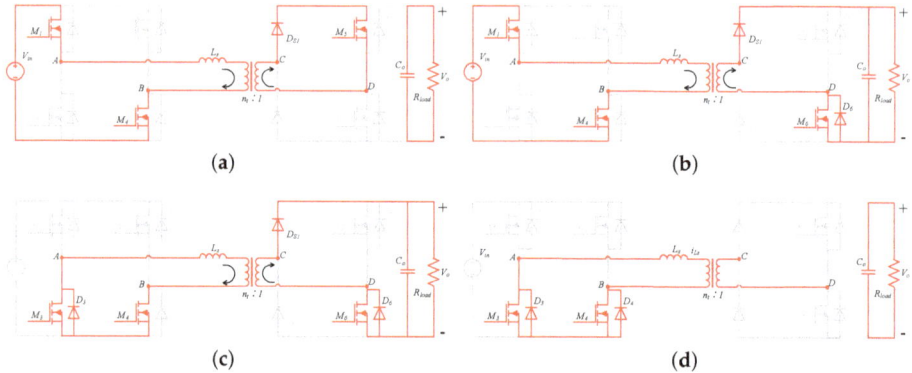

Figure 5. Equivalent circuits corresponding to the first four intervals in Mode B: (**a**) Interval 1 $[0, \phi - \alpha]$; (**b**) Interval 2 $[\phi - \alpha, \pi - \alpha]$; (**c**) Interval 3 $[\pi - \alpha, \gamma - \alpha]$; (**d**) Interval 4 $[\gamma - \alpha, \pi]$.

Mode C [Figure 4b]

Similarly, the equivalent circuits corresponding to first four intervals in Mode C are shown in Figure 6, respectively. It can be seen that the first second intervals in Mode C are almost the same as Intervals 2 and 3 in Mode A, except that the inductor current at end of Interval 2 (Mode C) can arrive again at zero.

Interval 3 [Figure 6c] in Mode C is different from those aforementioned intervals. Although the input DC source V_{in} is applied on the primary-side of transformer by the switches M_1 and M_4, there is no flowing current in the converter. The secondary-side of transformer is open-circuited since all secondary diodes are reversed biased. Thus, the secondary-side transformer voltage is clamped at $\frac{V_{in}}{n}$. Interval 3 ends up with M_3 turned on at zero current. This interval also belongs to the zero-current interval, and there is no power transferring. Interval 4 is the same as Interval 4 in Mode B.

Based on the steady-state analysis in each DCM, the instantaneous current values at the moment of transition can be calculated. Similarly, the output power and RMS current across inductor can be also obtained. These theoretical results are listed in Table 2.

Table 2. Theoretical values of inductor current and output power in DCM.

Value	Mode B	Mode C
$I_{0,pu}$	0	0
$I_{1,pu}$	$\phi - \alpha$	$\phi - \alpha$
$I_{2,pu}$	$\pi - \alpha + \phi M - \pi M$	0
$I_{3,pu}$	0	0
γ	$\frac{\pi - \alpha + \phi M}{M}$	$\frac{M\phi - \alpha}{M - 1}$
$P_{o,pu}$	$\frac{\alpha^2 + \pi^2 + 2\pi\phi M - 2\pi\alpha - \phi^2 M - \pi^2 M}{2\pi}$	$\frac{M(\phi - \alpha)^2}{2\pi(M-1)}$
$I_{Ls,rms,pu}$	$\frac{1}{\sqrt{3\pi M}}\sqrt{\begin{array}{l}3M^2\phi^2\pi - M^2\phi^3 - M^2\phi\pi^2 + M^2\pi^3 - M\alpha^3 \\ +3M\alpha^2\phi - 6M\alpha\phi\pi + 3M\alpha\pi^2 + 3M\phi\pi^2 - \\ 2M\pi^3 - \alpha^3 + 3\alpha^2\pi - 3\alpha\pi^2 + \pi^3\end{array}}$	$\sqrt{\frac{M(\phi - \alpha)^3}{3\pi(M-1)}}$

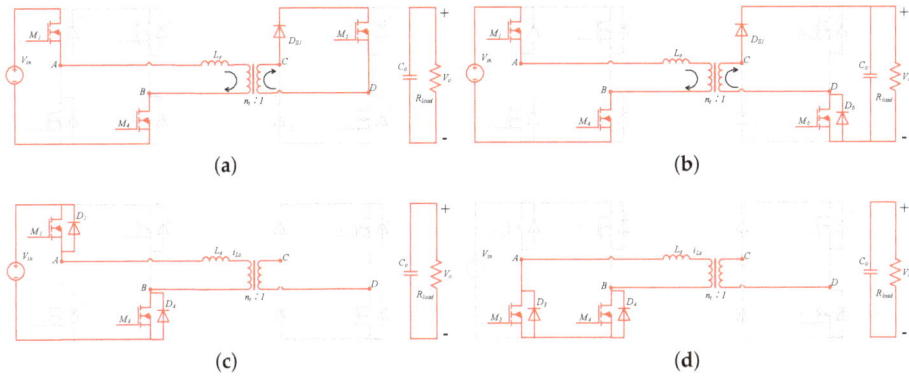

Figure 6. Equivalent circuits corresponding to the first four intervals in Mode C: (**a**) Interval 1 $[0, \phi - \alpha]$; (**b**) Interval 2 $[\phi - \alpha, \gamma - \alpha]$; (**c**) Interval 3 $[\gamma - \alpha, \pi - \alpha]$; (**d**) Interval 4 $[\pi - \alpha, \pi]$.

2.4. Switching Behaviour

Since the converter may work in three different steady-state modes, the switching behaviour of all switches and diodes also vary with the operation modes. According to the current polarity at switching moment of all switches and diodes, the switching behaviour in each mode are concluded in Table 3. First of all, the diodes in the secondary H-bridge can be turned on/off at zero current in any mode. In Mode A, all switches operate with ZVS and each diode is turned on/off with zero-current. Compared with Mode A, the switching loss in two DCMs is slightly increased due to the partial switch losing ZVS, and the switching loss in Mode C is higher than those in Mode B. Thus, Mode A should be selected as the main operation mode.

Table 3. Switching behavior in different modes.

Mode	M_1, M_3	M_2, M_4	M_5, M_6	D_{S1}, D_{S2}
A	ZVS	ZVS	ZVS	zero-current-on/off
B	ZVS	zero-current-on/off	ZVS	zero-current-on/off
C	zero-current-on/off	zero-current-on/off	ZVS	zero-current-on/off

2.5. Operation Range of Each Mode

Through comparing three steady-state modes, it can be found that Mode B is an in-between mode, and there are two boundary conditions existing between Mode B and the other two modes. Thus, the operating range of two controllable phase-shifts α and ϕ will be different in each mode. Knowing these conditions and range is helpful for the design of the converter.

When $\gamma = \pi$ in Mode B, the converter works at the boundary condition (6) between Mode A and B. At this boundary, the secondary H-bridge works in synchronous rectification mode:

$$\phi = \frac{\alpha + \alpha M + \pi M - \pi}{M}. \tag{6}$$

When $\gamma = \pi - \alpha$ in Mode B, it works at another boundary condition (7) between those two DCMs, in which the second zero-crossing happens at the moment of the switch M_3 being turned on:

$$\phi = \frac{\alpha + \pi M - \pi}{M}. \tag{7}$$

According to both boundary conditions, the operating range of each mode is shown in Figure 7 for $M = 1.5$ and $M = 2$, respectively.

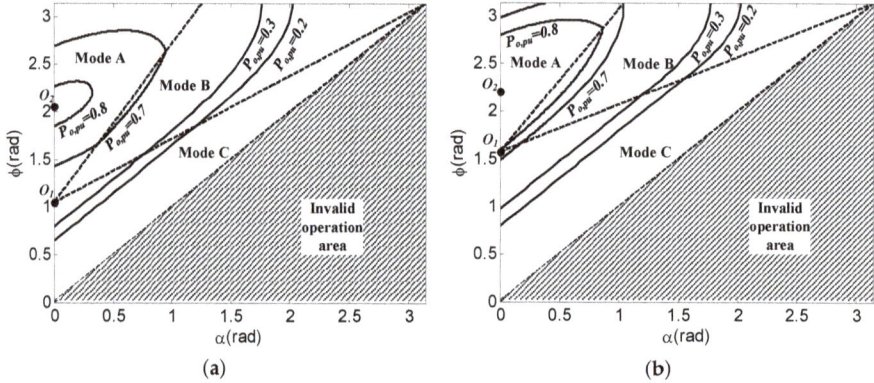

Figure 7. Operating range of each boost mode with suspended power contours for $M = 1.5$ and $M = 2$: (a) $M = 1.5$; (b) $M = 2$.

In Figure 7, the shaded area $\alpha \geq \phi$ represents an invalid operation area. The boundary of the neighboring modes is plotted by two dashed lines using (6) and (7), respectively, and these two boundary lines intersect at the point O_1. Meanwhile, the power contours with the same normalized power $P_{o,pu}$ value are shown by solid curves, in which the maximum load capacity of the converter is located at the point O_2, $P_{o,pu,max} = \frac{\pi M(M+1)}{2(M^2+2M+2)}$. The operating range with the conventional secondary phase-shifted control is only along the ϕ-axis. Compared with conventional control, PPWM + SPS can expand regulating range of output power and enhance flexibility of phase-shift control.

3. Proposed Control Route of an S-DAB Converter with PPWM + SPS Control

It is obvious that countless control routes exist from full power at O_2 to zero power at $\phi = \alpha$ for an S-DAB converter with PPWM + SPS control in Figure 7. Therefore, in order to select a reasonable control route, theoretical analysis of the inductor current is carried out to achieve lower conducting loss.

Based on Tables 1 and 2, the relationship between normalized inductor RMS current $I_{Ls,rms,pu}$ and phase-shift α at different power contours are shown in Figure 8, with $M = 1.5$ as an example. It can be found that, at the high power levels, the converter may work in Mode A. $I_{Ls,rms,pu}$ values can arrive at the minimum values when the converter is operated at conventional secondary phase-shifted control from O_2 to O_1 along the ϕ-axis. At the low power levels from O_1 to zero power O_0, the converter may operate in Mode C, in which $I_{Ls,rms,pu}$ is minimum and constant for the same power level.

In practical application, each switch and diode has its own snubber/parasitic capacitor. It is possible to get the voltage ringing on the transformer secondary-side when the converter is working in a zero-current interval. In Mode C, there are two zero-current Intervals (3 and 4) in the half period. Taking a capacitor into account, Interval 3 can be equivalent to the new circuit as Figure 9. In this interval, a resonance circuit is formed by a power inductor and the snubber/parasitic capacitor of the diode leg. Thus, the voltage ringing will be introduced into the transformer secondary-side. However, there is no voltage ringing in Interval 4. The main reason is that the DC source V_{in} in Figure 9 is short-circuited in interval 4. Compared with Mode C, there is only one zero-current interval in Mode B and it is free of voltage ringing, which is same as Interval 4 in Mode C. Considering that voltage ringing will potentially bring up system instability and damage the semiconductor devices, the control route for low power is put on the boundary line between Mode B and C. Thus, the selected route from full power to zero power with PPWM + SPS control is given as Equations (8) and (9), which is a piecewise function. Under the proposed route control, the converter can achieve at the minimum RMS current for the full power range and is free of voltage ringing across the transformer.

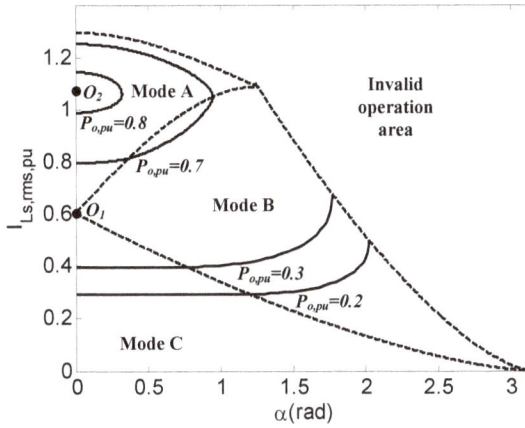

Figure 8. Normalized inductor RMS current vs. phase-shift α at different power contours for $M = 1.5$.

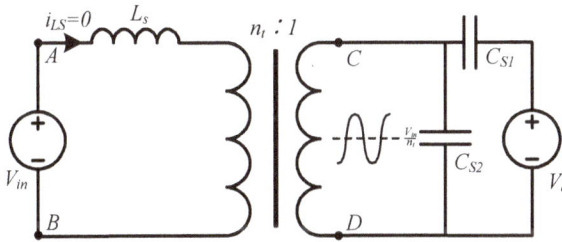

Figure 9. New equivalent circuit corresponding to interval 3 in Mode C.

When $P_{o,pu} \in \left[\frac{\pi(M-1)}{2M}, P_{o,pu,max} \right]$,

$$
\begin{cases}
\phi = \pi - \dfrac{(2+M)\sqrt{2\pi M\left(\pi M^2 + \pi M - 2M^2 P_{o,pu} - 4MP_{o,pu} - 4P_{o,pu}\right)} + X_1}{2M^3 + 4M^2 + 4M} \\
\alpha = 0.
\end{cases}
\tag{8}
$$

When $P_{o,pu} \in \left[0, \frac{\pi(M-1)}{2M} \right]$,

$$
\begin{cases}
\phi = \pi - \dfrac{X_2 \sqrt{P_{o,pu}}}{M} \\
\alpha = \pi - X_2 \sqrt{P_{o,pu}}
\end{cases}
\tag{9}
$$

where $X_1 = 2\pi M^2 + 2\pi M$, $X_2 = \frac{\sqrt{2\pi M(M-1)}}{M-1}$.

4. Experimental Verifications

To verify the theoretical analysis above, a 200 W S-DAB prototype is built, as shown in Figure 10. The specifications of the lab-scale converter are listed in Table 4. The gating signals of S-DAB are implemented using a TMS320F28335 DSP from TI (Texas Instruments, Dallas, TX, USA) and the switching frequency is set at 100 kHz.

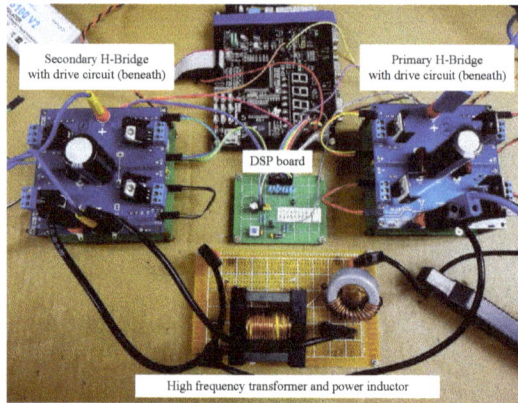

Figure 10. The layout of a 200 W S-DAB laboratory prototype.

Table 4. Specifications of a 200 W S-DAB converter.

Component	Parameter
Input DC voltage V_{in}	80 V
Load DC voltage V_o	120 V
Power inductance L_s	38 µH, CM400125/MPPcore
Transformer turns ratio n_t	15:15, ETD49/N97
Filter capacitor C_o	470 µF, 1 electrolytic cap
Switch $M_1 \sim M_6$	STP40NF20, $R_{ds} = 38$ mΩ
Diode D_{s1}, D_{s2}	MBR40250TG, $V_F = 0.86$ V

Based on the analysis of each operation mode in Section 2, a set of experimental waveforms corresponding to three effective modes are obtained and shown in Figure 11, respectively. These experimental results match the theoretical prediction closely. In Figure 11c, the voltage ringing shows up on the transformer secondary-side v_{CD} after i_{Ls} decreases to zero and remains a small duration until $v_{AB} = 0$. As expected, the voltage ringing is not identified in Figure 11b.

A series of experimental tests are then performed along the proposed minimized rms current route. The boundary power between the two stages of the control route is calculated to be 52% load, i.e., 104 W. Thus, the converter works in CCM (Mode A) at the high power of 200 W and 150 W. Two phase-shift angles are calculated as $\alpha = 0°$, $\phi = 90.25°$ (200 W) and $\alpha = 0°$, $\phi = 63.76°$ (150 W) according to (8). Experimental results of 200 W and 150 W are shown in Figure 12a,b, respectively. It can be seen from those two figures that the inductor current i_{Ls} is continuous. The related waveforms satisfy the operation condition of $\phi > \beta > \alpha = 0$ and all switches can operate at ZVS.

When the output power is lower than 52% load, the converter is operated at the boundary between Mode B and C. Using (9), two phase-shift angles are obtained as follows: (1) 100 W, $\alpha = 28.06°$, $\phi = 78.71°$; (2) 50 W, $\alpha = 72.46°$, $\phi = 108.3°$. Experimental results of 100 W and 50 W are shown in Figure 12c,d, respectively. It can be seen that the transition moment of v_{AB} from V_{in} to zero happens at the zero-crossing point of the inductor current and there is a small zero-current duration. The operation conditions of two experimental results match the boundary feature between both DCMs. In addition, the voltage ringing on the waveform v_{CD} is prevented in comparison to DCM under conventional secondary phase-shifted control (Figure 13). Based on these experimental results, the values of the RMS current, peak current, and efficiency are listed in Table 5, where the highest efficiency can arrive at 95.53% for 150 W. For 50 W operation with conventional control in Figure 13, the measured efficiency is 88.46%. It is seen that the efficiency using the proposed control is improved slightly since the current values and the switching behaviour under the two control methods are almost the same and the loss due to the ringing accounts for a small portion of total loss. However, the removal of the ringing phenomenon depresses EMI so that the risk of distortion in gating signals is reduced and the operation stability is improved consequently.

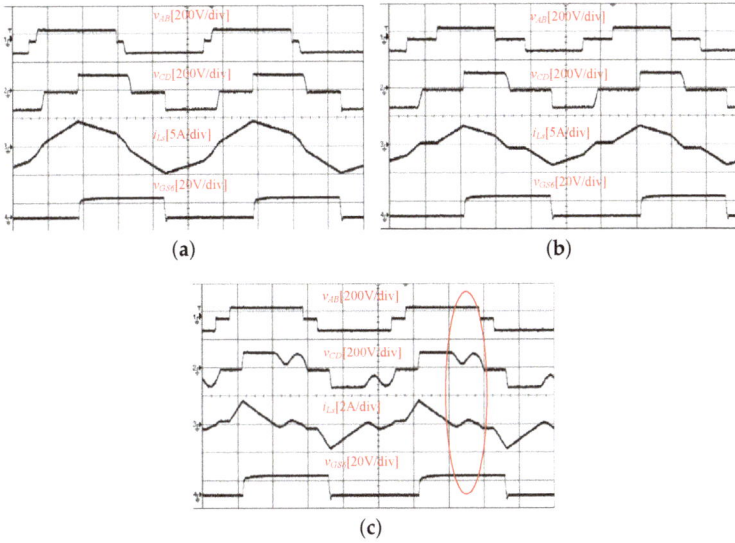

Figure 11. Experimental waveforms of v_{AB}, v_{CD}, i_{Ls} and V_{GS6} (from **top–bottom**), time scale: (2 μs/div), (**a**) Mode A; (**b**) Mode B; (**c**) Mode C.

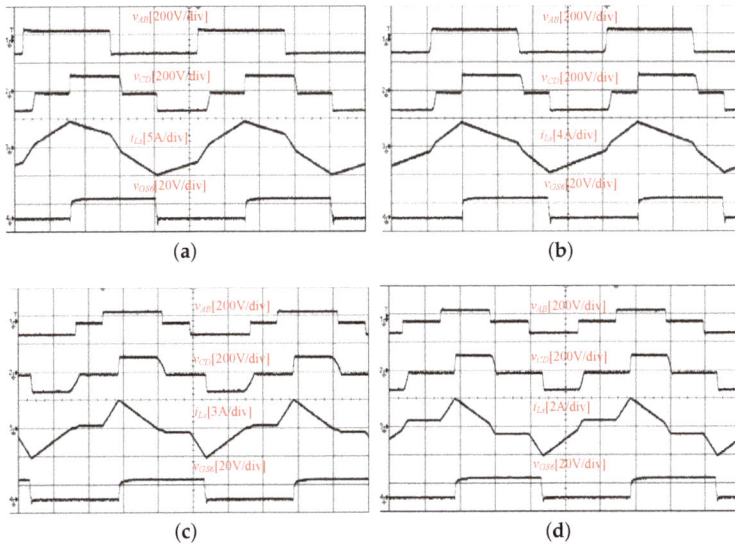

Figure 12. Experimental waveforms at four different power levels under the proposed route for $V_{in} = 80$ V and $V_o = 120$ V, time scale: (2 μs/div), (**a**) 200 W; (**b**) 150 W; (**c**) 100 W; (**d**) 50 W.

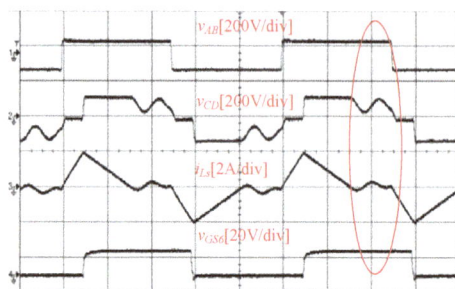

Figure 13. Experimental waveforms at 50 W with conventional secondary phase-shifted control.

Table 5. Measured results at $V_{in} = 80$ V and $V_o = 120$ V.

Power		$I_{Ls,rms}$ (A)	$I_{Ls,peak}$ (A)	η (%)
200 W	theor.	2.9	4.52	-
	exp.	3.08	4.7	94.97%
150 W	theor.	2.14	3.63	-
	exp.	2.19	3.63	95.53%
100 W	theor.	1.57	2.96	-
	exp.	1.58	2.92	92.13%
50 W	theor.	0.94	2.1	-
	exp.	0.95	2.07	88.95%

5. Conclusions

In this work, PPWM + SPS control with two controllable phase-shifts is applied on an S-DAB converter for boost operation and all effective steady-state modes are identified. Based on the characteristics of each mode, a reasonable control route is developed and implemented on a lab-scale S-DAB prototype. The experimental results show the consistency with the theoretical analysis results. Compared with conventional secondary phase-shifted control, the proposed control route not only makes the converter operate with the minimized RMS current for the whole power range, but also eliminates the voltage ringing on the secondary-side of the HF transformer completely. More importantly, the proposed hybrid control can be also applied on the other S-DAB converters in [22–24] to prevent the voltage ringing and improve stability. In addition, other optimization objectives could be developed according to various application requirements.

Author Contributions: M.L. did theoretical analysis, derivation, circuit implementation, experimental test and paper writing. X.L. was responsible for planning, coordination and proofreading.

Funding: This research was funded by Fundo para o Desenvolvimento das Ciências e da Tecnologia under Grant No. 004/2015/A1.

Acknowledgments: The authors would like to acknowledge Song Hu, Guo Chen for their support in the preparation of circuit implementation.

Conflicts of Interest: The authors declare no conflict of interest.

References

1. Blaabjerg, F.; Chen, Z.; Kjaer, S.B. Power electronics as efficient interface in dispersed power generation systems. *IEEE Trans. Power Electron.* **2004**, *19*, 1184–1194. [CrossRef]
2. Tahir, S.; Wang, J.; Baloch, M.; Kaloi, G. Digital control techniques based on voltage source inverters in renewable energy applications: A review. *Electronics* **2018**, *7*, 18. [CrossRef]
3. Chen, Z.; Guerrero, J.M.; Blaabjerg, F. A review of the state of the art of power electronics for wind turbines. *IEEE Trans. Power Electron.* **2009**, *24*, 1859–1875. [CrossRef]

4. Abdelsalam, A.K.; Massoud, A.M.; Ahmed, S.; Enjeti, P.N. High-performance adaptive perturb and observe MPPT technique for photovoltaic-based microgrids. *IEEE Trans. Power Electron.* **2011**, *26*, 1010–1021. [CrossRef]

5. Almalaq, Y.; Matin, M. Three topologies of a non-isolated high gain switched-inductor switched-capacitor step-up cuk converter for renewable energy applications. *Electronics* **2018**, *7*, 94. [CrossRef]

6. Lee, W.J.; Kim, C.E.; Moon, G.W.; Han, S.K. A new phase-shifted full-bridge converter with voltage-doubler-type rectifier for high-efficiency PDP sustaining power module. *IEEE Trans. Ind. Electron.* **2008**, *55*, 2450–2458.

7. Cha, H.; Chen, L.; Ding, R.; Tang, Q.; Fang, Z.P. An alternative energy recovery clamp circuit for full-bridge PWM converters with wide ranges of input voltage. *IEEE Trans. Power Electron.* **2008**, *23*, 2828–2837.

8. Chen, W.; Ruan, X.; Zhang, R. A novel zero-voltage-switching PWM full bridge converter. *IEEE Trans. Power Electron.* **2008**, *23*, 793–801. [CrossRef]

9. Jang, Y.; Jovanovic, M.M.; Chang, Y.-M. A new ZVS-PWM full-bridge converter. *IEEE Trans. Power Electron.* **2003**, *18*, 1122–1129. [CrossRef]

10. Yoon, H.K.; Han, S.K.; Choi, E.-S.; Moon, G.-W.; Youn, M.-J. Zero-voltage switching and soft-commutating two-transformer full-bridge PWM converter using the voltage-ripple. *IEEE Trans. Ind. Electron.* **2008**, *55*, 1478–1488. [CrossRef]

11. Lee, I.O.; Moon, G.W. Phase-shifted PWM converter with a wide ZVS range and reduced circulating current. *IEEE Trans. Power Electron.* **2013**, *28*, 908–919. [CrossRef]

12. Jang, Y.; Jovanovic, M.M. A new family of full-bridge ZVS converters. *IEEE Trans. Power Electron.* **2004**, *19*, 701–708. [CrossRef]

13. Ruan, X.; Yan, Y. A novel zero-voltage and zero-current-switching PWM full-bridge converter using two diodes in series with the lagging leg. *IEEE Trans. Ind. Electron.* **2001**, *48*, 777–785. [CrossRef]

14. Lo, Y.K.; Lin, C.Y.; Hsieh, M.T.; Lin, C.Y. Phase-shifted full-bridge series-resonant DC-DC converters for wide load variations. *IEEE Trans. Ind. Electron.* **2011**, *58*, 2572–2575. [CrossRef]

15. Gautam, D.S.; Bhat, A.K.S. A comparison of soft-switched DC-to-DC converters for electrolyzer application. *IEEE Trans. Power Electron.* **2013**, *28*, 54–63. [CrossRef]

16. Ali, K.; Das, P.; Panda, S.K. Analysis and design of APWM half-bridge series resonant converter with magnetizing current assisted ZVS. *IEEE Trans. Ind. Electron.* **2017**, *64*, 1993–2003. [CrossRef]

17. Mumtahina, U.; Wolfs, P.J. Multimode optimization of the phase shifted LLC eries resonant converter. *IEEE Trans. Power Electron.* **2018**, *1*. [CrossRef]

18. Lee, I.O.; Moon, G.W. The *k*-Q Analysis for an LLC series resonant converter. *IEEE Trans. Power Electron.* **2014**, *29*, 13–16. [CrossRef]

19. Fang, X.; Hu, H.; Chen, F.; Somani, U.; Auadisian, E.; Shen, J.; Batarseh, I. Efficiency-oriented optimal design of the LLC resonant converter based on peak gain placement. *IEEE Trans. Power Electron.* **2013**, *28*, 2285–2296. [CrossRef]

20. Zhang, J.; Zhang, F.; Xie, X.; Jiao, D.; Qian, Z. A novel ZVS DC/DC converter for high power applications. *IEEE Trans. Power Electron.* **2004**, *19*, 420–429. [CrossRef]

21. Mishima, T.; Nakaoka, M. Practical evaluations of a ZVS-PWM DC-DC converter with secondary-side phase-shifting active rectifier. *IEEE Trans. Power Electron.* **2011**, *26*, 3896–3907. [CrossRef]

22. Kulasekaran, S.; Ayyanar, R. Analysis, design, and experimental results of the semidual-active-bridge converter. *IEEE Trans. Power Electron.* **2014**, *29*, 5136–5147. [CrossRef]

23. Li, W.; Zong, S.; Liu, F.; Yang, H.; He, X.; Wu, B. Secondary-side phase-shift-controlled ZVS DC/DC converter with wide voltage gain for high input voltage applications. *IEEE Trans. Power Electron.* **2013**, *28*, 5128–5139. [CrossRef]

24. Wu, H.; Chen, L.; Xing, Y. Secondary-side phase-shift-controlled dual-transformer-based asymmetrical dual-bridge converter with wide voltage gain. *IEEE Trans. Power Electron.* **2015**, *30*, 5381–5392. [CrossRef]

25. Hu, S.; Li, X. Performance evaluation of a semi-dual-bridge resonant DC/DC converter with secondary phase-shifted control. *IEEE Trans. Power Electron.* **2017**, *32*, 7727–7738. [CrossRef]

26. Hu, S.; Li, X.; Lu, M.; Luan, B.-Y. Operation modes of a secondary-side phase-shifted resonant converter. *Energies* **2015**, *8*, 12314–12330. [CrossRef]

27. Park, K.B.; Kim, C.E.; Moon, G.W.; Youn, M.J. Voltage oscillation reduction technique for phase-shift full-bridge converter. *IEEE Trans. Ind. Electron.* **2007**, *54*, 2779–2790. [CrossRef]
28. Garabandic, D.; Dunford, W.G.; Edmunds, M. Zero-voltage-zero-current switching in high-output-voltage full-bridge PWM converters using the interwinding capacitance. *IEEE Trans. Power Electron.* **1999**, *14*, 343–349. [CrossRef]

electronics

MDPI

Review

Digital Control Techniques Based on Voltage Source Inverters in Renewable Energy Applications: A Review

Sohaib Tahir [1,2], Jie Wang [1,*], Mazhar Hussain Baloch [1,3] and Ghulam Sarwar Kaloi [1,4]

[1]   School of Electronic, Information and Electrical Engineering, Shanghai Jiao Tong University,
     Shanghai 200000, China; sohaibchauhdary@sjtu.edu.cn (S.T.); mazhar.hussain08ele@gmail.com (M.H.B.);
     Sarwar.kaloi59@gmail.com (G.S.K.)
[2]   Department of Electrical Engineering, COMSATS Institute of Information Technology, Sahiwal 58801, Pakistan
[3]   Department of Electrical Engineering, Mehran University of Engineering & Technology,
     Khairpur Mirs 67480, Pakistan
[4]   Department of Electrical Engineering, Quaid e Awam University of Engineering & Technology,
     Larkana 77150, Pakistan
*   Correspondence: jiewangxh@sjtu.edu.cn

Received: 4 December 2017; Accepted: 2 February 2018; Published: 7 February 2018

Abstract: In the modern era, distributed generation is considered as an alternative source for power generation. Especially, need of the time is to provide the three-phase loads with smooth sinusoidal voltages having fixed frequency and amplitude. A common solution is the integration of power electronics converters in the systems for connecting distributed generation systems to the stand-alone loads. Thus, the presence of suitable control techniques, in the power electronic converters, for robust stability, abrupt response, optimal tracking ability and error eradication are inevitable. A comprehensive review based on design, analysis, validation of the most suitable digital control techniques and the options available for the researchers for improving the power quality is presented in this paper with their pros and cons. Comparisons based on the cost, schemes, performance, modulation techniques and coordinates system are also presented. Finally, the paper describes the performance evaluation of the control schemes on a voltage source inverter (VSI) and proposes the different aspects to be considered for selecting a power electronics inverter topology, reference frames, filters, as well as control strategy.

Keywords: voltage source inverters (VSI); voltage control; current control; digital control; predictive controllers; advanced controllers; stability; response time

1. Introduction

Nowadays, energy demand is getting increased with the passage of time and distributed generation (DG) power systems especially through wind, solar and fuel cells as well as their related power conversion systems are conferred immensely. Many problems like grid instability, low power factor and power outage etc. for power distribution have also been increased with increase in energy demand [1]. However, DG power systems are found to be a sensible solution for such problems as they have relatively robust stability and causes additional flexibility balance. Moreover, their utilization can also improve the distribution networks management and carbon release is also reduced. VSIs are extensively necessitated for the commercial purpose as well as for the industrial applications as they play a key role in converting the DC voltage and current, usually produced by various DG applications, into AC before being discharged into the grid or consumed by the load. Several control systems are introduced, various schemes are proposed and numerous techniques are updated in order to facilitate the control of three-phase VSI. The objectives of these control schemes are to constrain the high and low-frequency electromagnetic pollution and to inject the active power with zero power factor into the grid [2]. The smooth and steady sinusoidal waveform can be a good input to a load for getting the most suitable response, therefore, the output of the inverter, which normally enjoys special standards and characteristics, should be controlled for providing an aforementioned waveform to load and grid.

Generally, it is observed that several problems are caused in linking the DG power system to a grid or grid to load in bidirectional inverters, i.e., grid instability, distortion in the waveform, attenuation as well as major and minor disturbances. Hence, in order to overcome these problems and to provide high-quality power, appropriate controllers with rapid response, compatible algorithm, ability to remove stable errors, less transit time, high tracking ability, less total harmonic distortion, THD value and smooth sinusoidal output should be designed. Various controllers are designed for achieving these qualities. The cascade technologies are introduced in the literature comprises of an inner current loop and outer voltage loop [3–12]. As the inner-loop current controller plays a fundamental role in closed-loop performance, various control approaches like PI [3–6], $H\infty$ [7,8], deadbeat [9–11,13] and μ-synthesis [11] are extensively applied. Outer voltage loop in the aforementioned cases refines the tracking ability and decreases the tracking error. In case of no input limitations, aforesaid PI controllers are the best choice for stabilizing the inner loop performance. However, input constraints restrict their performance and no optimization is usually observed by using PI controllers. The deadbeat control method is proposed in [9] to enhance the closed-loop performance but unfortunately, it was found highly sensitive to the disturbances, parameters mismatches and measurement noise. Later on, some observed based deadbeat controllers are introduced in order to provide compensation for these discrepancies, however, a trade-off was observed between phase margin and closed-loop performance [9,10]. Afterwards, $H\infty$ controllers in [7,8] are offering robust output response instead of input constraints, however, guaranteeing only the local stability like the μ-synthesis controller in [12].

Several other manuscripts are also amalgamated with literature for fulfilling the demand of electric power regarding fulfilling the environmental principles concerning green-house effects [14–18]. Various structures and topologies for interconnecting DGs are presented in [19–21] for parallel operation and in [22–24] for independent operation. For this reason, various control strategies are anticipated for stabilizing the system to control the voltage and frequency in case of unbalanced load and nonlinear loads. Many researchers have proposed several schemes for designing the controller in order to refine the quality of output voltage of DC to AC inverter. In [25], a control scheme is presented for a DG unit in islanded mode, this control technique is suitable for balanced load conditions for a DG unit when it is electronically coupled. However, this technique is constrained to small load variations and remain unable to stabilize the system in large load variations. A robust controller is proposed in [26] for balanced as well as unbalanced systems. However, it fails to address non-linear load properly. In [27], a repetitive control is implemented for controlling the inverters but the relatively slow response and absence of a systematical technique for stabilizing the error dynamics are the core problems. In [28], the uncomplicatedly designed controller is used to mitigate the load disturbances up to a significant extent through a feedforward compensation element, however, it is only restricted to balanced load conditions. In [29], a spatial repetitive control technique is implemented for controlling the current in a single-phase inverter. The results are satisfactory under non-linear load conditions; however, it is not guaranteeing the optimal tracking ability for a three-phase inverter. In [30], a discrete-time sliding mode current controller is proposed, it is optimally operating to control the system at a sudden load change, an unbalanced load and a nonlinear load, however, the system is quite intricate. In [31], the voltage and frequency controller is presented through a discrete-time mathematical model in order to operate the distributed resource units. This technique is achieving good voltage regulations under different load conditions but the results are not verified through the experimental setup. In [32], a controller is proposed having an adaptive feedforward compensation method applied through a Kalman filter for estimating the variation in parameters, the response was robust; however, tuning of covariance matrices are not appropriately described in the paper. In [33], a corresponding controller is recommended for distributed generation systems in grid applications, the anticipated controller is good in handling the grid disturbances and handling the nonlinearities, however, it is not suitable in stand-alone mode due to the nonexistence of voltage loop. In [34,35], the adaptive controller is used and voltage tracking is achieved precisely. The system is guaranteed under systems parameter variations, however, complexity in computation exists and a certain pre-defined value is needed for parameters. In [36], an output voltage controller based on the resonant harmonic filters is presented. It measures the capacitor current and load current in the same sensor. Unbalanced voltage condition and harmonic distortion are compensated in this controller. However, THD value is not defined appropriately, therefore, it is complicated to assess the quality of the controller. An adaptive control technique based proportional derivative controller is presented in [37], for a pulse width modulated inverter operation in islanded distributed generation system, voltage regulation under numerous load conditions is evaluated, though it is not easy to achieve the suitable control gains as par the designing procedure specified in the paper. Moreover, voltage and frequency are optimally controlled, active and reactive power unbalancing is aptly compensated through small signal modeling of inverters in [38].

The key purpose of this study is to provide a comprehensive review of the digital control strategies for different types of three-phase inverters in stand-alone as well as grid-connected modes. Correspondingly, explanation, discussion and comparison of the various control strategies are described in this manuscript in detail.

The manuscript is organized as: classification of voltage source inverters is described in Section 2. Section 3 discusses the characteristics of control systems, followed by a depiction of reference frames in Section 4. The control strategy in decoupled dq frame and time-delay sampling scheme for VSI are depicted in Sections 5 and 6 respectively. An overview of the most commonly used filters and damping techniques is illustrated in Sections 7 and 8 respectively. The grid synchronization techniques followed by modulation techniques are described in Sections 9 and 10 of the manuscript, respectively. Moreover, control Techniques along with their pros. & cons. are described in Section 11. In Section 12, comparative analysis and future goals for the researchers are elaborated. Whereas, conclusions are drawn in Section 13.

2. Classification of VSIs

There are various types, in which the inverters are categorized. Figure 1 shows the complete detail of categories in which voltage source inverters are classified.

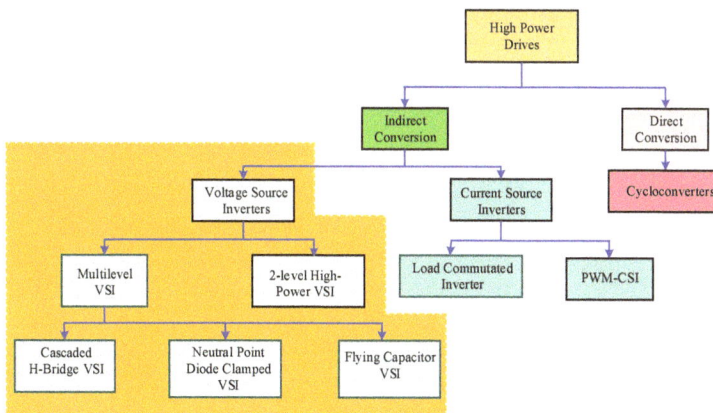

Figure 1. Classification of voltage source inverters (VSIs) in high power drives.

2.1. Multilevel Diode Neutral-Point Clamped Inverter

Multilevel inverter (MLI) was proposed in 1975, its design was like a cascade inverter with diodes facing the source. This inverter was later transformed into a Diode Clamped Multilevel Inverter, which is also named as a Neutral-Point Clamped Inverter (NPC) [39]. In this type of multilevel inverters, the integration of voltage clamping diodes is indispensable. An ordinary DC-bus is separated by an even number of bulk capacitors connected in series with a neutral point in the middle of the line that is dependent on the voltage levels of the inverter. In Figure 2, a five-level NPC-MLI is shown, here the clamping diodes are interlinked to M-1 regulatory pairs if M is considered as voltage levels of the inverter.

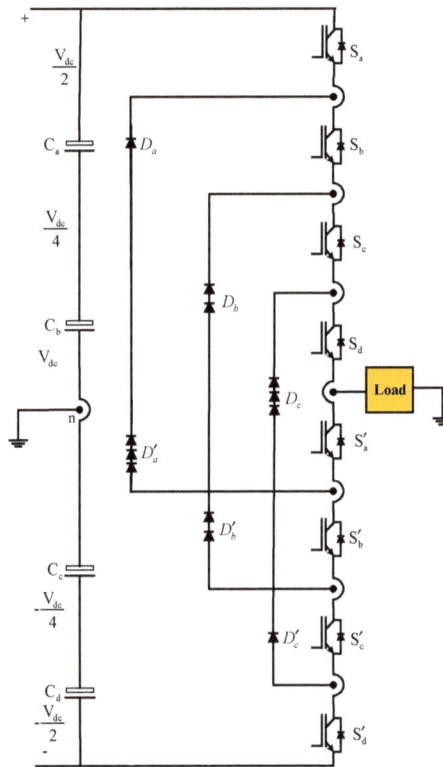

Figure 2. Five-level diode neutral-point clamped inverter.

The neutral point converter was designed by Nabae, Takahashi and Akagi in 1981, this was basically a three-level diode-clamped inverter [40]. A three-phase Three-level diode-clamped inverter is shown in Figure 3.

The NPC-MLI is considered as an important device in conventional high-power ac motor drive applications like mills, fans, pumps and conveyors, moreover, it also offers solutions for industries including chemicals, gas, power, metals, oil, marine, water and mining. The back-to-back configuration of inverters for reformative applications is also considered as a major plus point of this topology, used, for example, in regenerative conveyors, mining industry and grid interfacing of renewable energy sources like wind power [41].

There are several benefits as well as drawbacks of multilevel diode-clamped [39,42]. A common dc bus is shared by all the phases, this results in the reduction of capacitance requirements of the inverter.

Due to this reason, implementation of a back-to-back topology is not only credible but can also be applied practically for performing different operations in an adjustable speed drive and a high-voltage back-to-back inter-connection. The capacitors can be recharged as a group. On fundamental frequency, switching efficacy is relatively higher. However, real power flow is problematic in case of a single inverter as the intermediate dc levels will tend to overcharge or discharge due to inappropriate monitoring and control. The number of clamping diodes are quadratically associated with the number of levels, which can be unwieldy for units with a high number of levels.

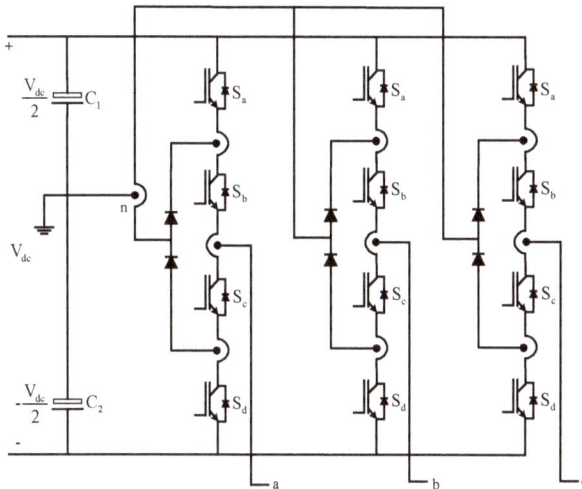

Figure 3. Three-level diode neutral-point clamped inverter.

2.2. Multilevel Capacitor Clamped/Flying Capacitor Inverter

A corresponding topology for the NPC-MLI topology is the Flying Capacitor (FC), or Capacitor Clamped, MLI topology, it is depicted in Figure 4. As an alternative to clamping diodes, capacitors are used for holding the voltages to the referred values. In the NPC-MLI, M − 1 number of capacitors are integrated on a shared DC-bus, where M is the level number of the inverter and 2(M − 1) switch-diode regulatory pairs are used. Though, for the FC-MLI, instead of clamping diodes, one or more capacitors are used to produce the output voltages depends upon the position and the level of the inverter. They are coupled to the midpoints of two regulatory pairs on the same position on each side of a midpoint [42], see capacitors C_a, C_b and C_c in Figure 5.

The basic difference is the usage of clamping capacitors in place of clamping diodes, as using them increases the number of switching combinations as capacitors do not block reverse voltages [42]. Numerous switching states would be able to produce the same voltage level and the redundant switching states would also be available.

DC side capacitors in this topology have a ladder-like structure and the voltage on each capacitor deviates from that of the other capacitor. The voltage increment between two adjacent legs of the capacitors provides the size of the voltage steps in the output waveform. One advantage of the flying-capacitor-based inverter is the redundancies for inner voltage levels; i.e., two or more effective switching amalgamations can produce an output voltage.

Unlike the diode-clamped inverter, the flying-capacitor inverter never requires all of the switches to be on (conducting state) in a consecutive series. Moreover, the flying-capacitor inverter has phase redundancies, while the diode-clamped inverters have only the line-line redundancies [40]. These redundancies provide selective charging and discharging of specific capacitors and it can be incorporated in the control system for the voltage balancing across the various levels.

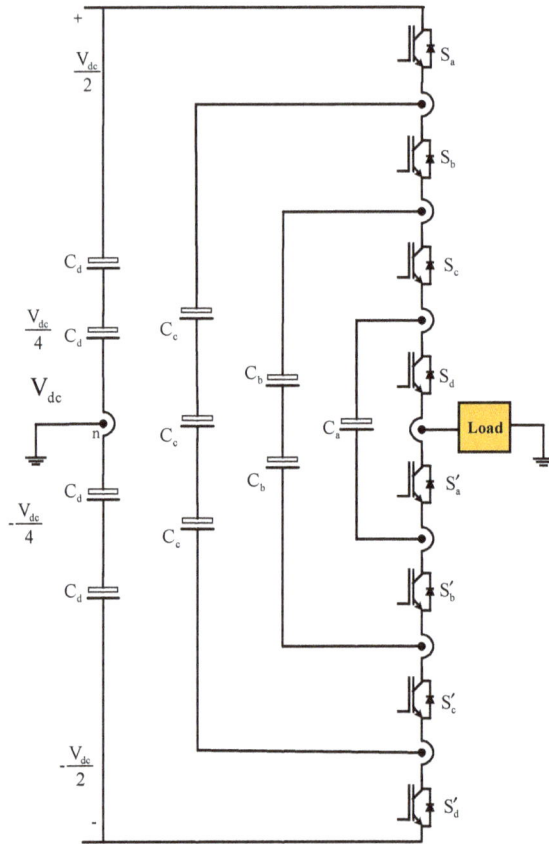

Figure 4. Multilevel (Five-level) capacitor clamped/flying capacitor inverter.

There are several advantages and disadvantages of multilevel flying capacitor inverters [41,43]. Phase redundancies are offered for balancing the voltage levels between the capacitors. Active and reactive power flow can be regulated. The presence of various capacitors allows the inverter to ride through outages for short duration and deep voltage sags. However, the control system is complex for tracking the voltage levels for all of the capacitors. Correspondingly, recharging all the capacitors to the same voltage level and startup are complex. Switching operation and efficacy are poor for real power transmission. The installation of large numbers of capacitors is not much economical and it also makes the system bulky as compared to the clamping diodes in multilevel diode-clamped converters. Likewise, packing is also tougher in the inverters with a higher number of levels. The five-level and three-level FC-MLIs are represented in Figures 4 and 5 respectively.

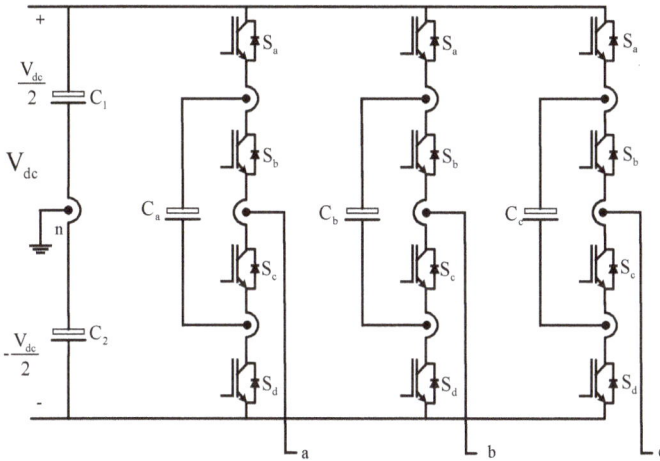

Figure 5. Three-level capacitor clamped/flying capacitor inverter.

2.3. Cascaded H-Bridge Inverter

There are minimum three voltage levels for a multilevel inverter using cascaded topologies. In order to attain a three-level waveform, a single full-bridge or H-bridge inverter is considered. Each inverter is provided with a separate DC source. A three-level cascaded inverter is shown in Figure 6.

By using different combinations of the four switches, S_a, S_b, S_c and S_d, each inverter level can produce three different outputs of voltage, i.e., V_{dc}, 0 and $-V_{dc}$ by connecting the dc source to the ac output. $-V_{dc}$ can be obtained by turning on switches S_b and S_c whereas for obtaining V_{dc}, switches S_a and S_d can be turned on. However, for achieving the output voltage on 0 level either S_a and S_b or S_c and S_d can be turned on. The different full-bridge inverters must be connected in series in the way that the finally produced voltage waveform should be the sum of the inverter outputs. Multilevel cascaded inverters are proposed for the applications such as static VAR generation (reactive power control), an interface with renewable energy sources and for battery-based applications. The main reasons for preferring a cascaded multilevel H-bridge inverter are the availability of possible output levels more than twice the number of dc sources [42–44]. The series of H-bridges enables the manufacturing and packaging process more easy, quick and economical. However, the requirement of a separate dc source for each H-bridge constrains the applications of these inverters to the products having multiple separate DC sources already or readily available.

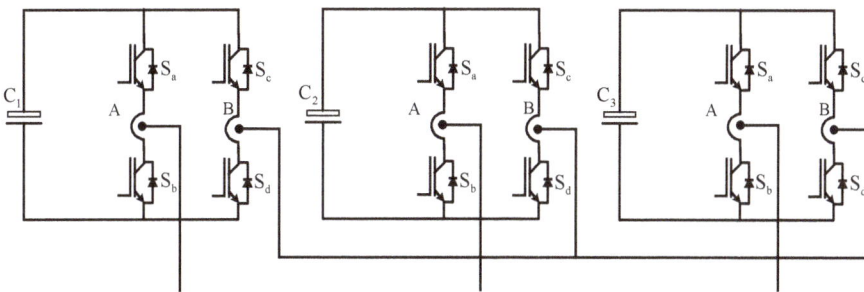

Figure 6. Cascaded H-bridge multilevel inverter.

2.4. Two-Level Three Phase VSI with an Output Filter

A simple two-level inverter is used to convert dc to ac output. It consists of six switches, IGBTs and MOSFETs are the two most suitable switching components for these inverters. Due to simplicity in their structure and ability to handle the voltage by keeping the system stable, they are preferred utmost in the industry and for commercial purpose due to their support in uninterruptible power supply applications. These are usually connected to the load or the grid by using LC or LCL filter. Various types of control systems are implemented by the researchers to improve their performance, robustness and stabilization, compensating the power losses and lowering the THD value. SPWM or SVPWM are mostly applied to these types of inverters for getting appropriate values. Two level three phase VSI is shown in Figure 7. In Figure 7, the S_1 to S_3 and S'_1 to S'_3 shows the switches of the inverter. Whereas, u_c represents the voltage across the capacitors, C.

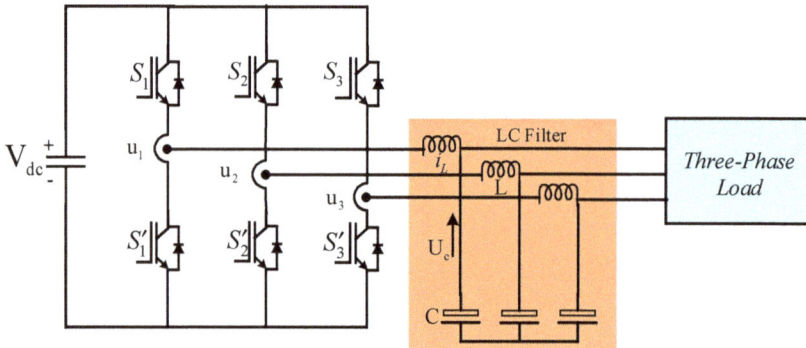

Figure 7. Two-level three phase VSI with an output LC filter.

2.5. Three Phase Four-Leg VSI with an Output Filter

Nowadays, a growing interest in using the three-phase four-leg inverters is observed from the researchers' due to their ability to handle the unbalanced loads efficaciously in four-wire systems [45,46]. In this topology, the neutral point is proposed by connecting the neutral path to the mid-point of the additional fourth leg, as shown in Figure 8. In Figure 8, u_o represents the output voltage of the LC filter, whereas M represents the point neutral point between two switches, S_M and S'_M. Even though the configuration in this topology does not need expensive and large capacitors and produces lower ripple on the DC link voltage, however, using two extra switches lead to a complex control system [47]. Additionally, the split DC-link voltage is about 15% less as compared to the AC voltage in this configuration [48].

Another topology can be using split DC link, which is the most common way of providing a neutral point to three-phase VSIs. This configuration can be provided by using two capacitors i.e., splitting the DC-bus into two parts by using a pair of capacitors and by connecting a neutral path to the mid-point of these capacitors, as shown in Figure 9. Both these configurations have several advantages and disadvantages, however, the split dc-link is found unsuitable for handling the unbalanced loads, whereas, three-phase four leg inverter is found most appropriate for handling the non-linear and unbalanced load conditions. A comparison of different types of VSIs with respect to their characteristics, control contents and complexity is described in Table 1.

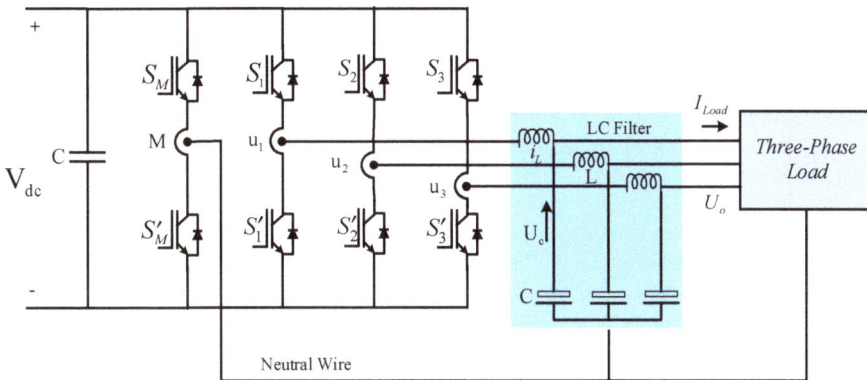

Figure 8. Three-phase four-leg inverter with an output LC filter.

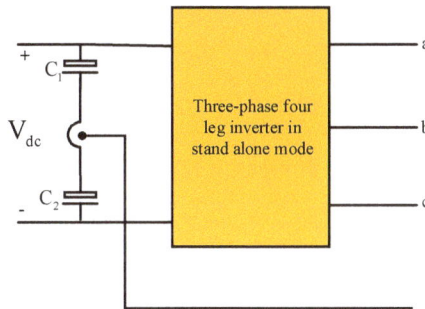

Figure 9. Schematic of a Three-phase four-leg inverter with split DC-link.

Table 1. Comparison of different types of VSI in terms of design, implementation & complexity.

Characteristic	Cascaded H-Bridge VSI	NP-Diode Clamped VSI	Flying Capacitor VSI	2L-VSI
Design & implementation complexity	High	Low	Medium	Low
Specific Requirements	Separate DC sources	Clamping diodes	Additional capacitors	IGBTs/MOSFETs
Control Concerns	Power Sharing	Voltage balancing	Voltage Setup	Voltage/current regulation
Modularity	High	Low	High	Low
Fault tolerance ability	Easy	Difficult	Easy	Easy
Reliability	Medium	Medium	Medium-High	High
Converter Complexity	Medium	Medium	Low-Medium	High
Controller Complexity	Medium-High	Medium-High	Medium-High	Medium
Power Quality	Good	Good	Good	Medium
Operational Power (MW)	3–6	3–7	3–6	3
Switching devices	MV-IGBT, IGCT	MV-IGBT, IGCT	MV-IGBT, IGCT	LV-IGBT

3. Characteristics of Control Systems

There are several parameters and characteristics through which a particular control system is identified. Mainly, there are two characteristics of a control system are found i.e., analog or digital control systems. Both are having some advantages and disadvantages, described as follows:

3.1. Analog Control System

The control systems in which the input and output are designed and analyzed by continuous time analysis or Laplace transform (in s-domain) using state-space formulations. In analog control systems, the representation of the time domain variable is assumed to have infinite precision. Hence, the equations of state space model are differential equations. These systems can be designed without using a computer, microcontrollers or a programmable logic control (PLC). Implementation of analog signals is generally done by using Op-amps, capacitors etc. Robustness against crash or breakdown, having a wide dynamic range, analytical composition accessibility and continuous processing indicate numerous advantages of the analog control systems. However, slow processing speed, interference, complicated implementation in comparative logic, intelligent control systems, neural networks and MIMO are several disadvantages of analog control systems.

3.2. Digital Control System

In digital control systems, modeling, designing, implementation and analysis is carried out in discrete-time or z-transformation domain. In digital control systems, as the name depicts that digital signals are analyzed. Therefore, time is sampled and quantized for state space equations. Additionally, as a digital computer has finite precision, extra attention is needed to ensure that error in coefficients, i.e., A/D conversion, D/A conversion etc. are not producing any disturbances or inadequate effects. In a digital controller, the output is a weighted sum of current as well as previous input and output samples, therefore, its implementation requires the storage of relevant values in a digital controller.

Mostly, a digital controller is implemented via a computer, so, found most economical to control the plants. Moreover, it is relatively easier to constitute and reconstitute through software. Likewise, programs can be leveled to the confines of storage without any additional cost. Correspondingly, digital controllers are compliant with constraints of the program can be changed. Furthermore, the digital controllers are less responsive to the changes in environmental conditions, unlike the analog controllers. Flexibility, swift expansion, uncomplicated implementation in comparative logic, intelligent systems and MIMO, high accuracy as well as robustness against interference are several advantages of these systems. Though, low processing speed, low dynamic range and non-user-friendly interface are the several drawbacks of the digital control systems. The digital controllers are implemented with various technologies which are classified into three categories expressed as follows:

1. Microcontroller Based implementation (MC) [49–51]
2. Digital Signal processing-based implementation (DSP) [52–54]
3. Field programmable gate array-based implementation (FPGA) [55–57]

In reliable scientific research, generally, DSP is used. Fixed point arithmetic and floating-point algorithms are mostly used in implementing the digital control technique by DSP. A traditional slow microprocessor is used normally in slow applications. However, an FPGA is found adequate in fast controllers, due to its abilities of bug fixing and to be reprogrammed in complex structures.

A general structure of a closed loop grid connected digital control system, with an inner current loop and an outer voltage loop, is depicted in Figure 10. In this figure, a voltage source inverter with an output filter is considered. An AC bus is connected to point of common coupling, PCC. Moreover, coordinates transformation from abc to dq is achieved by a phase angle, PH. However, PLL represents the phase locked loop. The symbols S_1, S_2, S_3, S_1', S_2' and S_3' represents the switches, responsible for positive and negative sequences of the inverter output.

The $v_{dref.}$ and $v_{qref.}$ represents the reference voltages in dq frame. SVPWM shows the space vector pulse width modulation technique for generating drive signals for a voltage source inverter.

The voltage across capacitors, u_c and current across inductors, i_L are measured and transformed into a synchronized dq reference frame. The input voltage is computed in the dq frame on the basis of $v_{ref.}$ in the three-phase reference frame. The computed data is then transformed from rotating dq to abc reference frame. Afterward, the PWM technique would be selected accordingly.

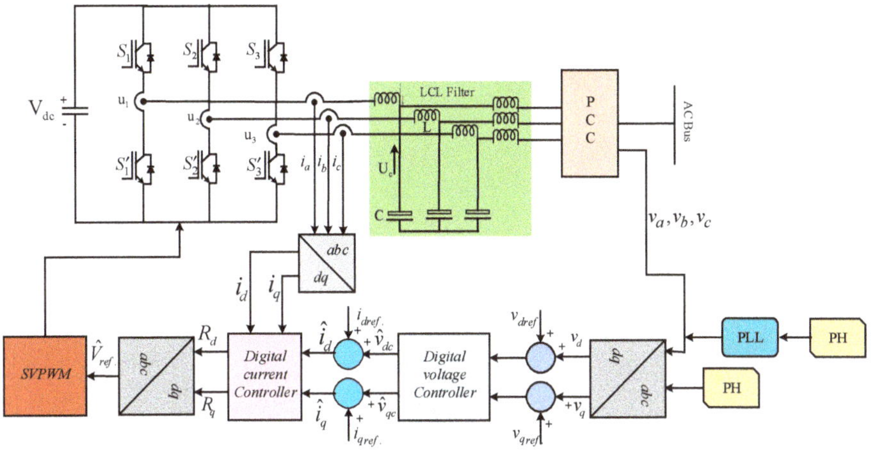

Figure 10. Schematic diagram of a controlled three-phase grid connected VSI with a digital controller.

4. Reference Frames

Control systems are implemented in either a single phase or a three-phase synchronous reference frame. These frames are synchronized with each other through special formulation in order to be compatible for facilitating the modeling, design, analysis and transformation of one phase and three phase systems into other systems. Complex structures, especially for multi-level converters, can be simplified by using these reference frames describes as follows [1,58].

4.1. abc Reference Frame

A general three-phase system is said to be applied to *abc* frame without any transformation. An individual controller is to be used for each phase current in *abc* frame but Delta and star connection has to be considered for designing a control system. Non-linear controllers are used in this system due to their rapid dynamic response.

4.2. dq Reference Frame

This frame is used in three-phase systems. Park's Transformation is used for transforming the *abc* frame into *dq* frame. This transformation causes the current and voltage waveforms to be converted into a frame that rotates synchronously with the grid voltage. As a result, the variables are converted into DC variables and they can easily be controlled and filtered if required.

4.3. αβ Reference Frame

This frame is used in three-phase systems and sometimes sensationally in single phase systems too. Grid current is transformed into a stationary reference frame from *abc* frame or single-phase frame by using Clark's transformation. Therefore, by using this transformation control variable can be transformed into sinusoidal quantities.

5. The Control Strategy in Decoupled dq Frame

In a digital control scheme in dq reference frame, decoupling is the most important issue to be discussed. Generally, a balanced and interrupted sinusoidal waveform can be obtained by adopting ac voltage control in an inverter station. Therefore, the fundamental requirement is to simplify the control design [59]. The controller in an inverter station is based on a mathematical steady-state model in the synchronous reference frame. Moreover, during a balanced network state, the direction of the current injected into the loads is assumed as the reference direction. The mathematical representation of a steady-state model is expressed as following:

$$\begin{cases} u_{bd} = \omega L i_{sq} + u_{sd} \\ u_{bq} = -\omega L i_{sd} + u_{sq} \end{cases} \tag{1}$$

In Equation (1), the terms u_{bd} and u_{bq} represents the voltages in dq frame under balanced network conditions. Likewise, k_p and k_i represents the proportional and integrated controllers and the equation by using aforementioned coefficients represents a PI controller. Correspondingly, u_{sd} and u_{sq} represents the bus voltages in *dq* axis. However, i_{sd} and i_{sq} represents the active and reactive current respectively. Commonly, the *d*-axis is fixed to the voltage source space vector, i.e., the amplitude of the desired ac voltage space vector is kept constant and the value of $u_{sq} = 0$. Then Equation (1) can be simplified as:

$$\begin{cases} u_{bd} = \omega L i_{sq} + u_{sd} \\ u_{bq} = -\omega L i_{sd} \end{cases} \tag{2}$$

According to Equation (2), the control structure of the inverter station is shown in Figure 11, where a PI controller is employed in the ac voltage control [60]. Moreover, $u_{sref.}$ is the reference voltage which can be set accordingly for the desirable amplitude of AC bus voltage.

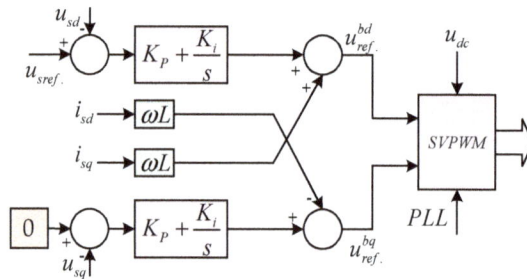

Figure 11. The decoupling control strategy in dq reference frame.

6. Time Delay Sampling Scheme for VSI

Time sampling for digital controllers is done by using a discrete time-domain analysis, i.e., z-domain. Two fundamental advantages of using z-domain analysis over s-domain (continuous time domain) analysis for designing a current controller are: First the control implementation is achieved on a computer-based system, i.e., the control calculation, sampling measurements and PWM signals sequence are updated in discrete time steps. Although, this sample and hold feature is a characteristic of a control system and effects its dynamics as per the referred sampling frequency. Secondly, the multiple time delays can be modeled by using a backshift operator, which affords no simplifications in linear control design, unlike continuous time domain, where the multiple time delays were sampled using an exponential term, which is approximated generally by applying Taylor-series expansion. The sampling effect is a most critical requirement to handle model uncertainties, issues in power supplies and relative disturbances. Therefore, in order to deal with aforementioned issues, zero order hold, ZOH should be incorporated in the control system. In ZOH, a pole or a zero is added into the existed controller through the compensator. The fundamental advantage of this technique is its uncomplicated structure to be implemented on a system, though, it only affects a limited share of the overall delay.

There are two basic sampling routines generally employed in the digital control systems, i.e., single updated sampling and double updated sampling [61]. A single-update sampling method comprises of the measurement samplings, in which calculated modulation indexes are updated once in every switching period. Whereas, a double-update sampling concept conferred to a PWM concept in which the measurement sampling and therefore, the calculated modulation index are updated twice in every switching period [61]. The detailed single-update and double-update sampling are shown in Figure 12, where, T(k) represents the switching period of the present time slot. However, T(k − 1) and T(k − 2) shows the switching period of the former time slots.

A single-update PWM-technique with sampling at the beginning of a switching period is depicted in Figure 12a. In this technique, the modulation index is updated once in beginning of a switching period. A time domain of one sampling time is introduced in the control loops. This effect is modeled with a backshift operator while taking discrete time domain into the account.

Figure 12b shows another scheme of a single-update PWM sampling in which the modulation index is updated in middle of a switching period. Therefore, the time delay due to sampling and updating routine is the mean value of the two converter voltage reference values, i.e., actual and former control cycles. Therefore, the transfer function of a single-update PWM technique with sampling in middle of a switching period is determined.

In the double-update sampling concept, sampling and updating occurs twice in each sampling period. In this technique, the modulation index is updated on the basis of former control cycle's measurements. According to this behavior, the time-delay is one control cycle. The pattern of a double-update sampling is presented in Figure 12c.

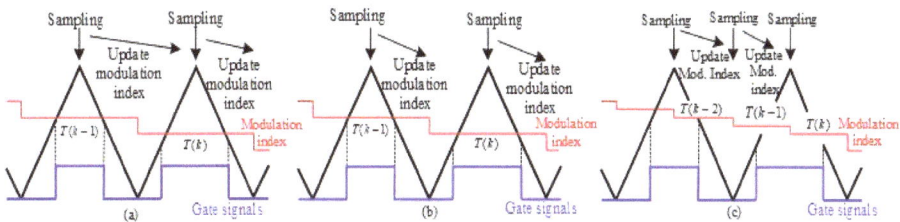

Figure 12. Time delay model of a VSI (**a**) single-update time delay model (sampling at beginning) (**b**) single-update time delay model (sampling at middle) (**c**) double-update time delay model for a VSI.

7. Output Filters for Inverters

The harmonics reduction is the foremost priority of the researchers while designing a power electronics or an electrical system. Therefore, an output filter is used for this purpose. An output filter uses the controlled phenomenon of switching the semiconductor devices for harmonics reduction. There are numerous topologies of such filters introduced in the literature by combining the inductor (L) and capacitor (C) i.e., L, LC and LCL filters unified with the inverters to their output.

7.1. L-Filter

In high switching frequency inverters, the first order L-filter is considered as the most suitable filter. However, inductance decreases the dynamics of the whole system.

7.2. LC-Filter

An LC filter is a second-order filter having substantially sophisticated damping behavior as compared to an L-filter. This filter topology is relatively easier to design and it is a compromise between the values of inductance and capacitance. The cut-off frequency needs the relatively higher value of inductance whereas the voltage quality can be improved through the higher value of capacitance. The value of resonant frequency is dependent on the impedance of the grid when the system is connected to the grid supply. An LC-filter is mostly preferred in standalone mode. The three-phase two-level and three-phase four legs voltage source inverters with an integrated LC filter are shown in Figures 7 and 8 respectively.

7.3. LCL-Filter

An LCL filter is a third order filter, mostly used for the grid-tied inverters. The lower frequency is preferable in presence of aforementioned filters. This filter supports the comparatively healthier decoupling between the filter and the grid impedance. This filter should be precisely designed by taking into consideration the parameters of the inverters. Otherwise even the smaller values of inductance can bring resonance and unstable states into the system. However, the smaller inductance can provide optimized current ripple diminishing values. A three-phase VSI with an LCL filter is shown in Figure 13. Where, V_{th} and Z_{th} represents the Thevenin voltage and Thevenin impedance respectively. However, the complexity of the control system inflated significantly and the dynamic performance of the inverter can perhaps be affected when relatively complex filter structures are employed. Thus, these topologies are most suitable for high power applications, which employ low switching frequencies. However, Figures 14 and 15 show the one-leg block diagram of a single-phase and three-phase grid-connected systems, respectively. Where, K_{pwm} represents the pulse width modulation characteristic of the system, whereas, u_g, u_i, u_c, L_g, L_i, i_i and i_g represents the grid side voltage, inverter side voltage, voltage across capacitor, grid side inductance, inverter side inductance, inverter side current and grid side current respectively.

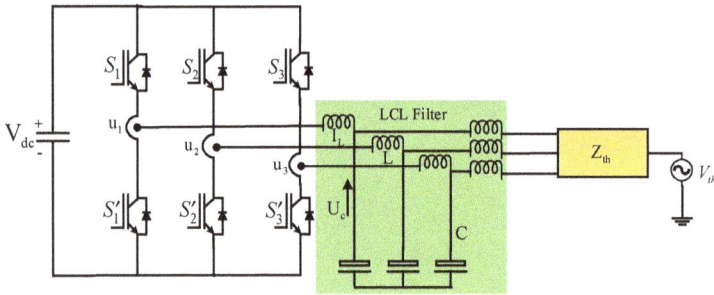

Figure 13. A Three-phase voltage source inverter in grid-connected mode.

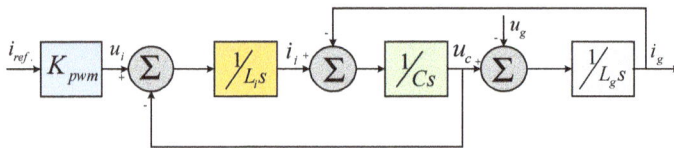

Figure 14. One-leg block diagram of a single-phase grid-connected system.

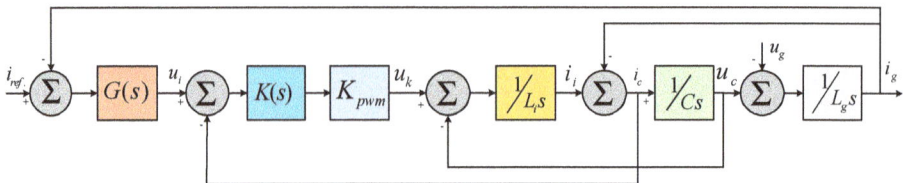

Figure 15. One-leg block diagram of a dual-loop current control strategy for VSI.

8. Damping Techniques for Grid-Connected VSIs

In the grid-connected applications, LCL filter is highly preferred due to its harmonic suppressing capability. In this case, the voltage across the point of common coupling PCC is controlled in synchronism with the current. Therefore, it becomes possible to regulate the active and reactive power injected into the grid according to the requirement. The LCL filter offers a resonance frequency which can be a source of instability in the closed-loop system. This problem is stated by various researchers in the literature and numerous damping strategies are proposed to solve it [62–65]. Damping methods can be classified into two groups. (i) Passive damping and (ii) Active damping.

8.1. Passive Damping

Passive damping is to inserting passive elements in the filter for reduction of the resonant peak in the system [32]. Generally, passive damping schemes never desire any amendments in the control strategy. Though, these approaches change attenuation of the filter, as a result of which losses increases [18,32,34]. The passive damping techniques, presented generally in the literature, results in the addition of a simple resistor in series with the filter capacitor [63]. The major drawback of this technique is a reduction in filter attenuation, increasing power losses and large filter volume [62]. A general schematic of passive damping control strategy for a grid connected VSI is shown in Figure 16.

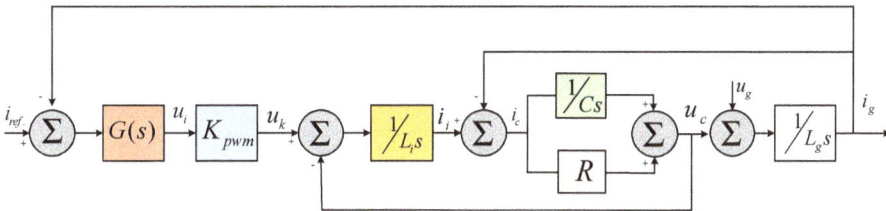

Figure 16. One-leg block diagram of a passive damping control strategy for a grid-connected VSI.

8.2. Active Damping

The active damping methods are proposed to overcome the drawbacks associated with the passive damping techniques. Active damping techniques offer modifications in the control policy in order to afford closed loop damping [65,66]. The active damping techniques are classified into 3 groups, i.e., single loop, multi-loop and complex controllers. Single loop methods are incorporated to damp the LCL filter resonance, without supplementary measurement. These methods comprise of low pass filter-based method, virtual flux estimation method, sensor-less method, splitting capacitor method, notch-filter method and grid current feedback method. Generally, single-loop methods are found relatively robust during uncertainty in parameters and variation in grid inductance [62]. Multiloop methods explore additional measurements. This group comprises of capacitor current feedback, capacitor voltage feedback and weighted average current control techniques. However, the third group of active damping methods is based on complex control structures. This outcome of these techniques is usually a suitable and a robust dynamic response [67]. These techniques include predictive control, state-space controllers, adaptive controllers, sliding mode controller and vector control. Additionally, when LCL filter is selected, there are two options for current control: grid current or converter current. Various techniques are proposed but there exists a disagreement in the literature about the suitable solution of these issues and it is agreed that the current control strategy should be carefully selected. An active damping technique with a damping resistance as well as a harmonic compensator are described in [65]. A general schematic of active damping control strategy for a grid connected VSI is shown in Figure 17.

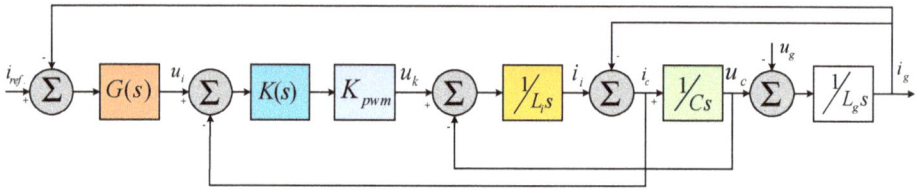

Figure 17. One-leg block diagram of an active damping control strategy for a grid-connected VSI.

9. Grid Synchronization Techniques

The grid voltage must be synchronized with the injected current in a utility network for a significant output. In synchronization algorithm, phase of the grid voltage vector is considered and control variables i.e., grid voltages and grid currents are synchronized by using it. Various methods are introduced in literature for extracting the phase angle [68]. Some commonly used techniques found in credible research articles are discussed as:

9.1. Zero-Crossing Technique

The simplest method to implement is Zero-Crossing method. However, it is not considered on a larger scale due to poor performances reported in the literature. Especially, during voltage variations, ample values of harmonics and notches are observed.

9.2. Filtering of Grid Voltages

The grid voltages can be filtered in the *dq* frame as well as in the αβ reference frame. The performance of zero-crossing method is improved by voltage filtering [68]. However, it is a complicated process to extract the phase angle out of utility voltage, especially during a fault condition. This method uses the arctangent function to realize the phase angle. Generally, a delay is observed in processing a signal while using the filtering method. Therefore, designing of the filter must be considered critically.

9.3. Phase Locked Loop Technique

The phase locked loop, PLL technique is considered as the state-of-the-art method to obtain the phase angle of the grid voltages. The PLL is implemented in dq-synchronous reference frame. In this case, the coordinates transformation from abc to dq is preferred and reference voltage, $\hat{u}_d$ would be set to zero for realizing the lock. A general schematic of PLL technique is depicted in Figure 18. A PI regulator is generally used to control the reference variable. Afterward, the grid frequency is integrated in the system and utility voltage angle is acquired after passing through a voltage-controlled oscillator, VCO. This voltage angle is then fed into the αβ to dq transformation module for transforming into the synchronous reference frame.

This technique is found the most suitable for rejecting notches, grid harmonics and other disturbances. However, additional improvements are needed to handle the unsymmetrical voltage faults. Especially, filtering techniques to filter the negative sequence should be proposed in case of unsymmetrical voltage faults, as second-order harmonics are propagated by the PLL system and reflected in the obtained phase angle. Moreover, it should be assured to estimate the phase angle of the positive sequence of the grid voltages during unbalanced grid voltages [68].

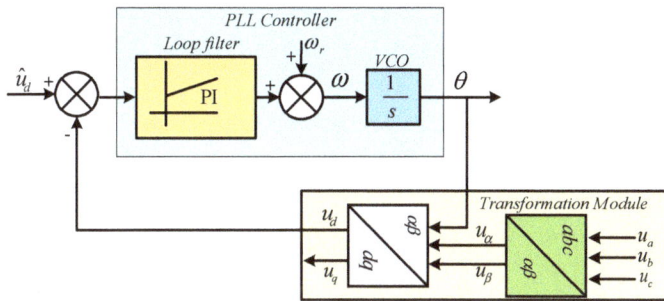

Figure 18. General schematic of a three-phase PLL technique.

10. Modulation Schemes

In power electronics converters, the major problem is the reduction in harmonics. PWM control techniques provide the most suitable solution for harmonics reduction. A sinusoidal output having controlled values of frequency and magnitude is the core purpose for using these PWM techniques. Primarily, PWM techniques are classified into three major categories i.e., Triangular Comparison-based PWM (TC-PWM), Space Vector-based PWM (SV-PWM) and Voltage look-up table-based PWM (VLUT-PWM).

10.1. Triangular Comparison Based PWM

In Triangular Comparison based PWM (TC-PWM) techniques, PWM waves are produced by the combination of an ordinary triangular carrier and a fundamental modulating reference signal. The triangular carrier signal has relatively very higher frequency than that of a fundamental modulating reference signal. The magnitude and frequency of the fundamental modulating reference signal control the magnitude and frequency of the central module in the grid side. PWM and Synchronous PWM (SPWM) are the core techniques to be mentioned in TC-PWM [69].

10.2. Space Vector Based PWM

In SVPWM techniques, the revolving reference vectors provide the reference signals. The magnitude and frequency of central module in grid side are controlled by the frequency and magnitude of the revolving reference vectors respectively. This technique was first introduced to generate vector based PWM in the three-phase inverters. However, nowadays it is expanded to various other newly introduced inverters. SV-PWM is considered to be the more advanced technique for PWM generation for getting qualified sinusoidal output with low THD values [69].

10.3. Voltage Look-Up Table-Based PWM

In VLUT-PWM, a new method is introduced to obtain the voltage reference based on the current reference for an inverter. The major advantage of this technique is its compatibility and simplicity with the load conditions. The switching frequency in this technique is usually taken significantly lower as compared to various other presented techniques [52].

11. Control Techniques

Connecting the grid to the distributed generation system plays a key role and if bit negligence is shown in implementing this procedure, a number of problems can arise i.e., the grid uncertainty and disturbance, so in order to overcome this situation, a suitable controller must be designed for it. In this section, the most appropriate control techniques are described according to their applications. Various single loop and multiloop control systems are discussed in the literature for power droop control, voltage and current control. In which inner loop is for current regulation and outer loop is for voltage regulation [13,70,71]. In Figure 19, the categorization of classical an advanced control technique is depicted clearly.

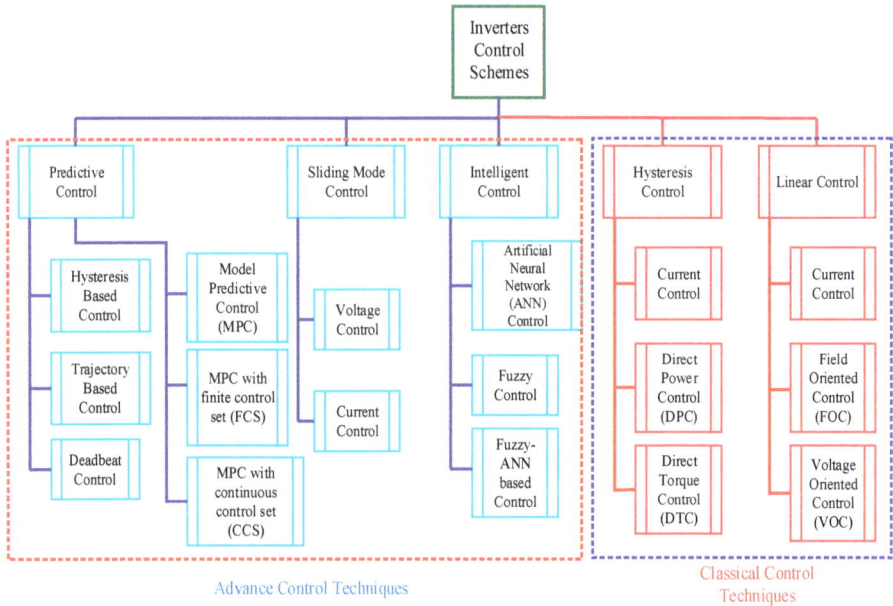

Figure 19. Classification of control techniques for VSIs.

11.1. Classical Control Techniques

The classical controllers include the category of controllers for adding or subtracting a proportion and adjusting the system accordingly. These controllers involve proportional (P), proportional integration (PI), proportional integral derivative (PID) and proportional derivative (PD) controllers. These controllers are considered as the most fundamental controllers in the industry for controlling linear systems and considered as the base of control theory. Lot of work in literature is being done on these controllers [49–52,72–78]. The fundamental benefits of implementing these controllers are their ability to tune themselves according to the requirement of the plant and their simple structure. Moreover, they are the most commonly used controllers on commercial levels, so easily available. However, their tracking ability, response time and ability to handle stable error are relatively lower as compared to modern state-of-the-art controller. The schematic of a digital PI controller for controlling a three-phase VSI with an LC filter in stand-alone mode is shown in Figure 20. In Figure 20, i_{af}, i_{bf} and i_{cf} represents the filter current across phase a, b and c respectively. Likewise, v_a, v_b and v_c characterizes the voltage across phase a, b and c respectively. Likewise, i_d and i_q represents the current across the d and q axis respectively. Moreover, S_a, S_b and S_c represents the switching commands across phase a, b and c respectively. Correspondingly, $V_{ref.}^d$ and $V_{ref.}^q$ symbolizes the reference voltages along d and q axis respectively.

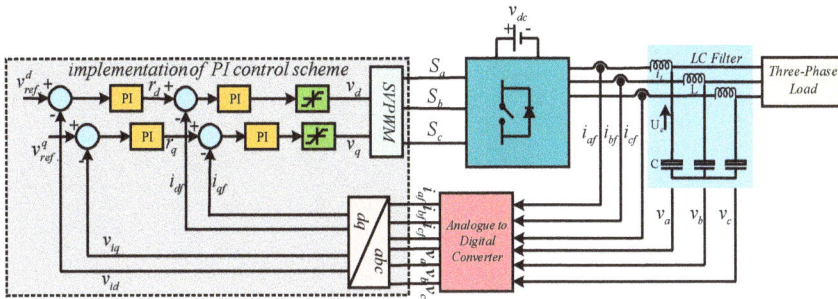

Figure 20. Schematic of a PI control algorithm for a VSI.

11.2. PR Controllers

PR controllers are the combination of proportional and resonant controllers. The frequencies closer to resonant frequency are integrated by the integrator. Therefore, phase shift or stationary error do not occur. This controller can be applied in both *ABC* and αβ frames. Due to high gain near resonant frequencies, this controller has the ability to eliminate the steady-state errors of electrical quantities. The resonant controller maintains the network frequency equal to the resonant frequency. It is capable of adjusting the frequency according to changes in grid frequency. However, an accurate tuning is always needed for optimal results and this technique is found sensitive to the frequency variations [30,31]. These controllers are relatively better than PI controllers in terms of their tracking ability and response time. If used with a harmonic compensator, they can optimally handle THD. Their capability to handle current in grid-connected inverters is also remarkable. However, damping issues still exist. The active and passive damping adjustments and integration in a system with a harmonic compensator are somehow, the complicated issues. Moreover, they do not have outstanding ability to handle stable error and phase shift. The limitation to handle specific frequencies i.e., closer to resonant frequencies is also a drawback of these controllers. A PR controller with a harmonic compensator, HC, in stand-alone mode for a VSI is shown in Figure 21. Structures of a simple PR controller and a discrete PR controller are shown in Figures 22 and 23, respectively. Where, T_s represents the sampling period, ω represents the grid angular frequency. However, K_p and K_r denotes the proportional and resonant coefficients respectively. The PR controller with a harmonic compensator is proposed in [65].

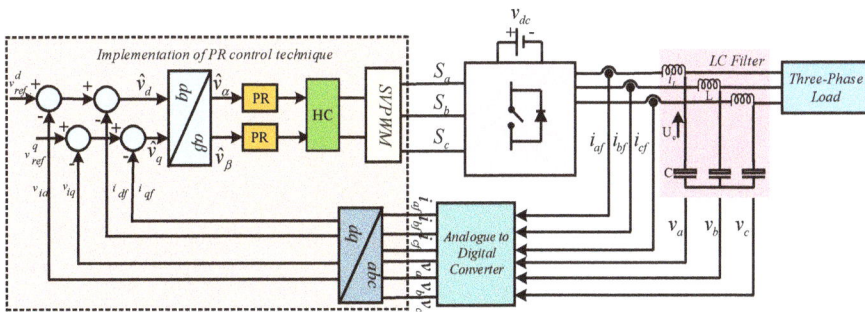

Figure 21. General schematic of a PR Controller and HC for a three-phase VSI.

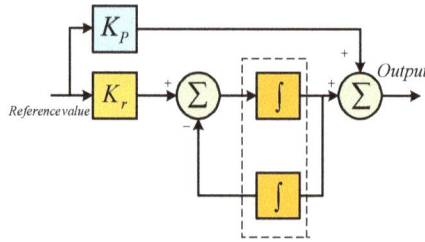

Figure 22. General structure of a PR Controller.

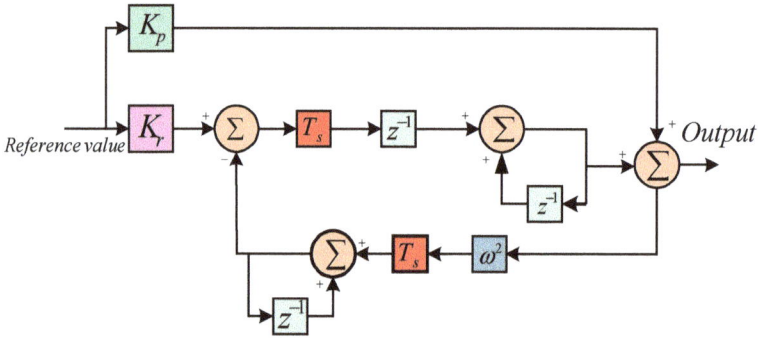

Figure 23. Structure of a discrete PR Controller.

11.3. LQG Control Technique

The integration of Kalman filter with an LQR controller gives rise to an LQG controller. In this technique, Kalman Filter, as well as an LQG controller, can be designed independently of each other. This control scheme is valid for both linear time-invariant systems as well as for linear time-varying systems. LQG control technique facilitates the designing of a linear feedback controller for an uncertain nonlinear control system [79–81]. An LQG control structure with a Kalman estimator is shown in Figure 24. Where, u_e represents the known input and y_c is the estimated noise/disturbance. The Kalman estimator provides the optimal solution to the continuous or discrete estimation problems.

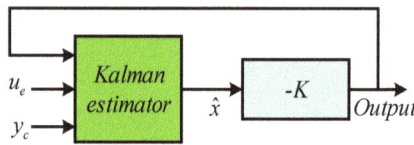

Figure 24. Structure of an LQG controller.

11.3.1. Linear Quadratic Regulator

The linear quadratic regulator (LQR) technique is found optimal for steady as well as transient states [82–84]. As the name depicts, this control technique is a combination of linear and quadratic functions, where the dynamics of the system are described by a set of linear equations and the cost of the system is a quadratic function. The cost function parameters are considered critically while designing the controller. LQR algorithm is an automatic approach for finding a suitable state-feedback controller. Pole placement with state feedback controller provides the system with a high degree of freedom and makes it simpler to implement. This method is characteristically steady and it can be employed even if some of the system parameters are unknown. However, exertion in finding the exact weighting factors limits the applications of LQR control scheme. Moreover, it has a discrepancy of tracking accuracy during load changes [83–85].

11.3.2. Linear Quadratic Integrator

In linear quadratic integrator (LQI) scheme, cost minimization is considered critically. This technique is implemented for nullifying the steady-state error between actual grid voltage and reference grid voltage during load variations [82]. An integral term used with LQ control is for minimizing the tracking error produced by uncertain disturbances in instantaneous reference voltage. Optimal gains for providing adequate tracking with zero steady-state error are relatively simpler to attain by using this technique. The rapid dynamic response, accurate tracking ability and relatively simpler designing procedure provide this technique a benefit over other techniques. However, complications in extracting the model and phase shift in voltage tracking even in normal operative condition are the major drawbacks of this scheme.

11.4. Hysteresis Control Technique

Hysteresis control is considered as a nonlinear method [86–93]. The hysteresis controllers are used to track the error between the referred and measured currents. Therefore, the gating signals are generated on the basis of this reference tracking. Hysteresis bandwidth is adjusted for error removal in reference tracking. This is an uncomplicated concept and has been used since analog control platforms were intensively used. This technique does not require a modulator; therefore, the switching frequency of an inverter is dependent on the hysteresis bandwidth operating conditions and filter parameters [94]. The major drawback of hysteresis controller is its uncontrolled switching frequency; however, researchers are working on improving this controller and several works are presented and several techniques are proposed in the literature. Main advances in this technique are direct torque control (DTC) [87,88,95,96] and direct power control (DPC) [97–99]. In DPC, active and reactive powers are directly controlled, however, in DTC torque and flux of the system are controlled. Error signals are produced by hysteresis controllers and drive signals are generated by the look-up according to the magnitude of the error signals. Hysteresis controllers require very high frequency for constraining the variables in hysteresis band limits, whenever implemented on a digital platform as shown in Figure 25. Moreover, switching losses are very high in this type of controllers. So, Hysteresis controllers are found inappropriate for high power applications.

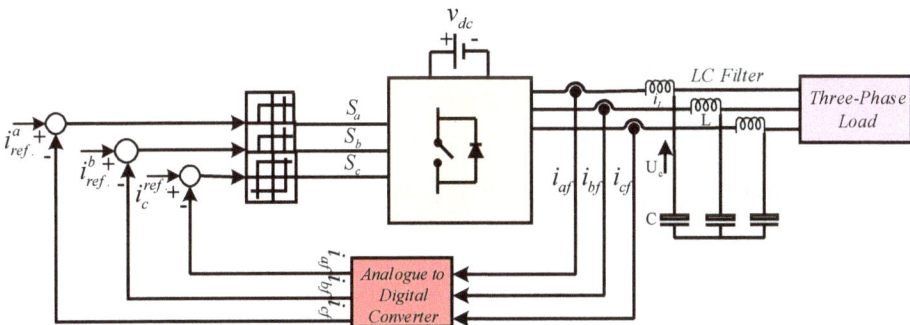

Figure 25. A Hysteresis control technique for VSI.

11.5. Sliding Mode Control

The sliding mode control is considered to be an advanced power control technique for the power converters. It fits into the family of adaptive control and variable structure control [100–104].

Sliding mode control is a non-linear technique, whereas it can be instigated to both non-linear as well as linear systems [100]. In Figure 26, a sliding mode control along with SVM/PWM is presented. Where, β_v represents the gain, λ is a strictly positive constant and ϕ is a trade-off between the tracking error and smoothing of the control discontinuity. The sliding controller produces the voltage references in a converter for generating the drive signals. A predefined trajectory is executed and the control variable is forced to slide along it [102–104]. The robust and stable response is achieved even in the system parameters variation or load disturbances by implementing sliding mode control technique. This controller is more robust and capable of removing the stable error as compared to the classical controllers. However, some drawbacks in implementing a sliding mode control are difficulty in finding a suitable sliding surface and limitation of sampling rate that degrades the performance of SMC will be degraded. Whenever tracking a variable reference, the chattering phenomenon is another drawback of SMC technique. As a result, overall system efficacy is reduced [105,106].

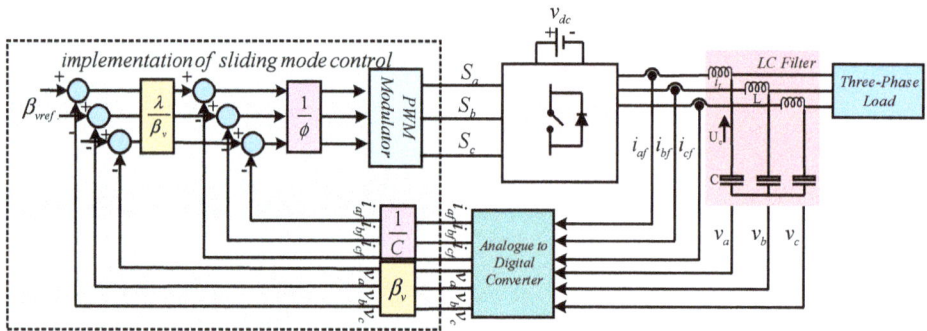

Figure 26. A Sliding mode control technique on a VSI.

11.6. Partial Feedback Controllers

There are several techniques presented for conversion of non-linear systems to linear systems for their uncomplicated computation. Partial feedback controllers are one of the most effective and forthright techniques for transforming the non-linear systems into the linear ones. By this technique, a system can be converted either partially or fully into a linear system, depends on the system constraints. Linearity in a system is attained by the cancellation of the nonlinearities inside the system. So, these systems can be controlled by using the linear controllers whenever a non-linear system is fully transformed into a linear system i.e., exact feedback linearization method. However, if it is partially converted into a linear system then it is known to be partial feedback linearization. PFL controller is implemented in [104,106–109]. In PFL, it is difficult to ensure the stability of complicated renewable energy system applications. However, an independent subsystem can be obtained from PFL for constraining the extensive use of this method. Moreover, in order to deal with these problems, exact feedback linearization (EFL) is a forthright and model-based technique for scheming nonlinear control techniques. EFL receipts the built-in nonlinearity characteristic of the system under deliberation and consents the conversion of a nonlinear structure into a linear one, algebraically. EFL removes nonlinearities of a system through nonlinear feedback, as a result, the transformed system is not reliant on an operating point.

11.7. Repetitive Control

The plug-in scheme (PIS) and internal model (IM) principle are the basic concepts of repetitive control (RC). RC uses an IMP which is in correspondence to the model of a periodic signal. In order to derive this model, trigonometric Fourier series expansion is used. If the model of reference is fed into the closed loop path, optimal reference tracking can be obtained. Moreover, it is found robust against disturbances and has the ability to reject them. RC mostly deals with periodic signals. Closed loop behavior of the system and Magnitude response of the IM are the core factors used for analyzing the performance of the repetitive controller in case of frequency variation or any other uncertainty in the system. Both these factors indicate the performance sagging in case of variation or uncertainty in the reference signal. In presence of a periodic disturbance, RC intends to attain zero tracking error when a periodic or a constant command is referred to it. RC has an ability to locate an error, a time-period before and fine-tunes the next command according to the feedback control system for eliminating the observed error. However, it lacks the ability to handle physical noise. For this purpose, an LPF can be used. Kalman's filtering approach is also noticeable to remove this noise [27,110–113]. The general structure of a repetitive controller is shown in Figure 27.

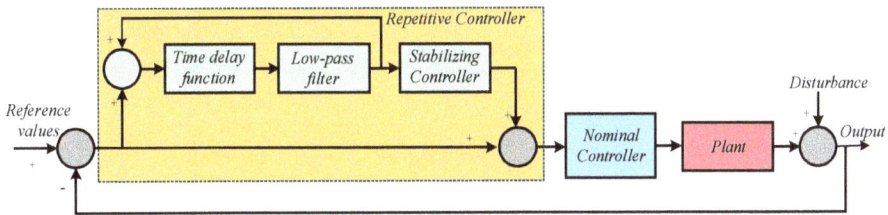

Figure 27. Block diagram of Repetitive control algorithm.

11.7.1. Fuzzy Control

This control technique belongs to the family of intelligent control systems. The PI controller is replaced by a fuzzy logic controller in this technique as shown in Figure 28. Where, $v_{fz.}$ is the fuzzified output voltage. However, its block diagram is shown in Figure 29. In a fuzzy controller, the tracking error of load current and its derivative are given as the input. This controller design is dependent on the awareness, knowledge, skills and experience of the converter designer in terms of functions involvement. Due to non-linear nature of the power converters, the system can be stabilized in case of parameters variation even if the exact model of the converter is unknown. Fuzzy logic controllers are also categorized as non-linear controllers and probably the best controllers amongst the repetitive controllers [113–116]. However, strong assumptions and adequate experience are required in fuzzification of this controller. As it is dependent on the system input and draw conclusions according to the set of rules assigned to them during the process of their modeling and designing.

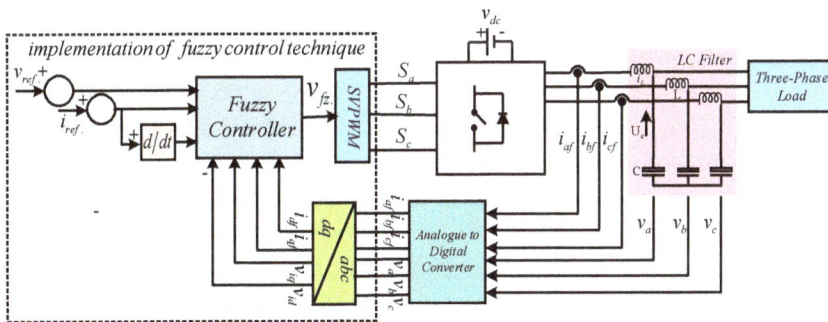

Figure 28. A Fuzzy control algorithm topology on a VSI.

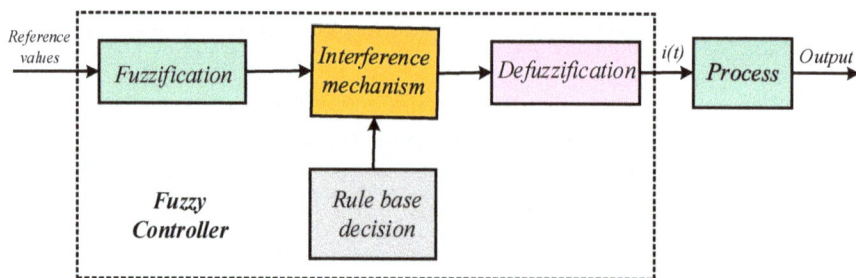

Figure 29. Block diagram of Fuzzy control algorithm.

11.7.2. Artificial Neural Network Control

The Artificial neural network (ANN) controllers are the fundamental form of the controllers based on the human-thinking mode. It consists of a number of artificial neurons to behave as a biological human brain. The reference tracking error signals are given through a suitable gain or a scaling factor (S) as input to the ANN for generating the switching signals into the power converters. This approach is used for achieving the constant switching operation in power converters. ANN can be used in both online as well offline modes while operating it on system control. It has high tolerance level to faults because of its ability to estimate the function mapping. Its topology is shown in Figure 30.

Fuzzy and ANN can be combined to achieve an optimal control performance in a power converter [113–115]. ANN does not need a converter model for its operation, however, the operational behavior of a power converter should be precisely known to the designer/operator while designing the ANN control system.

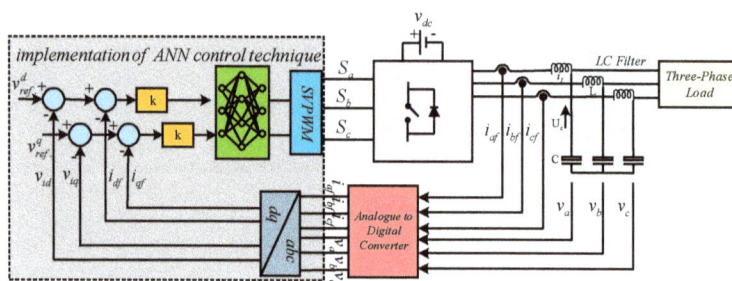

Figure 30. Schematic of Artificial Neural Network control for VSIs.

11.8. Robust Controllers

In robust control theory, a control system vigorous against uncertainties and disturbances is offered. The basic aim is to attain the stability in case of inadequate modeling. All the descriptions, criteria and limitations should be appropriately defined in order to get robust control. This controller guarantees the stability and high performance of closed-loop system even in multivariable systems [117].

11.8.1. H-Infinity Controllers

The expression H∞ control originates from the term mathematical space on which the optimization takes place: H∞ is considered as a space of matrix-valued functions that are investigative and confined in the open right-half of the complex plane. In this type of control system, first of all, the control problem is formulated and then mathematical optimization is implemented i.e., selection of the best element according to criterion from the set of obtainable alternatives. H-infinity control techniques are generally pertinent for the multivariable systems. The impact of a perturbation can be reduced by using H-infinity control techniques in a closed loop system subject to the problem formulation. The impact can be measured either in terms of performance or stabilization of the system. However, modeling of the system should be well-defined for implementation of these control techniques. Moreover, H-infinity control techniques have another discrepancy of high computational complications. In case of non-linear systems limitations, the control system cannot handle them well and response time also increases [118]. However, these controllers are implemented and well defined in [111,112,119].

11.8.2. μ-Synthesis Controllers

Mu-synthesis is based on the multivariable feedback control technique, which is used to handle the structured as well as unstructured disturbances in the system. Where μ mentions the singular value that is reciprocal of the multivariable stability margin. The basic purpose is to mechanize the synthesis of multivariable feedback controllers that are insensitive to uncertainties of the plant and be able to attain the anticipated performance objectives. This method is well described in [120,121].

11.9. Adaptive Controllers

An adaptive controller is designed to have the ability of self-tuning, i.e., to regulate itself spontaneously according to variations in the system parameters. It does not require initial conditions, system parameters or limitations for its implementation due to its ability to modify the control law according to system requirements. Recursive least squares and Gradient descent are two most commonly known technique for parameters estimation in adaptive controllers. The structure of a typical adaptive controller is shown in Figure 31. In the literature, some credible research articles and state-of-the-art techniques for adaptive controllers are found in [14,37,53,55,113,122–125]. These controllers are applicable for both dynamic as well as static processes. However, the complicated computational process leads to exertion in its implementation.

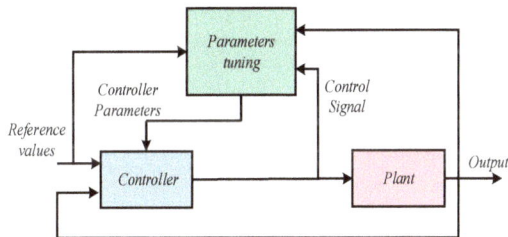

Figure 31. Block diagram of Adaptive control algorithm.

11.10. Predictive Controllers

Predictive controllers are commenced as a propitious control technique for electronics inverters. The system model is considered critically and then imminent behavior of the control variables is predicted conferring to the specified criterion. It is an uncomplicated technique and can handle multivariable systems efficiently. Moreover, it can handle the system with several limitations or non-linearities. It is generally preferred due to its prompt static as well as dynamic response and ability to handle stable errors. However, its computational analysis is complex as compared to classical controllers. It is further categorized into Deadbeat control and Model Predictive control. It can refer to literature [105,125–127] for predictive controllers. A comparison of predictive control techniques on basis of their pros. and cons. is described in Table 2.

11.10.1. Deadbeat Control

Deadbeat control technique is the most authentic, competent and attractive technique in terms of low THD value, frequency as well as rapid transient response. Differential equations are derived and discretized in this type of control system for controlling the dynamic behavior of the system. The control signal is predicted for the new sampling period for attaining the reference value. Its effective dynamic performance and high bandwidth simplify the current control for this type of controller. Error compensation is a specialty of a deadbeat controller. However, its major discrepancy is its sensitivity for network parameters and accurate mathematical filter modeling [13,54,56,128–135]. Its topology is shown in Figure 32, where a disturbance observer, a state estimator and a digital deadbeat controller are used to control the voltage and current of a VSI. The coefficient $\hat{d}$ represents the output of disturbance observer comprises of current and voltage. However, $\hat{v}_d$ and $\hat{v}_q$ represents the controlled voltage across d-axis and q-axis respectively.

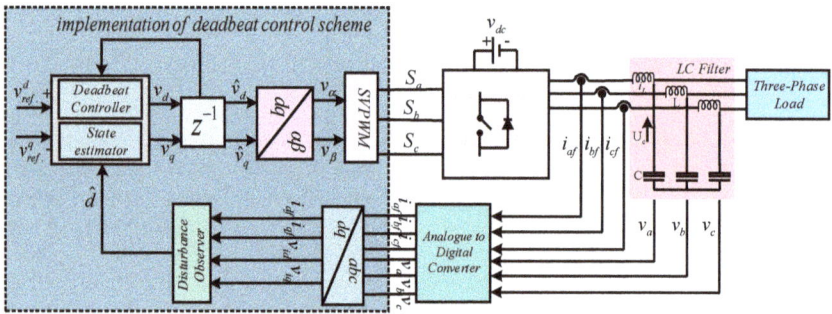

Figure 32. Deadbeat control topology for VSIs.

11.10.2. Model Predictive Control

As the name depicts, a model of the system is used to predict the behavior of the system in model predictive control (MPC) technique. A cost function criterion is defined in this type of control system, which can be minimized for optimal control actions. The controller adapts the optimal switching states according to the cost function criterion. Forecast error can be lessened for current tracking implementing. Moreover, system limitations and non-linearities, as well as multiple inputs and output systems, are handled well by MPC. Control actions of the present state are considered in order to predict the control actions of the system in the next state. Like deadbeat control, it is also found sensitive to system parameter variations [136–147]. The topology for implementation of MPC on VSI is shown in Figure 33, whereas, its control schematic is depicted in Figure 34.

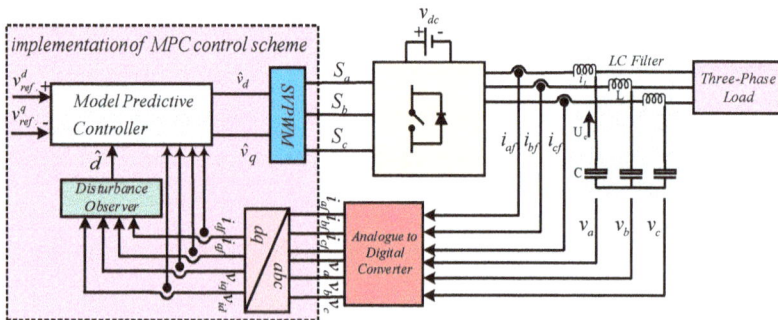

Figure 33. Model Predictive Control topology for VSIs.

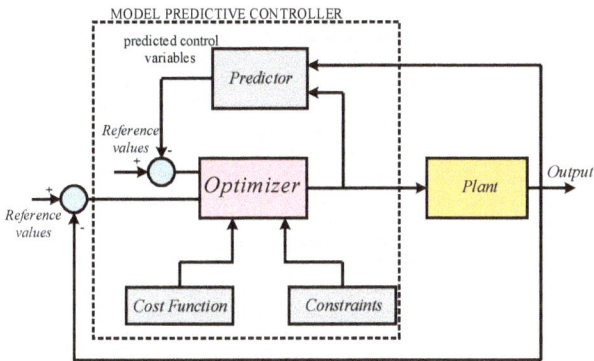

Figure 34. Block Diagram of Model Predictive Control algorithm.

11.11. Iterative Learning Scheme

Iterative Learning Scheme (ILS) is a complicated but authentic technique for attaining zero tracking error. In this scheme, each control command is executed and the system is examined and then adjusted accordingly before each repetition. Highly accurate modeling of the system is essential for the implementation of ILS; therefore, its designing technique is relatively more complicated than other schemes. ILS is capable of removing the tracking error caused by the periodic disturbances. The next cycle is predicted by considering the learning gain, system adjustment in z-transform, tracking error at each repetition, control function of the designed controller and the error function between two consecutive iterations.

12. Comparative Analysis and Future Research Goals

VSIs are specially designed for converting DC to three-phase AC, therefore, control strategies must be according to the three-leg three-phase power inverters. However, for MLIs, the control strategies must be inherited from three-leg three-phase power inverters. The control policies of VSIs in stand-alone mode can be categorized into numerous categories depending upon similar and dissimilar considerations. Considering the PWM, VSIs can be classified into two categories i.e., carrier-based modulation and carrier-less modulation. The carrier-based modulation schemes such as Selective Harmonic Elimination (SHE), 3D SVPWM, Sinusoidal Pulse Width Modulation (SPWM) and Minimum-Loss Discontinuous PWM (MLDPWM) based PWM techniques have attained significant consideration due to their constant switching frequency.

SPWM offers constant switching frequency and flexible control schemes; nevertheless, one major disadvantage of this technique is the compact efficacy of the DC voltage [148]. The 3D-SVPWM delivers an adequate DC bus utilization and a standardized load harmonic curvature as compared to the SPWM technique. However, it is complex in nature to be implemented on the digital devices. Correspondingly, the SHE-PWM suggests a flexible controller by considering the switching angle. However, the real-time enactment of this carrier-based modulation is quite difficult. The capability of the MLDPWM under nonlinear and unbalanced conditions is found relatively admissible; however, its real-time execution is found very much circuitous. However, the carrier-less modulation approaches such as flux vector and hysteresis provide a rapid-dynamic response [149]. But, they suffer from variable switching frequency [91]. Additionally, they require composite switching tables for their implementation.

The conventional PI controllers encounter problems to eliminate the steady-state error. In order to solve this problem, a PR controller is commonly used in the stationary reference frame for regulating the output voltages of the VSIs due to its sophisticated explication in eliminating the steady-state error, while controlling sinusoidal signals. Additionally, it is competent in eliminating selective harmonic uncertainties.

It is also taken into consideration that the PR controller perceives the resonant frequency to offer gains at specific frequencies. Thus, the resonant frequency should be synchronized with the frequency of the microgrid. Hence, it can be said that it is very sensitive with respect to the variations in system frequency. The PI controller is also extensively used in the *dqo* frame and performs robustly with pure DC signals. Though, in order to allocate the control variables from the *abc* to the *dqo* frame, the phase angle of the microgrid should be known. Likewise, using cross-coupling and voltage feedforward terms are the secondary problems in implementing this method.

In the stand-alone operating mode, VSIs primarily controls the power transfer, voltage and frequency of the system. Nevertheless, power quality can be enhanced by offering a suitable control technique in the inverter-based type DGs. As in VSIs, the auxiliary services for improvement in power quality embedded in the control assembly. In case of VSIs, compensation of unbalanced voltage, a lower value for total harmonic distortion, harmonic power-sharing schemes, power sharing between active/reactive powers, imbalance power, active/reactive power control and augmentation in power quality are critically considered and embedded in control schemes. VSIs are also applied on several applications in microgrid systems, extensively, for improving the power quality. This power control strategy is presented in [38]. However, a comprehensive review of various control strategies for microgrids is described in [150]. Moreover, using modular multilevel inverters can improve the modularity and scalability to meet reference voltage levels, efficiency in high power applications, reduction in harmonics in high voltage applications and size of passive filters as well as no requirement for dc-link capacitors [151].

In Table 3, different types of filters are suggested by various researchers based on the control systems. However, it is significant to use L-filters for low power applications having a simple design, nevertheless, L-filters are not found suitable in resonance state as well as for high power applications. So, LCL-filter is highly preferred in aforementioned system characteristics. The designing of this filter is comparatively complicated due to a few constrictions related to the system stability. The accuracy in designing and modeling of the system leads to better performance against resonance and harmonics. Nevertheless, the choice should be made according to the customer's demand. The prime parameters should be chosen on the basis of system condition and intended tasks to be performed by the system. Afterward, the designing of power system and control system parameters should be finalized.

This corresponding study incorporates the advantages and disadvantages of each controller in terms of stability, rapid response time, harmonic elimination, the nonlinearity of the system, unbalanced compensation and robustness against parameter variation. Various suitable control schemes for different types of VSIs are documented in this paper. However, their implementations for power generation and power quality improvement are still not perfect simultaneously. Moreover, each controller has its own benefits and obstructions. Therefore, it is not easy to decide that which control scheme is better than the others. These are significant subjects for the future research. On the basis of the analysis of former publications, appended research is suggested to be carried out in the aforementioned area.

Regardless of the several investigations in this field, none of the proposed control techniques can be selected as an immaculate solution to meet al.l the requirements of power quality, i.e., harmonic/reactive/imbalance power-sharing and voltage unbalanced/harmonic/swell/sag and Interruption compensation at the same time. Therefore, further research should be focused on the novel power-electronics topologies to fulfill all aforementioned necessities simultaneously.

Three-phase three-wire VSIs are now a well-developed and mature research topic with respect to their hierarchical control. But on the other hand, control hierarchies are not as well established for ML-VSIs, as for three-phase three-wire VSIs. It may be beneficial to consider ML-VSI system, as well as the primary, secondary and tertiary stages, whenever a control scheme is to be designed.

Substantially, a lot of work is to be done for exploiting the new control approaches for ML-VSIs. In order to achieve enhanced performance, it is compulsory to use some innovative techniques such as robust, MPC and LQR control techniques.

It is also observed through a number of studies that coupling among the phases is neglected whenever controlling an ML-VSI by means of a conventional PI controller, which results in a reduction of the system's robustness. Hence, it can also be beneficial to implement decoupled phase voltage control to realize the referred response in time domain. A comparison of the credible research articles found in literature with respect to their control techniques, modulation schemes, control parameters, loop characteristics, employed filters and applications is described in Table 3.

13. Conclusions

On the basis of research, conventional multilevel inverter topologies given in the previous sections, general and asymmetrically constituted ML-VSIs have been also reviewed in this paper. Several new hybrid topologies can be designed through the combinations of three main MLI topologies. Besides the combination of topologies, the trade-offs in MLI structures can be dealt by using H-MLIs that is formed using different DC source levels in inverter cells. PWM strategies that generate switching frequency at fundamental frequency are also introduced for H-MLIs for the switching devices of the higher voltage modules to operate at high frequencies only during some inverting instants. Due to numerous applications of conventional MLIs and flexibility to design the hybrid MLI topologies, this paper cannot cover all utilizations with MLIs but the authors intend to provide a useful basis to define the most proper control schemes and applications. In addition to these, the fundamental design and control principles of MLIs have been introduced as a result of a detailed literature survey. This paper has been destined to provide a reference to readers and the results given in this paper can also be extended with experimental studies.

Table 2. Description of predictive controllers on the basis of their pros. & cons.

	Deadbeat Control
	• -Modulator required
	• -Fixed switching frequency
	• -Low Computations
	• -Limitations not undertaken
	Trajectory Based control
	• -Modulator not required
	• -Variable frequency
	• -No cascaded structure
Predictive Control	Hysteresis Based predictive control
	• -Modulator not required
	• -Variable frequency
	• -Uncomplicated structure
	Model Predictive Control
	• -Modulator required in case of continuous control set (CCS) and not required in case of finite control set (FCS).
	• -Likewise, fixed switching frequency (CCS) and variable switching frequency exists in (FCS).
	• -Online optimization and simple designing is included in case of FCS.
	• -Constraints are considered in both cases

Table 3. Digital control system characteristics in numerous credible scientific proposals.

Application	Controller	Filter	Ref. Frame	Feedback	Modulation	Ctrl. Parameter	Ref.
General	adaptive	LCL	Single Phase	Multi-loop	SPWM	V, I	[51]
General	Classic, PR	LCL	Single Phase	Multi-loop	PWM	I	[152]
General	Adp., Rpt.	L	Single Phase	Single-loop	SPWM	I	[105]
UPS	DB	LC	Single Phase	Multi-loop	PWM	V, I	[153]
General	Rpt.	LC	Single-Phase	Single-loop	PWM	V	[101]
General	Rpt C	LCL	Single Phase	Single-loop	PWM	I	[104]
DG	Classic	LCL	abc, αβ	Single-loop	PWM	V, P	[47]
DG	Classic	LC	abc, αβ	Multi-loop	SVPWM	V, I	[48]
DG	Classic	LC	abc, αβ	Multi-loop	SPWM	V, I	[49]
DG	Classic	L	abc, αβ	Single-loop	VLUT	V	[50]
General	DB	L	abc, αβ	Single-loop	PWM	I	[52]
APF	Adp., Rpt.	LC	abc, dq	Multi-loop	SVPWM	I	[53]
General	DB	LCL	abc, dq	Multi-loop	PWM	I	[54]
DG	Adp., MPC	LCL	abc, αβ	Multi-loop	SVPWM	I	[55]
General	LQG	LCL	abc, dq	Single-loop	PWM	I	[66]
PV	PR, LQG	L	abc, αβ	Single-loop	SVPWM	I	[64]
General	Adp.	L	abc, αβ	Multi-loop	PWM	I	[107]
UPS	Pred.	LC	abc, dq	Single-loop	SVPWM	V	[109]
PV, APF	Pred., Fuzzy	L	abc, αβ	Multi-loop	PWM	P	[108]
PV, APF	SMC, Pred.	L	abc, αβ	Multi-loop	PWM	P	[111]
General	DB	L	abc, dq	Single-loop	SVPWM	I	[113]
General	Adp., DB	L	abc, dq	Single-loop	PWM	I	[112]
DG	DB	L	abc, dq	Single-loop	SVPWM	I	[116]
UPS	DB, Rpt.	LC	abc, αβ	Single-loop	PWM	V	[115]
General	MPC	LCL	abc, abc	Single-loop	PWM	V, I	[122]
General	MPC	L	abc, αβ	Single-loop	PWM	I	[125]
General	MPC	L	abc, dq	Single-loop	SVPWM	I	[128]
PV	MPC	L	abc, dq	Single-loop	SVPWM	I	[130]
PV	Classic, Rpt.	L	abc, dq	Single-loop	SVPWM	I	[97]

Acknowledgments: The National Natural Science Foundation of China supported this research work under Grant No. 61374155. Moreover, the Specialized Research Fund for the Doctoral Program of Higher Education PR China under Grant No. 20130073110030 is highly acknowledged.

Author Contributions: Sohaib proposed the idea for writing the manuscript. Wang suggested the literature and supervised in writing the manuscript. Mazhar helped Sohaib in writing and formatting. Sarwar helped in modifying the figures and shared the summary of various credible articles to be included in this manuscript.

Conflicts of Interest: The authors declare no conflict of interest.

References

1. Blaabjerg, F.; Teodorescu, R.; Liserre, M.; Timbus, A.V. Overview of control and grid synchronization for distributed power generation systems. *IEEE Trans. Ind. Electron.* **2006**, *53*, 1398–1409. [CrossRef]

2. European Commission Directorate-General for Energy. *DG ENER Work in the Paper—The Future Role and Challenges of Energy Storage*; European Commission Directorate-General for Energy: Brussel, Belgium, 2013.

3. Loh, P.C.; Newman, M.J.; Zmood, D.N.; Holmes, D.G. A comparative analysis of multiloop voltage regulation strategies for single and three-phase UPS systems. *IEEE Trans. Power Electron.* **2003**, *18*, 1176–1185.

4. Kassakian, J.G.; Schlecht, M.F.; Verghese, G.C. *Principles of Power Electronics*; Addison-Wesley: Reading, PA, USA, 1991; Volume 1991.

5. Mohan, N.; Undeland, T.M. *Power Electronics: Converters, Applications, and Design*; John Wiley & Sons: New Delhi, India, 2007.

6. Abdel-Rahim, N.M.; Quaicoe, J.E. Analysis and design of a multiple feedback loop control strategy for single-phase voltage-source UPS inverters. *IEEE Trans. Power Electron.* **1996**, *11*, 532–541. [CrossRef]

7. Lee, T.-S.; Chiang, S.-J.; Chang, J.-M. H/sub/spl infin//loop-shaping controller designs for the single-phase UPS inverters. *IEEE Trans. Power Electron.* **2001**, *16*, 473–481.

8. Willmann, G.; Coutinho, D.F.; Pereira, L.F.A.; Libano, F.B. Multiple-loop H-infinity control design for uninterruptible power supplies. *IEEE Trans. Ind. Electron.* **2007**, *54*, 1591–1602. [CrossRef]

9. Kawabata, T.; Miyashita, T.; Yamamoto, Y. Dead beat control of three phase PWM inverter. *IEEE Trans. Power Electron.* **1990**, *5*, 21–28. [CrossRef]

10. Ito, Y.; Kawauchi, S. Microprocessor based robust digital control for UPS with three-phase PWM inverter. *IEEE Trans. Power Electron.* **1995**, *10*, 196–204. [CrossRef]

11. Cho, J.-S.; Lee, S.-Y.; Mok, H.-S.; Choe, G.-H. Modified deadbeat digital controller for UPS with 3-phase PWM inverter. In Proceedings of the Thirty-Fourth IAS Annual Meeting, Conference Record of the 1999 IEEE Industry Applications Conference, Phoenix, AZ, USA, 3–7 October 1999; Volume 4.

12. Lee, T.S.; Tzeng, K.S.; Chong, M.S. Robust controller design for a single-phase UPS inverter using μ-synthesis. *IEE Proc. Electr. Power Appl.* **2004**, *151*, 334–340. [CrossRef]

13. Tahir, S.; Wang, J.; Kaloi, S.G.; Hussain, M. Robust digital deadbeat control design technique for 3 phase VSI with disturbance observer. *IEICE Electron. Express* **2017**, *14*, 20170351. [CrossRef]

14. Jung, J.-W.; Vu, N.T.-T.; Dang, D.Q.; Do, T.D.; Choi, Y.S.; Choi, H.H. A three-phase inverter for a standalone distributed generation system: Adaptive voltage control design and stability analysis. *IEEE Trans. Energy Convers.* **2014**, *29*, 46–56. [CrossRef]

15. Bogosyan, S. Recent advances in renewable energy employment. *IEEE Ind. Electron. Mag.* **2009**, *3*, 54–55. [CrossRef]

16. Liserre, M.; Sauter, T.; Hung, J.H. Future energy systems: Integrating renewable energy sources into the smart power grid through industrial electronics. *IEEE Ind. Electron. Mag.* **2010**, *4*, 18–37. [CrossRef]

17. Kim, M.-Y.; Song, Y.-U.; Kim, K.-H. The advanced voltage regulation method for ULTC in distribution systems with DG. *J. Electr. Eng. Technol.* **2013**, *8*, 737–743. [CrossRef]

18. Mokhtarpour, A.; Shayanfar, H.; Bathaee, S.M.T.; Banaei, M.R. Control of a single phase unified power quality conditioner-distributed generation-based input output feedback linearization. *J. Electr. Eng. Technol.* **2013**, *8*, 1352–1364. [CrossRef]

19. He, J.; Li, Y.W. An enhanced microgrid load demand sharing strategy. *IEEE Trans. Power Electron.* **2012**, *27*, 3984–3995. [CrossRef]

20. Marwali, M.N.; Jung, J.-W.; Keyhani, A. Stability analysis of load sharing control for distributed generation systems. *IEEE Trans. Energy Convers.* **2007**, *22*, 737–745. [CrossRef]

21. Zhang, Y.; Yu, M.; Liu, F.; Kang, Y. Instantaneous current-sharing control strategy for parallel operation of UPS modules using virtual impedance. *IEEE Trans. Power Electron.* **2013**, *28*, 432–440. [CrossRef]

22. Vechiu, I.; Curea, O.; Camblong, H. Transient operation of a four-leg inverter for autonomous applications with unbalanced load. *IEEE Trans. Power Electron.* **2010**, *25*, 399–407. [CrossRef]

23. Kasal, G.K.; Singh, B. Voltage and frequency controllers for an asynchronous generator-based isolated wind energy conversion system. *IEEE Trans. Energy Convers.* **2011**, *26*, 402–416. [CrossRef]

24. Nian, H.; Zeng, R. Improved control strategy for stand-alone distributed generation system under unbalanced and non-linear loads. *IET Renew. Power Gener.* **2011**, *5*, 323–331. [CrossRef]

25. Karimi, H.; Nikkhajoei, H.; Iravani, R. Control of an electronically-coupled distributed resource unit subsequent to an islanding event. *IEEE Trans. Power Deliv.* **2008**, *23*, 493–501. [CrossRef]

26. Karimi, H.; Yazdani, A.; Iravani, R. Robust control of an autonomous four-wire electronically-coupled distributed generation unit. *IEEE Trans. Power Deliv.* **2011**, *26*, 455–466. [CrossRef]

27. Escobar, G.; Valdez, A.A.; Leyva-Ramos, J.; Mattavelli, P. Repetitive-based controller for a UPS inverter to compensate unbalance and harmonic distortion. *IEEE Trans. Ind. Electron.* **2007**, *54*, 504–510. [CrossRef]

28. Yazdani, A. Control of an islanded distributed energy resource unit with load compensating feed-forward. In Proceedings of the 2008 IEEE Power and Energy Society General Meeting-Conversion and Delivery of Electrical Energy in the 21st Century, Pittsburgh, PA, USA, 20–24 July 2008.

29. Dasgupta, S.; Sahoo, S.K.; Panda, S.K. Single-phase inverter control techniques for interfacing renewable energy sources with microgrid—Part I: Parallel-connected inverter topology with active and reactive power flow control along with grid current shaping. *IEEE Trans. Power Electron.* **2011**, *26*, 717–731. [CrossRef]

30. Dai, M.; Marwali, M.N.; Jung, J.-W.; Keyhani, A. A three-phase four-wire inverter control technique for a single distributed generation unit in island mode. *IEEE Trans. Power Electron.* **2008**, *23*, 322–331. [CrossRef]

31. Delghavi, M.B.; Yazdani, A.N. Islanded-mode control of electronically coupled distributed-resource units under unbalanced and nonlinear load conditions. *IEEE Trans. Power Deliv.* **2011**, *26*, 661–673. [CrossRef]

32. Delghavi, M.B.; Yazdani, A.N. An adaptive feedforward compensation for stability enhancement in droop-controlled inverter-based microgrids. *IEEE Trans. Power Deliv.* **2011**, *26*, 1764–1773. [CrossRef]

33. Prodanovic, M.; Timothy, C.; Green, T.C. Control and filter design of three-phase inverters for high power quality grid connection. *IEEE Trans. Power Electron.* **2003**, *18*, 373–380. [CrossRef]

34. Mattavelli, P.; Escobar, G.; Stankovic, A.M. Dissipativity-based adaptive and robust control of UPS. *IEEE Trans. Ind. Electron.* **2001**, *48*, 334–343. [CrossRef]
35. Valderrama, G.E.; Stankovic, A.M.; Mattavelli, P. Dissipativity-based adaptive and robust control of UPS in unbalanced operation. *IEEE Trans. Power Electron.* **2003**, *18*, 1056–1062. [CrossRef]
36. Escobar, G.; Mattavelli, P.; Stankovic, A.M.; Valdez, A.A.; Leyva-Ramos, J. An adaptive control for UPS to compensate unbalance and harmonic distortion using a combined capacitor/load current sensing. *IEEE Trans. Ind. Electron.* **2007**, *54*, 839–847. [CrossRef]
37. Do, T.D.; Leu, V.Q.; Choi, Y.-S.; Choi, H.H.; Jung, J.-W. An adaptive voltage control strategy of three-phase inverter for stand-alone distributed generation systems. *IEEE Trans. Ind. Electron.* **2013**, *60*, 5660–5672. [CrossRef]
38. Rasheduzzaman, M.; Mueller, J.; Kimball, J.W. Small-signal modeling of a three-phase isolated inverter with both voltage and frequency droop control. In Proceedings of the 2014 Twenty-Ninth Annual IEEE Applied Power Electronics Conference and Exposition (APEC), Fort Worth, TX, USA, 16–20 March 2014.
39. Rodriguez, J.; Lai, J.-S.; Peng, F.Z. Multilevel inverters: A survey of topologies, controls, and applications. *IEEE Trans. Ind. Electron.* **2002**, *49*, 724–738. [CrossRef]
40. Nabae, A.; Takahashi, I.; Akagi, H. A new neutral-point-clamped PWM inverter. *IEEE Trans. Ind. Appl.* **1981**, *5*, 518–523. [CrossRef]
41. Franquelo, L.G.; Rodriguez, J.; Leon, J.I.; Kouro, S.; Portillo, R.; Prats, M.A.M. The age of multilevel converters arrives. *IEEE Ind. Electron. Mag.* **2008**, *2*, 28–39. [CrossRef]
42. Lai, J.-S.; Peng, F.Z. Multilevel converters—A new breed of power converters. *IEEE Trans. Ind. Appl.* **1996**, *32*, 509–517.
43. Patel, H.S.; Hoft, R.G. Generalized techniques of harmonic elimination and voltage control in thyristor inverters: Part I—Harmonic elimination. *IEEE Trans. Ind. Appl.* **1973**, *3*, 310–317. [CrossRef]
44. Khomfoi, S.; Tolbert, L.M. Multilevel power converters. In *Power Electronics Handbook*; Elsevier/Academic Press: Burlington, VT, USA, 2007; pp. 451–482.
45. Zhang, R.; Prasad, V.H.; Boroyevich, D.; Lee, F.C. Three-dimensional space vector modulation for four-leg voltage-source converters. *IEEE Trans. Power Electron.* **2002**, *17*, 314–326. [CrossRef]
46. Lohia, P.; Mishra, M.K.; Karthikeyan, K.; Vasudevan, K. A minimally switched control algorithm forthree-phase four-leg VSI topology tocompensate unbalanced and nonlinear load. *IEEE Trans. Power Electron.* **2008**, *23*, 1935–1944. [CrossRef]
47. Zhong, Q.-C.; Liang, J.; Weiss, G.; Feng, C.M.; Green, T.C. H∞ Control of the Neutral Point in Four-Wire Three-Phase DC–AC Converters. *IEEE Trans. Ind. Electron.* **2006**, *53*, 1594–1602. [CrossRef]
48. Liang, J.; Green, T.C.; Feng, C.; Weiss, C. Increasing voltage utilization in split-link, four-wire inverters. *IEEE Trans. Power Electron.* **2009**, *24*, 1562–1569. [CrossRef]
49. Miret, J.; Camacho, A.; Castilla, M.; de Vicuña, L.G.; Matas, J. Control scheme with voltage support capability for distributed generation inverters under voltage sags. *IEEE Trans. Power Electron.* **2013**, *28*, 5252–5262. [CrossRef]
50. Liu, Z.; Liu, J.; Zhao, Y. A unified control strategy for three-phase inverter in distributed generation. *IEEE Trans. Power Electron.* **2014**, *29*, 1176–1191. [CrossRef]
51. Li, Y.; Jiang, S.; Cintron-Rivera, J.G.; Peng, F.Z. Modeling and control of quasi-Z-source inverter for distributed generation applications. *IEEE Trans. Ind. Electron.* **2013**, *60*, 1532–1541. [CrossRef]
52. Ebadi, M.; Joorabian, M.; Moghani, J.S. Voltage look-up table method to control multilevel cascaded transformerless inverters with unequal DC rail voltages. *IET Power Electron.* **2014**, *7*, 2300–2309. [CrossRef]
53. Eren, S.; Pahlevani, M.; Bakhshai, A.; Jain, P. An adaptive droop DC-bus voltage controller for a grid-connected voltage source inverter with LCL filter. *IEEE Trans. Power Electron.* **2015**, *30*, 547–560. [CrossRef]
54. Abu-Rub, H.; Guzinski, J.; Krzeminski, Z.; Toliyat, H.A. Predictive current control of voltage-source inverters. *IEEE Trans. Ind. Electron.* **2004**, *51*, 585–593. [CrossRef]
55. Espi, J.M.; Castello, J.; Garcia-Gil, R.; Garcera, G.; Figueres, E. An adaptive robust predictive current control for three-phase grid-connected inverters. *IEEE Trans. Ind. Electron.* **2011**, *58*, 3537–3546. [CrossRef]
56. Moreno, J.C.; Espi Huerta, J.M.; Gil, R.G.; Gonzalez, S.A. A robust predictive current control for three-phase grid-connected inverters. *IEEE Trans. Ind. Electron.* **2009**, *56*, 1993–2004. [CrossRef]

57. Ahmed, K.H.; Massoud, A.M.; Finney, S.J.; Williams, B.W. A modified stationary reference frame-based predictive current control with zero steady-state error for LCL coupled inverter-based distributed generation systems. *IEEE Trans. Ind. Electron.* **2011**, *58*, 1359–1370. [CrossRef]

58. Miveh, M.R.; Rahmat, M.F.; Ghadimi, A.A.; Mustafa, M.W. Control techniques for three-phase four-leg voltage source inverters in autonomous microgrids: A review. *Renew. Sustain. Energy Rev.* **2016**, *54*, 1592–1610. [CrossRef]

59. Chen, H. Research on the control strategy of VSC based HVDC system supplying passive network. In Proceedings of the 2009 PES'09 IEEE Power & Energy Society General Meeting, Calgary, AB, Canada, 26–30 July 2009.

60. Qian, C.; Tang, G.; Hu, M. Steady-state model and controller design of a VSC-HVDC converter based on dq0-axis. *Autom. Electr. Power Syst.* **2004**, *16*, 015.

61. Hoffmann, N.; Fuchs, F.W.; Dannehl, J. Models and effects of different updating and sampling concepts to the control of grid-connected PWM converters—A study based on discrete time domain analysis. In Proceedings of the 2011 14th European Conference on Power Electronics and Applications (EPE 2011), Birmingham, UK, 30 August–1 September 2011.

62. Büyük, M.; Tan, A.; Tümay, M.; Bayındır, K.Ç. Topologies, generalized designs, passive and active damping methods of switching ripple filters for voltage source inverter: A comprehensive review. *Renew. Sustain. Energy Rev.* **2016**, *62*, 46–69. [CrossRef]

63. Peña-Alzola, R.; Liserre, M.; Blaabjerg, F.; Ordonez, M.; Yang, Y. LCL-filter design for robust active damping in grid-connected converters. *IEEE Trans. Ind. Inform.* **2014**, *10*, 2192–2203. [CrossRef]

64. Zhang, C.; Dragicevic, T.; Vasquez, J.C.; Guerrero, J.M. Resonance damping techniques for grid-connected voltage source converters with LCL filters—A review. In Proceedings of the 2014 IEEE International Energy Conference (ENERGYCON), Cavtat, Croatia, 13–16 May 2014; pp. 169–176.

65. Jia, Y.; Zhao, J.; Fu, X. Direct grid current control of LCL-filtered grid-connected inverter mitigating grid voltage disturbance. *IEEE Trans. Power Electron.* **2014**, *29*, 1532–1541.

66. Teodorescu, R.; Liserre, M.; Rodriguez, P. *Grid Converters for Photovoltaic and Wind Power Systems*; John Wiley & Sons: Chichester, UK, 2011; Volume 29.

67. Jalili, K.; Bernet, S. Design of LCL filters of active-front-end two-level voltage-source converters. *IEEE Trans. Ind. Electron.* **2009**, *56*, 1674–1689. [CrossRef]

68. Kazmierkowski, M.P.; Malesani, L. Current control techniques for three-phase voltage-source PWM converters: A survey. *IEEE Trans. Ind. Electron.* **1998**, *45*, 691–703. [CrossRef]

69. Kumar, K.V.; Michael, P.A.; John, J.P.; Kumar, S.S. Simulation and comparison of SPWM and SVPWM control for three phase inverter. *ARPN J. Eng. Appl. Sci.* **2010**, *5*, 61–74.

70. Lim, J.S.; Park, C.; Han, J.; Lee, Y.I. Robust tracking control of a three-phase DC–AC inverter for UPS applications. *IEEE Trans. Ind. Electron.* **2014**, *61*, 4142–4151. [CrossRef]

71. Huerta, J.M.E.; Castello, J.; Fischer, J.R.; García-Gil, R. A synchronous reference frame robust predictive current control for three-phase grid-connected inverters. *IEEE Trans. Ind. Electron.* **2010**, *57*, 954–962. [CrossRef]

72. Samui, A.; Samantaray, S.R. New active islanding detection scheme for constant power and constant current controlled inverter-based distributed generation. *IET Gener. Transm. Distrib.* **2013**, *7*, 779–789. [CrossRef]

73. Yuan, X.; Merk, W.; Stemmler, H.; Allmeling, J. Stationary-frame generalized integrators for current control of active power filters with zero steady-state error for current harmonics of concern under unbalanced and distorted operating conditions. *IEEE Trans. Ind. Appl.* **2002**, *38*, 523–532. [CrossRef]

74. Miret, J.; Castilla, M.; Matas, J.; Guerrero, J.M.; Vasquez, J.C. Selective harmonic-compensation control for single-phase active power filter with high harmonic rejection. *IEEE Trans. Ind. Electron.* **2009**, *56*, 3117–3127. [CrossRef]

75. Beza, M.; Bongiorno, M. Improved discrete current controller for grid-connected voltage source converters in distorted grids. In Proceedings of the 2012 IEEE Energy Conversion Congress and Exposition (ECCE), Raleigh, NC, USA, 15–20 September 2012.

76. Kandil, M.S.; El-Saadawi, M.M.; Hassan, A.E.; Abo-Al-Ez, K.M. A proposed reactive power controller for DG grid connected systems. In Proceedings of the 2010 IEEE International Energy Conference and Exhibition, Manama, Bahrain, 18–22 December 2010; pp. 446–451.

77. Radwan, A.A.A.; Abdel-Rady, I.M.Y. Power Synchronization Control for Grid-Connected Current-Source Inverter-Based Photovoltaic Systems. *IEEE Trans. Energy Convers.* **2016**, *31*, 1023–1036. [CrossRef]

78. Chilipi, R.; Sayari, N.A.; Hosani, K.A.; Beig, A.R. Control scheme for grid-tied distributed generation inverter under unbalanced and distorted utility conditions with power quality ancillary services. *IET Renew. Power Gener.* **2016**, *10*, 140–149. [CrossRef]

79. Busada, C.; Jorge, S.G.; Leon, A.E.; Solsona, J. Phase-locked loop-less current controller for grid-connected photovoltaic systems. *IET Renew. Power Gener.* **2012**, *6*, 400–407. [CrossRef]

80. Athans, M. The role and use of the stochastic linear-quadratic-Gaussian problem in control system design. *IEEE Trans. Autom. Control* **1971**, *16*, 529–552. [CrossRef]

81. Huerta, F.; Pizarro, D.; Cobreces, S.; Rodriguez, F.J.; Giron, C.; Rodriguez, A. LQG servo controller for the current control of *LCL* grid-connected voltage-source converters. *IEEE Trans. Ind. Electron.* **2012**, *59*, 4272–4284. [CrossRef]

82. Hossain, M.A.; Azim, M.I.; Mahmud, M.A.; Pota, H.R. Primary voltage control of a single-phase inverter using linear quadratic regulator with integrator. In Proceedings of the 2015 Australasian Universities Power Engineering Conference (AUPEC), Wollongong, Australia, 27–30 September 2015; pp. 1–6.

83. Ahmed, K.H.; Massoud, A.M.; Finney, S.J.; Williams, B.W. Optimum selection of state feedback variables PWM inverters control. In Proceedings of the IET Conference on Power Electronics, Machines and Drives, York, UK, 2–4 April 2008; pp. 125–129.

84. Xue, M.; Zhang, Y.; Kang, Y.; Yi, Y.; Li, S.; Liu, F. Full feedforward of grid voltage for discrete state feedback controlled grid-connected inverter with LCL filter. *IEEE Trans. Power Electron.* **2012**, *27*, 4234–4247. [CrossRef]

85. Lalili, D.; Mellit, A.; Lourci, N.; Medjahed, B.; Boubakir, C. State feedback control of a three-level grid-connected photovoltaic inverter. In Proceedings of the 2012 9th International Multi-Conference on Systems, Signals and Devices (SSD), Chemnitz, Germany, 20–23 March 2012; pp. 1–6.

86. Jaen, C.; Pou, J.; Pindado, R.; Sala, V.; Zaragoza, J. A linear-quadratic regulator with integral action applied to PWM DC-DC converters. In Proceedings of the IECON 2006-32nd Annual Conference on IEEE Industrial Electronics, Paris, France, 6–10 November 2006; pp. 2280–2285.

87. Bose, B.K. *Power Electronics and Motor Drives: Advances and Trends*; Academic Press: Oxford, UK, 2010.

88. Kaźmierkowski, M.P.; Krishnan, R.; Blaabjerg, F. (Eds.) *Control in Power Electronics: Selected Problems*; Academic Press: New York, NY, USA, 2002.

89. Shukla, A.; Ghosh, A.; Joshi, A. Hysteresis modulation of multilevel inverters. *IEEE Trans. Power Electron.* **2011**, *26*, 1396–1409. [CrossRef]

90. Prabhakar, N.; Mishra, M.K. Dynamic hysteresis current control to minimize switching for three-phase four-leg VSI topology to compensate nonlinear load. *IEEE Trans. Power Electron.* **2010**, *25*, 1935–1942. [CrossRef]

91. Zhang, X.; Wang, J.; Li, C. Three-phase four-leg inverter based on voltage hysteresis control. In Proceedings of the 2010 International Conference on Electrical and Control Engineering (ICECE), Wuhan, China, 25–27 June 2010.

92. Verdelho, P.; Marques, G.D. Four-wire current-regulated PWM voltage converter. *IEEE Trans. Ind. Electron.* **1998**, *45*, 761–770. [CrossRef]

93. Ali, S.M.; Kazmierkowski, M.P. Current regulation of four-leg PWM/VSI. In Proceedings of the 1998 IECON'98 24th Annual Conference of the IEEE Industrial Electronics Society, Aachen, Germany, 31 August–4 September 1998; Volume 3.

94. Rodriguez, J.; Cortes, P. *Predictive Control of Power Converters and Electrical Drives*; John Wiley & Sons: Chichester, UK, 2012; Volume 40.

95. Wu, B.; Narimani, M. *High-Power Converters and AC Drives*; John Wiley & Sons: Hoboken, NJ, USA, 2017.

96. Martins, C.A.; Roboam, X.; Meynard, T.A.; Carvalho, A.S. Switching frequency imposition and ripple reduction in DTC drives by using a multilevel converter. *IEEE Trans. Power Electron.* **2002**, *17*, 286–297. [CrossRef]

97. Hu, J.; Zhu, Z.Q. Investigation on switching patterns of direct power control strategies for grid-connected DC–AC converters based on power variation rates. *IEEE Trans. Power Electron.* **2011**, *26*, 3582–3598. [CrossRef]

98. Bouafia, A.; Gaubert, J.-P.; Krim, F. Predictive direct power control of three-phase pulsewidth modulation (PWM) rectifier using space-vector modulation (SVM). *IEEE Trans. Power Electron.* **2010**, *25*, 228–236. [CrossRef]

99. Kazmierkowski, M.P.; Jasinski, M.; Wrona, G. DSP-based control of grid-connected power converters operating under grid distortions. *IEEE Trans. Ind. Inform.* **2011**, *7*, 204–211. [CrossRef]

100. Hung, J.Y.; Gao, W.; Hung, J.C. Variable structure control: A survey. *IEEE Trans. Ind. Electron.* **1993**, *40*, 2–22. [CrossRef]

101. Tsang, K.M.; Chan, W.L. Adaptive control of power factor correction converter using nonlinear system identification. *IEE Proc. Electr. Power Appl.* **2005**, *152*, 627–633. [CrossRef]

102. Massing, J.R.; Stefanello, M.; Grundling, H.A.; Pinheiro, H. Adaptive current control for grid-connected converters with LCL filter. *IEEE Trans. Ind. Electron.* **2012**, *59*, 4681–4693. [CrossRef]

103. Herran, M.A.; Fischer, J.R.; Gonzalez, S.A.; Judewicz, M.G.; Carrica, D.O. Adaptive dead-time compensation for grid-connected PWM inverters of single-stage PV systems. *IEEE Trans. Power Electron.* **2013**, *28*, 2816–2825. [CrossRef]

104. Mohamed, Y.A.-R.I. Mitigation of converter-grid resonance, grid-induced distortion, and parametric instabilities in converter-based distributed generation. *IEEE Trans. Power Electron.* **2011**, *26*, 983–996. [CrossRef]

105. Athari, H.; Niroomand, M.; Ataei, M. Review and Classification of Control Systems in Grid-tied Inverters. *Renew. Sustain. Energy Rev.* **2017**, *72*, 1167–1176. [CrossRef]

106. Niroomand, M.; Karshenas, H.R. Hybrid learning control strategy for three-phase uninterruptible power supply. *IET Power Electron.* **2011**, *4*, 799–807. [CrossRef]

107. Mahmud, M.A.; Hossain, M.J.; Pota, H.R.; Roy, N.K. Robust nonlinear controller design for three-phase grid-connected photovoltaic systems under structured uncertainties. *IEEE Trans. Power Deliv.* **2014**, *29*, 1221–1230. [CrossRef]

108. Baloch, M.H.; Wang, J.; Kaloi, G.S. Dynamic Modeling and Control of Wind Turbine Scheme Based on Cage Generator for Power System Stability Studies. *Int. J. Renew. Energy Res.* **2016**, *6*, 599–606.

109. Baloch, M.H.; Wang, J.; Kaloi, G.S. Stability and nonlinear controller analysis of wind energy conversion system with random wind speed. *Int. J. Electr. Power Energy Syst.* **2016**, *79*, 75–83. [CrossRef]

110. Baloch, M.H.; Wang, J.; Kaloi, G.S. A Review of the State of the Art Control Techniques for Wind Energy Conversion System. *Int. J. Renew. Energy Res.* **2016**, *6*, 1276–1295.

111. Hornik, T.; Zhong, Q.-C. A Current-Control Strategy for Voltage-Source Inverters in Microgrids Based on H^∞ and Repetitive Control. *IEEE Trans. Power Electron.* **2011**, *26*, 943–952. [CrossRef]

112. Hornik, T.; Zhong, Q.-C. H∞ repetitive current controller for grid-connected inverters. In Proceedings of the 2009 IECON'09 35th Annual Conference of IEEE Industrial Electronics, Porto, Portugal, 3–5 November 2009.

113. Guo, Q.; Wang, J.; Ma, H. Frequency adaptive repetitive controller for grid-connected inverter with an all-pass infinite impulse response (IIR) filter. In Proceedings of the 2014 IEEE 23rd International Symposium on Industrial Electronics (ISIE), Istanbul, Turkey, 1–4 June 2014.

114. Bose, B.K. *Modern Power Electronics and AC Drives*; Bose, B.K., Ed.; Prentice Hall PTR: Upper Saddle River, NJ, USA, 2002.

115. Cirstea, M.; Dinu, A.; McCormick, M.; Khor, J.G. *Neural and Fuzzy Logic Control of Drives and Power Systems*; Elsevier: Oxford, UK, 2002.

116. Vas, P. *Artificial-Intelligence-Based Electrical Machines and Drives: Application of Fuzzy, Neural, Fuzzy-Neural, and Genetic-Algorithm-Based Techniques*; Oxford University Press: New York, NY, USA, 1999; Volume 45.

117. Damen, A.; Weiland, S. *Robust Control*; Measurement and Control Group Department of Electrical Engineering, Eindhoven University of Technology: Eindhoven, the Netherlands, 2002.

118. Zames, G. Feedback and optimal sensitivity: Model reference transformations, multiplicative seminorms, and approximate inverses. *IEEE Trans. Autom. Control* **1981**, *26*, 301–320. [CrossRef]

119. Yang, S.; Lei, Q.; Peng, F.Z.; Qian, Z. A robust control scheme for grid-connected voltage-source inverters. *IEEE Trans. Ind. Electron.* **2011**, *58*, 202–212. [CrossRef]

120. Chhabra, M.; Barnes, F. Robust current controller design using mu-synthesis for grid-connected three phase inverter. In Proceedings of the 2014 IEEE 40th Photovoltaic Specialist Conference (PVSC), Denver, CO, USA, 8–13 June 2014.

121. Chen, S.; Malik, O.P. Power system stabilizer design using/SPL MU/synthesis. *IEEE Trans. Energy Convers.* **1995**, *10*, 175–181. [CrossRef]

122. Mascioli, M.; Pahlevani, M.; Jain, P.K. Frequency-adaptive current controller for grid-connected renewable energy systems. In Proceedings of the 2014 IEEE 36th International Telecommunications Energy Conference (INTELEC), Vancouver, BC, Canada, 28 September–2 October 2014.

123. Jorge, S.G.; Busada, C.A.; Solsona, J.A. Frequency-adaptive current controller for three-phase grid-connected converters. *IEEE Trans. Ind. Electron.* **2013**, *60*, 4169–4177. [CrossRef]

124. Timbus, A.V.; Ciobotaru, M.; Teodorescu, R.; Blaabjerg, F. Adaptive resonant controller for grid-connected converters in distributed power generation systems. In Proceedings of the 2006 APEC'06. Twenty-First Annual IEEE Applied Power Electronics Conference and Exposition, Dallas, TX, USA, 19–23 March 2006; p. 6.

125. Zeng, Q.; Chang, L. Improved current controller based on SVPWM for three-phase grid-connected voltage source inverters. In Proceedings of the 2005 PESC'05 36th IEEE Power Electronics Specialists Conference, Recife, Brazil, 16 June 2005.

126. Ouchen, S.; Betka, A.; Abdeddaim, S.; Menadi, A. Fuzzy-predictive direct power control implementation of a grid connected photovoltaic system, associated with an active power filter. *Energy Convers. Manag.* **2016**, *122*, 515–525. [CrossRef]

127. Ouchen, S.; Abdeddaim, S.; Betka, A.; Menadi, A. Experimental validation of sliding mode-predictive direct power control of a grid connected photovoltaic system, feeding a nonlinear load. *Sol. Energy* **2016**, *137*, 328–336. [CrossRef]

128. Mohamed, Y.A.-R.I.; El-Saadany, E.F. An improved deadbeat current control scheme with a novel adaptive self-tuning load model for a three-phase PWM voltage-source inverter. *IEEE Trans. Ind. Electron.* **2007**, *54*, 747–759. [CrossRef]

129. Bode, G.H.; Loh, P.C.; Newman, M.J.; Holmes, D.G. An improved robust predictive current regulation algorithm. *IEEE Trans. Ind. Appl.* **2005**, *41*, 1720–1733. [CrossRef]

130. Zeng, Q.; Chang, L. An advanced SVPWM-based predictive current controller for three-phase inverters in distributed generation systems. *IEEE Trans. Ind. Electron.* **2008**, *55*, 1235–1246. [CrossRef]

131. Mattavelli, P. An improved deadbeat control for UPS using disturbance observers. *IEEE Trans. Ind. Electron.* **2005**, *52*, 206–212. [CrossRef]

132. Kim, J.; Hong, J.; Kim, H. Improved Direct Deadbeat Voltage Control with an Actively Damped Inductor-Capacitor Plant Model in an Islanded AC Microgrid. *Energies* **2016**, *9*, 978. [CrossRef]

133. Zhang, X.; Zhang, W.; Chen, J.; Xu, D. Deadbeat control strategy of circulating currents in parallel connection system of three-phase PWM converter. *IEEE Trans. Energy Convers.* **2014**, *29*, 406–417.

134. Timbus, A.; Liserre, M.; Teodorescu, R.; Rodriguez, P.; Blaabjerg, F. Evaluation of current controllers for distributed power generation systems. *IEEE Trans. Power Electron.* **2009**, *24*, 654–664. [CrossRef]

135. Song, W.; Ma, J.; Zhou, L.; Feng, X. Deadbeat predictive power control of single-phase three-level neutral-point-clamped converters using space-vector modulation for electric railway traction. *IEEE Trans. Power Electron.* **2016**, *31*, 721–732. [CrossRef]

136. Hu, J.; Zhu, J.; Dorrell, D.G. Model predictive control of grid-connected inverters for PV systems with flexible power regulation and switching frequency reduction. *IEEE Trans. Ind. Appl.* **2015**, *51*, 587–594. [CrossRef]

137. Cortés, P.; Kazmierkowski, M.P.; Kennel, R.M.; Quevedo, D.E.; Rodríguez, J. Predictive control in power electronics and drives. *IEEE Trans. Ind. Electron.* **2008**, *55*, 4312–4324. [CrossRef]

138. Mariéthoz, S.; Morari, M. Explicit model-predictive control of a PWM inverter with an LCL filter. *IEEE Trans. Ind. Electron.* **2009**, *56*, 389–399. [CrossRef]

139. Sosa, J.M.; Martinez-Rodriguez, P.R.; Vazquez, G.; Serrano, J.P.; Escobar, G.; Valdez-Fernandez, A.A. Model based controller for an LCL coupling filter for transformerless grid connected inverters in PV applications. In Proceedings of the IECON 2013 39th Annual Conference of the IEEE Industrial Electronics Society, Vienna, Austria, 10–13 November 2013; pp. 1723–1728.

140. Tan, K.T.; So, P.L.; Chu, Y.C.; Chen, M.Z.Q. Coordinated control and energy management of distributed generation inverters in a microgrid. *IEEE Trans. Power Deliv.* **2013**, *28*, 704–713. [CrossRef]

141. Rodriguez, J.; Pontt, J.; Silva, C.A.; Correa, P.; Lezana, P.; Cortés, P.; Ammann, U. Predictive current control of a voltage source inverter. *IEEE Trans. Ind. Electron.* **2007**, *54*, 495–503. [CrossRef]

142. Tan, K.T.; Peng, X.Y.; So, P.L.; Chu, Y.C.; Chen, M.Z.Q. Centralized control for parallel operation of distributed generation inverters in microgrids. *IEEE Trans. Smart Grid* **2012**, *3*, 1977–1987. [CrossRef]

143. Ayad, A.F.; Kennel, R.M. Model predictive controller for grid-connected photovoltaic based on quasi-Z-source inverter. In Proceedings of the 2013 IEEE International Symposium on Sensorless Control for Electrical Drives

and Predictive Control of Electrical Drives and Power Electronics (SLED/PRECEDE), München, Germany, 17–19 October 2013.

144. Trabelsi, M.; Ghazi, K.A.; Al-Emadi, N.; Ben-Brahim, L. An original controller design for a grid connected PV system. In Proceedings of the IECON 2012 38th Annual Conference on IEEE Industrial Electronics Society, Montreal, QC, Canada, 25–28 Octorber 2012; pp. 924–929.

145. Lee, K.-J.; Park, B.-C.; Kim, R.-Y.; Hyun, D.-S. Robust predictive current controller based on a disturbance estimator in a three-phase grid-connected inverter. *IEEE Trans. Power Electron.* **2012**, *27*, 276–283. [CrossRef]

146. Krishna, R.; Kottayil, S.K.; Leijon, M. Predictive current controller for a grid connected three level inverter with reactive power control. In Proceedings of the 2010 IEEE 12th Workshop on Control and Modeling for Power Electronics (COMPEL), Boulder, CO, USA, 28–30 June 2010.

147. Sathiyanarayanan, T.; Mishra, S. Synchronous reference frame theory based model predictive control for grid connected photovoltaic systems. *IFAC-PapersOnLine* **2016**, *49*, 766–771. [CrossRef]

148. Zeng, Z.; Yang, H.; Zhao, R.; Cheng, C. Topologies and control strategies of multi-functional grid-connected inverters for power quality enhancement: A comprehensive review. *Renew. Sustain. Energy Rev.* **2013**, *24*, 223–270. [CrossRef]

149. Patel, D.C.; Sawant, R.R.; Chandorkar, M.C. Three-dimensional flux vector modulation of four-leg sine-wave output inverters. *IEEE Trans. Ind. Electron.* **2010**, *57*, 1261–1269. [CrossRef]

150. Andishgar, M.H.; Gholipour, E.; Hooshmand, R.-A. An overview of control approaches of inverter-based microgrids in islanding mode of operation. *Renew. Sustain. Energy Rev.* **2017**, *80*, 1043–1060. [CrossRef]

151. Debnath, S.; Qin, J.; Bahrani, B.; Saeedifard, M.; Barbosa, P. Operation, control, and applications of the modular multilevel converter: A review. *IEEE Trans. Power Electron.* **2015**, *30*, 37–53. [CrossRef]

152. Mahmud, M.A.; Pota, H.R.; Hossain, M.J. Nonlinear controller design for single-phase grid-connected photovoltaic systems using partial feedback linearization. In Proceedings of the 2012 2nd Australian Control Conference (AUCC), Sydney, Australia, 15–16 November 2012.

153. Vas, P.; Stronach, A.F.; Neuroth, M. DSP-controlled intelligent high-performance ac drives present and future. In Proceedings of the IEE Colloquium on Vector Control and Direct Torque Control of Induction Motors, London, UK, 27 October 1995.

electronics

MDPI

Article

SHIL and DHIL Simulations of Nonlinear Control Methods Applied for Power Converters Using Embedded Systems

Arthur H. R. Rosa *, Matheus B. E. Silva, Marcos F. C. Campos, Renato A. S. Santana, Welbert A. Rodrigues, Lenin M. F. Morais and Seleme I. Seleme Jr.

Graduate Program in Electrical Engineering, Universidade Federal de Minas Gerais, Av. Antônio Carlos 6627, Belo Horizonte 31270-901, MG, Brazil; matbeiras@gmail.com (M.B.E.S.); engemarcoscampos@gmail.com (M.F.C.C.); rass.eletrica@gmail.com (R.A.S.S.); welbertalves@gmail.com (W.A.R.); lenin@cpdee.ufmg.br (L.M.F.M.); seleme@cpdee.ufmg.br (S.I.S.J.)
* Correspondence: arthurcpdee@gmail.com

Received: 17 August 2018; Accepted: 30 September 2018; Published: 6 October 2018

Abstract: In this work, a new real-time Simulation method is designed for nonlinear control techniques applied to power converters. We propose two different implementations: in the first one (Single Hardware in The Loop: SHIL), both model and control laws are inserted in the same Digital Signal Processor (DSP), and in the second approach (Double Hardware in The Loop: DHIL), the equations are loaded in different embedded systems. With this methodology, linear and nonlinear control techniques can be designed and compared in a quick and cheap real-time realization of the proposed systems, ideal for both students and engineers who are interested in learning and validating converters performance. The methodology can be applied to buck, boost, buck-boost, flyback, SEPIC and 3-phase AC-DC boost converters showing that the new and high performance embedded systems can evaluate distinct nonlinear controllers. The approach is done using matlab-simulink over commodity Texas Instruments Digital Signal Processors (TI-DSPs). The main purpose is to demonstrate the feasibility of proposed real-time implementations without using expensive HIL systems such as Opal-RT and Typhoon-HL.

Keywords: real-time simulation; power converters; nonlinear control; embedded systems; high level programing; SHIL; DHIL

1. Introduction

The rapid advance of digital and embedded systems has enabled the use of such systems in different applications [1]. Although still little explored, one of these utilities includes Hardware in The Loop (HIL) Simulations, in which both software and hardware are tested.

Real-time simulation (RTS) methods can be a feasible way to verify controllers performance and stability of dynamic systems. Commercial platforms, such as OPAL-RT Technologies Inc. (Montreal, QC, Canada), that implemented sophisticate and expensive test bench, are widely available [2]. Examples of Digital real-time simulator (DRTS) with applications attaining high accuracy results are: TYPHON HIL [2], OPAL-RT [3], dSPACE [4] and RTDS [5].

On the other hand, a real-time simulation platform with less complexity than those previously mentioned may be desirable. On these terms, the employment of powerful computational devices does not justify increased costs. Along these lines, a Processor in the Loop (PIL) applying the SimCoder platform of PSIM (Power System Simulator) is designed in [6], where a F28335 Texas Instruments micro-controller is employed to embed a PFC (Power Factor Correction) and motor drive circuits via software simulation. Also, Ref. [7] presents a simple and interesting real-time implementation.

In view of these concerns, an RTS based method is proposed in order to verify the power converters dynamics and validate the stability of their implemented control equations. The approach is made in a way that justifies the required computational power needed to simulate the elementary converters in real-time, with lower cost and spent time. In the proposed Single Hardware in The Loop (SHIL), both control and state equations implemented on Matlab/Simulink development environment are directly embedded in a C2000 F28377 Texas Instruments device, through Simulink Coder and Embedded Coder packages. In Double Hardware in The Loop (DHIL), the equations are embedded in distinctive DSPs, as illustrated in Figure 1. In the first DSP, the converter model is embedded (usually described by state-space equations or switched model). In the second DSP, the equation of the duty cycle *d* is calculated for the input control of the switch.

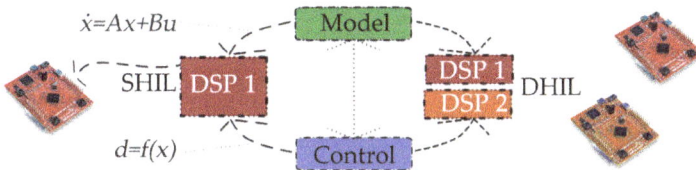

Figure 1. SHIL and DHIL.

As illustrated in Figure 1, the methodology of this work differs from the concepts found in the literature about Software in The Loop (SIL), Processor in The Loop (PIL) [7] and HIL [8] implementation. That is the reason we call it "Single Hardware In The Loop" (SHIL), since it contains hybrid characteristics of these methodologies. Given this, we can easily test the control law without needing a desktop computer and a real plant. In addition, the data transfer occurs directly and more quickly when both model and control are inserted in the same DSP or different DSPs.

The main objective of this work is to validate nonlinear control laws in embedded systems using the proposed real-time simulation methods. When dealing with unconventional control equations the following question appears: are these new methods feasible? To achieve this goal, it is not necessary to use complex models, since such models will be replaced by the real prototype. In fact, it is worth highlighting here that the controllers are usually performed by embedded systems in power electronics applications. This is the preponderant trend.

So, this work presents two different RTS approaches, where model and control equations are executed on DSP processors. The models and control equations are demonstrated in Section 2 and Appendix A. The proposed SHIL and DHIL simulation methods and their experimental results are explained in Section 3. Furthermore, the additional contribution of this work is the comparison of nonlinear control techniques (SFL, PBC and IDAPBC) applied to static power converters. As a whole, three converters models (Table 1) and nine control equations (Table 2) are validated using the proposed methods. Finally, results and conclusions are presented in Sections 4 and 5.

Table 1. Converters models.

	Boost	Buck	Buck-Boost
SS	$\dot{x}_1 = -(1-d)\frac{1}{L}x_2 + \frac{E}{L}$ $\dot{x}_2 = (1-d)\frac{1}{C}x_1 - \frac{G}{C}x_2$	$\dot{x}_1 = -\frac{1}{L}x_2 + d\frac{E}{L}$ $\dot{x}_2 = \frac{1}{C}x_1 - \frac{G}{C}x_2$	$\dot{x}_1 = (1-d)\frac{1}{L}x_2 + d\frac{E}{L}$ $\dot{x}_2 = -(1-d)\frac{1}{C}x_1 - \frac{G}{C}x_2$

EL: $x = \begin{bmatrix} x_1 \\ x_2 \end{bmatrix}; D_B = \begin{bmatrix} L & 0 \\ 0 & C \end{bmatrix}; R_B = \begin{bmatrix} 0 & 0 \\ 0 & G \end{bmatrix}; F = \begin{bmatrix} E \\ 0 \end{bmatrix}$

Boost	Buck	Buck-Boost
$D_B\dot{x} + (1-d)J_Bx + R_Bx = F$ $J_B = \begin{bmatrix} 0 & 1 \\ -1 & 0 \end{bmatrix};$	$D_B\dot{x} + (J_B + R_B)x = dF$ $J_B = \begin{bmatrix} 0 & 1 \\ -1 & 0 \end{bmatrix};$	$D_B\dot{x} + (1-d)J_Bx + R_Bx = dF$ $J_B = \begin{bmatrix} 0 & -1 \\ 1 & 0 \end{bmatrix};$

PCH: $x = \begin{bmatrix} x_1 \\ x_2 \end{bmatrix}; \dot{x} = [J_H(d) - R_H]\frac{\partial H}{\partial x}(x) + g_H E; H(x) = \frac{1}{2}Lx_1^2 + \frac{1}{2}Cx_2^2$

Boost	Buck	Buck-Boost
$J_H = \begin{bmatrix} 0 & -\frac{1-d}{LC} \\ \frac{1-d}{LC} & 0 \end{bmatrix};$	$J_H = \begin{bmatrix} 0 & -\frac{1}{LC} \\ \frac{1}{LC} & 0 \end{bmatrix};$	$J_H = \begin{bmatrix} 0 & \frac{1-d}{LC} \\ -\frac{1-d}{LC} & 0 \end{bmatrix};$
$R_H = \begin{bmatrix} 0 & 0 \\ 0 & \frac{1}{RC^2} \end{bmatrix};$	$R_H = \begin{bmatrix} 0 & 0 \\ 0 & \frac{1}{RC^2} \end{bmatrix};$	$R_H = \begin{bmatrix} 0 & 0 \\ 0 & -\frac{1}{RC^2} \end{bmatrix};$
$g_H = \begin{bmatrix} \frac{1}{L} \\ 0 \end{bmatrix}$	$g_H = \begin{bmatrix} \frac{d}{L} \\ 0 \end{bmatrix};$	$g_H = \begin{bmatrix} \frac{d}{L} \\ 0 \end{bmatrix};$

Table 2. Control equations.

	SFL	PBC	IDA-PBC
Boost	$d = 1 - \frac{[E + Lk_1(x_1 - x_{1d}) - L\dot{x}_{1d}]}{x_2}$	$d = 1 - \frac{[E + R_{1damp}(x_1 - x_{1d}) - L\dot{x}_{1d}]}{x_{2d}}$ $\dot{x}_{2d} = \frac{(1-d)x_{1d} - Gx_{2d} + R_{2damp}(x_2 - x_{2d})}{C}$	$\bar{d}_1 = 1 - \frac{E}{V_d}$ $d = 1 - (1-\bar{d}_1)\left(\frac{x_2}{V_d}\right)^\alpha$
Buck	$d = \frac{L\dot{x}_{1d} - Lk_1(x_1 - x_{1d}) + x_2}{E}$	$d = \frac{L\dot{x}_{1d} - R_{1damp}(x_1 - x_{1d}) + x_{2d}}{E}$ $\dot{x}_{2d} = \frac{x_{1d} - Gx_{2d}}{C}$	$\bar{d}_1 = 1 - \frac{E - V_d}{E}$ $d = 1 - (1-\bar{d}_1)\left(\frac{x_2}{V_d}\right)^\alpha$
Buck-Boost	$d = \frac{-L\dot{x}_{1d} + Lk_1(x_1 - x_{1d}) + x_2}{x_2 - E}$	$d = \frac{-L\dot{x}_{1d} + R_{1damp}(x_1 - x_{1d}) + x_{2d}}{x_{2d} - E}$ $\dot{x}_{2d} = \frac{-(1-d)x_{1d} - Gx_{2d}}{C}$	$\bar{d}_1 = 1 - \frac{E}{E - V_d}$ $d = 1 - (1-\bar{d}_1)\left(\frac{x_2}{V_d}\right)^\alpha$

2. Modeling and Control Equations

The basic power converters, such as boost, buck and buck-boost (shown in Figure 2), are typical switching-mode nonlinear systems, which customarily adopt conventional linear control method. These classic linear controllers, as mentioned in [9], exhibit some natural inconsistencies (for example, the intrinsic non-minimum phase characteristic related in [10]) and cannot satisfy the meaningful prerequisites of high performance control. The boost has inductor positioned in the input to reducing spikes in grid voltage, so is recommended to power factor (PFC) systems. Buck-boost inverts the polarity of the output voltage signal relative to the input signal and allows up-down output voltages. Note that the state variables are those related to energy store elements, i.e., capacitors and inductors.

In this context, there is a growing demand for new controllers to deal with this problem. Some nonlinear methods, such as SFL [11,12], PBC [13,14], IDA-PBC [15–17], fuzzy logic control [18], backstepping approach [19], predictive control [20], piecewise affine (PWA) [21] and repetitive control [22] have been designed and implemented in power converters.

This section presents the relevant models and control equations used in this work, collected in a literature review [10–17]. Notice in Figure 3, the Euler Lagrange (EL) is the base model to find the others. With the EL model, the PBC control equations are obtained. But the SFL control uses the model description in state space (SS). In turn, the IDA-PBC control requires the Port-controlled Hamiltonian model (PCH) system. Note that each model is associated with a control technique. Despite having specific mathematical and physical interpretations, the Euler-Lagrange and Hamiltonian models are mathematically similar to the models described in state space. It should be noted that the controllers are designed for continuous mode operation [23].

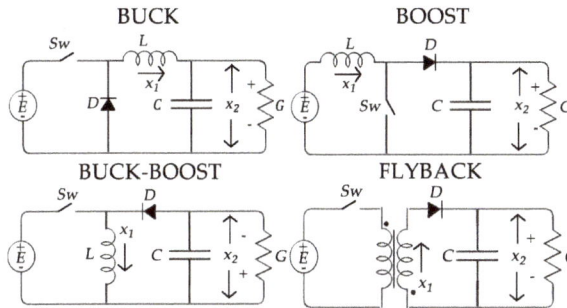

Figure 2. Basic power converters circuits.

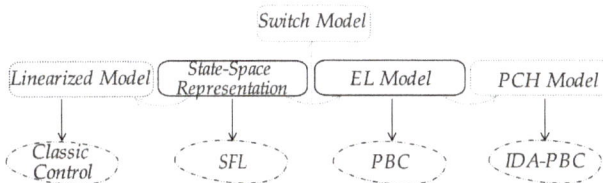

Figure 3. Models associated with non-linear control techniques.

As case studies, the control methods used in this work are SFL, PBC and IDA-PBC. A study and comparison of these methods are presented in [14]. The SFL control uses state space equations. PBC and IDAPBC include passivity properties, applying Lagrangian and Hamiltonian approaches, respectively. It should be noted that nonlinear control methods are currently widely discussed in the literature. However, another important trend in the design of these new controllers is the practical implementation.

To make the paper self-contained, we recall non linear concepts, including and intercalating different control equations associated with specific converter (see Appendix A). Further details and concepts of the applied methodology can be seen in the Section 3.

2.1. Buck-Boost and Flyback Examples

In the following paragraphs, the control equations (summarized in Table 3) and models of the Buck-Boost and Flyback converters will be described. Figure 2 presents the converters topologies. The readers that are familiar with nonlinear control can go directly to Section 3.

Table 3. Control equations.

	SFL	PBC	IDA-PBC
Buck-Boost	$d = \frac{-L\dot{x}_{1d} + Lk_1(x_1 - x_{1d}) + x_2}{x_2 - E}$	$d = \frac{-L\dot{x}_{1d} + R_{1damp}(x_1 - x_{1d}) + x_{2d}}{x_{2d} - E}$ $\dot{x}_{2d} = \frac{-(1-d)x_{1d} - Gx_{2d}}{C}$	$d = 1 - \left(\frac{E}{E - V_d}\right)\left(\frac{x_2}{V_d}\right)^{\alpha}$.
Flyback	$d = \frac{-L\dot{x}_{1d} + L_{eq}k_1(x_1 - x_{1d}) + x_2}{x_2 + E_{eq}}$	$d = \frac{-L_{eq}\dot{x}_{1d} + R_{1damp}(x_1 - x_{1d}) + x_{2d}}{x_{2d} + E_{eq}}$ $\dot{x}_{2d} = \frac{(1-d)x_{1d} - Gx_{2d}}{C}$	$d = 1 - \left(\frac{E_{eq}}{E_{eq} + V_d}\right)\left(\frac{x_2}{V_d}\right)^{\alpha}$.

According to [24] the average Buck-Boost converter circuit can be written by equivalent state space equations:

$$\dot{x}_1 = (1-d)\frac{1}{L}x_2 + d\frac{E}{L}, \tag{1}$$

$$\dot{x}_2 = -(1-d)\frac{1}{C}x_1 - \frac{G}{C}x_2. \tag{2}$$

where, d is the converter duty cycle, $0 \leq d < 1$. As it can be seen from (1) and (2), there are two state variables, x_1 and x_2 and an input (control) variable, the duty cycle d.

2.2. SFL Control

Summarily, the procedure to obtain the state feedback linearization [25] includes the steps:

1. Select the state variable to be controlled. Two possibilities: indirect control (current x_1) or direct control (voltage x_2);
2. Derivation of the output (n) times until an explicit relation between output (y) and the input (E) is achieved;
3. Determine $d = d(v, x)$ in order to perform the feedback linearization;
4. Investigate the stability of internal dynamics.

Defining L_f as the derivative of Lie [25] and consider that:

$$x_1 = h(x), \quad y = x_1, \quad x_2 = L_f h,$$
$$\dot{y} = \dot{x}_1, \quad \dot{x}_1 = (1 - d)\frac{1}{L}x_2 + d\frac{E}{L}. \tag{3}$$

Since we have to accomplish one derivation to obtain a relation between the input and output, the relative degree is unitary ($n = 1$). On these terms, the general expression for the duty cycle equation is:

$$d_{SFL} = \frac{L\left[\dot{x}_1 d - k_1\left(x_1 - x_{1d}\right)\right] - x_2}{E - x_2}. \tag{4}$$

A literature review of the main stability analysis methods applied to power converters is presented in [26]. PBC and IDAPBC control techniques, reported in [27,28], are demonstrated in details on appendices.

2.3. Flyback Modelling and Control Equations

Derived from a mathematical formulation, the flyback converter can be interpreted as an isolated buck-boost converter. As shown by [29], the average state-space model of the circuit illustrated in Figure 2 are given by:

$$\dot{x}_1 = \frac{L_1}{L_1 L_2 - L_M^2}(1 - d)x_2 - \frac{L_M}{L_1 L_2 - L_M^2}dE. \tag{5}$$

$$\dot{x}_2 = (1 - d)\frac{1}{C}x_1 - \frac{G}{C}x_2. \tag{6}$$

where L_1 and L_2 are the primary and secondary inductances, respectively, and L_M is the mutual inductance. After replacing:

$$L_{eq} = \frac{L_1 L_2 - L_M^2}{L_1}, \quad E_{eq} = \frac{L_M}{L_1}E. \tag{7}$$

the flyback converter Equations (8) and (9) become similar to the buck-boost Equation:

$$\dot{x}_1 = \frac{1}{L_{eq}}(1 - d)x_2 - d\frac{E_{eq}}{L_{eq}}. \tag{8}$$

$$\dot{x}_2 = (1 - d)\frac{1}{C}x_1 - \frac{G}{C}x_2. \tag{9}$$

Thus, the adapted mathematical models from buck-boost are evaluated to represent and withdraw the flyback control equations. All control equations (detailed in [30]) are summarized in Table 3.

By collecting and manipulating the terms, it is possible to obtain the general expression for the duty cycle(by considering $\dot{x}_{1d} = 0$), defined by:

$$d_{SFL} = \frac{x_2 + R_{1damp}(x_1 - x_{1d})}{E_{eq} + x_2}. \tag{10}$$

One of the lessons learned from previous researches [31] is that the nonlinear controllers need an integral action to achieve voltage regulation. Therefore, in order to improve stead-state performance and assure the convergence of error between the output voltage and desired value V_d, a proportional integrative term is recommended, given by:

$$G_{int} = -k_{int} \int_0^t [x_2(s) - V_d] ds. \tag{11}$$

3. SHIL and DHIL Proposed Methods

In order to validate a simulation, a modeling or a controller design it is necessary to obtain experimental results through hardware implementation. In the context of Power Electronics, as systems complexity increases [3,32]:

- Costs with semiconductor devices and power components rise significantly.
- Implementation of controller conditioning and communication systems complexity increase.
- Time spent for concluding the hardware implementation may become a problem.

In this session, we present a procedure for high-level programming of a DSP (Digital Signal Processor) using SHIL and DHIL Simulations. The HIL based method simulation is a technique that mixes both virtual and real elements. Currently, this technique is often used to test embedded control systems, where both the hardware and system software are tested. Also, we can verify the control and the system operations without the need of a physical circuit.

Besides the independence of the physical prototype, the proposed methodology has other advantages:

- There is no need for costly real-time Simulators (RTS) systems, such as those offered by OPAL RT, Typhon HIL, dSPACE and RTDS. In the same way, it is possible to use the method remotely, in residences, in the laboratory, using desktop pc, laptop, without being conditioned to a complex system—which involves both hardware and software—previously installed;
- It is possible to emulate only the converter model and perform several tests, regardless of control;
- The control of the system is embedded and its proper functionality can be evaluated in DSP; therefore, the determination of the processing time of each step of the algorithm can also be achieved. It is possible to monitor, make initial parameter updates, controller gains, input and load disturbances, etc., through the friendly interface offered by Matlab/Simulink.
- In addition, there is the possibility of the DSP to emulate the model or the control independently of the pc/laptop. In other words, its possible to upload the codes into the flash memory of the embedded system (tests are limited to DSP input/output capabilities, for example, DAC and other digital/analog ports).
- Simple and complex converters can be evaluated;

The major drawbacks are also listed:

- It needs Matlab installation.
- The approach depends on the mathematical model of the converter.

3.1. SHIL

Figure 4 shows the overview of SHIL proposed methodology. The first step is obtaining the plant and the control equation models. After that, such models can be simulated using commonly softwares as Matlab or PSIM. As an example, Figure 4 illustrates the Flyback's converter model and the SFL control equations, designed with Matlab/Simulink tools. Figure 5 shows the simulation in PSIM software. After the implementation, it is necessary to run the simulation and verify if the control and state variables are converging to the desired state. Once this is achieved, the next step is to embed the simulated system in the DSP. By using the *external mode* of Simulink and a compiler for the DSP, the model will be converted in code and then embedded to the target. Finally, the target will run the code, emulating the converter and control models. It is possible to verify desired signals in an oscilloscope by programming the DSP pins through DAC (Digital Analog Converter) blocks available in the C2000 *Texas Instruments* package. An important detail is the need of a scale adjustment for voltage compatibility between the simulation and target.

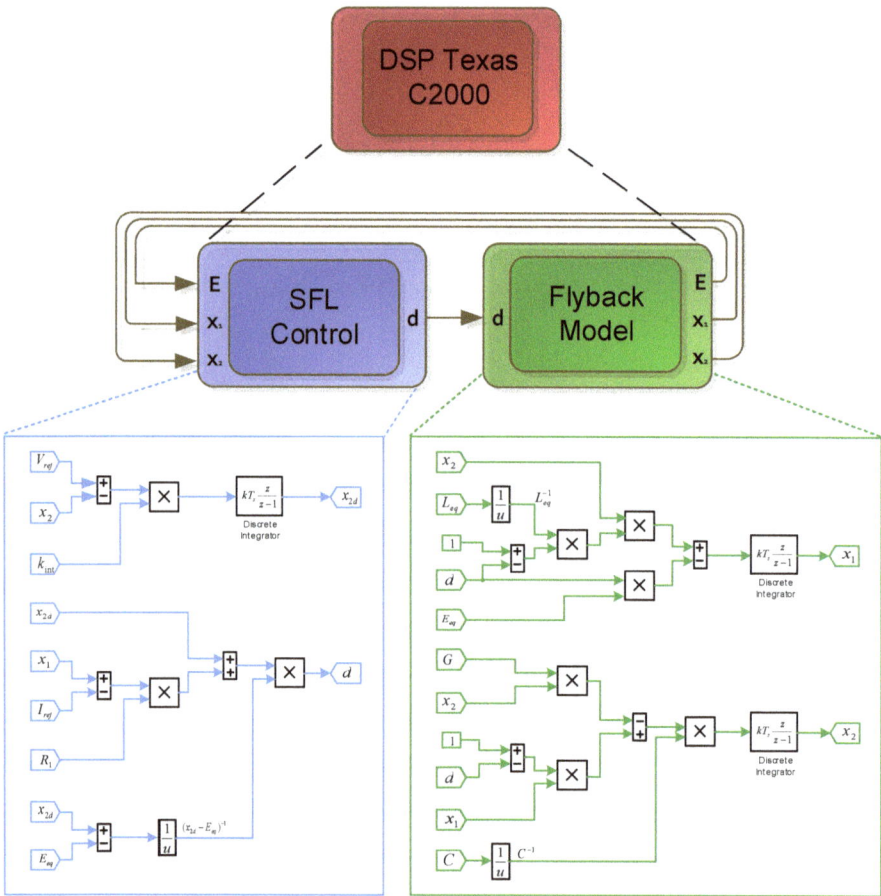

Figure 4. Flyback model, Equations (8) and (9), and SFL control Equation (10) in block diagrams (high level programing) that is embedded in DSP C2000.

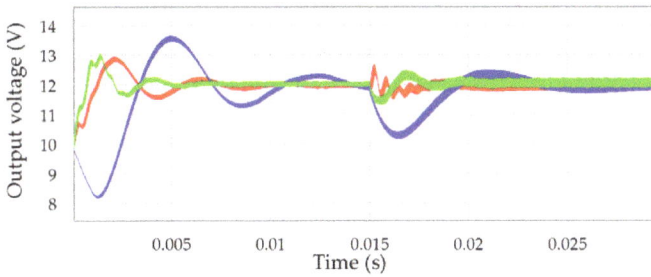

Figure 5. Simulation of flyback switched model using PSIM software. Normalized output voltage x_2 for load perturbation with a fixed time step 100 times smaller than the switching period. SFL (red), PBC (green) and IDAPBC (blue).

3.2. DHIL

With the proposed SHIL and DHIL methodologies, we can easily test the control law without needing a desktop computer or a real plant. In SHIL, the data transfer occurs directly and faster when both model and control are inserted in the same Digital Signal Processor. However, DHIL is best suited for synchronization, measurement and data communication tests between different systems.

The supporting package for C2000 microcontrollers available in Matlab/Simulink can be found in the Simulink library. Basic information about block functions, simulation configurations for real-time simulation or external mode and examples of systems implementation using the fundamental blocks such as PWM, ADC, DAC and interruptions can be found in Mathworks [33] website, or in Matlab "Help" area. Once selected, a list of C2000 DSP family will be displayed. By choosing the corresponding DSP, available blocks for the microcontroller are displayed. it is possible to build block systems with other Simulink blocks, by simply dragging them to the model window.

In DHIL, the use of a PWM (Pulse Width Modulation) block is necessary in order to control the converter switching sequence and also to synchronize 3 ADCs (Analog Digital Converter) available for measuring the inductor current (x_1), capacitor voltage (x_2) and the input voltage E. As seen in Figure 6, the control laws and the converter model are embedded in different microcontrollers. For computing the control laws, the inputs of the model are (x_1), (x_2) and E. Since those inputs are originated from the converter model computation—configured as analog signals type—it is necessary to convert those signals to a digital one, through analog-digital converter (ADC). After conversion, the output of the control law is the duty cycle. Since this control variable is digital, it will be converted to analog (DAC), for reading in the ADC of the DSP embedded with the converter model. For closing the loop, the variables (x_1), (x_2) and E are calculated and consequently converted from digital to analog type. It is important to notice that the conversions are based on the PWM sample rate, then requires the synchronization between both DSPs for a correct computation of control laws and converter model.

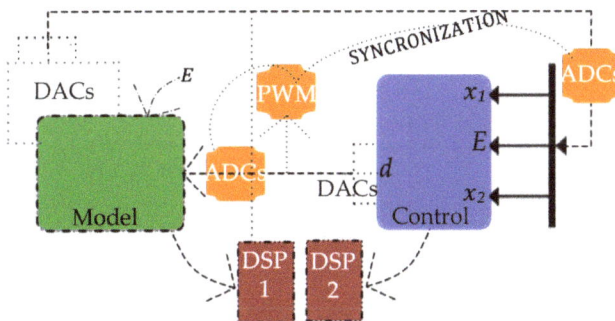

Figure 6. DHIL.

3.3. Details of the Implementation

Also, the approach used for programming the microcontroller diverges from the conventional one, being unnecessary the development of code lines. By using code generation tools and software libraries it is possible to resort the implementation of converter models and controllers through Matlab/Simulink blocks. The evident advantages of this approach are the clear visualization of the programming process and the time spared for development. Figure 7 illustrates the proposed steps for this methodology.

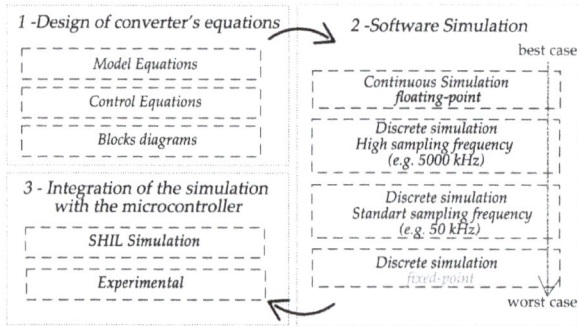

Figure 7. Methodology steps for SHIL simulation. The optional steps are highlighted in gray.

The first stage consists in developing an equation model for the system, demonstrating the relation between the state and control variables. It is up to the user to consider or not the nonlinearities of the system. Next, it is necessary to choose a control technique for actuating in the variable of interest. The nature of the control technique is wide, and can include since classical techniques, as PID controllers, to nonlinear control approaches. This stage ends with the implementation of the model and control equations in the simulation software. Figure 8 presents the implemented model and control equations of a buck-boost converter, as an example. Figure 9 shows the general SHIL simulation scheme for any converter. The main control goal is to calculate the duty cycle *d* (used in mathematical model), then the corresponding PWM signal is generated as an input control to command the switch of the physical converter. The state variables (inductor current x_1 and voltage capacitor x_2) and input voltage E are the required measures. An integral action is added to better regulate the output voltage.

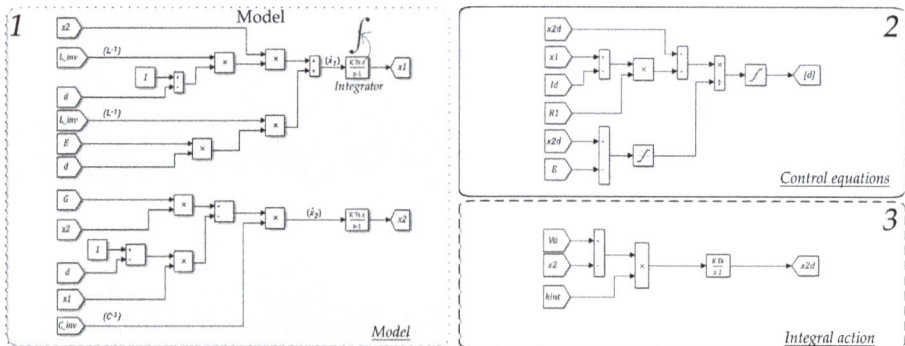

Figure 8. Buck-boost model and SFL control equations in block diagrams. Model (1), control Equation (2) and integral action (3).

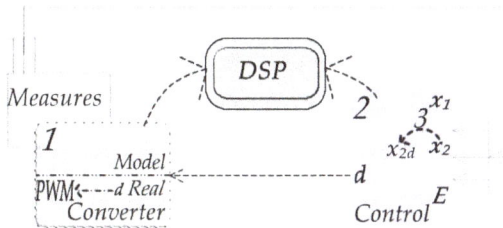

Figure 9. General SHIL Simulation Scheme.

The second phase consists of simulating the implemented system. Aiming at minimizing the errors, it is advised to simulate the system in continuous time and afterly in discrete time. For the continuous system it is important to work with float data type, by programming the operations of multiplication, division and constant blocks to "single/double" types. Although the increasing in processing time, this data type conversion configures one less source of error in the continuous time simulation.

After that, it is necessary to discretize the system, by changing the controllers structure and continuous operators to the corresponding discrete blocks (defining a sample time in which the blocks will be sampled). In a way to approach the results of the first discretized system to the continuous simulation it is proposed the use of a high sample frequency (or a small sample time).

The next task is to simulate the system, considering a standard sample frequency (e.g., nominal 50 kHz), in such way to obtain a discrete model that is less approached by the continuous model. The selection of this frequency must be cautiously chosen, since the system can converge to instability. Another common problem associated to a bad choice of the sample frequency is the signal aliasing (Nyquist rule). The final objective of this stage consists in transforming the floating point data in fixed point data. This conversion can be done by programming the operations of multiplication, division and constant blocks to fixed-point type, or by using operational blocks offered in specific libraries for microcontrollers (libraries that are offered by Mathworks for users of C2000's microcontrollers family, by Texas Instruments, for example), as shown in Figure 10. Some recomendations: avoid operations with floating data and divisions that increase the processing time. Always when possible, use multiplication operations instead of division operations (ex: when the denominator is a constant). Declare the variables as fixed point, preventing calculus with float and optmizing the code's execution. Discrete models can be embedded for HIL simulation, since the user compiler can convert the blocks in code. However, a discrete model that converges, when working with fixed point data, makes the compilation and the processing time of the microcontroller smaller (and also the memory used smaller). In this way, the last step brings a discrete model more appropriated for a HIL procedure.

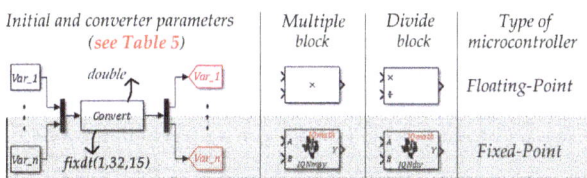

Figure 10. Fixed-point and Floating-point different implementations.

The final stage is the SHIL simulation itself. Once the compiler has generated the code and the system is embedded to the microcontroller (also called target) the communication between the target and the computer, in which Matlab/Simulink is running, begins. Usually microcontrollers of C2000 family communicates to the computer through USB or ethernet cable. In this application it is proposed the use of a USB cable. For running the system as a real-time simulation it is proposed to set the simulation as "external mode" simulation on Simulink. Also, it is necessary to set on the simulation configurations which target the connection must occur.

Since we are interested in plotting or viewing the gathered data, it is necessary to use specific blocks in the simulation for real-time plotting or setting DSP pins as analog outputs [33]. By doing that the user can see the data generated in the target in a Simulink "scope" or in an oscilloscope (by analog output reading). Each case can be achieved by using the RTDX (real-time Data Exchanged) or the DAC blocks, as seen in Figure 11. Since the R2016b version of Matlab there is a DAC block for DSP28377S of C2000 family (Texas Instruments). This block configures 3 digital inputs as 3 analog outputs (also called channels A, B and C). In this way it is possible to use 3 channels of an oscilloscope and view the curves in real-time.

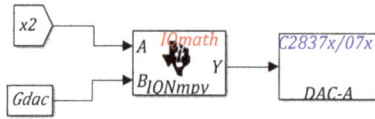

Figure 11. DAC Block: Used for showing state and internal variables in a scale between 0 V–3.3 V (12-bit resolution).

More details of the control algorithms implementation in block diagrams can be seen in the Figures 12 and 13. Since the didactical background available in the literaute lacks of information, details of the functional blocks are shown, providing a development base for future works. Although the approach of this chapterdeals with a specific study of case, the available content makes it simple to adapt the program to other applications. General files, containing the control methods for Buck-Boost, Boost and Buck converters are shown in Figures 8, 12 and 13, respectively.

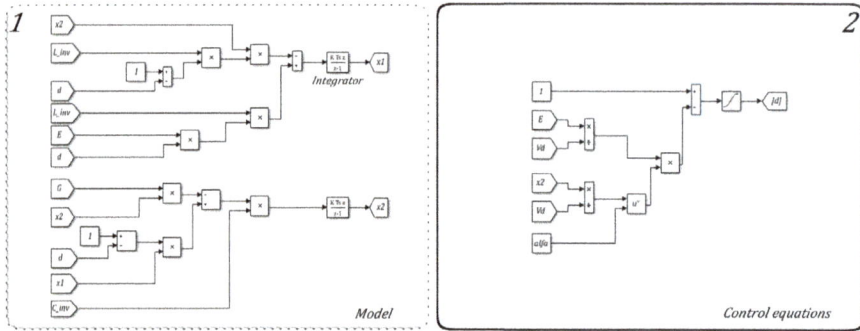

Figure 12. Boost model and IDA-PBC control equations in block diagrams.

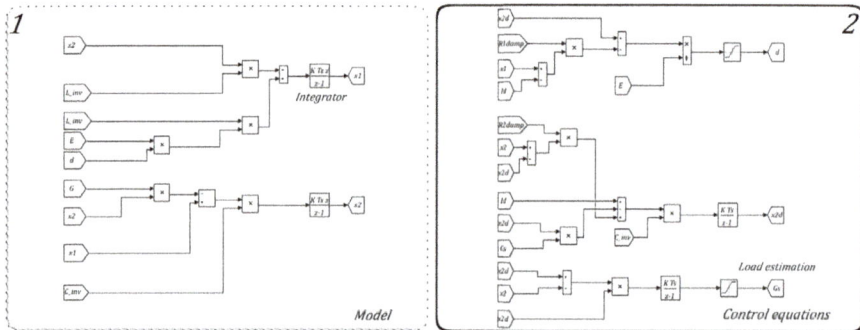

Figure 13. Buck model and PBC control equations in block diagrams.

4. SHIL and DHIL Results for Buck-Boost and Flyback

This section shows the digital simulations using Matlab (Model in The Loop: MIL) and SHIL/DHIL experimental results. The converters are designed according to specifications listed on Table 4 and the three control laws studied in this work. The experimental setup is sketched in Figure 14. The plots displayed in Figure 15a,c present the capacitor voltage response when an input voltage and a load variations, respectively, are included in the Buck-Boost converter simulated in software. The same effect is reported for the Flyback converter in Figure 15b,d. In both input and load variation, consecutive steps of 70% to 100% are applied in the simulated systems. An open-loop control test is also presented in Figure 16. Figures 17 and 18 show the output capacitor voltage of Flyback and Buck-Boost converters, respectively, for the SHIL and DHIL applications. A load variation (70–100%) is applied for evaluate the three control laws.

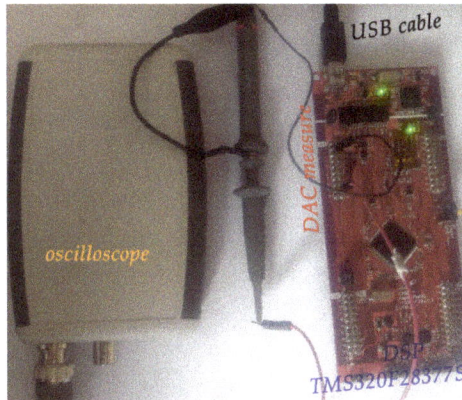

Figure 14. Experimental setup.

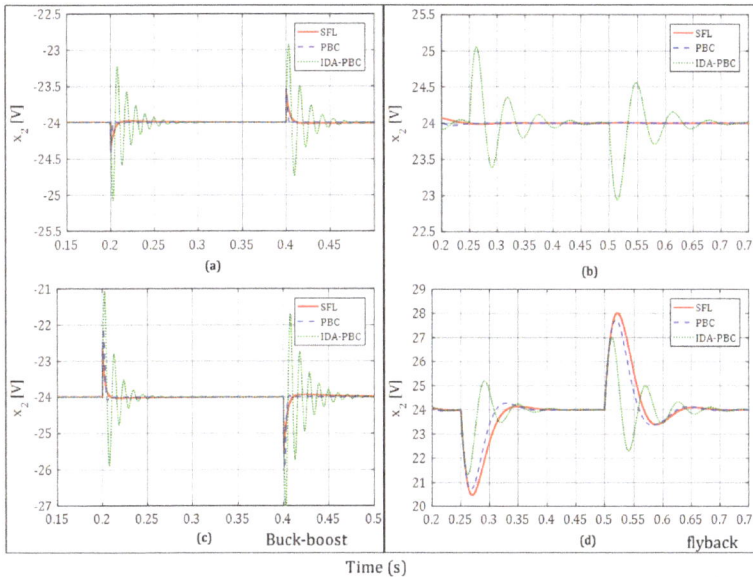

Figure 15. Software simulation result for Buck-Boost (**a,c**) and Flyback (**b,d**) (using control techniques SFL (red), PBC (blue) and IDA-PBC (green). Output voltage x_2 for input voltage (**a,b**) and for load variation (**c,d**).

Table 4. Initial and converters parameters.

Parameters	Buck-Boost	Flyback
V_d	−24 V	24 V
E	50 V	50 V
R	10 Ω	11.5 Ω
L	0.6 mH	146 μH, 35 μH
C	470 μF	470 μF
f_{sw}	50 kHz	100 kHz
$R_{1damp} = Lk_1$	100	15
k_{int}	200	−5000

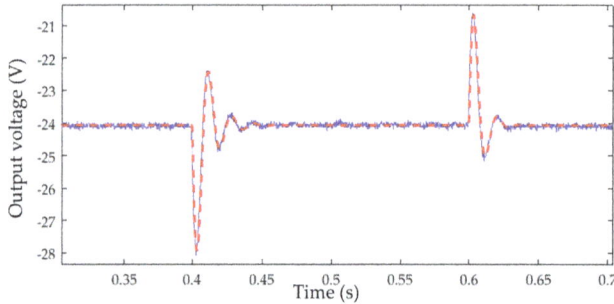

Figure 16. Open loop control comparing SHIL (blue line) with Software Simulation (red line) results for Buck-Boost converter. Normalized output voltage x_2 for load perturbation (70–100%) using fixed $d = 0.325$.

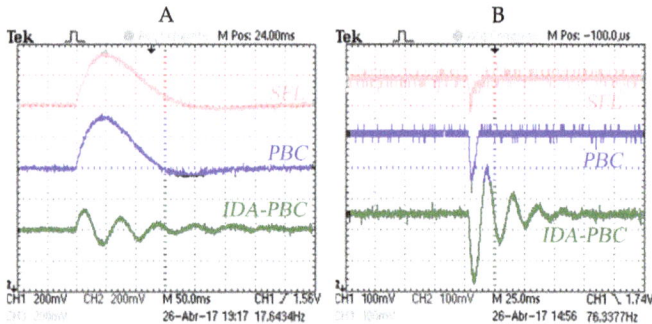

Figure 17. SHIL experimental result for Flyback (**A**) and Buck-boost (**B**) converters. Normalized output voltage x_2 for load perturbation (70–100%).

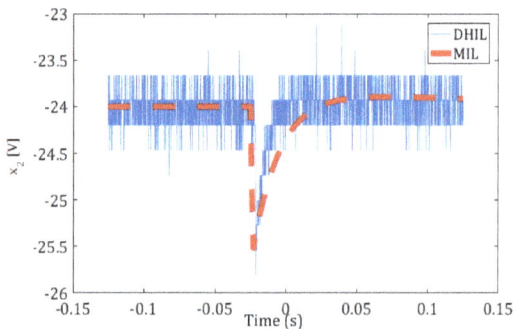

Figure 18. DHIL experimental compared with Software Simulation (MIL) result for Buck-Boost converter (SFL control). Normalized output voltage x_2 for load perturbation (70–100%).

It is possible to appreciate in the figures how the waveforms in both software and real-time simulations are compatibles. Although consecutive perturbations can be applied, the convergence of the variables to steady-state values is assured. In addition, similar transient dynamics can be observed, none of the implemented systems in software or in HIL present instability. Therefore, the embedded models and control equations are validated. The processing time for the DSP to compute the control law and converter state, and therefore to run a real-time simulation, is 1.2 µs.

5. SHIL Results for Second Order Power Converters and Comparison

This section presents the digital simulation results using Matlab and SHIL method for buck, boost and buck-boost. The converters are implemented according to design specifications of Table 5 and the three control laws studied in this work. Figure 19 shows the capacitor voltage response to a load voltage variation (70% G to 100% G), respectively, in the converters simulated in software. Figure 20 shows the capacitor voltage of the converters for the SHIL application. A load variation (70–100%) is applied to evaluate the three control laws. Figure 21 shows the PWM signal generated in HIL simulation for the buck-boost converter in steady state operation.

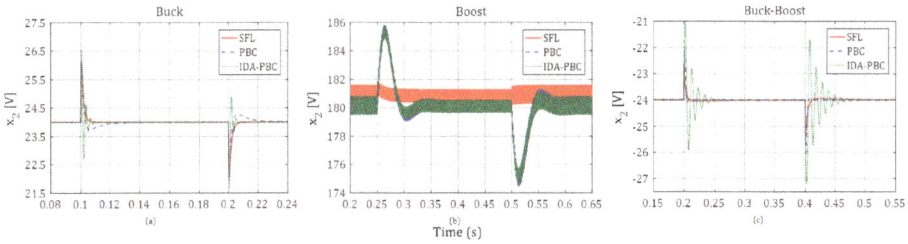

Figure 19. Software simulation result using control techniques SFL (red), PBC (blue) and IDA-PBC (green). Output voltage x_2 for buck (**a**) and boost PFC (**b**); buck-boost (**c**) for load variation in 0.25 s and 0.75 s.

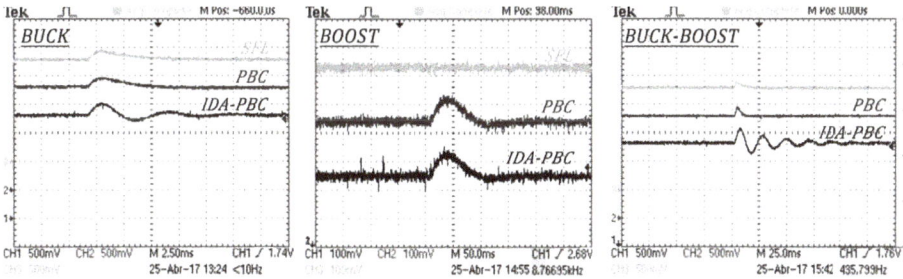

Figure 20. SHIL experimental result. Normalized output voltage x_2 for load perturbation (70–100%).

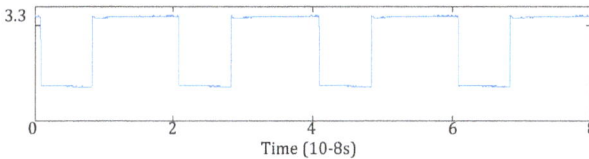

Figure 21. PWM signal generated in SHIL Simulation for buck-boost permanent condition ($d = 0.37$).

Table 5. Initial and converters parameters.

Parameters	Boost	Buck	Buck-Boost
I_d	$\frac{G}{E}V_d^2$	GV_d	$GV_d(\frac{V_d}{E}-1)$
R	52.5 Ω	10 Ω	10 Ω
L	0.6 mH	0.6 mH	0.6 mH
C	2800 μF	470 μF	470 μF
E	100 V	50 V	50 V
V_d	180 V	24 V	−24 V
$Pout$	630 W	57.6 W	57.6 W
f	50 kHz	50 kHz	50 kHz
R_{1damp}	33	500	100
R_{2damp}	50	0	0
k_g	0.0356	2.5	14
k_{int}	−150	2000	200
α	0.8	−10	0.8

As seen in the figures, both software simulations and SHIL results converged to the steady state value after the consecutive step applications. It should be noticed that the dynamical response of the systems is compatible, since the same transient dynamics is seen, even for the IDA-PBC's oscillatory dynamics. None of the implemented systems in software or in SHIL presented instability. Therefore, it means that the embedded models and control equations are capable in controlling the systems, and therefore, validating the control techniques.

In general, the SFL, PBC and IDA-PBC control laws show satisfactory results for the three types of converters studied: boost, buck and buck-boost. It was observed that the inductor current and output voltage in the capacitor, the main variables, follow the reference set points, reaching the control objectives. Since one of the main objectives of this work is the comparison of control methods, Table 6 shows, from an implementation point of view, the advantages and disadvantages of each method.

It is seen that SFL and PBC present similar results. Reminding that, for both control techniques the indirect control is the only possibility [10]. On one hand the SFL is a didactic and easier solution than the PBC. On the other hand, the control complexity of PBC is justified for load estimation and better voltage output regulation. However, there is a trade-off, since the overshoot and undershoot increase with the adaptative control law and the integral gain used.

SFL and PBC control present small error in steady state for the output voltage of the capacitor, without the integral action. On the other hand, the IDA-PBC control presented larger overshoot/ undershoot and accommodation times, notably for input voltage variations. This is because this technique is even more dependent on the exact knowledge model parameters. Thus, to improve the results it is necessary to include other non-modeled effects like parasitic resistances, diode and switches voltage drops [34].

Once IDA-PBC control is a direct control, it does not require the measurement of the current x_1, which is an advantage in terms of implementation. In general terms, we can verify:

- SFL: simpler, didactic, effective, dispenses load estimation for DC-DC systems when using integrative gain.
- PBC: has medium complexity, needs more measurements and control parameters to better estimate the load and regulate the output capacitor voltage. It is the most recommended technique for PFC systems, since it offers lower THD levels [13]. It has the same processing time as the SFL control because of the same amount of division operations, which effectively determine the total processing time (sums and multiplications offer irrelevant contributions).
- IDA-PBC: non-trivial control solution, allows direct control, which exempts current measurement and eventual problems. By the same direct nature of the control, does not work correctly for AC-DC systems, since the objective is to impose the current in phase with grid input voltage.

Table 6. Comparison of non linear control methods.

Comparison	SFL		PBC		IDA-PBC	
	Advantage	Disadvantage	Advantage	Disadvantage	Advantage	Disadvantage
Control law	simple			medium	open loop	
Solution	easy			medium complexity		Nontrivial
Parameters	2			4	1	
Measurements		3		3	2	
AC-DC System	low THD levels and high PF		low THD levels and high PF			incorrectly, high THD rates
Control Type		direct control is only for the buck		direct control is only for the buck	Direct control is possible for all	
Integral term	Soft integration		Soft integration			
Dependence on realistic model		medium		medium		very sensitive very dependent on realistic mode
Parametric dependence control	Does not depend on C			It depends on the G, L and C	Control law does not depend on G, L or C	
Processing Time (TMS320F2812 fixed-point DSP)	Fast (2.4 µs)		Fast (3.2 µs)		Fast (6 µs)	
Processing Time (TMS320F28377S floating-point DSP)	Fast (1.2 µs)		Fast (1.2 µs)		Fast (1.2 µs)	
sum	4			8	3	
div	1			1	0	
mult	3			6	3	
other operation						1 (exp)
Load Estimation		Cannot estimate by the output voltage error.	Can estimate the load by voltage error		It is not necessary estimate the load	

6. Conclusions

This work showed the buck, boost, buck-boost, flyback, SEPIC and 3-phase AC/DC boost converters modeling as well as the development of nonlinear control techniques using SHIL and DHIL implementations. In addition, the control and converter models were implemented in a DSP, resulting in a quick and cheap HIL realization of the proposed systems, ideal for students and engineers interested in learning and validating converters performance. Using a switching frequency of 50 kHz (20 μs), the processing time of the model/control equations (1.2 μs) demands 6% of the bandwidth (for buck, boost, buck-boost, flyback and SEPIC). According to the data in Table 6, there is a clear preference for new embedded systems with floating-point operation. As illustrated in Figures 16 and 18 and despite measurement noises, the SHIL/DHIL results remained close to the model-simulated one. The advantages provided by the proposed method are: security, saving development time, facilitating the understanding of the programming process, standardization, concurrent simulation, rapid prototyping and, mainly, an easy and cheap way to validate linear and nonlinear controllers.

Author Contributions: Conceptualization, A.H.R.R., L.M.F.M. and S.I.S.J.; Data curation, M.F.C.C., R.A.S.S. and W.A.R.; Formal analysis, A.H.R.R., W.A.R. and L.M.F.M.; Investigation, A.H.R.R., M.B.E.S. and L.M.F.M.; Methodology, A.H.R.R.; Project administration, S.I.S.J.; Resources, A.H.R.R., L.M.F.M., W.A.R.; Software, A.H.R.R., M.B.E.S., M.F.C.C. and R.A.S.S.; Supervision, A.H.R.R. and L.M.F.M.; Validation, A.H.R.R., M.B.E.S., M.F.C.C., R.A.S.S. and W.A.R.; Visualization, A.H.R.R. and S.I.S.J.; Writing—original draft, A.H.R.R. and M.B.E.S.; Writing—review & editing, A.H.R.R., L.M.F.M. and S.I.S.J.

Funding: This research was funded by Coordenacao de Aperfeicoamento de Pessoal de Nivel Superior (CAPES) grant number [086/2013].

Conflicts of Interest: The authors declare no conflicts of interest.

Nomenclature

E	Input voltage.
d	Duty cycle.
x_1	Inductor current.
x_{1d}	Desired inductor current.
I_d	Constant desired inductor current.
x_2	Capacitor voltage.
x_{2d}	Desired capacitor voltage.
V_d	Constant desired capacitor voltage.
L	Converters inductance.
L_1	Flyback primary-side inductance.
L_2	Flyback secondary-side inductance.
L_M	Flyback mutual inductance.
C	Converters capacitance.
G	Load conductance.
R_{damp}	Nonlinear PBC gain.
k_g	Load estimation gain.
k_{int}	Integral gain.
α	IDAPBC control gain.
k	SFL control gain.

Appendices

These appendices are optionals for those who are familiar with non-linear control applied to converters. So, we exemplify how to obtain SFL control for the buck, PBC for the buck-boost and IDA-PBC equations for the boost. Notice that we mix the three control laws and the three converters distinctly, to cover the maximum information in a smaller space. In time, the boost converter will be analyzed with power factor correction (PFC), since it is the most suitable for this specific application. The other systems are analyzed as voltage regulators (DC-DC) in which the input voltage comes up to a constant value. We also included SEPIC (Appendix B) and 3-phase AC-DC boost converter (Appendix C).

Appendix A. Nonlinear Controllers

The state feedback linearization control is used in this work mainly because it represents a didactic and effective procedure. This method facilitates the understanding of the system being useful for an initial contact with nonlinear control techniques and leads to a change Of coordinates that shows an interesting structure and mathematical properties. Moreover, it allows dynamic change of a nonlinear system into a linear dynamics through a nonlinear feedback of the output state conveniently chosen. For this purpose, it is necessary to perform a change of state variable input and an auxiliary input variable. Then, it is possible to use familiar linear techniques to effect control of the proposed system.

Appendix A.1. SFL Control Equations of the Buck Converter

According to [35,36] the average Buck converter circuit can be written as:

$$\dot{x}_1 = -\frac{1}{L}x_2 + d\frac{E}{L}, \tag{A1}$$

$$\dot{x}_2 = \frac{1}{C}x_1 - \frac{G}{C}x_2, \tag{A2}$$

where, d is the converter duty cycle, $0 \leq d < 1$. As it can be seen from (A1) and (A2), there two state variables, x_1 and x_2 and an input (control) variable, the duty cycle d. Defining L_f as the derivative of Lie [25] and choosing:

$$x_1 = h(x), y = x_1, x_2 = L_f h, \dot{y} = 1\dot{x}_1,$$
$$\dot{x}_1 = -\frac{1}{L}x_2 + d\frac{E}{L}. \tag{A3}$$

Since we have to derive $g_r = 1$ times to obtain a relation between the input and output, the relative degree is $g_r = 1$. In this way, the new coordinate system is:

$$[\dot{z}_1] = [z_2] = [v],$$
$$\begin{bmatrix} z_1 \\ z_2 \end{bmatrix} = \begin{bmatrix} x_1 \\ v \end{bmatrix}. \tag{A4}$$

Using the control law $v = r^{(g_r)} - k^T e$, with k and e given by:

$$k = [k_1],$$
$$e = [e_1] = [x_1 - r] \tag{A5}$$

obtains:

$$v = \dot{r} - k_1 e, \tag{A6}$$

$$-\frac{1}{L}x_2 + d\frac{E}{L} = \dot{r} - k_1(x_1 - r). \tag{A7}$$

Isolating d and considering the reference $r = x_{1d}$, the general expression for the duty cyclic equation is:

$$d_2 = d_{SFL},$$
$$d_{SFL} = \frac{L[\dot{r} - k_1(x_1 - r)] + x_2}{E}. \tag{A8}$$

We can observe that as the system relative degree is one (there is only one switch to control two variables), we need to perform only one branch, which is already inferred directly from (A1). Thus, we need only control equation given by (A8).

Appendix A.2. Passivity-Based Control (PBC)

The goal of the passivity-based control is to modify the dissipative structure since the inputs and store elements are constant. The basic premise is to keep the energy stored in the capacitors and inductors less than injected by the source. This effect is achieved by the addition of "virtual" resistors in parallel or in series with the load. Such resistances are emulated by the controller through the duty cycle signal conditioning.

Other definitions about passivity, as well as the equations necessary to control in view of this method, can be visualized in [10,13].

Appendix A.3. PBC Control of Buck-Boost Converter

According to [35,37] the average Buck-Boost converter circuit can be written by Euler-Lagrange equations, as:

$$D_B \dot{x} + (1-d) J_B x + R_B x = dF,$$ (A9)

with

$$x = \begin{bmatrix} x_1 \\ x_2 \end{bmatrix}, D_B = \begin{bmatrix} L & 0 \\ 0 & C \end{bmatrix},$$

$$R_B = \begin{bmatrix} 0 & 0 \\ 0 & G \end{bmatrix}, F = \begin{bmatrix} E \\ 0 \end{bmatrix}, J_B = \begin{bmatrix} 0 & -1 \\ 1 & 0 \end{bmatrix}.$$ (A10)

The equivalent state space equations are:

$$\dot{x}_1 = (1-d)\frac{1}{L}x_2 + d\frac{E}{L},$$ (A11)

$$\dot{x}_2 = -(1-d)\frac{1}{C}x_1 - \frac{G}{C}x_2,$$ (A12)

For PBC control, let us consider the state error in function of desired vector x_d:

$$\tilde{x} = e,$$
$$\tilde{x} = x - x_d.$$ (A13)

The error equation formulated as in (A11) and (A12) becomes:

$$D_B \dot{\tilde{x}} + (1-d) J_B \tilde{x} + R_B \tilde{x} + R_{damp}\tilde{x} = \psi,$$
$$\psi = F - [D_B \dot{x}_d + (1-d) J_B x_d + R_B x_d] + R_{damp}\tilde{x}$$ (A14)

In order to guarantee the error vector to converge to zero, one has to impose $\Psi = 0$, which can be written as:

$$L\dot{x}_{1d} - (1-d)x_{2d} - R_{1damp}\tilde{x}_1 = dE,$$
$$C\dot{x}_{2d} + (1-d)x_{1d} + Gx_{2d} = 0.$$ (A15)

where R_{damp} is the damping matrix defined as:

$$R_{damp} = \begin{bmatrix} R_{1damp} & 0 \\ 0 & R_{2damp} \end{bmatrix}.$$ (A16)

R_{damp} is the damping added to the system which shapes its energy. Some fundamental definitions regarding passivity, and the derivation of the control equations in view of this method, can be found in [10,38]. Aiming at rendering the system passive, via the condition established by (A15), one has:

$$d_3 = d_{PBC},$$

$$\dot{x}_{2d} = \frac{-(1-d)x_{1d} - Gx_{2d}}{C}, \tag{A17}$$

$$d_{PBC} = \frac{R_{1damp}(x_1 - x_{1d}) + x_{2d}}{x_{2d} - E} \tag{A18}$$

The load estimation is given by (A19):

$$\dot{G}_s = -k_g x_{2d}(x_2 - x_{2d}). \tag{A19}$$

Equation (A19) can be used for all four converters.

Appendix A.4. IDA-PBC Control

The IDA-PBC control methodology provides a clear separation between elements of the system in terms of their energy functions, enabling the controllers design with a clear physical interpretation [17]. Based on the Hamiltonian model, in which the term $H(x)$ is represented explicitly, describes how the energy flows within the system and between the subsystems interconnections, represented by the H matrix, and energy dissipation elements, represented by R_H matrix. The IDA-PBC controller design is to find the solution that leads to the stabilization of the system in closed loop, by modifying the matrix interconnection and system damping. Thus, it is necessary to solve partial differential equations from the interconnected subsystems, to enter the desired damping energy function.

Based on [15,17,36,39] the IDA-PBC control equations are obtained for the boost, buck and buck-boost converters.

Appendix A.5. IDA-PBC Control for Boost Converter

The average boost converter circuit can be written by equivalent state space equations, as:

$$\dot{x}_1 = -(1-d)\frac{1}{L}x_2 + \frac{E}{L}, \tag{A20}$$

$$\dot{x}_2 = (1-d)\frac{1}{C}x_1 - \frac{G}{C}x_2, \tag{A21}$$

The modeling and IDA-PBC control of boost converter is presented in [39]. Consecutively, PCH can be obtained by EL model:

$$x = \begin{bmatrix} x_1 \\ x_2 \end{bmatrix}, H(x) = \frac{1}{2}Lx_1^2 + \frac{1}{2}Cx_2^2,$$

$$J_H = \begin{bmatrix} 0 & \frac{-1-d}{LC} \\ \frac{1-d}{LC} & 0 \end{bmatrix}, R_H = \begin{bmatrix} 0 & 0 \\ 0 & \frac{1}{RC^2} \end{bmatrix}, g_H = \begin{bmatrix} \frac{1}{L} \\ 0 \end{bmatrix},$$

$$\dot{x} = [J_H(d) - R_H]\frac{\partial H}{\partial z}(z) + g_H E \tag{A22}$$

The equilibrium points of the boost converter system obtained when $\dot{x}_1 = 0$ and $\dot{x}_2 = 0$ on Equations (A20) and (A21) are:

$$\bar{x}_1 = \frac{EG}{(1-d)^2}, \bar{x}_2 = \frac{E}{(1-d)} \tag{A23}$$

Considering the desired output capacitor voltage as $x_{2d} = \bar{x}_2 = V_d$, the equilibrium point to stabilize $\bar{x}$ and the constant input control $\bar{d}$ given by:

$$\bar{d} = 1 - \frac{E}{V_d}, \bar{x} = [\bar{x}_1, \bar{x}_2]^T = \left[GV_d\left(\frac{V_d}{E}\right), V_d \right]^T. \tag{A24}$$

The main objective of IDA-PCB control is to find a static function through space state feedback, $d = v(x)$. In this way the closed loop dynamics becomes a Port Controlled Hamiltonian, given by:

$$\dot{x} = [J_d(x, d) - R_d] \frac{\partial H_d}{\partial x}(x) \tag{A25}$$

Given this, the IDA-PBC control equation:

$$d = 1 - (1 - \bar{d}) \left(\frac{x_2}{V_d}\right)^\alpha. \tag{A26}$$

Substituting (A24) in (A26) derives:

$$d = 1 - \left(\frac{E}{V_d}\right) \left(\frac{x_2}{V_d}\right)^\alpha. \tag{A27}$$

Appendix A.6. Boost PFC

For Boost PFC converter, considering a rectified sinusoidal input voltage, the inductor desired current, x_{1d}, must be sinusoidal and in phase with the input voltage E. So, if:

$$E = E_{max} |sin(wt + \mathcal{E})|, I_d = \frac{2V_d^2 G}{E_{max}}, \tag{A28}$$

then

$$x_{1d} = I_d |sin(wt + \mathcal{E})|. \tag{A29}$$

Note that for a system with input E constant, we obtain some simplifications

$$\dot{x}_{1d} = 0, I_d = \frac{G}{E} V_d^2, x_{1d} = I_d. \tag{A30}$$

By deriving from a direct control, IDA-PBC control Equation (A27) does not correct the power factor. Thus, we make the following adaptation based on PBC control law:

$$d_3 = d_{PBC},$$
$$\dot{x}_{2d} = \frac{(1 - d)x_{1d} - Gx_{2d} + R_{2damp}(x_2 - x_{2d})}{C},$$
$$\bar{d}_3 = 1 - \frac{\left[E + R_{1damp}(x_1 - x_{1d}) - L\dot{x}_{1d}\right]}{x_{2d}}, \tag{A31}$$
$$d = 1 - (1 - \bar{d}_3) \left(\frac{x_2}{V_d}\right)^\alpha$$

Appendix A.7. Integral Action

In order to minimize errors in steady state of the output voltage at a desired value V_d, it is useful to add a proportional integrative term in the control law, given by:

$$G_{Int} = -k_{int} \int_0^t [x_2(s) - V_d]ds. \tag{A32}$$

Equation (A32) can be used for all converters and SFL, PBC and IDA-PBC control laws. For SFL control and boost converter:

$$\mu = 1 - \frac{[E + Lk_1(x_1 - x_{1d}) - L\dot{x}_{1d}]}{x_{2d}},$$
$$x_{2d} = -k_{int} \int_0^t [x_2(s) - V_d]ds \tag{A33}$$

Appendix B. SHIL of SEPIC Converter

The state equation describing the behaviour of the CCM SEPIC converter [40], shown in Figure A1, is given by:

$$\begin{cases} L_1 \frac{dx_1}{dt} = E - (1-d) \cdot (x_4 + x_2) \\ C_o \frac{dx_2}{dt} = (1-d) \cdot (x_1 + x_3) - Gx_2 \\ L_2 \frac{dx_3}{dt} = d \cdot x_4 - (1-d).x_2 \\ C_1 \frac{dx_4}{dt} = (1-d) \cdot x_1 - d \cdot x_3 \end{cases} \tag{A34}$$

where d is the duty cycle of the semiconductor switch.

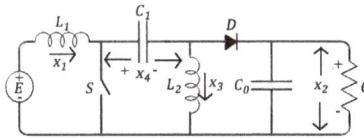

Figure A1. SEPIC converter.

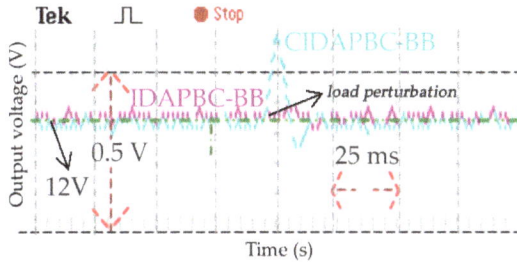

Figure A2. SHIL result for SEPIC: IDAPBC-BB (magenta) and CIDAPBC-BB (cyan). Normalized output voltage x_2 for load perturbation (30–100%).

Now, let us illustrate applications of SHIL using nonlinear equations, previously and recently found in the literature. Firstly, we consider two control laws based on IDA-PBC. In [39], Classic IDAPBC, which will refer as CIDAPBC, is applied to boost converters attaining a simplified control equation described by:

CIDAPBC-BB (CIDAPBC—Based on Boost converter):

$$\bar{d} = 1 - \frac{E}{V_d}, \quad d = 1 - (1 - \bar{d}) \left(\frac{x_2}{V_d} \right)^{k_\alpha}. \tag{A35}$$

Yet, Ref. [41] accomplish an evolution of (A35) given by:

IDAPBC-BB (IDAPBC—Based on Boost converter):

$$d = 1 - \frac{k_z E}{2Ex_2 + (k_z - 2E)x_{2d}} \tag{A36}$$

For more details, refer to [42]. In Figure A2 is sketched the output voltage x_2 in view of load change and nonlinear controllers IDAPBC-BB and CIDAPBC-BB. The following nominal conditions are: R = 10 Ω, L_1 = 146 µH, L_2 = 35 µH, C = 470 µF, E = 50 V, k_α = −0.77, k_z = −150, V_d = 12 V.

Appendix C. 3-Phase AC/DC Boost

Based on the work of [43,44], the average model(a-b-c) of the three-phase boost converter, shown in Figure A3, are given by:

$$\begin{cases} L\frac{di_a}{dt} = E_a - R_p i_a - \frac{1}{2} d_a v_o, \\ L\frac{di_b}{dt} = E_b - R_p i_b - \frac{1}{2} d_b v_o, \\ L\frac{di_c}{dt} = E_c - R_p i_c - \frac{1}{2} d_c v_o, \\ C\frac{dv_o}{dt} = \frac{1}{2}(d_a i_a + d_b i_b + d_c i_c) - G v_o. \end{cases} \tag{A37}$$

where E_a, E_b and E_c represent the line input voltages, L, C and R_p denote the inductance, the filter capacitance and the line resistance, respectively. The bipolar functions that control the semiconductor switches are d_a, d_b and d_c.

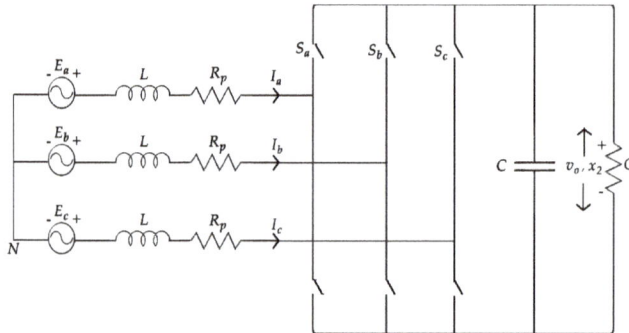

Figure A3. 3-phase AC-DC boost converter circuit.

The space-state variables of (A38) are already represented on the dq axis:

$$\begin{cases} L\dot{x}_1 = E - R_p x_1 - \frac{1}{2} x_2 \mu_d - wL x_3, \\ \frac{2}{3} C\dot{x}_2 = \frac{1}{2} x_1 \mu_d + \frac{1}{2} x_3 \mu_q - \frac{2}{3} G x_2, \\ L\dot{x}_3 = wL x_1 - -\frac{1}{2} x_2 \mu_q - R_p x_3 \end{cases} \tag{A38}$$

where w is the angular frequency of the sinusoidal voltage, x_1 is the mean current in the d-axis, x_2 is the output voltage in the capacitor, x_3 is the average current in the q-axis, μ_d and μ_q are the duty cycles and E_m is the amplitude of input voltage:

Thereby, the control goal is always to find the equations for μ_d and μ_q. With dq/abc transformations, the signals of the duty cycle $d_{a,b,c}$ are synthesized and then the corresponding PWM signal is produced for input to the converter. The converter state variables (currents on the dq axis of the inductor x_1 and x_3 and the voltage on the capacitor x_2) and the references (desired values of the output voltage V_d and the currents on the inductor $x_{1d} = I_d$ and $x_{3d} = 0$) feedback the nonlinear controller, given by:

$$\mu_d = \frac{2}{V_d}\left[-R_p x_{1d} + k_1(x_1 - x_{1d}) + E_m\right], \tag{A39}$$

$$\mu_q = \frac{2}{V_d}\left[wL x_{1d} + k_2 x_3\right], \tag{A40}$$

$$x_{1d} = \frac{1}{2}\left[\frac{E_m}{R_p} - \sqrt{\frac{E_m^2}{R_p^2} - \frac{4V_d^2}{RR_p}}\right]. \tag{A41}$$

For demonstrating the evaluation of the method applied to 3-phase boost converter, Figure A4 shows the SHIL experimental results in view of the nominal conditions: R = 40 Ω, R_p = 0.1 Ω, L= 5 mH, C = 2200 μF, E_m = 80 V, k_1 = 50, k_2 = 20, V_d = 200 V. For further information, refer to [43].

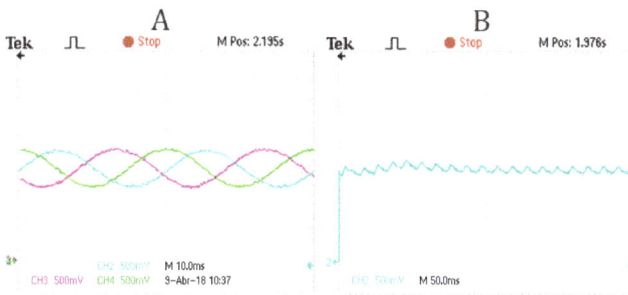

Figure A4. SHIL experimental result of 3-phase boost converter). Input voltage ((**A**): E_a—blue, E_a—magenta, E_c—green) and normalized output voltage x_2 (1.5 V DAC output voltage correspond to V_d = 200 V) using nonlinear control (A39-19)—(**B**).

Appendix D. Other Applications

The proposed methodology is used to evaluate another study cases: a PFC (Power Factor Correction) Boost converter [45] and the SST (Solid-State Transformer) [46]. In the last case, 32 control loops are embedded in the same DSP, which also justifies the use of the approach in more complex applications.

References

1. Shoushtari, M.; Dutt, N. SAM: Software-Assisted Memory Hierarchy for Scalable Manycore Embedded Systems. *IEEE Embed. Syst. Lett.* **2017**, *9*, 109–112. [CrossRef]
2. Noureen, S.; Shamim, N.; Roy, V.; Bayne, S. real-time Digital Simulators: A Comprehensive Study on System Overview, Application, and Importance. *Int. J. Res. Eng.* **2017**, *4*, 266–277. [CrossRef]
3. Grégoire, L.; Cousineau, M.; Seleme, S., Jr.; Ladoux, P. real-time Simulation of Interleaved Converters with Decentralized Control. In Proceedings of the International Conference on Renewable Energies and Power Quality (ICREPQ 2016), Madrid, Spain, 4–6 May 2016; pp. 1–6.
4. Li, Y.; Xu, X.; Sun, X.; Xue, H.; Jiang, H.; Qu, Y. Theoretical and experimental analytical study of powertrain system by hardware-in-the-loop test bench for electric vehicles. *Int. J. Veh. Syst. Model. Test.* **2017**, *12*, 44–71. [CrossRef]
5. Fernández-Álvarez, A.; Portela-García, M.; García-Valderas, M.; López, J.; Sanz, M. HW/SW Co-Simulation System for Enhancing Hardware-in-the-Loop of Power Converter Digital Controllers. *IEEE J. Emerg. Sel. Top. Power Electron.* **2017**, *5*, 1779–1786. [CrossRef]
6. Vardhan, H.; Akin, B.; Jin, H. A Low-Cost, High-Fidelity Processor-in-the Loop Platform: For Rapid Prototyping of Power Electronics Circuits and Motor Drives. *IEEE Power Electron. Mag.* **2016**, *3*, 18–28. [CrossRef]
7. Motahhir, S.; El Ghzizal, A.; Sebti, S.; Derouich, A. MIL and SIL and PIL tests for MPPT algorithm. *Cogent Eng.* **2017**, *4*, 1378475. [CrossRef]
8. Bélanger, J.; Venne, P.; Paquin, J. The What, Where and Why of Real-Time Simulation. Available online: https://www.opal-rt.com/wp-content/themes/enfold-opal/pdf/L00161_0436.pdf (accessed on 1 June 2018).
9. Erickson, R.W.; Maksimovic, D. *Fundamentals of Power Electronics*; Springer Science & Business Media: New York, NY, USA, 2007.
10. Sira-Ramirez, H.; Perez-Moreno, R.A.; Ortega, R.; Garcia-Esteban, M. Passivity-based controllers for the stabilization of DC-to-DC power converters. *Automatica* **1997**, *33*, 499–513. [CrossRef]
11. Sanders, S.R. Nonlinear Control of Switching Power Converters. Ph.D. Thesis, Massachusetts Institute of Technology, Cambridge, MA, USA, 1989.
12. Rosa, A.H.R. Estudo e Comparação de Técnicas de Controle não Lineares Aplicadas a Conversores Estáticos de Potência. Ph.D. Thesis, UFMG, Belo Horizonte, Brazil, 2015.

13. Seleme, S.I.; Rosa, A.H.R.; Morais, L.M.F.; Donoso-Garcia, P.F.; Cortizo, P.C. Evaluation of adaptive passivity-based controller for power factor correction using a boost converter. *IET Control Theory Appl.* **2012**, *6*, 2168–2178. [CrossRef]

14. Rosa, A.; Morais, L.M.; Seleme, I. A study and comparison of nonlinear control techniques apply to second order power converters using HIL simulation. In Proceedings of the 2016 12th IEEE International Conference on Industry Applications (INDUSCON), Curitiba, Brazil, 20–23 November 2016; pp. 1–8.

15. Wang, Y.; Yu, H.; Yu, J. The modeling and control of buck-boost converter based on energy-shaping theory. In Proceedings of the IEEE International Conference on Industrial Technology (ICIT 2008), Chengdu, China, 21–24 April 2008; pp. 1–6.

16. Ortega, R.; Garcia-Canseco, E. Interconnection and damping assignment passivity-based control: A survey. *Eur. J. Control* **2004**, *10*, 432–450. [CrossRef]

17. Ortega, R.; Van Der Schaft, A.; Maschke, B.; Escobar, G. Interconnection and damping assignment passivity-based control of port-controlled Hamiltonian systems. *Automatica* **2002**, *38*, 585–596. [CrossRef]

18. Raviraj, V.; Sen, P.C. Comparative study of proportional-integral, sliding mode, and fuzzy logic controllers for power converters. *IEEE Trans. Ind. Appl.* **1997**, *33*, 518–524. [CrossRef]

19. Fu, J.; Jin, Y.; Zhao, J. Nonlinear control of power converters: A new adaptive backstepping approach. *Asian J. Control* **2009**, *11*, 653–656. [CrossRef]

20. Beccuti, A.G.; Papafotiou, G.; Morari, M. Explicit model predictive control of the boost dc-dc converter. *IFAC Proc. Vol.* **2006**, *39*, 315–320. [CrossRef]

21. Almér, S.; Mariéthoz, S.; Morari, M. Piecewise affine modeling and control of a step-up DC-DC converter. In Proceedings of the American Control Conference (ACC), Baltimore, MD, USA, 30 June–2 July 2010; pp. 3299–3304.

22. Morais, L.M.F.; Santos Filho, R.M.; Cortizo, P.C.; Seleme, S.I.; Garcia, P.F.D.; Seixas, P.F. Pll-based repetitive control applied to the single-phase power factor correction using boost converter. In Proceedings of the 35th Annual Conference of IEEE Industrial Electronics (IECON'09), Porto, Portugal, 3–5 November 2009; pp. 737–742.

23. Kancherla, S.; Tripathi, R. Nonlinear average current mode control for a DC-DC buck converter in continuous and discontinuous conduction modes. In Proceedings of the 2008 IEEE Region 10 Conference (TENCON 2008), Hyderabad, India, 19–21 November 2008; pp. 1–6.

24. He, W.; Ortega, R.; Machado, J.E.; Li, S. An Adaptive Passivity-Based Controller of a Buck-Boost Converter with a Constant Power Load. *arXiv* **2017**, arXiv:1712.07792.

25. Khalil, H.K. *Noninear Systems*; Prentice-Hall: Upper Saddle River, NJ, USA, 1996.

26. El Aroudi, A.; Giaouris, D.; Iu, H.H.; Hiskens, I. A review on stability analysis methods for switching mode power converters. *IEEE J. Emerg. Sel. Top. Circuits Syst.* **2015**, *5*, 302–315. [CrossRef]

27. Sira-Ramírez, H.; Ortega, R.; García-Esteban, M. Adaptive passivity-based control of average dc-to-dc power converter models. *Int. J. Adapt. Control Signal Proc.* **1998**, *12*, 63–80. [CrossRef]

28. Hosseinzadeh, M.; Yazdanpanah, M.J. Robust adaptive passivity-based control of open-loop unstable affine non-linear systems subject to actuator saturation. *IET Control Theory Appl.* **2017**, *11*, 2731–2742. [CrossRef]

29. Seker, M.; Zergeroglu, E. Nonlinear control of flyback type DC to DC converters: An indirect backstepping approach. In Proceedings of the 2011 IEEE International Conference on Control Applications (CCA), Denver, CO, USA, 28–30 September 2011; pp. 65–69.

30. Rosa, A.; Silva, M.; Campos, M.; Santana, R.; Cortizo, P.; Mendes, M.; Morais, L.; Seleme, I. Hil simulation of non linear control methods applied for buck-boost and flyback converters. In Proceedings of the 2017 Brazilian Power Electronics Conference (COBEP), Juiz de Fora, Brazil, 19–22 November 2017; pp. 1–6.

31. Zhang, M.; Borja, P.; Ortega, R.; Liu, Z.; Su, H. PID Passivity-Based Control of Port-Hamiltonian Systems. *IEEE Trans. Autom. Control* **2018**, *63*, 1032–1044. [CrossRef]

32. Balluchi, A.; Benvenuti, L.; Engell, S.; Geyer, T.; Johansson, K.H.; Lamnabhi-Lagarrigue, F.; Lygeros, J.; Morari, M.; Papafotiou, G.; Sangiovanni-Vincentelli, A.L.; et al. Hybrid control of networked embedded systems. *Eur. J. Control* **2005**, *11*, 478–508. [CrossRef]

33. Parameter Tuning and Signal Logging with Serial External Mode. Document Description. Available online: https://www.mathworks.com/help/supportpkg/texasinstrumentsc2000/examples/parameter-tuning-and-signal-logging-with-serial-external-mode.html (accessed on 30 May 2017).

34. Márquez-Contreras, R.; Rodríguez-Cortés, H.; Spinetti-Rivera, M. Revisiting IDA-PBC, Open-Loop Control, and Modeling for the Boost DC-DC Power Converter. In Proceedings of the Latin American Congress of Automatic Control, Rio de Janeiro, Brazil, 5–6 November 2008.

35. Sira-Ramirez, H.; deNieto, M.D. A Lagrangian approach to average modeling of pulsewidth-modulation controlled DC-to-DC power converters. *IEEE Trans. Circuits Syst. I Fundam. Theory Appl.* **1996**, *43*, 427. [CrossRef]

36. Stadlmayr, R.; Schlacher, K. An energy-based control strategy for DC/DC power converters. In Proceedings of the 2009 European Control Conference (ECC), Budapest, Hungary, 23–26 August 2009; pp. 3967–3972.

37. Yildiz, H.A.; Goren-Sumer, L. Lagrangian modeling of DC-DC buck-boost and flyback converters. In Proceedings of the European Conference on Circuit Theory and Design (ECCTD 2009), Antalya, Turkey, 23–27 August 2009; pp. 245–248.

38. Seleme, S.I., Jr.; Morais, L.M.F.; Rosa, A.H.R.; Torres, L.A.B. Stability in passivity-based boost converter controller for power factor correction. *Eur. J. Control* **2013**, *19*, 56–64. [CrossRef]

39. Rodriguez, H.; Ortega, R.; Escobar, G. A new family of energy-based non-linear controllers for switched power converters. In Proceedings of the IEEE International Symposium on Industrial Electronics (ISIE 2001), Pusan, Korea, 12–16 June 2001; Volume 2, pp. 723–727.

40. Ma, H.; Li, Y.; Lai, J.; Zheng, C.; Xu, J. An Improved Bridgeless SEPIC Converter without Circulating Losses and Input Voltage Sensing. *IEEE J. Emerg. Sel. Top. Power Electron.* **2018**, *6*, 1447–1455. [CrossRef]

41. Zhang, M.; Ortega, R.; Liu, Z.; Su, H. A new family of interconnection and damping assignment passivity-based controllers. *Int. J. Robust Nonlinear Control* **2017**, *27*, 50–65. [CrossRef]

42. Rosa, A.; de Souza, T.; Morais, L.; Seleme, S., Jr. Adaptive and Nonlinear Control Techniques Applied to SEPIC Converter in DC-DC, PFC, CCM and DCM Modes Using HIL Simulation. *Energies* **2018**, *11*, 602. [CrossRef]

43. Lee, T.S. Lagrangian modeling and passivity-based control of three-phase AC/DC voltage-source converters. *IEEE Trans. Ind. Electron.* **2004**, *51*, 892–902. [CrossRef]

44. Rodriguez, L.; Jones, V.; Oliva, A.R.; Escobar-Mejía, A.; Balda, J.C. A new SST topology comprising boost three-level AC/DC converters for applications in electric power distribution systems. *IEEE J. Emerg. Sel. Top. Power Electron.* **2017**, *5*, 735–746. [CrossRef]

45. Rosa, A.; Morais, L.; Seleme, S., Jr. Practical hybrid solutions based on nonlinear controllers applied to PFC boost converter. *Przeglkad Elektrotech.* **2018**, *1*, 10–16. [CrossRef]

46. Rodrigues, W.; Oliveira, T.; Morais, L.; Rosa, A. Voltage and Power Balance Strategy without Communication for a Modular Solid State Transformer Based on Adaptive Droop Control. *Energies* **2018**, *11*, 1802. [CrossRef]

electronics

MDPI

Article

Exploring the Limits of Floating-Point Resolution for Hardware-In-the-Loop Implemented with FPGAs

Alberto Sanchez [1,*], **Elías Todorovich** [2,3] **and Angel de Castro** [1]

1   HCTLab Research Group, Universidad Autonoma de Madrid, 28049 Madrid, Spain; angel.decastro@uam.es
2   Facultad de Ciencias Exactas, Universidad Nacional del Centro de la Provincia de Buenos Aires,
    Tandil B7001BBO, Argentina; etodorov@exa.unicen.edu.ar
3   Faculty of Engineering, FASTA University, Mar del Plata B7600, Argentina
*   Correspondence: alberto.sanchezgonzalez@uam.es; Tel.: +34-914-97-3614

Received: 7 September 2018 ; Accepted: 26 September 2018; Published: 27 September 2018

Abstract: As the performance of digital devices is improving, Hardware-In-the-Loop (HIL) techniques are being increasingly used. HIL systems are frequently implemented using FPGAs (Field Programmable Gate Array) as they allow faster calculations and therefore smaller simulation steps. As the simulation step is reduced, the incremental values for the state variables are reduced proportionally, increasing the difference between the current value of the state variable and its increments. This difference can lead to numerical resolution issues when both magnitudes cannot be stored simultaneously in the state variable. FPGA-based HIL systems generally use 32-bit floating-point due to hardware and timing restrictions but they may suffer from these resolution problems. This paper explores the limits of 32-bit floating-point arithmetics in the context of hardware-in-the-loop systems, and how a larger format can be used to avoid resolution problems. The consequences in terms of hardware resources and running frequency are also explored. Although the conclusions reached in this work can be applied to any digital device, they can be directly used in the field of FPGAs, where the designer can easily use custom floating-point arithmetics.

Keywords: hardware-in-the-loop; floating-point; fixed-point; real-time emulation; field programmable gate array

---

1. Introduction

Digital control for power converters has been growing during the past two decades [1–5]. Despite all the advantages of digital control, the debugging process of this type of control is more complex because the power converter is an analog system while the control is digital. Hardware-in-the-loop (HIL) is a technique that consists in the hardware implementation of mathematical models that represent a real system. HIL simulation presents numerous advantages such as having a safe environment to test controllers, allowing the use of the controller in its final implementation, even before building the real plant to be controlled. HIL techniques are being increasingly implemented using computers [6–11] and also digital devices like FPGAs (Field Programmable Gate Array) [12–17]. The latter make it possible to perform complex calculations faster. Thus, it is not surprising that several companies have released commercial HIL products [18–20].

Arithmetics used in HIL systems have a noteworthy impact in speed, hardware resources needed for the model, the complexity to design the model, and the accuracy of the system. Fixed-point arithmetics provide optimized operations in terms of area and speed. In [21], a comparison between fixed-point and floating-point arithmetics, in the context of FPGA-based HIL systems, was presented. Results showed that floating-point required ten times as many logic resources as well as it ran 10 times slower than fixed-point. For that reason, many HIL systems are based on fixed-point arithmetics when there are hard temporal restrictions [22–26].

The main drawback of fixed-point is that the implementation is more complex because the designer has to define the number of bits of the integer and fractional parts. Thus, the maximum representable value and the required resolution need to be calculated for every signal. However, in floating-point arithmetics, the designer does not take this definition into consideration, as an IEEE-754 single-precision floating-point number can store values up to $\pm 2^{127}$, and the resolution is optimized in every calculation. This is accomplished by the floating-point libraries which automatically adapt the point location through the exponent field. Because of this remarkable advantage of floating-point, most HIL models actually use floating-point arithmetics [9,27,28], including commercial implementations [18–20].

Floating-point arithmetics for FPGAs were not viable in the past as there were no support libraries, and all the logic had to be implemented by the designer. However, with the release of floating-point support libraries, such as float_pkg of the VHDL-2008 Standard, it is easy to include floating-point arithmetics in a VHDL design. Lucia et al. [27] presented one of the first examples of a HIL model using floating-point in an FPGA.

In the literature not many cases of floating-point numerical issues for HIL systems have been reported, as the earliest purposes of HIL was to simulate complex systems with relatively low natural frequencies and integration steps of tens or hundreds of microseconds. With the advances in FPGAs, HIL technique started to be used for new applications, such as power electronics. Firstly, it was applied to converter models with low switching frequencies (kHz or tens of kHz). However, to simulate converters with medium to high switching frequency (hundreds of kHz or MHz), the integration step should be reduced accordingly and the system may present numerical problems, and as a result, obtain wrong simulations. The numerical issues are not related to overflows, as the exponent is automatically adapted. The problem is that, as the integration step is reduced, the increments of every step are smaller, and resolution issues may arise.

This paper explores the limits of floating-point arithmetics for HIL systems, and how to predict the floating-point format needed for accurate simulations. It explores not only the standard formats but also custom formats that can be used thanks to the VHDL-2008 standard libraries or any other libraries.

The rest of the paper is organized as follows. Section 2 shows the application example to illustrate the resolution problems for floating-point HIL simulations. Section 3 explains where the limits for single precision floating-point arithmetics are and how many bits would be necessary to increase the accuracy if needed. Sections 4 and 5 show the experimental and synthesis results respectively. Finally, Section 6 gives the conclusions.

2. Application Example

In this paper a PFC (Power Factor Correction) boost converter is used as an application example. The PFC technique allows regulating the output voltage while reducing the input current harmonics, so the converter behaves as a resistor emulator to the mains. The schematic of a boost converter is shown in Figure 1, excluding the previous diode bridge for ac/dc operation. The parameters of this plant are shown in Table 1. This boost configuration is proposed in an Infineon Design Note [29].

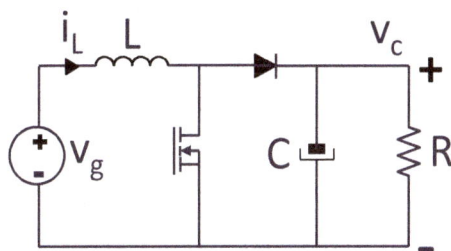

Figure 1. Boost converter topology.

Table 1. Boost converter parameters used in Section 4.

Parameter	C	L	V_{in}	V_{out}	*Power*
Value	540.5 μF	416.5 μH	230 V	400 V	400 W

The converter can be modeled using the state variables of the system: the inductor current and the capacitor voltage. Therefore, both variables should be updated every simulation step.

The behavior of the inductor and the capacitor can be described using the following equations:

$$v_L = L \cdot \frac{di_L}{dt} \tag{1}$$

$$i_C = C \cdot \frac{dv_{out}}{dt} \tag{2}$$

These equations can be discretized using different numerical methods but the simplest method to be used is explicit Euler [30]. While this method presents several disadvantages such as greater local and global error and risk of instability, these problems are negligible whenever very small integration steps (below microseconds) are used, so it is frequently used for HIL systems in power electronics [13,31–33]. Therefore, the previous Equations (1) and (2) can be discretized and the state variables can be defined as:

$$i_L(n) = i_L(n-1) + \frac{\Delta t}{L} \cdot v_L(n-1)$$

$$v_c(n) = v_c(n-1) + \frac{\Delta t}{C} \cdot i_C(n-1) \tag{3}$$

where dt has been converted into Δt, which is the simulation step, i.e. the time between two calculations of the model.

As can be seen in the previous equations, the state variables depend on the inductor voltage and the capacitor current. These values depend on the conduction state of the switch and the diode, so several states should be considered.

If the switch is closed, the inductor voltage is $v_g - 0$, while the capacitor current is $-i_R$. In this case, the state variables are defined as:

$$i_L(n) = i_L(n-1) + \frac{\Delta t}{L} \cdot v_g(n-1)$$

$$v_c(n) = v_c(n-1) - \frac{\Delta t}{C} \cdot i_R(n-1) \tag{4}$$

If the switch is open, the conduction state of the diode depends on the inductor current. If the current is positive, the diode is conducting (called CCM or Continuous Current Mode) and the inductor voltage is $v_g - v_c$, while the capacitor current is $i_L - i_R$, so the state variables are:

$$i_L(n) = i_L(n-1) + \frac{\Delta t}{L} \cdot (v_g(n-1) - v_c(n-1))$$

$$v_c(n) = v_c(n-1) + \frac{\Delta t}{C} \cdot (i_L(n-1) - i_R(n-1)) \tag{5}$$

Finally, if the inductor is fully discharged, the diode stops conducting (called DCM or Discontinuous Current Mode), so the capacitor current is $-i_R$:

$$i_L(n) = 0$$

$$v_c(n) = v_c(n-1) - \frac{\Delta t}{C} \cdot i_R(n-1) \tag{6}$$

The explicit Euler approach has been chosen in order to simplify the equations and get minimum simulation step. As the proposed system is based on an FPGA, both equations (inductor current and capacitor voltage) can be evaluated and updated in parallel, simplifying the resolution of the system. This is an important advantage compared with software-based HIL systems, in which equations need to be solved sequentially, one after another. Therefore, parallelization is one of the main reasons for the acceleration obtained using FPGAs. Taking advantage of this, every time step (Δt), all the state variables are updated. The accuracy of the system relies on the small value of the simulation step. However, a small time step implies tiny increments for the state variables, as both are proportional. This can lead to resolution issues as it is explained in the next section.

3. Numerical Resolution

As it was explained in the previous section, the model updates its state variables every simulation step. There is a relation between the simulation step and the switching frequency, because the model should be updated with enough intermediate steps inside a switching period. This update is necessary to detect accurately the state of the switch and the diode of the converter so, the smaller the switching period is, the smaller the simulation step should be. In other words, the relation between the switching period and the simulation step is the resolution of the duty cycle. Therefore, the simulation step, and then the increments for the state variables, should be quite small. As HIL systems were firstly applied to low switching frequency converters, numerical issues have not been thoroughly studied in the literature. However, as HIL systems are being used for higher frequency converters, resolution problems arise because the variables width is limited to be able to run the model in real-time [34].

Obtaining low numerical resolution leads to poor accuracy simulations or even an unpredictable simulation behavior. For example, if the system is in steady state and it suffers any small change in the input voltage or load, it is possible that the output voltage changes with the opposite sign rather than the expected one. Besides, the problem is difficult to detect because the system may be able to detect bigger changes in the input conditions, but not the smaller ones.

The optimal solution is to increase the variables width. However, if the width is increased, the calculations are more complex, more hardware resources must be used, and the combinational paths between the flip-flops inside the FPGA get longer. In conclusion, increasing the width leads to longer simulation steps, which have a negative impact on the accuracy of the simulation.

Taking all the previous considerations into account, a trade-off between the simulation step and resolution should be reached. The minimum number of bits can be estimated considering the relation between the maximum expected value of a variable x, $max(x)$, and the increment that should be added, Δx [34]. Besides, some extra bits, n, should be included to store the increment value with more than 1 bit of resolution:

$$width_x = \lceil \lceil \log_2 max(x) \rceil - \log_2 \Delta x \rceil + n \tag{7}$$

The first log_2 operation calculates the number of integer bits needed to store $max(x)$, that is, the exponent. The second log_2 gives the bits needed to store the increment. That second term may be negative, as the increment is usually below 0, indicating that fractional digits should be included. Therefore the subtraction calculates the number of bits needed for the significand field, as it has to store both the value of x and its increments simultaneously.

The maximum values of the state variables are easy to calculate because they are defined by the limits of the converter design. Regarding the increments, it is also easy if they are stable, e.g., in the case of a dc-dc converter in steady state, but otherwise further analysis must be done.

For a boost-based PFC configuration, which is the example of this paper, the minimum incremental values for the output capacitor voltage are reached when i_R is similar to i_L, as shown in Equations (4)–(6). Likewise, the minimum incremental values for the inductor current are reached when v_g is near 0 (in the case of closed switch).

As the number of bits is limited, infinitesimal incremental values will not be properly computed, but the designer can estimate the number of bits that will be necessary to obtain accurate resolutions. In [34], a fixed-point based HIL model for a PFC converter was presented. In that article, the model presented high accuracy when it had enough bits to store the incremental values during 95% of the ac period. We have to take into account that the remaining 5% has a minimum impact on the overall simulation because during that time the increments are smaller than the rest of the time and, therefore, almost negligible.

Following the aforementioned rule of 95%, and given the characteristics of the PFC converter proposed in Table 1 and the current and voltage waveforms from Figure 2, it is possible to calculate the minimum incremental values considered for the state variables. Figure 2 shows points A1 and A2 which correspond to the minimum considered input voltage. Likewise, points B1-4 correspond to the minimum considered difference between currents. Using these points, the minimum considered increments are as follows:

$$\frac{\Delta t}{L} \cdot v_g(n-1) = \frac{50 \text{ ns}}{416.5 \text{ } \mu\text{F}} \times 25.52 \text{ V} = 3.064 \text{ mA}$$

$$\frac{\Delta t}{C} \cdot (i_L(n-1) - i_R(n-1)) = \frac{50 \text{ ns}}{540.5 \text{ } \mu\text{F}} \times 0.0848 \text{ A} = 7.844 \text{ } \mu\text{V} \tag{8}$$

In the previous equations, a value of 50 ns was used as the simulation step (Δt). With the results of the previous equations, it is possible to calculate the number of bits needed for both variables, which are included in Section 4.

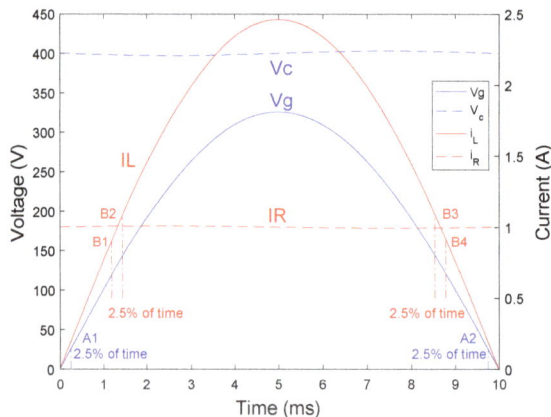

Figure 2. Current and voltages waveforms of the PFC boost converter.

Once the required significand width is obtained, it is possible to estimate the needed floating-point format. IEEE-754 floating-point format [35] defines three binary-based floating-point formats with 32, 64, and 128 bits, also known as Single, Double, and Quadruple precision, respectively. In the field of FPGAs, most cases use single-precision floating-point, as the hardware cost of Double and Quadruple precision formats, in terms of area and speed, makes it inviable to use them.

Custom floating-point formats can be used in order to reach a trade-off between speed, area and numerical resolution. The floating-point formats include separated fields for the sign, exponent, and significand, as can be seen in Figure 3. The problem in power converter HIL models is not the number of exponent bits, as the magnitudes are not extremely big or small, but the number of bits of the significand. Therefore, only the significand field should be enlarged reaching the number of bits calculated using Equations (7) and (8).

In the case of VHDL, this custom floating-point format can be defined using the floating-point package included in the VHDL-2008 standard [36]. Using this package, the codification of Equations (4)–(6) is trivial, so the format of the model variables only has to be taken into account while declaring the signals.

Section 4 compares the accuracy of the PFC boost model using different floating-point formats, and Section 5 shows the hardware results once the model is synthesized in an FPGA.

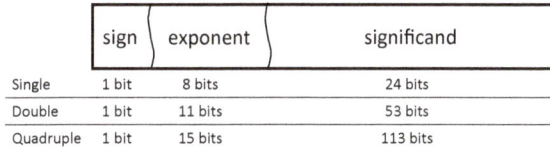

	sign	exponent	significand
Single	1 bit	8 bits	24 bits
Double	1 bit	11 bits	53 bits
Quadruple	1 bit	15 bits	113 bits

Figure 3. IEEE-754 floating-point formats.

4. Simulation Results

The previous section showed that the designer must reach a trade-off between speed, hardware resources, and accuracy. Therefore, the number of bits cannot be increased as desired. This section compares the accuracy of different floating-point formats regardless of speed and hardware resources.

It is important to note that the designed model has to be tested in open loop, without using any feedback from the control loop. If closed loop were used, the regulator would compensate the numerical errors of the model, so the whole system would probably get the desired current and voltage values at steady state. Simulations in open loop for power factor correction can be done using pre-calculated duty cycles for the PWM signal. This technique has been used previously in the literature because of its low cost, since it gets rid of the current sensor in the case of PFC converters [37–39]. Although using pre-calculated duty cycles also presents disadvantages, such as sensitivity to non-nominal conditions, it can be perfectly used to quantitatively measure the accuracy of the model, as any drift of the model will not be compensated, because it allows open-loop operation for PFC.

Table 2 presents the different experimental scenarios that have been tested, including different output loads, and cases starting at nominal steady state (400 V) and also with small capacitor voltage transients. In the case of the transients, the system will move slowly towards the nominal state following the dynamics of the chosen PFC/Boost converter, as the duty cycles are not modified in these simulations. All the scenarios have been simulated during 100 ms (10 ac semi-cycles) in order to allow the evolution of the output voltage, especially in the case of the small transients (cases 2, 4 and 6). The models have been compared with a double-precision floating-point model (53 bits for the significand field), which implements the same equations. This model should not present resolution issues and, therefore, it is used as our reference model.

Table 2. Experimental scenarios for the boost model.

	Case 1	Case 2	Case 3	Case 4	Case 5	Case 6
Output load	100%	100%	20%	20%	10%	10%
Starting capacitor voltage	400 V	410 V	400 V	410 V	400 V	410 V

Table 3 summarizes the theoretically minimum floating-point format needed for every load. These widths have been calculated using Equations (7) and (8) and the scenarios of Table 2. Regardless of the calculated widths, all scenarios have been simulated with significand widths between 24 and 32 bits.

Table 3. Significand length needed for optimal simulations using Equation (7).

	100% Load Case 1 & 2	20% Load Case 3 & 4	10% Load Case 5 & 6
i_L	$7 + n$	$5 + n$	$4 + n$
v_C	$26 + n$	$29 + n$	$30 + n$

In order to compare the simulation results quantitatively, some figures of merit should be defined. The Mean Absolute Error (MAE) between the state variables and their references (double-precision model) offers an overview of the precision of the model. The main drawback is that it does not take into account whether the error is spread out along the simulation or condensed in a small zone.

The RMSE (Root Mean Square Error) considers the square of the errors, so the main advantage of using RMSE is that it gives a high weight to large errors, and therefore it is much more sensitive to outliers. It is important to note that RMSE is the square root of the average squared error, so the results will be given directly in volts and amperes, and therefore will be directly compared with MAE.

Figure 4 shows the MAE and RMSE for the inductor current and capacitor voltage in every scenario, relative to the RMS current and RMS voltage respectively. It can be seen that the voltage calculation is more sensitive to the variable width, which is consistent with Table 3, as the width for the current variable is less restrictive. As Table 3 predicted, the scenarios with 100% of load improve when the capacitor voltage is stored using 26 bits or more. Likewise, scenarios with 20% and 10% of load improve over 30 bits.

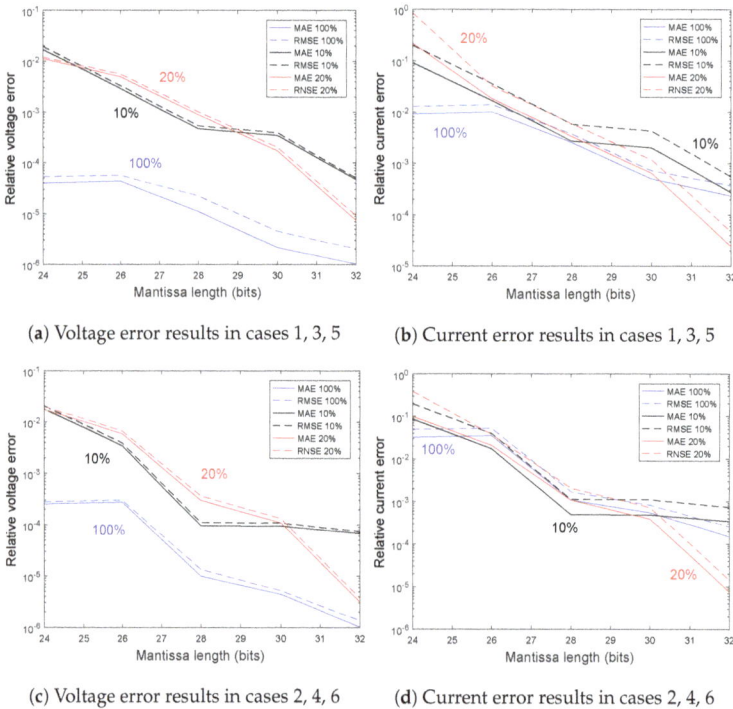

(**a**) Voltage error results in cases 1, 3, 5   (**b**) Current error results in cases 1, 3, 5

(**c**) Voltage error results in cases 2, 4, 6   (**d**) Current error results in cases 2, 4, 6

Figure 4. MAE and RMSE results of every scenario.

There are several cases that are worth mentioning. For instance, Figure 5 shows the capacitor voltage of case 6 (10% of load and voltage transient). It can be seen that, using 24 bits for the significand field, the capacitor voltage not only does not decrease but even increases. This is due to the insufficient resolution of the output voltage. As Table 3 shows that at least 30 bits are needed for the voltage state variable. The capacitor voltage increments are $(i_L - i_R) \cdot \frac{\Delta t}{C}$ when it is increasing and $-i_R \cdot \frac{\Delta t}{C}$ when it is decreasing. Taking into account that the voltage is around 400 V and the output current is around 0.1 A, the negative increments are around $-0.1 \times \frac{50 \text{ ns}}{540.5 \mu\text{F}} \approx -9 \times 10^{-6}$ V and $400 - 9 \times 10^{-6}$ V is rounded to 400 V when using 24 bits for the significand. When the switch is on, the current inductor reaches 1 A and, after switch-off, positive increments are around $(1 - 0.1) \times \frac{50 \text{ ns}}{540.5 \mu\text{F}} \approx 8 \times 10^{-5}$ V, and $400 + 8 \times 10^{-5}$ V is rounded to 400.00009 V using 24 bits for the significand. Therefore, the problem is that $-i_R \cdot \frac{\Delta t}{C}$ is so small that it is rounded to 0 when it is compared with the actual capacitor voltage. However, $(i_L - i_R) \cdot \frac{\Delta t}{C}$ is numerically bigger, so it is not rounded to 0, and the capacitor voltage only increases until the model reaches steady state.

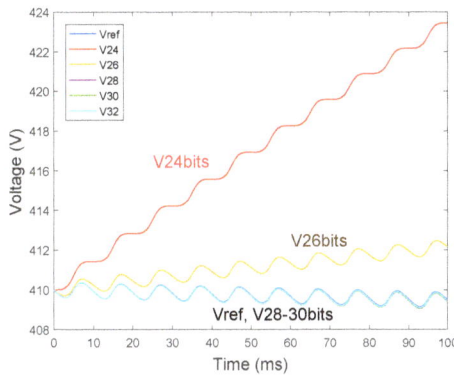

Figure 5. Simulation of 10% of load and a voltage transient (Case 6).

Case 3 is another interesting simulation (20% of load without voltage transient), which can be seen in Figure 6. In this scenario, the system is in DCM as the load is low. When using 24 bits, the output voltage decreases due to resolution problems, until it crosses the limit between DCM and CCM. However, the duty cycles are calculated to operate in DCM and the simulation is in open loop, so they are not modified. As the DCM mode needs higher duty cycles than CCM for the same values of input and output voltages, when the model enters CCM mode, the inductor suffers a short but pronounced transient. The system does not become unstable because the current transient is followed by a growth of the capacitor voltage, and the model comes back to the DCM mode.

As stated, depending on the scenarios and the variable widths, some simulations offer completely wrong waveforms. The previous statistics — MAE and RMSE — give an idea of the simulation error but do not provide clear information about the similarity of the waveforms in terms of tendency. Therefore, another statistic could be found to achieve that. The PCC (Pearson Correlation Coefficient) measures the correlation between a model and its reference, so it also offers a quick test to know if the signs of a state variable and its reference match (positive when matching and negative otherwise). The Pearson correlations for all scenarios are presented in Table 4.

As almost all the simulations present relatively similar waveforms, the PCC is around 1 in almost all cases. In fact, the case of Figure 6 presents a PCC of 0.8311, because the waveforms tendency is similar most of the time, but the current transient worsens this similarity. The case of Figure 5 gives a negative PCC because the signs of the capacitor voltage tendency are opposite. It can be seen that, only when the PCC is over 0.999, the errors of Figure 4 may be acceptable. It is also important to note that the MAE and RMSE statistics make sense only when the tendencies of the tested simulation and its reference are similar. In other words, the similarity in the tendency is reached before the error reaches acceptable values. A comparison of Tables 3 and 4 shows that the results are coherent. The theoretical widths are 26+n, 29+n and 30+n bits for 100%, 20% and 10% of load respectively, while the PCC results show that the simulations are sufficiently similar to their references (in terms of tendency) with 24, 28 and 32 bits.

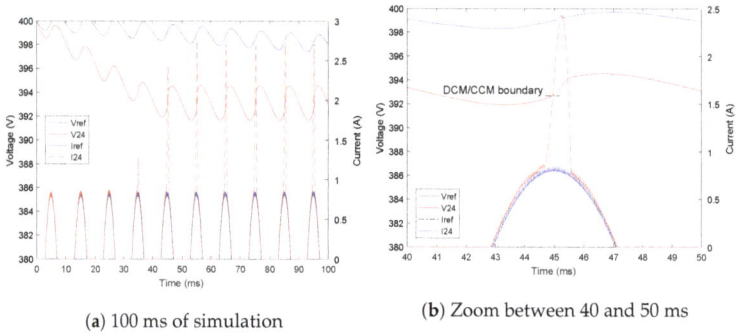

(a) 100 ms of simulation

(b) Zoom between 40 and 50 ms

Figure 6. Simulation of 20% of load (Case 3). (**a**) 100 ms of simulation. (**b**) Zoom between 40 and 50 ms.

Table 4. Pearson correlation taking the capacitor voltage.

	Case 1 100% Steady	Case 2 100% Trans	Case 3 20% Steady	Case 4 20% Trans	Case 5 10% Steady	Case 6 10% Trans
24 bits	1.0000	0.9991	0.8311	0.9832	0.8708	0.9712
26 bits	0.9999	0.9990	−0.0555	0.7582	0.9496	−0.0685
28 bits	1.0000	1.0000	1.0000	1.0000	0.9928	0.9980
30 bits	1.0000	1.0000	0.9993	0.9998	0.9976	0.9980
32 bits	1.0000	1.0000	1.0000	1.0000	1.0000	0.9994

As Table 3 shows, the necessary width for the current is much smaller than the voltage width. However, the current error of Figure 4 is similar to the voltage error using the same widths. The reason for such similarity is that both state variables depend on each other, and the accuracy issues of one variable affect the other one, so the worst case — the maximum width — should be considered.

Taking all the results into account, some conclusions can be reached regarding the necessary variable widths. First of all, the waveforms of the state variables should be similar to their references. This similarity can be measured using the PCC and selecting only the widths that obtain a PCC above 0.999, and taking the worst case – the most sensitive state variable. The previous step may choose insufficient widths. For instance, the case of 28 bits and 20% of load has a PCC of 1.0, but the MAE and RMSE errors are relatively high (around 0.5%), however, with 30 bits, the PCC is 0.993, while the MAE and RMSE errors drop to 0.1%. Therefore, once PCC has chosen a reasonable width, RMSE should also be considered. Errors below 0.1% may be sufficient for almost any application. It is possible to increase even more the width to reduce the numerical resolution error, but it is important to note that the model inherently has other error sources, such as non-idealities that have not been modeled, or tolerances in the values of C, R, etc.

For example, for this application we should look for widths that produce a PCC over 0.999 and MAE or RMSE between 0.1% and 0.5% in both state variables. These constraints would imply 28–30 bits for 100% load, 30–32 for 20% load and also 30–32 bits for 10% load. Table 3 predicts these results when using $n = 2$ or $n = 4$, verifying that the method proposed in Section 3 is a good approximation to determine the necessary width of state variables without having to run long simulations.

5. Synthesis Results

In this work, the synthesis is targeted to a device of the Arria 10 FPGA family using the Quartus Prime version 17.0 Standard Edition tool configured with the default parameters and automatic constraints except for the required clock period.

Table 5 shows area and time results for several significand widths, s. On the one hand, the area occupied in the target device for the converter is studied in terms of the logic utilization in ALMs (Adaptive Logic Module), total registers, number of DSP Blocks inferred by the synthesizer, and number of required pins. On the other hand, the maximum clock frequency for each converter configuration is evaluated. The constraint for the CLK period was set such that the synthesis and fitter (place and route) tools generate the fastest circuits.

Concerning time results, the circuit speed worsens at a rate of 1.5% per additional bit in the significand, while the area does so at the higher rate of 4%. It means that this architecture is more tolerant, in terms of speed, to resolution improvements than it is in terms of area, which is good for real-time simulations. Furthermore, the selected device, one of the largest in this family, has 427,200 ALMs.

Area and speed are closely related in digital circuits. In this case, although DSP blocks in the Arria 10 family have dedicated single-precision floating-point operators implemented in silicon, these resources are not inferred by the HDL synthesizer. Instead, the synthesizer configures the DSP block to compute the significand-part fixed-point operations. Therefore, similar results should be obtained using other FPGA families.

As the results have shown, the significand-part growth has not influenced the hardware usage or the maximum achievable frequency significantly. Therefore, the estimation method explained in Section 3, using a value between 4 and 6 for n, is valid. The simulations done in Section 4 are not necessary to estimate the state variable widths, but they were accomplished to demonstrate the validity of the method.

Table 5. Post place & route area and time results.

Significand Width	24	26	28	30	32
ALMs	4998	5413	5522	6146	6606
Regs	64	68	72	76	80
DSP	3	3	9	9	9
Pins	131	139	147	155	163
CLK const. [ns]	52	52	53	55	57
Fmax [MHz]	19.38	19.21	18.6	17.86	17.23

6. Conclusions

Thanks to the improvement in the performance of digital devices, HIL systems are starting to be used in applications that require small simulation steps (below 1 μs). The reduction of the simulation step allows more accurate simulations or make it possible to apply the technique to systems with higher natural frequencies, but the integration increments are inherently reduced. This can cause resolution problems if the arithmetics cannot handle values which are so small compared with the actual values of the state variables. In FPGA-based HIL applications, 32-bit floating-point is the most widely used arithmetic because of its simplicity from the designer point of view, along with its good performance compared with 64-bit floating-point. However, 32-bit floating-point numerical resolution is not suitable for all applications as it was observed in this work. Instead of using 64-bit arithmetics, intermediate widths can be chosen. This work has shown the limits of 32-bit floating-point for HIL simulations, and it has also provided a method to calculate the optimal width, taking into account the accuracy and the performance of the HIL system. Results have proven that the addition of few bits can dramatically improve the accuracy of the simulation but, once the numerical resolution is better than the increments, it is unproductive to increase the width.

Author Contributions: Conceptualization, A.S. and A.d.C.; methodology, A.S. and A.d.C.; software, E.T.; validation, E.T.; writing—original draft preparation and writing—review and editing, A.S., E.T. and A.d.C.

Funding: This research was funded by Spanish Ministerio de Economía y Competitividad grant number TEC2013-43017-R.

Conflicts of Interest: The authors declare no conflict of interest.

References

1. Patella, B.J.; Prodic, A.; Zirger, A.; Maksimovic, D. High-frequency digital PWM controller IC for DC-DC converters. *IEEE Trans. Power Electron.* **2003**, *18*, 438–446. [CrossRef]
2. Peterchev, A.V.; Xiao, J.; Sanders, S.R. Architecture and IC implementation of a digital VRM controller. *IEEE Trans. Power Electron.* **2003**, *18*, 356–364. [CrossRef]
3. Albatran, S.; Smadi, I.A.; Ahmad, H.J.; Koran, A. Online Optimal Switching Frequency Selection for Grid-Connected Voltage Source Inverters. *Electronics* **2017**, *6*, 110. [CrossRef]
4. Nguyen, T.D.; Hobraiche, J.; Patin, N.; Friedrich, G.; Vilain, J. A Direct Digital Technique Implementation of General Discontinuous Pulse Width Modulation Strategy. *IEEE Trans. Ind. Electron.* **2011**, *58*, 4445–4454. [CrossRef]
5. Güvengir, U.; Çadırcı, I.; Ermiş, M. On-Line Application of SHEM by Particle Swarm Optimization to Grid-Connected, Three-Phase, Two-Level VSCs with Variable DC Link Voltage. *Electronics* **2018**, *7*, 151. [CrossRef]
6. Champagne, R.; Dessaint, L.A.; Fortin-Blanchette, H. Real-time simulation of electric drives. *Math. Comput. Simul.* **2003**, *63*, 173–181. [CrossRef]
7. Short, M.; Abugchem, F.; Abrar, U. Dependable Control for Wireless Distributed Control Systems. *Electronics* **2015**, *4*, 857–878. [CrossRef]
8. Dennetière, S.; Saad, H.; Clerc, B.; Mahseredjian, J. Setup and performances of the real-time simulation platform connected to the INELFE control system. *Electr. Power Syst. Res.* **2016**, *138*, 180–187. [CrossRef]
9. Barragan, L.A.; Urriza, I.; Navarro, D.; Artigas, J.I.; Acero, J.; Burdio, J.M. Comparing simulation alternatives of FPGA-based controllers for switching converters. In Proceedings of the 2007 IEEE International Symposium on Industrial Electronics, Vigo, Spain, 4–7 June 2007; pp. 419–424. [CrossRef]
10. Short, M.; Abugchem, F. A Microcontroller-Based Adaptive Model Predictive Control Platform for Process Control Applications. *Electronics* **2017**, *6*, 88. [CrossRef]
11. Viola, F.; Romano, P.; Miceli, R. Finite-Difference Time-Domain Simulation of Towers Cascade Under Lightning Surge Conditions. *IEEE Trans. Ind. Appl.* **2015**, *51*, 4917–4923. [CrossRef]
12. Aiello, G.; Cacciato, M.; Scarcella, G.; Scelba, G. Failure analysis of AC motor drives via FPGA-based hardware-in-the-loop simulations. *Electr. Eng.* **2017**, *99*, 1337–1347. [CrossRef]
13. Herrera, L.; Li, C.; Yao, X.; Wang, J. FPGA-Based Detailed Real-Time Simulation of Power Converters and Electric Machines for EV HIL Applications. *IEEE Trans. Ind. Appl.* **2015**, *51*, 1702–1712. [CrossRef]
14. Sandre-Hernandez, O.; Rangel-Magdaleno, J.; Morales-Caporal, R.; Bonilla-Huerta, E. HIL simulation of the DTC for a three-level inverter fed a PMSM with neutral-point balancing control based on FPGA. *Electr. Eng.* **2018**, *100*, 1441–1454. [CrossRef]
15. Morales-Caporal, M.; Rangel-Magdaleno, J.; Peregrina-Barreto, H.; Morales-Caporal, R. FPGA-in-the-loop simulation of a grid-connected photovoltaic system by using a predictive control. *Electr. Eng.* **2018**, *100*, 1327–1337. [CrossRef]
16. Waidyasooriya, H.M.; Takei, Y.; Tatsumi, S.; Hariyama, M. OpenCL-Based FPGA-Platform for Stencil Computation and Its Optimization Methodology. *IEEE Trans. Parallel Distrib. Syst.* **2017**, *28*, 1390–1402. [CrossRef]
17. Fernández-Álvarez, A.; Portela-García, M.; García-Valderas, M.; López, J.; Sanz, M. HW/SW Co-Simulation System for Enhancing Hardware-in-the-Loop of Power Converter Digital Controllers. *IEEE J. Emerg. Sel. Top. Power Electron.* **2017**, *5*, 1779–1786. [CrossRef]
18. Typhoon HIL. Available online: https://www.typhoon-hil.com (accessed on 11 December 2017).
19. dSPACE. Available online: https://www.dspace.com (accessed on 11 December 2017).
20. OPAL-RT. Available online: https://www.opal-rt.com (accessed on 11 December 2017).

21. Sanchez, A.; de Castro, A.; Garrido, J. A Comparison of Simulation and Hardware-in-the-Loop Alternatives for Digital Control of Power Converters. *IEEE Trans. Ind. Inform.* **2012**, *8*, 491–500. [CrossRef]

22. Razzaghi, R.; Mitjans, M.; Rachidi, F.; Paolone, M. An automated FPGA real-time simulator for power electronics and power systems electromagnetic transient applications. *Electr. Power Syst. Res.* **2016**, *141*, 147–156. [CrossRef]

23. MacCleery, B.; Trescases, O.; Mujagic, M.; Bohls, D.M.; Stepanov, O.; Fick, G. A new platform and methodology for system-level design of next-generation FPGA-based digital SMPS. In Proceedings of the 2012 IEEE Energy Conversion Congress and Exposition (ECCE), Raleigh, NC, USA, 15–20 September 2012; pp. 1599–1606. [CrossRef]

24. Parma, G.; Dinavahi, V. Real-Time Digital Hardware Simulation of Power Electronics and Drives. *IEEE Trans. Power Deliv.* **2007**, *22*, 1235–1246. [CrossRef]

25. Matar, M.; Iravani, R. Massively Parallel Implementation of AC Machine Models for FPGA-Based Real-Time Simulation of Electromagnetic Transients. *IEEE Trans. Power Deliv.* **2011**, *26*, 830–840. [CrossRef]

26. Myaing, A.; Dinavahi, V. FPGA-Based Real-Time Emulation of Power Electronic Systems with Detailed Representation of Device Characteristics. *IEEE Trans. Ind. Electron.* **2011**, *58*, 358–368. [CrossRef]

27. Lucia, O.; Urriza, I.; Barragan, L.A.; Navarro, D.; Jimenez, O.; Burdio, J.M. Real-Time FPGA-Based Hardware-in-the-Loop Simulation Test Bench Applied to Multiple-Output Power Converters. *IEEE Trans. Ind. Appl.* **2011**, *47*, 853–860. [CrossRef]

28. Sanchez, A.; Todorovich, E.; de Castro, A. Impact of the hardened floating-point cores on HIL technology. *Electr. Power Syst. Res.* **2018**, *165*, 53–59. [CrossRef]

29. Infineon Technologies AG. *Design Note: CCM PFC Boost Converter Design (DN 2013-01)*; Rev. 1.0.; Infineon Technologies AG: Neubiberg, Germany, 2013.

30. Butcher, J.C. *The Numerical Analysis of Ordinary Differential Equations: Runge-Kutta and General Linear Methods*; Wiley-Interscience: New York, NY, USA, 1987.

31. Ibarra, L.; Rosales, A.; Ponce, P.; Molina, A.; Ayyanar, R. Overview of Real-Time Simulation as a Supporting Effort to Smart-Grid Attainment. *Energies* **2017**, *10*, 817. [CrossRef]

32. Saralegui, R.; Sanchez, A.; Martínez-García, M.S.; Novo, J.; de Castro, A. Comparison of Numerical Methods for Hardware-In-the-Loop Simulation of Switched-Mode Power Supplies. In Proceedings of the 2018 IEEE 19th Workshop on Control and Modeling for Power Electronics (COMPEL), Padova, Italy, 25–28 June 2018; pp. 1–6. [CrossRef]

33. Sutikno, T.; Idris, N.R.N.; Jidin, A.Z.; Daud, M.Z. FPGA based high precision torque and flux estimator of direct torque control drives. In Proceedings of the 2011 IEEE Applied Power Electronics Colloquium (IAPEC), Johor Bahru, Malaysia, 18–19 April 2011; pp. 122–127. [CrossRef]

34. Goñi, O.; Sanchez, A.; Todorovich, E.; de Castro, A. Resolution Analysis of Switching Converter Models for Hardware-in-the-Loop. *IEEE Trans. Ind. Inf.* **2014**, *10*, 1162–1170. [CrossRef]

35. *IEEE Standard for Floating-Point Arithmetic*; IEEE Std 754-2008; IEEE Standards: Piscataway, NJ, USA, 2008; pp. 1–70. [CrossRef]

36. *IEEE Standard VHDL Language Reference Manual*; IEEE Std 1076-2008 (Revision of IEEE Std 1076-2002); IEEE Standards: Piscataway, NJ, USA, 2009; pp. 1–626. [CrossRef]

37. Merfert, I. Analysis and application of a new control method for continuous-mode boost converters in power factor correction circuits. In Proceedings of the PESC97, Record 28th Annual IEEE Power Electronics Specialists Conference, Formerly Power Conditioning Specialists Conference 1970-71, Power Processing and Electronic Specialists Conference 1972, Saint Louis, MO, USA, 27 June 1997; Volume 1, pp. 96–102. [CrossRef]

38. Merfert, I.W. Stored-duty-ratio control for power factor correction. In Proceedings of the Fourteenth Annual Applied Power Electronics Conference and Exposition, APEC '99, Dallas, TX, USA, 14–18 March 1999; Volume 2, pp. 1123–1129. [CrossRef]

39. Sanchez, A.; de Castro, A.; López, V.M.; Azcondo, F.J.; Garrido, J. Single ADC Digital PFC Controller Using Precalculated Duty Cycles. *IEEE Trans. Power Electron.* **2014**, *29*, 996–1005. [CrossRef]

electronics

MDPI

Article

Robust DC-Link Voltage Tracking Controller with Variable Control Gain for Permanent Magnet Synchronous Generators

Seok-Kyoon Kim [1] and Kyo-Beum Lee [2,*]

[1] Seok-Kyoon Kim is with the Department of Creative Convergence Engineering, Hanbat National University, Daejeon 341-58, Korea; lotus45kr@gmail.com
[2] Kyo-Beum Lee is with the Department of Electrical and Computer Engineering, Ajou University, Suwon 443-749, Korea
*   Correspondence: kyl@ajou.ac.kr; Tel.: +82-031-219-2376

Received: 9 October 2018; Accepted: 16 November 2018; Published: 21 November 2018

Abstract: This study develops a robust DC-link voltage tracking controller with variable control gain for permanent magnet synchronous generators. The first feature is to suggest an auto-tuning algorithm to drive the control gain to update the closed-loop cut-off frequency. The second one is to prove that the proposed controller incorporating auto-tuner and disturbance observer (DOB) coerces the closed-loop system to achieve the desired voltage tracking behavior, exponentially, with the steady-state rejection property. The control performance is demonstrated by emulating a wind-turbine power system using the powerSIM (PSIM) software.

Keywords: PMSG; DC-link voltage control; variable control gain; disturbance observer

1. Introduction

Nowadays, owing to major advantages such as high power density and efficiency, the permanent magnet synchronous machines (PMSMs) have rapidly replaced induction machines (IMs) for a wide range of industrial applications, including wind power systems [1–7]. The elimination of rotor excitation results in a considerable simplification of the machine structure and control algorithm.

PMSMs can be used as generators in various industrial applications, such as wind power systems and electric vehicles, and these generators are called permanent magnet synchronous generators (PMSGs). A PMSG connected to an external mechanical system acts as a power source with variable magnitude and frequency. The three-phase inverter has to be controlled to convert the AC power coming from a PMSG to the desired DC power, which can be viewed as an AC/DC conversion system with a several power source. The previous control techniques for AC/DC converters can be utilized for PMSG DC-link voltage control applications with a slight modification. Cascade-type controllers are commonly adopted for regulating the DC-link voltage of AC/DC converters as they provide better closed-loop performance than single-loop type controllers [8,9]. A cascade-type control system has current and voltage regulators in the inner- and outer-loops, respectively. Conventionally, both the inner- and outer-loops have been realized using the proportional-integral (PI) regulators with well-tuned PI gains through trial-and-error procedures. Bode and Nyquist methods have also been used to find a reasonable PI gain to achieve the desired specification given in the frequency domain, for a specified operating point. To cover an operating region, these techniques must be repeatedly applied for each operating point. The resulting PI gains need to be assigned to a closed-loop system through an additional gain scheduling algorithm as in [10]. The feedback-linearization (FL) controller was devised to overcome this drawback; it introduces a parameter-dependent additional feed-forward compensator with PI gains [8,11,12]. The resulting closed-loop transfer function is obtained in the form of a first-order low-pass filter (LPF) with the desired cut-off frequency in the absence of model-plant mismatches. The reduction in parameter dependence was accomplished through passivity approaches [13–15], which inject a fixed damping effect to the closed-loop after shaping the desired energy function using partial converter parameter information. The same advantages can be obtained by using the adaptive [16] and sliding mode techniques [17]. There have been several attempts to incorporate the disturbance observers (DOBs) into classical PI controllers [18,19] for a better transient performance. A novel proportional-type controller embedding DOBs was suggested with a fixed closed-loop cut-off frequency for PMSG output voltage control applications [20]. The predictive techniques seek to achieve optimal control command for each control period by predicting the future state variable behavior, where the discretized dynamical equation with converter parameter values is used to predict the converter state [21,22]. The closed-loop performance driven by the extant parameter-dependent controllers could be improved by embedding an additional parameter identification mechanism as in [23–25] into the controller. Moreover, it is desirable to update the control gain automatically for a desirable cut-off frequency leading to a better closed-loop performance during transient periods.

This paper provides an auto-tuner-based robust DC-link voltage tracking controller for PMSGs driven by external mechanical systems. The parameter and load variation problems are handled by considering the perturbed dynamical model and adopting properly designed DOBs. The contributions are twofold: (a) the introduction of a closed-loop cut-off frequency update mechanism by the use of the variable control gain from the proposed auto-tuner (b) a rigorous closed-loop analysis for convergence and performance recovery without steady-state errors in the absence of tracking error integrators. Numerical verifications are conducted to demonstrate the effectiveness of the proposed controller by simulating the DC power supply system driven by a PMSG with wind turbines. The powerSIM (PSIM) software is used to emulate the wind power system, with the controller implemented using a dynamic link library (DLL).

2. PMSG Dynamics in Rotating *d-q* Axis

The time-varying coordinate transformation aligned to the electrical speed of PMSGs leads to the dynamical equations described in rotating *d-q* axis as [26]:

$$\mathbf{L}_{dq}\dot{\mathbf{i}}_{dq}(t) = -R_s\mathbf{i}_{dq}(t) + \mathbf{p}(\mathbf{i}_{dq}(t))\omega_r(t) + \mathbf{u}(t), \tag{1}$$

$$J\dot{\omega}(t) = -B\omega(t) + T_m(t) - T_e(i_d(t), i_q(t)), \ \forall t \geq 0, \tag{2}$$

where $\mathbf{i}_{dq}(t) := \begin{bmatrix} i_d(t) & i_q(t) \end{bmatrix}^T$ and $\mathbf{u}(t) := \begin{bmatrix} u_d(t) & u_q(t) \end{bmatrix}^T$ represent the state and control input vectors whose component correspond to the *d-q* axis current and terminal voltages. The nonlinearity of $\mathbf{p}(\mathbf{i}_{dq}(t))\omega_r(t)$ acts as a disturbance to the current dynamics of (1) where $\mathbf{p}(\mathbf{i}_{dq}(t)) := \begin{bmatrix} L_q i_q(t) & -(L_d i_d(t) + \lambda_{PM}) \end{bmatrix}^T$ and $\omega_r(t) := P\omega(t)$ with P being the pole pair. In the mechanical dynamics of (2), the mechanical speed is represented as $\omega(t)$ in rad/s, and external mechanical and electrical torques are denoted as $T_m(t)$ and $T_e(t)$ where $T_e(i_d(t), i_q(t)) := \frac{3}{2}P\left(\Delta L_{dq}i_d(t)i_q(t) + \lambda_{PM}i_q(t)\right), \forall t \geq 0$, with $\Delta L_{dq} := L_d - L_q$. The electrical and mechanical machine parameters are given as follows: the *d-q* inductance of L_x, $x = d, q$, stator resistance of R_s, magnet flux of λ_{PM}, viscous damping of B, and rotor inertia of J.

Figure 1 depicts a DC power supply system driven by a PMSG with an external mechanical torque where $P_{in}(t)$, $P_{cap}(t)$, and $P_{grid}(t)$ denote the input power, output capacitor power, and grid power, respectively. These power signals are related as

$$P_{cap}(t) = Cv_{dc}(t)\frac{dv_{dc}(t)}{dt} = P_{in}(t) - P_{grid}(t), \ \forall t \geq 0. \tag{3}$$

By combining the relationship of $P_{in}(t) = P_G(t) - P_{loss,inv}(t)$, $\forall t \geq 0$, with the PMSG power of $P_G(t)$ and inverter power loss of $P_{inv,loss}(t)$, the equation of (3) gives

$$C\dot{v}_{dc}(t) = \frac{\omega(t)}{v_{dc}(t)}T_e(t) - \frac{P_{loss,inv}(t)}{v_{dc}(t)} - i_{grid}(t), \; \forall t \geq 0, \tag{4}$$

with $i_{grid}(t)$ denoting the load current toward the grid.

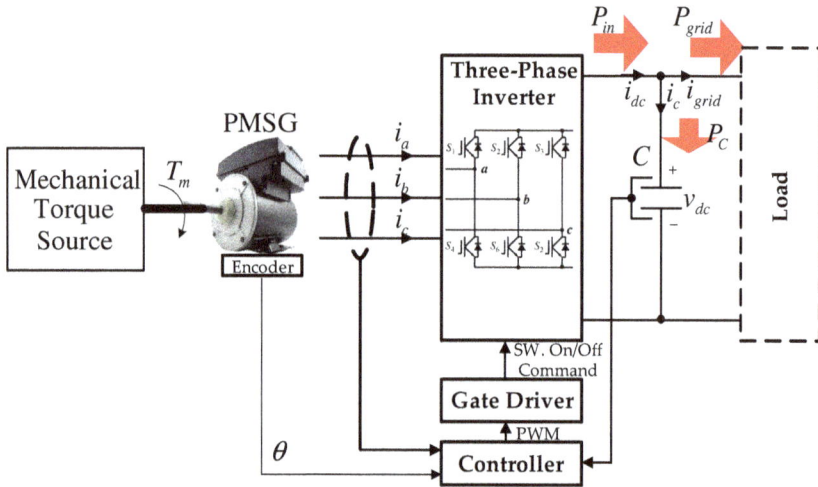

Figure 1. PMSG power system configuration.

For the system depicted in Figure 1, the *d-q* axis current of $i_x(t)$, $x = d, q$, and the DC-link voltage of $v_{dc}(t)$ are treated as state-variables for feedback, and the *d-q* axis PMSG terminal voltage of $u_x(t)$, $x = d, q$, correspond to the control input to be designed later. The system parameters of L_x, $x = d, q$, λ_{PM}, C, inverter power loss of $P_{loss,inv}(t)$, and load current of $i_{grid}(t)$ are assumed to be unknown because they can be varied significantly depending on the operating conditions.

3. DC-Link Voltage Controller Design

The goal of this section is to develop a control algorithm such that

$$\lim_{t \to \infty} v_{dc}(t) = v_{dc}^*(t), \tag{5}$$

exponentially, where the target trajectory of $v_{dc}^*(t)$ satisfies the LPF:

$$\dot{v}_{dc}^*(t) = \omega_{vc}(v_{dc,ref}(t) - v_{dc}^*(t)), \; \omega_{vc} > 0, \; \forall t \geq 0, \tag{6}$$

for a given reference signal of $v_{dc,ref}(t)$. This study investigates the tracking performance improvement by modifying the LPF dynamics of (6) as

$$\dot{v}_{dc}^*(t) = \hat{\omega}_{vc}(t)(v_{dc,ref}(t) - v_{dc}^*(t)), \; \forall t \geq 0, \tag{7}$$

where $\hat{\omega}_{vc}(t)$ denotes the time-varying cut-off frequency associated with the control gain to be designed later. The time-varying LPF of (7) is called the target dynamics in this study. Section 3.1 presents a control algorithm with the classical cascade structure, and Section 3.2 analyzes the closed-loop properties.

3.1. Controller Design

This section develops the DC-link voltage-loop, including an auto-tuner, and the current-loop in a separated manner. To this end, rewrite the DC-link voltage and current dynamics of (1) and (4) with respect to the nominal parameter values of $L_{x,0}$, $x = d, q$, $R_{s,0}$, $\lambda_{PM,0}$, and C_0 as

$$
\begin{aligned}
C_0 \dot{v}_{dc}(t) &= \frac{\omega(t)}{v_{dc}(t)} T_{e,0}(t) + d_v(t) \\
&= \frac{\omega(t)}{v_{dc}(t)} b i_q(t) + \frac{\omega(t)}{v_{dc}(t)} \frac{3}{2} P \Delta L_{dq,0} i_d(t) i_q(t) + d_v(t),
\end{aligned} \tag{8}
$$

$$
L_{dq,0} \dot{\mathbf{i}}_{dq}(t) = -R_{s,0} \mathbf{i}_{dq}(t) + \mathbf{p}_0(\mathbf{i}_{dq}(t)) \omega_r(t) + \mathbf{u}(t) + \mathbf{d}_0(t), \quad \forall t \geq 0, \tag{9}
$$

where $\quad b: \quad = \quad \frac{3}{2} P \lambda_{PM,0}, \quad T_{e,0}(t): \quad = \quad T_e(i_d(t), i_q(t)) \Big|_{L_x = L_{x,0}, \lambda_{PM} = \lambda_{PM,0}},$

$\mathbf{p}_0(\mathbf{i}_{dq}(t)) := \begin{bmatrix} L_{q,0} i_q(t) & -(L_{d,0} i_d(t) + \lambda_{PM,0}) \end{bmatrix}^T,$ and $\mathbf{L}_{dq,0}: = \operatorname{diag}\{L_{d,0}, L_{q,0}\}, \quad \forall t \geq 0.$ The disturbances of $d_v(t)$ and $\mathbf{d}_0(t): = \begin{bmatrix} d_{d,0}(t) & d_{q,0}(t) \end{bmatrix}^T$ represent the model-plant mismatches and load variations.

3.1.1. DC-Link Voltage-Loop

This section handles the q-axis current reference of $i_{q,ref}(t)$ as a design variable. First, consider the DC-link voltage dynamics of (8) as

$$
\begin{aligned}
C_0 \dot{v}_{dc}(t) &= \frac{\omega(t)}{v_{dc}(t)} b i_q(t) + \frac{\omega(t)}{v_{dc}(t)} \frac{3}{2} P \Delta L_{dq,0} i_d(t) i_q(t) + d_v(t) \\
&= \frac{\omega(t)}{v_{dc}(t)} b i_{q,ref}(t) - \frac{\omega(t)}{v_{dc}(t)} b \tilde{i}_q(t) + \frac{\omega(t)}{v_{dc}(t)} \frac{3}{2} \Delta L_{dq,0} i_d(t) i_q(t) + d_v(t), \quad \forall t \geq 0, \tag{10}
\end{aligned}
$$

with $\tilde{i}_q(t): = i_{q,ref}(t) - i_q(t), \forall t \geq 0$. Then, the q-axis current reference is proposed as

$$
i_{q,ref}(t) = \frac{v_{dc}(t)}{b\omega(t)} (C_0 \hat{\omega}_{vc}(t) \tilde{v}_{dc}(t) - \frac{\omega(t)}{v_{dc}(t)} \frac{3}{2} \Delta L_{dq,0} i_d(t) i_q(t) - \hat{d}_v(t)), \quad \forall t \geq 0, \tag{11}
$$

where the DC-link voltage tracking error is defined as $\tilde{v}_{dc}(t): = v_{dc,ref}(t) - v_{dc}(t), \forall t \geq 0$, and the variable control gain of $\hat{\omega}_{vc}(t)$ comes from the proposed auto-tuning mechanism:

$$
\dot{\hat{\omega}}_{vc}(t) = \gamma_{at}(\tilde{v}_{dc}^2(t) + \rho_{at} \tilde{\omega}_{vc}(t)), \quad \gamma_{at} > 0, \rho_{at} > 0 \, \forall t \geq 0, \tag{12}
$$

with $\tilde{\omega}_{vc}(t): = \omega_{vc} - \hat{\omega}_{vc}(t), \hat{\omega}_{vc}(0) = \omega_{vc} > 0, \forall t \geq 0$. The dynamical compensator of $\hat{d}_v(t)$ is updated as

$$
\begin{aligned}
\dot{z}_v(t) &= -l_v z_v(t) - l_v^2 C_0 v_{dc}(t) - l_v T_{e,0}(t) \frac{\omega(t)}{v_{dc}(t)}, \tag{13} \\
\hat{d}_v(t) &= z_v(t) + l_v C_0 v_{dc}(t), \, l_v > 0, \forall t \geq 0, \tag{14}
\end{aligned}
$$

with $z_v(t)$ being the state-variable, which is the DOB for the DC-link voltage loop. It is easy to see that the proposed DC-link voltage loop controller produces the closed-loop dynamics by substituting (11) in (10) as

$$
\begin{aligned}
\dot{v}_{dc}(t) &= \hat{\omega}_{vc}(t)\tilde{v}_{dc}(t) - \frac{\omega(t)}{C_0 v_{dc}(t)} b \tilde{i}_q(t) + \frac{1}{C_0} \tilde{d}_v(t) \\
&= \omega_{vc}\tilde{v}_{dc}(t) - \tilde{\omega}_{vc}(t)\tilde{v}_{dc}(t) - \frac{\omega(t)}{C_0 v_{dc}(t)} b \tilde{i}_q(t) + \frac{1}{C_0} \tilde{d}_v(t), \ \forall t \geq 0,
\end{aligned}
\tag{15}
$$

with $\tilde{d}_v(t) := d_v(t) - \hat{d}_v(t), \ \forall t \geq 0$.

Remark 1. *Unlike [20], the voltage-loop controller of (11) feedbacks the tracking error with the time-varying gain of $C_0 \hat{\omega}_{vc}(t)$ and the proposed auto-tuning mechanism of (12) updates the cut-off frequency of $\hat{\omega}_{vc}(t)$. Moreover, the resulting closed-loop behavior is also analyzed in Section 3.2 in a different way.* ◇

3.1.2. Current-Loop

Defining the current error of $\tilde{i}_{dq}(t) := i_{dq,ref}(t) - i_{dq}(t)$ with the current reference of $i_{dq,ref}(t) := \begin{bmatrix} i_{d,ref}(t) & i_{q,ref}(t) \end{bmatrix}^T$, it follows from (9) that

$$
\begin{aligned}
\mathbf{L}_{dq,0}\dot{\tilde{i}}_{dq}(t) &= \mathbf{L}_{dq,0}\dot{i}_{dq,ref}(t) - \mathbf{L}_{dq,0}\dot{i}_{dq}(t) \\
&= R_{s,0}i_{dq}(t) - \mathbf{p}_0(i_{dq}(t))\omega_r(t) - \mathbf{u}(t) + \mathbf{d}(t), \ \forall t \geq 0,
\end{aligned}
\tag{16}
$$

with $\mathbf{d}(t) := \mathbf{L}_{dq,0}\dot{i}_{dq,ref}(t) - \mathbf{d}_0(t), \ \forall t \geq 0$. A controller is suggested to stabilize the error dynamics of (16) as follows:

$$
\mathbf{u}(t) = R_{s,0}i_{dq}(t) - \mathbf{p}_0(i_{dq}(t))\omega_r(t) + \hat{\mathbf{d}}(t) + \mathbf{L}_{dq,0}\omega_{cc}\tilde{i}_{dq}(t), \ \omega_{cc} > 0, \ \forall t \geq 0,
\tag{17}
$$

with the dynamical compensator of $\hat{\mathbf{d}}(t)$ updating as

$$
\begin{aligned}
\dot{\mathbf{z}}(t) &= -l\mathbf{z}(t) - l^2\mathbf{L}_{dq,0}\tilde{i}_{dq}(t) + l(-R_{s,0}i_{dq}(t) + \mathbf{p}_0(i_{dq}(t))\omega_r(t) + \mathbf{u}(t)), \\
\hat{\mathbf{d}}(t) &= \mathbf{z}(t) + l\mathbf{L}_{dq,0}\tilde{i}_{dq}(t), \ l > 0, \ \forall t \geq 0,
\end{aligned}
\tag{18}
\tag{19}
$$

using the state-variable of $\mathbf{z}(t)$, which is called the DOB for the current-loop. It is also easy to see that the proposed controller of (17) produces closed-loop current error dynamics by combining (16) and (17) as

$$
\dot{\tilde{i}}_{dq}(t) = -\omega_{cc}\tilde{i}_{dq}(t) + \mathbf{L}_{dq}^{-1}\tilde{\mathbf{d}}(t), \ \forall t \geq 0,
\tag{20}
$$

with $\tilde{\mathbf{d}}(t) := \mathbf{d}(t) - \hat{\mathbf{d}}(t), \ \forall t \geq 0$.

3.2. Closed-Loop Properties

This section presents the useful properties of the closed-loop system and shows that the closed-loop system guarantees the control objective of (5) by analyzing the closed-loop system behaviors. Firstly, Theorem 1 provides the convergence property.

Theorem 1. *The closed-loop system controlled by the proposed control law of (11)–(14), (17)–(19) gives the output voltage convergence property. i.e.,*

$$
\lim_{t \to \infty} v_{dc}(t) = v_{dc,ref}(t)
\tag{21}
$$

as $\dot{v}_{dc,ref}(t), \dot{d}_v(t) \to 0$ and $\dot{\mathbf{d}}(t) \to \mathbf{0}$, exponentially. ◇

Proof. First, rewrite the DOBs for the DC-link voltage- and current-loops of (13), (14), (18), and (19) as

$$\dot{\hat{d}}_v - l_v C_0 \dot{v}_{dc} = -l_v(\hat{d}_v - l_v C_0 v_{dc}) - l_v^2 C_0 v_{dc} - l_v T_{e,0} \frac{\omega}{v_{dc}},$$

$$\dot{\hat{\mathbf{d}}} - l\mathbf{L}_{dq,0}\dot{\mathbf{i}}_{dq} = -l(\hat{\mathbf{d}} - l\mathbf{L}_{dq,0}\mathbf{i}_{dq}) - l^2\mathbf{L}_{dq,0}\mathbf{i}_{dq} + l(-R_{s,0}\mathbf{i}_{dq} + \mathbf{p}_0\omega_r + \mathbf{u}), \ \forall t \geq 0,$$

which gives their error dynamics (by (8) and (16)):

$$\dot{\tilde{d}}_v = -l_v\tilde{d}_v + \dot{d}_v, \ \dot{\tilde{\mathbf{d}}} = -l\tilde{\mathbf{d}} + \dot{\mathbf{d}}, \ \forall t \geq 0. \tag{22}$$

Defining the DC-link voltage error of $\tilde{v}_{dc}(t) := v_{dc,ref}(t) - v_{dc}(t)$, $\forall t \geq 0$, it follows from (15) that

$$\dot{\tilde{v}}_{dc} = \dot{v}_{dc,ref} - \dot{v}_{dc}$$

$$= -\omega_{vc}\tilde{v}_{dc} + \tilde{\omega}_{vc}\tilde{v}_{dc} + \frac{\omega}{C_0 v_{dc}}b\mathbf{e}_2^T\tilde{\mathbf{i}}_{dq} - \frac{1}{C_0}\tilde{d}_v + \dot{v}_{dc,ref}, \ \forall t \geq 0, \tag{23}$$

with $\mathbf{e}_2 := \begin{bmatrix} 0 & 1 \end{bmatrix}^T$. Then, the time-derivative along (12), (20), (22), and (23) of the positive definite function defined as

$$V := \frac{1}{2}\tilde{v}_{dc}^2 + \frac{\kappa_1}{2}\|\tilde{\mathbf{i}}_{dq}\|^2 + \frac{1}{2\gamma_{at}}\tilde{\omega}_{vc}^2 + \frac{\kappa_2}{2}\tilde{d}_v^2 + \frac{\kappa_3}{2}\|\tilde{\mathbf{d}}\|^2, \ \kappa_i > 0, \ i = 1,2,3, \ \forall t \geq 0, \tag{24}$$

is obtained as

$$\dot{V} = \tilde{v}_{dc}\dot{\tilde{v}}_{dc} + \kappa_1\tilde{\mathbf{i}}_{dq}^T\dot{\tilde{\mathbf{i}}}_{dq} - \frac{1}{\gamma_{at}}\tilde{\omega}_{vc}\dot{\tilde{\omega}}_{vc} + \kappa_2\tilde{d}_v\dot{\tilde{d}}_v + \kappa_3\tilde{\mathbf{d}}^T\dot{\tilde{\mathbf{d}}}$$

$$= \tilde{v}_{dc}\left(-\omega_{vc}\tilde{v}_{dc} + \tilde{\omega}_{vc}\tilde{v}_{dc} + \frac{\omega}{C_0 v_{dc}}b\mathbf{e}_2^T\tilde{\mathbf{i}}_{dq} - \frac{1}{C_0}\tilde{d}_v + \dot{v}_{dc,ref}\right)$$

$$+ \kappa_1\tilde{\mathbf{i}}_{dq}^T\left(-\omega_{cc}\tilde{\mathbf{i}}_{dq} + \mathbf{L}_{dq}^{-1}\tilde{\mathbf{d}}\right) - \tilde{\omega}_{vc}(\tilde{v}_{dc}^2 + \rho_{at}\tilde{\omega}_{vc})$$

$$- \kappa_2 l_v\tilde{d}_v^2 + \kappa_2\dot{d}_v\tilde{d}_v - \kappa_3 l\|\tilde{\mathbf{d}}\|^2 + \kappa_3\dot{\mathbf{d}}^T\tilde{\mathbf{d}}$$

$$= -\omega_{vc}\tilde{v}_{dc}^2 - \kappa_1\omega_{cc}\|\tilde{\mathbf{i}}_{dq}\|^2 - \rho_{at}\tilde{\omega}_{vc}^2 - \kappa_2 l_v\tilde{d}_v^2 - \kappa_3 l\|\tilde{\mathbf{d}}\|^2$$

$$+ \tilde{v}_{dc}\frac{\omega}{C_0 v_{dc}}b\mathbf{e}_2^T\tilde{\mathbf{i}}_{dq} - \frac{1}{C_0}\tilde{v}_{dc}\tilde{d}_v + \kappa_1\tilde{\mathbf{i}}_{dq}^T\mathbf{L}_{dq}^{-1}\tilde{\mathbf{d}} + \mathbf{w}^T\mathbf{y}, \ \forall t \geq 0,$$

where $\mathbf{w} := \begin{bmatrix} \dot{v}_{dc,ref} & \kappa_2\dot{d}_v & \kappa_3\dot{\mathbf{d}}^T \end{bmatrix}^T$ and $\mathbf{y} := \begin{bmatrix} \tilde{v}_{dc} & \tilde{d}_v & \tilde{\mathbf{d}}^T \end{bmatrix}^T$. Applying the Young's inequality of $\mathbf{x}^T\mathbf{y} \leq \frac{\epsilon}{2}\|\mathbf{x}\|^2 + \frac{1}{2\epsilon}\|\mathbf{y}\|^2$, $\forall \epsilon > 0$, $\forall \mathbf{x}, \mathbf{y} \in \mathbb{R}^n$ to the indefinite terms of $\dot{V}$, it holds that

$$\dot{V} \leq -\frac{\omega_{vc}}{3}\tilde{v}_{dc}^2 - \left(\kappa_1\omega_{cc} - \frac{\omega_{max}^2 b^2}{4\omega_{vc}C_0^2 v_{dc,min}^2} - \frac{1}{2}\right)\|\tilde{\mathbf{i}}_{dq}\|^2 - \rho_{at}\tilde{\omega}_{vc}^2$$

$$- \left(\kappa_2 l_v - \frac{1}{4\omega_{vc}C_0^2}\right)\tilde{d}_v^2 - \left(\kappa_3 l - \frac{\kappa_1^2\|\mathbf{L}_{dq}^{-1}\|^2}{2}\right)\|\tilde{\mathbf{d}}\|^2 + \mathbf{w}^T\mathbf{y}, \ \forall t \geq 0,$$

with ω_{max} and $v_{dc,min}$ being the maximum and minimum values of ω and v_{dc}, respectively, whose upper bound can be obtained by the constants of $\kappa_1 := \frac{1}{\omega_{cc}}\left(\frac{\omega_{max}^2 b^2}{4\omega_{vc}C_0^2 v_{dc,min}^2} + 1\right)$, $\kappa_2 := \frac{1}{l_v}\left(\frac{1}{4\omega_{vc}C_0^2} + \frac{1}{2}\right)$, and $\kappa_3 := \frac{1}{l}\left(\frac{\kappa_1^2\|\mathbf{L}_{dq}^{-1}\|^2}{2} + \frac{1}{2}\right)$ as

$$\dot{V} \leq -\frac{\omega_{vc}}{3}\tilde{v}_{dc}^2 - \frac{1}{2}\|\tilde{\mathbf{i}}_{dq}\|^2 - \rho_{at}\tilde{\omega}_{vc}^2 - \frac{1}{2}\tilde{d}_v^2 - \frac{1}{2}\|\tilde{\mathbf{d}}\|^2 + \mathbf{w}^T\mathbf{y}$$

$$\leq -\alpha V + \mathbf{w}^T\mathbf{y}, \ \forall t \geq 0, \tag{25}$$

with $\alpha := \min\{\frac{2\omega_{vc}}{3}, \frac{1}{\kappa_1}, 2\rho_{at}\gamma_{at}, \frac{1}{\kappa_2}, \frac{1}{\kappa_3}\}$, which completes the proof. $\square$

Lemma 1 presents the boundedness property of the auto-tuning algorithm of (12), which simplifies the proof of performance recovery property of Theorem 2.

Lemma 1. *The variable cut-off frequency of $\hat{\omega}_{vc}(t)$ coming from the auto-tuner of* (12) *satisfies*

$$\hat{\omega}_{vc}(t) \geq \omega_{vc}, \forall t \geq 0. \tag{26}$$

$\diamondsuit$

Proof. The auto-tuner update rule of (12) is equivalent to the expression of

$$\dot{\hat{\omega}}_{vc} = -\gamma_{at}\rho_{at}\hat{\omega}_{vc} + \gamma_{at}\rho_{at}\omega_{vc} + \gamma_{at}\tilde{v}_{dc}^2,$$

which indicates that (by integrating both sides)

$$
\begin{aligned}
\hat{\omega}_{vc} &= e^{-\gamma_{at}\rho_{at}t}\omega_{vc} + \int_0^t e^{-\gamma_{at}\rho_{at}(t-\tau)}(\gamma_{at}\rho_{at}\omega_{vc} + \gamma_{at}\tilde{v}_{dc}^2)d\tau \\
&\geq e^{-\gamma_{at}\rho_{at}t}\omega_{vc} + \gamma_{at}\rho_{at}\omega_{vc}e^{-\gamma_{at}\rho_{at}t}\int_0^t e^{\gamma_{at}\rho_{at}\tau}d\tau \\
&= \omega_{vc}, \forall t \geq 0.
\end{aligned}
$$

$\square$

Theorem 2 asserts that the proposed controller establishes the control objective of (5), that is the performance recovery property, using the result of Lemma 1.

Theorem 2. *The closed-loop system controlled by the proposed control law of* (11)–(14), (17)–(19) *ensures the performance recovery property. i.e.,*

$$\lim_{t\to\infty} v_{dc}(t) = v_{dc}^*(t) \tag{27}$$

as $\dot{d}_v(t) \to 0$ and $\dot{\mathbf{d}}(t) \to \mathbf{0}$, exponentially.

$\diamondsuit$

Proof. Using (7) and (15), the dynamics of the tracking error $\tilde{v}_{dc}^* := v_{dc}^* - v_{dc}$ is obtained as

$$\dot{\tilde{v}}_{dc}^* = -\hat{\omega}_{vc}\tilde{v}_{dc}^* + \frac{\omega}{C_0 v_{dc}}b\mathbf{e}_2^T\tilde{\mathbf{i}}_{dq} - \frac{1}{C_0}\tilde{d}_v, \forall t \geq 0, \tag{28}$$

with $\mathbf{e}_2 = \begin{bmatrix} 0 & 1 \end{bmatrix}^T$. The time-derivative along (20), (22), and (28) of the positive definite function given by

$$V^* := \frac{1}{2}(\tilde{v}_{dc}^*)^2 + \frac{c_1}{2}\|\tilde{\mathbf{i}}_{dq}\|^2 + \frac{c_2}{2}\tilde{d}_v^2 + \frac{c_3}{2}\|\tilde{\mathbf{d}}\|^2, \ c_i > 0, \ i = 1,2,3, \ \forall t \geq 0, \tag{29}$$

is given by

$$
\begin{aligned}
\dot{V}^* &= \tilde{v}_{dc}^*(-\hat{\omega}_{vc}\tilde{v}_{dc}^* + \frac{\omega}{C_0 v_{dc}}b\mathbf{e}_2^T\tilde{\mathbf{i}}_{dq} - \frac{1}{C_0}\tilde{d}_v) \\
&\quad + c_1\tilde{\mathbf{i}}_{dq}^T(-\omega_{cc}\tilde{\mathbf{i}}_{dq} + \mathbf{L}_{dq}^{-1}\tilde{\mathbf{d}}) - c_2 l_v\tilde{d}_v^2 - c_3 l\|\tilde{\mathbf{d}}\|^2 + c_2 l_v \dot{d}_v\tilde{d}_v + c_3 l\dot{\mathbf{d}}^T\tilde{\mathbf{d}} \\
&= -\hat{\omega}_{vc}(\tilde{v}_{dc}^*)^2 - c_1\omega_{cc}\|\tilde{\mathbf{i}}_{dq}\|^2 - c_2 l_v\tilde{d}_v^2 - c_3 l\|\tilde{\mathbf{d}}\|^2 \\
&\quad + \tilde{v}_{dc}^*\frac{\omega}{C_0 v_{dc}}b\mathbf{e}_2^T\tilde{\mathbf{i}}_{dq} - \tilde{v}_{dc}^*\frac{1}{C_0}\tilde{d}_v + c_1\tilde{\mathbf{i}}_{dq}^T\mathbf{L}_{dq}^{-1}\tilde{\mathbf{d}} + c_2 l_v\dot{d}_v\tilde{d}_v + c_3 l\dot{\mathbf{d}}^T\tilde{\mathbf{d}}, \ \forall t \geq 0.
\end{aligned}
$$

Applying the result of Lemma 1 and the Young's inequality to the indefinite terms of $\dot{V}^*$, it holds that

$$
\dot{V}^* \leq -\frac{\omega_{vc}}{3}(\tilde{v}_{dc}^*)^2 - (c_1\omega_{cc} - \frac{\omega_{max}^2 b^2}{4\omega_{vc}C_0^2 v_{dc,min}^2} - \frac{1}{2})\|\tilde{\mathbf{i}}_{dq}\|^2
$$
$$
-(c_2 l_v - \frac{3}{4\omega_{vc}C_0^2})\tilde{d}_v^2 - (c_3 l - \frac{c_1^2\|\mathbf{L}_{dq}^{-1}\|^2}{2})\|\tilde{\mathbf{d}}\|^2 + c_2 l_v \dot{d}_v \tilde{d}_v + c_3 l \dot{\mathbf{d}}^T \tilde{\mathbf{d}}, \ \forall t \geq 0,
$$

whose upper bound can be obtained by the constants of $c_1 := \frac{1}{\omega_{cc}}(\frac{\omega_{max}^2 b^2}{4\omega_{vc}C_0^2 v_{dc,min}^2} + 1)$, $c_2 := \frac{1}{l_v}(\frac{1}{4\omega_{vc}C_0^2} + \frac{1}{2})$,
and $c_3 := \frac{1}{l}(\frac{c_1^2\|\mathbf{L}_{dq}^{-1}\|^2}{2} + \frac{1}{2})$ as

$$
\dot{V}^* \leq -\frac{\omega_{vc}}{3}(\tilde{v}_{dc}^*)^2 - \frac{1}{2}\|\tilde{\mathbf{i}}_{dq}\|^2 - \frac{1}{2}\tilde{d}_v^2 - \frac{1}{2}\|\tilde{\mathbf{d}}\|^2 + c_2 l_v \dot{d}_v \tilde{d}_v + c_3 l \dot{\mathbf{d}}^T \tilde{\mathbf{d}}
$$
$$
\leq -\alpha^* V^* + c_2 l_v \dot{d}_v \tilde{d}_v + c_3 l \dot{\mathbf{d}}^T \tilde{\mathbf{d}}, \ \forall t \geq 0, \tag{30}
$$

with $\alpha^* := \min\{\frac{2\omega_{vc}}{3}, \frac{1}{c_1}, \frac{1}{c_2}, \frac{1}{c_3}\}$, which completes the proof. $\square$

It is not obvious for the proposed controller to ensure the offset-free property in actual implementations due to the absence of integral actions of tracking errors in the controller law of (17) and (11). Theorem 3 addresses this issue.

Theorem 3. *The control system controlled by the proposed control law of (11)–(14), (17)–(19) establishes the tracking objective of (5) without offset-errors. i.e.,*

$$
v_{dc}(\infty) = v_{dc,ref}(\infty), \tag{31}
$$

where $\lim_{t\to\infty} v_{dc}(t) = v_{dc}(\infty)$ *and* $\lim_{t\to\infty} v_{dc,ref}(t) = v_{dc,ref}(\infty)$. $\diamond$

Proof. The closed-loop dynamics of (15), (20), and (22) give the steady-state equations as

$$
0 = \hat{\omega}_{vc}(\infty)\tilde{v}_{dc}(\infty) - \frac{\omega(\infty)}{C_0 v_{dc}(\infty)}b\mathbf{e}_2^T\tilde{\mathbf{i}}_{dq}(\infty) + \frac{1}{C_0}\tilde{d}_v(\infty), \tag{32}
$$
$$
0 = -\omega_{cc}\tilde{\mathbf{i}}_{dq}(\infty) + \mathbf{L}_{dq}^{-1}\tilde{\mathbf{d}}(\infty), \tag{33}
$$
$$
0 = \tilde{d}_v(\infty), \ 0 = \tilde{\mathbf{d}}(\infty). \tag{34}
$$

The equation of (33) implies $\tilde{\mathbf{i}}_{dq}(\infty) = \mathbf{0}$ from the equation of (34), which shows that $\tilde{v}_{dc}(\infty) = 0$ by the combination of (32) and (34). Therefore, the proposed controller removes the offset errors as long as the closed-loop system reaches a steady-state. $\square$

4. Simulations

In this section, numerical verifications were carried out to demonstrate the effectiveness of the proposed scheme, and the FL controller is used for comparison. A wind power system driven by a PMSG was emulated by using the function blocks provided in the PSIM software. The control algorithms were built by C-language, which results in the DLL block. The control output signals of $u_x(t)$, $x = d, q$, were synthesized using the three-phase inverter with the pulse-wide modulation (PWM). The control and PWM periods were selected as 0.1 ms. The system parameters were chosen as

$$
R_s = 0.099 \ \Omega, \ L_x = 4.07 \ mH, \ x = d, q, \ \lambda_{PM} = 0.3166 \ Wb, \ P = 40,
$$
$$
J = 0.02 \ kgm^2, \ B = 0.000425 \ Nm/rad/s, \ C = 2350 \ \mu F, \tag{35}
$$

and the control algorithms were constructed using the nominal system parameters:

$$R_{s,0} = 0.7R_s, \ L_{x,0} = 1.5L_x, \ \lambda_{PM,0} = 1.2\lambda_{PM}, \ C_0 = 0.6C, \ x = d,q, \tag{36}$$

instead of the use of true system parameters to consider the model-plant mismatches. The wind turbine parameters were set as follows; nominal output power: 15 kW, inertia: 1.5×10^{-3} kgm^2, base rotational speed: 55 rpm, and initial rotational speed: 15 rpm. The wind pattern was emulated using the wind model based-on Weibull distribution [27]. The structure of the implemented wind power system is shown in Figure 2.

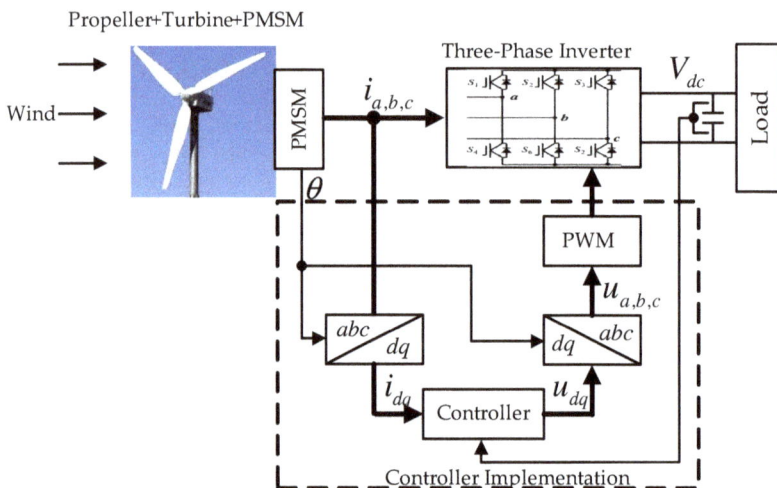

Figure 2. Closed-loop system implementation.

The control law of FL method is described as

$$u_d(t) = L_{d,0}\omega_{cc}\tilde{i}_d(t) + R_{s,0}\omega_{cc}\int_0^t \tilde{i}_d(\tau)d\tau - L_{q,0}\omega_r(t)i_q(t), \tag{37}$$

$$u_q(t) = L_{q,0}\omega_{cc}\tilde{i}_q(t) + R_{s,0}\omega_{cc}\int_0^t \tilde{i}_q(\tau)d\tau + L_{d,0}\omega_r(t)i_d(t)$$
$$+ \lambda_{PM,0}\omega_r(t), \ \forall t \geq 0, \tag{38}$$

with

$$i_{q,ref}(t) = \frac{v_{dc}(t)}{b\omega(t)}\left(2C_0\omega_{vc}\tilde{v}_{dc}(t) + C_0\omega_{vc}^2\int_0^t \tilde{v}_{dc}(\tau)d\tau\right), \ \forall t \geq 0, \tag{39}$$

where $\tilde{v}_{dc}(t) = v_{dc,ref}(t) - v_{dc}(t), \forall t \geq 0$, which gives the closed-loop transfer functions for the voltage- and current-loops:

$$\frac{I_x(s)}{I_{x,ref}(s)} = \frac{\mathcal{L}\{i_x(t)\}}{\mathcal{L}\{i_{x,ref}(t)\}} = \frac{\omega_{cc}}{s + \omega_{cc}}, \ x = d,q,$$

$$\frac{V_{dc}(s)}{V_{dc,ref}(s)} = \frac{\mathcal{L}\{v_{dc}(t)\}}{\mathcal{L}\{v_{dc,ref}(t)\}} = \frac{\omega_{vc}}{s + \omega_{vc}},$$

approximately, via pole-zero cancellation in the absence of model-plant mismatches, where $\mathcal{L}(\cdot)$ denotes the Laplace transform operator. The design parameters commonly used for the two controllers were set to $f_{cc} = 200$ Hz and $f_{vc} = 4$ Hz for $\omega_{cc} = 2\pi f_{cc} = 1256$ rad/s and $\omega_{vc} = 2\pi f_{vc} = 25.1$ rad/s.

The proposed controller was tuned as $l_v = l = 50$, $\gamma_{at} = 0.05$, and $\rho_{at} = 15/\gamma_{at}$. Note that the d-axis current reference was set to zero for simplicity.

The first verification was carried out to demonstrate the robustness improvement for several loads under the voltage tracking control mode. The DC-link voltage reference was given in the form of a pulse from 300 V to 500 V, and the closed-loop tracking behavior changes were observed for three resistive loads, $R_L = 30, 60, 120\ \Omega$. Figure 3 shows the comparison results of the DC-link voltage response, which implies that the proposed controller effectively improves the closed-loop robustness by preventing closed-loop performance variation in spite of load changes. From Figure 4, it can be seen that the proposed controller drives the q-axis current more rapidly than the FL controller, resulting in better closed-loop robustness. The corresponding cut-off frequency and DOB behaviors are presented in Figure 5, and Figure 6 shows the wind velocity pattern from the Weibull distribution.

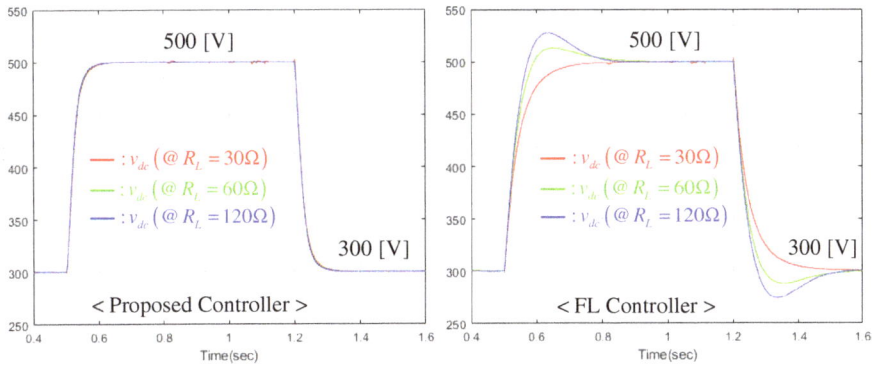

Figure 3. DC-link voltage tracking behavior changes for three loads, $R_L = 30, 60, 120\ \Omega$.

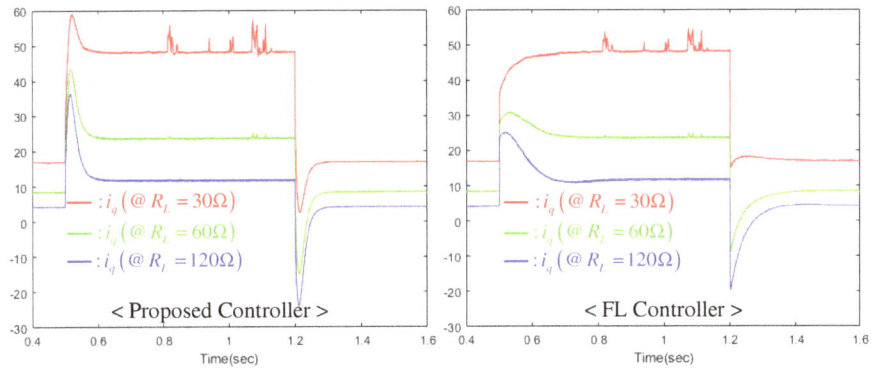

Figure 4. q-axis current response comparison result under DC-link voltage tracking control mode.

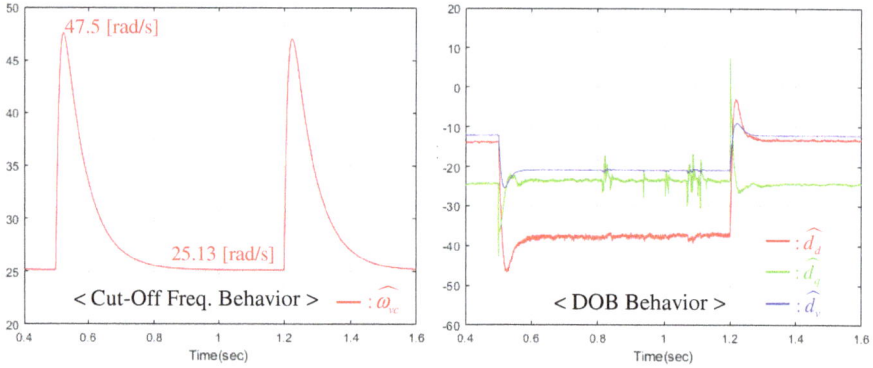

Figure 5. Cut-off frequency and DOB responses of proposed controller under DC-link voltage tracking control mode.

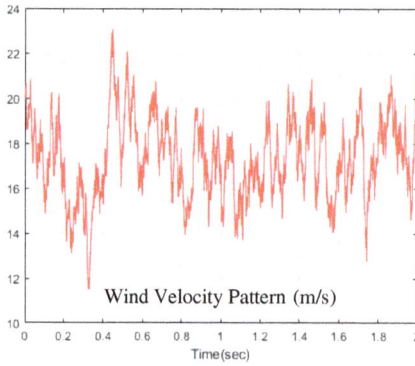

Figure 6. Wind velocity pattern from Weibull distribution.

The second verification was carried out to observe the closed-loop robustness under the voltage regulation mode with several sudden load change scenarios. The DC-link voltage reference was fixed to 300 V, and the closed-loop regulation behavior changes were observed for three resistive load change scenarios: (a) restoring the resistive load after increasing it from $R_L = 60\ \Omega$ to $R_L = 24\ \Omega$; (b) restoring the resistive load after increasing it from $R_L = 120\ \Omega$ to $R_L = 30\ \Omega$; and (c) restoring the resistive load after increasing it from $R_L = 30\ \Omega$ to $R_L = 17\ \Omega$. Figure 7, which shows the comparison result of the closed-loop DC-link voltage response, indicates that the closed-loop robustness improvement is achieved by the proposed technique, as it decreases the overshoots/undershoots considerably. The corresponding q-axis current response is given in Figure 8, which indicates that the proposed controller leads to a rapid current dynamics for a better DC-link voltage regulation performance.

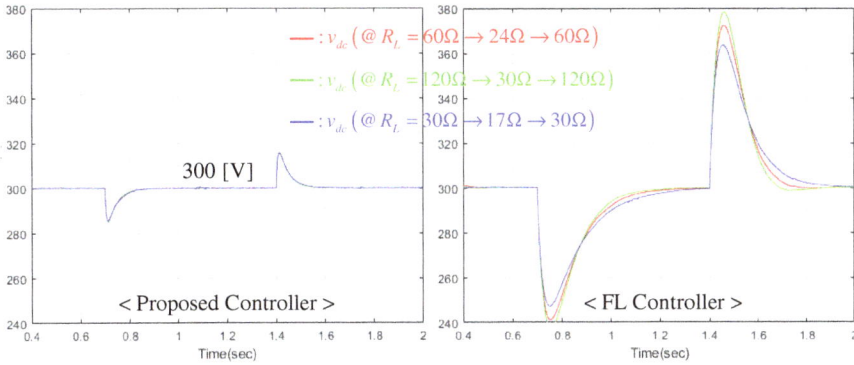

Figure 7. DC-link voltage regulation behavior changes under three resistive load change scenarios: (a) $R_L = 60\,\Omega \to 24\,\Omega \to 60\,\Omega$; (b) $R_L = 120\,\Omega \to 30\,\Omega \to 120\,\Omega$; and (c) $R_L = 30\,\Omega \to 17\,\Omega \to 30\,\Omega$.

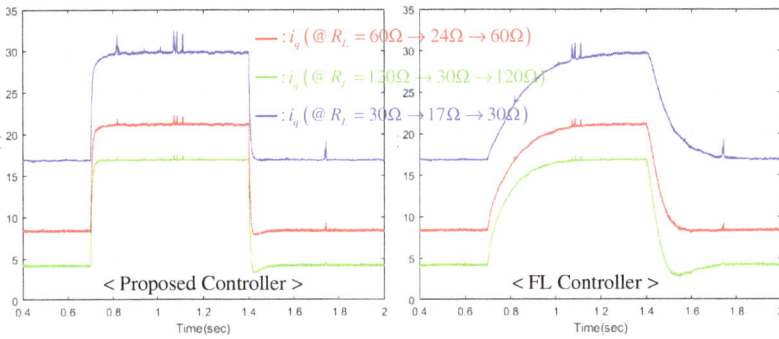

Figure 8. q-axis current response comparison under DC-link voltage regulation mode.

The third verification shows the output voltage tracking performance improvement at the resistive load $R_L = 60\,\Omega$, compared with a recent DOB-based technique introduced in [20], which is given by

$$u_d(t) = R_{s,0}i_d(t) - L_{q,0}\omega_r(t)i_q(t) + L_{d,0}\lambda_{cc}\tilde{i}_d(t) + \hat{d}_d(t), \tag{40}$$

$$u_q(t) = R_{s,0}i_q(t) + L_{d,0}\omega_r(t)i_d(t) + \lambda_{PM,0}\omega_r(t) + \frac{L_{q,0}\omega(t)}{C_0 v_{dc}(t)}b\tilde{v}_{dc}^*$$
$$+ L_{q,0}\lambda_{cc}\tilde{i}_q(t) + \hat{d}_q(t), \ \lambda_{cc} > 0, \ \forall t \geq 0, \tag{41}$$

with the DOBs of

$$\hat{d}_x(t) = \zeta_x(t) + l_x L_{x,0}\tilde{i}_x(t), \ x = d, q,$$

$$\dot{\zeta}_d(t) = -l_d\zeta_d(t) - l_d^2 L_{d,0}\tilde{i}_d(t) + l_d\left(-R_{s,0}i_d(t) + L_{q,0}\omega_r(t)i_q(t) + u_d(t)\right), \ l_d > 0, \tag{42}$$

$$\dot{\zeta}_q(t) = -l_q\zeta_q(t) - l_q^2 L_{q,0}\tilde{i}_q(t)$$
$$+ l_q\left(-R_{s,0}i_q(t) - L_{d,0}\omega_r(t)i_d(t) - \lambda_{PM,0}\omega_r(t) + u_q(t)\right), l_q > 0, \ \forall t \geq 0. \tag{43}$$

The corresponding q-axis current reference is updated as

$$i_{q,ref}(t) = \frac{v_{dc}(t)}{b\omega(t)}\left(C_0\lambda_{vc}\tilde{v}_{dc}^*(t) - \frac{\omega(t)}{v_{dc}(t)}\frac{3}{2}P\Delta L_{dq,0}i_d(t)i_q(t) + \hat{d}_v(t)\right), \ \lambda_{vc} > 0, \ \forall t \geq 0, \tag{44}$$

along with the output voltage tracking error of $\tilde{v}_{dc}^*(t) = v_{dc}^*(t) - v_{dc}(t)$ where the desired trajectory of $v_{dc}^*(t)$ comes from

$$\dot{v}_{dc}^*(t) = \omega_{vc}\left(v_{dc,ref}(t) - v_{dc}^*(t)\right), \ \forall t \geq 0, \tag{45}$$

and the DOB for (44) is given by

$$
\begin{aligned}
\hat{d}_v(t) &= \zeta_v(t) + l_v C_0 \tilde{v}_{dc}^*(t), \\
\dot{\zeta}_v(t) &= -l_v\zeta_v(t) - l_v^2 C_0 \tilde{v}_{dc}^*(t) + l_v \frac{\omega(t)}{v_{dc}(t)}\left(bi_q(t) + \frac{3}{2}P\Delta L_{dq,0}i_d(t)i_q(t)\right), \ l_v > 0, \ \forall t \geq 0.
\end{aligned}
$$

The cut-off frequency of ω_{vc} in (45) was set to the same as the initial cut-off frequency of proposed auto-tuning algorithm. The rest of design parameters were adjusted as $\lambda_{vc} = 125.6$, $\lambda_{cc} = 1256$, $l_v = l_d = l_q = 314$. This stage used the three-kinds of sinusoidal reference signals given as $v_{dc,ref}(t) = 500 + 100\sin(2\pi f t)$ with $f = 3, 6, 12$ Hz. Figure 9 shows the tracking performance comparison results, which clearly observes a frequency response performance improvement thanks to the proposed auto-tuning algorithm.

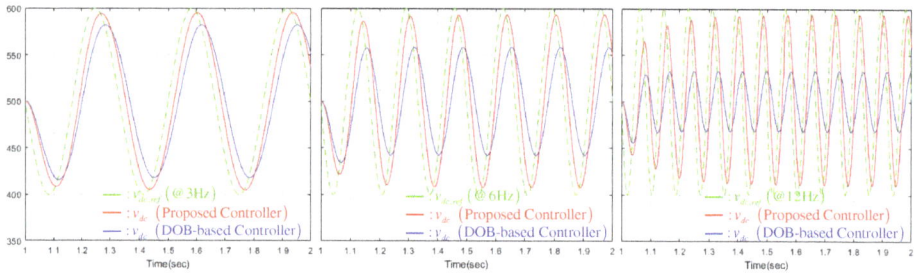

Figure 9. DC-link voltage tracking performance comparison under sinusoidal reference signals at resistive load $R_L = 60\ \Omega$.

In the last verification, the efficacy of the proposed auto-tuner was investigated under the DC-link voltage tracking control mode with a pulse reference from 300 V to 500 V and a resistive load of $R_L = 60\ \Omega$. The initial cut-off frequency was decreased to $f_{vc} = 2$ Hz for $\omega_{cc} = 2\pi f_{vc} = 12.56$ rad/s to demonstrate the effectiveness of the proposed auto-tuner clearly. The resulting DC-link voltage tracking behavior changes are shown in Figure 10, including the variable cut-off frequency dynamics from the auto-tuner. This shows that the proposed auto-tuner effectively boosts the closed-loop tracking performance during transient periods.

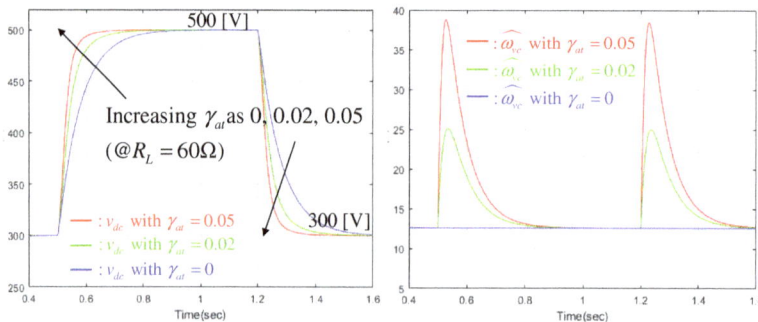

Figure 10. DC-link voltage tracking performance improvement tendency as increasing $\gamma_{at} = 0, 0.02, 0.05$.

From these numerical data, it is seen that the useful closed-loop properties that were proven in Section 3.2 result in practical advantages compared with the FL controller, which depends on the true system parameters. Thus, the proposed controller can be considered as a promising solution for several industrial applications.

5. Conclusions

This study offers a novel DC-link voltage tracking control algorithm with convincing numerical data from realistic simulations. The proposed controller automatically updates the cut-off frequency using the embedded auto-tuning algorithm. Rigorous closed-loop analysis was also presented for the performance recovery and convergence properties. The closed-loop performance improvement was confirmed by simulating a wind power system controlled by the proposed controller. A guideline for systematic and optimal design parameter determination will be provided in a future study with experimental data.

Author Contributions: S.-K.K. surveyed the backgrounds of this research, designed the whole control and estimation algorithms, and performed the simulations so as to show the novelties of the proposed technique. K.-B.L. supervised and financially supported this study.

Funding: This research was supported by a grant (17TLRP-C135446-01,Development of Hybrid Electric Vehicle Conversion Kit for Diesel Delivery Trucks and its Commercialization for Parcel Services) from Transportation & Logistics Research Program (TLRP) funded by Ministry of Land, Infrastructure and Transport of Korean government, and was supported by "Human Resources Program in Energy Technology" of the Korea Institute of Energy Technology Evaluation and Planning (KETEP), granted financial resource from the Ministry of Trade, Industry & Energy, Republic of Korea. (No. 20174030201660)

Conflicts of Interest: The authors declare no conflict of interest.

References

1. Tiwari, R.; Krishnamurthy, K.; Neelakandan, R.B.; Padmanaban, S. Neural Network Based Maximum Power Point Tracking Control with Quadratic Boost Converter for PMSG-Wind Energy Conversion System. *Electronics* **2018**, *7*, 20. [CrossRef]
2. Chen, M.H. Use of Three-Level Power Converters in Wind-Driven Permanent-Magnet Synchronous Generators with Unbalanced Loads. *Electronics* **2015**, *4*, 339–358. [CrossRef]
3. Kamel, T.; Abdelkader, D.; Said, B.; Padmanaban, S.; Iqbal, A. Extended Kalman Filter Based Sliding Mode Control of Parallel-Connected Two Five-Phase PMSM Drive System. *Electronics* **2018**, *7*, 14. [CrossRef]
4. Kim, S.K.; Lee, K.B. Robust Offset-Free Speed Tracking Controller of Permanent Magnet Synchronous Generator for Wind Power Generation Applications. *Electronics* **2018**, *7*, 48. [CrossRef]
5. Asensio, A.P.; Gomez, S.A.; Rodriguez-Amenedo, J.L.; Plaza, M.G.; Carrasco, J.E.G.; de las Morenas, J.M.A.M. A Voltage and Frequency Control Strategy for Stand-Alone Full Converter Wind Energy Conversion Systems. *Energies* **2018**, *11*, 474. [CrossRef]
6. Liu, Z.; Li, K.; Sun, Y.; Wang, J.; Wang, Z.; Sun, K.; Wang, M. A Steady-State Analysis Method for Modular Multilevel Converters Connected to Permanent Magnet Synchronous Generator-Based Wind Energy Conversion Systems. *Energies* **2018**, *11*, 461. [CrossRef]
7. Kim, S.K.; Song, H.; Lee, J. Adaptive Disturbance Observer-Based Parameter-Independent Speed Control of an Uncertain Permanent Magnet Synchronous Machine for Wind Power Generation Applications. *Energies* **2015**, *8*, 4496–4512. [CrossRef]
8. Kazmierkowski, M.P.; Krishnan, R.; Blaabjerg, F. *Control in Power Electronics- Selected Problems*; Academic Press: Cambridge, MA, USA, 2002.
9. Dixon, J.W.; Ooi, B.T. Indirect current control of a unity power factor sinusoidal current boost type three phase rectifier. *IEEE Trans. Ind. Electron.* **1988**, *35*, 508–515. [CrossRef]
10. Matausek, M.R.; Jeftenic, B.I.; Miljkovic, D.M.; Bebic, M.Z. Gain scheduling control of DC motor drive with field weakening. *IEEE Trans. Ind. Electron.* **1996**, *43*, 153–162. [CrossRef]
11. Lee, T.S. Input-output linearization and zero-dynamics control of three-phase AC/DC voltage-source converters. *IEEE Trans. Power Electron.* **2003**, *31*, 11–22.

12. Lee, D.C.; Lee, G.M.; Lee, K.D. DC-bus voltage control of three-phase AC/DC PWM converters using feedback linearization. *IEEE Trans. Ind. Appl.* **2000**, *36*, 826–833.
13. Lee, T.S. Lagrangian Modeling and Passivity Based Control of Three Phase AC to DC Voltage Source Converters. *IEEE Trans. Ind. Electron.* **2004**, *51*, 892–902. [CrossRef]
14. Gomez, M.H.; Ortega, R.; Lagarrigue, F.L.; Escobar, G. Adaptive PI Stabilization of Switched Power Converters. *IEEE. Trans. Control Syst. Technol.* **2010**, *18*, 688–698. [CrossRef]
15. Flores, D.D.P.; Scherpen, J.M.A.; Liserre, M.; de Vries, M.M.J.; Kransse, M.J.; Monopoli, V.G. Passivity-Based Control by Series/Parallel Damping of Single-Phase PWM Voltage Source Converter. *IEEE. Trans. Control Syst. Technol.* **2012**, *3*, 459–471.
16. De Araujo Ribeiro, R.L.; de Oliveira Alves Rocha, T.; de Sousa, R.M.; dos Santos, E.C.; Lima, A.M.N. A Robust DC-Link Voltage Control Strategy to Enhance the Performance of Shunt Active Power Filters Without Harmonic Detection Schemes. *IEEE Trans. Ind. Electron.* **2015**, *62*, 803–813. [CrossRef]
17. Vazquez, S.; Sanchez, J.A.; Carrasco, J.M.; Leon, J.I.; Galvan, E. A model based direct power control for three-phase power converters. *IEEE. Trans. Ind. Electron.* **2008**, *55*, 1647–1657. [CrossRef]
18. Salomonsson, D.; Sannino, A. Direct Power Control Based on Natural Switching Surface for Three-Phase PWM Rectifiers. In Proceedings of the 2007 IEEE 42nd IAS Annual Meeting, Conference Record of the Industry Applications Conference, New Orleans, LA, USA, 23–27 September 2007.
19. Wang, C.; Li, X.; Guo, L.; Li, Y.W. A Nonlinear-Disturbance-Observer-Based DC-Bus Voltage Control for a Hybrid AC/DC Microgrid. *IEEE Trans. Power Electron.* **2014**, *29*, 6162–6177. [CrossRef]
20. Kim, S.K. Proportional-Type Performance Recovery DC-Link Voltage Tracking Algorithm for Permanent Magnet Synchronous Generators. *Energies* **2017**, *10*, 1387. [CrossRef]
21. Rodriguez, J.; Cortes, P. *Predictive Control of Power Converters and Electrical Drives*; Wiley-IEEE Press: Hoboken, NJ, USA, 2012.
22. Bouafia, A.; Gaubert, J.P.; Krim, F. Predictive Direct Power Control of Three-Phase Pulsewidth Modulation (PWM) Rectifier Using Space-Vector Modulation (SVM). *IEEE Trans. Power Electron.* **2010**, *25*, 228–236. [CrossRef]
23. Gan, C.; Wu, J.; Hu, Y.; Yang, S.; Cao, W.; Kirtley, J.L. Online Sensorless Position Estimation for Switched Reluctance Motors Using One Current Sensor. *IEEE Trans. Power Electron.* **2016**, *31*, 7248–7263. [CrossRef]
24. Zhao, L.; Huang, J.; Chen, J.; Ye, M. A Parallel Speed and Rotor Time Constant Identification Scheme for Indirect Field Oriented Induction Motor Drives. *IEEE Trans. Power Electron.* **2016**, *31*, 6494–6503. [CrossRef]
25. Perez, J.N.H.; Hernandez, O.S.; Caporal, R.M.; de J R Magdaleno, J.; Barreto, H.P. Parameter Identification of a Permanent Magnet Synchronous Machine based on Current Decay Test and Particle Swarm Optimization. *IEEE Lat. Am. Trans.* **2013**, *11*, 1176–1181. [CrossRef]
26. Krause, P.; Wasynczuk, O.; Sudhoff, S. *Analysis of Electric Machinery*; IEEE Press: Baltimore, MD, USA, 1995.
27. Mathew, S. *Wind Energy: Fundamentals, Resource Analysis and Economics*; Springer: Berlin, Germany, 2006.

electronics

MDPI

Article

Comparative Analysis of Two and Four Current Loops for Vector Controlled Dual-Three Phase Permanent Magnet Synchronous Motor

Muhammad Ahmad [1], Zhixin Wang [1,*], Sheng Yan [2], Chengmin Wang [1], Zhidong Wang [3], Chenghzi Zhu [4] and Hua Qin [5]

[1] School of Electronics, Information and Electrical Engineering (SEIEE), Shanghai Jiao Tong University, Shanghai 200240, China; muhammad.ahmad14@hotmail.com (M.A.); wangchengmin@sjtu.edu.cn (C.W.)
[2] State Grid Corporation of China, Beijing 100031, China; sheng-yan@sgcc.com.cn
[3] State Grid Economic and Technological Research Institute Co., LTD, Beijing 102209, China; wangzhidong@chinasperi.sgcc.com.cn
[4] State Grid Zhejiang Electric Power Co., Ltd., Hangzhou 310007, China; zhu_chengzhi@zj.sgcc.com.cn
[5] Shanghai Huacheng Elevator Technology Co., Ltd., Shanghai 201111, China; info@sh-huacheng.com
* Correspondence: wangzxin@sjtu.edu.cn; Tel.: +86-21-3420-4527

Received: 19 September 2018; Accepted: 18 October 2018; Published: 23 October 2018

Abstract: Dual three-phase (DTP) permanent magnet synchronous motors (PMSMs) are specialized machines which are commonly used for high power density applications. These machines offer the merits of high efficiency, high torque density, and superior supervisor fault tolerant capability compared to conventional three-phase AC-machines. However, the electrical structure of such machines is very complicated, and as such, control becomes challenging. In conventional vector controlled DTP-PMSMs drives, the components of the dq-subspace are associated with electromechanical energy conversion, and two currents, i.e., I_d and I_q belonging to this subspace, are used in feedback-loops for control. Such orthodox control methods can cause some anomalies e.g., the voltage source inverter's (VSI) dead time effect and other nonlinear factors, and can induce large harmonics. These glitches can be greatly alleviated by the introduction of the two-extra current loops to directly control the currents in Z_1Z_2-subspace in order to suppress the insertion of harmonics. In this paper, two approaches—one with two-current loops and other with four-current loops—for vector controlled DTP-PMSMs are investigated with the aid of different MATLAB-based simulations. Furthermore, in the paper, the influence of additional current loops is quantified using simulation-based results.

Keywords: current control loops; dual three-phase (DTP) permanent magnet synchronous motors (PMSMs); space vector pulse width modulation (SVPWM); vector control; voltage source inverter

---

1. Introduction

In recent years, governments across the globe have raised the issues of climate changes due to the emission of harmful gases into the atmosphere [1]. The foremost source of these emissions is automobiles, which are running mainly on hydrocarbon fuels. Owing to this fact, the greener aspects of automobiles have increasingly received the attention of governments and the commercial sector. All these aspects have led, in recent times, to important technological advancements in the field of hybrid-electric road vehicles, electric ship propulsion, more electric aircraft, and electric locomotive traction [2,3]. By and large, all transport-related applications necessitate high power density motors, where these motors are principally used as a means of propulsion. Conventional drives systems, which utilize three-phase AC-machinery, are not inevitably the appropriate solution for such applications;

due to high power density, dual-three phase PMSMs are used for these purposes [4,5]. Moving forward, these days, AC-machines are fed by inverters, which practically decouple the machine from the mains; thereby, the number phase then cannot be reserved to three phases anymore. Certainly, electrical machines with more than three phases are named *multiphase machines*, and possess several advantages over three-phase AC-machines. Some of these are discussed in [2,3] and [6–9], and are briefly reiterated below.

For multi-phase AC-machines, power can be separated into more than three phases, which leads to a reduction in switching current stress. Therefore, the rating requirements of power semiconductor devices can be reduced to a great extent. Thus, the cost of machine drives will be reduced. It must be noted not only that high rating power semiconductor devices are expensive, but also, that they are sometimes not easy to access. Other potential advantages of DTP AC-machines over typical three-phase AC-machines include superior fault tolerance capability, reduced torque pulsations, and improved reliability etc.

Two different topologies of DTP-PMSMs, i.e., symmetric and asymmetric, have frequently been discussed in the literature. A DTP-PMSM utilizes a pair of three-phase windings shifted by 30 degrees and 60 degrees is termed as asymmetric and symmetric winding configurations respectively [10]; both formations are illustrated in Figure 1. The difference between symmetric and asymmetric winding configurations is explained in [11]. Since asymmetric winding arrangements for DTP-PMSMs possess low iron losses and low torque ripple, they have widely been used by researchers. Therefore, for this study, asymmetric winding configuration is considered. Furthermore, there are two circuit topologies which can be used for inverter-fed DTP-PMSMs: one with a single natural (where all six-phases of the machine connected to a single natural point), and another with two isolated natural (where two sets of windings i.e., ACE and BDF are connected separately with two naturals, ACE and BDF represents the six-phases of machine), as depicted in Figure 2.

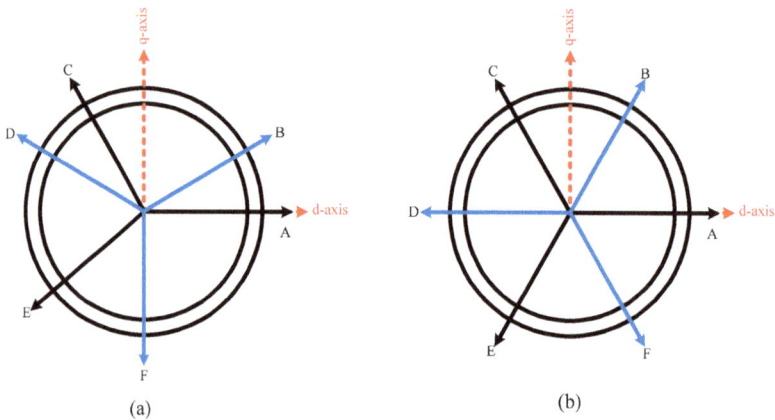

(a)   (b)

Figure 1. This figure represents the Winding configuration dual three-phase (DTP) AC-machines: (**a**) Asymmetric configuration; (**b**) Symmetric configuration.

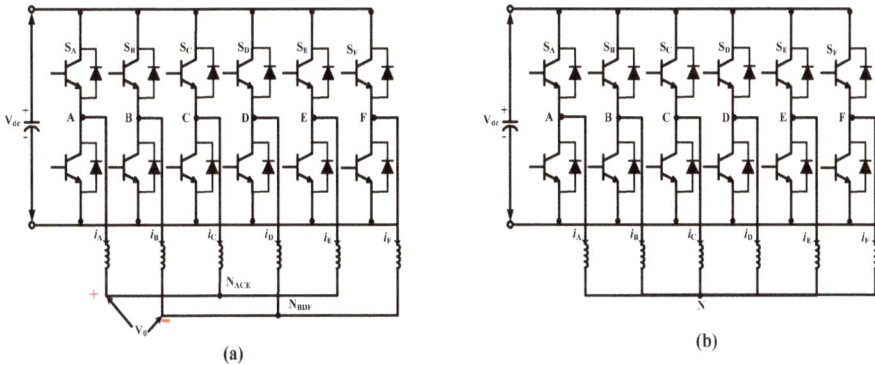

Figure 2. This figure symbolizes the winding configurations of inverter fed DTP-PMSM (permanent magnet synchronous motor): (**a**) Double neutral topology; (**b**) Single neutral topology.

Theoretical and implementation aspects of both these approaches have been comprehensively explained in [10]. Double natural topology is used for the simulation analyses in this work. Although DTP AC-machines offer several advantages over conventional three-phase AC-machines, their complicated electrical structures can be considered as one of the disadvantages. For VSI-fed DTP drive systems, in practice, six independent currents flow to the machine, which can be viewed as six-dimensional electromechanical structures. Therefore, developing any control structure in the original nature frame for DTP AC-machine is almost impossible. In practice, the control topologies for DTP AC-machines are usually developed in two-dimensional, stationary $\alpha\beta$-frames or two-dimensional, rotating dq-frames instead of reference. Comprehensive details concerning transformations are given in the following section. It is important to mention at this stage that once the machine parameters are altered to an $\alpha\beta$-frame (orthogonal in nature), these parameters are separated into three different subspaces.

These are named the $\alpha\beta$, Z_1Z_2, and O_1O_2 subspaces. Among these, only the components linked with $\alpha\beta$-subspaces contribute to electromechanical energy transfiguration, whereas components of the remaining two subspaces can induce large harmonics into machine currents. In conventional vector control for DTP-PMSMs, two currents associated with the $\alpha\beta$-subspace (i.e., I_α and I_β) are used in feedback loops. But, the inverter dead time effect and other nonlinear factors (e.g., magnetic saturation) can give rise to currents in the Z_1Z_2 subspace of the machine, resulting in poor harmonics of the phase currents.

In recent investigations by different researchers, a method to suppress the harmonic inclusion implicates the direct control of current in Z_1Z_2 subspace. For that purpose, two additional current loops for I_{Z1} and I_{Z2} are introduced for control [12]. Two approaches of vector control for DTP-PMSMs, one with conventional two current loops and other with unorthodox four current loops, will be the focus of this paper. At first stage, a simulation-based analysis was conducted for both approaches. In the later stages of this paper, brief compression will be done to understand the effectiveness of extra loops and their practical implications for vector control of DTP-PMSMs. Moreover, in [12], an FFT-based (FFT stands for Fast Fourier Transform) analysis current was ignored, which helped us to quantify the influence of additional current loops. In this paper, once the results for both control approaches were obtained, FFT-based analyses were carried out, so that the influence of additional current loops for the purpose of control could be quantified.

The different sections of this paper are organized as follows: Section 2 deals with the theory of space vector decomposition and analytical model of DTP-PMSMs. In Section 3, the theory and implementation of two and four current-loops for vector control are discussed with a brief theoretical background of DTP-SVPWM (space vector pulse width modulation). Comparative analysis of performance for both approaches with the help of simulation-based results are the subject of Section 4; Section 5 concludes this paper.

2. Space Vector Decomposition and Machine Model

2.1. Space Vector Decomposition

In VSI feed DTP-PMSM, six independent currents can flow into the machine. Owing to this fact, six-phase AC-machines can be viewed as a six-dimensional plane. Consequently, modelling and control of such machines in their original reference frame become a little more complicated. These challenges were greatly alleviated by the introduction of space vector decomposition theory [10]. According to this concept, six-dimensional machine parameters can be transformed into a stationary frame ($\alpha\beta$-frame) which is orthogonal in nature. This transformation procedure chose six vectors from the original nature frame and formed a new basis for six-dimensional space. This transformation can be accomplished by Equation (1).

$$T_{6S/2S} = \frac{1}{\sqrt{3}} \begin{bmatrix} 1 & \frac{\sqrt{3}}{2} & -\frac{1}{2} & -\frac{\sqrt{3}}{2} & -\frac{1}{2} & 0 \\ 0 & \frac{1}{2} & \frac{\sqrt{3}}{2} & \frac{1}{2} & -\frac{\sqrt{3}}{2} & -1 \\ 1 & -\frac{\sqrt{3}}{2} & -\frac{1}{2} & \frac{\sqrt{3}}{2} & -\frac{1}{2} & 0 \\ 0 & \frac{1}{2} & -\frac{\sqrt{3}}{2} & \frac{1}{2} & \frac{\sqrt{3}}{2} & -1 \\ 1 & 0 & 1 & 0 & 1 & 0 \\ 0 & 1 & 0 & 1 & 0 & 1 \end{bmatrix} \tag{1}$$

Once machine parameters have been transformed into the orthogonal frame, they can then be alienated in three subspaces which are referred to as the $\alpha\beta$, Z_1Z_2, and O_1O_2 subspaces. Moreover, it is important to mention that the terms linked with $\alpha\beta$-subspace are responsible for electromechanical conversion, while those associated with other subspaces can give rise to harmonics and can cause losses. The transformation matrix retains the following properties: [10]

- Fundamental machine constituents and harmonics of order k = 12m ± 1 where (m = 1, 2, 3......) are mapped on the $\alpha\beta$-plane.
- Harmonic components with order k = 6m ± 1 where (m = 1, 3, 5......) are mapped into Z_1Z_2-subspace or on Z_1Z_2-plane.
- Harmonic components with order k = 3m where (m = 1, 3, 5......) are mapped into O_1O_2-subspace or on O_1O_2-plane. The harmonics components mapped into this plane are considered as non-electromechanical energy conversion components and form the zero-sequence components.

2.2. Machine Model in Stationary Frame

Since the fundamental machine model in the original reference frame is beyond the scope of this paper, it is purposely omitted. Only the machine model in the stationary frame is reported, which is pivotal for developing the vector control of six-phase AC-machines. A set of equations which are used to express the machine model are presented below [6–8,13]:

$$u_\alpha = R_s i_\alpha + L_\alpha (di_\alpha/dt) + L_{\alpha\beta}(di_\beta/dt) + k_e w_e \cos\theta \tag{2}$$

$$u_\beta = R_s i_\beta + L_\beta (di_\beta/dt) + L_{\alpha\beta}(di_\alpha/dt) + k_e w_e \sin\theta \tag{3}$$

$$u_{z1} = R_s i_{z1} + L_{z1}(di_{z1}/dt) + L_{z1z2}(di_{z2}/dt) \tag{4}$$

$$u_{z2} = R_s i_{z2} + L_{z2}(di_{z2}/dt) + L_{z1z2}(di_{z1}/dt) \tag{5}$$

$$u_{o1} = R_s i_{o1} + L_{o1}(di_{o1}/dt) \tag{6}$$

$$u_{o2} = R_s i_{o2} + L_{o2}(di_{o2}/dt) \tag{7}$$

$$T_e = 3P + \left[\psi_\alpha i_\beta - \psi_\beta i_\alpha \right] \tag{8}$$

The specifics of coefficients attached to the above equations are as follows: $\begin{bmatrix} u_\alpha & u_\beta \end{bmatrix}^T$, $\begin{bmatrix} u_{z1} & i_{z2} \end{bmatrix}^T$ and $\begin{bmatrix} u_{o1} & u_{o2} \end{bmatrix}^T$ are voltage vectors, likewise $\begin{bmatrix} i_\alpha & i_\beta \end{bmatrix}^T$, $\begin{bmatrix} i_{z1} & i_{z2} \end{bmatrix}^T$ and $\begin{bmatrix} i_{o1} & i_{o2} \end{bmatrix}^T$ are the currents vectors in the $\alpha\beta$, Z_1Z_2, and O_1O_2 subspaces respectively; R_s is the stator resistance. Other variables include $L_\alpha, L_\beta, L_{z1}, L_{z2}, L_{o1}$, and L_{o2}, the self-inductance in different subspaces; $L_{\alpha\beta}, L_{z1z2}$ and L_{o1o2} are the mutual inductance in different subspaces. Moving forward, "*P*" represents the number of pole pairs and "*ψ*" represents the flux linkages in the $\alpha\beta$-subspace. It must be observed that the machine model utilized for this project involves stator windings with two isolated neutrals and fed by VSI. Consequently, the vectors of O_1O_2-subspace are mapped at the origin. As a result, no current flows in the O_1O_2-subspace, and so the fundamental machine model can be broken down into two sets of decoupled equations which are associated with the $\alpha\beta$ and Z_1Z_2 subspaces. Based on these aspects, it can be concluded that torque and flux components involve only $\alpha\beta$-subspace components, which makes the controlling of DTP machines similar to common three-phase machines. Theoretical explanations of vector control for three-phase motors can be found in [14,15].

3. Vector Control of Dual-Three Phase PMSMs

To understand the vector control for DTP-PMSMs, it is obligatory to explain the space vector pulse width modulation (SVPWM). In this section of the paper, a modulation scheme for a DTP-SVPWM is comprehensively explained. In a conventional, two-level dual three-phase inverter as shown in Figure 2a, the machine is fed with $2^6 = 64$ inverter voltage vectors; out of those, 60 vectors are active, and 4 vectors, i.e., 0, 21, 42 and 63, are inactive, and are mapped at the origin. These voltage vectors are projected onto the $\alpha\beta$- and Z_1Z_2-planex; the mapping of voltage vectors on the $\alpha\beta$-plane is described in Figure 3.

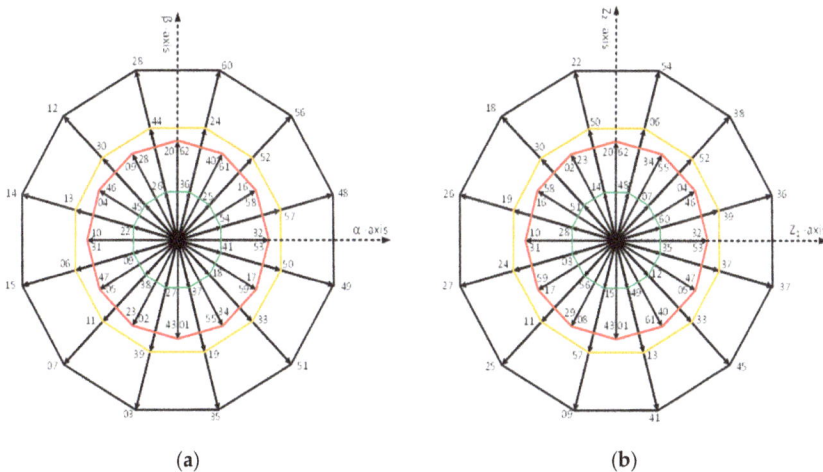

(a)   (b)

Figure 3. Projection of voltage vectors: (**a**) On $\alpha\beta$-plane; (**b**) On Z_1Z_2-plane.

Moreover, it must be observed that inverter voltage vectors have different lengths, and the combination of these voltage vectors can be utilized in different modulation schemes. One essential feature for the realization of the modulation scheme is the calculation of vector projections, which are used to determine the dwell time of vectors. The procedure to estimate these projections is highlighted below.

Each vector (shown in Figure 3) is represented by a decimal number, which is converted to a binary number. The most significant bit (MSB) represents the ON or OFF state of IGBT's associated with phase A; likewise, the second MSB represents phase B and so on. "1's" indicate that the upper IGBTs associated with any of the phases are ON, and "0's" indicate that lower IGBTs associated with either phase are ON.

Based on these switching states, the line-to-line voltages against any of the vectors can easily be determined. The line-to-line voltages are then required to be converted into phase voltages using a transformation matrix. Double natural topology for the inverter is used for this project, and the transformation matrix to transform line voltages into phase voltages for this scheme is explained in [10], and reproduced below.

$$
\begin{bmatrix} u_0 \\ u_a \\ u_b \\ u_c \\ u_d \\ u_e \\ u_f \end{bmatrix} = \frac{1}{6} \begin{bmatrix} 2 & 0 & 2 & 0 & 2 & -2 & 2 \\ 4 & 4 & 2 & 2 & 0 & 2 & 0 \\ 0 & 4 & 4 & 2 & 2 & 0 & 2 \\ -2 & -2 & 2 & 2 & 0 & 2 & 0 \\ 0 & -2 & -2 & 2 & 2 & 0 & 2 \\ -2 & -2 & -4 & -4 & 0 & 2 & 0 \\ 0 & -2 & -2 & -4 & -4 & 0 & 2 \end{bmatrix} \begin{bmatrix} u_{ab} \\ u_{bc} \\ u_{cd} \\ u_{de} \\ u_{ef} \\ 0 \\ 0 \end{bmatrix} \tag{9}
$$

Space Vector PWM Control Strategy

When a machine's parameters are transferred to orthogonal frame current, voltages vectors in the $\alpha\beta$-subspace will participate in electromechanical energy conversion, as pointed out earlier. So, the objective of PWM control is to synthesize the voltage vectors to fulfil the torque control requirements for the machine while keeping the average volt-sec balance to zero in the other two subspaces.

$$
\begin{bmatrix} T_1 \\ T_2 \\ T_3 \\ T_4 \end{bmatrix} = \begin{bmatrix} u_\alpha^1 & u_\alpha^2 & u_\alpha^3 & u_\alpha^4 \\ u_\beta^1 & u_\beta^2 & u_\beta^3 & u_\beta^4 \\ u_{z_1}^1 & u_{z_1}^2 & u_{z_1}^3 & u_{z_1}^4 \\ u_{z_2}^1 & u_{z_2}^2 & u_{z_2}^3 & u_{z_2}^4 \end{bmatrix}^{-1} \begin{bmatrix} u_\alpha^* T_s \\ u_\beta^* T_s \\ 0 \\ 0 \end{bmatrix} \tag{10}
$$

The space vector topology used for this project was based on the double neutral arrangement, which is explained in the former portion of the paper. It is worthy to note that when stator winding is connected with double neutrals, the projection of vectors on the O_1O_2-planes is zero. That's why space vector PWM is performed only on $\alpha\beta$ and Z_1Z_2-planes. The DTP-SVPWM scheme is implemented using Equation (10); for details, see [10]. In Equation (10), V_x^k is the projection of kth voltage vector on the α-axis on the voltage vector plane in particular subspaces i.e., $\alpha\beta$ and Z_1Z_2, T_k is the dwell time of active vectors, k = 1, 2, 3, 4., and u_α^* and u_β^* are reference voltages in $\alpha\beta$-subspace.

The modulation arrangement is described as, in each PWM cycle set of four active voltage vectors must be chosen to ensure that Equation (10) has a unique and positive result. There are many methods to select and place these vectors; in this study, four adjacent voltage vectors are chosen, and have a maximum length in $\alpha\beta$-plane. Suppose that the reference voltage vector resides in sector III as displayed in Figure 4, in this sector, four active vectors, 48, 56, 60, and 28, are used for modulation purposes; these vectors are colored red. It must be observed that together with the four active vectors, two inactive 0 and 63 vectors are also placed in each switching cycle. In this way, six vectors are used for modulation purposes in each PMW cycle. The PWM waveform for a few sectors is exemplified in Figure 5 to clarify the placement of the active and inactive vectors in each switching cycle.

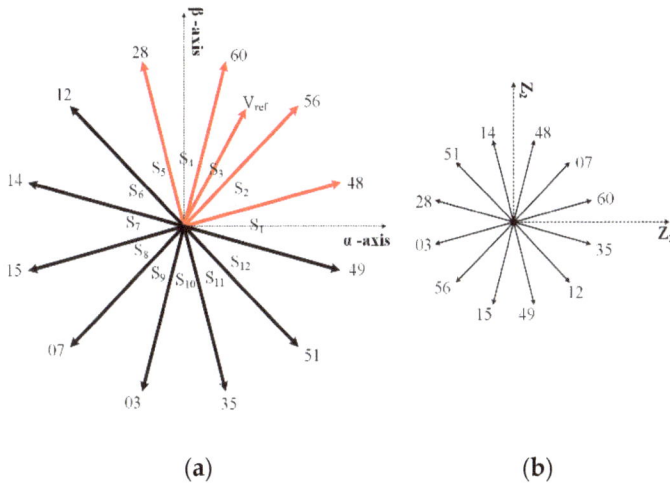

Figure 4. Projection of vectors used for modulation. (**a**) On $\alpha\beta$-plane and (**b**) On Z_1Z_2-plane.

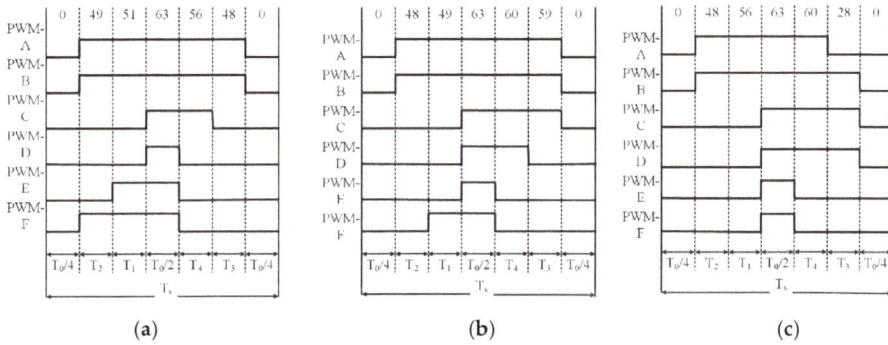

Figure 5. PWM waveforms for all phases: (**a**) Sector I; (**b**) Sector II and (**c**) Sector III.

4. The Result of Vector Control for DTP-PMSMs

4.1. Two-Loops Current Control of DTP-PMSMs

In usual practice, vector control for DTP-PMSMs is developed and simulated in a two-dimensional stationary (or $\alpha\beta$) frame or two dimensional rotating (or dq) frame, instead of in a nature reference; for particulars see [10]. As discussed earlier, once the machine parameters are transferred to a two dimensional static or rotating frame, only the components of $\alpha\beta$-subspace or dq-subspace participate in electromechanical energy conversion. So, for typical vector control of DTP-PMSMs, the currents of dq-subspace i.e., I_d and I_q are used in a feedback loop for good dynamic performance of the machine. The conversion process of parameters from the $\alpha\beta$ to dq frame is elaborated in [10].

The schematic diagram for typical vector control with two axes current control is shown in Figure 6. The simulation model was developed based on the block diagram shown in Figure 6, and with consideration of the machine parameters, as discussed in the paper [14], and reproduced in Appendix A, to investigate the performance of the machine under different working conditions. In the first simulation-based experiment, two currents were used in feedback loops under the following conditions: a machine was tested at a rated speed and full load. In this sense, different results were obtained to understand and evaluate the dynamic performance of the machine. In Figure 7, the load currents are shown. In this result, transients at 0.1 s can be observed; the details are as follows. Before reaching to this point, the machine was running at no load. At 0.1 s, a rated load of 16 Nm was applied to the machine, and it started consuming its rated current. In Figure 8 phase currents of winding set ACE are represented. Some supporting results include electromechanical torque and speed, which are portrayed in Figures 9 and 10 respectively. It must be observed that these results were obtained at a rated speed of i.e., 3000 r/min. Moreover, these phase currents elucidated in the Figure 10 were obtained after the application of load and zooming between 0.15 s to 0.2 s.

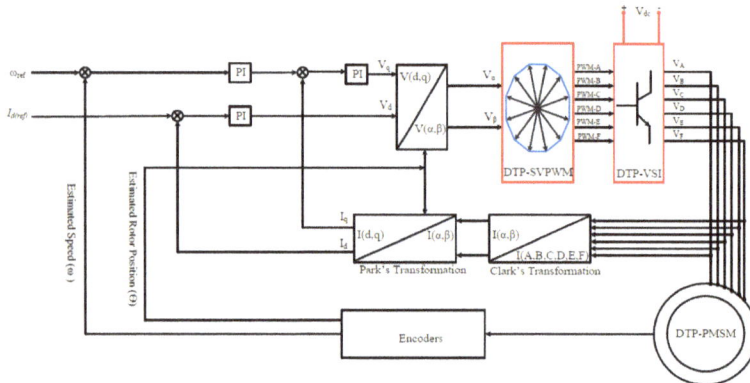

Figure 6. Block diagram of vector control with two-axis current control.

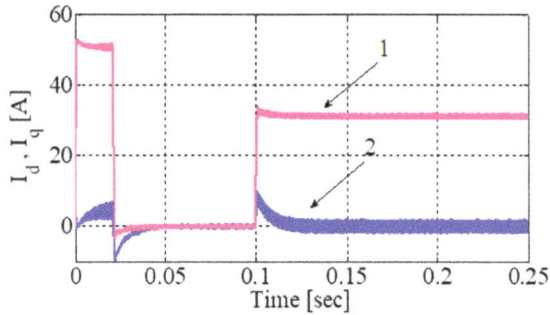

Figure 7. Load current; (**1**) Iq and (**2**) I$_d$ (measured in amperes).

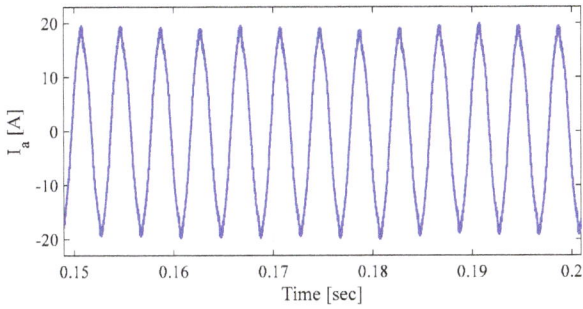

(a)

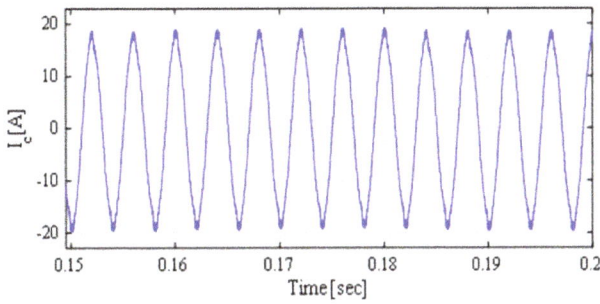

(b)

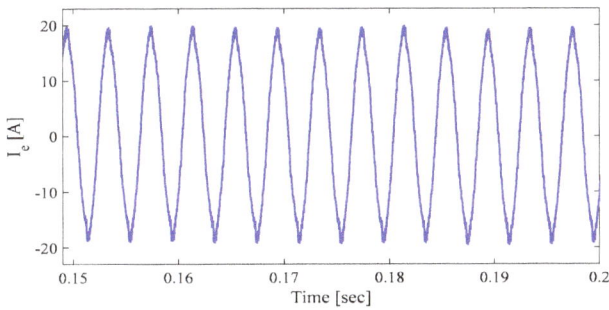

(c)

Figure 8. Phase currents of winding set ACE; (**a**) Phase current "A" measured in amperes; (**b**) Phase current "C" measured in amperes and (**c**) Phase current E measured in amperes.

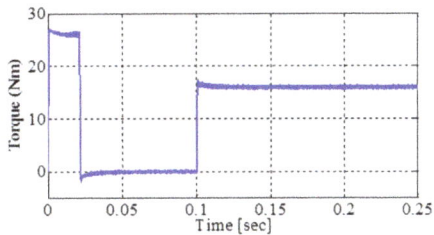

Figure 9. Electromechanical torque.

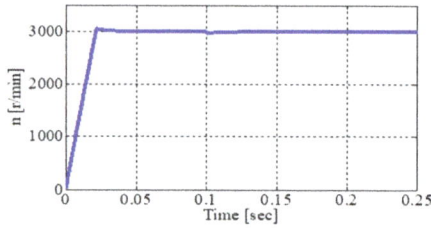

Figure 10. Speed (measured in r/min).

4.2. Four-Loops Current Control of DTP-PMSMs

In two current loop, vector-controlled DTP-PMSMs, the problems associated with harmonics is encountered indirectly by ensuring that the average current of Z_1Z_2-subspace is zero. But due to the inverter's or the machine's asymmetry, this problem cannot be fully eradicated, and certainly can cause a large harmonic current to flow through the machine; it could be the cause of increased losses as well [12]. Some of the problems due to Z_1Z_2-subspace are briefly discussed in [16].

In recent years different authors have proposed different approaches to mitigate these challenges. One such solutions introducing two extra current loops is reported in [12]. These extra current loops are introduced for the currents of Z_1Z_2-subspace, i.e., I_{Z1} and I_{Z2} are put into the closed feedback loops and controlled by the conventional PI controllers. The control diagram for four loops control is described in Figure 11. Based on the arrangement depicted in Figure 11, a Matlab-based simulation was built, and four loop control was verified. Similar parameters (as discussed for two-loops) under the influence of four loop current control were used, and the results are highlighted in this section of the paper. Figure 12 represents the load currents; it can be observed from the result that Id was set to zero, and also, at 0.1 s, rated torque was applied to the machine and it started taking the rated current. Some supporting results include the electromechanical torque and speed represented by Figures 13 and 14 respectively. Likewise, in Figure 15, the phase currents of winding set ACE were obtained. It must also be observed that these results are obtained at full load and rated speed.

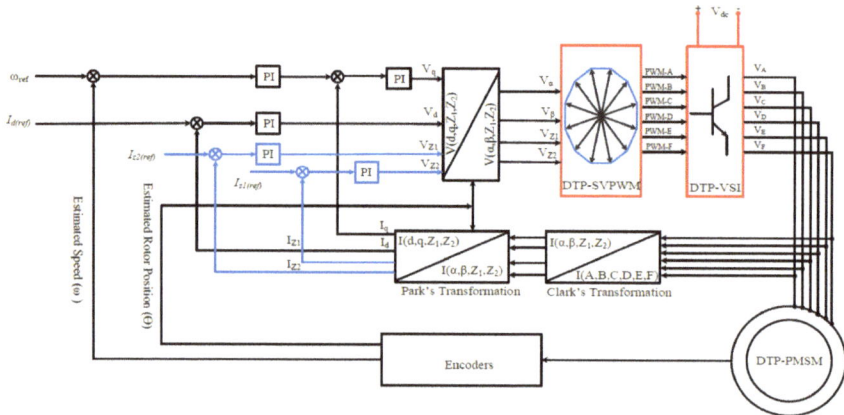

Figure 11. Block diagram of vector control with four-axis current control.

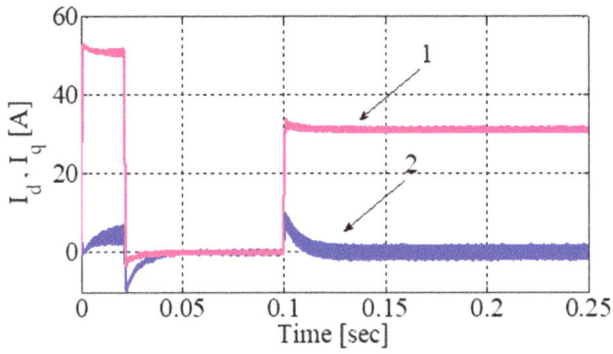

Figure 12. Load current (**1**) Iq and (**2**) I_d.

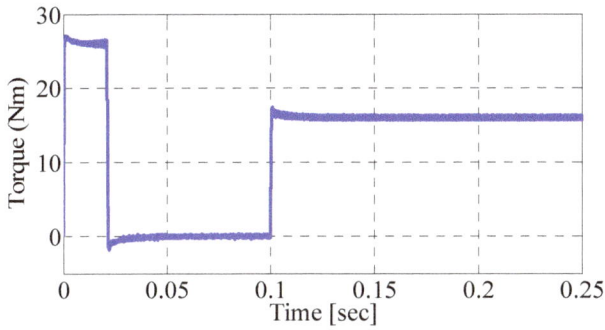

Figure 13. Electromechanical torque.

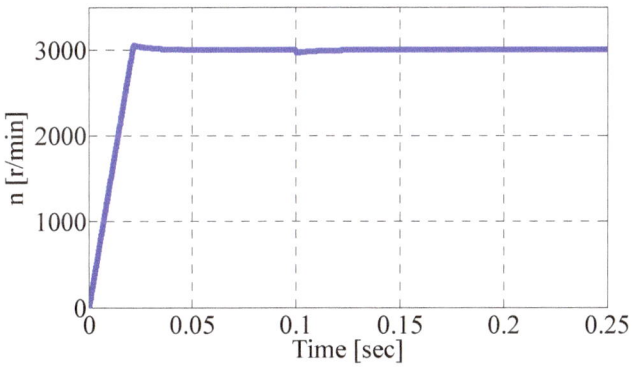

Figure 14. Speed (measured in r/min).

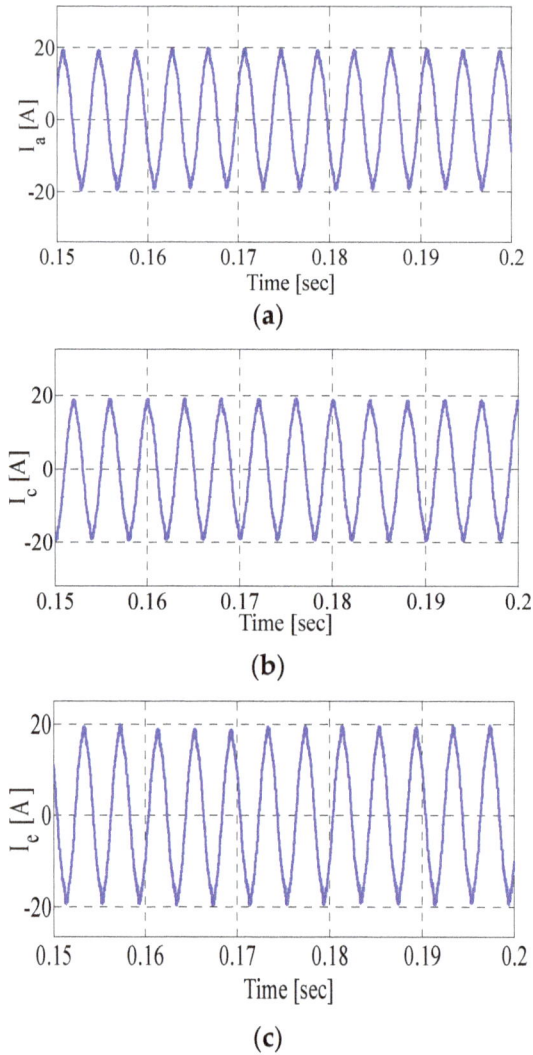

Figure 15. Phase currents of winding set ACE; (**a**) Phase current "A" measured in amperes; (**b**) Phase current "C" measured in amperes and (**c**) Phase current E measured in amperes.

Finally, an FFT analysis of phase current A was conducted under the influence of two and four current loops, to observe the harmonics performance of machine. The FFT analysis of phase current A is depicted in Figure 16.

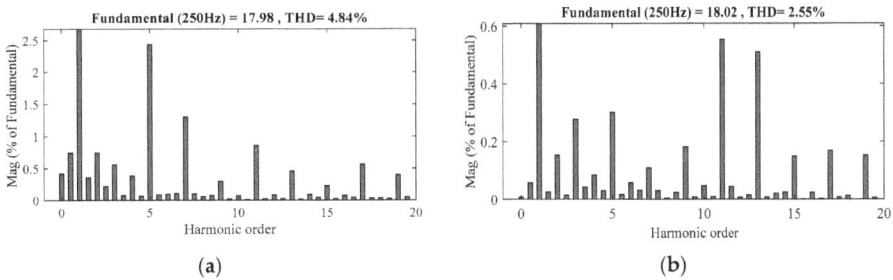

Figure 16. FFT Analysis of phase current A: (**a**) with two loops current control (**b**) with four loops current control.

5. Conclusions

Conventional two current loops vector control and recently-developed four current loops vector control for DTP-PMSMs are discussed in this paper. The theoretical and implementation aspects of both approaches were elaborated in comprehensive detail. Different results for both control structures at the rated speeds of the machines were obtained to assess the accuracy of the control algorithms. The following points conclusions on the simulation analysis of both control methodologies can be made: In conventional, two-current control loops methods, the modulation scheme is adopted in such a way that the average volt/sec balance in each PWM cycle remains zero on the $Z_1 Z_2$ and $O_1 O_2$ subspaces. However, the $Z_1 Z_2$-subspace of the machine contains stator resistance and leakage inductance, for to this reason, low-level voltage harmonics can be induced to the machine, and as a result, high-level of harmonics can be induced in the phase currents. Therefore, instead of controlling the currents indirectly for the $Z_1 Z_2$-subspace by making sure the volt/sec balance equal to zero, an additional two loops were added so that direct control of currents in $Z_1 Z_2$-subspaces can be achieved. The additional two-current loops for $Z_1 Z_2$-subspaces of machines can be used to reduce the level of harmonics in the phase currents. Harmonics analyses of both approaches were presented in the results, which indicated that the addition of two extra currents loops can improve the harmonics in phase currents. The inclusion of additional current loops also improves torque ripple, but will have no influence on the average value of torque. In terms of hardware implementation, four current loops have some drawbacks as well. The addition of two extra loops can increase the overall cost of the prototype. Moreover, extra loops also complicate the overall hardware structure.

Author Contributions: M.A., Z.W. (Zhixin Wang) and C.W. proposed the idea for this paper. The author M.A. was involved in designing this study and also done the simulation work. The simulation work was analyzed and verified by Z.W. (Zhixin Wang) and C.W. The authors S.Y., Z.W. (Zhidong Wang), C.Z., and H.Q. provided their valuable input for relevant literature need to be discussed in the paper and also helped in writing the introductory part of paper. Major portion of this paper was written by the author M.A. under the supervision of Z.W. (Zhixin Wang) and C.W. and at the final stage, paper was comprehensively reviewed by all authors.

Funding: Authors would like to acknowledge the following agencies for their support: Project Supported by State Grid Science & Technology: Research on Morphologies and Pathways of Future Power Systems(B3440818K010); Special Funding of Technical Standard of Shanghai Science and Technology Commission (18DZ2205700); Project of Science and Technology Project of Minhang District of Shanghai (2017MH271).

Conflicts of Interest: The authors declare no conflicts of interest.

Appendix A

Table A1. DTP-PMSM Specifications.

Parameter	Value
Power Rating	5 kW
Speed Rating	3000 r/min
Torque Rating	16 Nm
Pole Pairs	5
Stator Resistance	0.0495 Ω
L_d, L_q	2.4633 mH, 2.4733mH
Leakage Inductance	1.5207 mH

References

1. Tong, C.; Zheng, P.; Wu, Q.; Bai, J.; Zhao, Q. A brushless claw-pole double-rotor machine for power-split hybrid electric vehicles. *IEEE Trans. Ind. Electron.* **2014**, *61*, 4295–4305. [CrossRef]
2. Levi, E. Multiphase electric machines for variable-speed applications. *IEEE Trans. Ind. Electron.* **2008**, *55*, 1893–1909. [CrossRef]
3. Levi, E.; Barrero, F.; Duran, M.J. Multiphase machines and drives-revisited. *IEEE Trans. Ind. Electron.* **2016**, *63*, 429–432. [CrossRef]
4. Pisek, P.; Stumberger, B.; Marcic, T.; Virtic, P. Design analysis and experimental validation of a double rotor synchronous PM machine used for HEV. *IEEE Trans. Magn.* **2013**, *49*, 152–155. [CrossRef]
5. Dalal, A.; Ansari, M.N.; Kumar, P. A novel steady-state model of a hybrid dual rotor motor comprising electrical equivalent circuit and performance equations. *IEEE Trans. Magn.* **2014**, *50*, 1–11. [CrossRef]
6. Ren, Y.; Zhu, Z.-Q. Enhancement of steady-state performance in direct-torque-controlled dual three-phase permanent-magnet synchronous machine drives with modified switching table. *IEEE Trans. Ind. Electron.* **2015**, *62*, 3338–3350. [CrossRef]
7. Almarhoon, A.H.; Ren, Y.; Zhu, Z. Sensorless switching-table-based direct torque control for dual three-phase PMSM drives. In Proceedings of the 2014 17th International Conference on Electrical Machines and Systems (ICEMS), Hangzhou, China, 22–25 October 2014.
8. Bojoi, R.; Lazzari, M.; Profumo, F.; Tenconi, A. Digital field-oriented control for dual three-phase induction motor drives. *IEEE Trans. Ind. Appl.* **2003**, *39*, 752–760. [CrossRef]
9. Parsa, L. On advantages of multi-phase machines. In Proceedings of the 2005 31st Annual Conference of IEEE Industrial Electronics Society (IECON 2005), Raleigh, NC, USA, 6–10 November 2005.
10. Zhao, Y.; Lipo, T.A. Space vector PWM control of dual three-phase induction machine using vector space decomposition. *IEEE Trans. Ind. Appl.* **1995**, *31*, 1100–1109. [CrossRef]
11. Zheng, P.; Wu, F.; Lei, Y.; Sui, Y.; Yu, B. Investigation of a novel 24-slot/14-pole six-phase fault-tolerant modular permanent-magnet in-wheel motor for electric vehicles. *Energies* **2013**, *6*, 4980–5002. [CrossRef]
12. Changpan, Z.; Jianyong, S.; Guijie, Y.; Nianwei, X. Four-dimension current vector control for dual three-phase PMSM. In Proceedings of the 2014 17th International Conference on Electrical Machines and Systems (ICEMS), Hangzhou, China, 22–25 October 2014.
13. Ahmad, M.; Zhang, W.; Gao, Q. Low and zero speed position estimation of dual three-phase PMSMs based on the excitation of PWM waveforms. In Proceedings of the 2017 IEEE 3rd International Future Energy Electronics Conference and ECCE Asia (IFEEC 2017—CCE Asia), Kaohsiung, Taiwan, 3–7 June 2017.
14. Yano, M.; Abe, S.; Ohno, E. History of power electronics for motor drives in Japan. In Proceedings of the 2004 IEEE Conference on the History of Electronics, Bletchley Park, Bletchley Town, UK, 28–30 June 2004.
15. Vas, P. *Sensorless Vector and Direct Torque Control*; Oxford University Press: Oxford, UK, 1998.
16. Che, H.S.; Levi, E.; Jones, M.; Hew, W.-P.; Rahim, N.A. Current control methods for an asymmetrical six-phase induction motor drive. *IEEE Trans. Power Electron.* **2014**, *29*, 407–417. [CrossRef]

![electronics logo] *electronics*

MDPI

Article

Extended Kalman Filter Based Sliding Mode Control of Parallel-Connected Two Five-Phase PMSM Drive System

Tounsi Kamel [1], Djahbar Abdelkader [1], Barkat Said [2], Sanjeevikumar Padmanaban [3] and Atif Iqbal [4],*

[1] Department of Electrical Engineering, LGEER laboratory, U.H.B.B-Chlef University, Chlef 02000, Algeria; t_kamel@outlook.com (T.K.); a_djahbar@yahoo.fr (D.A.)
[2] Laboratoire de Génie Électrique, Faculté de Technologie, Université de M'Sila, M'Sila 28000, Algeria; sa_barkati@yahoo.fr
[3] Department of Energy Technology, Aalborg University, 6700 Esberg, Denmark; sanjeevi_12@yahoo.co.in
[4] Department of Electrical Engineering Qatar University, Doha, Qatar
* Correspondence: atif.iqbal@qu.edu.qa; Tel.: +97-433-276-330

Received: 19 December 2017; Accepted: 19 January 2018; Published: 26 January 2018

Abstract: This paper presents sliding mode control of sensor-less parallel-connected two five-phase permanent magnet synchronous machines (PMSMs) fed by a single five-leg inverter. For both machines, the rotor speeds and rotor positions as well as load torques are estimated by using Extended Kalman Filter (EKF) scheme. Fully decoupled control of both machines is possible via an appropriate phase transposition while connecting the stator windings parallel and employing proposed speed sensor-less method. In the resulting parallel-connected two-machine drive, the independent control of each machine in the group is achieved by controlling the stator currents and speed of each machine under vector control consideration. The effectiveness of the proposed Extended Kalman Filter in conjunction with the sliding mode control is confirmed through application of different load torques for wide speed range operation. Comparison between sliding mode control and PI control of the proposed two-motor drive is provided. The speed response shows a short rise time, an overshoot during reverse operation and settling times is 0.075 s when PI control is used. The speed response obtained by SMC is without overshoot and follows its reference and settling time is 0.028 s. Simulation results confirm that, in transient periods, sliding mode controller remarkably outperforms its counterpart PI controller.

Keywords: five-phase permanent magnet synchronous machine; five-leg voltage source inverter; multiphase space vector modulation; sliding mode control; extended Kalman filter

---

1. Introduction

Recently, five-phase AC machine drives have gained an increasing attention for a wide variety of industrial applications such as electric vehicles, aerospace applications, naval propulsion systems and paper mills. Major advantages of using a five-phase machine over three-phase machine are better fault tolerant, higher torque density, reduced torque pulsations, improvement of the drive noise characteristic and decrease in the required rating per inverter leg [1–3]. In addition, there are three possible connections for the windings, which is able to enlarge the speed operation range compared with three-phase machines. [4].

Five-phase machines include either induction or synchronous machines. However, compared with induction machine, under the synchronous machines category, the permanent magnet synchronous machine possesses many advantages such as high-power density, better torque generating capability and high conversion efficiency [2]. The rotor excitation of the PMSM is provided by PMs. The PMSM do not need extra DC power supply or field windings in order to provide rotor excitations. So, the power losses related to the filed windings are eliminated in the PMSM. In addition, the magnets and redundant teeth in stators allow magnetic decoupling from the different groups of windings [5,6]. Therefore, more and more multiphase permanent magnet synchronous machines are addressed in a variety of specialized literatures [7–10]. Fortunately, with the increasing development of the technology in traction and industrial applications such as for electrical railways and steel processing, the parallel/series-connected multi-machine systems fed by a single supply become strongly suggested. The reasons for that are: low cost drive, compactness and lightness [11,12]. However, the series-connected system suffers from some serious drawbacks compared with parallel-connected system. In such connection, both beginnings and ending of each phase should be brought out to the terminal box of each multiphase machine. Connecting the phase endings into the star point within the machine can eliminate this disadvantage, as it is the case for parallel connection [13]. Further, the series-connected machines suffer from drawback of poor efficiency because of higher losses.

This paper therefore exploits the maturity of the control ideas proposed for series-connected multiphase multi-machine drives [14–16], as the starting point and extends them to parallel-connected multiphase multi-machine drives.

Actually, to control the torque and flux of any multiphase only direct and quadrature current components are used. The remaining components can be used to control the other machines which are fed by a single multi-leg inverter. This constitutes the main idea behind the concept of parallel-connected multiphase multi-machine drive system fed by a single multi-leg inverter supply. This idea has been developed for all induction machines with even and odd supply phase numbers as pointed in [1,17], respectively.

Usually, in order to control the multiphase drive, the standard controllers have been widely used. However, neglected dynamics, parameter variations, friction forces and load disturbances are the main disturbances and uncertainties that can affect the effective functioning of the drive system. So, it will be very difficult to limit these disturbances effectively if linear control methods like PI controllers are adopted [18]. To overcome the aforementioned problems, other advanced methods have been proposed [1,19–21]. These approaches include, among others, the sliding mode control (SMC). The SMC is a nonlinear control method known to have robust control characteristics under restricted disturbance conditions or when there are limited internal parameter modeling errors as well as when a there are some nonlinear behaviors [22]. The robustness of the SMC is guaranteed usually by using a switching control law. Unfortunately, this switching strategy often leads to a chattering phenomenon. In order to mitigate the chattering phenomenon, a common method is to use the smooth function instead of the switching function [18,23,24].

The five-phase PMSM is invariably supplied by a five-phase voltage source inverter (VSI). There are many techniques to control the five-phase inverter such as carrier based PWM (CBPWM) and space vector modulation (SVM). However, SVM has become the most popular due to its ease of digital implementation and higher DC bus utilization [25–27]. To develop the SVM technique for the parallel-connected two five-phase PMSMs configuration, the concept of multiple 2-D subspaces is used. The idea is to select in each of the two planes, completely independently, a set of four active space vectors neighboring the corresponding reference. So, it can be possible to create two voltage space vector references independently, by using the same approach and the same analytical expressions as for the case of purely sinusoidal output voltage generation. However, in the first switching period, the space vector modulator will apply α-β voltage reference. In the next switching period the space vector modulator will apply x-y voltage reference [27].

In the most electric drives, an accurate knowledge on rotor position is crucial for feedback control. It can be achieved from some types of shaft sensors such as an optical encoder or resolver connected to the rotor shaft [11,28]. However, the use of these sensors will increase the cost and reduce the reliability of the drive and may suffer from some restrictions such as temperature, humidity and vibration.

In order to overcome these shortcomings, a number of researchers have developed the well-known sensorless control technology. Various sensorless algorithms have been investigated and reported in many publications [29–31]. The main idea of sensorless control of parallel-connected two five-phase machines is to estimate the rotor positions and their corresponding speeds through an appropriate way using measurable quantities such as five-phase currents and voltages. However, few applications deal with the sensorless control of multiphase machines such as, model reference system [32], Kalman filtering technique [33] and sliding mode observer of five-phase induction motor [10]. Unlike the other approaches, EKF is more attractive because it delivers rapid, precise and accurate estimation. The feedback gain used in EKF achieves quick convergence and provides stability for the observer [34]. For stochastic systems, the extended Kalman filter is the preferable solution capable to provide states estimation or of both the states and parameters estimation.

The purpose of this paper is to study a sensorless sliding mode control scheme using the extended Kalman filter of parallel connected two five-phase PMSMs fed by a single five-leg inverter. To meet this end, the SMC is implemented for speeds and currents control and EKF is used for sensorless operation purposes. The resulting control scheme combines the features of the robust control and the stochastic observer to enhance the performances of the proposed two-machine drive. The performance of the estimation and control scheme is tested with challenging variations of the load torque and velocity reference. The obtained results prove that the two machines are totally decoupled under large speeds and loads variations, although they are connected in parallel and supplied by a single inverter. In addition to that, a comparison between SMC and the traditional PI for sensorless operation is also considered.

2. Modeling of Multiphase AC Drive System

The two-machine drive system under consideration is shown in Figure 1. It consists of a five-leg inverter feeding two five-phase PMSMs. The five-phase PMSM has five-phase windings spatially shifted by 72 electrical degrees. In Figure 1 each stator is star-connected with isolated neutral point which eliminates the zero sequence voltages. It can be seen from Figure 1 that the phase transposition rules of parallel-connected two five-phase PMSMs system are as follows [17]: as_1-as_2, bs_1-cs_2, cs_1-es_2, ds_1-bs_2, es_1-ds_2. Where indices 1 and 2 identify the two machines as indicated in Figure 1. So, the relationships between voltages and currents are given as:

$$v_{abcde} = \begin{bmatrix} v_a^{inv} \\ v_b^{inv} \\ v_c^{inv} \\ v_d^{inv} \\ v_e^{inv} \end{bmatrix} = \begin{bmatrix} v_{as1} = v_{as2} \\ v_{bs1} = v_{cs2} \\ v_{cs1} = v_{es2} \\ v_{ds1} = v_{bs2} \\ v_{es1} = v_{ds2} \end{bmatrix} \quad i_{abcde} = \begin{bmatrix} i_a^{inv} \\ i_b^{inv} \\ i_c^{inv} \\ i_d^{inv} \\ i_e^{inv} \end{bmatrix} = \begin{bmatrix} i_{as1} + i_{as2} \\ i_{bs1} + i_{cs2} \\ i_{cs1} + i_{es2} \\ i_{ds1} + i_{bs2} \\ i_{es1} + i_{ds2} \end{bmatrix} \quad (1)$$

The main five dimensional systems can be decomposed into five dimensional uncoupled subsystems (d-q-x-y-0). Let the correlation between the original phase variables and the new (d-q-x-y-0) variables are given by $f_{dqxy} = Cf_{abcde}$, where C is the following invariant transformation matrix:

$$[C] = \frac{2}{5} \begin{bmatrix} \cos(\theta) & \cos(\theta - 2\pi/5) & \cos(\theta - 4\pi/5) & \cos(\theta + 4\pi/5) & \cos(\theta + 2\pi/5) \\ \sin(\theta) & \sin(\theta - 2\pi/5) & \sin(\theta - 4\pi/5) & \sin(\theta + 4\pi/5) & \sin(\theta + 2\pi/5) \\ \cos(\theta) & \cos(\theta + 4\pi/5) & \cos(\theta - 2\pi/5) & \cos(\theta + 2\pi/5) & \cos(\theta - 4\pi/5) \\ \sin(\theta) & \sin(\theta + 4\pi/5) & \sin(\theta - 2\pi/5) & \sin(\theta + 2\pi/5) & \sin(\theta - 4\pi/5) \\ 1/2 & 1/2 & 1/2 & 1/2 & 1/2 \end{bmatrix} \quad (2)$$

By applying the transformation matrix (2) on Equation (1), the voltage and current components of the five-phase VSI become:

$$\begin{bmatrix} v_d^{inv} \\ v_q^{inv} \\ v_x^{inv} \\ v_y^{inv} \\ v_0^{inv} \end{bmatrix} = [C]v_{abcde} = \begin{bmatrix} v_{ds1} = v_{xs2} \\ v_{qs1} = -v_{ys2} \\ v_{xs1} = v_{ds2} \\ v_{ys1} = v_{qs2} \\ 0 \end{bmatrix}, \begin{bmatrix} i_d^{inv} \\ i_q^{inv} \\ i_x^{inv} \\ i_y^{inv} \\ i_0^{inv} \end{bmatrix} = [C]i_{abcde} = \begin{bmatrix} i_{ds1} + i_{xs2} \\ i_{qs1} - i_{ys2} \\ i_{xs1} + i_{ds2} \\ i_{ys1} + i_{qs2} \\ 0 \end{bmatrix} \quad (3)$$

From (3), it is evident that the inverter voltage d-q components can control the first machine (PMSM1), while the second machine (PMSM2) can be controlled separately using the inverter voltage x-y components.

The model of each five-phase PMSM is presented in a rotating d-q-x-y frame as:

$$\begin{aligned} v_{dsj} &= r_{sj}i_{dsj} + L_{dj}\frac{di_{dsj}}{dt} - \omega_j L_{qsj}i_{qsj} \\ v_{qsj} &= r_{sj}i_{qsj} + L_{qj}\frac{di_{qsj}}{dt} + \omega_j L_{dsj}i_{dsj} + \omega_j\Phi_{fj} \\ v_{xsj} &= r_{sj}i_{xsj} + L_{lsj}\frac{di_{xsj}}{dt} \\ v_{ysj} &= r_{sj}i_{ysj} + L_{lsj}\frac{di_{ysj}}{dt} \end{aligned} \quad (4)$$

where $j = 1,2$. $v_{dj}, v_{qj}, v_{xj}, v_{yj}$ are the stator voltages in the d, q, x, y axes, respectively. $i_{dj}, i_{qj}, i_{xj}, i_{yj}$ are the stator currents in d, q, x, y axes, respectively. L_{dj}, L_{qj}, L_{lsj} are inductances in the rotating frames. r_{sj} is the stator resistance.

The torques equations for the first and the second machines are given by:

$$\begin{aligned} T_{em1} &= \frac{5p_1}{2}(\Phi_{f1}i_{qs} + (L_{d1} - L_{q1})i_{qs}i_{ds}) \\ T_{em2} &= \frac{5p_2}{2}(\Phi_{f2}i_{ys} + (L_{d2} - L_{q2})i_{ys}i_{xs}) \end{aligned} \quad (5)$$

where p_j are pole pairs, Φ_{fj} are the permanent magnet fluxes.

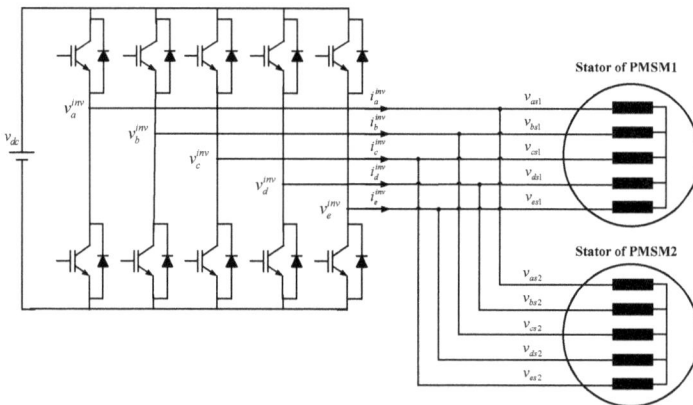

Figure 1. A parallel connected five-phase two-motor drive.

The proposed sensor-less control of the parallel-connected two five-phase permanent magnet synchronous machines is presented in Figure 2, where the two main parts EKF and SMC are considered. The EKF is designed to estimate the rotor position, speed and load torque of each machine by using a current observer. The feedback actual speed, estimated speed and load torques are the inputs of the speeds SMCs to determine the q_1-y_2 axes reference current components. The other current components are maintained to zero. The measured currents are processed in the current SMCs to obtain as outputs the *dqxy* axes reference voltages components. These reference voltages are transformed into the *abcde* frame and transformed again to *αβxy* frame to become input signals to the SVM blocks. The SVM transmits the signals to the inverter to drive the two five-phase PMSMs connected in parallel.

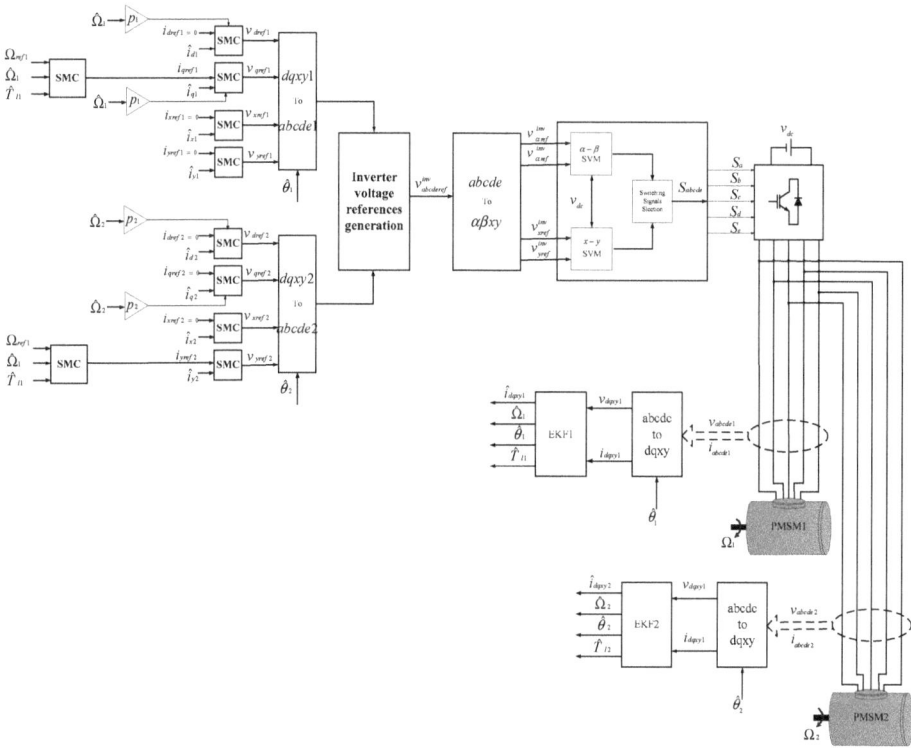

Figure 2. Sensor-less SMC of parallel-connected two five-phase PMSMs drive system.

3. Sliding Mode Controller (SMC)

The design of a sliding mode controller requires mainly two stages. The first stage is choosing an appropriate sliding surface. The second stage is designing a control law, which will drive the state variables to the sliding surface and will keep them there.

3.1. Sliding Surfaces Choice

In order to prescribe the desired dynamic characteristics of the controlled system, the following general form of sliding surface can be adopted [35].

$$S(x) = \left(\frac{d}{dt} + \lambda\right)^{r-1} e(x) \tag{6}$$

With: $e(x) = x_{ref} - x$. λ: is a positive coefficient. r: is the relative degree, which is the number of times required to differentiate the surface before the input u {\display style u} appears explicitly.

3.2. Controller Design

In order to drive the state variables to the sliding surface, the following control law is defined as:

$$u = u_{eqc} + u_{dic} \tag{7}$$

The equivalent control u_{eqc} is capable to keep the state variables on the switching surface, once they reach it and to achieve the desired performance under nominal model. It is derived as the solution of the following equation:

$$S(x) = \dot{S}(x) = 0 \tag{8}$$

The discontinuous control u_{dic} is needed to assure the convergence of the system states to sliding surfaces in finite time and it should be designed to eliminate the effect of any unpredictable perturbation. The discontinuous control input can be determined with the help of the following Lyapunov function candidate:

$$V = \frac{1}{2}S(x)^2 \tag{9}$$

The stability is shown under two conditions as:

- The Lyapunov function V is positive definite.
- The derivative of the sliding function should be negative $\dot{V} = \dot{S}(x)S(x) < 0 \ (\forall S)$.

The so-called reaching stability condition $(\dot{V} = S\dot{S} < 0)$ is fulfilled using the following discontinuous control:

$$u_{dic} = G \, sign(S(x)) \tag{10}$$

where G is a control gain.

In order to reduce the chattering phenomenon, a saturation function instead of the switching one can be used. The saturation function depicted in Figure 3 is expressed as follows:

$$sat(S(x)) = \begin{cases} sgn(S(x)) & if \quad |S(x)| > \delta \\ \frac{S(x)}{\delta} & if \quad |S(x)| < \delta \end{cases} \tag{11}$$

With δ is the boundary layer width.

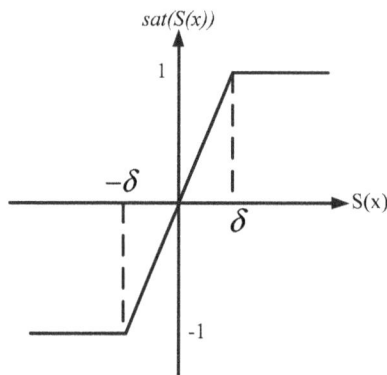

Figure 3. Saturation function.

4. Sliding Mode Control of the Five-Phase Two-Machine AC Drive System

4.1. Speeds SMC Design

The first task in the speeds SMC design process is to select suitable sliding surfaces $S(\Omega_j)$. Since the relative degree is one, the following sliding surfaces are adopted:

$$S(\Omega_j) = \Omega_{refj} - \Omega_j \tag{12}$$

By taking the derivative of sliding surfaces (12) with respect to time and using the machines motion equations, it yields:

$$
\begin{aligned}
\dot{S}(\Omega_1) &= \dot{\Omega}_{ref1} - \frac{5p_1(L_{d1}-L_{q1})i_{ds1}+5p_1\Phi_{f1}}{2J_1}i_{qs1} + \frac{T_{l1}}{J_1} + \frac{f_1\Omega_1}{J_1} \\
\dot{S}(\Omega_2) &= \dot{\Omega}_{ref2} - \frac{5p_2(L_{d2}-L_{q2})i_{xs2}+5p_2\Phi_{f2}}{2J_2}i_{ys2} + \frac{T_{l2}}{J_2} + \frac{f_2\Omega_2}{J_2}
\end{aligned}
\tag{13}
$$

where J_j, f_j and T_{lj} are moment of inertia, damping coefficient and load torque of each machine. The currents controls i_{qsref1} and i_{ysref2} are defined by:

$$
\begin{aligned}
i_{qs1} &= i_{qseqc1} + i_{qsdic1} \\
i_{ys2} &= i_{yseqc2} + i_{ysdic2}
\end{aligned}
\tag{14}
$$

where:

$$
\begin{aligned}
i_{qseqc1} &= -\frac{J_1\dot{\Omega}_{ref1}+T_{l1}+f_1\Omega_1}{\frac{5}{2}p_1(L_{d1}-L_{q1})i_{ds1}+\frac{5}{2}p_1\Phi_{f1}}; i_{yseqc2} = -\frac{J_2\dot{\Omega}_{ref1}+T_{l2}+f_2\Omega_2}{\frac{5}{2}p_2(L_{d2}-L_{q2})i_{xs2}+\frac{5}{2}p_2\Phi_{f2}} \\
i_{qsdic1} &= G_{\Omega1}sat(S(\Omega_1)); i_{ysdic2} = G_{\Omega2}sat(S(\Omega_2))
\end{aligned}
$$

During the convergence mode, the condition $\dot{V} = \dot{S}(x)S(x) < 0\,(\forall S)$ must be verified. By replacing (14) into (13), we get:

$$
\begin{aligned}
\dot{S}(\Omega_1) &= -\frac{5p_1(L_{d1}-L_{q1})i_{ds1}+5p_1\Phi_{f1}}{2J_1}i_{qsdic1} < 0 \\
\dot{S}(\Omega_2) &= -\frac{5p_2(L_{d2}-L_{q2})i_{xs2}+5p_2\Phi_{f2}}{2J_2}i_{ysdic2} < 0
\end{aligned}
\tag{15}
$$

4.2. Currents SMC Design

The control objectives are to track the desired currents trajectories. So, the sliding surfaces can be calculated as follows:

$$
\begin{aligned}
S(i_{dsj}) &= i_{dsrefj} - i_{dsj} \\
S(i_{qsj}) &= i_{qsrefj} - i_{qsj} \\
S(i_{xsj}) &= i_{xsrefj} - i_{xsj} \\
S(i_{ysj}) &= i_{ysrefj} - i_{ysj}
\end{aligned}
\tag{16}
$$

The time derivative of (16) is:

$$
\begin{aligned}
\dot{S}(i_{dsj}) &= \dot{i}_{ds\,refj} - \dot{i}_{dsj} \\
\dot{S}(i_{qsj}) &= \dot{i}_{qs\,refj} - \dot{i}_{qsj} \\
\dot{S}(i_{xsj}) &= \dot{i}_{xs\,refj} - \dot{i}_{xsj} \\
\dot{S}(i_{ysj}) &= \dot{i}_{ys\,refj} - \dot{i}_{ysj}
\end{aligned}
\tag{17}
$$

Using (4), the Equation (17) can be rewritten as:

$$\dot{S}(i_{dsj}) = \dot{i}_{dsrefj} + \frac{r_{sj}}{L_{dj}}i_{dsj} - \frac{L_{qj}}{L_{dj}}\omega_j i_{qsj} - \frac{1}{L_{dj}}v_{dsj}$$
$$\dot{S}(i_{qsj}) = \dot{i}_{qsrefj} + \frac{r_{sj}}{L_{qj}}i_{qsj} + \frac{L_{dj}}{L_{qj}}\omega_j i_{dsj} + \frac{\omega_j \Phi_{fj}}{L_{qj}} - \frac{1}{L_{qj}}v_{qsj}$$
$$\dot{S}(i_{xsj}) = \dot{i}_{xsrefj} + \frac{r_{sj}}{L_{lsj}}i_{xsj} - \frac{1}{L_{lsj}}v_{xsj}$$
$$\dot{S}(i_{ysj}) = \dot{i}_{ysrefj} + \frac{r_{sj}}{L_{lsj}}i_{ysj} - \frac{1}{L_{lsj}}v_{ysj}$$

(18)

So, it is possible to choose the control laws for stator voltages as follows:

$$v_{ds\,refj} = v_{dseqcj} + v_{dsdicj}$$
$$v_{qs\,refj} = v_{qseqcj} + v_{qsdicj}$$
$$v_{xs\,refj} = v_{xseqcj} + v_{xsdicj}$$
$$v_{ys\,refj} = v_{yseqcj} + v_{ysdicj}$$

(19)

where:

$$v_{dseqcj} = \left(\dot{i}_{dsrefj} + \frac{r_{sj}}{L_{dj}}i_{dsj} - \frac{L_{qj}}{L_{dj}}\omega_j i_{qsj}\right)L_{dj}; v_{qseqcj} = \left(\dot{i}_{qsrefj} + \frac{r_{sj}}{L_{qj}}i_{qsj} + \frac{L_{dj}}{L_{qj}}\omega_j i_{dsj} + \frac{\omega_j \Phi_{fj}}{L_{qj}}\right)L_{qj}$$
$$v_{xseqcj} = \left(\dot{i}_{xsrefj} + \frac{r_{sj}}{L_{lsj}}i_{xsj}\right)L_{lsj}; v_{yseqcj} = \left(\dot{i}_{ysrefj} + \frac{r_{sj}}{L_{lsj}}i_{ysj}\right)L_{lsj}$$

$$v_{dsdicj} = G_{dsj}sat(S(i_{dsj})); v_{qsdicj} = G_{qsj}sat(S(i_{qsj})); v_{xsdicj} = G_{xsj}sat(S(i_{xsj})); v_{ysdicj} = G_{ysj}sat(S(i_{ysj}))$$

During the convergence mode, the condition $\dot{V} = \dot{S}(x)S(x) < 0 \ (\forall\, S)$ must be verified. By replacing (19) into (18), we get:

$$\dot{S}(i_{dsj}) = -\frac{1}{L_{dj}}v_{dsdicj}$$
$$\dot{S}(i_{qsj}) = -\frac{1}{L_{qj}}v_{qdicsj}$$
$$\dot{S}(i_{xsj}) = -\frac{1}{L_{lsj}}v_{xsdicj}$$
$$\dot{S}(i_{ysj}) = -\frac{1}{L_{lsj}}v_{ysdicj}$$

(20)

The control voltages given by (19) are transformed in *abcde* frame and then the inverter phase voltage references are calculated according to the following expression:

$$v_a^{inv} = v_{as\,ref1} + v_{as\,ref2}$$
$$v_b^{inv} = v_{bs\,ref1} + v_{cs\,ref2}$$
$$v_c^{inv} = v_{cs\,ref1} + v_{es\,ref2}$$
$$v_d^{inv} = v_{ds\,ref1} + v_{bs\,ref2}$$
$$v_e^{inv} = v_{es\,ref1} + v_{ds\,ref2}$$

(21)

5. Space Vector Modulation Technique for Parallel Connected Multiphase AC Drive System

5.1. Five-Leg VSI Modeling

The five-leg inverter output phase-to-neutral voltages can be expressed as:

$$v_a^{inv} = \frac{v_{dc}}{5}(4S_a - S_b - S_c - S_d - S_e)$$
$$v_b^{inv} = \frac{v_{dc}}{5}(-S_a + 4S_b - S_c - S_d - S_e)$$
$$v_c^{inv} = \frac{v_{dc}}{5}(-S_a - S_b + 4S_c - S_d - S_e)$$
$$v_d^{inv} = \frac{v_{dc}}{5}(-S_a - S_b - S_c + 4S_d - S_e)$$
$$v_e^{inv} = \frac{v_{dc}}{5}(-S_a - S_b - S_c - S_d + 4S_e)$$

(22)

where v_{dc} denotes DC-link voltage and S_i, $i = a, b, c, d, e$ refer to switching functions.

The five-phase inverter has totally thirty-two space voltage vectors, thirty non-zero voltage vectors and two zero voltage vectors. These space vectors can be projected on α-β subspace as well as on x-y subspace as shown in Figure 4. Every plane is divided in ten sectors, each occupying a 36° angle around the origin by means of the following two space vectors [27]:

$$v_{\alpha\beta}^{inv} = \frac{2}{5}(v_a^{inv} + v_b^{inv}e^{j\alpha} + v_c^{inv}e^{j2\alpha} + v_d^{inv}e^{j3\alpha} + v_e^{inv}e^{j4\alpha})$$
$$v_{xy}^{inv} = \frac{2}{5}(v_a^{inv} + v_b^{inv}e^{j2\alpha} + v_c^{inv}e^{j4\alpha} + v_d^{inv}e^{j6\alpha} + v_e^{inv}e^{j8\alpha}) \tag{23}$$

where $\alpha = \frac{2\pi}{5}$.

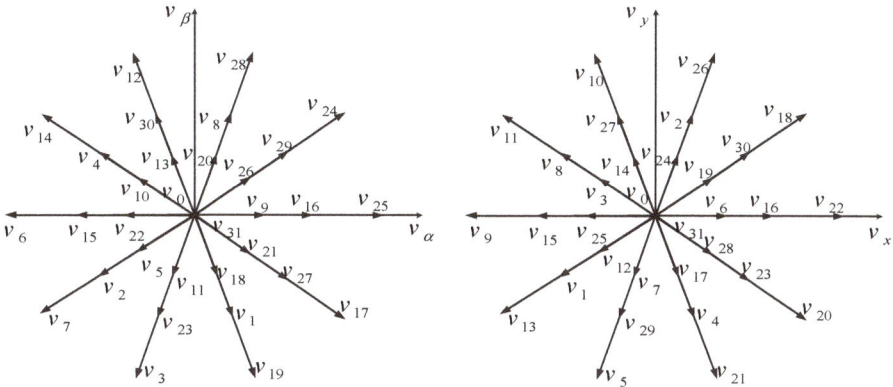

Figure 4. Space vectors of a five-phase inverter in two 2-D subspaces.

From Figure 3 the space vectors are divided into three groups in accordance with their magnitudes: small, medium and large space vector groups. The magnitudes are identified with indices s, m and l and are given as: $|V_s| = 4/5\cos(2\pi/5)v_{dc}$, $|V_m| = 2/5v_{dc}$ and $|V_l| = 4/5\cos(\pi/5)v_{dc}$, respectively [27,35–38].It can be observed from Figure 3 that medium length space vectors of the α-β plane are mapped into medium length vectors in the x-y plane and large vectors of the α-β plane are mapped into small vectors in the x-y plane and vice-versa.

5.2. SVM Method for Five-Leg VSI

The reference voltage can be obtained by averaging a certain number of active space vectors for adequate time intervals, without saturating the VSI. Four active space vectors are required to reconstruct the reference voltage vector [27,36–38].

The dwell times for active space vectors T_{1m}, T_{1l}, T_{2m}, T_{2l} are:

$$T_{1l} = \frac{|v_{ref}|\sin(s\pi/5-\vartheta)}{(|V_l|+|V_s|)\sin(\pi/5)}T_s$$
$$T_{2m} = \frac{|V_s||v_{ref}|\sin(s\pi/5-\vartheta)}{|V_m|(|V_l|+|V_s|)\sin(\pi/5)}T_s$$
$$T_{1l} = \frac{|v_{ref}|\sin(s\pi/5-\vartheta)}{(|V_l|+|V_s|)\sin(\pi/5)}T_s \tag{24}$$
$$T_{2m} = \frac{|V_s||v_{ref}|\sin(s\pi/5-\vartheta)}{|V_m|(|V_l|+|V_s|)\sin(\pi/5)}T_s$$
$$T_o = T_s - (T_{1l} + T_{1m} + T_{2l} + T_{2m})$$

where: T_s is the switching period and ϑ is the voltage reference vector position.

The control strategy adopted herein is based on the approach proposed in [27,36]. Indeed, in the first switching period, the space vector modulator will apply α-β voltage reference. In the next switching period the space vector modulator will apply x-y voltage reference as shown in Figure 5. The selection of switching signals is depicted in Figure 6. So, two independent space vector modulators are further utilized to realize the required two voltage space vector references, with dwell times calculated independently in the two planes using (24).

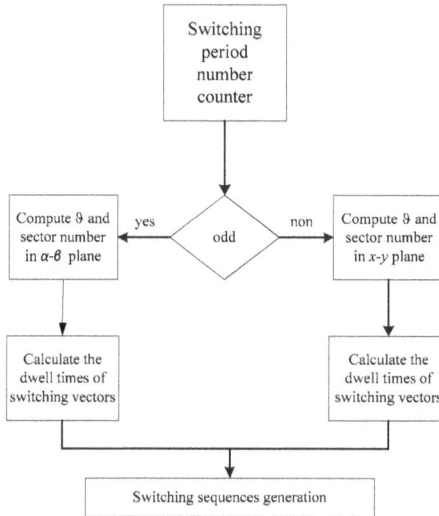

Figure 5. Steps of SVM technique for parallel connected two-machine drive.

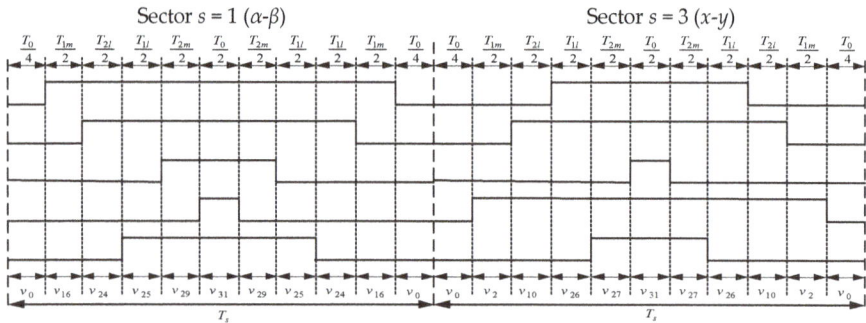

Figure 6. Switching pattern obtained with SVM.

6. Extended Kalman Filter Based Speed Estimator for Parallel Connected Two Motor Drivel

Normally, speed observers used for three-phase machines can be easily extended to multi-phase multi-machine drives. For each machine, the speed estimator requires only stator voltages and currents components. The SMC block diagram based on extended kalman filter of parallel-connected two five-phase machines drive system is shown in Figure 2. The main task of EKF is to find the best estimate of state variables and the unknown load torques since the knowledge of the load torque is necessary for each speed SMC implementation.

In the five-phase PMSM control case, d, q, x, y currents and voltages are measured and the Equation (4) is sampled to obtain a discrete state space representation to be used in the observer synthesis. Assuming that the sampling interval T_e is very short compared to the system dynamics, the augmented discrete-time of each five-phase PMSM model is given as follows:

$$x_{j(k+1)} = A_{jk}x_{jk} + B_{jk}u_{kj} + w_{jk}$$
$$y_{jk} = C_{jk}x_{jk} + v_{jk}$$

(25)

With:

$$x_{jk} = [i_{dsjk}\ i_{qsjk}\ i_{xsjk}\ i_{ysjk}\ \Omega_{jk}\ \theta_{jk}\ T_{ljk}]^t$$

$$u_{jk} = \begin{bmatrix} 0 & v^{inv}_{qk} & 0 & v^{inv}_{yk} \\ v^{inv}_{dk} & 0 & v^{inv}_{xk} & 0 \end{bmatrix}^t$$

$$y_{jk} = [i_{dsjk}\ i_{qsjk}\ i_{xsjk}\ i_{ysjk}]^t$$

$$A_{jk} = \begin{bmatrix} 1 - T_e\left(\frac{r_{sj}}{L_{dj}}\right) & p_j\Omega_j\frac{L_{qj}}{L_{dj}} & 0 & 0 & 0 & 0 & 0 \\ -p_j\Omega_j\frac{L_{qj}}{L_{dkj}} & 1 - T_e\left(\frac{r_{sj}}{L_{qj}}\right) & 0 & 0 & p_j\Phi_{ff}\frac{T_e}{L_{qj}} & 0 & 0 \\ 0 & 0 & 1 - T_e\left(\frac{r_{sj}}{L_{lsj}}\right) & 0 & 0 & 0 & 0 \\ 0 & 0 & 0 & 1 - T_e\left(\frac{r_{sj}}{L_{lsj}}\right) & 0 & 0 & 0 \\ T_e\frac{5P_j}{2J_j}\left(L_{dj}-L_{qj}\right)i_{qsj} & T_e\frac{5P_j}{2J_j}\Phi_{ff} & 0 & 0 & 1 - T_e\left(\frac{f_j}{J_j}\right) & 0 & 0 \\ 0 & 0 & 0 & 0 & T_eP_j & 1 & 0 \\ 0 & 0 & 0 & 0 & 0 & 0 & 1 \end{bmatrix}\quad B_{jk} = \begin{bmatrix} T_e\frac{1}{L_{dj}} & 0 \\ 0 & T_e\frac{1}{L_{qj}} \\ T_e\frac{1}{L_{lsj}} & 0 \\ 0 & T_e\frac{1}{L_{lsj}} \\ 0 & 0 \\ 0 & 0 \\ 0 & 0 \end{bmatrix}\quad C_{jk} = \begin{bmatrix} 1 & 0 & 0 & 0 & 0 & 0 & 0 \\ 0 & 1 & 0 & 0 & 0 & 0 & 0 \\ 0 & 0 & 1 & 0 & 0 & 0 & 0 \\ 0 & 0 & 0 & 1 & 0 & 0 & 0 \end{bmatrix}$$

where x_{jk}, u_{jk} and y_{jk} are the augmented state vector and input vector and output vector at the sampling instant k of machine j, respectively. A_{jk} and B_{jk} are discrete system matrix and discrete input matrix for each machine, respectively. w_{jk} and v_{jk} are the system noise and measurement noise, respectively.

The added white-noise vectors are Gaussian and uncorrelated from each other with zero mean and covariance Q_j and R_j, respectively. The covariance matrices Q_j and R_j of these noises are defined as:

$$Q_j = E\left\{w_{jk}w^T_{jk'}\right\}, \quad \text{for}\quad k \neq k'$$
$$R_j = E\left\{v_{jk}v^T_{jk'}\right\}, \quad \text{for}\quad k \neq k'$$

(26)

In a first main stage the state $x_{j(k+1)}$ is predicted using discrete matrices and previous state. In a second main stage, the feedback correction weight matrix K_j (filter coefficients) is used to have an accurate prediction of the state $x_{j(k+1/k)}$. This is obtained by computing K_j depending not only on the error made but also with an adjustment using weight P_j (covariance state matrix). This allows estimating accurately x_j with respect to Q_j and R_j covariance matrices corresponding respectively to state noise and measurement noise levels [30]. Using Equation (25), the rotor speeds and load torques can be estimated by the extended Kalman filter algorithm described as follows:

- Sate prediction:

$$\hat{x}_{j(k+1/k)} = A_{jk}x_{j(k/k)} + B_{jk}u_{jk}$$

(27)

- Estimation of the matrix of the covariance error:

$$\hat{P}_{j(k+1/k)} = A_{jk}P_{j(k/k)}A^T_{jk} + Q_j$$

(28)

- Kalman coefficient update:

$$K_{j(k+1)} = \hat{P}_{j(k+1/k)}C_{jk}^T[C_{jk}\hat{P}_{j(k+1/k)}C_{jk}^T + R_j]^{-1}$$

(29)

- State estimation:

$$\hat{x}_{j(k+1/k+1)} = \hat{x}_{j(k+1/k)} + K_{j(k+1)}(y_{j(k+1)} - C_{jk}\hat{x}_{j(k+1/k)})$$

(30)

- Covariance error matrix update:

$$\hat{P}_{j(k+1/k+1)} = \hat{P}_{j(k+1/k)} - K_{j(k+1)}C_{jk}\hat{P}_{j(k+1/k)} \tag{31}$$

where $\hat{x}$ is the system state, u_{jk} is the system input vector, y is the system output vector, P, Q and R are the covariance matrices, C is the transformation matrix.

7. Numerical Simulation Results

In order to verify the applicability of the proposed control scheme for the two-machine drive system of Figure 2, the following simulations are performed using two identical 2-pole, 50 Hz five-phase PMSM. The parameters of each machine are listed in Table 1. The performance of the SMC controller is compared with that of the conventional controller. The tuning parameters for the PI controllers and SMC controllers are also given in Table 2. Many simulation tests are performed in order to verify the independence of the control of the two machines in sensor-less mode.

Table 1. Five-phase PMSM parameters.

p_j	L_{dj}	L_{qj}	L_{lsj}	Φ_{fj}	J_j	r_{sj}	f_j
2	8.5 mH	8 mH	0.2 mH	0.175 Wb	0.004 kg m^2	1 Ω	0

Table 2. PI and SMC parameters.

	Speed Controller	i_{sd} Controller	i_{sq} Controller	i_{sx} Controller	i_{sy} Controller
PI	$k_p = 0.8$ $k_i = 40$	$k_p = 33$ $k_i = 32{,}000$	$k_p = 33$ $k_i = 32{,}000$	$k_p = 33$ $k_i = 32{,}000$	$k_p = 33$ $k_i = 32{,}000$
SMC	$G_{\Omega j} = 5$	$G_{idj} = 4000$	$G_{iqj} = 7000$	$G_{ixj} = 4000$	$G_{iyj} = 7000$

The behavior of the overall drive system is presented in Figures 7–11 at different test conditions. Figure 7 shows then estimated speeds, currents and torques of the unloaded two machines for many different speeds references. At the beginning, the first machine is running at 100 rad/s, at $t = 0.7$ s it decelerated to -10 rad/s, after that, it is accelerated again to the speed 60 rad/sat $t = 1.4$ s. For the second machine the speed reference is set at 50 rad/s, 25 rad/s, 100 rad/s, -100 rad/s and 80 rad/s at $t = 0$ s, 0.4 s, 0.9 s, 1.2 s, 1.7 s, respectively. Effect of the speed rotation reversion of one machines on the other machine performance is investigated Figure 8. In this test, most of the time when one machine is rotating at +100 rad/s the other is running at the opposite speed.

Some additional reversing tests are conducted next to further verify decoupling of the control of the two machines. Figure 9 displays results for the case when the speed Ω_2 is kept at standstill, while Ω_1 is reversed from: +100 to -100 rad/s at $t = 0.5$ s and returns to zero at $t = 1$ s. At the subsequent test, the speed Ω_1 is held at zero, while Ω_2 is reversed from 100 to -100 rad/s at 1.5 s.

Figure 10 shows the speeds, torques and currents of the two-machine drive controlled by both PI and SMC controller in the presence of load torques variations. The reference of the first speed is fixed at 100 rad/s, while the speed reference of the second machine is fixed at 50 rad/s. Load torques are applied on the two machines at $t = 0.5$ s and $t = 0.7$ s, respectively.

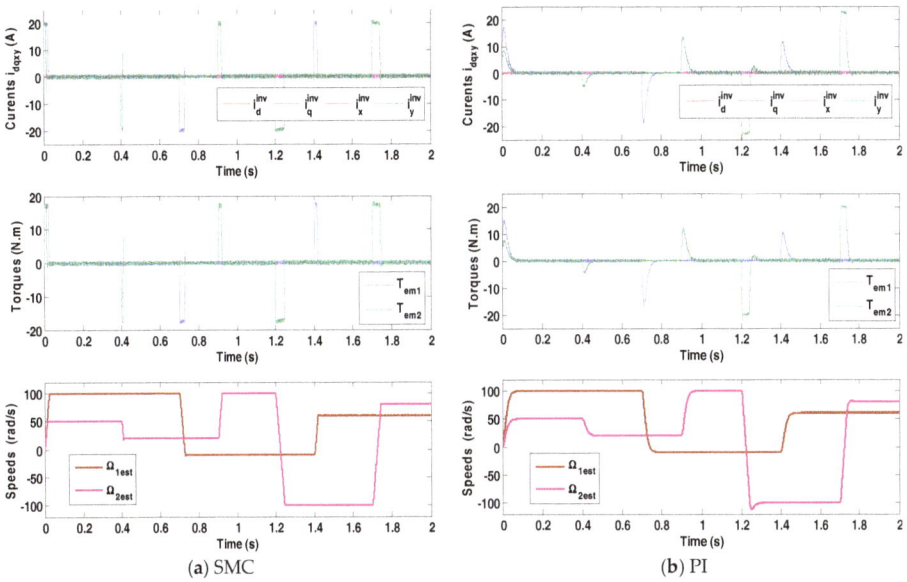

Figure 7. Dynamic responses of parallel-connected two five-phase PMSMs system at different reference speeds values.

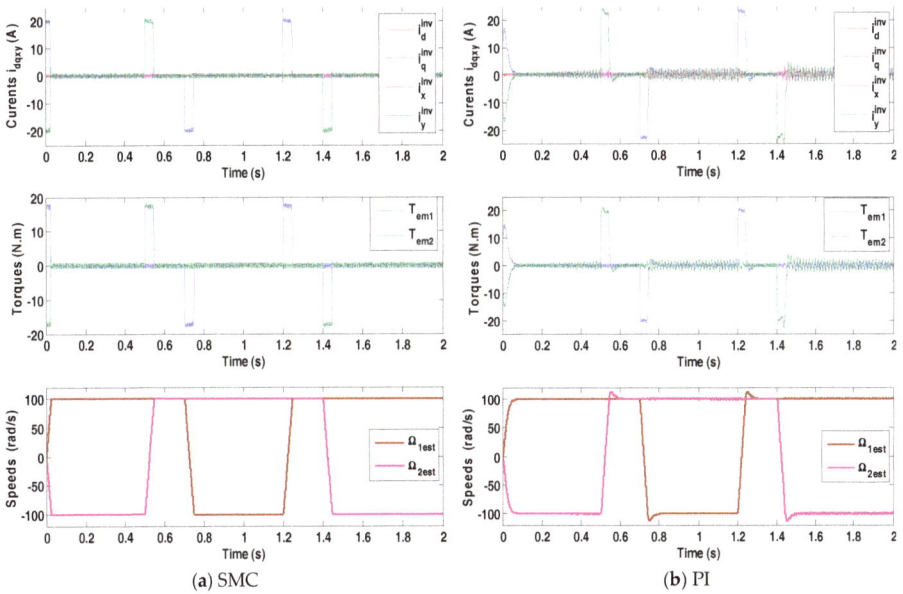

Figure 8. Dynamic responses of parallel-connected two five-phase PMSMs system: when the two motors are operating in the opposite directions.

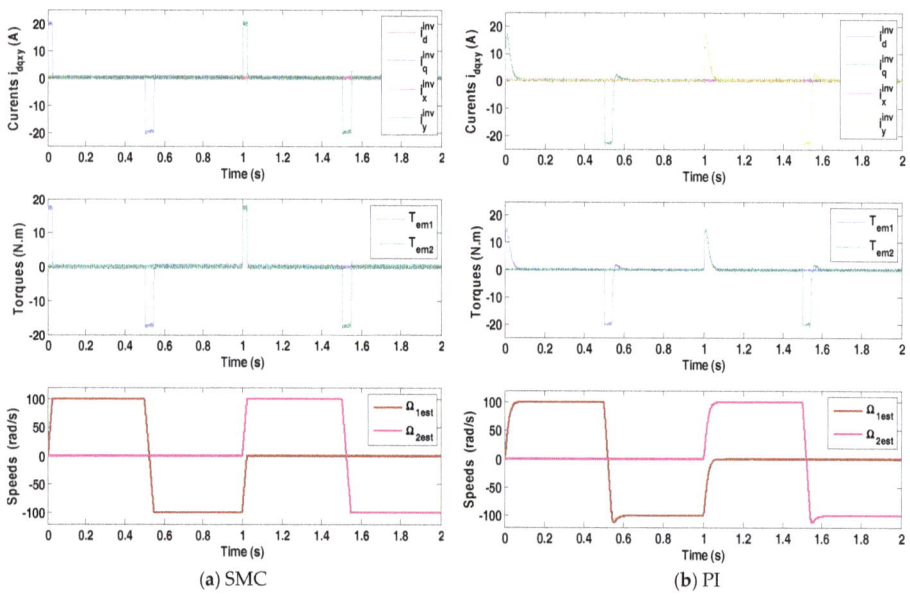

Figure 9. Dynamic responses of parallel-connected two five-phase PMSMs system: when one machine is at standstill and the other is still running.

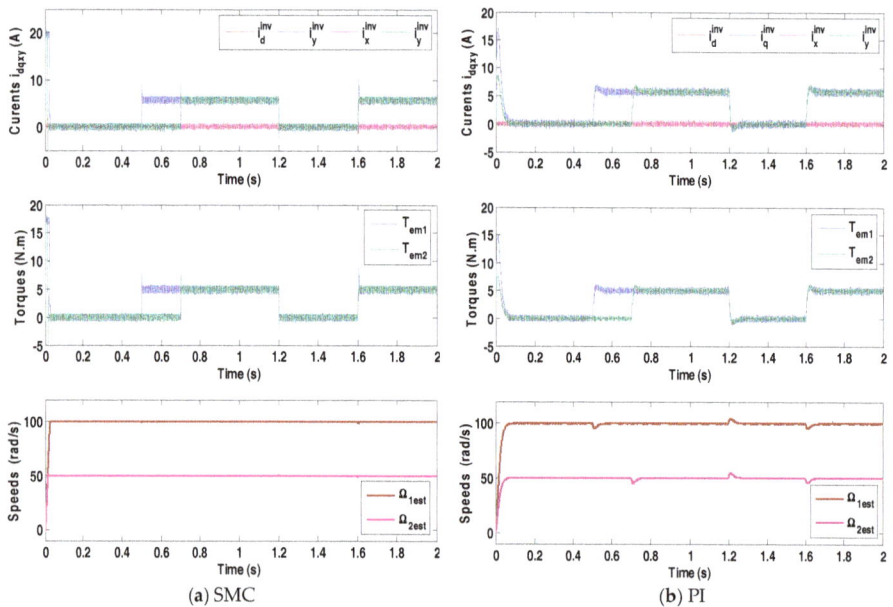

Figure 10. Dynamic responses of parallel-connected two five-phase PMSMs system at different loading conditions.

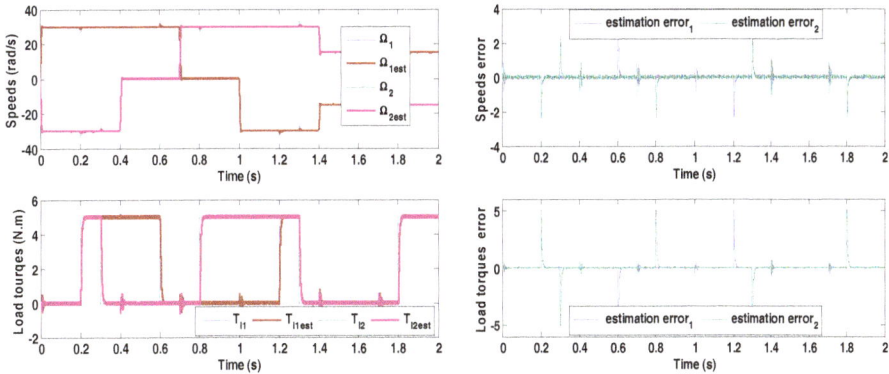

Figure 11. Actual and estimated speeds and load torques and their corresponding estimation errors (SMC case).

It is clear from all estimated speeds characteristics that in every test the speed estimators provide accurate speed estimations. These results also prove that both speeds machines are independently controlled even in sensor-less mode. Indeed, the speed variation of the first machine in the two-machine drive system does not affect the behavior of the speed of the second machine even in reversal conditions.

The electromagnetic torque generated by each machine during the simulated speed step response is shown in Figures 7–10. Note that the generated torques are directly proportional to the q-x axes currents and fully decoupled from d-y axes currents.

Comparison of results in Figure 7 shows once more that the control of the two machines is completely decoupled. There is hardly any evidence of torque disturbance of one machine during the reversal of the other one. Furthermore, the direct axis currents responses remain completely unaffected during these transients.

As shown from Figure 8, the starting and reversing transients of one machine do not have any tangible consequence on the operation of the second machine. The decoupled control is preserved and the characteristics of both machines are unaffected.

Figure 9 illustrates results for the case when the speed of one machine is kept at zero, while the second is reversed. Speed of one machine and its electromagnetic torque remain completely undisturbed during the reversion of the other machine, indicating a complete decoupling of the control.

Figure 10 shows inverter current characteristic, motor torques and estimated speed of motors at different loading conditions for parallel-connected two five-phase machines drive system. It is clear from Figure 10 that when one machine is loaded or unloaded, the second machine performance is unaffected; which proves once again that both motors connected in parallel are totally decoupled. In case of sliding mode control, no variation whatsoever can be observed in the speeds responses of the both machines during these transients.

The estimated and actual values of speeds and load torques as well as their estimations errors are reported in Figure 11. The EKF algorithms give accurate and fast speeds estimations over entire speed range including low speed and standstill operations with low speed errors, even in transients. Furthermore, the estimated values of loads torques are very close to their applied ones. Consequently, the load torque estimation errors are almost zeros; this reflects the stability of EKF during load torques variations. These results confirm that the extended Kalman filter is very suitable for two-machine drive system.

It is worth to notice that there is no impact on the speed and electromagnetic torque of one machine when the speed or the load of the other machine in parallel-connected system changes. Thus, through proper phase transformation rules, the decoupled control of two five-phase PMSMs connected in parallel can be achieved with a single supply from a five-phase voltage source inverter. Furthermore, measured and estimated speeds are in excellent agreement in both steady state and transient operations.

Figures 7–10 illustrate the behavior of the *dq*-axes and *xy*-axes inverter currents for both controllers. In case of the PI controller, the stator currents i_q^{inv} and i_y^{inv} peak above 17 A, then decay exponentially to the steady-state while the currents i_d^{inv} and i_x^{inv} are maintained at zero as illustrated in Figures 7b, 8b, 9b and 10b. Figures 7a, 8a and 9a show currents in case of sliding mode control. In contrast, the i_q^{inv} and i_y^{inv} currents peak slightly above 20 A and continue on this value, until the speed reaches its reference value, this leads to a short settling time, as shown in Figures 7a, 8a, 9a and 10a.

Figures 7–9 show the behavior of the two-machine drive under different speeds step variations and without load torque. In Figures 7b, 8b and 9b, the system comportment using PI controller exhibits the expected step response characteristics of a second order system. The response has a short rise time, an overshoot of approximately 18% during reverse modes and settling times close to 0.075 s. Figures 7a, 8a and 9a show comparable dynamic behavior using SMC. However, it is clear from these figures that the system reaches steady-state at 0.028 s without overshoot.

From Figure 10b and by analyzing the transient of two-machine drive controlled by PI controller, it is easy to observe speeds drops taken place at the moments of loads changes. These speed drops are compensated by the PI controller after a necessary recovery time. Figure 10a shows the drive responses in the same load conditions with PI control. At the moments of load variations, the SMCs keep the speeds close to their references without overshoots and without drops. Therefore, the SMC can be considered as more robust under loads variations.

A general comparison between SMC and PI is given in Table 3. Compared to PI controller, SMC shows superiority in terms of settling time and overshoot. However, it needs more energy in starting transient then that needed by the conventional controller.

Table 3. Comparison between SMC and PI.

Comparison Criterion	SMC	PI
Settling time (s)	0.028	0.075
Recovery time (at abrupt load) (s)	0.0045	0.05
Overshoot in reversal mode (%)	0	18%
Starting current (A)	20	17
Starting torque (Nm)	18	15
Speeds drops (%)	0	5

8. Conclusions

In this paper, sensor-less non-linear sliding mode control based on the Lyapunov theory of parallel-connected two five-phase PMSMs drive fed by a single inverter has been developed in order to make the system asymptotically stable. In the proposed control scheme, the Extended Kalman Filter is used for rotor speeds, positions and load torques estimations, while a sliding mode controller is used for speeds and currents control. The sliding mode control has several advantages such as, robustness, high precision, stability and simplicity, very low Settling time. The added value of EKF based sensor-less control is the improvement in system dynamics through the accuracy in speeds, rotor positions and load torques estimations. The effectiveness of the control approach has also been verified through extensive computer simulations and compared with PI controller. The response has a short rise time, an overshoot during reverse modes in PI controller and settling times close to 0.075 s. However, the speed response obtained by SMC is without overshoot and follows its reference and settling times close to 0.028 s. The results also show that the torque obtained by the PI control decreases progressively, while the torque obtained by the SMC is maintained longer at its maximum value, until the speed reaches its reference value. Speeds drops taken place at the moments of loads changes in PI controller. The SMC keep the speeds close to their references without overshoots and without drops.

Therefore, the SMC can be considered as more robust under loads variations. SMC shows better speed tracking performance at both dynamic and steady state than conventional PI controller in the situation reverse modes and load torque variations. Thus, simulation results have verified the proposed whole system has great robust to external disturbances. The simulation of the two-machine drive under various test conditions confirmed that the control of the parallel-connected two five-phase machines is truly decoupled even in sensor-less mode. These results affirm also the ability of the observer to guarantee good estimations in steady state and transients as well. Simulation results point out also that using sliding mode control the dynamic performance of the two-machine drive is further improved compared with the conventional PI controller.

Acknowledgments: No source of funding for this research investigation.

Author Contributions: All authors contributed equally for the final decimation of the research investigation as a full article.

Conflicts of Interest: The authors declare no conflict of interest.

References

1. Navid, R.A. Sliding-mode control of a six-phase series/parallel connected two induction motors drive. *ISA Trans.* **2014**, *53*, 1847–1856.

2. Chen, H.H.; Chong, X.S. Current control for single-inverter-fed series-connected five-phase PMSMS. In Proceedings of the IEEE International Symposium on Industrial Electronics, Taipei, Taiwan, 28–31 May 2013; pp. 1–6.

3. Liliang, G.; John, E.F. A space vector switching strategy for three-level five-phase inverter drives. *IEEE Trans. Ind. Electron.* **2010**, *57*, 2332–2343. [CrossRef]

4. Sneessens, C.; Labbe, T.; Baudart, F.; Matagne, E. Position sensorless control for five-phase permanent-magnet synchronous motors. In Proceedings of the International Conference on Advanced Intelligent Mechatronics, Besançon, France, 8–11 July 2014; pp. 794–799.

5. Wang, Z.; Wang, X.J.C.; Cheng, M.; Hu, Y. Direct torque control of T-NPC inverters fed double-stator-winding PMSM drives with SVM. *IEEE Trans. Power Electron.* **2018**, *33*, 1541–1553. [CrossRef]

6. Wang, X.; Wang, Z.; Cheng, M.; Hu, Y. Remedial strategies of T-NPC three-level asymmetric six-phase PMSM drives based on SVM-DTC. *IEEE Trans. Ind. Electron.* **2017**, *64*, 6841–6853. [CrossRef]

7. Lei, Y.; Ming-liang, C.; Jian-qing, S.; Fei, X. Current Harmonics Elimination Control Method for Six-Phase PM Synchronous Motor Drives. *ISA Trans.* **2015**, *59*, 443–449.

8. Leila, P.; Hamid, A.T. Sensorless Direct Torque Control of Five-Phase Interior Permanent-Magnet Motor Drives. *IEEE Trans. Ind. Appl.* **2007**, *43*, 952–959.

9. Siavash, S.; Lusu, G.; Hamid, A.T.; Leila, P. Wide Operational Speed Range of Five-Phase Permanent Magnet Machines by Using Different Stator Winding Configurations. *IEEE Trans. Ind. Electron.* **2012**, *59*, 2621–2631.

10. Hosseyni, A.; Ramzi, T.; Faouzi, M.M.; Atif, I.; Rashid, A. Sensorless Sliding Mode Observer for a Five-Phase Permanent Magnet Synchronous Motor Drive. *ISA Trans.* **2015**, *58*, 462–473. [CrossRef] [PubMed]

11. Ahmad, A.A.; Dahaman, I.; Pais, S.; Shahid, I. Speed-Sensorless Control of Parallel-Connected PMSM Fed By A Single Inverter Using MRAS. In Proceedings of the IEEE International Power Engineering and Optimization Conference, Melaka, Malaysia, 6–7 June 2012; pp. 35–39.

12. Zhang, H.; Luo, S.; Yu, Y.; Liu, L. Study On Series Control Method For Dual Three-Phase PMSM Based On Space Vector Pulse Width Modulation. *Int. J. Control Autom.* **2015**, *8*, 197–210. [CrossRef]

13. Martin, J.; Emil, L.; Slobodan, N.V. Independent Control of Two Five-Phase Induction Machines Connected In Parallel To A Single Inverter Supply. In Proceedings of the IEEE Industrial Electronics Conference, Paris, France, 6–10 November 2006; pp. 1257–1262.

14. Levi, E.; Jones, M.; Vukosavic, S.N.; Iqbal, A.; Toliyat, H.A. Modeling, control, and experimental investigation of a five-phase series-connected two-motor drive with single inverter supply. *IEEE Trans. Ind. Electron.* **2007**, *54*, 1504–1516. [CrossRef]

15. Levi, E.; Jones, M.; Slobodan, N.V.; Hamid, A.T. Steady-State Modeling of Series-Connected Five-Phase And Six-Phase Two-Motor Drives. *IEEE Trans. Ind. Appl.* **2008**, *44*, 1559–1568. [CrossRef]

16. Mekri, F.; Charpentier, J.F.; Semail, E. An Efficient Control of A Series Connected Two-Synchronous Motor 5-Phase with Non-Sinusoidal EMF Supplied By A Single 5-Leg VSI: Experimental And Theoretical Investigations. *Electr. Power Syst. Res.* **2012**, *92*, 11–19. [CrossRef]

17. Jones, M.; Vukosavic, S.N.; Levi, E. Parallel-Connected Multiphase Multidrive Systems with Single Inverter Supply. *IEEE Trans. Ind. Electron.* **2019**, *56*, 2047–2057. [CrossRef]

18. Zhang, X.; Lizhi, S.; Zhao, K.; Sun, L. Nonlinear Speed Control for PMSM System Using Sliding-Mode Control and Disturbance Compensation Techniques. *IEEE Trans. Power Electron.* **2013**, *28*, 1358–1365. [CrossRef]

19. Fatemi, F.S.M.J.R.; Navid, R.A.; Jafar, S.; Saeed, A. Speed Sensorless Control of a Six-Phase Induction Motor Drive Using Backstepping Control. *IET Power Electron.* **2014**, *7*, 114–123. [CrossRef]

20. Anissa, H.; Ramzi, T.; Atif, I.; Med, F.M. Backstepping Control for a Five-Phase Permanent Magnet Synchronous Motor Drive. *Int. J. Power Electron. Drive Syst.* **2015**, *6*, 842–852.

21. Ahmed, M.; Karim, F.M.; Abdelkader, M.; Abdelber, B. Input Output Linearization And Sliding Mode Control of a Permanent Magnet Synchronous Machine Fed by a Three Levels Inverter. *J. Electr. Eng.* **2006**, *57*, 205–210.

22. Le-Bao, L.; Ling-Ling, S.; Sheng-Zhou, Z.; Qing-Quan, Y. PMSM Speed Tracking And Synchronization of Multiple Motors Using Ring Coupling Control and Adaptive Sliding Mode Control. *ISA Trans.* **2015**, *58*, 635–649.

23. Lin, S.F.J.; Hung, Y.C.; Tsai, M.T. Fault-Tolerant Control for Six-Phase PMSM Drive System via Intelligent Complementary Sliding Mode Control Using TSKFNN-AMF. *IEEE Trans. Ind. Electron.* **2013**, *60*, 5747–5762. [CrossRef]

24. Chen, S.; Luo, Y.; Pi, Y.G. PMSM Sensorless Control with Separate Control Strategies and Smooth Switch from Low Speed to High Speed. *ISA Trans.* **2015**, *58*, 650–658. [CrossRef] [PubMed]

25. Dujic, D.; Jones, M.; Levi, E.; Prieto, J.; Barrero, F. Switching Ripple Characteristics of Space Vector PWM Schemes for Five-Phase Two-Level Voltage Source Inverters—Part 1: Flux Harmonic Distortion Factors. *IEEE Trans. Ind. Electron.* **2011**, *58*, 2789–2798. [CrossRef]

26. Iqbal, A.; Levi, E. Space Vector Modulation Schemes for a Five-Phase Voltage Source Inverter. In Proceedings of the European Conference on Power Electronics and Applications, Dresden, Germany, 11–14 September 2005; pp. 1–12.

27. Dujic, D.; Grandi, G.; Jones, M.; Levi, E. A Space Vector PWM Scheme for Multi Frequency Output Voltage Generation with Multiphase Voltage-Source Inverters. *IEEE Trans. Ind. Electron.* **2008**, *55*, 1943–1955. [CrossRef]

28. Quang, N.K.; Hieu, N.T.; Ha, Q.P. FPGA-Based Sensorless PMSM Speed Control Using Reduced-Order Extended Kalman Filters. *IEEE Trans. Ind. Electron.* **2014**, *12*, 6574–6582. [CrossRef]

29. Zhuang, X.; Rahman, M.F. Comparison of a Sliding Observer and a Kalman Filter for Direct-Torque-Controlled IPM Synchronous Motor Drives. *IEEE Trans. Ind. Electron.* **2012**, *59*, 4179–4188.

30. Dong, X.; Zhang, S.; Liu, J. Very-Low Speed Control of PMSM Based on EKF Estimation with Closed Loop Optimized Parameters. *ISA Trans.* **2013**, *52*, 835–843.

31. Abbas, N.K.; Jafar, S. MTPA Control of Mechanical Sensorless IPMSM Based on Adaptive Nonlinear Control. *ISA Trans.* **2016**, *61*, 348–356.

32. Khan, M.R.; Iqbal, A. MRAS Based Sensorless Control of a Series-Connected Five-Phase Two-Motor Drive System. *J. Electr. Eng. Technol.* **2008**, *3*, 224–234. [CrossRef]

33. Khan, M.R.; Iqbal, A. Extended Kalman Filter Based Speeds Estimation of Series-Connected Five-Phase Two-Motor Drive System. *Simul. Model. Pract. Theory* **2009**, *17*, 1346–1360. [CrossRef]

34. Ali, W.H.; Gowda, M.; Cofie, P.; Fuller, J. Design of a Speed Controller Using Extended Kalman Filter for PMSM. In Proceedings of the IEEE 57th International Midwest Symposium on Circuits and Systems, College Station, TX, USA, 8–11 July 2014; pp. 1101–1104.

35. Slotine, J.J. *Applied Nonlinear Control*; Tice Hall: Englewood Cliffs, NJ, USA, 1991.

36. Iqbal, A.; Levi, E. Space Vector PWM for a Five-Phase VSI Supplying Two Five-Phase Series-Connected Machines. In Proceedings of the 12th International Power Electronics and Motion Control Conference, Portoroz, Slovenia, 30 August–1 September 2006; pp. 222–227.

37. Dujic, D.; Jones, M.; Levi, E. Generalised Space Vector PWM for Sinusoidal Output Voltage Generation with Multiphase Voltage Source Inverters. *Int. J. Ind. Electron. Drives* **2009**, *1*, 1–13. [CrossRef]

38. Jones, M.; Dordevic, O.; Bodo, N.; Levi, E. PWM Algorithms for Multilevel Inverter Supplied Multiphase Variable-Speed Drives. *Electronics* **2012**, *16*, 22–31. [CrossRef]

electronics

MDPI

Article

Improving Performance of Three-Phase Slim DC-Link Drives Utilizing Virtual Positive Impedance-Based Active Damping Control

Ahmet Aksoz [1,*], Yipeng Song [2], Ali Saygin [1], Frede Blaabjerg [2] and Pooya Davari [2,*]

[1] Department of Electrical and Electronics Engineering, Faculty of Technology, Gazi University, 06500 Ankara, Turkey; asaygin@gazi.edu.tr (A.S.)

[2] Department of Energy Technology, Aalborg University, AAU 9220 Aalborg East, Denmark; yis@et.aau.dk (Y.S.); fbl@et.aau.dk (F.B.)

* Correspondence: ahmetaksoz@gazi.edu.tr (A.A.); pda@et.aau.dk (P.D.); Tel.: +90-554-2774428 (A.A.); +45-3147-8845 (P.D.)

Received: 3 September 2018; Accepted: 1 October 2018; Published: 4 October 2018

Abstract: In this paper, a virtual positive impedance (VPI) based active damping control for a slim DC-link motor drive with 24 section space vector pulse width modulation (SVPWM) is proposed. Utilizing the proposed control and modulation strategy can improve the input of current total harmonic distortion (THD) while maintaining the cogging torque of the motor. The proposed system is expected to reduce the front-end current THD according to international standards, as per IEC 61000 and IEEE-519. It is also expected to achieve lower cost, longer lifetime, and fewer losses. A permanent magnet synchronous motor (PMSM) is fed by the inverter, which adopts the 24 section SVPWM technique. The VPI based active damping control for the slim DC-link drive with/without the 24 section SVPWM are compared to confirm the performance of the proposed method. The simulation results based on MATLAB are provided to validate the proposed control strategy.

Keywords: slim DC-link drive; VPI active damping control; total harmonic distortion; cogging torque

1. Introduction

In many industrial applications, slim DC-link drives have become increasingly favored day by day. A classical driver consists of a 6-pulse diode bridge rectifier, an intermediate circuit with a big capacitor, an inductor, and an inverter. To maintain stable DC-link voltage, the DC-link capacitor needs to be carefully selected. Although the big size capacitor with large capacitance is at a higher cost and shorter lifetime, it has strong robustness against the stability problem. However, cost, lifetime, and loss must be taken into consideration for industrial applications. Thus, using a film capacitor as the slim DC-link capacitor in drivers is preferred, in spite of the stability problem. A diode rectifier based slim DC-link drive is shown in Figure 1. This grid-connected driver has a diode rectifier, a slim DC capacitor, and a 6-switches three-phase inverter. Additionally, point of common coupling (PCC) phase currents for stiff grid and weak grid are given in Figure 2 at different operation speeds. The i_pcc simulated waveforms show that stiff grid and high operation speed (c) is the best current waveform. However, it can be improved with control methods, especially for weak grid conditions.

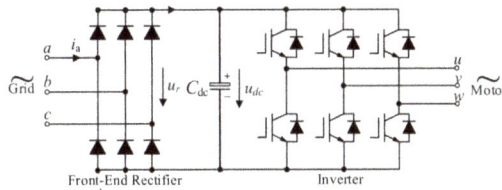

Figure 1. Diode rectified slim DC-link drive.

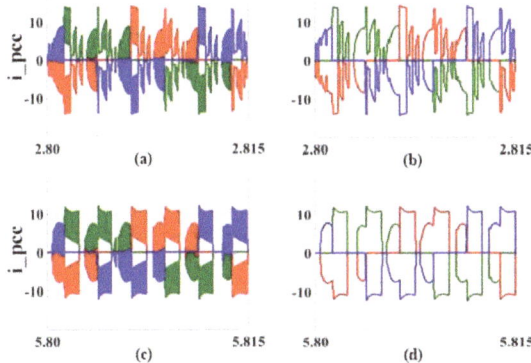

Figure 2. PCC phase current. (**a**) Stiff grid at 1500 rpm. (**b**) Weak grid at 1500 rpm. (**c**) Stiff grid at 3000 rpm. (**d**) Weak grid at 3000 rpm.

To solve the stability problem, Virtual Positive Impedance (VPI) based active damping control has been effectively implemented [1,2]. Active damping control for slim DC-link drive ensures that no extra passive damping component is needed. The negative impedance instability is described as:

$$\frac{-P_L}{v_{dc}^2} = \frac{-i_{dc}v_{dc0}}{v_{dc}(v_{dc0} - \tilde{v})} = \frac{-(i_{dc0} + \widetilde{i_{dc}})v_{dc0}}{v_{dc}(v_{dc0} - \tilde{v})} \tag{1}$$

where P_L is the load power, v_{dc0} is the DC component of the DC-link voltage, $\tilde{v}$ is the AC part of the DC-link voltage, and v_{dc} is the DC-link voltage [1]. i_{dc} is the DC-link current, $\widetilde{i_{dc}}$ is the AC part of the DC-link current, and i_{dc0} is the DC component of the DC-link current. In contrast to the case using the big capacitor [3], the constant power load behavior of the motor with a slim-DC-link capacitor causes the larger ripple on the DC voltage. Both ripples on the DC-link voltage and the front-end current harmonics are higher when using a small capacitor [4,5]. In order to reduce the input current harmonics, VPI based active damping control decreases the ripple on the DC-link voltage [1–4]. A virtual positive impedance block diagram is illustrated in Figure 3 [1]. This model can be used for the generation of the DC-link voltage reference. In order to control the AC component of the DC-link voltage, the 1st-order High Pass Filter is represented by a high pass filter (HPF) block. g_v is the gain on the accompaniment of the DC-link voltage. In spite of the fact that VPI contains an HPF block, it resembles the 1st-order Low Pass Filter and harmonic detection block. Additionally, this block diagram can be used for detecting the harmonic [5]. The harmonic detection is also ensured by the VPI method, providing the lower ripples on the DC-link voltage. Harmonic mitigation, harmonic cancellation, or generally a harmonic problem is an important issue for motor drivers [6,7]. This problem can deteriorate grid voltage quality, as well as the performance of both the driver and the load. Although it cannot be completely removed, it needs to be mitigated as much as possible. For this purpose, several control techniques and PWM techniques have been studied [8–10].

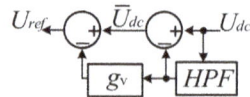

Figure 3. Virtual positive impedance block diagram [1].

At the same time, not only ripple on the DC-link voltage on a small capacitor, but also a motor cogging torque due to interactions between core and magnet result in both grid input current harmonics and motor current harmonics. Owing to the harmonic problem, active damping control (ADC) and VPI can be used to decrease THD. Even through these methods achieve harmonics suppression, the cogging torque also needs to be solved by harmonic effect, because motor current harmonics cause the higher cogging torque [11–16]. When the 3DSVPWM aims to optimize switching waveforms, it can achieve a lower cogging torque. The algorithm of the 3DSVPWM was based on four steps. Firstly, the reference vector was transformed into 2D. In addition, the length of the reference vector was described according to the length of the basis vectors. Secondly, the closest three vectors were found. When they were detected, finding high–low values of the reference vector coordinates could be facilitated. Duty cycles were calculated in the third step. Lastly, the best switching states were selected when 2D coordinates are transformed to 3D coordinates [9,12].

In this study, the DC-link voltage, the grid input current THD, the VPI bode results, the cogging torque, and the THD of motor currents were simulated in MATLAB (R2016b, MathWorks, Natick, MA, USA), where four simulation models were developed: (1) Weak grid without VPI without 3DSVPWM (wOVPIwO3D), (2) weak grid with VPI without 3DSVPWM (wVPIwO3D), (3) weak grid with VPI with 3DSVPWM (wVPIw3D), and (4) stiff grid with VPI with 3DSVPWM (stiffwVPIw3D).

Simulation results of these models are compared and discussed. Section 2 analyzes the interaction between the cogging torque and the current harmonic. The 3DSVPWM, the PMSM model, and the input admittance are described. Then, virtual positive impedance based active damping control is given in Section 3. In Section 4, the performance analysis of the DC-link current THD and the motor current THD of the slim DC-link capacitor is shown. In addition, the stability analysis, the control structure, and the control impedance Y_{ctrl} are explained in the same section. Additionally, the simulation validation of the grid input current THD_i and the cogging torque are obtained. Lastly, the study is summarized in Section 5.

2. Interaction between Cogging Torque and Harmonic

2.1. The Cogging Torque Reduction Methods

In order to decrease the cogging torque, some methods are used. These are mainly:

- Skewing stator stack or magnets;
- Modulation drive current waveform;
- Using fractional slots per pole;
- Optimizing the magnet pole arc or width [9,17].

The schema of the cogging torque reduction is illustrated in Figure 4. In order to obtain a better modulation drive current waveform, there are three main methods. They are decreasing harmonics, switching at high frequency, and using advanced PWM techniques [9,17].

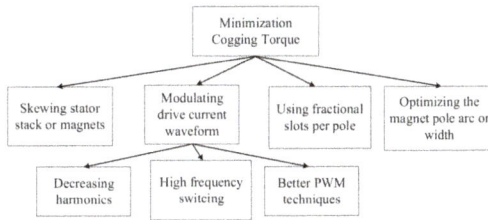

Figure 4. Reducing cogging torque schema [17].

On the other hand, the kth harmonic is related to the cogging torque, as expressed in (2). T_{ck} is the amplitude of the kth harmonic component of the cogging torque, θ is the angle of rotation, and k is the order of cogging harmonics.

$$T_{ck}(\theta) = \sum_{n=-\infty}^{\infty} T_n e^{2ni(\theta - k\theta_s)} \tag{2}$$

where T_n is the Fourier series coefficient and θ_s is the electrical angle slot pitch. It is expressed as:

$$\theta_s = \pi N_m / N_s \tag{3}$$

where N_m is the number of the magnet pole and N_s is the number of the slot. Accordingly, the cogging torque T_{cog} can be written as the Fourier series as:

$$T_{cog}(\theta) = \sum_{k=0}^{N_s - 1} T_{ck}(\theta) \tag{4}$$

The proposed models are applied not only for achieving a reduced cogging torque, but also improving system stability thanks to decreased harmonics. Given the fact that the T_{ck} is decreased, the cogging torque can be reduced.

2.2. 3DSVPWM Technique

In the proposed modulation technique, there are 24 sectors, including zero voltage vectors [9]. The modulation space is divided into 6 sections (S1–S6), each section consisting of 4 delta sectors ($\Delta 1$, $\Delta 2$, $\Delta 3$, $\Delta 4$).

This proposed modulation technique is adopted in the 3-phase 3-level or multilevel inverter. However, it is used in this study in the 3-phase 2-level inverter. Thus, modulation angles are made smaller and the number of the sector is increased in sections. Here, using the definition of the vector norm, the vectors of the inverter are defined in a plane as given in (5):

$$v_{ab} + v_{bc} + v_{ca} = 0 \tag{5}$$

where v_{ab}, v_{bc}, and v_{ca} are the vectors of the inverter in the 3D coordinate system. The switching state vectors are shown in Figure 5 [18,19]. Additionally, the numbers of 0, 1, and 2 in Figure 5 represent V_{ab}, V_{bc}, or V_{ca}/V_{dc}.

$$\overrightarrow{V}_{15}\ (-2,2) \quad \overrightarrow{V}_8\ (-1,2) \quad \overrightarrow{V}_{14}\ (0,2)$$

(Space vector diagram of 3DSVPWM, hexagon with vectors $\overrightarrow{V}_0$–$\overrightarrow{V}_{18}$ and coordinates: $\overrightarrow{V}_{15}(-2,2)$, $\overrightarrow{V}_8(-1,2)$, $\overrightarrow{V}_{14}(0,2)$, $\overrightarrow{V}_9(-2,1)$, $\overrightarrow{V}_3(-1,1)$, $\overrightarrow{V}_2(0,1)$, $\overrightarrow{V}_7(1,1)$, $\overrightarrow{V}_{16}(-2,0)$, $\overrightarrow{V}_4(-1,0)$, $\overrightarrow{V}_0$, $\overrightarrow{V}_6$, $\overrightarrow{V}_1(1,0)$, $\overrightarrow{V}_{13}(2,0)$, $\overrightarrow{V}_5(0,-1)$, $(1,-1)$, $\overrightarrow{V}_{10}(-1,-1)$, $\overrightarrow{V}_{12}(2,-1)$, $\overrightarrow{V}_{17}(0,-2)$, $\overrightarrow{V}_{11}(1,-2)$, $\overrightarrow{V}_{18}(2,-2)$)

$\overrightarrow{V}_0$	$\overrightarrow{V}_1$	$\overrightarrow{V}_2$	$\overrightarrow{V}_3$	$\overrightarrow{V}_4$	$\overrightarrow{V}_5$	$\overrightarrow{V}_6$
00 00 00	00 01 01	10 10 00	01 00 01	00 10 10	01 01 00	10 00 10
00 00 11	11 01 01	10 10 11	01 11 01	11 10 10	01 01 11	10 11 10
00 11 00	10 00 00	00 00 01	00 10 01	01 00 00	00 00 10	00 01 00
00 11 11	10 00 11	00 11 01	00 11 11	01 00 11	00 11 10	00 01 11
11 00 00	10 11 00	11 00 01	11 10 00	01 11 00	11 00 10	11 01 00
11 00 11	10 11 11	11 11 01	11 10 11	01 11 11	11 11 10	11 01 11
11 11 00						
11 11 11	$\overrightarrow{V}_7$	$\overrightarrow{V}_8$	$\overrightarrow{V}_9$	$\overrightarrow{V}_{10}$	$\overrightarrow{V}_{11}$	$\overrightarrow{V}_{12}$
01 01 01	10 00 01	00 10 01	01 10 00	01 00 10	00 01 10	10 01 00
10 10 10	10 11 01	11 10 01	01 10 11	01 11 10	11 01 10	10 01 11

$\overrightarrow{V}_{13}$	$\overrightarrow{V}_{14}$	$\overrightarrow{V}_{15}$	$\overrightarrow{V}_{16}$	$\overrightarrow{V}_{17}$	$\overrightarrow{V}_{18}$
10 01 01	10 10 01	01 10 01	01 10 10	01 01 10	10 01 10

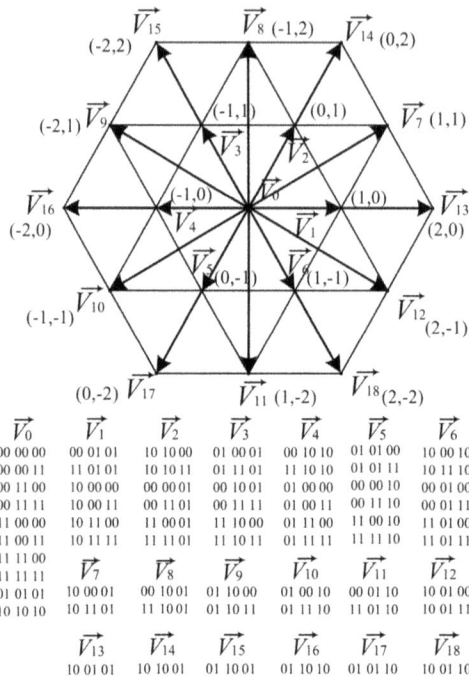

Figure 5. Space vector diagram of 3DSVPWM in 2D and 3D vectors after 3D transformation.

According to (5) and Figure 5, the delta vectors can be expressed in the 2D coordinate system as:

$$\overrightarrow{V}_{1_{(v_{ab},v_{bc},v_{ca})}},\ \overrightarrow{V}_{2_{(v_{ab},v_{bc},v_{ca})}} = \begin{bmatrix} V_{dc} \\ 0 \\ -V_{dc} \end{bmatrix}, \begin{bmatrix} 0 \\ V_{dc} \\ -V_{dc} \end{bmatrix} \tag{6}$$

where, $\overrightarrow{V}_{1(vab,\ vbc,\ vbc)}$ and $\overrightarrow{V}_{2(vab,\ vbc,\ vbc)}$ are the delta vectors. They are the transformed vectors from 3D to 2D. The reference vector can be placed in a sector and the switching-state vector is shown at the corner of each sector. It is able to produce switching-state vectors in 2D, which are then transformed into 3D switching-state vectors.

This 24 sectors SVPWM technique can eliminate the need for dead-time protection and allow the upper and lower switches to switch at the same time. In this case, the dead-time effect is removed, an additional midpoint voltage is generated, and the effective output switching frequency is doubled. Thus, the current harmonics in the output current waveform are significantly suppressed by applying the three-level voltage output and doubling the effective switching frequency. In order to reduce the current harmonics, the adjacent three vectors and the reference vector must be defined in the best way. V_x, V_y, and V_z are the adjacent three vectors as follow:

$$T_s = d_x + d_y + d_z \tag{7}$$

$$\overrightarrow{V}_{ref} = d_x \overrightarrow{V}_x + d_y \overrightarrow{V}_y + d_z \overrightarrow{V}_y \tag{8}$$

where the dwell time of vectors are d_x, d_y, and d_z, respectively. The reference vector is determined in the hexagon to state which triangle will be used. The biggest difference between 3DSVPWM and classical SVPWM is dwell times: The 3DSVPWM provides better dwell times for switching angles.

2.3. PMSM Model

The PMSM motor is modeled in the dq reference frame, which relies on the field oriented control (FOC), and the mathematical equations are given below:

$$\begin{bmatrix} v_{sd} \\ v_{sq} \end{bmatrix} = \begin{bmatrix} R + sL_{sd} & -\omega_r L_{sq} \\ \omega_r L_{sd} & R + sL_{sq} \end{bmatrix} \begin{bmatrix} i_{sd} \\ i_{sq} \end{bmatrix} + \begin{bmatrix} 0 \\ \omega_r \lambda \end{bmatrix} \tag{9}$$

$$T_e = \frac{3}{2} P(\lambda i_{sq} + (L_{sd} - L_{sq}) i_{sd} i_{sq}) \tag{10}$$

where the R, λ, P, L_{sd}, L_{sq}, T_e, and ω_r represent the stator resistor, the flux produced by the permanent magnets, the number of pole pairs, the stator inductances in the *dq*-frame, the electrical torque, and the rotor speed individually. In addition, v_{sd}, v_{sq}, i_{sd}, and i_{sq} represent the stator voltages and the stator currents in the *dq*-frame, respectively. Due to the fact that the speed and current loops force the stator current i_{sd} to be 0, (9) can be rewritten as:

$$V_{sd} = -\omega_r L_{sq} i_{sq} \tag{11}$$

$$V_{sq} = R i_{sq} + s L_{sq} i_{sq} + \omega_r \lambda \tag{12}$$

2.4. Input Admittance

Using the above equations, the input admittance of the control block and the constant power load are specified as follow:

$$G_{iq} = \frac{1}{Z_q + F_{iq}} \frac{V_q}{V_{dc}} (1 + g_v D A) \tag{13}$$

$$G_{id} = \frac{1}{Z_d + F_{id}} \frac{V_q}{V_{dc}} (1 + g_v D A) \tag{14}$$

$$G_{vd} = \frac{3}{2} (Z_d I_d + V_d + \omega_r L_d I_q \\ + \frac{3}{2} \frac{(L_d - L_q)^2 I_d I_q^2 N_{pp}^2}{Js}) \tag{15}$$

$$G_{vq} = \frac{3}{2} (Z_q I_q + V_q - \omega_r L_q I_d \\ + \frac{3}{2} \frac{(L_d - L_q)^2 I_q I_q^2 N_{pp}^2}{Js}) \tag{16}$$

$$Y_{in} = \frac{1}{Z_{in}} = \frac{G_{vd} G_{id} + G_{vq} G_{iq}}{V_{dc}} + \frac{-P_L}{V_{dc}^2} \tag{17}$$

$$1/Y_{ctrl} = \frac{V_{dc}}{G_{vd} G_{id} + G_{vq} G_{iq}} \tag{18}$$

Y_{ctrl} is the admittance of the control part with VPI based active damping control and Y_{cpl} $(-P_L/V_{dc}^2)$ is the admittance of the constant power load behavior. D, A, N_{pp}, J, Z_{dq}, F_{id}, and F_{iq} relate the PWM delay, the 1st-order HPF, the pole-pairs, the inertia, the *dq*-axis impedance of the PMSM, and the current controller of the *dq*-axis separately. Two input admittances in (17) are illustrated with an equivalent DC-link circuit in Figure 6, which is on simplified equivalent circuit model of the diode rectified based slim DC-link drive.

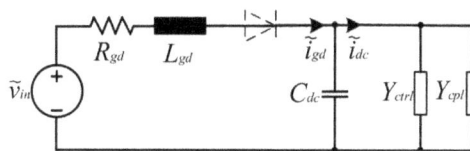

Figure 6. Simplified equivalent DC-link in a drive unit.

3. Virtual Positive Impedance Based Active Damping Control

The parameters of the drive and the PMSM are given in Table 1. Moreover, the sample period is Ts, the reference torque of 3 ph trapezoidal motor is Tm, C_{dc} is the slim DC-link capacitor value, SCR is the short circuit ratio, and the stator resistive and inductive values are R and Ld–Lq. In addition, the current loop, the speed loop, and PWM block are shown in Figure 7.

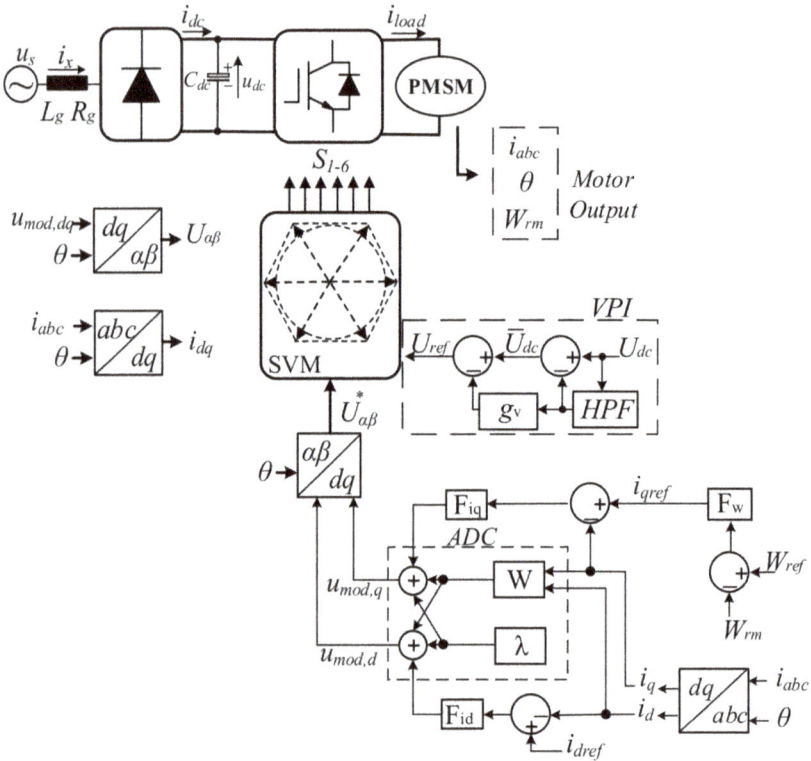

Figure 7. Control system block diagram with Virtual Positive Impedance (VPI) based active damping control (ADC).

Table 1. The parameters of the drive and the PMSM.

Parameters	Values
V_{rms}	400 V/3 ph
C_{dc}	14 uF
SCR (weak grid-stiff grid)	35–350
T_s	50 us
T_m	4 Nm
R	150 mΩ
L_d–L_q	8.5–8.8 mH
ω_r	1500–3000 rpm

The control diagram includes the Space Vector Modulation (SVM) block, together with a VPI and an ADC block, the Park and the inverse Park transformation, the speed controller F_w, the current controller F_{iq} and F_{id} together with the decoupling block W. All the components are assumed as ideal. The power loss and the saturation effects in the drive system are neglected. The VPI is illustrated in Figure 3; U_{dc} is the voltage feedback from the slim DC-link capacitor, which is used for voltage ripple elimination. Then, U_{ref} is the control reference of the V_{dc}, which is calculated based on the reference voltages is 400 V, and the sum of the reversed voltage is $-U_{dc}$.

Furthermore, the speed closed-loop control and the current closed-loop control are shown in Figure 7 in the dq-frame. W is the decoupling function block presented in Equation (19) and ω_r is the rotor speed. The current control equation is shown in Equation (20) and the speed control equation is shown in Equation (21) as follows:

$$W = \begin{bmatrix} 0 & -\omega_r L_q \\ \omega_r L_d & 0 \end{bmatrix} \tag{19}$$

$$\begin{aligned} u_{\text{mod},dq}(t) = K_p \left[i_{dq,ref}(t) - i_{dq}(t) \right] \\ + K_i \left[i_{dq,ref}(t) - i_{dq}(t) \right] + 2\lambda + i_{dq}(t)W \end{aligned} \tag{20}$$

$$i_q(t) = K_p \left[(\omega_{ref}(t) - \omega_r(t)) \right] + K_i T_s (\omega_{ref}(t) - \omega_r(t)) \tag{21}$$

4. Performance Analysis of The DC-Link Voltage THD and the Motor Current THD

According to the Equation (18), bode diagrams of the $1/Y_{ctrl}$ at 1500 rpm and 3000 rpm are given in Figures 8 and 9.

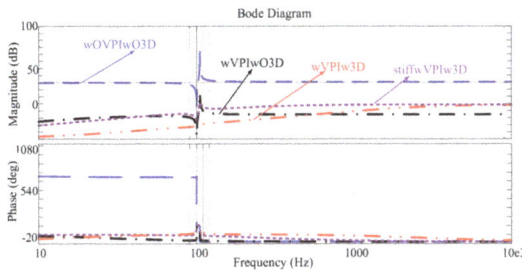

Figure 8. Bode plot of $1/Y_{ctrl}$ at 1500 rpm (Blue: wOVPIwO3D. Black: wVPIwO3D. Red: wVPIw3D. Purple: stiffwVPIwO3D).

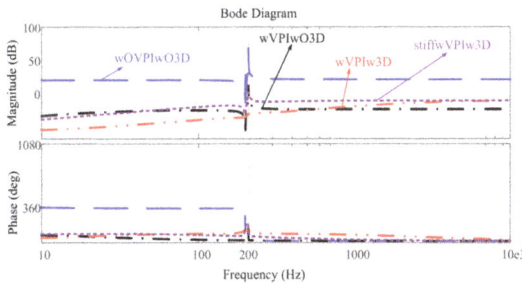

Figure 9. Bode plot of $1/Y_{ctrl}$ at 3000 rpm (Blue: wOVPIwO3D. Black: wVPIwO3D. Red: wVPIw3D. Purple: stiffwVPIwO3D).

The Bode diagrams of the control impedance of the four cases can be seen at 1500 rpm and 3000 rpm. In the case of not using VPI based ADC and 3DSVPWM that is wOVPIwO3D (blue), the impedance magnitude behaves flat, but it does not reach zero. By contrast, in the case of the wVPIwO3D (black), wVPIw3D (red), and stiff wVPIw3D (purple), the magnitude of the control impedance becomes lower than that without active damping in the frequency range. The resonance is named as Negative-Impedance (NI) resonance due to the frequency character decided by NI at the constant power load (CPL) situation. Its impedance characteristic behaves as an inductive plus negative-resistive impedance during [10, 5000] Hz. This is helpful in suppressing the harmonics, caused by the resonance between L_{gd} and C_{dc}. In order to improve the THD (lower ripple on magnitude), impedance is increased at the current controller bandwidth. Increased impedance with the bandwidth of the current controller is helpful for suppressing the current harmonics (100 Hz and 200 Hz). Additionally, control impedance always behaves as positive-resistive plus inductive at high frequency while capacitive at low frequency. This positive-resistive characteristic helps to damp the system into a stable state.

4.1. The Performance Analysis of the DC-Link Voltage THD and the Motor Current THD

According to the VPI based ADC, the performance analysis of the DC-link voltage THD and the motor current THD is presented. Owing to the fact that the big size capacitor or RLC components have a higher cost and shorter lifetime, using the film capacitor as the slim DC-link capacitor in drivers can be a good alternative [1]. In spite of the stability problem, the DC-link voltage of the slim capacitor is controlled well with VPI based ADC and 3DSVPWM. The DC-link voltage performances of the four cases are given below. The DC-link voltage when rotor speed is 1500 rpm is shown in Figure 10 and the DC-link voltage when rotor speed is 3000 rpm is shown in Figure 11. Firstly, the motor is operated at 1500 rpm from 0 s to 3 s, and then it is operated at 3000 rpm from 3 s to 6 s. However, time periods of the simulation are only 2.8–2.815 s and 5.8–5.815 s, because the results of the simulation are the same during 0s to 3 s and 3 s to 6 s. Thus, 2.8–2.815 s as TP1 (time period 1) and 5.8–5.815 as TP2 (time period 2) are used.

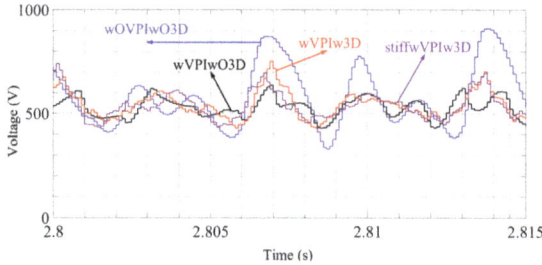

Figure 10. DC-link voltage when rotor speed is 1500 rpm.

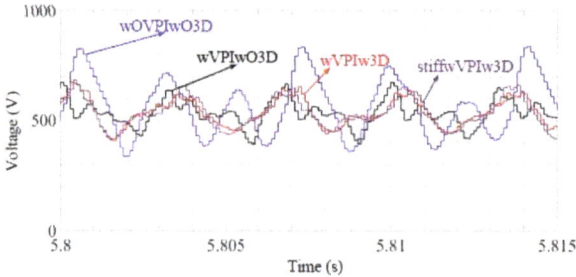

Figure 11. DC-link voltage when rotor speed is 3000 rpm.

The ripples on the DC-link voltage when rotor speed is 1500 rpm and 3000 rpm are given in Table 2.

Table 2. The ripples of the DC-link voltage.

Cases	Voltage (TP1)	Voltage (TP2)
wOVPIwO3D	240.3	190.8
wVPIwO3D	158.9	144.2
wVPIw3D	225.0	169.6
stiffwVPIwO3D	82.1	76.79

In addition, the fast Fourier transform (FFT) results of the DC-link voltage when rotor speed is 1500 rpm and 3000 rpm are given in Table 3.

Table 3. The FFT results of the DC-link voltage.

Cases	THD% (TP1)	THD% (TP2)
wOVPIwO3D	34.34	32.84
wVPIwO3D	13.15	15.80
wVPIw3D	15.93	17.01
stiffwVPIwO3D	15.93	17.02

As shown in Table 2, the case of stiffwVPIwO3D has the best performance, as expected, with the lowest ripples on the DC-link voltage for both operation speeds as 82.1 V and 76.79 V. However, wVPIwO3D with a weak grid has the best THD results for both operation speeds according to Table 3, as 13.15% at TP1 and 15.80% at TP2. Additionally, the ripples on DC-link voltage wVPIwO3D are obviously better than those on wVPIw3D (158.9–225.0 V and 144.2–169.6 V). The ripple on the DC-link voltage of the VPI based ADC with traditional SVPWM can oscillate. Lower ripples on DC-link voltage are obtained as 158.9 V at TP1 and 144.2 V at TP2. This means that using both the 3DSVPWM and the VPI based ADC does not provide better results of the ripple on DC-link voltage. Moreover, the FFT results of the motor current harmonics are displayed in Table 4.

Table 4. The FFT results of the motor current.

Cases	THD% (TP1)	THD% (TP2)
wOVPIwO3D	41.36%	29.82%
wVPIwO3D	27.83%	20.46%
wVPIw3D	10.74%	12.07%
stiffwVPIw3D	10.84%	11.95%

From Table 4, the motor current harmonics of wVPIw3D with a weak grid or stiff grid are acceptable. When the 3DSVPWM is enabled, the motor current harmonics are suppressed effectively as 10.74% at TP1 and 12.07% at TP2 for Case 3 and 10.84% at TP1 and 11.95% at TP2 for Case 4.

The motor current (i_abc) waveforms of 4 cases are given in Figure 12. As seen there, Case 4 supplies the best results (g and h), thanks to virtual positive impedance-based active damping control and 3DSVPWM under the stiff grid. Then, Case 3 gives good results (e and f), thanks to virtual positive impedance-based active damping control and 3DSVPWM under the weak grid. Motor currents without 3DSVPWM means are seen in Case 2 (c and d). Lastly, motor currents with classical SVPWM without VPI based ADC means (a and b) are given in Case 1.

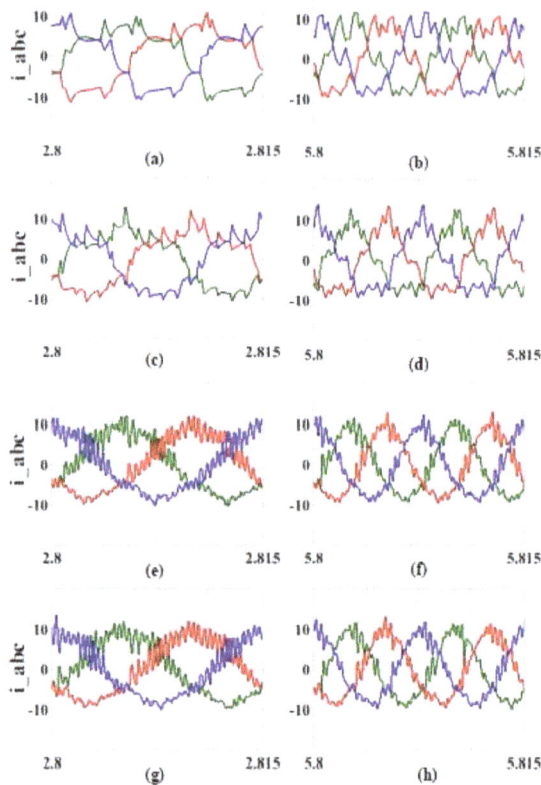

Figure 12. Motor current waveforms. (**a**) wOVPIwO3D at TP1. (**b**) wOVPIwO3D at TP2. (**c**) wVPIwO3D at TP1. (**d**) wVPIwO3D at TP2. (**e**) wVPIw3D at TP1. (**f**) wVPIw3D at TP2. (**g**) stiffwVPIw3D at TP1. (**h**) stiffwVPIw3D at TP2.

4.2. The Performance Analysis of the Grid Current THD and the Cogging Torque

Table 5 shows the analysis of the grid current THD according to the four cases of simulation results.

Table 5. The grid input current FFT results.

Cases	THD% (TP1)	THD% (TP2)
gridCurrent wOVPIwO3D	110.71%	98.45%
gridCurrent wVPIwO3D	54.53%	51.60%
gridCurrent wVPIw3D	47.74%	43.88%
gridCurrent stiffwVPIwO3D	48.60%	43.48%

As shown in Table 5, the grid input current FFT results are shown when the drive load is 3 kW. The FFT results of the grid input current of Case 1 are not as expected. When the 3DSVPWM is enabled, the THD_i decreases from 51.60% to 43.88% at TP2, and it also decreases from 54.53% to 47.74% at TP1. Although Case 4 has a better result than Case 3 at 3000 rpm, the result of Case 4 gives worse THD_i than Case 3 at 1500 rpm. The grid input current harmonics with the VPI based ADC and with the 3DSVPWM in Case 3 or Case 4 (with a weak grid or stiff grid) are acceptable. When the four cases are compared, the THD_i results of Case 3 and Case 4 are rather desirable for both operation speeds.

The cogging torque results are given in Table 6. As seen in Table 6, the cogging torque clearly decreases when adopting the 3DSVPWM. When Case 2 and Case 3 are compared, the cogging torque values get lower, from 0.25170 Nm to 0.15931 Nm at 3000 rpm and from 0.33270 Nm to 0.17871 Nm. Moreover, since the stiff grid is used, these results are 0.15285 Nm at 3000 rpm and 0.16720 Nm at 1500 rpm. In addition, the worst results are 0.28560 Nm at 3000 rpm and 0.44510 Nm at 1500 rpm from Case 1. These results are also seen in Figures 13 and 14.

Table 6. The cogging torque results.

	Cogging Torque (Nm) (TP1)	Cogging Torque (Nm) (TP2)
wOVPIwO3D	0.44510	0.29560
wVPIwO3D	0.33270	0.25170
wVPIw3D	0.17871	0.15931
stiffwVPIwO3D	0.16720	0.15285

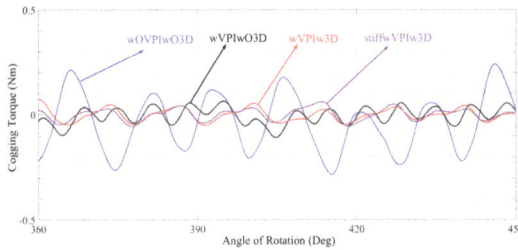

Figure 13. Cogging torque at 1500 rpm.

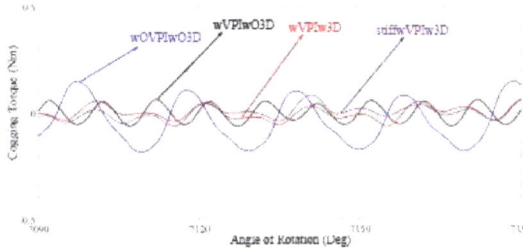

Figure 14. Cogging torque at 3000 rpm.

It can be seen that the cogging torque results are higher at a lower operation speed. The cogging torque results of Case 3 and Case 4 are more preferable than those of Case 1 and Case 2. Because the cogging torque is an important problem at low speed, the performance of the cogging torque in Case 3 and Case 4 are desired, especially at lower speed. At the same time, the cogging torque results of Case 3 and Case 4 at higher speed are better than those of the other cases. When both results in the tables and the figures are compared under either a weak or stiff grid, the adoption of the VPI based ADC and the 3DSVPWM together gives better results. The VPI based ADC ensures better harmonics, using a more advanced modulation technique, like 3DSVPWM (0.17871–0.15931 Nm and 0.16720–0.15285 Nm). Although the ripples on the DC-link voltage of the wVPIwO3D (Case 2) are lower without 3DSVPWM, the wVPIw3D (Case 3) and the stiffwVPIw3D (Case 4) are able to better suppress the grid current THD, the motor current THD, and the cogging torque.

After the results are obtained, all of them are given in Figure 15.

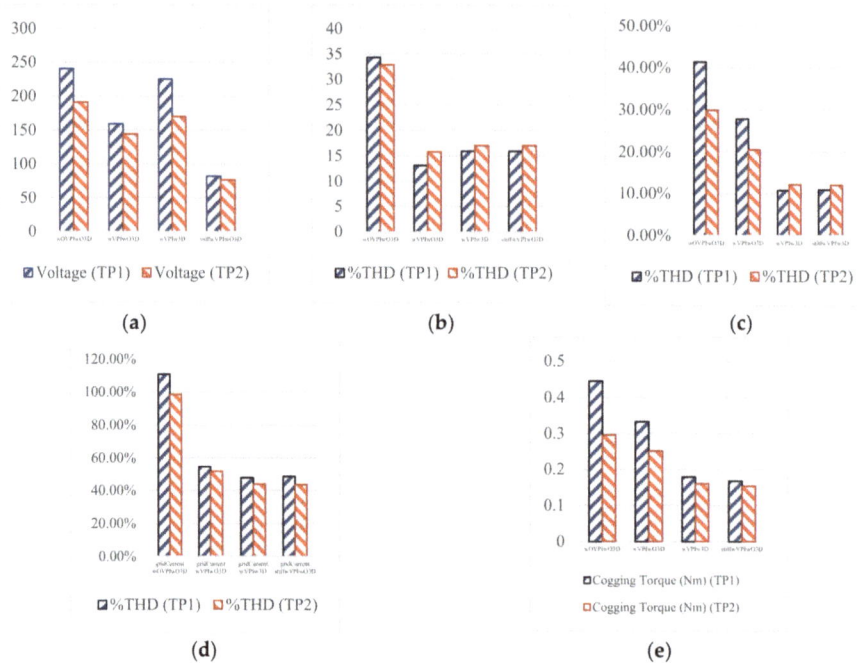

Figure 15. The performance difference between cases. (**a**) The ripples of the DC-link voltage. (**b**) The FFT results of the DC-link voltage. (**c**) The FFT results of the motor current. (**d**) The grid input current FFT results. (**e**) Cogging torque results.

The performance difference between cases is given in Figure 15. In this figure, the ripples of the DC-link voltage, the FFT results of the DC-link voltage, the FFT results of the motor current, the grid input current FFT results, and the cogging torque results are illustrated as bar graphics.

5. Conclusions

The motor drives equipped with the slim DC capacitor using the VPI based ADC and the 3DSVPWM are able to achieve lower input current harmonics and a lower cogging torque. In spite of the decreased ripples on DC-link voltages, the grid current harmonics and the motor current harmonics must be suppressed for the stable systems, because these harmonics result in a shorter lifetime for the driver and the cogging torque in the PMSM. However, a slim DC-link drive may not reach the expected performance, due to the weak grid condition. In this paper, the virtual positive impedance based active damping control and the 3DSVPWM are employed. Under either weak or stiff grid conditions, the four cases are clearly simulated. According to the simulation results, the performance of the VPI based ADC and the 3DSVPWM are investigated. In spite of the ripples on DC-link voltage, the 3DSVPWM with the VPI based ADC achieves harmonic suppression and a decreased cogging torque. Thus, a more stable and longer lifetime driver is obtained when the VPI based ADC and the 3DSVPWM are used.

Author Contributions: Conceptualization, F.B. and A.S.; Methodology, P.D; Software, A.A.; Validation, A.A., Y.S. and P.D.; Formal Analysis, A.A.; Investigation, P.D.; Resources, F.B. and P.D.; Data Curation, A.A.; Writing-Original Draft Preparation, A.A.; Writing-Review & Editing, P.D. and Y.S.; Visualization, Y.S.; Supervision, A.S. and F.B.; Project Administration, F. B. and A.S.; Funding Acquisition, A.S.

Funding: This research was funded by [TUBITAK] grant number [1059B141600864].

Conflicts of Interest: The authors declare no conflict of interest.

References

1. Yang, F.; Mathe, L.; Lu, K.; Blaabjerg, F.; Wang, X.; Davari, P. Analysis of harmonics suppression by active damping control on multi slim dc-link drives. In Proceedings of the IEEE IECON 2016, Florence, Italy, 24–27 October 2016; pp. 5001–5006.

2. Feng, Y.; Wang, D.; Blaabjerg, F.; Wang, X.; Davari, P.; Lu, K. Active damping control methods for three-phase slim DC-link drive system. In Proceedings of the IEEE IFEEC-ECCE 2017, Kaohsiung Taiwan, 3–7 June 2017; pp. 2165–2170.

3. Soltani, H.; Davari, P.; Kumar, D.; Zare, F.; Blaabjerg, F. Effects of DC-link filter on harmonic and interharmonic generation in three-phase adjustable speed drive systems. In Proceedings of the IEEE ECCE 2017, Cincinnati, OH, USA, 1–5 October 2017; pp. 675–681.

4. Maheshwari, R.; Munk-Nielsen, S.; Lu, K. An active damping technique for small dc-link capacitor based drive system. *IEEE Trans. Ind. Inf.* **2013**, *9*, 848–858. [CrossRef]

5. Wang, D.; Lu, K.; Rasmussen, P.O.; Mathe, L.; Feng, Y. Analysis of voltage modulation based active damping techniques for small dc-link drive system. In Proceedings of the IEEE ECCE 2015, Montreal, CA, USA, 20–24 September 2015; pp. 20–24.

6. Davari, P.; Yang, Y.; Zare, F.; Blaabjerg, F. A multi-pulse pattern modulation scheme for harmonic mitigation in three-phase multi-motor drives. *IEEE J. Emerg. Sel. Top. Power Electron.* **2016**, *4*, 174–185. [CrossRef]

7. Hansen, S.; Nielsen, P.; Blaabjerg, F. Harmonic cancellation by mixing nonlinear single-phase and three-phase loads. *IEEE Trans. Ind. Appl.* **2000**, *36*, 152–159. [CrossRef]

8. Malinowski, M.; Kazmierkowski, M.P.; Hansen, S.; Blaabjerg, F.; Marques, G.D. Virtual-flux-based direct power control of three-phase PWM rectifiers. *IEEE Trans. Ind. Appl.* **2001**, *37*, 1019–1027. [CrossRef]

9. Kerem, A.; Aksoz, A.; Saygin, A.; Yilmaz, E.N. Smart grid integration of micro hybrid power system using 6-switched 3-level inverter. In Proceedings of the IEEE ICSG 2017, Istanbul, Turkey, 19–21 April 2017; pp. 161–165.

10. Erol, C.; Sayan, H.H. A novel SSPWM controlling inverter running nonlinear device. *Electr. Eng.* **2018**, *100*, 39–46.

11. Vafakhah, B.; Salmon, J.; Knight, A.M. A New Space-Vector PWM with Optimal Switching Selection for Multilevel Coupled Inductor Inverters. *IEEE Trans. Ind. Electron.* **2010**, *57*, 2354–2364. [CrossRef]

12. Wang, Z.; Wang, Y.; Chen, J.; Hu, Y. Decoupled Vector Space Decomposition Based Space Vector Modulation for Dual Three-Phase Three-Level Motor Drives. *IEEE Trans. Power Electron.* **2018**, *33*, 10683–10697. [CrossRef]

13. Song, J.Y.; Kang, K.J.; Kang, C.H.; Jang, G.H. Cogging Torque and Unbalanced Magnetic Pull Due to Simultaneous Existence of Dynamic and Static Eccentricities and Uneven Magnetization in Permanent Magnet Motors. *IEEE Trans. Mag.* **2017**, *53*, 1–9. [CrossRef]

14. Yang, Y.; Zhou, K.; Wang, H.; Blaabjerg, F. Analysis and Mitigation of Dead Time Harmonics in the Single-Phase Full-Bridge PWM Converters with Repetitive Controllers. *IEEE Trans. Ind. Appl.* **2018**. [CrossRef]

15. Perez-Basante, A.; Ceballos, S.; Konstantinou, G.; Pou, J.; Kortabarria, I.; de Alegria, I.M. A Universal Formulation for Multilevel Selective Harmonic Elimination—PWM with Half-Wave Symmetry. *IEEE Trans. Power Electron.* **2018**, *1*. [CrossRef]

16. Lee, J.S.; Kwak, R.; Lee, K.B. Novel Discontinuous PWM Method for a Single-Phase Three-Level Neutral Point Clamped Inverter with Efficiency Improvement and Harmonic Reduction. *IEEE Trans. Power Electron.* **2018**, *33*, 9253–9266. [CrossRef]

17. Flankl, M.; Tüysüz, A.; Kolar, J.W. Cogging Torque Shape Optimization of an Integrated Generator for Electromechanical Energy Harvesting. *IEEE Trans. Ind. Electron.* **2017**, *64*, 9806–9814. [CrossRef]

18. Holmes, D.G.; Lipo, T.A. *Pulse Width Modulation for Power Converters*; IEEE Press: Piscataway, NJ, USA, 2003.

19. Celanovic, N.; Boroyevich, D. A fast space-vector modulation algorithm for multilevel three-phase converters. *IEEE Trans. Ind. Appl.* **2001**, *37*, 637–641. [CrossRef]

electronics

MDPI

Article

Performance Improvement for PMSM DTC System through Composite Active Vectors Modulation

Tianqing Yuan and Dazhi Wang *

School of Information Science and Engineering, Northeastern University, Shenyang 110819, China; tqyuan@stumail.neu.edu.cn
* Correspondence: noblefuture@163.com; Tel.: +86-133-2245-2013

Received: 19 September 2018; Accepted: 18 October 2018; Published: 22 October 2018

Abstract: In this paper, a novel direct torque control (DTC) scheme based on composite active vectors modulation (CVM) is proposed for permanent magnet synchronous motor (PMSM). The precondition of the accurate compensations of torque error and flux linkage error is that the errors can be compensated fully during the entire control period. Therefore, the compensational effects of torque error and flux linkage error in different operating conditions of the PMSM are analyzed firstly, and then, the operating conditions of the PMSM are divided into three cases according to the error compensational effects. To bring the novel composite active vectors modulation strategy smoothly, the effect factors are used to represent the error compensational effects provided by the applied active vectors. The error compensational effects supplied by single active vector or synthetic voltage vector are analyzed while the PMSM is operated in three different operating conditions. The effectiveness of the proposed CVM-DTC is verified through the experimental results on a 100-W PMSM drive system.

Keywords: direct torque control (DTC); composite active vectors modulation (CVM); permanent magnet synchronous motor (PMSM); effect factors

---

1. Introduction

Permanent magnet synchronous motors (PMSM) have a lot of merits such as high reliability, high efficiency, simple construction, and good control performance, and thus, it has been applied in various control systems including electrical drives, industrial applications, and medical devices in recent years [1–7]. Direct torque control (DTC) and field-oriented control (FOC) are two widely applied high-performance control strategies for the PMSM. Different from the decoupled-analyzing method in FOC, torque and flux linkage are controlled directly in DTC, and therefore, the quickest dynamic response can be obtained in the PMSM driven by DTC. However, as only six active vectors can be selected to compensate the errors of flux linkage and torque in conventional DTC (CDTC), the PMSM suffers from some drawbacks, such as large torque and flux linkage ripples. To improve the steady-state performance of the PMSM, many researchers have attempted to reduce these ripples by adding the amount of the active vectors through different methods.

With more appropriate active vectors selected in each control period, a novel DTC-fed PMSM system is proposed on the basis of a three-level inverter [8,9], and thus, the ripples of torque and flux linkage in PMSM can be suppressed effectively. In References [10,11], a novel DTC strategy using a matrix converter is proposed. Four enhanced switching tables are designed for the selection of switching states, and therefore, the ripples of the PMSM can be reduced effectively. Despite the fact that multiple active vectors can be supplied by three-level inverter or matrix converter, the cost of the DTC system is inevitably increased.

In fact, torque error and flux linkage error are tiny in most cases, and therefore, these errors will be over-compensated if the selected vector is applied over the whole control period. To solve these problems, duty ratio modulation strategy is introduced into the DTC-fed PMSM. Different duty ratio modulation methods for DTC (DDTC) are studied in References [12–16], and the ripples of torque and flux linkage can be reduced effectively without degrading the fast dynamic response in CDTC. In Reference [17], a new suboptimal control algorithm applying dynamic programming and a ramp trajectory method is proposed to DB-DTFC. In DB-DTFC, the maximum torque changes in one inverter switching period are used to determine the number of quantized stages for the minimum-time ramp trajectory method. It can be found that, the error compensational effects are considered in DB-DTFC, and therefore, torque and flux linkage command trajectories can be developed in different shapes according to the desired objectives, and fast dynamic responses can be achieved easily.

The stator flux linkage currents are decoupled in *d-q* axes and controlled independently in FOC, and thus, the outstanding operating performance of the PMSM can be obtained easily. The decoupled-analysis method is adopted in the novel DTC based on a space vector modulation (SVM) strategy with simple proportional-integral (PI) regulator or sliding mode observer [18–25]. With the independent control of torque and flux linkage in SVM-DTC, the amplitude and the phase of the wanted active vector can be determined accurately, and the errors of torque and flux linkage can be compensated precisely. However, the introduced PI regulator or sliding mode observer will degrade the dynamic response of the system.

To improve the operation performance of PMSM effectively, a novel DTC scheme utilizing composite active vectors modulation (CVM) strategy is presented in this paper. The compensational effects of torque error and flux linkage error in the PMSM driven by different control strategies are analyzed. Subsequently, the precondition of the accurate error compensation is obtained, and then, the precondition is adopted to determine the applied control strategy for the PMSM in different operation conditions, which is ignored in SVM-DTC and CDTC.

It should be noted that the most complicated control process is the transient-state. The large error component should be compensated fully, and the low error component should not be over-compensated; therefore, the duty ratio direct torque control strategy which, considering the active angles and the impact angles in Reference [1], can be used. The effectiveness of the proposed CVM-DTC scheme is validated through the experimental results. It should be noted that the steady-state performance and the dynamic response of the PMSM driven by CDTC, DDTC, and SVM-DTC are also studied in this paper.

The rest of this paper comprises the following sections. The principles of the conventional DTC are analyzed in Section 2. The compensational effects of torque error and flux linkage error in the SVM-DTC system are also illustrated in Section 2. The dividing process of the PMSM operation conditions and the error compensational effect supplied by different vectors in different operation conditions are described in Section 3 and the precondition of the accurate error compensations are also analyzed in Section 3. The description of experimental setup and discussions on experimental results are given in Section 4. The conclusion is analyzed in Section 5.

2. Principle of the Conventional DTC and SVM-DTC

2.1. Principle of the Conventional DTC

In the PMSM DTC system driven by a two-level voltage source inverter, eight voltage vectors can be applied to compensate the errors of torque and flux linkage, including six active vectors V_n (n = 1, 2, 3, 4, 5, 6) and two null vectors (V_0 and V_7). The spatial placements of the six active vectors in $\alpha\beta$-reference frames are shown in Figure 1. The whole rotation space of stator flux linkage φ_s can be divided into six sectors through the section boundary lines l_i (i = 1, 2, 3, 4, 5, 6), as shown in Figure 1. The six sectors are represented with number "N", and the sector vector vs. represents the active vector in every sector which the stator flux linkage φ_s is located in.

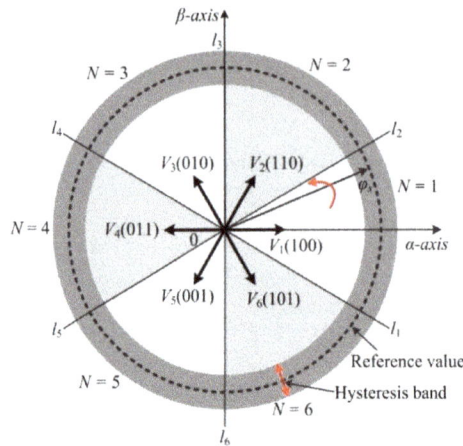

Figure 1. Active vectors in DTC system.

The torque error e_T is obtained by the comparison between the reference value T_{ref} and the real-time value T. The hysteresis comparator is used to determine the property ε_T of torque error e_T. The property ε_T value is 1 or -1, which indicates torque T needs to be increased if the value of property ε_T is 1, while the torque needs to be decreased if the value of property ε_T is -1. The determination methods of another parameter flux linkage φ is in the same way. The active vector selection rules in the SV-CDTC system are described in Table 1.

Table 1. Conventional switching table.

Sector number N		Torque (ε_T)	
		1	**−1**
Stator flux linkage (ε_F)	1	V_{N+1}	V_{N-1}
	−1	V_{N+2}	V_{N-2}

2.2. Compensations of Torque Error and Flux Linkage Error in SVM-DTC System

The number and the direction of the active vectors are fixed in CDTC, and therefore, the errors of torque and flux linkage are difficult to be compensated effectively, leading to large ripples. To improve the steady-state performance of the PMSM, the decoupling control strategy adopted in FOC is introduced into DTC. The PI controllers are used to obtain the amplitude of the torque vector and flux linkage vector on the basis of torque error and flux linkage error; then, the space vector modulation (SVM) is used to determine the precise vectors. The schematic diagram of SVM-DTC is shown in Figure 2.

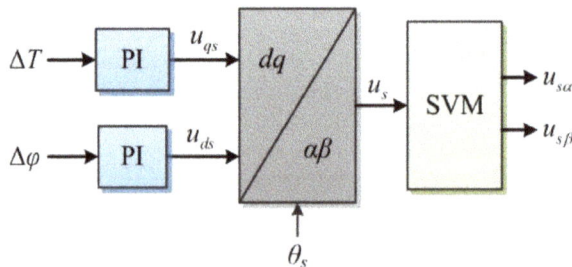

Figure 2. Schematic diagram of SVM-DTC.

Figure 2 shows the error compensations provided by a synthesis voltage vector. It can be found that torque error ΔT and flux linkage error $\Delta\varphi$ can be compensated through vectors u_{qs} and u_{ds}, respectively. Consequently, the synthesis voltage vector u_s can be obtained on the basis of the rotor position θ_s. As shown in Figure 3, the voltage vectors $u_{s\alpha}$ and $u_{s\beta}$ will be obtained through coordinate transformation based on u_s.

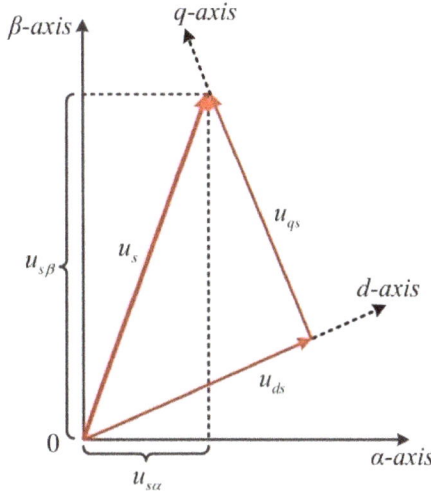

Figure 3. Synthesis voltage vectors in *d-q* reference frame.

From the aforementioned analyses, it can be observed that the switching table in CDTC is replaced by the PI controllers in SVM-DTC. Therefore, the active vectors used to compensate the errors of torque and flux linkage are not limited to the six basic active vectors. Additionally, the ideal steady-state performance of the PMSM can be obtained easily. Despite the fact that the precision of the wanted synthetic vector can be ensured with the using of SVM strategy in SVM-DTC, the dynamic response of the PMSM is affected inevitably.

3. Analysis of Error Compensations

The ripples of torque and flux linkage in the PMSM driven by SVM-DTC are relatively minor while the PMSM is operated in the steady-state condition. On the other hand, the dynamic performance will be affected by the complicated calculations of the synthesis voltage vector and the over-modulation process while the PMSM is operated in the dynamic response condition. This is the main reason that the fast dynamic performance of permanent magnet synchronous motor driven by SVM-DTC is degraded.

To improve the steady-state performance of PMSM, and maintain the fast dynamic response at the same time, appropriate control strategy should be selected and applied to the system according to the operation conditions, including CDTC, DDTC, and SVM-DTC. Therefore, the differences of the error compensational effects provided by the synthesis voltage vector and single active vector under different operation conditions should be analyzed firstly.

3.1. Operation Conditions

The stator flux linkage φ_s changes from φ_{s1} to φ_{s2} during one control period and the variation of the stator flux linkage φ_s is $\Delta\varphi_s$, which can be decoupled into $\Delta\varphi_{sd}$ and $\Delta\varphi_{sq}$ in the *d-q* axis. In this control period, the errors of torque and flux linkage are ΔT and $\Delta\varphi$, respectively. Hence, the torque component variation of the stator flux linkage is $\Delta\varphi_{sq}$, and the amplitude component variation of the stator flux linkage is $\Delta\varphi_{sd}$, as shown in Figure 4.

The torque component variation and the amplitude component variation of the stator flux linkage can be expressed as

$$\Delta\varphi_{sq} = \frac{2L_s}{3p} \cdot \frac{1}{\varphi_f} \cdot \Delta T \tag{1}$$

$$\Delta\varphi_{sd} = \Delta\varphi \tag{2}$$

where L_s is the stator inductance, p is the number of pole pairs, and φ_f is the permanent magnet flux linkage.

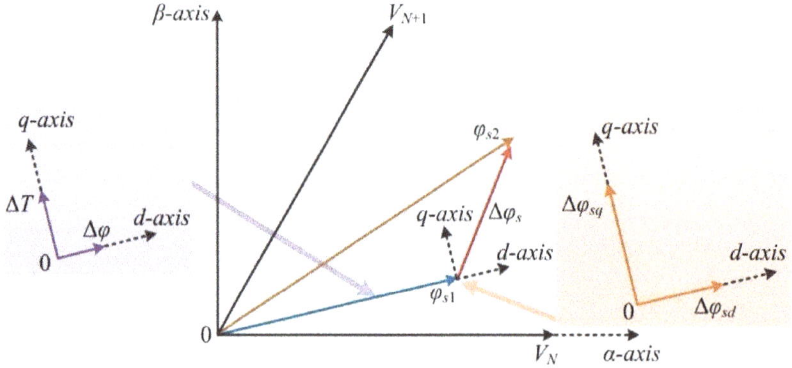

Figure 4. Analysis of error compensational effects.

The compensational effect of the stator flux linkage supplied by single active vector V_N and synthesis voltage vector u_s are $\Delta\varphi'_{s1}$ and $\Delta\varphi'_s$, respectively, as shown in Figure 5.

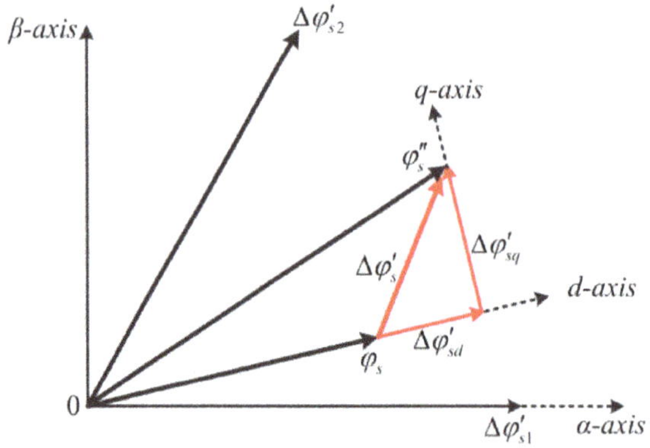

Figure 5. Analysis of error compensational effects.

The torque component compensation of the stator flux linkage supplied by synthesis voltage vector u_s is $\Delta\varphi'_{sq}$, and the amplitude component compensation of the stator flux linkage provided by synthesis voltage vector u_s is $\Delta\varphi'_{sd}$. It is obvious that the parameters of $\Delta\varphi'_s$, $\Delta\varphi'_{sq}$, and $\Delta\varphi'_{sd}$ are fixed during each control period in the system. While the torque error ΔT and the flux linkage error $\Delta\varphi$ will vary with the variation of the stator flux linkage location in different control period. Therefore, the real values of $\Delta\varphi_{sq}$ and $\Delta\varphi_{sd}$ are also different.

The relationships between the real error compensations of the stator flux linkage and the errors can be described in the following way.

First item, the actual compensations are greater than the errors:

$$\left(\begin{array}{l} \Delta\varphi'_{sd} > \Delta\varphi_{sd} \\ \Delta\varphi'_{sq} > \Delta\varphi_{sq} \end{array} \right. \tag{3}$$

Second item, the actual compensations are less than the errors:

$$\left(\begin{array}{l} \Delta\varphi'_{sd} < \Delta\varphi_{sd} \\ \Delta\varphi'_{sq} < \Delta\varphi_{sq} \end{array} \right. \tag{4}$$

Third item, the actual compensation of the amplitude component is less than the error while the actual compensation of torque component is greater than the error:

$$\left(\begin{array}{l} \Delta\varphi'_{sd} < \Delta\varphi_{sd} \\ \Delta\varphi'_{sq} > \Delta\varphi_{sq} \end{array} \right. \tag{5}$$

Fourth item, the actual compensation of the amplitude component is greater than the error while the actual compensation of torque component is less than the error:

$$\left(\begin{array}{l} \Delta\varphi'_{sd} > \Delta\varphi_{sd} \\ \Delta\varphi'_{sq} < \Delta\varphi_{sq} \end{array} \right. \tag{6}$$

The operation conditions of the PMSM can be divided into three items in accordance with the errors and the actual compensations, as shown in Table 2.

Table 2. Operation conditions

Operation Conditions	Reference Equation
Steady-state	(3)
Transient-state	(5) and (6)
Dynamic-state	(4)

3.2. Error Compensation Analysis in Steady-State Case

The values of torque error ΔT and flux linkage error $\Delta\varphi_s$ are relatively low in steady-state case [6]. The angle between the stator flux linkage φ_s and the active vector V_N is θ_1, as shown in Figure 6. It is also shown in Figure 6 that both the compensations of ΔT and $\Delta\varphi_s$ are bigger than the errors. Consequently, the errors will be over-compensated if the active vector or the synthesized voltage vector is applied during the entire control period.

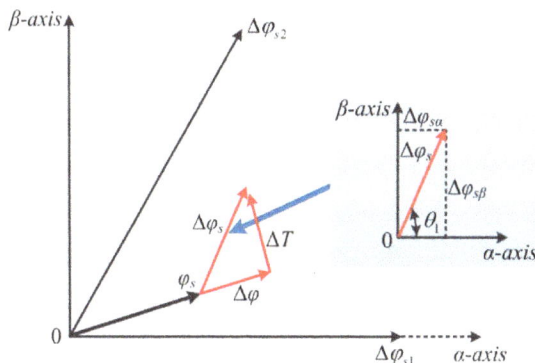

Figure 6. Analysis of error compensational effects in steady-state.

In DDTC-fed PMSM, the applied time of the active vector is modulated by duty ratio modulation strategy. As a result, the over-compensation of the errors can be avoided; nevertheless, the fixed active vectors limit the compensational effects.

In the PMSM driven by SVM-DTC, the adjacent active vectors V_N and V_{N+1} are selected as the benchmark vectors to obtain the synthesized voltage vector u_s. Furthermore, the applied time of V_N and V_{N+1} are T_1 and T_2, respectively. The error compensations can be evaluated by

$$\Delta \varphi_{s1} = V_N \cdot T_1 \tag{7}$$

$$\Delta \varphi_{s2} = V_{N+1} \cdot T_2 \tag{8}$$

The modulation process of the active vectors can be expressed as

$$\Delta \varphi_{s1} + \Delta \varphi_{s2} \cdot \cos \frac{\pi}{3} = \Delta \varphi_{s\alpha} = \Delta \varphi_s \cdot \cos \theta_1 \tag{9}$$

$$\Delta \varphi_{s2} \cdot \sin \frac{\pi}{3} = \Delta \varphi_{s\beta} = \Delta \varphi_s \cdot \sin \theta_1 \tag{10}$$

$$T_1 + T_2 + T_0 = T_s \tag{11}$$

where T_0 is the zero voltage vector applied time.

From the aforementioned analyses, it can be found that the errors of torque and flux linkage can be compensated accurately through SVM strategy while the PMSM is operated in steady-state.

3.3. Error Compensation Analysis in Dynamic-State Case

The errors of torque or flux linkage may become greater in the dynamic-state while the speed or the torque changes. As shown in Figure 7, the stator flux linkage error is $\Delta \varphi_s$; the angle between the stator flux linkage error $\Delta \varphi_s$ and the active vector V_N is θ_2. It can be found that the torque error and the flux linkage error are greater than the error compensations.

It can be found that the torque error ΔT and the flux linkage error $\Delta \varphi$ cannot be compensated fully by any single active vector or the synthesized voltage vector in the next control period. The differences of the error compensation effect supplied by the single active vector or the synthesized voltage vector are described in following parts.

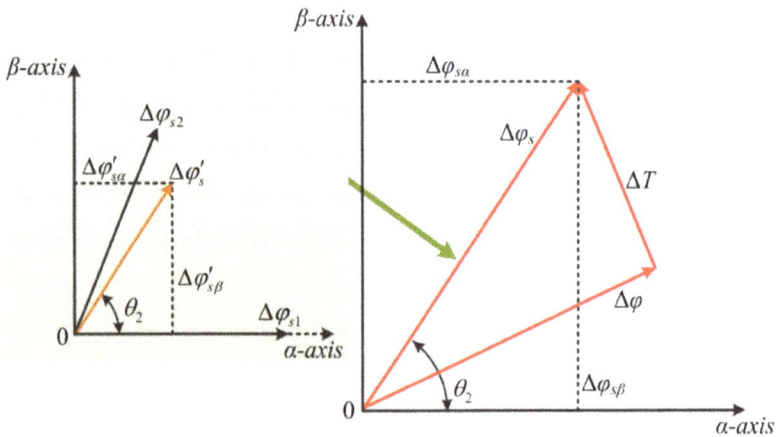

Figure 7. Analysis of error compensational effects in dynamic-state.

3.3.1. Synthetic Voltage Vector

The adjacent active vectors V_N and V_{N+1} are selected as the benchmark vectors. The applied time of V_N and V_{N+1} are T_1 and T_2, respectively. Therefore, the error compensations can be calculated as

$$\Delta\varphi'_{s1} = V_N \cdot T_1 \tag{12}$$

$$\Delta\varphi'_{s2} = V_{N+1} \cdot T_2 \tag{13}$$

$$\Delta\varphi'_{s1} + \Delta\varphi'_{s2} \cdot \cos\frac{\pi}{3} = \Delta\varphi_{s\alpha} = \Delta\varphi_s \cdot \cos\theta_2 \tag{14}$$

$$\Delta\varphi'_{s2} \cdot \sin\frac{\pi}{3} = \Delta\varphi_{s\beta} = \Delta\varphi_s \cdot \sin\theta_2 \tag{15}$$

Since the actual compensations are smaller than the errors, therefore

$$T_1 + T_2 > T_s \tag{16}$$

The applied time of the applied active vectors can be over-modulated as

$$T'_1 = \frac{T_1}{T_1 + T_2} \cdot T_s \tag{17}$$

$$T'_2 = \frac{T_2}{T_1 + T_2} \cdot T_s \tag{18}$$

The applied time of the applied active vectors can be rewritten as

$$T_1 = k_1 \cdot T_s \tag{19}$$

$$T_2 = k_2 \cdot T_s \tag{20}$$

where k_1 and k_2 are the duty ratio values of applied time of V_N and V_{N+1}, respectively.

Therefore, the actual compensation of the stator flux linkage in one control period is

$$\Delta\varphi'_{s\alpha} = V_N \cdot \frac{k_1}{k_1 + k_2} \cdot T_s + V_{N+1} \cdot \cos\frac{\pi}{3} \cdot \frac{k_2}{k_1 + k_2} \cdot T_s \tag{21}$$

$$\Delta\varphi'_{s\beta} = V_{N+1} \cdot \sin\frac{\pi}{3} \cdot \frac{k_2}{k_1 + k_2} \cdot T_s \tag{22}$$

which can be simplified as

$$\left(\begin{array}{l} \Delta\varphi'_{s\alpha} = V_N \cdot T_s \cdot \left(\frac{k_1}{k_1 + k_2} + \frac{1}{2} \cdot \frac{k_2}{k_1 + k_2} \right) \\ \Delta\varphi'_{s\beta} = V_N \cdot T_s \cdot \frac{\sqrt{3}}{2} \cdot \frac{k_2}{k_1 + k_2} \end{array} \right. \tag{23}$$

Therefore, the compensation of the stator flux linkage supplied by synthesized voltage vector u_s can be given as

$$|\Delta\varphi'_s| = \sqrt{\left(\Delta\varphi'_{s\alpha}\right)^2 + \left(\Delta\varphi'_{s\beta}\right)^2} = V_N \cdot T_s \cdot \sqrt{\frac{k_1^2 + k_1 k_2 + k_2^2}{(k_1 + k_2)^2}} < V_N \cdot T_s \tag{24}$$

3.3.2. Single Active Vector

The compensations of the stator flux linkage supplied by adjacent vectors V_N and V_{N+1} during the whole control period are

$$\Delta\varphi''_{s1} = V_N \cdot T_s \tag{25}$$

$$\Delta\varphi''_{s2} = V_{N+1} \cdot T_s \qquad (26)$$

Therefore, the compensations of the stator flux linkage provided by the single active vector can be expressed as

$$\Delta\varphi''_s = \Delta\varphi''_{s1} = \Delta\varphi''_{s2} = V_N \cdot T_s \qquad (27)$$

The comparison result of the stator flux linkage compensations supplied by the different vectors can be described as

$$\Delta\varphi'_s < \Delta\varphi''_s \qquad (28)$$

From the aforementioned analyses, it can be observed that the compensational effects of the stator flux linkage supplied by the synthesized voltage vector is weaker than the single active vector. Therefore, the SVM strategy is not required to compensate the errors of torque and flux linkage while the PMSM is operated in the dynamic state. To simplify the calculations of the system, the appropriate active vector can be selected from a conventional switching table and be used in the system over the entire control period.

In short, CDTC strategy should be used to reduce the ripples of torque and flux linkage in the PMSM when the PMSM is operated in a dynamic state. Hence, the delayed dynamic response caused by the PI controller can be eliminated, and the ripples' depressing effects of the PMSM driven by CDTC are the same as that driven by SVM-DTC.

3.4. Error Compensation Analysis in Transient-State Case

The operation condition of the PMSM may deviate the steady-state due to external disturbance. Therefore, the PMSM may operate in a transient-state if one parameter of torque error and flux linkage error is large while another parameter is relatively low.

The torque error ΔT is high and the flux linkage error $\Delta\varphi$ is relatively low as shown in Figure 8. The variation of the stator flux linkage is $\Delta\varphi_s$.

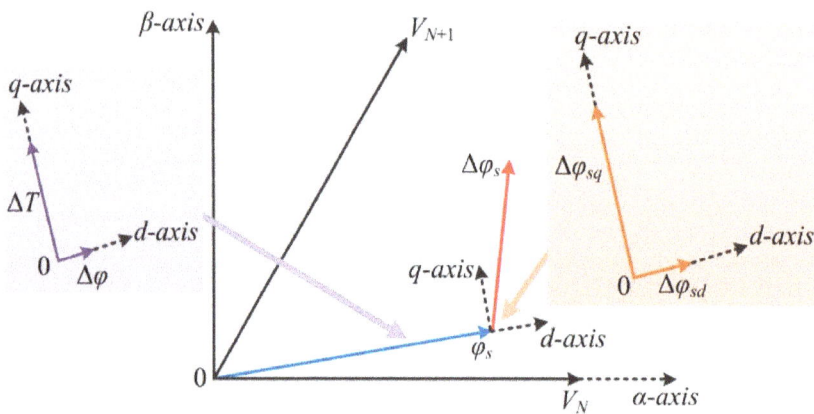

Figure 8. Analysis of error compensational effects in transient-state.

The differences of the error compensation effect supplied by single active vector or synthesized voltage vector are described in the following section.

3.4.1. Synthetic Voltage Vector

Figure 9 shows the error compensational effects provided by different active vectors.

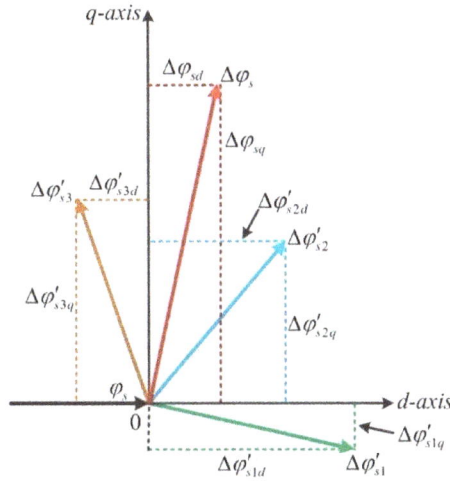

Figure 9. Analysis of error compensational effects provided by different active vectors.

As shown in Figure 9, the adjacent active vectors V_{N+1} and V_{N+2} are selected as the benchmark vectors. The applied time of V_{N+1} and V_{N+2} are T_1 and T_2, respectively. Therefore, the error compensations can be evaluated by

$$\Delta\varphi'_{s2q} \cdot T_1 + \Delta\varphi'_{s3q} \cdot T_2 = \Delta\varphi_{sq} \cdot T_s \tag{29}$$

$$\Delta\varphi'_{s2d} \cdot T_1 - \Delta\varphi'_{s3d} \cdot T_2 = \Delta\varphi_{sq} \cdot T_s \approx 0 \tag{30}$$

3.4.2. Single Active Vector V_n

The stator flux linkage error $\Delta\varphi_s$ is located in the middle of error compensations $\Delta\varphi'_{s2}$ and $\Delta\varphi'_{s3}$, as shown in Figure 9. To compensate the error $\Delta\varphi_{sq}$ effectively and avoid the over-compensation of the error $\Delta\varphi_{sd}$ at the same time, the adjacent vectors V_{N+1} and V_{N+2} can be selected and applied to half of the control period.

From the above analysis, it can be found that the torque error can be compensated fully supplied by a single active vector while the PMSM is operated in the transient-state; however, the flux linkage error cannot be compensated fully. Despite the fact that the torque error and the flux linkage error can be compensated fully by synthetic voltage vector, the calculations of the system are inevitably increased. It should be noted that the novel DDTC strategy based on the active angle in Reference [1] has solved the problem while one parameter is large and another parameter is relatively small. Therefore, the DDTC strategy can be used to improve the performance of the system while the PMSM is operated in a transient-state.

3.5. Novel Composite Active Vectors Modulation Strategy

To improve the operation performance of the PMSM effectively, a novel composite active vectors modulation DTC (CVM-DTC) strategy considering the precondition of the accurate errors compensations is presented in this section. The schematic diagram of the presented CVM-DTC system is shown in Figure 10. The parameters in CVM-DTC are defined by:

u_{abc}: Stator voltage;
i_{abc}: Stator currents;
U_{DC}: DC bus voltage;
n: Actual rotor speed;

n_{ref}: Reference rotor speed;
σ: Rotor position;
T_{ref}: Reference torque;
φ_{ref}: Reference flux linkage;
ΔT: *Reference torque compensation;
$\Delta\varphi$: *Reference flux linkage compensation;
V_n: Single active vector;
d: Duty ratio value of applied time;
u_s: Synthetic voltage vector.

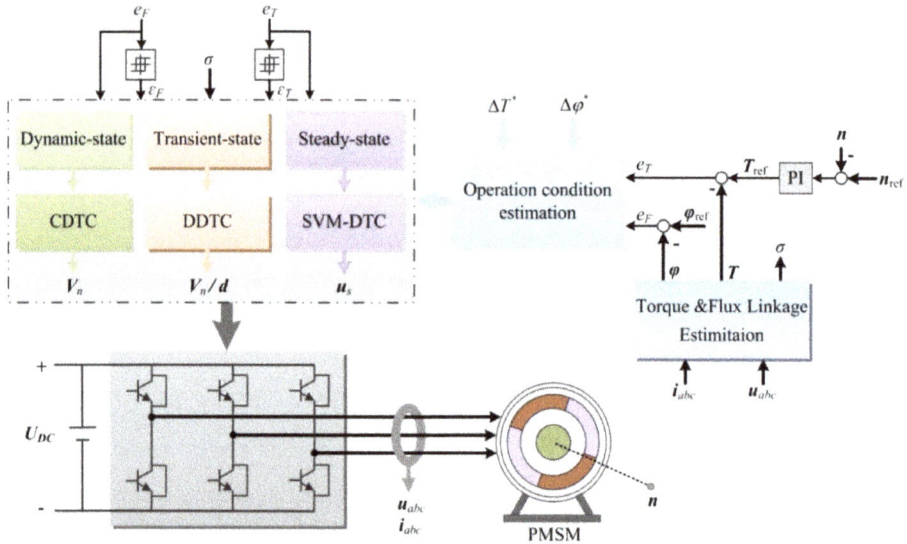

Figure 10. Schematic diagram of the CVM-DTC for PMSM.

In order to maintain the fast dynamic response in CDTC and obtain the minimum ripples of the system, the applied control strategy should adjust according to the operation conditions of the PMSM.

The precondition of the accurate compensations of torque error and flux linkage error is that the torque error and the flux linkage error can be compensated and fully supplied by the applied active vector in the whole control period. However, this precondition is ignored in the SVM-DTC system. Therefore, the torque error and the flux linkage error will be analyzed through decoupled calculations through PI controllers, while the compensational effects of the stator flux linkage in SVM-DTC and CDTC when the PMSM is operated in non-steady-state are nearly the same. As a result, the error compensational effects are not satisfied and the dynamic response will be affected without considering the operation conditions of the PMSM.

3.6. Determining of the Operation Condition through Effect Factors

The relationship between the active vector V_n and the stator flux linkage variation $\Delta\varphi_s$ in each control period is

$$\Delta\varphi_s = V_n \cdot T_s \tag{31}$$

During the whole control period, the max compensations of $\Delta\varphi_{sq}$ and $\Delta\varphi_{sd}$ can be expressed as

$$\Delta\varphi_{sq-max} = \Delta\varphi_s = V_n \cdot T_s \tag{32}$$

$$\Delta\varphi_{sd-max} = \Delta\varphi_s = V_n \cdot T_s \tag{33}$$

The max compensation of the torque is

$$\Delta T_{max} = \frac{3p}{2L_s} \cdot \varphi_f \cdot V_n \cdot T_s \tag{34}$$

And the max compensation of the flux linkage is

$$\Delta\varphi_{dmax} = V_n \cdot T_s \tag{35}$$

Defining the reference values of torque variation and flux linkage variation are ΔT^* and $\Delta\varphi^*$, respectively, which can be expressed as

$$\Delta T^* = \Delta T_{max} = \frac{3p}{2L_s} \cdot \varphi_f \cdot V_n \cdot T_s \tag{36}$$

$$\Delta\varphi^* = \Delta\varphi_{dmax} = V_n \cdot T_s \tag{37}$$

The effect factors of torque and flux linkage are k_T and k_φ, respectively, which can be given as

$$k_T = \frac{\Delta T}{\Delta T^*} \tag{38}$$

$$k_\varphi = \frac{\Delta\varphi}{\Delta\varphi^*} \tag{39}$$

The introduced effect factors can be obtained through the errors and the reference values of the variation in any control period. The operation conditions of the PMSM can be classified into three cases: steady-state, transient-state, and dynamic-state. The relationships between the effect factors and the operation conditions are shown in Table 3.

Table 3. Effect factors for different operation conditions.

Effect Factors		Operation Conditions
k_T	k_φ	
	$(-\infty, -1)$	Dynamic-state
$(-\infty, -1)$	$(-1, 1)$	Transient-state
	$(1, +\infty)$	Dynamic-state
	$(-\infty, -1)$	Transient-state
$(-1, 1)$	$(-1, 1)$	Steady-state
	$(1, +\infty)$	Transient-state
	$(-\infty, -1)$	Dynamic-state
$(1, +\infty)$	$(-1, 1)$	Transient-state
	$(1, +\infty)$	Dynamic-state

4. Experimental Analysis

4.1. Experimental System Setup

Experimental studies are carried out on a 100-W PMSM drive system to validate the feasibility and effectiveness of the proposed CVM-DTC strategy. The experimental hardware setup is illustrated in Figure 11. The parameters of the PMSM are given as follows: $R_s = 0.76\ \Omega$; $L_s = 0.00182$ H; the number of pole pairs $p = 4$. The DC voltage is 36 V. This study compares the steady-state and the dynamic response performance of CDTC, DDTC, SVM-DTC, and CVM-DTC. The experiments are implemented in a TMS320F28335 DSP control system with a sampling period of 100 μs.

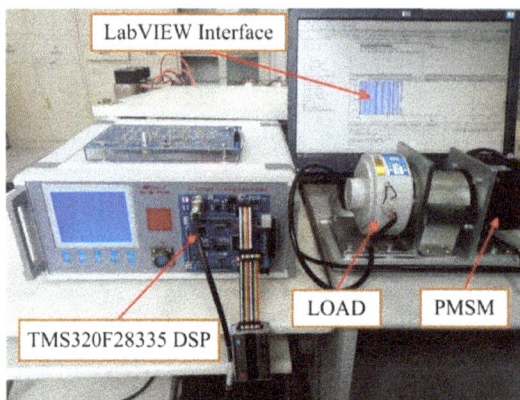

Figure 11. Experimental setup of control system.

4.2. Steady-State Performance

The steady-state performances of CDTC, DDTC, SVM-DTC, and CVM-DTC are compared under the same operating conditions. The PMSM is operated at 500 rpm and the reference values of torque and flux linkage are 0.8 N·m and 0.3 Wb, respectively. The torque and flux linkage waveforms of the PMSM are driven by different control strategies as shown in Figure 12.

From these experimental results, it can be found that the torque ripples of CDTC, SV-DDTC, SVM-DTC, and CVM-DTC are 0.56, 0.4, 0.32, and 0.34 N·m, respectively, and the flux linkage ripples of the four control system are 0.08, 0.06, 0.04, and 0.038 Wb, respectively. Therefore, compared with CDTC, DDTC and SVM-DTC can reduce the torque ripple by at least 28% and 42%, respectively, and reduce the flux linkage ripple at least 25% and 50%, respectively. While the steady-state performances of the PMSM driven by CVM-DTC in the setting operation conditions are nearly the same as SVM-DTC. The experimental results show that the errors of torque and flux should be compensated through SVM-DTC strategy, which indicates that the applied control strategy in CVM-DTC in the steady-state condition is appropriate.

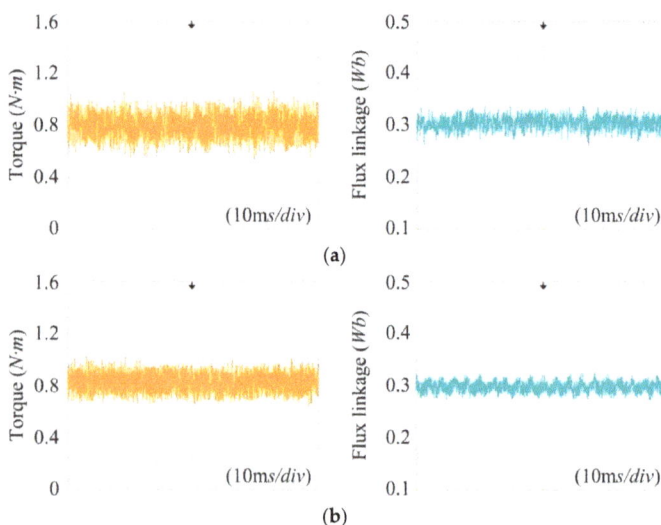

Figure 12. *Cont.*

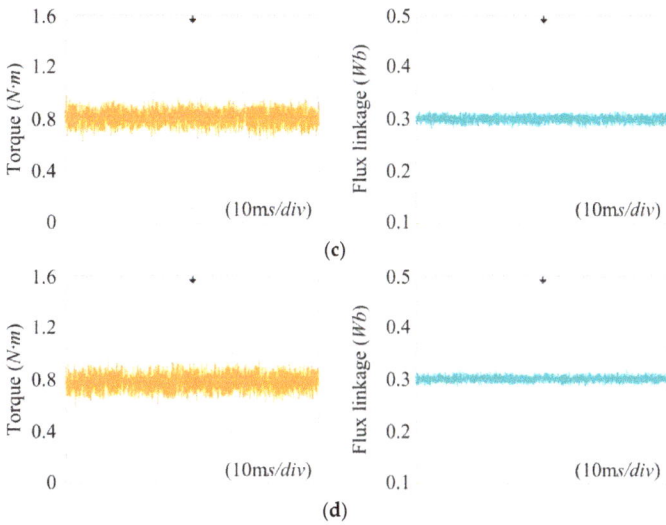

Figure 12. Experimental torque and flux linkage of the PMSM when using: (**a**) CDTC; (**b**) DDTC; (**c**) SVM-DTC; (**d**) CVM-DTC.

4.3. Dynamic Performance

To validate the fast dynamic response of the proposed novel CVM-DTC, the speed responses of the PMSM driven by the four control strategies are tested when the torque is set as 0.5 N·m. In these tests, a step change from 200 to 400 rpm is applied on the speed reference, as shown in Figure 13.

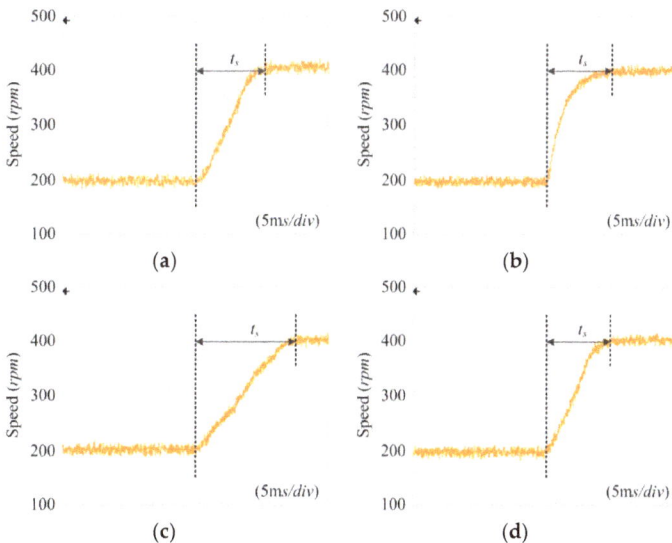

Figure 13. The speed trajectory from 200 rpm to 400 rpm when using: (**a**) CDTC; (**b**) DDTC; (**c**) SVM-DTC; (**d**) CVM-DTC.

It can be seen that the ripple of the rotor speed is 35 rpm when using CDTC, while the speed ripples of the PMSM can be reduced to 30, 25, and 24 rpm with the use of DDTC, SVM-DTC, and CVM-DTC.

Moreover, the settling times of the rotor speed using the four different control strategies are 0.013, 0.012, 0.019, and 0.012 s.

Therefore, the main advantage of CDTC, i.e., the fast dynamic response, is maintained in CVM-DTC. The experimental results show that dynamic response has a higher priority than ripples in dynamic-state condition, hence, DDTC or SVM-DTC should be abandoned. In short, the applied control strategy in CVM-DTC in the dynamic-state condition is appropriate.

5. Conclusions

The precondition of the accurate compensations of torque error and flux linkage error is considered in the proposed novel CVM-DTC scheme in this paper, which is ignored in CDTC and SVM-DTC. Therefore, the compensational effects of torque error and flux linkage error provided by the single active vector or synthetic voltage vector in different operation conditions are analyzed firstly, and then, the operating conditions of the PMSM are divided into three cases according to the compensational effects (effect factors). To improve the performance of the PMSM effectively, the applied control strategy for the PMSM in different sampling periods will vary on the basis of the introduced effect factors.

Experimental results clearly indicate that the novel CVM-DTC scheme exhibits excellent control of torque and flux linkage with lower steady-state ripples when compared to CDTC and DDTC, and faster transient response performances when compared to SVM-DTC.

Author Contributions: This paper was a collaborative effort between the authors. T.Y. and D.W. proposed the original idea; T.Y. wrote the full manuscript and carried out the experiments.

Funding: This research received no external funding.

Acknowledgments: This work was supported in part by National Natural Science Foundation of China under Grant 61,433,004 and 51467017, and in part by National Key Research and Development Program of China under Grant 2017YFB1300900.

Conflicts of Interest: The authors declare no conflict of interest.

References

1. Cheema, M.A.M.; Fletcher, J.E.; Xiao, D.; Rahman, M.F. A direct thrust control scheme for linear permanent magnet synchronous motor based on online duty ratio control. *IEEE Trans. Power Electron.* **2016**, *31*, 4416–4428. [CrossRef]
2. Abosh, A.; Zhu, Z.Q.; Ren, Y. Reduction of torque and flux ripples in space vector modulation-based direct torque control of asymmetric permanent magnet synchronous machine. *IEEE Trans. Power Electron.* **2017**, *32*, 2976–2986. [CrossRef]
3. Zhou, Y.Z.; Chen, G.T. Predictive DTC Strategy with fault-tolerant function for six-phase and three-phase PMSM series-connected drive system. *IEEE Trans. Ind. Electron.* **2018**, *65*, 9101–9112. [CrossRef]
4. Shinohara, A.; Inoue, Y.; Morimoto, S.; Sanada, M. Direct calculation method of reference flux linkage for maximum torque per ampere control in DTC-based IPMSM drives. *IEEE Trans. Power Electron.* **2016**, *32*, 2114–2122. [CrossRef]
5. Alsofyani, I.M.; Idris, N.R.N.; Lee, K.B. Dynamic hysteresis torque band for improving the performance of lookup-table-based DTC of induction machines. *IEEE Trans. Power Electron.* **2018**, *33*, 7959–7970. [CrossRef]
6. Yuan, T.Q.; Wang, D.Z.; Li, Y.L. Duty ratio modulation strategy to minimize torque and flux linkage ripples in IPMSM DTC system. *IEEE Access.* **2017**, *5*, 14323–14332. [CrossRef]
7. Putri, A.K.; Rick, S.; Franck, D.; Hameyer, K. Application of sinusoidal field pole in a permanent-magnet synchronous machine to improve the NVH behavior considering the MTPA and MTPV operation area. *IEEE Trans. Ind. Appl.* **2016**, *52*, 2280–2288. [CrossRef]
8. Tatte, Y.N.; Aware, M.V.; Pandit, J.K.; Nemade, R. Performance improvement of three-level five-phase inverter-fed DTC-controlled five-phase induction motor during low-speed operation. *IEEE Trans. Ind. Appl.* **2018**, *54*, 2349–2357. [CrossRef]
9. Payami, S.; Behera, R.K.; Iqbal, A. DTC of three-level NPC inverter fed five-phase induction motor drive with novel neutral point voltage balancing scheme. *IEEE Trans. Power Electron.* **2018**, *33*, 1487–1500. [CrossRef]

10. Xia, C.; Zhao, J.; Yan, Y.; Shi, T. A novel direct torque and flux control method of matrix converter-fed PMSM drives. *IEEE Trans. Power Electron.* **2014**, *29*, 5417–5430. [CrossRef]
11. Yan, Y.; Zhao, J.; Xia, C.; Shi, T. Direct torque control of matrix converter-fed permanent magnet synchronous motor drives based on master and slave vectors. *IET Power Electron.* **2015**, *8*, 288–296. [CrossRef]
12. Mohan, D.; Zhang, X.; Foo, G.H.B. A simple duty cycle control strategy to reduce torque ripples and improve low-speed performance of a three-level inverter fed DTC IPMSM drive. *IEEE Trans. Ind. Electron.* **2017**, *64*, 2709–2721. [CrossRef]
13. Niu, F.; Wang, B.; Babel, A.S.; Li, K.; Strangas, E.G. Comparative evaluation of direct torque control strategies for permanent magnet synchronous machines. *IEEE Trans. Power Electron.* **2016**, *31*, 1408–1424. [CrossRef]
14. Zhang, Y.; Zhu, J.; Xu, W.; Guo, Y. A simple method to reduce torque ripple in direct torque-controlled permanent-magnet synchronous motor by using vectors with variable amplitude and angle. *IEEE Trans. Ind. Electron.* **2011**, *58*, 2848–2859. [CrossRef]
15. Ren, Y.; Zhu, Z.Q.; Liu, J. Direct torque control of permanent-magnet synchronous machine drives with a simple duty ratio regulator. *IEEE Trans. Ind. Electron.* **2014**, *61*, 5249–5258. [CrossRef]
16. Niu, F.; Li, K.; Wang, Y. Direct torque control for permanent magnet synchronous machines based on duty ratio modulation. *IEEE Trans. Ind. Electron.* **2015**, *62*, 6160–6170. [CrossRef]
17. Lee, J.S.; Lorenz, R.D. Deadbeat direct torque and flux control of IPMSM drives using a minimum time ramp trajectory method at voltage and current limits. *IEEE Trans. Ind. Appl.* **2014**, *50*, 3795–3804. [CrossRef]
18. Mohan, D.; Zhang, X.; Foo, G. Generalized DTC strategy for multilevel inverter fed IPMSMs with constant inverter switching frequency and reduced torque ripples. *IEEE Trans. Energy Convers.* **2017**, *32*, 1031–1041. [CrossRef]
19. Zhang, Z.; Zhao, Y.; Qiao, W.; Qu, L. A space-vector-modulated sensorless direct-torque control for direct-drive PMSG wind turbines. *IEEE Trans. Ind. Appl.* **2014**, *50*, 2331–2341. [CrossRef]
20. Zhang, X.; Foo, G.H. Over-modulation of constant switching frequency based DTC for reluctance synchronous motors incorporating field-weakening operation. *IEEE Trans. Ind. Electron.* **2019**, *66*, 37–47. [CrossRef]
21. Berzoy, A.; Rengifo, J.; Mohammed, O. Fuzzy predictive DTC of induction machines with reduced torque ripple and high performance operation. *IEEE Trans. Power Electron.* **2018**, *33*, 2580–2587. [CrossRef]
22. Do, T.D.; Choi, H.H.; Jung, J.W. Nonlinear Optimal DTC Design and Stability Analysis for Interior Permanent Magnet Synchronous Motor Drives. *IEEE/ASME Trans. Mechatron* **2716**, *20*.
23. Choi, Y.-S.; Choi, H.H.; Jung, J.W. Feedback linearization direct torque control with reduced torque and flux ripples for IPMSM drives. *IEEE Trans. Power Electron.* **2016**, *31*, 3728–3737. [CrossRef]
24. Zhang, Z.; Wei, C.; Qiao, W.; Qu, L. Adaptive saturation controller-based direct torque control for permanent-magnet synchronous machines. *IEEE Trans. Power Electron.* **2016**, *31*, 7112–7122. [CrossRef]
25. Liang, D.L.; Li, J.; Qu, R.H.; Kong, W.B. Adaptive second-order sliding-mode observer for PMSM sensorless control considering VSI nonlinearity. *IEEE Trans. Power Electron.* **2018**, *33*, 8994–9004. [CrossRef]

electronics

MDPI

Article

Design of A Novel Line Start Synchronous Motor Rotor

Berkan Zöhra [1,*], Mehmet Akar [2] and Mustafa Eker [2]

[1]   Department of Electronics and Automation, Merzifon Vocational School, Amasya University, Amasya TR 05300, Turkey

[2]   Department of Mechatronics Engineering, Faculty of Engineering and Natural Sciences, Gaziosmanpaşa University, Tokat TR 60150, Turkey; mehmet.akar@gop.edu.tr (M.A.); mustafa.eker@gop.edu.tr (M.E.)

*   Correspondence: berkan.zohra@amasya.edu.tr; Tel.: +90-358-513-5103/2744

Received: 29 November 2018; Accepted: 21 December 2018; Published: 26 December 2018

Abstract: Line start permanent magnet synchronous motors (LS-PMSM) are preferred more and more in industrial applications, because they can start on their own and because of their high efficiency. In this study, a new LS-PMSM rotor typology is suggested, which is modelled using surface mount permanent magnets, in which two different slot types have been used together. The rotor of an asynchronous motor on the industrial market in the IE2 efficiency segment has been remodeled in the study, resulting in an increase in motor efficiency from 85% to 91.8%. A finite elements software was used for determining motor design and performance, in addition to analytical methods.

Keywords: line start; permanent magnet; synchronous motor; efficiency motor; rotor design

---

1. Introduction

Electrical motors have the highest share in electrical consumption for industrial and home applications. Therefore, there are many studies in literature that focus on making electrical motors more efficient. In addition to the electricity consumed by the electrical motors used in all fields, the resources used for motor manufacture are also another significant issue [1,2]. The use of a motor with a large core for obtaining the desired shaft power increases the material used for producing the motor, thereby directly affecting production cost [3].

There are many products that can respond to the operating conditions in the current electrical motors market. A necessity has emerged for regulating the electrical motor market with various standards, as there are many different products in the market with different brands that can replace one another. Many regulations are ongoing regarding the frame dimensions, operating characteristics, and operating conditions for these electrical motors, which are put on the market with standards that are also accepted by electrical motor producers [4]. Many standards, such as EPAct (1992), CEMEP (1998), and IEC 60034-2-1/60034-30-1 (2008), until now, have been suggested to arrange the market and increase product mobility [4,5]. Moreover, now, more efficient motors with a higher performance can be put on market, with the developing technology and the revisions that are made in the standards subject to these changes. According to the widely accepted IEC electrical motor standards, motor efficiencies are classified as IE1 for standard efficiency, IE2 for high efficiency, IE3 for premium efficiency, and IE4 for super premium efficiency [6].

Induction motors (IMs), which are preferred because of their robustness and requirement for less maintenance are put on the market with an efficiency of IE3 and below. It is not possible to produce these motors with an efficiency above IE3 because of technological and material limitations [7]. However, high-power electrical motors can be produced, which provide a higher efficiency and power factor by placing high performance permanent magnets on the IM rotor, the cost of which continues to decrease every day [8]. These motors with a permanent magnet in their rotor and an IM squirrel cage are known in the literature as line start permanent magnet synchronous motors (LS-PMSM). In addition to being robust and requiring less maintenance, like IMs, LS-PMSMs are also able to provide efficiencies of above IE3, thanks to the permanent magnets in their rotor [9].

While the LS-PMSM may start asynchronously via direct power from the line, like IMs, because of the squirrel cage in their rotors, contrary to IMs, which generate variable rpm subject to load, they can continue to operate at a constant speed after reaching a synchronous speed. A rotor current is not induced on the cage, as the squirrel cage does not cut off the stator magnetic field in the motor, which continues to rotate at a constant speed (synchronous speed) because of the effect of the extra magnetic flux generated by the permanent magnets, which leads to reducing the electromagnetic losses on the rotor to an almost non-existing level (when harmonics are considered) [10,11].

The LS-PMSMs, which can run up with the cage torque generated by the aluminum start winding (squirrel cage) in their rotor, are able to generate an opposite braking torque thanks to the permanent magnets placed on the rotor. Thus, the electromagnetic torque generated by the motor during the synchronization period is comprised of two torques, as the cage torque (T_c) and magnet torque, which is in the opposite direction (T_{pm}). The T_c and T_{pm} torque components generated at the time of starting are expressed by Equations (1) and (2), respectively [12–14].

$$T_c = \frac{p}{2}\frac{m}{w_s}\left\{ \begin{array}{c} (X_{2d} - X_{2q})I_{2d}I_{2q}+ \\ X_{md}I_dI_{2q} - X_{mq}I_qI_{2d} + E_0I_{2d} \end{array} \right\} \tag{1}$$

$$T_{pm} = \frac{p}{2}\frac{m}{w_s}\left\{ X_{md}I_{fm}I_{md} + (X_d - X_q)I_{md}I_{mq} \right\} \tag{2}$$

The total electromagnetic torque (T_e) can be determined using Equation (3).

$$T_e = \frac{p}{2}\frac{m}{w_s}\left\{ \begin{array}{c} (X_d - X_q)I_dI_q+ \\ X_{md}I_{2d}I_q - X_{mq}I_{2q}I_d + E_0I_q \end{array} \right\} \tag{3}$$

Figure 1 shows the torque components of the LS-PMSM. The dotted curve in the figure represents the torque generated by the start cage without a permanent magnet (T_c), and the dashed line represents the braking torque generated by the permanent magnets (T_{pm}). The asynchronous torque (T_{asyn}) generated by electromagnetic torque (T_e), which is the sum of T_c and T_{pm}, is represented by a normal line. The cage torque (T_c) should overcome the braking torque (T_{pm}) generated by the permanent magnets, the load torque (T_L), and motor inertia (J_m) in order for the synchronization process in LS-PMSM to be successful.

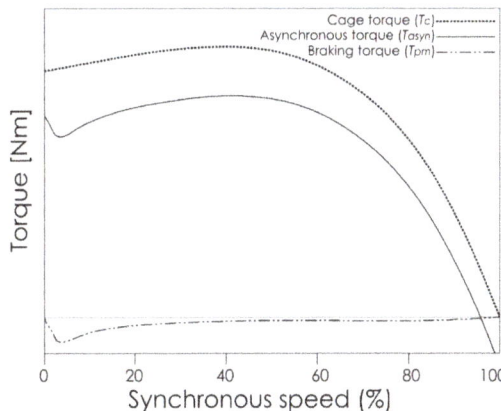

Figure 1. Torque components of the line start permanent magnet synchronous motors (LS-PMSM).

The successfully synchronized motor passes on to a steady state operation in order to continue rotating at a synchronous speed. The torque generated because of the currents during the steady state operation is as given in Equation (4).

$$T_{syn} = \frac{p}{2} \frac{m}{w_s} \{ E_0 I_q + I_d I_q (X_d - X_q) \} \tag{4}$$

In addition, if the angle between the EMF and terminal voltage is denoted by β, the synchronous torque when the resistances are neglected can be expressed as in Equation (5).

$$\begin{aligned} T_{syn} &= \frac{p}{2} \frac{m}{w_s} \left\{ \frac{E_0 V_{ph1}}{X_d} \sin \beta + \frac{V_{ph1}^2}{2} \left(\frac{1}{X_q} - \frac{1}{X_d} \right) \sin 2\beta \right\} \\ &= T_1 \sin \beta + T_2 \sin 2\beta \end{aligned} \tag{5}$$

There are many studies in the literature for improving the start performance of LS-PMSMs under different loads and line voltages, and for ensuring that the motor operates more efficiently after synchronization. It is observed that these studies have focused on improving the motor dynamic model, which expresses the operating characteristic of the motor better, as well as the determination and optimization of the design parameters with an impact on the synchronization performance of the motor [15–19].

A new LS-PMSM rotor design is presented in this study, with a high producibility, low cost, high efficiency, and with low maintenance requirement. A new squirrel cage using two different slot structures for improving motor run up and synchronization performance, along with surface mount permanent magnets, have been used in this suggested rotor design for increasing performance. The purpose of the study is to remodel the rotor of an already existing IM with IE2 1500 rpm 5.5 kW rated power. The ANSYS® Electromagnetics Suite finite elements software environment, which has proven itself in academic studies, was used for modelling the motor, in addition to the analytical models. The MathWorks Inc. MATLAB® development environment was also used in addition to Electromagnetics Suite for determining the motor performance characteristics, as well as for the optimization studies.

2. LS-PMSM Design

The IM design process is quite complex, but it has been discussed many times in the literature. LS-PMSMs are structurally similar to IMs to a great extent; however, their design process is more complex because of the inclusion of permanent magnets in their structure. It is a preferred method of LS-PMSM design to initially design the IM that will provide the desired power, and then placing the permanent magnets to the rotor [19–22]. Works for improving the slot and magnet design should be carried out in order to reach the desired motor design. Figure 2 shows the optimal LS-PMSM design algorithm [20,23].

It is very practical to start the design process by first determining the main dimensions of the motor (stator outer diameter, D_{so}; stator inner diameter, D_{si}; stack length, l; and air gap, δ). The air gap power of the motor can be determined subject to the output power of the motor, after the main dimensions of the motor have been determined. Stator winding and slot design can be carried out using the determined airgap power (S_{GAP}). Finally, the rotor design is carried out to complete the design process.

The primary goal in LS-PMSM design is to ensure that the motor can synchronize by itself. The synchronization performance should be reevaluated by updating the permanent magnet and starting the winding design of the motor, which failed in the synchronization process. The shaft torque characteristic and efficiency during the steady state operation are controlled for the motor, which provides a satisfactory synchronization performance. The design process should be repeated by updating the main dimensions of the motor and/or the permanent magnet design for the motor designs with an unsatisfactory torque characteristic or efficiency.

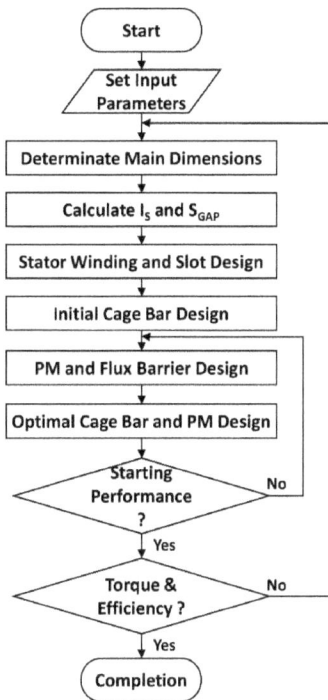

Figure 2. Optimal LS-PMSM design algorithm.

2.1. Determining the Main Dimensions of the Machine

Determining the main dimensions of the LS-PMSM should be carried out in an attentive manner. It becomes very easy to place the permanent magnets and slots to the main motor core when the motor core is selected as larger than necessary. However, this preference leads to the use of more production (lamination steel, aluminum, copper, etc.), thereby resulting in a motor design with a heavier core and a larger production cost. It may become a problem to fit a slot design and permanent magnet that will meet the desired airgap power if the motor core is selected as being smaller than necessary. When it is considered that the motor core is affected by the magnetic field in addition to the magnetic flux generated by the permanent magnets, especially the stator yoke, the saturation increases excessively, thereby leading to a significant decrease in motor efficiency. Finally, the accordance of the selected motor core dimensions with the standard frame dimensions is important for the compliance of the design with the standards.

In this study, the stator design was obtained from a current industrial IEC standard compliant IM. Thus, the main dimensions of the motor were determined by taking the stator design as reference. Therefore, it was ensured that the new design is compliant with the IEC motor standards. The main reference motor dimensions in the study have been given in Table 1.

Table 1. Main dimensions of the reference motor.

Frame	Output Power, P_n	Stack Length, l	Stator Outer Dia, D_{so}	Stator Inner Dia, D_{si}
132 s4	5.5 kW	125 mm	200 mm	125 mm

D_{so}, D_{si}, and l are already known in the study, as the stator design has been directly taken from a standard IM (Table 1). Additionally, the air gap has to be re-determined for the new rotor design.

Equation (6) is used in the literature [20] for determining the air gap in the machines with two or more poles.

$$\delta = \left(0.1 + 0.012\sqrt[3]{P_n}\right)10^{-3}m \tag{6}$$

The air gap, δ, stated here, is the ideal value required for the modeling of standard IM. As surface mount permanent magnets are used in this study, the risks for braking the torque generated by the magnets and the demagnetization of the permanent magnets increase when the air gap that is selected is very small. Therefore, it may be necessary to select this value as greater than the analytical values.

Another parameter that should be set during the design process is the targeted flux density for the air gap (B_δ). It is suggested [24] that the air gap flux density should be between (0.7–0.9 T). This value is taken into consideration when modeling the rotor cores and determining the dimensions of the permanent magnets.

2.2. Stator Design

This study aims to make an industrial asynchronous motor more efficient only by redesigning its rotor. For this purpose, the stator design was taken from a 5.5 kW IE2 asynchronous motor. Details regarding the 36-slot stator design have been given in Table 2.

Table 2. Stator design details.

Winding Layer	Winding Type	Conductor per Slots	Coil Pitch	Number of Strands	Wire Size	Steel Type	Number of Slots
2	Whole-Coiled	28	7	3	0.9116	M330 50A	36

The label information for the standard IE2 IMs taken as reference have been given in Table 3.

Table 3. Label information for the reference motor.

Output Power	Rated Current	R.P.M. min^{-1}	Power Factor	Eff. %	Starting Current	Rated Torque	Starting Torque	Weight
5.5 kW	11.8 A	1440	0.83	85	7 p.u.	36.5 Nm	2.3 p.u.	63 kg

2.3. Rotor Design

The rotor design for LS-PMSM can be examined under the two main headings of slot design and permanent magnet design. Determining the preferred rotor topology first, during the design process, followed by the design of start windings, and finally the placement of the required permanent magnets to the rotor core, makes the design process very practical. Many different rotor designs have been presented for LS-PMSM development until today [25–28]. It can be observed in the studies carried out that surface PM or interior PM type designs may be preferred [29,30]. Placing the slots on the regions left over from the permanent magnet may be problematic in designs, such as in interior PMs, where permanent magnets are placed away from the air gap. It may be necessary to readjust the slot design so as to reduce the slot height, especially in designs where permanent magnets are positioned behind the slot. In addition, part of the magnetic field generated by the permanent magnets placed away from the air gap only close in the rotor, thereby resulting in losses in the magnet magnetic field leading to an overheating of the rotor. This undesired situation is overcome by placing a flux barrier at the magnet tip. This brings forth the necessity to leave an additional space on the rotor core for the flux barrier. Placing the permanent magnet on the rotor core without deforming is another problem that needs to be overcome. However, it has been observed in the studies carried out that interior PM typologies are preferred more, as there is a lower risk for the permanent magnets to be deformed while the motor is in operation [25–28].

A new surface permanent magnet topology has been preferred in this study, which is to place the desired number of permanent magnets on the rotor. Also, it is more flexible with regard to the slot design. Thanks to this preference, it has been possible to place the permanent magnets on the rotor core without any deformation. In addition, the permanent magnet magnetic flux losses have been minimized and they have been directed to the airgap. Of course, this has led to giving more attention to the rotor cage design in order to ensure that the permanent magnets that are closer to the air gap do not have an adverse impact on the motor start. The torque fluctuations due to the impact of the permanent magnets are another issue that should be taken into consideration [31,32].

The most important issue that requires attention in the design of the surface mount permanent magnets is to select the magnet dimensions (thickness and height) properly. While a very thin permanent magnet preference leads to an increased risk of demagnetization, a thick permanent magnet preference results in the braking torque generated by the permanent magnet to have an adverse impact on the motor start [22,33].

The leakage flux and fringing effects may be neglected when it is taken into consideration that the flux generated by the permanent magnets in the surface mount permanent magnet motors directly passes onto the air gap, thus it is accepted that $\phi_\delta = \phi_{pm}$. Thereby, Equations (7) and (8) can be written [33].

$$h_{mag} = -\frac{(H_\delta \delta)}{H_{mag}} \tag{7}$$

$$t_{mag} = \frac{\frac{B_\delta}{\mu_r \mu_0} \delta}{H_{mag}} \tag{8}$$

The rotor bar and end ring circuits to be used in dimensioning the motor start windings can be determined by way of Equations (9) and (10) [24].

$$I_r = \frac{z_{Q_s}}{a} \frac{Q_s}{Q_r} I_s \cos \varphi \tag{9}$$

$$I_{ring} = \frac{I_r}{2 \sin\left(\frac{\pi p}{Q_r}\right)} \tag{10}$$

where

$$I_s = \frac{P_n}{m \eta V_{ph1} \cos \varphi} \tag{11}$$

As a result, the acquired rotor bar and end ring area can be calculated via Equations (12) and (13), respectively.

$$S_{cr} = \frac{I_r}{J_r} \tag{12}$$

$$S_{cring} = \frac{I_{ring}}{J_{ring}} \tag{13}$$

where J_r and J_{ring} indicate the rotor bar and end ring current densities, respectively. It is suggested in the literature [24] that this value should remain between the 3–6.5 A/mm^2 interval for aluminum rotor cages.

In light of the acquired data, the rotor tooth width, b_{tr}, can be calculated as given in Equation (14) [20].

$$b_{tr} \approx \frac{B_\delta}{K_{Fe} B_{tr}} \tau_r \tag{14}$$

The LS-PMSM's may start by overcoming the motor inertia and the braking torque generated by the permanent magnets. This puts forth the necessity for a high-performance cage design in order to prevent the motor from having synchronization problems. In addition, the start windings have to re-synchronize the motor in case the motor speed is below the synchronous speed at the time of starting.

The slot structures that determine the motor torque characteristic are standardized by NEMA with class A, B, C and D [34]. According to the NEMA standard classes A and B are used in general purpose motors. Also, Class C and D are preferred in applications requiring higher starting torque. As a result, a new rotor typology making use of both of the slot types is suggested in this study, to ensure that both the motor start and synchronization performances are high (Figure 3).

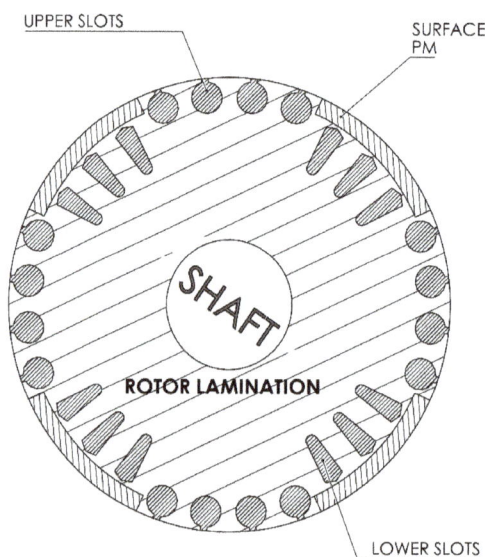

Figure 3. Suggested rotor design.

A total of 16 NEMA D class upper slots have been placed near the permanent magnets in the suggested model, by taking into consideration the speeding characteristic of LS-PMSM. In addition, 12 NEMA B class lower slots have been placed under the permanent magnets, in order to help re-synchronization in case the motor drops below the synchronous speed during steady state operation, and also in order to support the motor start. Care was also given to ensure that the cross-sections of both slot types are as close to each other as possible, for ensuring the homogeneous distribution of the rotor current induced on the start windings to the rotor bars. As was the case for the stator, M330 50A steel material was used for the rotor as well during the design procedure. Also, N45SH neodymium magnets are preferred for PM design.

2.4. Optimization of the Suggested Design

The motor operating characteristic in the LS-PMSM design depends to a large extent on the permanent magnet and slot design. So much so that a significant increase in the motor performance can be attained with a permanent magnet design with the correct geometrical position and geometry, in addition to reducing the cogging. In addition, determining the proper rotor slot geometry is another important issue to ensure that the motors start themselves.

It can be seen in Figure 4a that the analytically designed model has been synchronized successfully at 150 ms. The shaft torque generated by the motor during starting (0–150ms) and steady state (150–200 ms) periods can be seen in Figure 4b. The T_{ripple} = 19.98 (SI) generated during the steady state operation indicates that an excessively vibratory shaft torque has been generated (the SI unit corresponds to T_{RMS}/T_{mean}). The stator current and motor efficiency were determined during the steady state operation as $I_s = 9.6$ A and $\eta = 91.14\%$, respectively. When this efficiency value is compared with the 85% value obtained with the standard IM rotor, it can be observed that an efficiency increase of 6.14% has been attained. IE3 efficiency levels have been reached with this efficiency value, obtained using the analytically modelled design without any optimization.

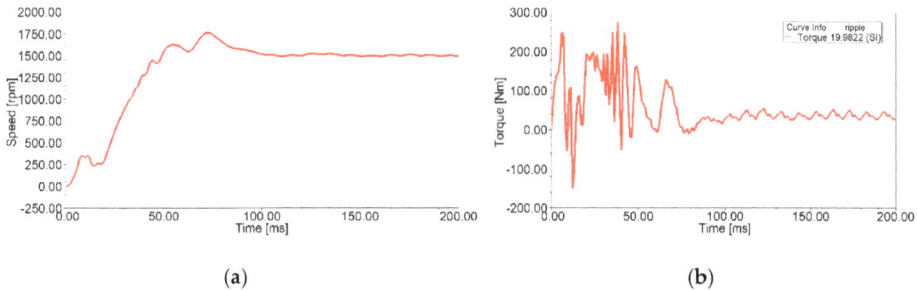

(a)

(b)

Figure 4. $T_L = 35$ Nm; $J_m = 1$ p.u. Starting process of the analytical model: (**a**) speed characteristic and (**b**) torque characteristic.

The objective with the optimization studies carried out was to ensure that the motor efficiency exceeds 91.9% (IE4), by determining the proper design parameters. The optimized design parameters can be seen in Figure 5.

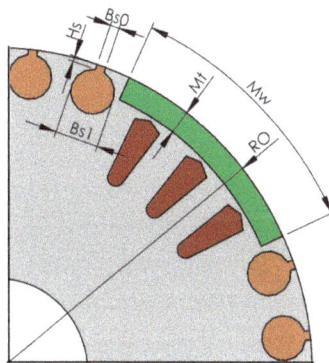

Figure 5. Design optimization.

M_t and M_w in Figure 5 stand for the magnet thickness and width, respectively; $Bs0$ denotes the slot gap for the rotor upper slots; H_s represents the slot depth; and $Bs1$ denotes the slot diameter. Finally, the rotor diameter (RO) was also included in the optimization studies for determining the optimum air gap value. The parameters used in the optimization study are shown in Table 4 together with their min. and max. values.

Table 4. Design parameters used for the optimization of the model.

Design Parameters	Initial Value	Min.	Max.
M_w	41°	36°	42°
M_t	4.7 mm	4 mm	5 mm
$Bs0$	1.5 mm	1 mm	2 mm
$Bs1$	9 mm	8 mm	9 mm
H_s	0.5 mm	0.25 mm	1.2 mm
RO	61.5	61 mm	62 mm

2.4.1. Optimization of the Model Suggested via Genetic Algorithm

It can be observed when studies on geometric optimization are examined, that, in general, the genetic algorithm (GA) is used for determining the optimum data from among large data sets. GA has also been preferred in this study during the optimization work carried out for determining the motor design parameters with the targeted efficiency characteristics.

The cost function obtained as a result of optimization via the genetic algorithm has been presented in Figure 6 as a graph. As can be seen from the graph, the targeted efficiency (92%) was reached after 383 evolutions.

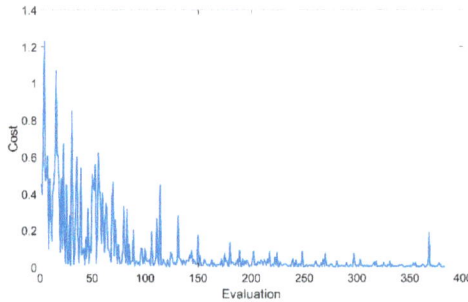

Figure 6. Cost function.

Figure 7 shows the design parameters processed via GA and the distribution of the parameters. It can be understood upon examining the graph that the values of $Bs1$, H_s, RO, M_t, and M_w do not have a homogeneous distribution, and the optimization study has focused more on certain value intervals. In addition, the graph presented indicates that the impact of the $Bs0$ value on motor performance is quite low. In conclusion, the targeted efficiency was reached as a result of GA optimization at around $Bs1 = 8.5$ mm, $H_s = 1$ mm, $Bs0 = 1.79$ mm, $M_t = 4.48$ mm, $M_w = 40.5°$, and $RO = 61.30$ mm. Moreover, the torque ripples generated by the motor have been decreased down to about $T_{ripple} = 7.09$ (SI), and the motor rated current was determined as $I_A = 9.26$ A. Table 5 summarizes the change in the design parameters and motor operating characteristic as a result of the GA optimization.

Table 5. Change in design parameters and motor efficiency after genetic algorithm (GA) optimization.

Design Parameters	Before GA Optimization	After GA Optimization
M_w	41°	40.5°
M_t	4.7 mm	4.48 mm
$Bs0$	1.5 mm	1.79 mm
$Bs1$	9 mm	8.5 mm
H_s	0.5 mm	1 mm
RO	61.5	61.30 mm
η	91.14%	92%
T_{ripple}	19.98 (SI)	7.09 (SI)
I_A	9.6 A	9.26 A

A significant number of analyses were carried out during the optimization studies for reaching an optimum rotor design with a high efficiency. The motor design was evaluated as being two dimensional during the analyses carried out, as it delivers faster solutions, and the analyses were finalized with a Maxwell 2D Transient solver. A Maxwell 3D Transient solver was used during the next stages of the study for a more detailed examination of the model obtained with a high efficiency via GA.

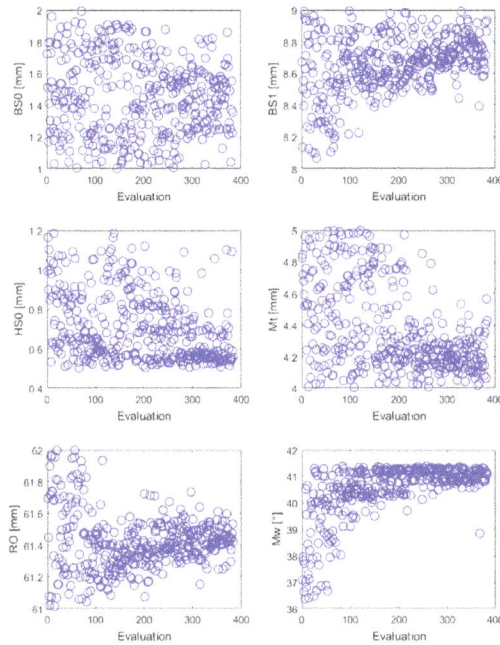

Figure 7. Distribution of genetic algorithm (GA) analysis parameters according to the number of evolutions.

The torque ripples in the motor shaft torque resulted in $T_{ripple} = 19.98$ (SI) in the model, whereas this value was decreased to $T_{ripple} = 7.08$ (SI) in the new design. Optimization works were carried out for the optimization of this value, which was still quite high in the later stages of the study. The component generated especially in the pole transitions generates ripples in the torque, because of the interactions between the permanent magnets and the stator tooth in LS-PMSM's [35]. It was observed that when the studies for reducing the ripples in the torque were examined, the focus was on the optimization of the stator tooth and/or permanent magnets, production of the skewed stator, or rotors [36–40].

Embedding the permanent magnets to the rotor core in order to decrease their interaction with the stator tooth is a widely used method for optimizing the permanent magnets so as to reduce torque ripples. However, this method cannot be applied on rotor topologies with a surface mount PM. In addition, it is not possible to update the stator slot and tooth structure, as the stator design was taken from an already existing IM.

On the other hand, producing skewed permanent magnet rotors makes it more difficult to place the magnets inside the rotor core, which increases the production costs. Therefore, the focus was on producing a skewed stator core for reducing ripples in torque. Thus, it was aimed at reducing the ripples in the torque by only producing a skewed core without any changes in the stator slot and tooth design.

2.4.2. Reducing Ripples in the Torque Using Skewed Stator

The graph in Figure 8 shows the T_{ripple} values generated by the motors modelled with different stator skew angles. It can be seen in the graph that there is no linear relationship between the T_{ripple} values, however the lowest T_{ripple} values could be generated as $T_{ripple}(\text{skew9}) = 4.48$ (SI), $T_{ripple}(\text{skew13}) = 2.9$ (SI), and $T_{ripple}(\text{skew15}) = 4.39$ (SI). It can also be seen from the graph that the highest efficiency values were obtained as η (skew7) = 91.33%, η(skew8) = 91.29%, and η (skew13) = 91.28%.

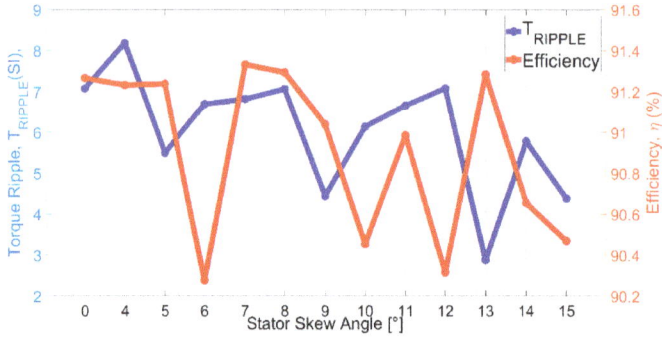

Figure 8. Ripple and efficiency obtained for different stator skew angles.

As can be seen in Figure 9, showing the impact of s stator skew angle on the motor synchronization performance, the motor has the highest synchronization performance at angles of skew0 and skew8. However, the high torque ripples of around T_{ripple} = 7 (SI) are generated for both skew angles, which results in an unacceptable decrease in the motor shaft torque quality. In addition, it was also observed that motor efficiency and performance decreased significantly for skew15 at which a satisfactory decrease in the torque ripple was obtained. In conclusion, skew13 was preferred as the stator skew angle, because of its satisfactory synchronization performance and lowest torque ripples.

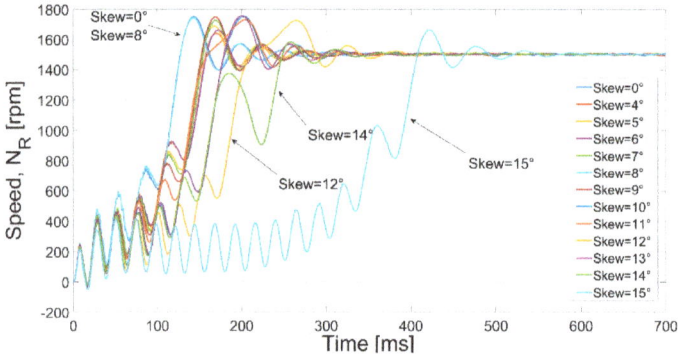

Figure 9. Motor synchronization performance for different stator skew angles.

Cogging torque is a torque effect that is generated in the motor shaft, which prevents the rotation of the motor shaft—the impact of which is felt when the motor is idle. This torque component generated because of the permanent magnets on the rotor core is a factor that results in the decrease in the quality of the shaft torque generated by the motor. Figure 10 shows the cogging torque for the skew0 and skew13 angles, which is generated during the 30° rotation of the motor. As can be seen in the graph, the cogging torque is reduced significantly (67.7%) with a skewed stator motor design. The graph in Figure 11 shows the electromagnetic torque generated by the motor for skew angles of skew0 and skew13. As can be seen from the graph a significant improvement in motor torque quality has been attained by preferring skew13 angle.

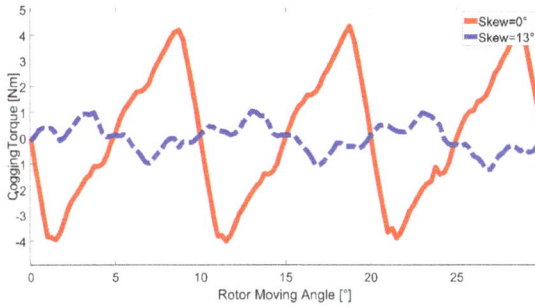

Figure 10. The cogging torque generated during a 30° rotation of the motor for skew0 and skew13 skew angles.

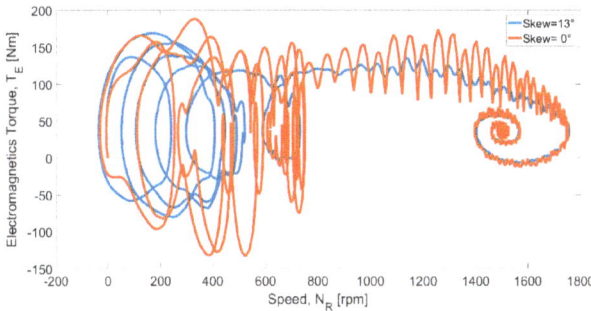

Figure 11. Change in electromagnetic torque subject to rotor revolution generated by the motor for skew0 and skew13 angles.

2.5. Analysis of the Suggested Model

At this stage of the study, the performance characteristic will be examined for the motor design with a stator having a skew angle of skew13, modelled using the suggested rotor topology. The most important problem with LS-PMSM's is the braking torque generated during operation, because of the permanent magnets. This results in issues such as the LS-PMSM's becoming too sensitive to circumstances, such as the overloading of the motor. Figure 12 shows the starting performance under different motor loads for the skewed stator model suggested. As can be seen from the graph, the motor can reach a synchronous speed of 1500 rpm in 0.3 s at T_L = 35 Nm (Full Load). An overloading of 14% (T_L = 40 Nm) results in the failure of the motor synchronization process.

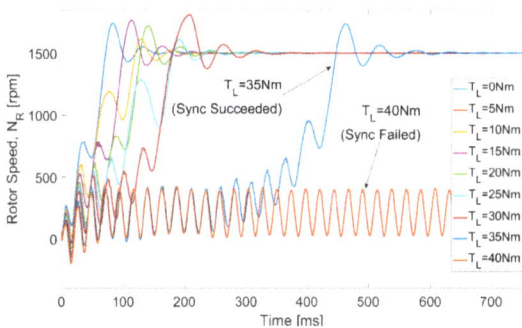

Figure 12. Starting performance for the suggested model with skewed stator under different motor loads.

The graph in Figure 13 shows the current and efficiency values subject to load torque. It can clearly be seen from the graph that the current drawn when the motor is idle is around $I_A = 2.3$ A. This value increases up to $I_A = 9.84$ A at an overloading of 40 Nm. The current drawn at $\frac{1}{4}$ of the load is 2.8 A and 4.3 A at $\frac{1}{2}$ loading. Motor efficiency at $\frac{1}{4}$ loading increased up to 86% and up to 91% at $\frac{1}{2}$, whereas a maximum value of 91.28% was reached at full load.

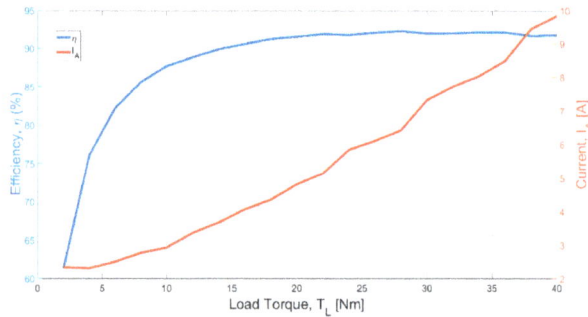

Figure 13. Current and efficiency values obtained from the motor subject to load torque.

Figure 14 shows the meshed model of the $\frac{1}{4}$ suggested LS-PMSM. The Tetrahedra mesh structure was used in the simulation analyses. The number of small regions formed in this structure is directly related to the closeness to the correct result. A total of 173,665 regions were created as a result of the mesh processing. While a total of 35,569 regions existed in each stator; 26,133 regions were consisted in the rotor, bar, and magnets. The rest of the regions are located in the defined windings, band, inner-outer regions, and shaft regions.

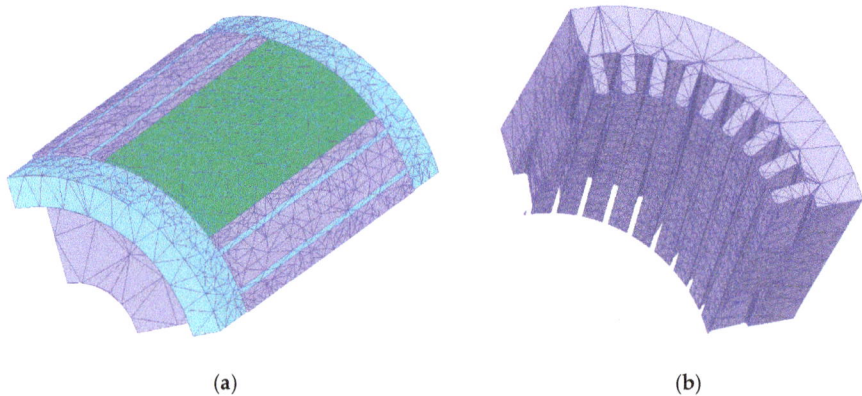

(a)   (b)

Figure 14. The meshed model of the LS-PMSM with suggested rotor design: (a) rotor and (b) stator.

Figure 15a shows the flux densities in the stator and rotor core during the steady state operation of the motor. As can be understood from the presented figure, the magnetic flux can penetrate deep into the rotor core at high speeds, because of the low frequency of the rotor, which results in an active motor synchronization at lower slots. In addition, it also makes it more difficult for the motor to decrease from a synchronous speed. Even though it is observed in the graph that the flux densities remain in safe limits at the rotor and stator yokes, it can also be observed that the flux accumulations that develop during the steady state period do not exceed the value of 1.9–2 T.

(**a**)   (**b**)

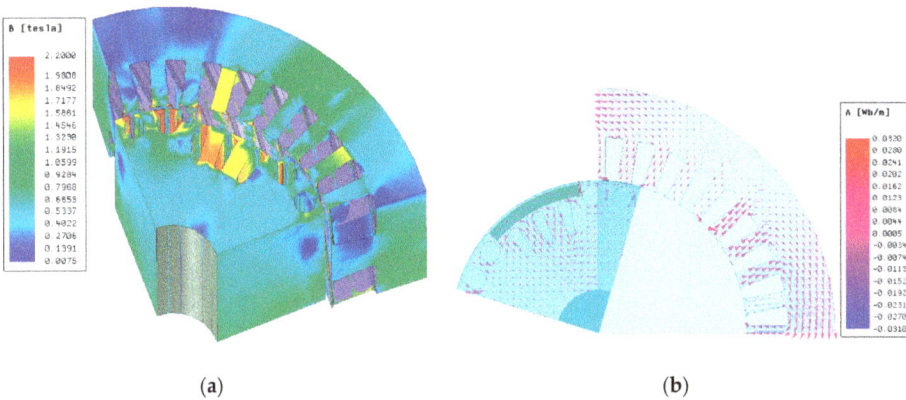

Figure 15. Field reports of the LS-PMSM with suggested rotor design: (**a**) flux distributions at the stator and rotor cores in steady state operation and (**b**) vector distribution.

Also, Figure 15b shows the vector distribution of the $\frac{1}{4}$ suggested model in 2D. As can be seen in the figure, the magnetic flux around the upper slots, lower slots, permanent magnet, and the rotor yoke changes easily.

2.6. Analysis of the Suggested Model Under Quadratic Load

The starting performance of the model, which is suggested with Figure 12 in the previous section, was discussed under a constant load; it has been observed that the suggested motor cannot be synchronized under a constant load of $T_L = 40$ Nm. This result, which has been acquired from analysis studies, is quite low if compared to the 230% of the starting torque generated by the reference motor.

The load torque increases with the proportion of the square of the rotation speed in the applications where the flow power is required, such as centrifugal pumps, fans, and blowers (Equation (15))

$$T_L = k\omega^2 \tag{15}$$

A superior performance can be obtained in the applications that have a square load profile, by using LM-PMSM, which offers a low starting performance under a constant load. The starting performance under the $T_L = 35$ Nm constant and square load of the suggested model in Figure 16 was compared. As it can be understood from the figure, the motor with a low performance under a constant load can offer superior performance under a square load. It can be seen the suggested model in Figure 17 can be synchronized in 200 ms up to $T_L = 87.50$ Nm (250% rated torque) loads successfully.

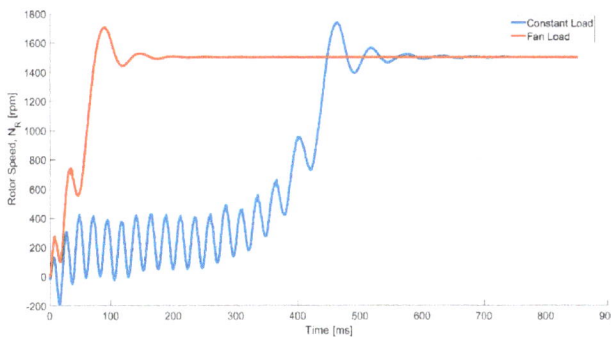

Figure 16. Starting performance under $T_L = 35$ Nm constant and square load of the suggested model.

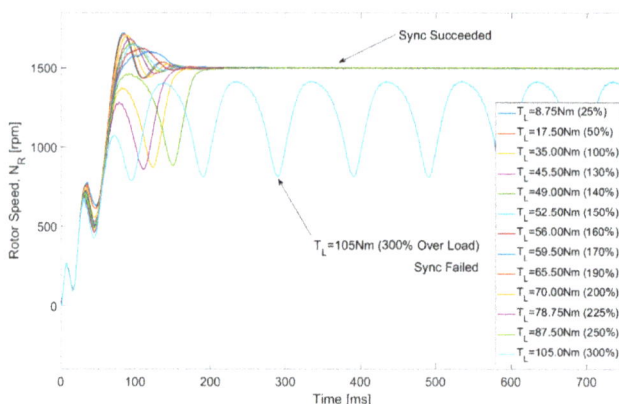

Figure 17. Starting performance under different square loads of the suggested model.

3. Conclusions

In this study, the rotor of a motor with a 5.5 kW power 4–pole IE2 efficiency class is remodeled using PM. Analytical methods and finite element analysis software environment were used in the modeling. As the stator section of the designed motor was obtained from an industrially marketed product, the rotor was designed using an accordant stator inner diameter and stack length values, thereby resulting in the designed motor operating in accordance with the other asynchronous motors with the same power. Works were carried out during the study for improving the torque ripples generated by the motor in LS-PMSM, which resulted in a significant decrease in the shaft torque quality. The best design parameters were determined via GA. An increase in the motor efficiency (of 6.28%) was obtained by redesigning the rotor of a motor in the moderate efficiency segment. A rotor design was suggested that can operate with IM, with the same power, as a result of using a compatible stator inner diameter and stack length. A new rotor topology with two different slot types as an upper and lower was suggested as a result of the design studies carried out. A satisfactory starting performance was obtained with the suggested topology. A new rotor design was introduced in the industrial pump and fan applications. In addition, a surface mount PM was used in the design of the suggested model, thereby eliminating problems such as the placement of the PM in the rotor core, which poses a difficult problem for motors with interior PMs.

Author Contributions: Formal analysis, B.Z. and M.E.; investigation, B.Z.; methodology, B.Z. and M.A.; resources, M.E.; supervision, M.A.; validation, B.Z.; writing (review and editing), B.Z.

Funding: This study was supported by the Gaziosmanpaşa University Scientific Research Projects Unit, project number 2017/90.

Conflicts of Interest: The authors declare no conflict of interest.

Nomenclature

m, p: Number of phase and pole
P_n, w_s: Output power and stator angular speed
X_d, X_q, X_{2q}, X_{2q}: Leakage reactances (the suffix "2" refers to the rotor)
I_d, I_q, I_{2d}, I_{2q}: Axis current (the suffix "2" refers to the rotor)
X_{md}, X_{qm}: Magnetizing reactances
I_{md}, I_{qm}: RMS of d- and q-axis currents
E_0, I_{fm}, V_{ph1}: Excitation voltage, excitation current, and phase voltage
T_c, T_{pm}, T_e: Cage torque, permanent magnet torque, and total electromagnetic torque
T_{asyn}, T_{syn}: Asynchronous torque and synchronous torque

J_m, T_L, k: Motor inertia, load torque, and square load profile coefficient

β: Angle between EMF and terminal voltage

D_{so}, D_{si}, RO: Outer and inner diameter of stator, rotor diameter

l, δ: Stack length and air gap

B_δ, S_{GAP}: Air gap flux density and air gap power

ϕ_δ, ϕ_{pm}: Air gap flux and permanent magnet flux

h_{mag}, t_{mag}: Permanent magnet height and thickness (in calculations)

M_t, M_w: Magnet thickness and width (in optimization study)

H_δ, H_{mag}: Field strength of the air gap and field intensity of the permanent magnet

μ_0, μ_r: Permeability of air and relative recoil permeability

I_s, I_r, I_{ring}: Stator, rotor, and end ring current

Q_s, Q_r: Number of stator and rotor slots

z_{Q_s}, a: Number of conductors in a slot and parallel paths in a stator winding

η, $\cos \varphi$, I_A: Efficiency, power factor, and rated current

S_{cr}, S_{cring}: Cross-sectional area of rotor bar and end ring

J_r, J_{ring}: Rotor and end ring current densities

b_{tr}, τ_r, K_{Fe}: The rotor tooth width, rotor slot pitch, and the stacking factor

T_{ripple}, T_{RMS}, T_{mean}: Torque ripple, RMS of motor torque, and mean of motor torque

$Bs0$, $Bs1$, H_s: Slot gap for upper slot, slot diameter, and rotor slot depth

References

1. Riba, J.R.; Torres, C.L.; Romeral, L.; Garcia, A. Rare-earth-free propulsion motors for electric vehicles: A technology review. *Renew. Sustain. Energy Rev.* **2016**, *57*, 367–379. [CrossRef]
2. Koch, S.F.; Peter, M.; Fleischer, J. Lightweight Design and Manufacturing of Composites for High-performance Electric Motors. *Procedia CIRP* **2017**, *66*, 283–288. [CrossRef]
3. Kampker, A.; Burggräf, P.; Nee, C. Costs, quality and scalability: Impact on the value chain of electric engine production. In Proceedings of the 2nd International Electric Drives Production Conference (EDPC), Nuremberg, Germany, 15–18 October 2012; pp. 1–6.
4. De Almeida, A.T.; Ferreira, F.J.T.E.T.E.; Fong, J.A.C. Standards for Efficiency of Electric Motors. *IEEE Ind. Appl. Mag.* **2011**, *17*, 12–19. [CrossRef]
5. De Almeida, A.T.; Ferreira, F.J.; Fong, J.A.C.; Brunner, C.U. Electric motor standards, ecodesign and global market transformation. In Proceedings of the IEEE/IAS Industrial and Commercial Power Systems Technical Conference, Clearwater Beach, FL, USA, 4–8 May 2008; pp. 1–9.
6. Brunner, C.U.; Niederberger, A.A.; De Almeida, A.T.; De Keulenaer, H. Standards for efficient electric motor systems SEEEM building a worldwide community of practice. In Proceedings of the European Council for an Energy Efficient Economy (ECEEE) Summer Studies, La Colle sur Loup, France, 4–9 June 2007; pp. 1443–1455.
7. Isfahani, A.H.; Zadeh, S.V. Line start permanent magnet synchronous motors: Challenges and opportunities. *Energy* **2009**, *34*, 1755–1763. [CrossRef]
8. Behbahanifard, H.; Sadoughi, A. Line Start Permanent Magnet Synchronous Motor Performance and Design; a Review. *J. World Electr. Eng. Tech.* **2015**, *4*, 58–66.
9. Honsinger, V. Permanent Magnet Machines: Asynchronous Operation. *IEEE Trans. Power Appar. Syst.* **1980**, *PAS-99*, 1503–1509. [CrossRef]
10. Stoia, D.; Chirila, O.; Cernat, M.; Hameyer, K.; Ban, D. The Behaviour of The LSPMSM in Asynchronous Operation. In Proceedings of the 14th International Power Electronics and Motion Control Conference (EPE-PEMC 2010), Ohrid, Macedonia, 6–8 September 2010; pp. T4-45–T4-50.
11. Miller, T.J.E. Synchronization of Line-Start Permanent-Magnet AC Motors. *IEEE Trans. Power Appar. Syst.* **1984**, *PAS-103*, 1822–1828. [CrossRef]
12. Soulard, J.; Nee, H.P. Study of the synchronization of line-start permanent magnet synchronous motors. In Proceedings of the Thirty-Fifth IAS Annual Meeting and World Conference on Industrial Applications of Electrical Energy, Rome, Italy, 8–12 October 2000; Volume 1, pp. 424–431.

13. Isfahani, A.H.; Zadeh, S.V.; Rahman, M.A. Evaluation of Synchronization Capability in Line Start Permanent Magnet Synchronous Motors. In Proceedings of the IEEE International Electric Machines & Drives Conference (IEMDC), Niagara Falls, ON, Canada, 15–18 May 2011; pp. 1346–1350.

14. Rabbi, S.F.; Rahman, M.A. Determination of the synchronization criteria of line start IPM motors. In Proceedings of the International Electric Machines & Drives Conference, Chicago, IL, USA, 12–15 May 2013; pp. 1218–1224.

15. Nedelcu, S.; Tudorache, T.; Ghita, C. Influence of design parameters on a line start permanent magnet machine characteristics. In Proceedings of the 13th International Conference on Optimization of Electrical and Electronic Equipment (OPTIM), Brasov, Romania, 24–26 May 2012; pp. 565–571.

16. Yang, Y.; Wang, X.; Zhang, R.; Ding, T.; Tang, R. The optimization of pole arc coefficient to reduce cogging torque in surface-mounted permanent magnet motors. *IEEE Trans. Magn.* **2006**, *42*, 1135–1138. [CrossRef]

17. Kang, G.H.; Hur, J.; Nam, H.; Hong, J.P.; Kim, G.T. Analysis of irreversible magnet demagnetization in line-start motors based on the finite-element method. *IEEE Trans. Magn.* **2003**, *39*, 1488–1491. [CrossRef]

18. Liu, X.; Lin, Q.; Fu, W. Optimal Design of Permanent Magnet Arrangement in Synchronous Motors. *Energies* **2017**, *10*, 1700. [CrossRef]

19. Rabbi, S.F.; Rahman, M.A. Critical Criteria for Successful Synchronization of Line-Start IPM Motors. *IEEE J. Emerg. Sel. Top. Power Electron.* **2014**, *2*, 348–358. [CrossRef]

20. Boldea, I.; Nasar, S.A. *The Induction Machines Design Handbook*, 2nd ed.; CRC Press/Taylor & Francis: Boca Raton, FL, USA, 2010; ISBN 978-1-4200-6668-5.

21. Elistratova, V. Optimal Design of Line-Start Permanent Magnet Synchronous Motors of High Effiency. Ph.D. Thesis, Ecole Centrale de Lille, Villeneuve-d'Ascq, France, 2015.

22. Dosiek, L.; Pillay, P. Cogging Torque Reduction in Permanent Magnet Machines. In Proceedings of the IEEE Industry Applications Conference Forty-First IAS Annual Meeting, Tampa, FL, USA, 8–12 October 2006; pp. 44–49.

23. Jędryczka, C.; Knypiński, Ł.; Demenko, A.; Sykulski, J.K. Methodology for Cage Shape Optimization of a Permanent Magnet Synchronous Motor Under Line Start Conditions. *IEEE Trans. Magn.* **2018**, *54*, 1–4. [CrossRef]

24. Pyrhonen, J.; Jokinen, T.; Hrabovcova, V. *Design of Rotating Electrical Machines*, 2nd ed.; Wiley: Chichester, UK, 2014; ISBN 978-1-118-58157-5.

25. Kumar, A.; Srivastava, A. Performance Comparison of Two Different Rotor Topologies of Line Start Permanent Magnet Synchronous Motors. *Int. J. Res. Appl. Sci. Eng. Technol.* **2017**, *5*, 2313–2318.

26. Ugale, R.T.; Chaudhari, B.N.; Pramanik, A. Overview of research evolution in the field of line start permanent magnet synchronous motors. *IET Electr. Power Appl.* **2014**, *8*, 141–154. [CrossRef]

27. Dinh, B.M. Optimal Rotor Design of Line Start Permanent Magnet Synchronous Motor by Genetic Algorithm. *Adv. Sci. Technol. Eng. Syst. J.* **2017**, *2*, 1181–1187. [CrossRef]

28. Ding, T.; Takorabet, N.; Sargos, F.; Wang, X. Design and Analysis of Different Line-Start PM Synchronous Motors for Oil-Pump Applications. *IEEE Trans. Magn.* **2009**, *45*, 1816–1819. [CrossRef]

29. Huang, P.W.; Mao, S.H.; Tsai, M.C.; Liu, C.T. Investigation of line start permanent magnet synchronous motors with interior-magnet rotors and surface-magnet rotors. In Proceedings of the International Conference on Electrical Machines and Systems, Wuhan, China, 17–20 October 2008; pp. 2888–2893.

30. XuXiaozhuo, X. Performance of Line-start Permanent Magnet Synchronous Motor with Novel Rotor Structure. *Int. J. Digit. Content Technol. Its Appl.* **2013**, *7*, 1217–1225. [CrossRef]

31. Bianchi, N.; Bolognani, S. Design techniques for reducing the cogging torque in surface-mounted PM motors. In Proceedings of the Thirty-Fifth IAS Annual Meeting and World Conference on Industrial Applications of Electrical Energy (Cat. No.00CH37129), Rome, Italy, 8–12 October 2000; pp. 179–185.

32. Ortega, A.J.P.; Xu, L. Analytical Prediction of Torque Ripple in Surface-Mounted Permanent Magnet Motors Due to Manufacturing Variations. *IEEE Trans. Energy Convers.* **2016**, *31*, 1634–1644. [CrossRef]

33. Garner, K.; Grobler, A.J. Rotor Design of a Retrofit Line Start Permanent Magnet Synchronous Motor. In Proceedings of the 23rd Southern African Universities Power Engineering Conference (SAUPEC), Johannesburg, African, 28–30 January 2015; pp. 221–226.

34. National Electrical Manufacturers Association. *Motors and Generators*; MG1-1993; NEMA: Washington, DC, USA, 1993.

35. Cetin, E.; Daldaban, F. Analyzing the Profile Effects of the Various Magnet Shapes in Axial Flux PM Motors by Means of 3D-FEA. *Electronics* **2018**, *7*, 13. [CrossRef]

36. Islam, R.; Ortega, A.P. Practical aspects of implementing skew angle to reduce cogging torque for the mass-production of permanent magnet synchronous motors. In Proceedings of the 20th International Conference on Electrical Machines and Systems (ICEMS), Sydney, NSW, Australia, 11–14 August 2017; pp. 1–5.

37. Sheth, N.K.; Sekharbabu, A.R.C.; Rajagopal, K.R. Torque ripple minimization in a doubly salient permanent magnet motors by skewing the rotor teeth. *J. Magn. Magn. Mater.* **2006**, *304*, e371–e373. [CrossRef]

38. Kim, B.; Kim, D.; Know, B.; Lipo, T.A. Optimal Skew Angle for Improving of Start-Up Performance of a Single-Phase Line-Start Permanent Magnet Motor. In Proceedings of the IEEE Industry Applications Society Annual Meeting, Edmonton, AB, Canada, 5–9 October 2008; pp. 1–6.

39. Hsiao, C.-Y.; Yeh, S.-N.; Hwang, J.-C. A Novel Cogging Torque Simulation Method for Permanent-Magnet Synchronous Machines. *Energies* **2011**, *4*, 2166–2179. [CrossRef]

40. Zhu, Z.Q.; Howe, D. Influence of design parameters on cogging torque in permanent magnet machines. *IEEE Trans. Energy Convers.* **2000**, *15*, 407–412. [CrossRef]

electronics

MDPI

Article

Analysis of Equivalent Inductance of Three-Phase Induction Motors in the Switching Frequency Range

Milan Srndovic [1,*]**, Rastko Fišer** [2] **and Gabriele Grandi** [1]

[1]   Department of Electrical, Electronic, and Information Engineering, University of Bologna,
     40136 Bologna, Italy; gabriele.grandi@unibo.it
[2]   Department of Mechatronics, Faculty of Electrical Engineering, University of Ljubljana,
     1000 Ljubljana, Slovenia; rastko.fiser@fe.uni-lj.si
*   Correspondence: milan.srndovic2@unibo.it

Received: 31 December 2018; Accepted: 19 January 2019; Published: 22 January 2019

Abstract: The equivalent inductance of three-phase induction motors is experimentally investigated in this paper, with particular reference to the frequency range from 1 kHz to 20 kHz, typical for the switching frequency in inverter-fed electrical drives. The equivalent inductance is a basic parameter when determining the inverter-motor current distortion introduced by switching modulation, such as rms of current ripple, peak-to-peak current ripple amplitude, total harmonic distortion (THD), and synthesis of the optimal PWM strategy to minimize the THD itself. In case of squirrel-cage rotors, the experimental evidence shows that the equivalent inductance cannot be considered constant in the frequency range up to 20 kHz, and it considerably differs from the value measured at 50 Hz. This frequency-dependent behaviour can be justified mainly by the skin effect in rotor bars affecting the rotor leakage inductance in the considered frequency range. Experimental results are presented for a set of squirrel-cage induction motors with different rated power and one wound-rotor motor in order to emphasize the aforesaid phenomenon. The measurements were carried out by a three-phase sinusoidal generator with the maximum operating frequency of 5 kHz and a voltage source inverter operating in the six-step mode with the frequency up to 20 kHz.

Keywords: equivalent inductance; leakage inductance; switching frequency modelling; induction motor; current switching ripple

1. Introduction

A well pronounced frequency-dependent behaviour of an equivalent inductance in a three-phase induction motor (IM) has been analysed several times so far. Generally, the equivalent inductance of an IM, i.e., leakage inductance seen from the stator side at a stand-still, is at times determined as the sum of the stator and rotor inductances, both constant [1,2]. Such a simplification is not acceptable for higher order harmonics due to the frequency-dependent character of the rotor leakage inductance as a result of a skin effect in squirrel-cage rotor bars [3–5].

There are just a few analyses where the high-frequency behaviour of three-phase IMs was carried out over the actual switching-frequency range, but using impedance meters with an insufficient voltage level, and occasionally supplying just one out of three phases. The low power supply voltage might cause a significant limitation on the validity and applicability of proposed measurements [6,7]. Furthermore, supplying just one phase is not an entirely accurate approach and it might lead to a mismatch between estimated parameters when comparing them with the ones obtained at symmetrically supplied three phases.

The phenomenon of varying the rotor leakage inductance with frequency, and consequently the equivalent inductance, contributes considerably to motors' characteristics. The estimation of inverter current harmonic distortion, current ripple rms, and peak-to-peak current ripple amplitude [1–4,8,9]

is strongly affected as well. Especially, in the case of multiphase IMs, the equivalent inductance of each α-β plane seen from the stator side may additionally vary due to the different arrangements of machine connections [4].

On the whole, variation of the rotor leakage inductance over the frequency range is firmly connected to motor parameters, such as the power range, pole-pairs number, rotor bar shape, and depth [7]. In [10], a comprehensive equation for calculating the rotor leakage inductance considering some of the aforesaid parameters was introduced. The equation was derived with respect to the DC value of the rotor leakage inductance. The rotor leakage inductance decreases following a reciprocal value of the frequency square root over the whole frequency range. On the other hand, in [7], it was noted that the rotor leakage behaviour for frequencies lower than 1 kHz is reciprocally dependent on the square of frequency, and for frequencies higher than 1 kHz, the behaviour is the same as it was given by the equation in [10]. The explanation is that for lower frequencies, the skin depth and conductor size of squirrel-cage IM are of the same order of magnitude. Considering this, the ratio of leakage inductances at 20 kHz compared to the rated motor frequency is about 0.5 to 0.6. In [7], it was experimentally noted that the equivalent inductance of one IM had an increasing behaviour above a certain given frequency value, but without a proper explanation. Since a similar phenomenon was observed also in one of the tested motors, a capacitive effect was introduced, considering it as an adequate explanation. Similar effects were analysed in [11,12].

This paper gives a comprehensive set of experimental results for three different IMs with squirrel-cage rotors and one IM with a wound-rotor. The comparison between theoretical developments and the experiments are given over the considered frequency range, resulting in good correspondence. Apart from this, an additional experiment was carried out in order to show the effect of the total inductance frequency dependence on the current ripple estimation in case of one IM supplied by a three-phase inverter. Based on this, the graphical evidence of the analytically calculated current ripple envelopes at 50 Hz and at the switching frequency of 3 kHz are presented.

2. Theoretical Background and Basic Assumptions

In order to analyse the high-frequency behaviour of IM, the basic electrical scheme at a standstill is presented in Figure 1.

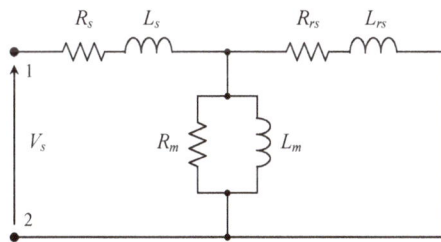

Figure 1. Basic per-phase equivalent circuit of the IM at a standstill.

The model consists of stator resistance, R_s, and stator leakage inductance, L_s, rotor resistance, R_{rs}, and rotor leakage inductance, L_{rs}, both seen from the stator side, and the magnetizing branch with parallel resistance, R_m, and inductance, L_m. For the higher order harmonics, the stator and rotor resistances can be neglected due to the fact that stator and rotor leakage inductances dominate [4]. By omitting these two parameters, the equivalent circuit is simplified to just four parameters. The parallel connection between the rotor leakage inductance and magnetizing inductance can be replaced with just one inductance with the proper equivalent value. If the magnetizing inductance is much higher than the rotor leakage inductance, which is usually the case in IMs, the parallel connection can be considered as the rotor leakage inductance per se, where $L_{rs} || L_m \cong L_{rs}$.

Such an assumption is made in this paper. After simplifying the proposed model, the equivalent impedance seen from the terminals 1 and 2 (Figure 1) can be derived:

$$Z' = R' + j\omega\,L' = \frac{\omega^2 L_{rs}{}^2 R_m}{R_m{}^2 + \omega^2\,L_{rs}{}^2} + j\omega\left(L_s + \frac{L_{rs} R_m{}^2}{R_m{}^2 + \omega^2\,L_{rs}{}^2}\right),\tag{1}$$

where f is the frequency, and $\omega = 2\pi f$ is the angular frequency. The frequency dependent behaviour of the rotor leakage inductance in case of rotor bars having a depth, d, is presented in [10,13]:

$$L_{rs}\,(f) = \frac{3\,L_{rs}^{dc}}{k\,d\,\sqrt{f}}\frac{\sinh\left(k\,d\,\sqrt{f}\right) - \sin\left(k\,d\,\sqrt{f}\right)}{\cosh\left(k\,d\,\sqrt{f}\right) - \cos\left(k\,d\,\sqrt{f}\right)},\tag{2}$$

where L_{rs}^{dc} denotes the rotor leakage inductance at DC (in our case, at the lowest initial measured frequency [5–7]), and:

$$k = \sqrt{\frac{4\pi\mu_0}{\rho}}.\tag{3}$$

The relative permeability of free space in (3) is $\mu_0 = 4\pi \times 10^{-7}$ H/m and the resistivity of the aluminium bars is $\rho = 2.65 \times 10^{-8}$ Ω·m. It can be noted that the behaviour of the rotor leakage inductance is inversely proportional to the square root of the frequency. In order to introduce the effect of stray capacitances for higher frequencies, the modified IM equivalent circuit at the standstill is presented in Figure 2.

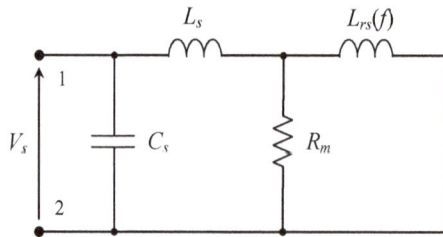

Figure 2. Proposed per-phase equivalent circuit of IM in the switching-frequency range, including stray capacitance and a simplified RL network.

The value of the capacitor, C_s, which is connected in parallel to the simplified circuit, is within the range of 0.5 nF to 5 nF [11,12]. Usually, the capacitive effect is not noticeable in the lower switching frequency range; however, in some particular cases, it has a relatively small contribution for higher frequencies. Considering the equivalent series resistance, R', and inductance, L', as defined in (1), and introducing the parallel capacitor, C_s, the overall equivalent inductance becomes:

$$L_{eq} = \frac{L'\left(1 - \omega^2 L' C_s\right) - C_s R'^2}{\left(1 - \omega^2 L' C_s\right)^2 + \left(\omega C_s R'\right)^2}.\tag{4}$$

3. Experimental Evaluation of Equivalent Inductance

Experimental analyses were carried out in order to estimate the frequency dependent behaviour of the equivalent inductance of three-phase IMs and to verify the analytical developments based on the simplified equivalent circuit presented in Figure 2. The set of motors under investigation consists of four IMs with different rated power (see Figure 3). Three out of four tested IMs have squirrel-cage rotors, where the skin effect in aluminium rotor bars must be considered. The fourth one has wound-rotor windings, mainly used experimentally to emphasise the considerably less pronounced skin effect in such rotor windings.

Figure 3. The considered set of IMs under testing.

The rated parameters of the tested IMs represented in Figure 2 are given in Table 1.

Table 1. Rated parameters of tested induction motors.

IM	Squirrel Cage Rotor			Wound Rotor	
Power (kW)	2.2	4.0	7.5	5.5	
Voltage (V)	400	400	400	Stator	Rotor
				380	186
Current (A)	5.2	9.2	15.3	14.0	19.5
Frequency (Hz)	50	50	50	50	
Speed (r/min)	1400	1425	1450	1400	

During measurements, each rotor shaft was mechanically locked to prevent its rotation. A three-phase sinusoidal power source, HP6834B (300 V, 4500 VA, 1Φ/3Φ, Keysight Technologies, Santa Rosa, CA, USA), was used for supplying the star-connected motors at a standstill. Due to its upper frequency limit of 5 kHz, in order to extend the test frequency range up to 20 kHz, a three-phase custom-made inverter operating in the six-step mode was used for the additional set of measurements, also guaranteeing almost sinusoidal motor currents. It consists of the three-phase IGBT Mitsubishi PS22A76 intelligent power module (1200 V, 25 A, Mitsubishi Electric Corporation, Tokyo, Japan) controlled by the Arduino Due microcontroller board (84 MHz Atmel, SAM3X83 Cortex-M3 CPU, Somerville, MA, USA). The Yokogawa DLM 2024 oscilloscope (Yokogawa Electric Corporation, Tokyo, Japan) with the PICO TA057 differential voltage probe (25 MHz, ±1400 V, ±2%, Pico Technology, Tyler, TX, USA) and LEM PR30 current probe (DC to 20 kHz, ±20 A, ±1%, LEM International SA, Plan-les-Ouates, Switzerland) were used to acquire motors' phase voltage and current. The whole experimental setup is shown in Figure 4.

The measurement determines the equivalent impedance for all the four IMs, calculated as the ratio between rms values of fundamental components of the voltage and current in the frequency range from 1 kHz to 20 kHz, with an equidistant step of 1 kHz. The equivalent reactance, and consequently the equivalent inductance (seen from the stator side), was calculated based on the phase displacement between the two fundamental components. In addition, the same quantities were also determined at the motor rated frequency of 50 Hz. The built-in digital filters of the oscilloscope were used to properly handle the current and voltage waveforms.

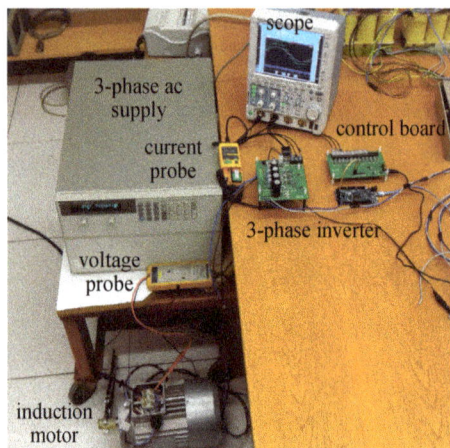

Figure 4. Experimental setup for measuring frequency dependence of L_{eq}.

To exclude the possible influence of voltage harmonics on the inductance measurements, a correlation between results acquired using the three-phase sinusoidal power source (up to 5 kHz) and the three-phase inverter with the six-step mode control was examined for all four motors. Comparing the results given in Table 2 for the 2.2 kW and 4.0 kW IMs clearly shows that the applied voltage source does not considerably influence the accuracy (within ±5%) of the determined equivalent inductance; therefore, only results for inverter supplied motors were presented to cover the whole frequency range from 1 kHz to 20 kHz. In Table 2, the parameter, L_{eq},~ represents the equivalent inductance measured in the case of the sinusoidal power source, $L_{eq,inv}$ denotes the equivalent inductance for the inverter supplied motor, and ε is the relative difference (in percent) between them.

Table 2. Variance of equivalent inductance with two different supplies.

IM	2.2 kW			4.0 kW		
f_{sw} (kHz)	$L_{eq\sim}$ (mH)	$L_{eq,inv}$ (mH)	ε (%)	$L_{eq\sim}$ (mH)	$L_{eq,inv}$ (mH)	ε (%)
1	19.43	18.95	+2.53	13.53	13.49	+0.29
2	17.95	18.37	−2.39	12.37	12.76	−3.06
3	17.63	17.44	+1.09	11.60	11.83	−1.94
4	17.09	17.11	−0.12	11.06	11.57	−4.41
5	16.79	16.46	+2.00	10.66	11.20	−4.82

Due to the voltage limitation of the supply source, it was not possible to perform the short circuit test at the rated current for the whole frequency range up to 20 kHz (the necessary voltage would also destroy the motor isolation). Therefore, the input inductance was determined at the same highest possible % of rated current for a particular IM. To evaluate the magnetizing conditions at lower currents than the rated stator currents and to eliminate possible mistakes in the determination of the equivalent inductance, verification tests with different currents were done for two motors at the rated frequency. The results presented in Table 3 show that the equivalent inductance, L_{eq}, practically, does not change (regularly within ±5%), considering different test currents (I_s).

The equivalent inductance, experimentally determined by the procedure described above, was compared with the value calculated by the proposed method (4), considering the motors' parameters given in Table 4. The results show an acceptable agreement in the whole considered frequency range from 1 kHz to 20 kHz.

Table 3. Variance of equivalent inductance at reduced stator currents.

	IM 2.2 kW					
I_s (A)	5.2 (*)	4	3	2	1	0.35
L_{eq} (mH)	23.56	23.44	24.04	24.64	24.04	25.02
ε (%)	0.00	−0.51	+2.04	+4.58	+2.04	+6.19
	IM 4.0 kW					
I_s (A)	9.2 (*)	7	5	3	1	0.33
L_{eq} (mH)	16.20	16.56	15.83	16.49	16.41	16.30
ε (%)	0.00	+2.22	−2.28	+1.79	+1.30	+0.62

* Rated current.

Table 4. Equivalent circuit parameters of tested induction motors.

IM	Squirrel-Cage Rotor			Wound-Rotor
Power (kW)	2.2	4.0	7.5	5.5
C_s (nF)	0.25	0.1	0.1	3.5
L_s (mH)	13	8.8	4	4
L_{rs}^{dc} (mH)	12	8	4.6	3.7
R_m (Ω)	500	150	80	350
d (mm)	6	5	5.5	—

Particularly, in Figure 5, the case of IMs with rotor bars is presented, showing the expected decreasing behaviour of the equivalent inductance over the frequency due to the rotor leakage inductance frequency dependence. In the case of the 2.2 kW IM, the effect of stray capacitances is noticeable after the frequency of 15 kHz (modelled by the capacitor connected parallel to the equivalent circuit, Figure 2). The capacitive effects regarding the other two IMs (4 kW and 7.5 kW) are not visible within the considered frequency range.

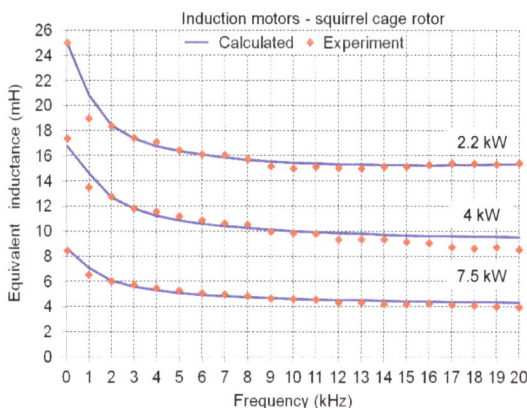

Figure 5. Model and experimental results for L_{eq} vs. supply frequency for three squirrel-cage IMs.

In the case of the IM with a wound rotor, as shown in Figure 6, the rotor leakage inductance is almost constant, with the skin effect being ineffective in such a rotor winding construction. The smooth decrease of the equivalent inductance is motivated by the resistive and reactive branches in the equivalent circuit (Figure 2), acting as a variable current divider as function of the frequency. A small increase of the L_{eq} after 15 kHz is also noticeable, well represented by the parallel capacitor.

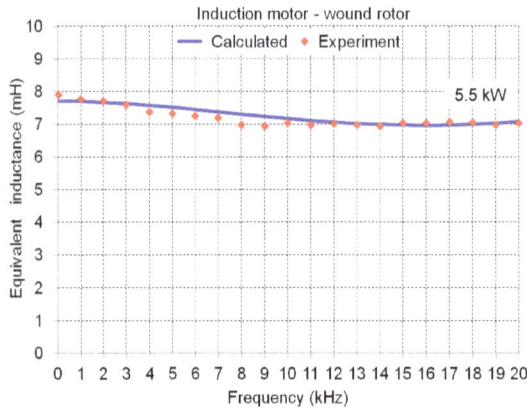

Figure 6. Model and experimental results for L_{eq} vs. supply frequency for the rotor-wound IM.

4. Application Example

When analysing the output current ripple regarding different PWM techniques and inverter configurations [9,14,15], the switching frequency and the load inductance are directly involved. In case of a squirrel-cage IM supplied by an inverter operating in the typical switching frequency range from 1 kHz to 20 kHz, and considering its equivalent inductance measured at a standstill at 50 Hz, this will result in a wrong current ripple estimation due to the frequency dependency of L_{eq}, which could generally decrease more than 50% (Figure 5).

Experimental results were carried out in order to demonstrate the influence of L_{eq} on the peak-to-peak current ripple amplitude in case of a three-phase squirrel-cage IM drive. The modulation technique used in this analysis was the space vector PWM technique, practically obtained by centering the three sinusoidal modulating signals of a carrier-based PWM with a common-mode injection (the so-called min/max injection).

The IM under test is the first given one in Table 2; Table 4 (2.2 kW). Figure 7 shows the line-to-neutral voltage (blue trace), instantaneous output current (red trace), and its fundamental component (grey trace) and ripple (orange trace), over one fundamental period (20 ms). The presented case corresponds to the DC bus voltage of V_{dc} = 300 V, switching frequency of 3 kHz, and modulation index, m = 0.5 (m_{max}= 0.577). The current ripple was obtained by subtracting the fundamental component from the instantaneous current [15]. All waveforms in Figure 7 are given in real scales.

In Figure 8, the following waveforms are presented: Measured current ripple (orange trace, obtained by downloading the experimental data with a high sample resolution and post-processing in Matlab/Simu-link), current ripple envelope analytically calculated by using L_{eq} = 17.36 mH, measured at f_{sw} = 3 kHz (blue trace), and the current ripple envelope, analytically calculated by using L_{eq} = 24.98 mH, measured at 50 Hz (red trace). The procedure of the current ripple envelope calculation is explained in detail in [9,14,15].

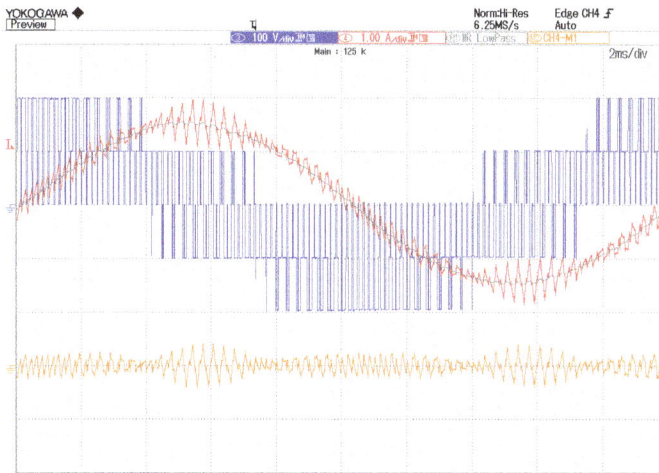

Figure 7. Details of experimentally obtained line-to-neutral voltage (blue trace), instantaneous current (red trace), fundamental current (grey trace), and current ripple (orange trace) for the 2.2 kW IM supplied by a three-phase inverter (centered PWM): $V_{dc} = 300$ V, $f_{sw} = 3$ kHz and $m = 0.5$.

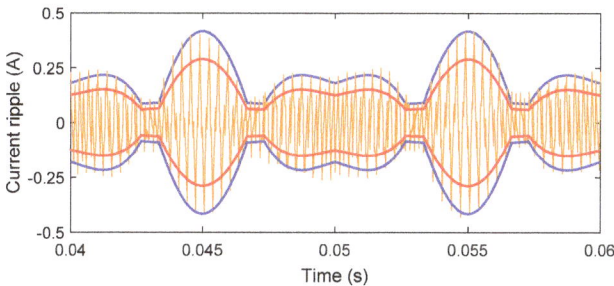

Figure 8. Measured current ripple (orange trace) and calculated current ripple envelopes (blue trace: L_{eq} measured at $f_{sw} = 3$ kHz, red trace: L_{eq} measured at 50 Hz).

The results clearly emphasise the case of an incorrect current ripple estimation when considering L_{eq} measured at the rated frequency instead of the switching frequency, giving a smaller current ripple amplitude (red envelope) compared with the correct one (blue envelope). L_{eq} further changes with the switching frequency according to Figure 5. Such an explicit discrepancy may lead to several mismatches in the procedures of current ripple minimization, modulation strategies' optimization, harmonic losses evaluation, EMC analysis, etc.

5. Conclusions

In this paper, the frequency-dependent behaviour of the IM equivalent inductance was presented in more detail compared to the existing literature. A simplified circuit model was proposed for induction motors in order to evaluate the equivalent inductance over the switching frequency range from 1 kHz to 20 kHz, which is the typical operating range of a PWM inverter in industrial ac motor drives.

Analytical results were compared with experimental ones, carried out by considering IMs with different rated powers, both squirrel-cage and wound-rotor types. The results show a good match, proving that the skin effect in rotor bars is mainly responsible for the equivalent inductance variation. There was huge variation in the frequency range up to 10 kHz, leading to a value that was 0.5 to

0.6 times more than the corresponding value measured at 50 Hz, depending on the IM design and power range. For higher frequencies, in the considered range up to 20 kHz, the equivalent inductance was almost constant. Thus, while selecting the switching frequency, it is very important to take into account the corresponding equivalent inductance, especially in the case of precise and sensitive motor control algorithms. The effect of stray capacitances was just noticeable, and it starts to be effective for frequencies higher than 20 kHz, which is already above the commonly used switching frequencies of mass-produced industrial inverters. More precise evaluation of the stray capacitance would require more complex circuits, instruments with higher resolution, and experiments in the capacitively-dominant working region; however, its influence is hardly noticeable in most cases. In the case of IM with wound rotor winding, the overall decrease of equivalent inductance was much less pronounced, and was not determined by the skin effect, and was well represented by the proposed circuit model. Apart from the presented analysis of the equivalent inductance, the evaluation of the equivalent resistance over the switching frequency range that is responsible for other relevant changes in some parameters could be performed in further research.

Author Contributions: Conceptualization, M.S. and G.G.; methodology, M.S., R.F., and G.G.; validation, M.S.; formal analysis, M.S., R.F. and G.G.; writing—original draft preparation, M.S.; writing—review and editing, R.F. and G.G.; funding acquisition, M.S.

Funding: This research received no external funding.

Conflicts of Interest: The authors declare no conflict of interest.

References

1. Casadei, D.; Serra, G.; Tani, A.; Zarri, L. Theoretical and experimental analysis for the RMS current ripple minimization in induction motor drives controlled by SVM technique. *IE* **2004**, *51*, 1056–1065. [CrossRef]
2. Kubo, H.; Yamamoto, Y.; Kondo, T.; Rajashekara, K.; Zhu, B. Current ripple analysis of PWM methods for open-end winding induction motor. In Proceedings of the 2014 IEEE Energy Conversion Congress and Exposition (ECCE), Pittsburgh, PA, USA, 14–18 September 2014.
3. Dujic, D.; Jones, M.; Levi, E. Analysis of output current-ripple RMS in multiphase drives using polygon approach. *PEL* **2010**, *25*, 1838–1849. [CrossRef]
4. Jones, M.; Dujic, D.; Levi, E.; Prieto, J.; Barrero, F. Switching ripple char-acteristics of space vector PWM schemes for five-phase two-level voltage source inverters—Part 2: Current ripple. *IA* **2011**, *57*, 2799–2808.
5. Calin, M.; Rezmerita, F.; Ileana, C.; Iordache, M.; Galan, N. Performance analysis of three phase squirrel cage induction motors with deep rotor bars in transient behavior. *Electr. Electron. Eng.* **2012**, *2*, 11–17. [CrossRef]
6. Potter, B.A.; Shirsavar, S.A.; Mcculloch, M.D. Study of the variation of the input impedance of induction machines with frequency. *IET Electr. Power Appl.* **2007**, *1*, 36–42. [CrossRef]
7. Novotny, D.W.; Nasar, S.A.; Jeftenic, B.; Maly, D. Frequency depend-ence of time harmonic losses in induction machines. In Proceedings of the ICEM, Boston, FL, USA, 13–15 August 1990.
8. Jiang, D.; Wang, F. Current ripple prediction for three-phase PWM converters. *IA* **2014**, *50*, 531–538. [CrossRef]
9. Grandi, G.; Loncarski, J.; Dordevic, O. Analysis and comparison of peak-to-peak current ripple in two-level and multilevel PWM inverters. *IE* **2015**, *62*, 2721–2730. [CrossRef]
10. Kwon, Y.S.; Lee, J.H.; Moon, S.H.; Kwon, B.K. Standstill parameter identification of vector-controlled induction motors using the frequency characteristics of rotor bars. *IA* **2009**, *45*, 1610–1618.
11. Hidaka, T.; Ishida, M.; Hori, T.; Fujita, H. High-frequency equivalent cir-cuit of an induction motor driven by a PWM inverter. *Electr. Eng. Jpn.* **2001**, *135*, 65–76. [CrossRef]
12. Grandi, G.; Casadei, D.; Reggiani, U. Equivalent circuit of mush wound AC windings for high frequency analysis. In Proceedings of the ISIE, Guimarães, Portugal, 7–11 July 1997.
13. Cho, K.R.; Seok, J.K. Induction motor rotor temperature estimation based on high-frequency model of a rotor bar. *IA* **2009**, *45*, 1267–1275.

14. Grandi, G.; Loncarski, J.; Srndovic, M. Analysis and minimization of output current ripple for discontinuous pulse-width modulation techniques in three-phase inverters. *Energies* **2016**, *9*, 380. [CrossRef]

15. Loncarski, J.; Leijon, M.; Srndovic, M.; Rossi, C.; Grandi, G. Comparison of output current ripple in single and dual three-phase inverters for electric vehicle motor drives. *Energies* **2015**, *8*, 3832–3848. [CrossRef]

electronics

MDPI

Article

Analyzing the Profile Effects of the Various Magnet Shapes in Axial Flux PM Motors by Means of 3D-FEA

Emrah Cetin * and Ferhat Daldaban

Engineering Faculty, Electrical and Electronics Engineering, Erciyes University, 38039 Kayseri, Turkey;
daldaban@erciyes.edu.tr
* Correspondence: emrahcetin@erciyes.edu.tr; Tel.: +90-532-660-5758

Received: 14 December 2017; Accepted: 23 January 2018; Published: 25 January 2018

Abstract: Axial flux machines have positive sides on the power and torque density profile. However, the price of this profile is paid by the torque ripples and irregular magnetic flux density production. To gather higher efficiency, torque ripples should close to the zero and the stator side iron should be unsaturated. Torque ripples mainly occur due to the interaction between the rotor poles and the stator teeth. In this study, different rotor poles are investigated in contrast to stator magnetic flux density and the torque ripple effects. Since the components of the axial flux machines vary by the radius, analysis of the magnetic resources is more complicated. Thus, 3D-FEA (finite element analysis) is used to simulate the effects. The infrastructure of the characteristics which are obtained from the 3D-FEA analysis is built by the magnetic equivalent circuit (MAGEC) analysis to understand the relationships of the parameters. The principal goal of this research is a smoother distribution of the magnetic flux density and lower torque ripples. As the result, the implementations on the rotor poles have interesting influences on the torque ripple and flux density profiles. The MAGEC and 3D-FEA results validate each other. The torque ripple is reduced and the magnetic flux density is softened on AFPM irons. In conclusion, the proposed rotors have good impacts on the motor performance.

Keywords: axial flux machines; magnetic equivalent circuit; torque ripple; back EMF; permanent-magnet machines

1. Introduction

Axial flux permanent magnet machines (AFPM) are one of the futuristic candidates for the higher performance aspiration. AFPM machines have high power/torque density, light mass/volume. It is applicable for many systems as researched in the literature. Mignot et al. designed an AFPM motor with magnetic equivalent circuit [1]. Kierstead et al. studied an in-wheel AFPM motor with non-overlapping windings [2]. Fei et al. researched an AFPM in-wheel motor with two air gap. They compared an approximation method with the 3D-FEA according to the calculation of the back EMF and cogging torque values [3]. Caricchi et al. suggested AFPM motors for direct-drive in-wheel applications with slotted windings. They considered the mitigation of the undesired effects, such as cogging torque and power loss due to flux pulsation in the core teeth, winding conductors, and rotor magnets [4]. Additionally, AFPM machines are investigated for many applications. Seo et al. studied robotic applications by using an analytic model and numeric analysis [5]. Parviainen et al. designed a generator in a small-scale wind-power applications [6]. Di Gerlando et al. focused on wind power generation after defining a general analysis of the model and design features of the AFPM machine [7]. De et al. proposed an ironless AFPM motor with low inductance for the aerospace industry in their paper [8]. One of the common points of these applications is sensitivity with the torque performance. Thus, torque ripples need to be as low as possible. Torque ripples mainly occur due to two main constituents, which are ripple torque and the cogging torque. The cogging torque is cultured by the mutual effects of the reluctance variation in stator and rotor magnetic flux. The ripple torque is mainly constituted by the coaction of the stator current magnetomotive force and rotor magnetic flux distribution in the surface permanent magnet (SPM) machines [9]. Both the cogging and ripple torques are related to rotor magnetic flux distribution, which is manipulated by shapes of PM in the SPM machines. In addition, the back EMF is one of the most crucial characteristics of the AFPM machine profile which is affected by the winding configurations as described by Saavedra et al. in [10].

Magnetic flux is the one basic characteristic of electric machines. There exist so many techniques to analyze the magnetic flux. One of the most prevailing methods is the MAGEC. Analyzing the magnetic flux by using MAGEC is one of the easiest analytical tools in comparison of the other methods given in the literature. Since magnetic flux paths simply turn into the circuit components and the problem is solved by the circuit analysis easily [11–14]. If the solution of the MAGEC is done, air gap magnetic flux density, permanent magnet axial length, the total permeance of the machine, back EMF value, winding resistance, self-inductance, and torque and output power can be composed according to the study of Mignot et al. [1]. The stator winding resistance, eddy current resistance, end winding resistance, power factor, phase voltages, output power, and steady-state torque are counted by the MAGEC perusal in the research of Wang et al. [15]. Parviainen et al. [11], Tiegna et al. [12], Bellara et al. [13], and Lubin et al. [14], mentioned different analytical calculation techniques for analyzing the characteristics of the AFPM machine in the literature.

In the literature, various topologies were investigated to reduce torque ripple including the shaping of rotor magnet pole and stator slots. Aydin et al. studied the shaping of rotor magnet pole which was realized by skewing or displacing magnet poles [16]. Sung et al. proposed the shaping of stator slots by recasting slot or teeth numbers in [17]. Saavedra et al. researched the effects of the magnet shaping under demagnetization fault conditions by means of 3D-FEA. The research aimed to determine a more efficient magnet geometry [18]. Kahourzade et al. summarized the torque ripple reducing methods by means of the classification of the AFPM machines [19].

This paper suggests analyzing the three-phase single air gap AFPM machine as given in Figure 1 by means of the 3D-FEA analysis. Five different rotor designs are investigated to achieve the goals. The first goal is to use MAGEC to observe the magnetic flux path and to describe the magnetic events in an analytic way. Another goal is to investigate the torque ripple, back EMF, and air gap magnetic flux distribution results of the proposed rotors in 3D-FEA. Two of the five designs are the novel magnet shapes, which are mainly developed for mitigating the torque ripples. The proposal is stepping and shifting the rotor magnets. Due to this action affects the magnetic flux path, back EMF, and torque waveforms are changed. All of the designs are simulated in 3D-FEA to compare the novel topologies and exciting results are obtained.

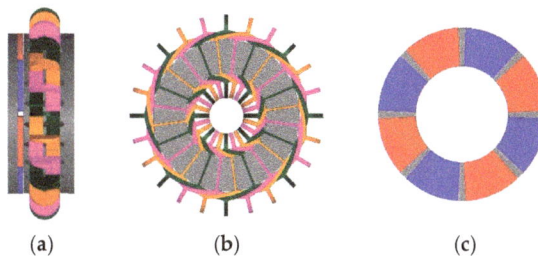

(a)   (b)   (c)

Figure 1. The single gap AFPM machine structure (**a**) total stack; (**b**) the stator with distributed windings; and (**c**) the rotor.

2. MAGEC Design and Analysis

MAGEC design is composed of the magnetic flux path at the machine given in Figure 2. Each definition of the flux sources and the permeance are situated by considering this path.

The describing of the elements of MAGEC eases the analytical solutions by obtaining the parameters easily.

An interesting specialty of the AFPM machine is that most of the parameters vary by the radius. The produced torque is defined by the radius, too. Due to this, the MAGEC is designed by considering the single air gap AFPM motor in this section. Conceptual 2D and 3D representations are seen in Figure 2.

The rotor and the stator back irons are produced from ferromagnetic steel, and these steels are designed from layers, like round strips, which are laminated in the circumference direction. Permanent magnets are placed on the surface of the rotor back iron. There are gaps between each pole to minimize the permanent magnet flux leakage. As seen from Figure 2, flux paths are the same for each permanent magnet pole. Since the flux divides by two for each pole, just one closed loop is modeled in the MAGEC Equations.

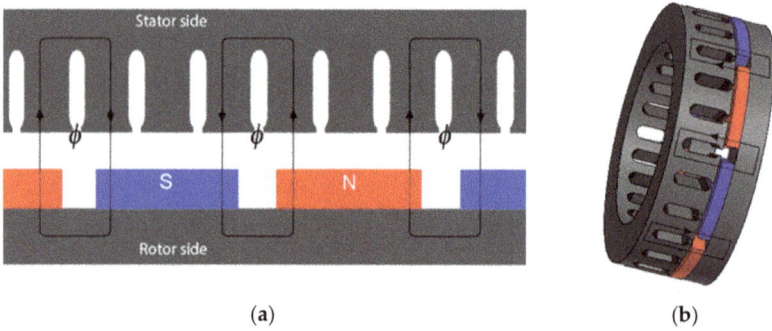

(a)   (b)

Figure 2. Magnetic flux path of the AFPM machine in (**a**) the 2D view and (**b**) the 3D view.

The developed MAGEC is shown in Figure 3, which is designed by minding the flux path illustrated in Figure 2a,b. The two permanent magnet halves, rotor and stator back iron, air gap, and the gap between poles are included in the MAGEC.

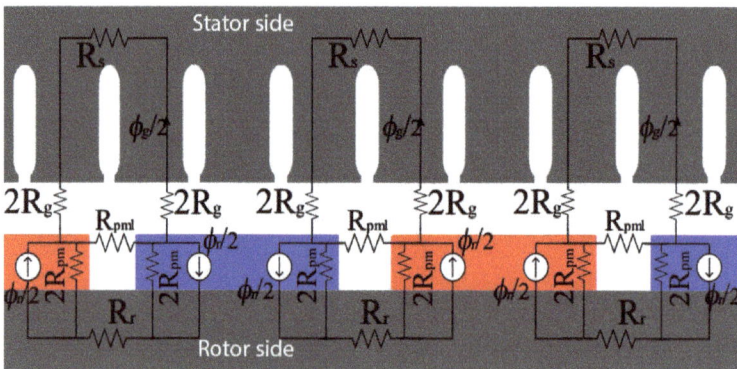

Figure 3. MAGEC design of the studied single air gap, slotted AFPM machine.

Three stator teeth per one pole are determined for the studied single air gap AFPM motor, as demonstrated in Figure 2. Thus, the stator has 24 teeth and the rotor has eight magnets. Other parameters are listed in Table 1.

Table 1. The parameters of the single air gap AFPM motor.

Parameter	Value	Unit
Inner radius (D_i)	40	mm
Outer radius (D_o)	75	mm
Slot/Pole	24/8	
Magnet height	5	mm
Magnet fill factor	0.8722	
Air gap	1	mm
Winding	Distributed, overlapping	
Turns	40	
Stator width	50	mm
Rotor back iron width	10	mm
Rated Speed	2200	rpm
Rated current	175	A

Since air gap permeability μ_0 is much lower than iron permeability, the air gap reluctance is much higher than the rotor and stator back iron reluctances. Due to this, the rotor and stator reluctances can be neglected to have an easier solution. Thus, the MAGEC can be simplified, as in Figure 4. In the end, the permeance values are taken into account instead of the reluctances.

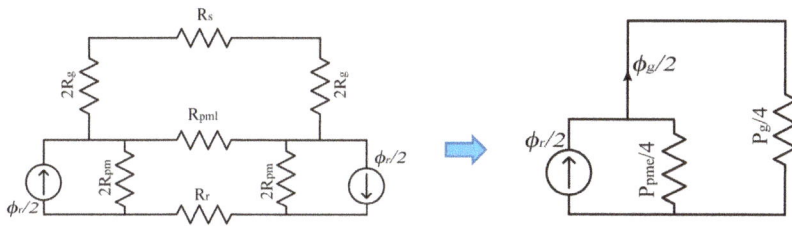

Figure 4. The simplified MAGEC.

The simplified MAGEC is located to the 2D design of the AFPM machine as given in Figure 5. The relationship between the air gap flux and the rotor flux is pointed out in the Equation (1).

$$\phi_g = \frac{1}{1 + P_{pme}/P_g}\phi_r \tag{1}$$

where P_{pme} is the effective permanent magnet permeance, and P_g is the air gap permeance.

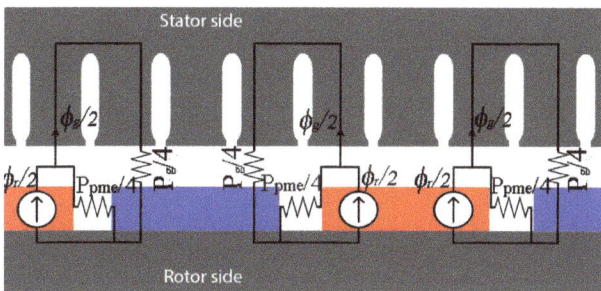

Figure 5. Simplified MAGEC design in the 2D view of the studied AFPM machine.

Permanent magnet and steel data are given in Table 2. Additionally, permeance of the magnet is defined in Equation (2):

$$P_{pm} = \frac{\mu_r \mu_0 A_{pm}}{L_{PM}},$$ (2)

Table 2. Data of the permanent magnet and the steel.

Permanent Magnet: NdFe–N35	
B_r (T)	1.17
μ_r	1.099
H_{cb} (kA/m)	868
H_{cj} (kA/m)	955
Steel: M250-35A	
B_{ref} (T)	1.5
μ_r	660
Loss (W/kg)	2.5
f (Hz)	50

Here, L_{PM} is the permanent magnet's height and the A_{PM} is the surface area of the permanent magnet. The height of the permanent magnet can be determined by Equation (3) [20]:

$$L_{PM} = \frac{\mu_r B_g}{B_r - \left(\frac{K_f}{K_d} B_g\right)} (gK_c),$$ (3)

The permanent magnet surface area is calculated in contrast to the inner and outer radii, as given by Equation (4):

$$A_{pm} = \alpha_{pm} \frac{\pi}{N_{pm}} \left(D_o^2 - D_i^2\right),$$ (4)

Here, N_{pm} is the number of the pole of the AFPM machine. The pole area A_p is necessary to find the magnet fill factor α_{pm}:

$$A_p = \frac{\pi}{N_{pm}} \left(D_o^2 - D_i^2\right)$$ (5)

$$\alpha_{pm} = \frac{A_{pm}}{A_p}$$ (6)

A magnetic flux leakage occurs between the adjacent magnets on the rotor. The path of this flux leakage draws an arc between two magnets. When this path is accounted, the obtained leakage permeance is calculated by Equation (7). P_{pml} is the permeance of the gap between the two adjacent permanent magnets. If the simplification of the MAGEC is taken into account by applying $P_{pme} = P_{pm} + 4P_{pml}$, Equation (8) can be derived to simplify the equation by a coefficient (K_{pml}) which is given in Equation (9). The effective permanent magnet permeance (P_{pme}) is defined by the multiplication of the permanent magnet permeance P_{pm} and K_{pml} in Equation (8):

$$P_{pml} = \frac{\mu_0 (D_o - D_i)}{\pi} \ln\left(1 + \pi \frac{g}{d_f}\right)$$ (7)

$$P_{pme} = K_{pml} P_{pm}$$ (8)

$$K_{pml} = 1 + 4 \frac{L_{PM} N_{pm}}{\pi^2 \mu_r \alpha_{pm} (D_o + D_i)} \ln\left(1 + \pi \frac{g}{d_f}\right)$$ (9)

Equations (7)–(9) allows simplifying the MAGEC, as seen in Figure 4. In addition, air gap permeance can be calculated correctly by defining the effective air gap $g_e = K_c g$, and the air gap area [21]. Thus, the interaction between the air gap flux and the rotor flux becomes as specified in Equation (11):

$$P_g = \frac{\mu_0 A_g}{g_e} \tag{10}$$

$$\phi_g [Wb] = \frac{1}{1 + 2\frac{\mu_r \alpha_{pm} K_{pml} K_c g}{(1 + \alpha_{pm}) L_{PM}}} \phi_r [Wb] \tag{11}$$

One of the main subjects to create the MAGEC is the defining the air gap magnetic flux density. In the light of the Equation (11), the magnetic flux density B_g can be calculated as stated in the Equation (12) where $K_{k\varphi} = A_{pm}/A_g$ and $C_p = L_{PM}/gK_{k\varphi}$:

$$B_g [T] = \frac{K_{k\phi}}{1 + \frac{\mu_r K_{pml} K_c}{C_p}} B_r [T] \tag{12}$$

As given in Equation (13), permanent magnet flux produces the air gap flux density and results in the voltage induction, called back EMF, in the stator windings. This can be seen from the MAGEC depicted in Figure 4.

$$e_{ind} = w N_{pm} N_w B_g (D_o - D_i) \tag{13}$$

The force Equation (15) is composed of the electric-magnetic loads from Equation (14) and the total area of the magnets from Equation (4). If these Equations are applied from the inner to the outer radius, the electromagnetic torque equation becomes that shown by Equation (16):

$$Q_{load} = B_g J_{in} \tag{14}$$

$$F_{emr_i} = \pi B_g J_{in} (D_o{}^2 - D_i{}^2) \tag{15}$$

$$T_{em} = \pi B_g J_{in} \int_{D_i}^{D_o} D_i \mathbf{r} d\mathbf{r} = \pi B_g J_{in} (D_o)^3 \lambda (1 - \lambda^2) \tag{16}$$

Here, J_{in}, is the current density at the inner radius D_i, and λ is the rate of the radiuses which is counted by D_i/D_o.

3. Analyzed Rotor Pole Designs

The single air gap, slotted AFPM motor is taken into account as the reference design structure, which is demonstrated in Figure 1. The studied motor parameters are given in Table 1. Five different rotor pole designs are investigated in this research. Design I is a conventional rotor pole design of an AFPM machine model, taken as the reference model for this study, which can be seen in Figure 6a. It has sharp edges. This type of magnet can be easily found on the market. Design II is an improved rotor pole model for an AFPM machines, as seen in Figure 6b. It has sinusoidal edges. This design is studied to reduce the cogging torque in the literature [22]. Design III is one of the proposed rotor pole models for this research, and is shown in Figure 6c. This design is the novel proposal for AFPM motors. It is studied to reduce the torque ripples. Design IV is another proposed rotor model for this research, as seen in Figure 6d. This design is a novel proposal for axial flux machines. Design V is developed for validating the FEA model. Aydin and Gulec proposed that cogging torque has minimum values when skewing angle is 18.75°, such as that used in this study for design V, given in Figure 6e. The FEA simulation of designs I and V prove the validation of the FEA model in comparison with [23].

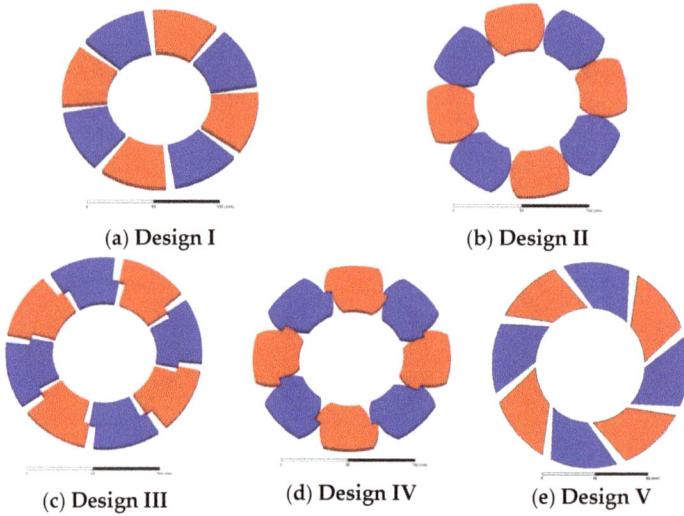

(a) Design I

(b) Design II

(c) Design III

(d) Design IV

(e) Design V

Figure 6. The researched designs of the magnets.

All of the permanent magnet poles are magnetized in the z-axis and the total value of the inner and outer diameters are the same for each pole designs. The MAGECs of each design do not change in majority due to the constant magnet fill factor α_{pm}, which is 0.8722 for each design. Additionally, all permanent magnet pole designs have symmetry in the radial direction.

4. 3D-FEA Analysis

The back EMF, torque, and flux density distribution waveforms are obtained from the three-dimensional finite element analysis. Both transient and static analyses are performed. 3D-FEA simulations are performed for the $\frac{1}{4}$ of the AFPM motor designs, as given in Figure 7, in order to shorten the simulation time. A runtime process of 10 milliseconds is chosen, thus, the motor turns more than one time during the simulation. M250-35A steel and NdFe magnet specifications are given in Table 2. The values are defined in 3D-FEA. Additionally, the cylindrical coordinate system is used to define the axial flux steel orientation.

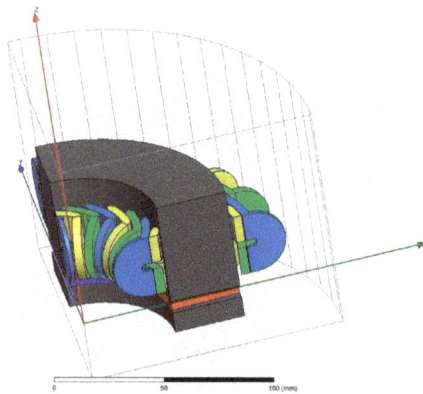

Figure 7. $\frac{1}{4}$ part of the simulated AFPM motor designs.

Before starting the comparison of five designs, the optimum shifting angles of the suggested rotor poles must be specified. The shifting angle means that inner rotor step magnets are displaced by an angle from the outer rotor magnets, as seen in Figure 6c,d. One of the aims of this shifting method is the mitigation of the torque ripples. There are some analytical methods to define the best shift angle in the literature. One of them is the cogging torque period method that is described in [11], but this method does not give the best results for the AFPM machines. In this research, parametric analysis with 3-D FEA is used to find the optimum shifting angle.

The shifting angle is defined as a variable and differs from 0° to 14° by 1° steps. The third magnet design is used to perform this analysis. The average torque and the torque ripple values are taken into account for each result in order to mitigate the total torque ripples. Figure 8 demonstrates the results of the parametric analysis. Table 3 shows each peak-to-peak torque ripple and the average torque value for each shifting angle.

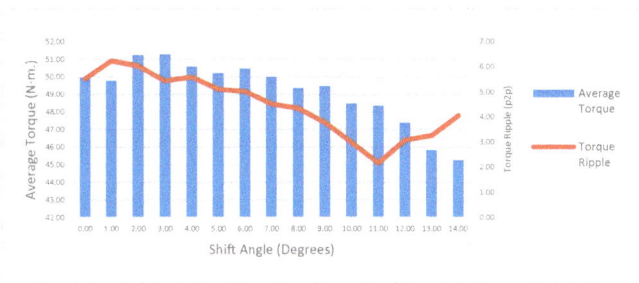

Figure 8. Defining the shifting angle by 3D-FEA parametric analysis.

The simulations gave interesting results from the total parametric analysis. If the torque ripple is the most important anchor of the application, the best result is the 11° shifting angle which gives 2.16 N·m. of peak-to-peak cogging torque. However, if the average torque value is the most valued parameter, the 3° shifting angle has the highest average torque of 51.27 N·m., which is 1.3 N·m. higher than the 0° shifting angle. Torque ripple drops from a shifting angle of 1° to 11°, but after 11° it starts to rise again.

Table 3. The results of the 3D-FEA parametric analysis of the shifting angle.

Shift Angle (Degree)	Average Torque (N·m.)	Torque Ripple (p2p) (N·m.)
0	49.95	5.52
1	49.77	6.23
2	51.24	6.04
3	51.27	5.44
4	50.58	5.59
5	50.21	5.11
6	50.46	5.02
7	50.00	4.51
8	49.35	4.34
9	49.47	3.77
10	48.46	2.98
11	**48.32**	**2.16**
12	47.37	3.06
13	45.81	3.25
14	45.25	4.04

After defining the shifting angle of the third and fourth designs, the magnetic simulations are completed for each design in both static and dynamic conditions. The stator was split into four identical parts and one of them was investigated due to the symmetrical geometry to reduce the simulation time.

5. Comparison of the Results of the Proposed Designs

In the 3D-FEA analysis, PMs have an 11° shifting angle in designs III and IV due to the seeking of the lowest torque ripple. Torque and back EMF waveforms are taken from the dynamic simulations. Figure 9 illustrates the electromagnetic torque results of the five designs that are shown in Figure 6. As seen from the torque results, the lowest torque ripple is in the third design, with 62.4% mitigation, despite a 4.3% reduction on the average torque compared to the design I. The table of the comparison is demonstrated in Table 4. Although it has step and shift on the magnets, design IV has some of the worst data in the view of the torque ripple in this study. This is because of the magnet edges. Since some arrays are sinusoidal, some arrays are sharp. Thus, magnetic flux distribution is unsteady. Additionally, design V has a 43.2% reduction in torque ripple with a 0.8722 magnet fill factor (pole-arc ratio), as validated by Aydin and Gulec, although with some different characteristics of the simulated motors, like magnet thickness, air gap, and the dimensions in [23]. As seen from the simulations, the skewing process has a lower reduction effect than the stepping and shifting process effect on the torque ripple, as demonstrated in Table 4.

Back EMF depends directly on the speed, number of turns, pole numbers, inner and outer radii, and the magnetic flux density, as given in the Equation (13). All of these parameters are stationary without the magnetic flux density in this study. Magnetic flux density depends on the permanent magnet magnetic flux. The average values of the parameters are given in Table 5. Analytical results are obtained by calculating the MAGEC Equations (12), (13), and (16) developed earlier. The constant values are given in Table 6. Figure 10 illustrates the back electromotive force waveforms of each design. The smoothness of these waveforms is crucial to have more constant torque. That means lower torque ripple. Hence, design III has smoother back EMF and lower torque ripple waveforms.

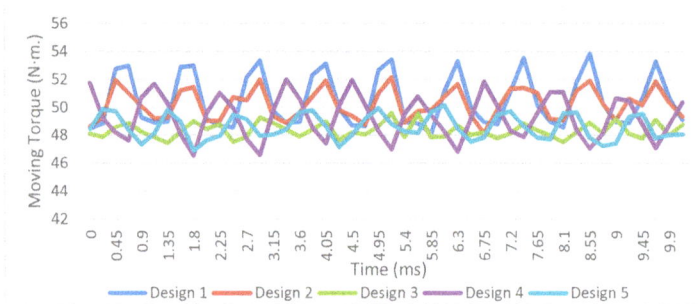

Figure 9. Electromagnetic torque results.

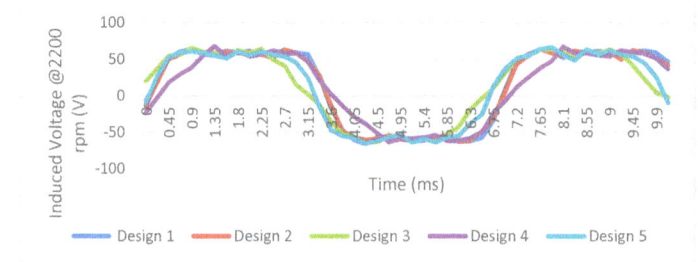

Figure 10. Back EMF results.

The flux density distributions are given in Figure 11 for five designs which demonstrate the radial components of the flux density between the magnets and the stator steel. Since the simulations are conducted for $\frac{1}{4}$ of the motor, the waveforms are produced in 90 degrees. The simulation has interesting results that have caused by the proposed rotor pole designs. The rotor permanent magnet flux does not drop under 0.2 T in designs III and IV, unlike designs I, II, and V. The geometry of the proposition provides these conclusions. Figure 12 shows the magnetic flux densities on the surfaces of the AFPM motor. The first and fifth designs have too high a magnetic flux leakage between the magnets, as seen in Figure 12. The high magnetic flux causes the saturation of the iron. Saturation is an undesired situation which may cause heat and unsteady inductance. Stepping and shifting of magnets allow resistance to the leakage flux. Designs I and V have strong magnetic saturation between the adjacent magnets due to the magnet shapes. Constant width and straight edges of the adjacent magnets ease the magnetic flux leakage. Due to the sinusoidal shape, design II has a lower magnetic saturation than designs I and V. Figure 13 shows the prototype of the AFPM machine.

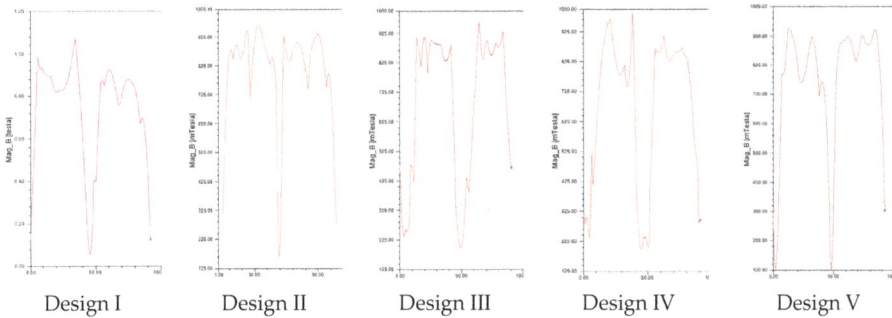

| Design I | Design II | Design III | Design IV | Design V |

Figure 11. Flux density distributions for $\frac{1}{4}$ machine (for 90 mechanical degrees).

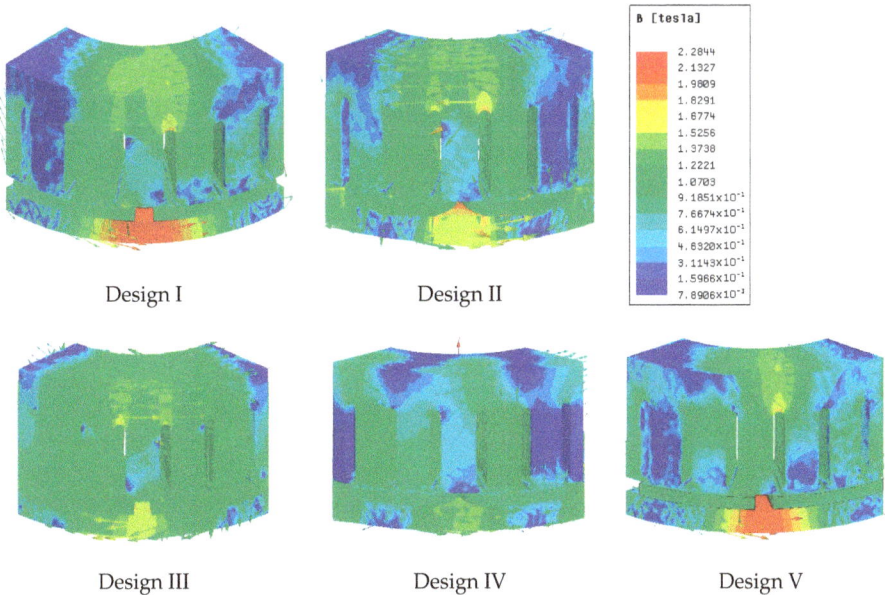

Design I   Design II

Design III   Design IV   Design V

Figure 12. Magnetic flux densities on the surfaces.

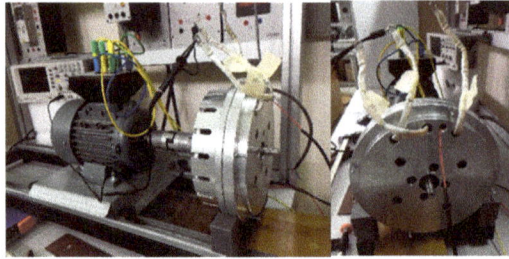

Figure 13. The prototype machine.

Table 4. The total comparison of the 3D-FEA results.

	Design I	Design II	Design III	Design IV	Design V
Average Torque (AT)	50.521	50.25	48.32	49.28	48.59
Torque Ripple (TR)	5.744	3.935	**2.159**	5.456	3.264
Rate (TR/AT)	0.114	0.078	**0.045**	0.111	0.067
AT Reduction	ref.	−%0.1	**−%4.3**	−%2.4	−%3.8
TR Reduction	ref.	−%31.5	**−%62.4**	−%5.1	−%43.2

Table 5. The average values of parameters by means of the MAGEC and 3D-FEA.

Average Values of	Simulation	Design I	Design II	Design III	Design IV	Design V
Magnetic Flux	MAGEC	0.7	0.69	0.62	0.59	0.61
Density, B_g (T)	3D-FEA	0.72	0.7	0.63	0.60	0.63
Back EMF (V)	MAGEC	17.24	16.58	14.97	14.24	14.23
	3D-FEA	17.643	16.878	15.1502	14.358	14.866
Torque (N·m.)	MAGEC	49.117	49.532	47.553	48.458	47.047
	3D-FEA	50.521	50.25	48.32	49.28	48.59

Table 6. Some of constant values of the AFPM motor counted by MAGEC.

Values of the Constants	Design I	Design II	Design III	Design IV	Design V
$K_{k\varphi}$	0.8386	0.8386	0.8386	0.8386	0.8386
K_{pml}	2.09	2.2	3.04	3.46	3.17
K_c	1.04	1.04	1.04	1.04	1.04
C_p	5.96	5.96	5.96	5.96	5.96
λ	0.533	0.533	0.533	0.533	0.533

6. Conclusions

Different rotor pole designs are investigated in this study by means of the MAGEC and FEA analyses. The MAGEC gives the understanding of the single air gap AFPM machine and FEA analyzes the characteristics of the AFPM machine. The MAGEC describes the infrastructure of the AFPM machine characteristics that are obtained from the 3D-FEA. The magnetic flux paths are illustrated by the MAGEC in Figures 2–5. Table 4 compares the results of the electromagnetic torque and torque ripples for all magnet shapes. TR/AT values prove that design III has the lowest rate and, hence, an average torque reduction of 4.3%. Transient analysis is performed by 3D-FEA for 10 milliseconds. Each design is discussed in contrast to the simulation results. Additionally, a parametric analysis is fulfilled to determine the best solution for the shifting angle. The electromagnetic torque and the back EMF waveforms are demonstrated in Figures 9 and 10, which are obtained from the transient analysis. Table 5 is demonstrated to prove the methods. The MAGEC and 3D-FEA results are compared in

Table 5 in terms of air gap magnetic flux density, back EMF, and torque characteristics. The 3D-FEA and MAGEC results validate each other.

Furthermore, the magnetic flux density distribution waveforms are given in Figure 11 and the surface magnetic flux density profiles are given in Figure 12. The air gap magnetic flux density is not collapsed under 0.2 T by the permanent magnets in designs III and IV, unlike design numbers I, II, and V. However, Figure 12 gives information for the saturation of the irons. Design I and V have strong flux leakage between the adjacent magnets, but the saturation points are mostly in the rotor iron, hence, the results are not affected much at the 3D-FEA simulation time as given in the Table 5. Additionally, the MAGEC results do not contain saturation effects. Thus, the heat effects are neglected in the 3D-FEA results in Table 5. If the permanent magnets are damaged by the heat caused by the saturation, all characteristics in Table 5 could be changed dramatically. Resultantly, the third design has the best results in contrast to the precision on the stability of moving torque. Additionally, the results show that stepping and shifting method has better results compared with the skewing method in the view of torque ripple mitigation. The magnets will be produced privately for designs III and IV. Hence, the costs may be higher for the prototype, but the magnet costs of each design will be the same for mass production since the magnet weights being the same. Moreover, a prototype machine can be seen in Figure 13 which is manufactured in the light of this paper for further studies.

Acknowledgments: This study researched by the supports of the TUBITAK (The Scientific and Technical Research Council of Turkey), and WEMPEC. (Wisconsin Electric Machines and Power Electronics Concorsium).

Author Contributions: E.C. and F.D. conceived and designed the MAGEC and 3D-FEA; E.C. performed the simulations; E.C. and F.D. analyzed the data; E.C. wrote the paper.

Conflicts of Interest: The authors declare no conflict of interest.

Nomenclature

R_s	The reluctance of the stator back iron,
R_r	The reluctance of the rotor back iron,
R_g	The reluctance of the air gap between the stator and the rotor,
R_{pm}	The reluctance of the permanent magnet,
R_{pml}	The reluctance of the air gap between the two permanent magnets,
φ_r	Magnetic flux flows from the rotor pole,
φ_g	Flux flow passed from the air gap into the stator
P_{pm}	The permeance of the permanent magnet
P_{pml}	The permeance of the gap between adjacent magnets
K_c	Carter's coefficient
K_f	The correction factor of the air gap magnetic flux density in radial direction
K_d	Flux leakage coefficient
K_{pml}	Leakage coefficient between the magnets
$K_{k\varphi}$	Flux density coefficient
d_f	The distance between adjacent magnets
N_{pm}	Number of the magnets
A_p	Area of a pole
C_p	Permeance factor
N_w	Number of turns
J_{in}	Current density
N_s	Slot number

References

1. Mignot, R.B.; Dubas, F.; Espanet, C.; Chamagne, D. Design of Axial Flux PM Motor for Electric Vehicle via a Magnetic Equivalent Circuit. In Proceedings of the First International Conference on REVET-2012 Renewable Energies and Vehicular Technology, Hammamet, Tunisia, 26–28 March 2012; pp. 212–217.
2. Kierstead, H.; Wang, R.; Kamper, M. Design optimization of a single sided axial flux permanent magnet in-wheel motor with non-overlap concentrated winding. In Proceedings of the 18th Southern African Universities Power Engineering Conference, Stellenbosch, South Africa, 28–29 January 2009; pp. 36–40.
3. Fei, W.; Luk, P.; Jinupun, K. A new axial flux permanent magnet segmented-armature-torus machine for in-wheel direct drive applications. In Proceedings of the Power Electronics Specialists Conference, Rhodes, Greece, 15–19 June 2008; pp. 2197–2202.
4. Carichi, F.; Capponi, F.G.; Crescimbini, F.; Solero, L. Experimental study on reducing cogging torque and no-load power loss in axial-flux permanent-magnet machines with slotted winding. *IEEE Trans. Ind. Appl.* **2004**, *40*, 1066–1075. [CrossRef]
5. Seo, J.M.; Rhyu, S.; Kim, J.; Choi, J.; Jung, I. Design of Axial Flux Permanent Magnet Brushless DC Motor for Robot Joint Module. In Proceedings of the IEEE International Power Electronics Conference, Sapporo, Japan, 21–24 June 2010; pp. 1336–1340.
6. Parviainen, A.; Pyrhönen, J.; Kontkanen, P. Axial flux permanent magnet generator with concentrated winding for small wind power applications. In Proceedings of the IEEE International Conference on Electric Machines and Drives, San Antonio, TX, USA, 15 May 2005; pp. 1187–1191.
7. Di Gerlando, A.; Foglia, G.; Iacchetti, M.F.; Perini, R. Axial flux pm machines with concentrated armature windings: Design analysis and test validation of wind energy generators. *IEEE Trans. Ind. Electron.* **2011**, *58*, 3795–3805. [CrossRef]
8. De, S.; Rajne, M.; Poosapati, S.; Patel, C.; Gopakumar, K. Low inductance axial flux BLDC motor drive for more electric aircraft. *IET Power Electron.* **2012**, *5*, 124–133. [CrossRef]
9. Jahns, T.M.; Soong, W.L. Pulsating torque minimization techniques for permanent magnet ac motor drives—A review. *IEEE Trans. Ind. Electron.* **1996**, *43*, 321–330. [CrossRef]
10. Saavedra, H.; Urresty, J.-C.; Riba, J.-R.; Romeral, L. Detection of interturn faults in PMSMs with different winding configurations. *Energy Convers. Manag.* **2014**, *79*, 534–542. [CrossRef]
11. Parviainen, A.; Niemela, M.; Pyrhonen, J. Modeling of axial flux permanent-magnet machines. *IEEE Trans. Ind. Appl.* **2004**, *40*, 1333–1340. [CrossRef]
12. Tiegna, H.; Bellara, A.; Amara, Y.; Barakat, G. Analytical modeling of the open-circuit magnetic field in axial flux permanent-magnet machines with semi-closed slots. *IEEE Trans. Magn.* **2012**, *48*, 1212–1226. [CrossRef]
13. Bellara, A.; Amara, Y.; Barakat, G.; Dakyo, B. Two-dimensional exact analytical solution of armature reaction field in slotted surface mounted pm radial flux synchronous machines. *IEEE Trans. Magn.* **2009**, *45*, 4534–4538. [CrossRef]
14. Lubin, T.; Mezani, S.; Rezzoug, A. 2-d exact analytical model for surface-mounted permanent-magnet motors with semi-closed slots. *IEEE Trans. Magn.* **2011**, *47*, 479–492. [CrossRef]
15. Wang, R.J.; Kamper, M.J.; Westhuizen, K.V.; Gieras, J.F. Optimal Design of a Coreless stator Axial Flux Permanent Magnet Generator. *IEEE Trans. Magn.* **2005**, *41*, 55–64. [CrossRef]
16. Aydin, M.; Zhu, Z.Q.; Lipo, T.A.; Howe, D. Minimization of Cogging Torque in Axial-Flux Permanent-Magnet Machines: Design Concepts. *IEEE Trans. Magn.* **2007**, *43*, 3614–3622. [CrossRef]
17. Sung, S.J.; Park, S.J.; Jang, G.H. Cogging torque of brushless DC motors due to the interaction between the uneven magnetization of a permanent magnet and teeth curvature. *IEEE Trans. Magn.* **2011**, *47*, 1923–1928. [CrossRef]
18. Saavedra, H.; Riba, J.-R.; Romeral, L. Magnet shape influence on the performance of AFPMM with demagnetization. In Proceedings of the 39th Annual Conference of the IEEE Industrial Electronics Society, Vienna, Austria, 10–13 November 2013; pp. 973–977.
19. Kahourzade, S.; Mahmoudi, A.; Ping, H.W.; Uddin, M.N. A Comprehensive Review of Axial-Flux Permanent-Magnet Machines. *Can. J. Electr. Comput. Eng.* **2014**, *37*, 19–33. [CrossRef]
20. Mahmoudi, A.; Kahourzade, S.; Abd Rahim, N.; Hew, W.P. Design, Analysis, and Prototyping of an Axial-Flux Permanent Magnet Motor Based on Genetic Algorithm and Finite-Element Analysis. *IEEE Trans. Magn.* **2013**, *49*, 1479–1492. [CrossRef]

21. Qishan, G.; Hongzhan, G. Effect of Slotting in PM Electric Machines. *Electr. Mach. Power Syst.* **1985**, *10*, 273–284. [CrossRef]
22. Shokri, M.; Rostami, N.; Behjat, V.; Pyrhönen, J.; Rostami, M. Comparison of performance characteristics of axial-flux permanent-magnet synchronous machine with different magnet shapes. *IEEE Trans. Magn.* **2015**, 51. [CrossRef]
23. Aydin, M.; Gulec, M. Reduction of Cogging Torque in Double-Rotor Axial-Flux Permanent-Magnet Disk Motors: A Review of Cost-Effective Magnet-Skewing Techniques With Experimental Verification. *IEEE Trans. Ind. Electron.* **2014**, *61*, 5025–5034. [CrossRef]

electronics

MDPI

Article

Series Active Filter Design Based on Asymmetric Hybrid Modular Multilevel Converter for Traction System

Muhammad Ali [1,*], Muhammad Mansoor Khan [1], Jianming Xu [2], Muhammad Talib Faiz [1], Yaqoob Ali [1], Khurram Hashmi [1] and Houjun Tang [1]

[1] School of Electronics, Information & Electrical Engineering (SEIEE), Smart Grid Research & Development Centre, Shanghai Jiao Tong University, Shanghai 200240, China; mansoor@sjtu.edu.cn (M.M.K.); talib_faiz@sjtu.edu.cn (M.T.F.); yaqoob.ali@uetpeshawar.edu.pk (Y.A.); khurram_hashmi@sjtu.edu.cn (K.H.); hjtang@sjtu.edu.cn (H.T.)

[2] Changzhou Power Supply Company, Changzhou 213176, Jiangsu, China; x_jianming@outlook.com

* Correspondence: engrmak.ee@sjtu.edu.cn; Tel.: +86-131-2215-7127

Received: 30 June 2018; Accepted: 30 July 2018; Published: 1 August 2018

Abstract: This paper presents a comparative analysis of a new topology based on an asymmetric hybrid modular multilevel converter (AHMMC) with recently proposed multilevel converter topologies. The analysis is based on various parameters for medium voltage-high power electric traction system. Among recently proposed topologies, few converters have been analysed through simulation results. In addition, the study investigates AHMMC converter which is a cascade arrangement of H-bridge with five-level cascaded converter module (FCCM) in more detail. The key features of the proposed AHMMC includes: reduced switch losses by minimizing the switching frequency as well as the components count, and improved power factor with minimum harmonic distortion. Extensive simulation results and low voltage laboratory prototype validates the working principle of the proposed converter topology. Furthermore, the paper concludes with the comparison factors evaluation of the discussed converter topologies for medium voltage traction applications.

Keywords: hybrid converter; multi-level converter (MLC); series active filter; power factor correction (PFC)

1. Introduction

An electrified ac railway system, being an energy-efficient and governmentally-friendly medium of mass transportation, achieved great demand in many countries. However, to maximize the traffic, it still requires developing highly efficient, reliable, and compact traction systems with reduced cost, minimum time delay and less vibrations to assure passengers comfort [1]. The efficiency of railway traction systems can be improved with a regenerative braking system which enables transforming the kinetic energy of the rail vehicle in slowing down the speed into electrical power. Using a bidirectional converter, the electric power generated due to the regenerative braking system can be harvested for reuse [2].

A high efficiency and low production cost is required mostly by industrial processes, which can be achieved by increasing the power rating of electrical components/equipment with the reduced installation size. The power can be increased either by developing semiconductor devices with a capability to withstand high voltage or by introducing multilevel converters which will allow to connect the converter system directly to medium-voltage line. Recently, continuous development in power semiconductors devices i.e., high-voltage insulated-gate bipolar transistors (IGBTs) and integrated-gate commutated thyristors (IGCTs) and their use in self-commutated converters increases the nominal voltage and power ratings of the converter system.

Lower medium voltage high power railway traction system multi-level converters (MLC) have significant advantages over classical two-level and three-level converters due to their lower current harmonics distortion, less electromagnetic interference, increased output voltage level, and high power density. Furthermore, due to the simple layout of classical two-level converters, connecting single power semiconductor switch to medium grid voltages directly is not appropriate [3,4]. The key conventional MLC topologies i.e., neutral point clamped (NPC), flying capacitors (FC), and the cascaded H-bridge (CHB) have been reported in [4–8]. The MLC contains numerous power semiconductor switches and a dc-link capacitor which are arranged according to the required output voltage level. However, increasing the number of components accounts for attaining a high voltage level to increase the control complexity, which will affect the efficiency and reliability of the converter [6]. The requirement for MLC includes the dc-link voltage value at reference voltage level, unity power factor on the ac side and low harmonics distortion in the injected current to ac grid [9]. Therefore, an appropriate control scheme is required to track the current signal and maintain the dc-link capacitor voltages of MLC at their respective desired reference. A pulse width modulation (PWM) technique which is used by various MLC drive systems has been investigated in [10,11].

Recently, arrangement of different conventional MLC topologies known as hybrid multilevel converters has been introduced in [12,13]. In medium and high voltage range application, numerous multilevel topologies share the market for industrial applications. Hybrid MLC topologies have been a focus of interest due to substantial advantages which include, a redundant converter design, wide operating range, minimum line harmonics, and improved power factor. Some of these arrangements exist in the literature, such as NPC-H module [14], FC-H module [15], NPC-FC module [16] and H-cascaded module [17]. Furthermore, unequal dc link voltages are used which minimizes the redundant switching states and increases the output voltage level of the converter. Such converters are known as asymmetric converters in the literature [17,18].

The non-linearity of the converter injects harmonic currents in the ac grid, which adversely affects stability and the power quality of the ac grid [19]. Various techniques are available to mitigate high harmonics; a modulation scheme based on selected harmonic elimination pulse width modulation is studied in [12]. Various passive filters are introduced in [18,20,21] to overcome the harmonics issues in converters, but due to its system parameters, the performance is limited and may cause a resonance problem [22]. A compensator based on an electronics converter known as series active power filter (APF) converter is studied in the literature [23–25] to tackle the limit imposed by the high harmonics. Moreover, a converter based compensator in the existing research needs improvement in a supply system for the electrified traction system.

This paper presents a comparative study of the recently proposed MLC converter for the medium voltage-high power traction system which is examined through simulation results. Among all converters, a new topology based on an asymmetric hybrid modular multilevel converter (AHMMC), which is a series arrangement of a classical H-bridge and FCCM voltage source PWM converters, is investigated in more detail. The design of AHMMC topology, its corrective current and its voltage control methods are demonstrated. The overall system's controllability, total harmonic distortion (voltage and current), power factor and voltage stability on the dc-link capacitors is analysed through extensive simulation results. The proposed topology is practically validated through low-power laboratory prototype.

The paper is organized as follows. Section 2 includes a comparative analysis of five-level converter topologies and their output results are examined. Section 3 first investigates the proposed converter model, afterwards, its modulation strategy and control scheme are discussed. The simulation and experimental results of the proposed converter are presented and discussed in Section 4. An evaluation of comparison factors of the discussed converter topologies are examined in Section 5. Finally, the study is concluded in Section 6.

2. Comparative Analysis of 5-Level Converter Topologies

A comparative analysis of a few MLC converter arrangements i.e., NPC, hybrid NPC with H-bridge module, CHB based NPC, and HNPC with a cascaded module for lower-medium voltage application are discussed in this section.

2.1. NPC Converter

NPC converters are most widely used in industrial application from the last two decades among the other high power converters [26,27] in the range of 2.3 kV, 4.16 kV and 6 KV applications [28]. Figure 1 shows a NPC voltage source converter (VSC), is the most commonly used topology for self-commutated medium voltage converters (MVCs). The distinct features of NPC VSC ensures a better output voltage quality compared to the conventional two-level converter.

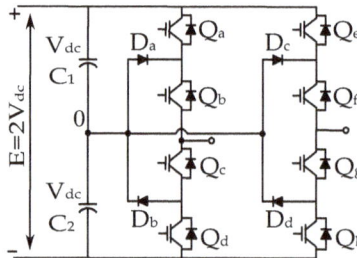

Figure 1. Five-level NPC converter topology.

The NPC converter circuit shown in Figure 1, contains eight power semiconductor switches, four clamping diodes $D_{(a-d)}$, a dc-bus voltage of E = 2Vdc and two dc-link capacitors, each of the dc-link capacitor is charged to half of the dc-bus voltage i.e., Vca = Vcb = $\frac{1}{2}$E. The output voltage of the NPC converter is set to five level $\pm 2V_{dc}$, $\pm V_{dc}$, and 0. The modulation of converter switches can be achieved by three methods: (1) Carrier Based Pulse Width Modulation (CBPWM); (2) Selective Harmonic Elimination (SHE); and (3) Space Vector Modulation (SVM). In this study a CBPWM based modulation is used for a 3L-NPC, for which the maximum current is achieved at the maximum modulation index (m = 1). Based on topological design, consider that the NPC converter half bridge is under voltage stress equal to half of the dc-bus voltage, whereas the half bridge of the H-bridge converter is stressed equal to dc-bus voltages. To correlate the system voltage and required semiconductor voltage in the three-level NPC converter, a 3.3 KV ratings devices are equipped in 2.3 kV converter without series connection of devices. Using 3L-NPC VSC, an output voltage of up to 4.16 kV can be achieved without any series connection of devices [29–31]. Currently, 3L-NPC converters are available in market with a variety of devices. Siemens^{TM}, a European manufacturer, utilizes 3.3 and 6.5 KV IGBT modules devices where higher voltage levels are attained through series connection of semiconductor devices, to cover a range of 2.3 to 7.2 kV converters [32–34]. ABB^{TM} use 4.5 kV IGCTs [35] to offer 3.3 KV converters and Converteam^{TM} recently launched a 3.3 KV NPC product using the press-pack 4.5 kV IGBT technology [36].

2.2. Hybrid NPC with H-Bridge

A hybrid modular converter based on the cascaded arrangement of NPC five level cell (having two three-level NPC leg) with a three-level H-bridge cell and H-NPC module with a H-bridge as shown in Figure 2, are investigated in [37–39] for different voltage ratios of the dc-link among the two power cells. In the study, a converter cell which gives a fundamental component support operates at low frequency, whereas a converter cell as a series active filter operates at high frequency for high harmonics minimization and coupling inductor size reduction.

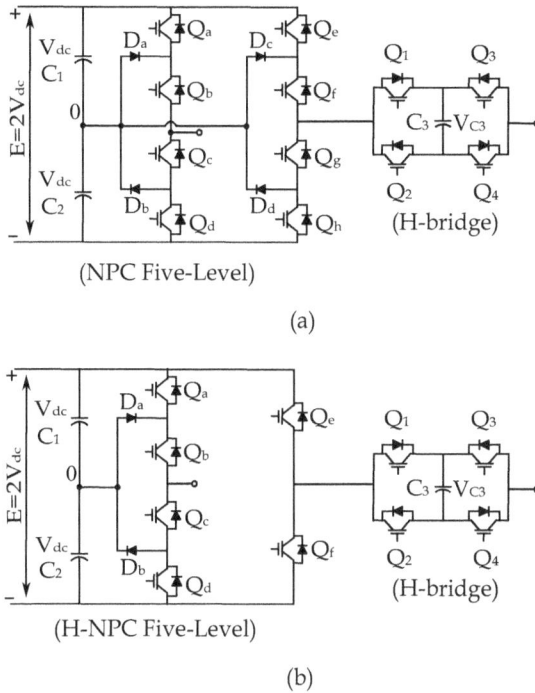

Figure 2. Five level NPC converter with H-bridge (**a**) NPC with H-bridge module, and (**b**) H-NPC with H-bridge module.

The power losses across the semiconductor switches are calculated and grouped by the power cell of each converter. The total power loss(P_t) includes, switching loss (P_{sw}) and conduction loss (P_c). The conduction losses of the H-NPC cell are less due to the minimum number of power semiconductor devices. The switching losses of the harmonic compensation converter will be higher because of the high voltage stress and high average switching frequency.

2.3. CHB Based NPC Converter

As discussed above, the hybrid NPC with H-bridge configuration will cause a high power loss in the voltage stressed power semiconductor switches because of the high average switching frequency. To overcome the said issues, the dc-link voltage is equally distributed by replacing an H-bridge cell with a five-level cascaded H-bridge (CHB) as shown in Figure 3. However, the topology with a five level CHB arrangement requires an increased number of power semiconductor switches which will cause high conduction losses and require a greater number of drive circuits.

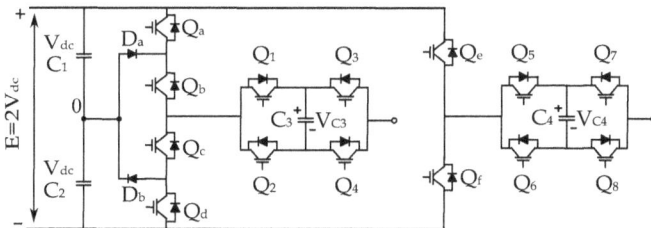

Figure 3. CHB based converter topology.

2.4. H-NPC with Cascaded Converter Module

Figure 4 depicts an arrangement for H-NPC with a cascaded module that is constructed in this study for lower medium voltage application. The topology has the benefit of a lower number of power semiconductor devices in comparison with the CHB based converter topology discussed in Section 2.3. The HNPC with cascaded module will minimise the total switch loss and number of drive circuits.

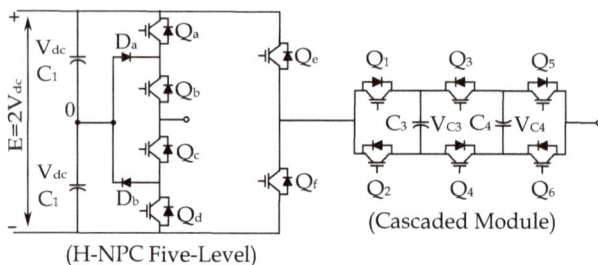

Figure 4. Five-level H-NPC with cascaded converter module.

2.5. Topological Analysis Based on Simulation Results

To verify the converter system performance, a few of these arrangements, including the NPC module, hybrid H-NPC with H-bridge module and H-NPC with cascaded converter module are simulated in Matlab/Simulink environment. A five-level and seven-level converter output voltage and the inductor current is obtained. The output voltage and inductor current of the converters are further analysed by fast Fourier transform (FFT). The simulation parameters of the NPC Converter are listed in Table 1.

Table 1. Simulation Parameters of NPC Converter.

Parameters	Value
Grid Voltage	1750 V (rms)
DC bus Voltage	3000 V
DC link Capacitor C_1, C_2	1500 V
Converter Current	70 Amp (rms)
L	0.6 mH
Modulation Depth	1
$Freq_{pwm}$	2 kHz

The simulation results of an NPC converter are shown in Figure 5. In Figure 5a,b the output voltage with its harmonics spectrum is shown using conventional PWM control technique with a modulation index m = 1. Figure 5c,d shows that the converter current is in phase with the grid voltage, ensuring the rectifier operation mode of the converter.

The simulation parameters of seven-level H-NPC with H-bridge module and H-NPC with cascaded converter module are presented in Table 2. By operating H-NPC with H-bridge module converter at maximum modulation index m = 1.06, a maximum seven-level of output voltage is achieved as shown in Figure 6a. The figure depicts the output voltage of the H-NPC, H-bridge and overall converter, whereas Figure 6b shows its respective harmonics spectrum by using conventional carrier based PWM control technique. Figure 7 shows H-NPC converter with H-bridge module operating in rectifier mode. It can be seen from Figure 7a that the converter current has the same phase with the grid voltage shown in Figure 7b. Figure 8 shows the simulation results of a 7-L HNPC with a cascaded module. Figure 8a depicts the output voltage of the H-NPC converter, cascaded module and overall converter, whereas Figure 8b shows its respective harmonics spectrum by using conventional carrier based PWM control technique with a modulation index m = 1.06. Figure 9 shows the converter operating in rectifier mode. It can be seen from Figure 9a that the converter current has the same phase with the grid voltage shown in Figure 9b.

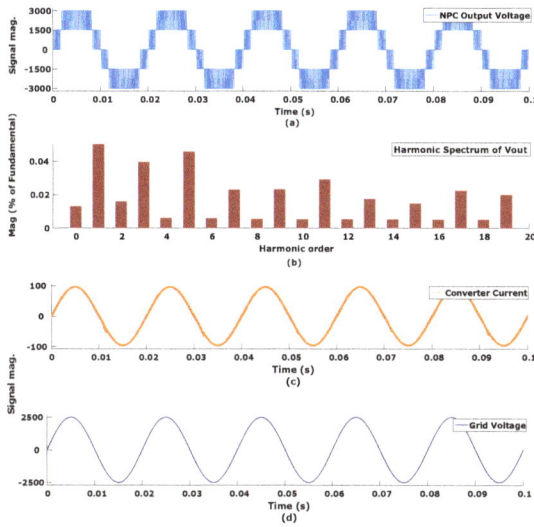

Figure 5. 5-L NPC converter topology (**a**) output voltage, (**b**) harmonic spectrum of output voltage, (**c**) converter current, and (**d**) grid voltage.

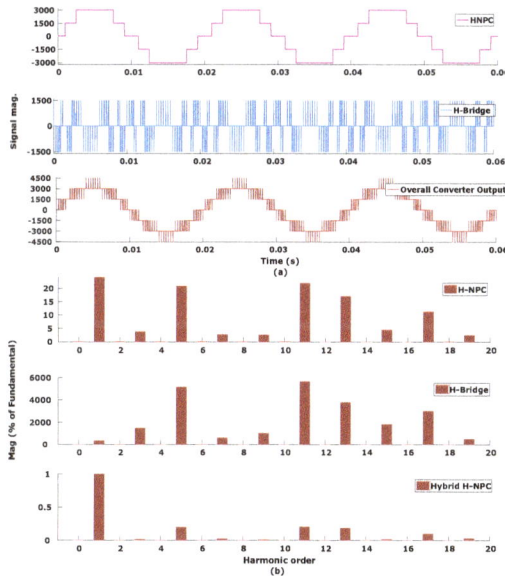

Figure 6. 7-L hybrid H-NPC with H-bridge converter topology (**a**) output voltage, and (**b**) harmonic spectrum of output voltage.

Table 2. Simulation parameters of Hybrid H-NPC converter.

Parameters	Value
Grid Voltage	1750 V (rms)
DC bus Voltage	3000 V
DC link Capacitor C_1, C_2, C_3, C_4	1500 V
Converter Current	70 Amp (rms)
L	0.6 mH
Modulation Depth	1.06
H-NPC converter $freq_{pwm}$	350 Hz
H-Bridge cell $freq_{pwm}$	2 kHz
Cascaded module converter $freq_{pwm}$	2 kHz

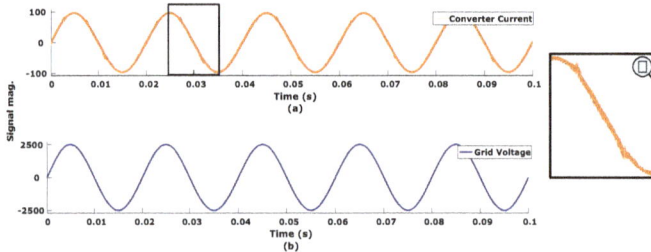

Figure 7. Converter operating in rectifier mode (**a**) converter current, and (**b**) grid voltage.

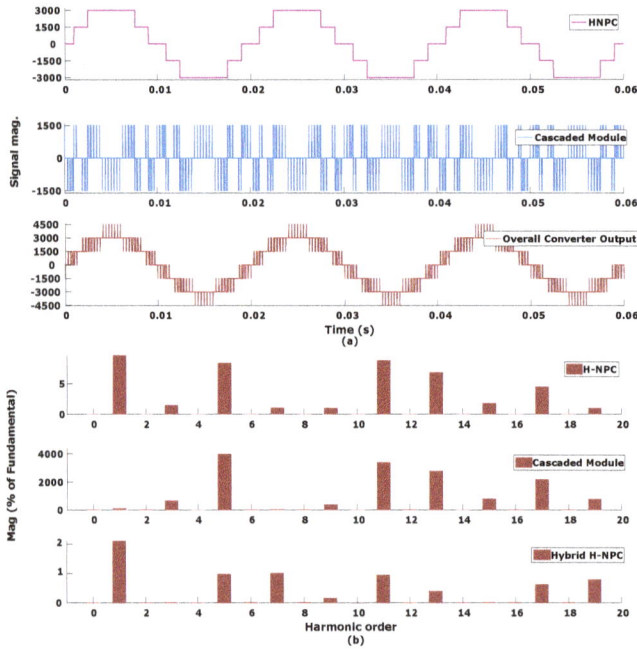

Figure 8. 7-L hybrid H-NPC with a cascaded converter topology (**a**) output voltage, and (**b**) harmonic spectrum of output voltage.

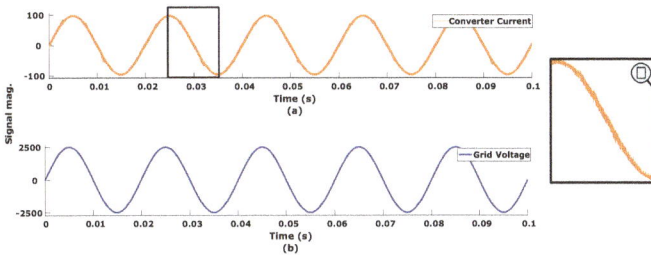

Figure 9. Converter operating in rectifier mode (**a**) converter current, and (**b**) grid voltage.

3. Proposed AHMMC Converter

3.1. Structure and Working Principle

Figure 10 shows the proposed AHMMC topology. The basic circuit configuration contains an FCCM in series with an H-bridge cell. An H-bridge converter contains four power semiconductor switches Q_{a-d}, a dc-bus voltage of $E = 2V_{dc}$, and a single dc-link capacitors C_a charged to the dc-bus voltage. The output voltage of the H-bridge converter $(V_{(AB)})$ is set to three level $\pm V_{dc}$, and 0. The FCCM which is the origin of the AHMMC topology contains six power semiconductor switches Q_{1-6} and two dc-link capacitors V_{c1} and V_{c2}, each of the dc-link is charged to $1/2\,E$ of the dc-bus voltage. The output voltage of FCCM $(V_{(BC)})$ is synthesized which generates 5-level that are $\pm 2V_{dc}, \pm V_{dc}$, and 0. The output voltage of the proposed AHHMC converter can be achieved by Equation (1):

$$V_{(AHMMC)} = V_{(AB)} + V_{(BC)} \tag{1}$$

The mathematical model of the proposed AHMMC grid tied converter can be expressed by the following set of Equations (2)–(4):

$$\frac{di_s}{dt} = \frac{1}{L}(S_x V_{c1} + S_y V_{c2} \mp S_z 2V_{dc} \pm V_g) \tag{2}$$

$$\frac{dV_{c1}}{dt} = \frac{S_x i}{C_1} \tag{3}$$

$$\frac{dV_{c2}}{dt} = \frac{S_y i}{C_2} \tag{4}$$

where $V_g = E\sin(wt)$ is a grid voltage, S_x, S_y and S_z are the switching functions of the AHMMC which can be expressed by the following Equations (5)–(7):

$$S_x = Q_1 Q_4 - Q_2 Q_3 \tag{5}$$

$$S_y = Q_4 Q_5 - Q_3 Q_6 \tag{6}$$

$$S_z = Q_a Q_d - Q_b Q_c \tag{7}$$

The eight distinct operating modes of FCCM are achieved by three complementary pairs of the power switches (Q_2, Q_4, Q_6 are the complement pair of Q_1, Q_3, Q_5, respectively) are listed in Table 3. Figure 11 depicts an output of seven-level $\pm 3V_{dc}, \pm 2V_{dc}, \pm V_{dc}$, and 0 which is achieved by operating the converter modules with the combination of switching states. Table 4 summarises the seven-level output voltage of the AHMMC converter.

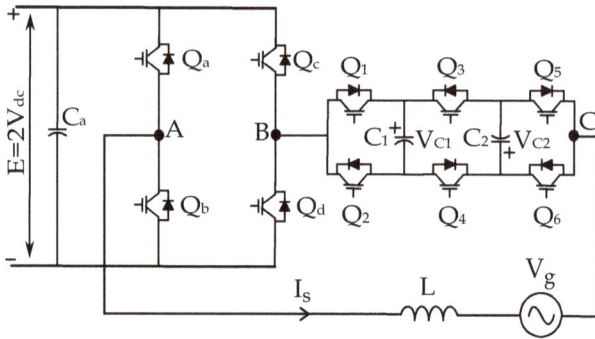

Figure 10. Proposed AHMMC converter topology.

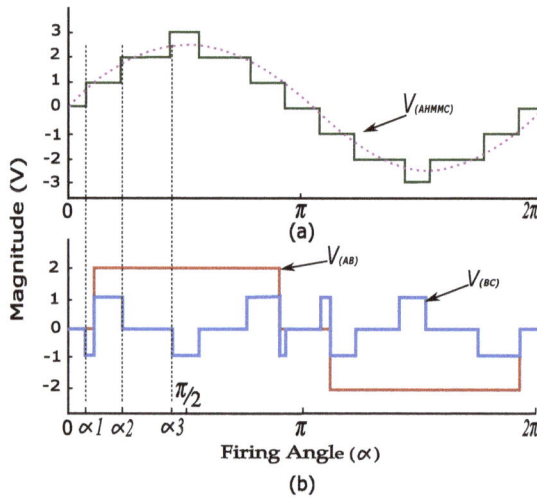

Figure 11. AHMMC operation mode (**a**) AHMM output voltage, and (**b**) H-bridge output voltage $(V_{(AB)})$ and FCCM output voltage $(V_{(BC)})$.

Table 3. Switching modes of FCCM module.

Modes	Switching States			Capacitor	Capacitor	FCCM Output Voltage
-	Q_1	Q_4	Q_5	C_1	C_2	$V_{(BC)}$
				Table (a) when $i_s > 0$		
1	1	1	1	discharge	discharge	$2V_{dc}$
2	1	1	0	discharge	by pass	V_{dc}
3	0	1	0	by pass	by pass	0
4	1	0	0	by pass	charge	$-V_{dc}$
5	0	0	0	charge	charge	$-2V_{dc}$
				Table (b) when $i_s > 0$		
1	1	1	1	discharge	discharge	$2V_{dc}$
2	0	1	1	by pass	discharge	V_{dc}
3	1	0	1	by pass	by pass	0
4	0	0	1	charge	by pass	$-V_{dc}$
5	0	0	0	charge	charge	$-2V_{dc}$

Table 4. AHMMC Converter Output Voltage.

Angle	$V_{(AB)}$	$V_{(BC)}$	$V_{(AHMMC)} = (V_{(AB)}) + (-V_{(BC)})$
$0 \leq \theta \leq \alpha_1$	0	0	0
$\alpha_1 \leq \theta \leq \alpha_2$	0	$-V_{dc}$	V_{dc}
$\alpha_1 \leq \theta \leq \alpha_2$	$2V_{dc}$	V_{dc}	V_{dc}
$\alpha_2 \leq \theta \leq \alpha_3$	$2V_{dc}$	0	$2V_{dc}$
$\alpha_3 \leq \theta \leq \pi$	$2V_{dc}$	$-V_{dc}$	$3V_{dc}$

3.2. Modulation Technique

A phase opposition disposition (POD) multi-carriers modulation strategy is implemented as a PWM technique for the proposed converter. In order to minimize the switching loss, the converter switches are categorised in two sets (i) high frequency switching of low voltage switches; (ii) low frequency switching of high voltage switches. Design of the heat dissipation of the system and switch type are the parameter for optimal selection of high frequency switches. In the proposed AHMMC converter, the switches Q_a and Q_c are under high voltage stress and therefore, they are operated with the fundamental switching frequency to minimise the switching loss. The voltage stress on the switches of the FCCM converter is not symmetric. Within the FCCM, the middle leg switches Q_3 and Q_4 are under high voltage stress and therefore, for the design consideration the PWM frequency of these switches is restricted below 1 kHz which will reduce the switching loss, whereas the PWM frequency for the outer leg switches Q_2 and Q_6 are set to 2 kHz (1 kHz effective switching frequency) due to the permissible operating range of high voltage devices up-to 2 kV. Moreover, the modulation scheme studied in this work has the benefit of a wide modulation index range. Figure 12 shows the graph of modulation index versus dc-link voltage value for series active filter (FCCM) at various operating angle.

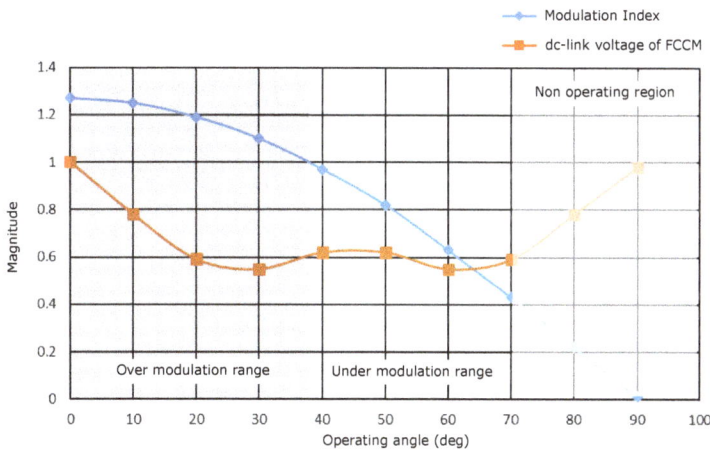

Figure 12. Modulation index versus dc-link voltage of FCCM.

3.2.1. Over Modulation Range (1~1.27)

In the proposed AHMMC converter, the H-bridge cell will provide a fundamental support whereas the FCCM module will cancel harmonics produced by the H-bridge cell. The overall output voltage level of the proposed AHMMC converter depends on the modulation index (m) which is expressed by Equation (8). Where θ is the initial phase angle of the h-bridge cell which is calculated using Equation (9).

$$m = \frac{4}{\pi}\cos(\theta) \tag{8}$$

$$\theta = arccos\frac{\pi}{4}\frac{V_g}{V_{dc}} \tag{9}$$

At $\theta = 0$, a 27% maximum modulation index is achieved which results in the higher output voltage of the proposed converter. Figure 12 shows the over modulation range along with the dc-link voltage required for harmonic compensation. When the theta lies between $(0 \leq \theta < 38)$ degree, maximum seven-level output voltage is achieved which will lower the current value for the same power rating of the converter.

3.2.2. Switch Stress and Modulation Range (0.4~1)

Operating the proposed converter in the modulation range of m = 0.4~1, a five level converter output voltage is achieved with approximately half of the dc-bus voltage required for series active filter (FCCM). This will lower the voltage stress on the semiconductor switches. The voltage stress on the middle leg of the FCCM will be $0.6 E$, whereas, voltage stresses on the upper legs will be $0.3E$. Therefore, the switching frequency can be varied according to the system design.

3.3. Switch Losses

The switch losses were divided into the conduction and switching power losses. The total power switch losses P_T are grouped by power converters. The proposed AHMMC topology studied in this paper consists of 10 power switches with anti-parallel diodes. The conduction loss of the switch depends on the power conduction through the switch due to direction of the current. The calculation of total conduction losses P_{cd} at any instant across the power switch (p_{sw}) and diode (d) can be given as Equations (10)–(12):

$$P_{c\_m(p_{sw})} = \frac{1}{2\pi}\int_0^{2\pi}\left[(m_{p_{sw}} \times v_{p_{sw}})i_s \cdot Q_l\right]d(\omega t) \tag{10}$$

$$P_{c\_m(d)} = \frac{1}{2\pi}\int_0^{2\pi}\left[(m_d \times v_d)i_s \cdot Q_l\right]d(\omega t) \tag{11}$$

$$P_{cd} = P_{c\_m(p_{sw})} + P_{c\_m(d)} \tag{12}$$

where $m_{p_{sw}}$ and m_d are switch and diode count during conducting period, $v_{p_{sw}}$ and v_d are voltage drops of the power switch and diode in conduction state, respectively, i_s is the conduction current of the power switch device, and Q_l is the switching command of the power switch.

The switching losses of active switch can be determined every *ON* and *OFF* state during a reference period. The overall switching losses can be calculated by Equations (13) and (14):

$$P_{sw} = P_{sw\_on} + P_{sw\_off} \tag{13}$$

$$= \sum_{j=1}^{2n+2}\left[\frac{1}{6}v_{b(j)} \cdot i \cdot (t_{on} + t_{off})f_j\right] \tag{14}$$

where, v_b, i and f_j is the blocking voltage, a conduction current, and the switching frequency of the active switch, respectively.

The total power switch losses across the converter can be computed as:

$$P_T = P_{cd} + P_{sw} \tag{15}$$

3.4. Control Scheme

For a stable operation of the proposed converter, corrective current and voltage control techniques shown in Figure 13 are implemented to track the reference current and reference voltage to ensure the unity power factor and balanced dc-link voltage, respectively. An inner current control loop is implemented at the FCCM module; since the operating frequency of this converter is high it will respond quickly to any change.

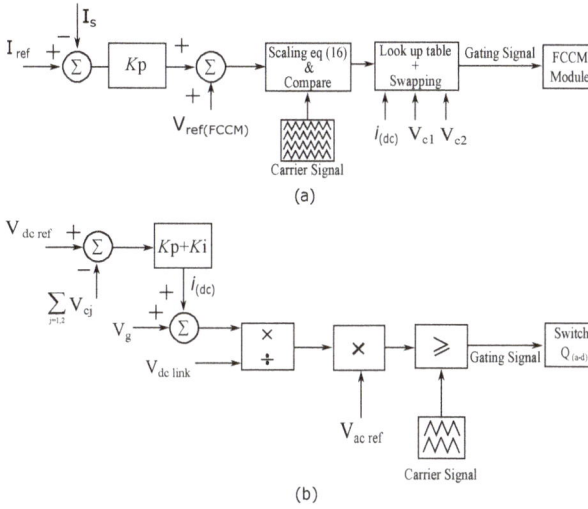

Figure 13. Corrective control scheme (**a**) current control loop, and (**b**) voltage control loop.

Figure 13a, shows a current control loop, in which an error signal is achieved by comparison of measured current I_s and I_{ref} which is further compensated by a proportional gain K_p and then added in the modulating signal for the FCCM converter. The dc link voltage variation impact in the current is minimized through dc voltage feed forward loop control by adjusting the PWM accordingly as described in Equation (16):

$$d = \left(p_1 \frac{|V_{out}|}{V_{c_1}} + p_2 \frac{|V_{out}|}{V_{c_2}} + p_3 \frac{|V_{out}| - V_{c_1}}{V_{c_2}} + p_4 \frac{|V_{out}| - V_{c_2}}{V_{c_1}}\right) \times Sign(V_{out}) \qquad (16)$$

where d and $p_{(1-4)}$ is the duty cycle and PWM output conditions for FCCM, respectively. Table 5 presents PWM output conditions, whereas the direction of PWM output is determined by $Sign(V_{out})$ which is expressed by Equation (17):

$$Sign(V_{out}) = \begin{cases} 1, if\ V_{out} \geq 0 \\ -1, if\ V_{out} < 0 \end{cases} \qquad (17)$$

Figure 13b shows the voltage control loop implemented for the proposed converter. In order to attain the desired output voltage level of the converter, The PWM allows to conduct through one node of the capacitor, which leads to the unbalancing of the capacitor V_{c1} and V_{c2}. Moreover, some of the power dissipates across the switches of the FCCM module which is fed by dc-link capacitors; therefore, a voltage control is needed to maintain balanced voltage on the dc-link. The voltage control loop is categorised into two parts (i) common mode voltage control (ii) voltage balancing among the capacitors. The common mode voltage is controlled by modifying the operating angle of the fundamental component of the H-bridge converter. For a stable operation of the converter, the difference among the dc-link capacitor voltages needs to be balanced, which is achieved by the swapping technique, using the redundant switching states given in Table 3. It is presented in Table 3 that the voltage level $\pm V$ and 0 can be achieved in multiple ways with their respective states of charging and discharging of capacitor. The charge swapping using redundant switching states is based on the following set of rules.

- When $(i_{(s)} \times V_{ref}) < 0$, if $V_{C1} < V_{C2}$, then Table 3b will be selected.
- When $(i_{(s)} \times V_{ref}) > 0$, if $V_{C1} > V_{C2}$, then Table 3a will be selected.

where $i_{(dc)}$ is a direction of the dc reference current. Using the above relation of selecting tables, the balancing of capacitor voltage is achieved.

Table 5. PWM operating conditions.

Operating Condition	p_1	p_2	p_3	p_4
V_{c1} is a lower capacitor				
$\mid V_{out} \mid \leq V_{c1}$	1	0	0	0
$V_{c1} < \mid V_{out} \mid < (V_{c1} + V_{c2})$	0	0	1	0
V_{c2} is a lower capacitor				
$\mid V_{out} \mid \leq V_{c2}$	0	1	0	0
$V_{c2} < \mid V_{out} \mid < (V_{c1} + V_{c2})$	0	0	0	1

4. Simulation and Experimental Results

The AHMMC converter is simulated in MATLAB/Simulink environment to demonstrate the system's effectiveness, and performance of the modulation and control scheme. The AHMMC converter is designed for lower medium voltage application, hence efficiency greater then 97% is expected. However, an experimental validation is performed at low voltage laboratory prototype in which the proposed converter efficiency is determined by conduction losses of the semiconductor switches, which is not a very good indicator of real power losses appearing on the system, operating at a voltage range greater than 1 KV. To verify the system results, simulation results are scaled accordingly to a low voltage laboratory prototype using the system's parameters presented in Table 6, which are the same for both simulation and experimental models. The selection of all component values has been done by extensive simulations. The criteria for capacitors selection are based on the voltage ripple and stability consideration of a converter i.e., the capacitor capacity should be high enough to withstand imbalance to a couple of cycles. Through simulation analysis, approximately 2% capacitor ripple is designed, whereas the inductor L value is selected for a current ripple of 0.5% (peak to peak) under full rated current.

Table 6. System parameters of AHMMC.

Parameter	Value
Inductance	0.6 mH
DC-link Capacitor	20 μF
FCCM Switching Frequency	2 kHz
Capacitor Voltage	50 V_{dc}
Grid Voltage	110 (rms)
Current I_s	14 A (rms)
DC-bus Voltage	100 V

Figure 14 depicts the simulation and practical results of the AHMMC converter output voltage at m = 1.26. The three angles per quarter cycle operation gives an output of seven-level. Figure 14a shows simulation results, whereas Figure 14b shows experimental results of the stable converter operation because of balanced capacitor voltages. An active power is transferred by compensating the harmonics produced by H-bridge through series active filter (FCCM).

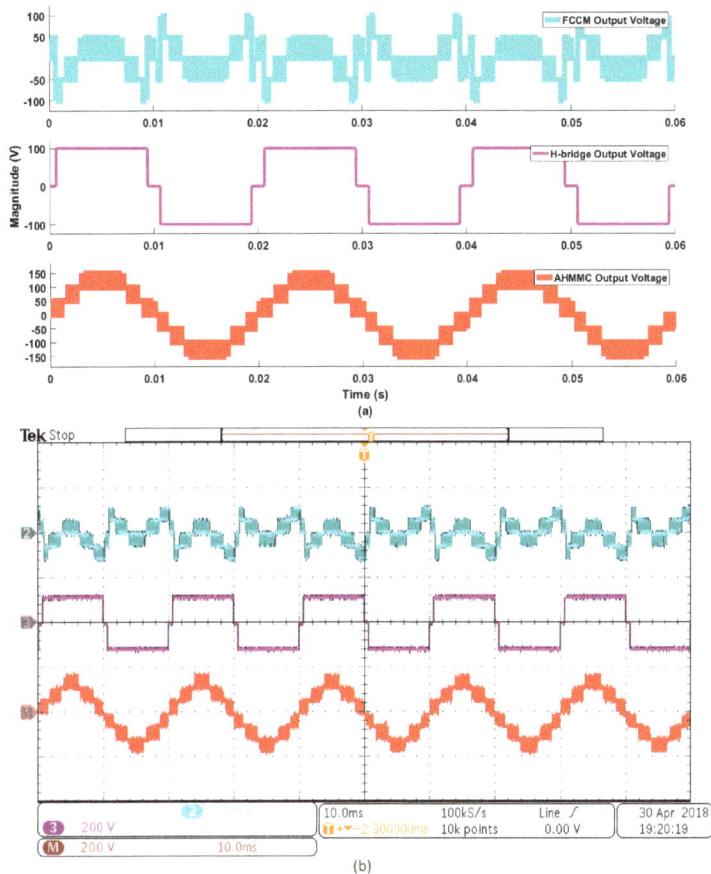

Figure 14. AHMMC converter output at m = 1.26, (**a**) simulation validation, and (**b**) experimental validation, traces channels 2 and 3 (200 V/div).

Figure 15 depicts the simulation and practical results of the AHMMC converter output voltage at m = 0.89. The two angles per quarter cycle operation gives an output of five-level. Figure 15a shows simulation results, whereas Figure 15b shows experimental results of the stable converter operation due to the balanced dc-link capacitor voltages. The situation is similar when an active power is transferred by compensating the harmonics produced by the H-bridge through series active filter (FCCM).

Figure 16 shows simulation and practical validation of the balance operation of the converter with balanced capacitors voltage due to voltage swapping technique by redundant switching state selection presented in Table 3 together with converter current. Figure 17 depicts the simulation and practical results of the grid voltage and converter current. The signals are in-phase which tells that the converter works as a rectifier. Figure 18 depicts the low voltage laboratory prototype of a proposed AHMMC converter. Table 7 presents the simulation and experimental current THD value of the converter operating at different power ratings.

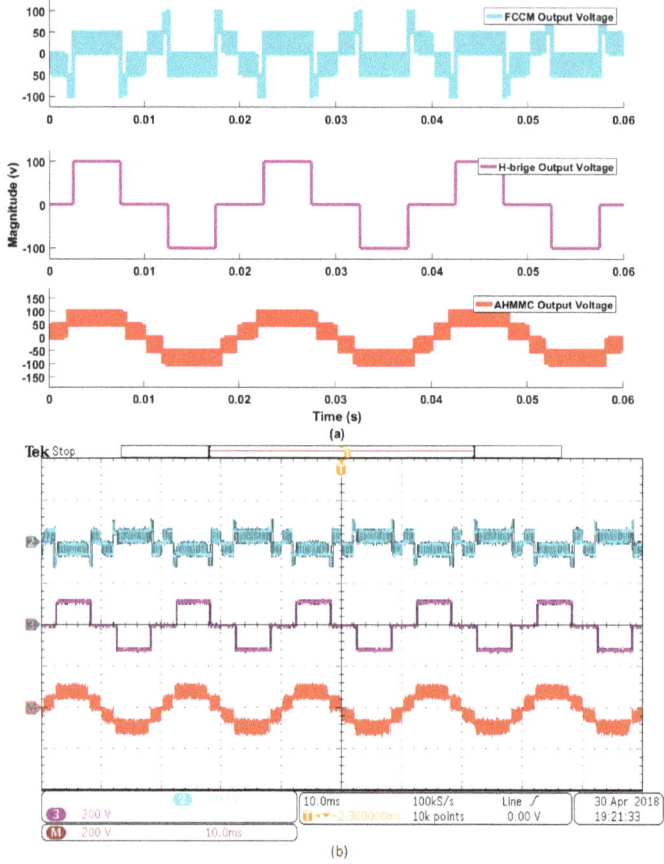

Figure 15. AHMMC converter output at m = 0.89, (**a**) simulation validation, and (**b**) experimental validation, traces chanels 2 and 3 (200 V/div).

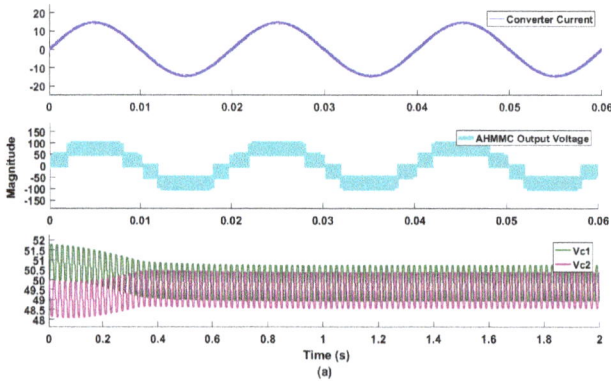

Figure 16. *Cont.*

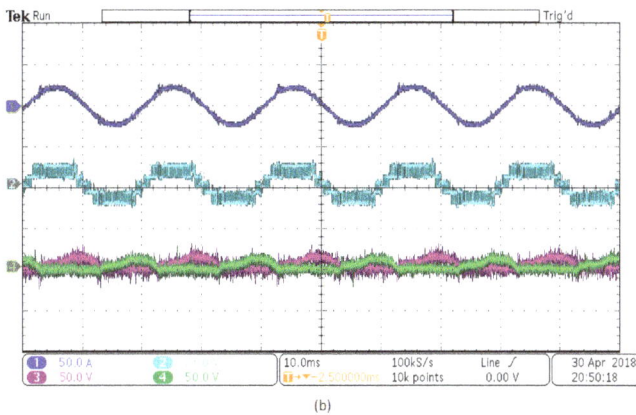

(b)

Figure 16. AHMMC converter output, (**a**) simulation validation of capacitor voltage balancing togather with a unity power factor, and (**b**) experimental validation, traces channels 1, 3, 4 (50 V/div), and 2 (200 V/div).

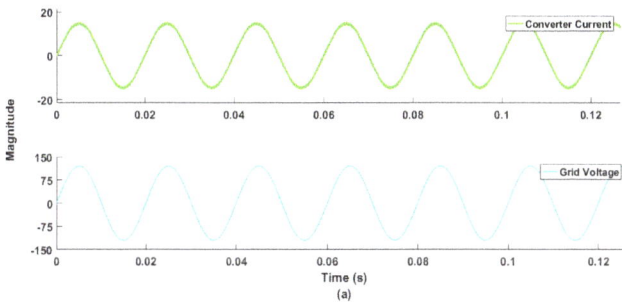

(a)

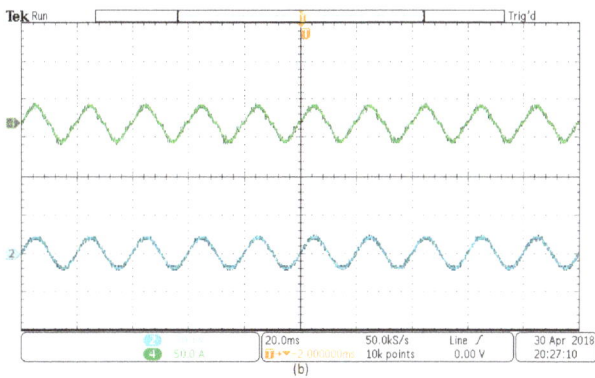

(b)

Figure 17. Low distortion in input current, (**a**) simulation validation of minimum distorion in current injected to ac main, and (**b**) experimental validation of minimum distortion in current injected to ac main. traces channels 2 (200 V/div), and 4 (50 A/div).

Table 7. Current THD of AHMMC operating at differnt power rating.

Current THD	Power Rating		
	25%	50%	100%
Simulation THD	6.55	3.31	1.66
Experiment THD	7.95	4.87	2.98

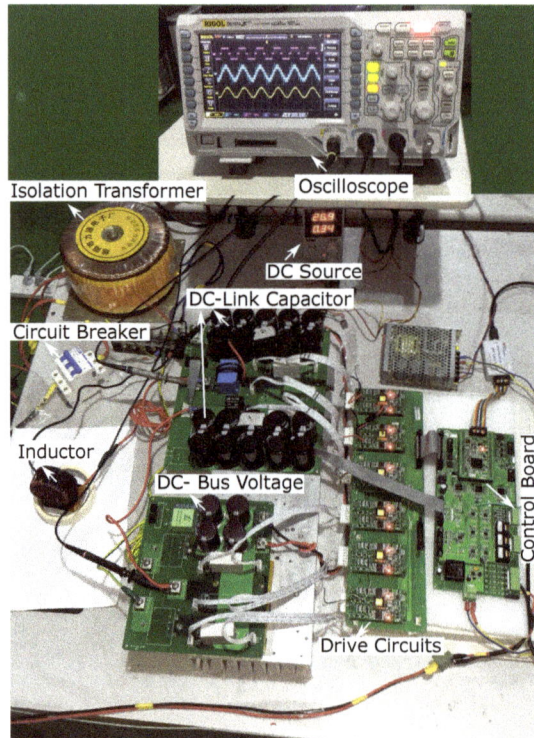

Figure 18. Experimental setup.

5. Comparision and Discussion

In this paper, the important characteristics of five different converter configurations were examined. A comparison is carried out among these topologies under power semiconductor devices (IGBT) available in different voltage levels (2.3, 3.3, and 4.16 kV) [30].

The NPC VSC has superior features compared to other hybrid converters presented here for upper voltage range up-to 3 kV for present devices operating at 1 kHz average switching frequency [40]. The key features include low conduction losses, better voltage and current THD, low components count and equal voltage stress on the active switches. However, due to the unequal loss distribution, the result is unsymmetrical temperature among the semiconductor devices.

Considering the dc bus voltage up-to 5 kV, the structural design of hybrid H-NPC with H-bridge has the benefit of seven-level output voltage by two more active switches which is inevitable. The topology can be used for higher voltage range as the dc link required for harmonic compensation is 0.35 *E* at maximum modulation index. Moreover, the switching losses is reduced by modulating the high voltage stressed switch at fundamental switching frequency. The structural drawback of the H-NPC converter is the high voltage stresses on half bridge leg of H-NPC converter. The voltage stresses can be higher across the H-bridge for the topology used in higher range voltage application.

A CHB based converter is introduced to tackle the limit imposed by high voltage stresses which is reduced by splitting the dc link voltage of the H-bridge converter using five-level CHB for higher voltage application. However, the high component count will lead to a high system cost. A topology based on a H-NPC and cascaded module achieves better output voltage with reduced active switches and minimum conduction loss in comparison with CHB based topology.

The proposed converter topology is designed for a lower medium voltage range up-to 5 kV with the maximum output voltage of seven-level. Considering the same system parameters, the proposed topology has certain advantages over other hybrid converter topologies, for example, the wide operating rang, low current THD presented in Table 11, and the lower component count in comparison with the hybrid converter topologies listed in Table 8. The switching losses are addressed by operating the high stressed switch with low switching frequency; the switching loss of the proposed converter is lower which results in higher efficiency in comparison with NPC VSC as shown in Figure 19 based on the converter parameters given in Table 9. However, the current THD of the NPC converter is less than the AHMMC converter.

Table 8. Component count of converter.

Converter	Components Count		
	Active Switches	Diode	Dc-Link Capacitor
NPC	8	4	2
HNPC with H-bridge	10	2	3
HNPC with cascaded module	12	2	4
Proposed AHMMC converter	10	-	3

Table 9. Converter parameter for efficiency comparison.

Parameter	Value			
DC bus voltage	3 kV			
Grid voltage	1.75 kV (rms)			
Modulation index	Maximum modulation index			
Reactor inductance	0.6 mH			
	V_{CES}	V_{CESat}	Rise time	Fall time
IGBT	1.7 kV	2.45 V	0.29 μs	0.29 μs
IGBT	4.5 kV	2.84 V	0.35 μs	0.35 μs

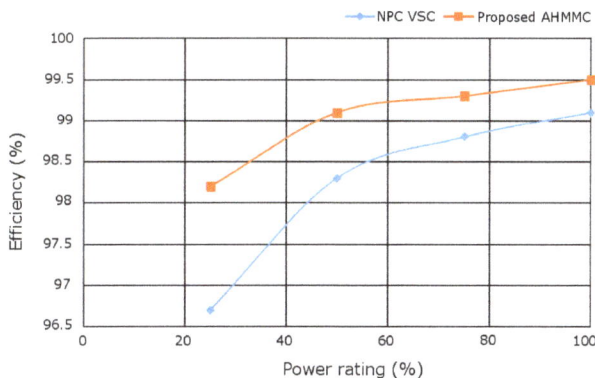

Figure 19. Converter's effiency.

Table 10 presents voltage stress on the semiconductor switches. The proposed topology has four active switches (in the outer leg of FCCM) which are under low voltage stress; thus, they are operated at high frequency to minimize the current ripple together with inductor size. The current total harmonic distortion (THD) of the converter operating at different power ratings is given in Table 11, which shows that the proposed converter current THD is lower than the other hybrid converter topologies discussed here.

Table 10. Voltage stress on semiconductor devices.

Semiconductor Devices	Voltage Stress	
	Active Switches	Diode
Half bridge NPC	E/2	E/2
Half bridge	E/2	–
Cascaded module centre cell	E	–
Cascaded module outer cell	E/2	–
Half Bridge (AHMMC)	E	–
Cascaded module centre cell	E	–
Cascaded module outer cell	E/2	–

Table 11. Converters current THD operating at different power ratings.

Converter	Power Rating		
	25%	50%	100%
NPC	5.31	2.83	1.38
HNPC with H-bridge	6.71	3.43	1.73
HNPC with cascaded module	6.59	3.39	1.71
Proposed AHMMC	6.55	3.31	1.66

6. Conclusions

In this paper, a comparative analysis of MLC topologies for medium voltage traction application has been demonstrated. The study will be helpful to select a suitable converter among the converters discussed for specific voltage ranges. Among all the converters, the proposed AHMMC configuration is studied in detail for medium voltage high power application due to its various features including a better output voltage, reduced switch losses by minimizing the switching frequency as well as the component count, and improved power factor with low THD. The proposed system performance is validated through simulation results using MATLAB/Simulink. Furthermore, a low voltage laboratory prototype is developed to test the performance of the proposed converter in practice.

Author Contributions: M.A. and M.M.K. proposed the idea for writing the manuscript. M.M.K. and H.T. suggested the literature and supervised in writing the manuscript. M.T.F. and K.H. helped M.A. in writing and formatting. Y.A. helped in modifying the figures and shared the summary of various credible articles to be included in this manuscript. X.J. helped in practical system design.

Funding: This research received no external funding.

Conflicts of Interest: The authors declare no conflict of interest.

References

1. Feng, J.; Chu, W.Q.; Zhang, Z.; Zhu, Z.Q. Power Electronic Transformer-Based Railway Traction Systems: Challenges and Opportunities. *IEEE J. Emerg. Sel. Top. Power Electron.* **2017**, *5*, 1237–1253. [CrossRef]
2. Flinders, F.; Oghanna, W. Energy efficiency improvements to electric locomotives using PWM rectifier technology. In Proceedings of the 1995 International Conference on Electric Railways in a United Europe, Amsterdam, The Netherlands, 27–30 March 1995; pp. 106–110. [CrossRef]

3. Rodriguez, J.; Lai, J.S.; Peng, F.Z. Multilevel inverters: A survey of topologies, controls, and applications. *IEEE Trans. Ind. Electron.* **2002**, *49*, 724–738. [CrossRef]

4. Nordvall, A. *Multilevel Inverter Topology Survey*; Chalmers University of Technology: Goteborg, Swenden, 2011.

5. Chauhan, N.; Jana, K.C. Cascaded multilevel inverter for underground traction drives. In Proceedings of the 2012 IEEE International Conference on Power Electronics, Drives and Energy Systems (PEDES), Bengaluru, India, 16–19 December 2012; pp. 1–5. [CrossRef]

6. Pulikanti, S.R.; Konstantinou, G.S.; Agelidis, V.G. Generalisation of flying capacitor-based active-neutralpoint-clamped multilevel converter using voltage-level modulation. *IET Power Electron.* **2012**, *5*, 456–466. [CrossRef]

7. Malinowski, M.; Gopakumar, K.; Rodriguez, J.; Perez, M.A. A Survey on Cascaded Multilevel Inverters. *IEEE Trans. Ind. Electron.* **2010**, *57*, 2197–2206. [CrossRef]

8. Salinas, F.; Gonzalez, M.A.; Escalante, M.F. Voltage balancing scheme for flying capacitor multilevel converters. *IET Power Electron.* **2013**, *6*, 835–842. [CrossRef]

9. Cecati, C.; Dell'Aquila, A.; Liserre, M.; Monopoli, V.G. A passivity-based multilevel active rectifier with adaptive compensation for traction applications. *IEEE Trans. Ind. Appl.* **2003**, *39*, 1404–1413. [CrossRef]

10. Tarisciotti, L.; Zanchetta, P.; Watson, A.; Bifaretti, S.; Clare, J.C.; Wheeler, P.W. Active DC Voltage Balancing PWM Technique for High-Power Cascaded Multilevel Converters. *IEEE Trans. Ind. Electron.* **2014**, *61*, 6157–6167. [CrossRef]

11. Konstantinou, G.; Capella, G.J.; Pou, J.; Ceballos, S. Single-Carrier Phase-Disposition PWM Techniques for Multiple Interleaved Voltage-Source Converter Legs. *IEEE Trans. Ind. Electron.* **2018**, *65*, 4466–4474. [CrossRef]

12. Pulikanti, S.R.; Konstantinou, G.; Agelidis, V.G. Hybrid Seven-Level Cascaded Active Neutral-Point-Clamped-Based Multilevel Converter Under SHE-PWM. *IEEE Trans. Ind. Electron.* **2013**, *60*, 4794–4804. [CrossRef]

13. Ren, L.; Gong, C.; He, K.; Zhao, Y. Modified hybrid modulation scheme with even switch thermal distribution for H-bridge hybrid cascaded inverters. *IET Power Electron.* **2017**, *10*, 261–268. [CrossRef]

14. Chattopadhyay, S.K.; Chakraborty, C.; Pal, B.C. Cascaded H-Bridge amp; neutral point clamped hybrid asymmetric multilevel inverter topology for grid interactive transformerless photovoltaic power plant. In Proceedings of the IECON 2012—38th Annual Conference on IEEE Industrial Electronics Society, Montreal, QC, Canada, 25–28 October 2012; pp. 5074–5079. [CrossRef]

15. Sneineh, A.A.; Wang, M.Y. A Novel Hybrid Flying-Capacitor-Half-Bridge Cascade 13-Level Inverter for High Power Applications. In Proceedings of the 2007 2nd IEEE Conference on Industrial Electronics and Applications, Harbin, China, 23–25 May 2007; pp. 2421–2426. [CrossRef]

16. Teymour, H.R.; Sutanto, D.; Muttaqi, K.M.; Ciufo, P. A Novel Modulation Technique and a New Balancing Control Strategy for a Single-Phase Five-Level ANPC Converter. *IEEE Trans. Ind. Appl.* **2015**, *51*, 1215–1227. [CrossRef]

17. Ali, M.; Mansoor, M.; Tang, H.; Rana, A. Analysis of a seven-level asymmetrical hybrid multilevel converter for traction systems. *IET Power Electron.* **2017**, *10*, 1878–1888. [CrossRef]

18. Carnielutti, F.; Pinheiro, H. Hybrid Modulation Strategy for Asymmetrical Cascaded Multilevel Converters Under Normal and Fault Conditions. *IEEE Trans. Ind. Electron.* **2016**, *63*, 92–101. [CrossRef]

19. Wang, S.; Song, W.; Zhao, J.; Feng, X. Hybrid single-carrier-based pulse width modulation scheme for single-phase three-level neutral-point-clamped grid-side converters in electric railway traction. *IET Power Electron.* **2016**, *9*, 2500–2509. [CrossRef]

20. Hu, H.; He, Z.; Gao, S. Passive Filter Design for China High-Speed Railway With Considering Harmonic Resonance and Characteristic Harmonics. *IEEE Trans. Power Deliv.* **2015**, *30*, 505–514. [CrossRef]

21. Faiz, M.T.; Khan, M.M.; Jianming, X.; Habib, S.; Tang, H. Parallel feedforward compensation based active damping of LCL-type grid connected inverter. In Proceedings of the 2018 IEEE International Conference on Industrial Technology (ICIT), Lyon, France, 20–22 February 2018; pp. 788–793. [CrossRef]

22. He, Z.; Hu, H.; Zhang, Y.; Gao, S. Harmonic Resonance Assessment to Traction Power-Supply System Considering Train Model in China High-Speed Railway. *IEEE Trans. Power Deliv.* **2014**, *29*, 1735–1743. [CrossRef]

23. Jianben, L.; Shaojun, D.; Qiaofu, C.; Kun, T. Modelling and industrial application of series hybrid active power filter. *IET Power Electron.* **2013**, *6*, 1707–1714. [CrossRef]

24. Pereda, J.; Dixon, J. 23-Level Inverter for Electric Vehicles Using a Single Battery Pack and Series Active Filters. *IEEE Trans. Veh. Technol.* **2012**, *61*, 1043–1051. [CrossRef]
25. Ortuzar, M.E.; Carmi, R.E.; Dixon, J.W.; Moran, L. Voltage-source active power filter based on multilevel converter and ultracapacitor DC link. *IEEE Trans. Ind. Electron.* **2006**, *53*, 477–485. [CrossRef]
26. Nabae, A.; Takahashi, I.; Akagi, H. A New Neutral-Point-Clamped PWM Inverter. *IEEE Trans. Ind. Appl.* **1981**, *IA-17*, 518–523. [CrossRef]
27. Baker, R.H. Bridge Converter Circuit. U.S. Patent 4270163, 26 May 1981.
28. Rodriguez, J.; Bernet, S.; Steimer, P.K.; Lizama, I.E. A Survey on Neutral-Point-Clamped Inverters. *IEEE Trans. Ind. Electron.* **2010**, *57*, 2219–2230. [CrossRef]
29. Sayago, J.A.; Bruckner, T.; Bernet, S. How to Select the System Voltage of MV Drives; A Comparison of Semiconductor Expenses. *IEEE Trans. Ind. Electron.* **2008**, *55*, 3381–3390. [CrossRef]
30. Bernet, S. State of the art and developments of medium voltage converters-an overview. *Przeglkad Elektrotechn.* **2006**, *82*, 1–10.
31. Rodriguez, J.; Bernet, S.; Wu, B.; Pontt, J.O.; Kouro, S. Multilevel Voltage-Source-Converter Topologies for Industrial Medium-Voltage Drives. *IEEE Trans. Ind. Electron.* **2007**, *54*, 2930–2945. [CrossRef]
32. D12, Siemens. Sinamics Catalogue. In *Automation and Drives*; Siemens AG: Nuremberg, Germany, 2006.
33. D12-Supplement 04/2007, Siemens. Sinamics Catalogue. In *Automation and Drives*; Siemens AG: Nuremberg, Germany, 2007.
34. Dietrich, C.; Gediga, S.; Hiller, M.; Sommer, R.; Tischmacher, H. A new 7.2kV medium voltage 3-Level-NPC inverter using 6.5kV-IGBTs. In Proceedings of the 2007 European Conference on Power Electronics and Applications, Aalborg, Denmark, 2–5 September 2007; pp. 1–9. [CrossRef]
35. ABB Industry. *Drive ACS6000 Brochure and Datasheet Rev. B*; ABB Switzerland Ltd.: Turgi, Switzerland, 2005.
36. Jakob, R.; Keller, C.; Mohlenkamp, G.; Gollentz, B. 3-Level high power converter with press pack IGBT. In Proceedings of the 2007 European Conference on Power Electronics and Applications, Aalborg, Denmark, 2–5 September 2007; pp. 1–7. [CrossRef]
37. Sousa, R.P.R.; Jacobina, C.B.; Bahia, F.A.C.; Barros, L.M. Comparative analysis of cascaded inverters based on 5-level and 3-level H-bridges. In Proceedings of the 2017 IEEE Energy Conversion Congress and Exposition (ECCE), Cincinnati, OH, USA, 1–5 October 2017; pp. 2161–2167. [CrossRef]
38. Xu, Y.; Zou, Y. Research on a novel hybrid cascade multilevel converter. In Proceedings of the 2007 International Power Engineering Conference (IPEC 2007), Singapore, 3–6 December 2007; pp. 1081–1085.
39. Kai, D.; Yunping, Z.; Lei, L.; Zhichao, W.; Hongyuan, J.; Xudong, Z. Novel Hybrid Cascade Asymmetric Inverter Based on 5-level Asymmetric inverter. In Proceedings of the 2005 IEEE 36th Power Electronics Specialists Conference, Recife, Brazil, 16 June 2005; pp. 2302–2306. [CrossRef]
40. Sanchez-Ruiz, A.; Mazuela, M.; Alvarez, S.; Abad, G.; Baraia, I. Medium Voltage–High Power Converter Topologies Comparison Procedure, for a 6.6 kV Drive Application Using 4.5 kV IGBT Modules. *IEEE Trans. Ind. Electron.* **2012**, *59*, 1462–1476. [CrossRef]

electronics

MDPI

Article

FPGA Implementation of a Three-Level Boost Converter-fed Seven-Level DC-Link Cascade H-Bridge inverter for Photovoltaic Applications

Nagaraja Rao Sulake [1,*], **Ashok Kumar Devarasetty Venkata** [2] **and Sai Babu Choppavarapu** [1]

[1] Department of Electrical and Electronics Engineering, Jawaharlal Nehru Technological University, Kakinada 533003, Andhra Pradesh, India; chs_eee@yahoo.co.in
[2] Department of Electrical and Electronics Engineering, Rajeev Gandhi Memorial College of Engineering and Technology, Nandyal 518501, Andhra Pradesh, India; rgmdad09@gmail.com
* Correspondence: nagarajraomtech@gmail.com or nagarajarao.ee.et@msruas.ac.in; Tel.: +91-974-2669617

Received: 3 October 2018; Accepted: 24 October 2018; Published: 29 October 2018

Abstract: This paper presents an optimized single-phase three-level boost DC-link cascade H-bridge multilevel inverter (TLBDCLCHB MLI) system to generate a seven-level stepped output voltage waveform for photovoltaic (PV) applications. The proposed TLBDCLCHB MLI system is obtained by integrating a three-level boost converter (TLBC) with a seven-level DC-link cascade H-bridge (DCLCHB) inverter. It consists of a TLBC, level generation unit (LGU) and phase sequence generation unit (PSGU). When compared with traditional boost converter-fed multilevel inverter systems, the proposed TLBDCLCHB MLI system requires a single DC source, fewer power switches and gate drivers. Reduction in the switch count and number of DC sources makes the system cost effective and requires a smaller installation area. Pulse generation for the power switches of an LGU in a DCLCHB inverter is accomplished by providing proper conducting angles that are generated by optimized conducting angle determination (CAD) techniques. In this paper two CAD techniques i.e., equal-phase CAD (EPCAD) and step pulse wave CAD (SPWCAD) techniques are proposed to evaluate the performance of the proposed system in terms of the total harmonic distortion (THD) and the quality of the stepped output voltage waveform. The proposed system has been modeled and simulated using MATLAB/SIMULINK software. Results are presented and discussed. Also, a prototype model of a single-phase TLBDCLCHB MLI system is developed using a field-programmable gate array (FPGA)-based pulse generation with a resistive load and its performance is analyzed for various operating conditions.

Keywords: three-level boost converter (TLBC); DC-link cascade H-bridge (DCLCHB) inverter; conducting angle determination (CAD) techniques; total harmonic distortion (THD)

1. Introduction

Development in power electronics lay down a widespread scope for the resourceful operation of power converters. A few setups of power converters are produced to do the sun-powered photovoltaic (PV) applications with enhanced adequacy [1,2]. PV power generation is an encouraging elective source of energy and has numerous focal points compared to the other elective energy sources like wind, ocean, biomass, fuel, geothermal, and so on. In PV power generation, boost converters and multilevel inverters (MLIs) are playing a major role in power conversion. These power converters are broadly being utilized as a connection between load and supply. As most of the renewable power source generation is DC in nature, the DC-DC boost converters are utilized to increase the voltage level, and the DC must be changed over to AC for grid connection. Therefore, MLIs are used for DC to AC conversion [3,4]. The power generation using a traditional boost converter and inverter consists of a greater number of components, requires a larger installation area, is bulky in size, and costly. Also, the traditional boost converters are unable to produce a high boost ratio [5,6]. This paper proposes a three-level boost converter (TLBC) with a high boost ratio, based on one switch, one inductor, (2N-1) capacitors, and (2N-1) diodes for 'N' levels. It is a pulse width modulation (PWM)-controlled boost converter capable of maintaining an equal voltage in all 'N' output levels and controlling the input current.

In this paper, the structure of single-phase three-level boost DC-link cascade H-bridge multilevel inverter (TLBDCLCHB MLI) system is proposed to generate a seven-level stepped output voltage waveform for PV applications. The proposed system is obtained by integrating a three-level boost converter (TLBC) with a seven-level DC-link cascade H-bridge (DCLCHB) [7,8]. Also, the objective of the proposed work is to investigate the performance of a single-phase seven-level DCLCHB inverter [9,10] using conducting angle determination (CAD) techniques [11] in terms of the total harmonic distortion (THD) and the quality of the stepped output voltage waveform. Here, equal-phase CAD (EPCAD) based on equal-area distribution and step pulse wave CAD (SPWCAD) based on volt-second area equal to step pulse wave techniques are proposed to evaluate the performance of the DCLCHB inverter [12]. The proposed TLBDCLCHB MLI system is modeled, simulated and validated through experimental setup using field-programmable gate array (FPGA)-based pulse generation.

2. Structure of TLBDCLCHB MLI System

The block diagram of the proposed TLBDCLCHB MLI system is shown in Figure 1. The proposed TLBDCLCHB MLI system consists of single DC voltage source, TLBC, and DCLCHB inverter to generate a seven-level stepped output waveform. The DCLCHB inverter is composed of level generation unit (LGU) and phase sequence generation unit (PSGU). LGU is used to generate the required number of levels and PSGU is used to generate positive and negative sequence voltage levels [13]. The equivalent structure of TLBDCLCHB MLI system is shown in Figure 2.

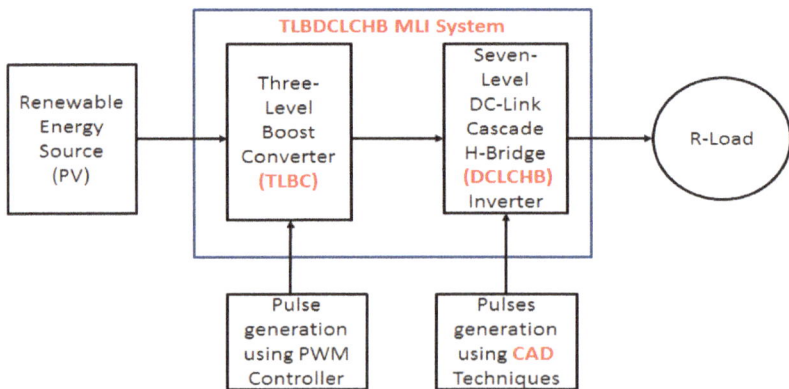

Figure 1. Block diagram of the proposed (TLBDCLCHB MLI) system.

The number of power switches '$N_{Switches}$' and the number of levels 'm' for a single-phase TLBDCLCHB MLI system are calculated using Equations (1) and (2), respectively.

$$m = (2C + 1)^H \tag{1}$$

$$N_{Switches} = (m - 1) + 5 \tag{2}$$

where 'C' is the number of DC link capacitors integrated to a DCLCHB inverter and 'H' is the number of H-Bridge circuits.

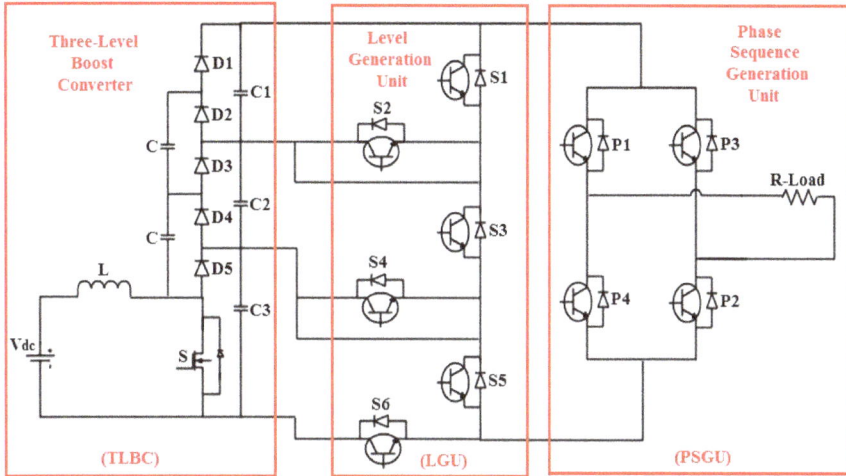

Figure 2. Equivalent structure of TLBDCLCHB MLI system.

Table 1 demonstrates the required number of power switches and DC sources for the traditional and proposed boost DC-link-based inverter to generate a stepped output waveform from seven levels to fifteen levels. From the investigation, it is gathered that the proposed TLBDCLCHB MLI system has reduced the switch count and requires just a single DC source compared to traditional boost multilevel inverter (MLI) systems [14,15].

Table 1. Component requirements for existing and proposed boost based MLI systems.

Number of Levels	Boost Cascade MLI (Conventional)		TLBDCLCHB MLI (Proposed)	
	Number of Power Switches	Number of DC Sources	Number of Power Switches	Number of DC Sources
7	15	3	11	1
9	20	4	13	1
11	25	5	15	1
13	30	6	17	1
15	35	7	19	1

3. Three-Level Boost Converter (TLBC)

The circuit configuration of DC-DC TLBC is represented in Figure 2. It consists of a traditional boost converter, (2N-1) capacitors, and (2N-1) diodes. The main advantages of using the TLBC topology are; it can be extended to any number of levels by adding only diodes and capacitors without changing the main circuit; no need of additional voltage balance circuit; and voltage gain can be increased without the use of a transformer by operating at minimum duty ratio. The TLBC circuit consists of three stages which are operated by varying duty cycles of 0.4, 0.5, and 0.6. The operation of the TLBC is explained in [16].

3.1. Analysis of DC-DC TLBC

Considering the presence and absence of inductor power loss for both traditional and proposed TLBCs gives important information to designers. From basic principles, the voltage gain or the boost factor of the traditional boost converter [14] is given by Equation (3):

$$\text{Voltage gain,} \ \frac{V_o}{V_{dc}} = \frac{1}{1-D} \tag{3}$$

where 'V_o' is the output voltage, 'V_{dc}' is the input voltage and 'D' is the duty cycle.

By considering the lossless system, the voltage gain for the N-level boost converter can be expressed as:

$$\frac{V_o}{V_{dc}} = \frac{N}{1-D} \tag{4}$$

For a lossless system, the input current, I_{dc}, can be expressed as:

$$I_{dc} = \frac{N\,I_o}{1-D} \tag{5}$$

From Equation (5), the input current, I_{dc}, can be controlled using duty cycle 'D' in the PWM. Now the voltage gain or boost factor expression for the N-level boost converter can be derived as follows:

Based on the condition that the average voltage across the inductor 'L' is equal to zero. The total inductor voltage during the ON—OFF condition can be expressed as,

$$V_L = D(V_{dc} - I_L R_L) + (1-D)(V_{dc} - V_C - I_L R_L) = 0 \tag{6}$$

where, 'I_L' is the inductor current which is equivalent to 'I_{dc}' and 'R_L' is the inductor resistance or parasitic resistance.

Here, the first term of Equation (6) is valid when the switch 'S' is turned ON, and the second term can write when the switch 'S' is turned OFF. From Equation (6) it can be written as,

$$V_{dc}(D+1-D) + I_L R_L(-D-1+D) = (1-D)V_C \tag{7}$$

From Equation (7),

$$V_{dc} = (1-D)V_C + I_L R_L \tag{8}$$

Therefore, from Equations (3)–(7), the input voltage, V_{dc} can be expressed as,

$$V_{dc} = (1-D)\frac{V_o}{N} + \frac{N\,V_o}{(1-D)R_O}R_L \tag{9}$$

Therefore, Equation (9) can be expressed as follows:

$$\frac{V_{dc}}{V_O} = \frac{1}{\frac{(1-D)}{N} + \frac{NR_L}{(1-D)R_O}} \tag{10}$$

Equation (10) is equal to Equation (3) if $N = 1$ and $R_L = 0$. From Equation (10), it can be observed that the voltage gain reaches a maximum before $D = 1$, and then becomes 0. The effect of parasitic resistance 'R_L' is responsible for the limitation in the boost factor. The actual boost factor or voltage gain against the duty cycle is analyzed by varying 'R_L' in Equation (10). Figures 3 and 4 describe the graph between voltage gain versus duty cycle for the traditional boost converter, i.e., $N = 1$ and for the proposed TLBC, i.e., $N = 3$ using different values of R_L/Ro.

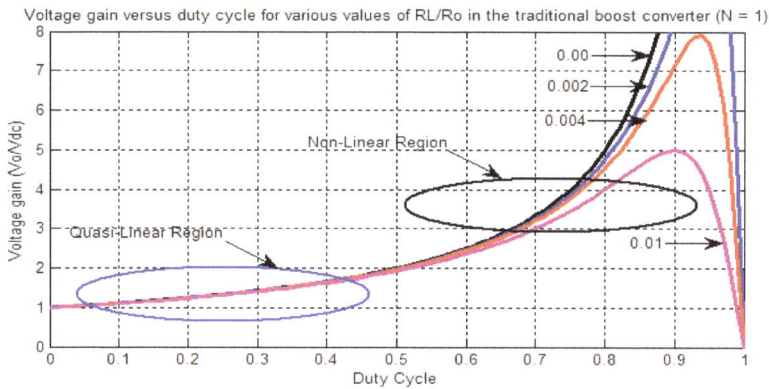

Figure 3. Duty cycle versus voltage gain of a traditional boost converter for various values of R_L/R_O ($N = 1$).

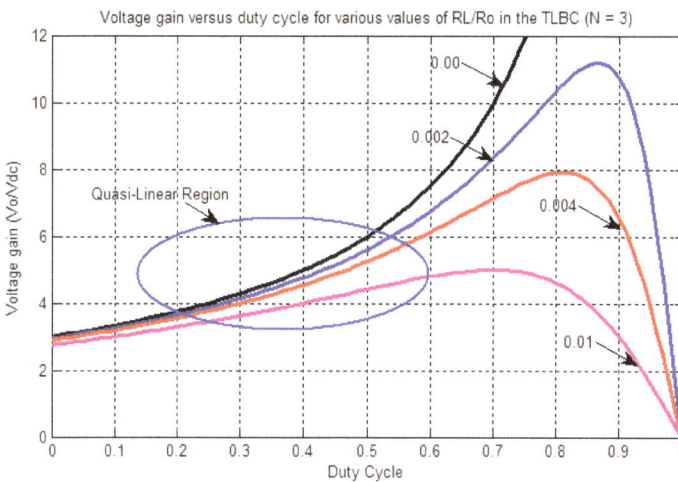

Figure 4. Duty cycle versus voltage gain of TLBC for various values of R_L/R_O ($N = 3$).

From the Figures 3 and 4, it can be noticed that the voltage gain of the traditional boost converter is quasi-linear when the duty cycle varies from 0 to 0.5, but beyond that, the boost factor of a traditional boost converter becomes non-linear; therefore, the control of a traditional boost converter is complicated.

Similarly, from Figure 4, i.e., when $N = 3$, it can be observed that the quasi-linear region is extended with a high voltage gain for the TLBC. Therefore, the TLBC achieves a high voltage gain compared to the traditional boost converter, and, also a better operating point of the duty cycle for the TLBC, which is from 0.4 to 0.6.

In the next section, the effect of the voltage drop across the switch and diodes is studied.

3.2. TLBC Voltage Drop Across Switch and Diodes

In actual operation of TLBC, the voltage drop across the switch and diodes must be considered since it avoids full charge across the capacitors [17,18]. Therefore, it reduces the conversion efficiency of the TLBC topology. In general, the voltage drop in the power switches and diodes can be around 2 V, and it can be neglected in medium- and high-voltage applications but must be considered in low-voltage applications. Here, the voltage drop across the switch and diodes is assumed to be equal to 'V_d'.

From Figure 5, it can be noticed that the voltage across the C_5 becomes,

$$V_{C5} = V_{C3} - V_{switch} - V_{diode} = V_{C3} - 2V_d \tag{11}$$

where, '$2V_d$' is the voltage drop across the switch 'S' and diode 'D_5'.

Similarly, V_{C2} and V_{C1} can be written as,

$$V_{C2} = 2V_{C3} - 4V_d \tag{12}$$

$$V_{C1} = 3V_{C3} - 8V_d \tag{13}$$

where, V_{C2} and V_{C1} are the expressions for the voltage output of two-level and three-level boost converters, respectively. The generalized output voltage expression for the N-level boost converter can be expressed as follows:

$$V_o = NV_C - (N-1)4V_d \tag{14}$$

where, 'V_C' is the lower capacitor output voltage, and follows Equation (3).

Equation (14) gives the output voltage for the multiple stages of the boost converter. The efficiency of the proposed converter for the N-level is given by the Equation (15).

$$Efficiency, \ \eta \ = \ \frac{V_o}{NV_C} \ = \ 1 - \frac{(N-1)4V_d}{NV_C} \tag{15}$$

Form Equation (15), the efficiency of the TLBC circuit can be reduced by considering the voltage drop across the switch and diode.

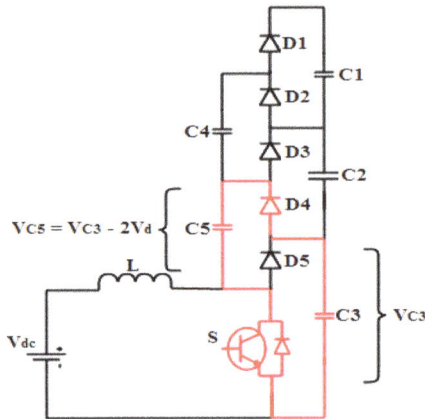

Figure 5. Charging C5 of a TLBC with switch and diode's voltage drop.

3.3. Closed-Loop Control of TLBC

From Figure 5 and Equation (10), it can be observed that the output voltage gain depends on the ratio of load resistance (R_O) and source resistance (R_L), i.e., if there is any variation between the load and source resistances, the output voltage of TLBC is not kept constant, and, from Equation (13), there will be a variation in the duty cycle to get the required amount of output voltage by considering the voltage drop across the switch and diodes. Therefore, the proposed TLBC circuit is modeled in closed-loop mode using an integral controller to maintain the constant output voltage.

Figure 6 represents the TLBC circuit in closed-loop mode. In the case of any variation in the load side or source side, the output changes, so a suitable controller is designed to change the duty cycle by comparing 'V_{ref}' with 'V_{out}' in order to maintain the required output voltage.

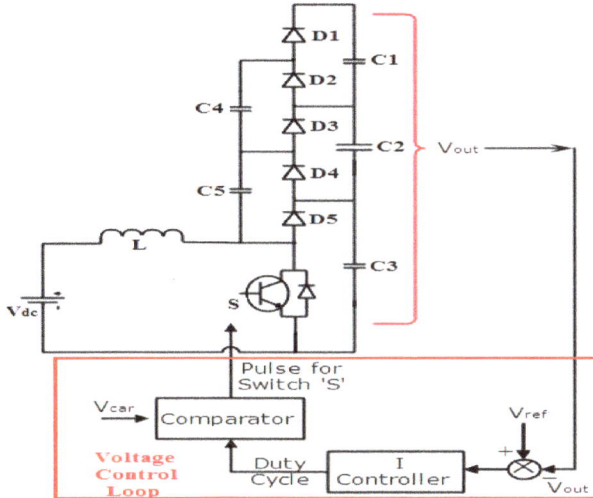

Figure 6. Model of TLBC in closed-loop mode.

4. DC-Link Cascade H-Bridge (DCLCHB) Inverter

Figure 7 depicts the DCLCHB inverter topology for the generation of a single-phase seven-level output voltage waveform. It is composed of LGU and PSGU [12].

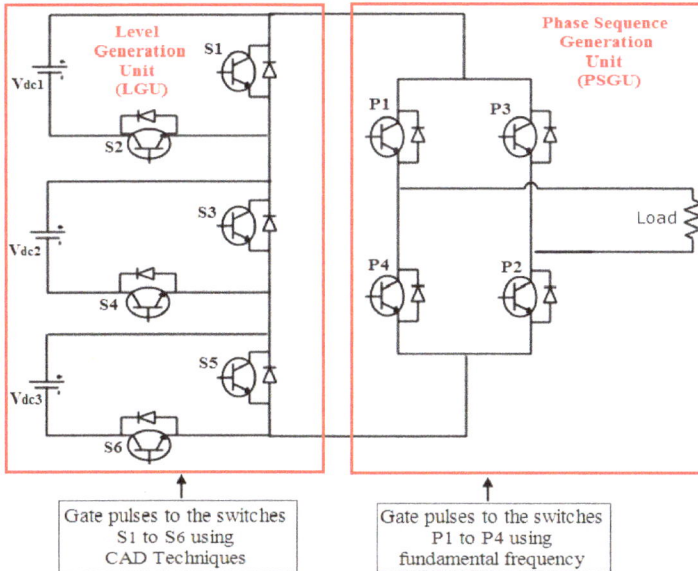

Figure 7. Single-phase seven-level DCLCHB inverter.

Switches in the LGU are used to generate the required number of levels and it is formed by connecting half-bridge cells in series. Each half-bridge cell consists of a DC source controlled by two switches. PSGU consists of an H-bridge circuit, which is used to generate the positive and negative sequence voltage levels. Table 2 gives a component requirement to generate a seven-level output voltage for the proposed and existing MLIs. It clearly shows a substantial component reduction when using a DCLCHB structure [7,8].

Table 2. Component requirements for existing and proposed cascade MLI systems.

Components	Traditional MLI	Proposed MLI
Switches	12	10
Clamping diodes	0	0
DC sources	3	3
Capacitors	0	0

In this DCLCHB inverter topology, all the magnitudes of DC voltage sources are equal $(V_{dc1} = V_{dc2} = V_{dc3})$.

$$\text{i.e., } V_{dci} = V_{dc}, \text{ where } i = 1, 2, \text{ and } 3 \tag{16}$$

The maximum value of the output phase voltage is obtained by using Equation (17).

$$V_{max} = \sum_{i=1}^{S} V_{dci} \tag{17}$$

The number of output phase voltage levels can be obtained from Equation (1). The number of power switches for the DCLCHB inverter can be calculated using Equation (18).

$$N_{Switches} = (m-1) + 4 \tag{18}$$

Equation (16) gives the output level of the LGU. By using PSGU, the positive and negative levels are obtained at the load (V_o), the synthesized stepped AC output phase voltage will be obtained by using the Equations (19) and (20).

$$V_{o,\ max} = \sum_{i=1}^{3} + V_{dci}, \text{ If } P_1, P_2 = 1 \tag{19}$$

$$V_{o,\ max} = \sum_{i=1}^{3} - V_{dci}, \text{ If } P_3, P_4 = 1 \tag{20}$$

For a single-phase seven-level DCLCHB inverter, the switching sequences to generate the required levels are given in Table 3.

Table 3. Switching sequence to generate seven-level output for a DCLCHB inverter.

S.No	LGU Switches						PSGU Switches				Voltage Levels (Volts)
	S_1	S_3	S_5	S_2	S_4	S_6	P_1	P_2	P_3	P_4	
1	1	1	1	0	0	0	0	0	0	0	0
2	0	1	1	1	0	0	1	1	0	0	100
3	0	0	1	1	1	0	1	1	0	0	200
4	0	0	0	1	1	1	1	1	0	0	300
5	0	1	1	1	0	0	0	0	1	1	−100
6	0	0	1	1	1	0	0	0	1	1	−200
7	0	0	0	1	1	1	0	0	1	1	−300

5. Conducting Angle Determination (CAD) Techniques

CAD techniques are a vital part of any inverter since they are directly related to the efficiency of the entire system [6,12]. It is used to control the proposed DCLCHB inverter output phase voltage and also for the calculation of the two main parameters such as %THD and V_{rms}. In this paper, a step pulse wave CAD (SPWCAD) technique has been employed to trigger the switches of LGU in the DCLCHB inverter and is compared with a conventional CAD technique, i.e., equal-phase CAD (EPCAD) technique. The generation of the seven-level stepped voltage waveform using CAD techniques is shown in Figure 8.

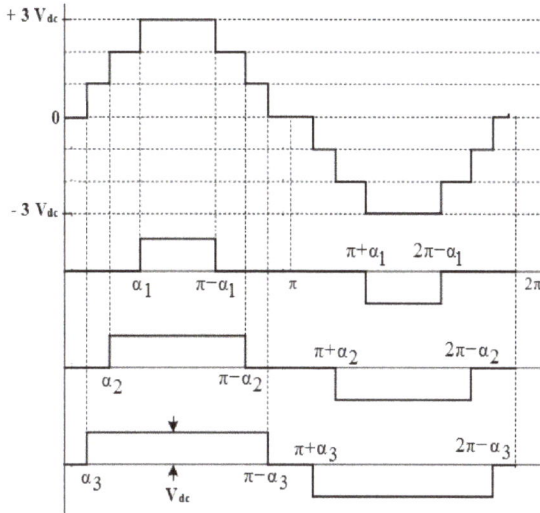

Figure 8. Generation of the seven-level stepped voltage waveform for DCLCHB inverter using CAD techniques.

In the presented EPCAD and SPWCAD techniques, for an m-level stepped waveform in the period of the first quadrant, i.e., 0 to 90°, $2(m-1)/2$ conducting angles need to be determined. From Figure 8, to generate a seven-level stepped waveform in the first quadrant, i.e., 0 to 90°, three conducting angles need to be determined. They are defined as the main conducting angles, i.e., α_1, α_2 and α_3 using the time-sequence. From Figure 8, it can be noticed that only the main conducting angles need to be determined; the rest of the conducting angles can be derived from the main conducting angles. The solution of the main conducting angles, i.e., α_1, α_2 and α_3 must satisfy the following condition:

$$0 \leq \alpha_1 \leq \alpha_2 \leq \alpha_3 \leq \frac{\pi}{2} \tag{21}$$

5.1. EPCAD Technique

In the EPCAD technique the main conducting angles are derived by taking an average distribution of the conducting angles from 0 to 180°. In this technique, the main conducting angles are obtained by using Equation (22).

$$\alpha_i = i\left(\frac{180°}{m}\right) \tag{22}$$

where $i = 1, 2, \ldots\ldots \left(\frac{m-1}{2}\right)$, $m = $ number of levels.

For a seven-level stepped waveform using the EPCAD technique, three main conducting angles need to be determined using Equation (22), i.e., α_1, α_2 and α_3, the values of which are 25.71°, 51.43°, and 77.14°, respectively.

5.2. SPWCAD Technique

In the proposed SPWCAD technique, conducting angles are acquired by computing the volt-second areas of the sine reference voltage waveform that is equivalent to the stepped output phase voltage waveform of the DCLCHB inverter. In the seven-level DCLCHB inverter, since three half-bridge cells are connected in series, the reference voltage 'V_{ref}' and output-phase voltage '$V_{out-phase}$' can be obtained by the Equations (23) and (24) respectively:

$$V_{ref} = 3 \left(\frac{4V_{dc}}{\pi} \right) (M_i \, Sin \, \omega t) \tag{23}$$

$$V_{out-phase} = i.V_{dc} \, (1 \le i \le 3) \tag{24}$$

where, 'M_i' is the modulation index, and 'i' is the integer number.

Figure 9 demonstrates the dummy conducting angles (α_i^1) in the case of $M_i = \pi/4$. The areas of A_1^1, A_2^1, and A_3^1 are encompassed by the sine reference voltage wave and the stepped output-phase voltage levels in the positive half-cycle of the seven-level DCLCHB inverter.

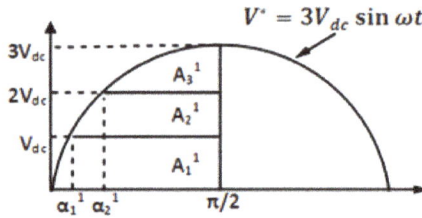

Figure 9. Reference voltage waveform with dummy conducting angles.

The generation of a step pulse wave in the DCLCHB inverter to meet the equivalent areas as A_1^1, A_2^1, and A_3^1. Here, main conducting angles (α_i) are defined as the switching timing angles of step pulse waves in the DCLCHB inverter. Figure 10 represents the main conducting angles and the stepped output phase voltage of the seven-level DCLCHB inverter during the positive half cycle.

Assuming that the areas A_1, A_2, and A_3 made by the main conducting angles in Figure 10 are equivalent to A_1^1, A_2^1, and A_3^1 made by dummy conducting angles which are obtained as follows:

Area, A_1^1 can be obtained from Equation (25):

$$A_1^1 = (A_1^1 + A_2^1 + A_3^1) - (A_2^1 + A_3^1) \tag{25}$$

Therefore, from Equation (25), A_1^1 can be written as:

$$A_1^1 = \int_0^{\frac{\pi}{2}} 3 \, V_{dc} \sin \omega t d(\omega t) - \left\{ \int_{\alpha_1^1}^{\frac{\pi}{2}} 3 V_{dc} \sin \omega t d(\omega t) - \left(\frac{\pi}{2} - \alpha_1^1 \right) V_{dc} \right\} \tag{26}$$

From Figure 10, Area, A_1 can be expressed as,

$$A_1 = \left(\frac{\pi}{2} - \alpha_1 \right) V_{dc} \tag{27}$$

Since $A_1{}^1$ is equal to A_1, the real conducting angle α_1 of the proposed SPWCAD technique is obtained as,

$$\alpha_1 = \frac{\pi}{2} - \left\{ \int_0^{\frac{\pi}{2}} 3\, V_{dc} \sin \omega t d(\omega t) - \left(\int_{\alpha_1^1}^{\frac{\pi}{2}} 3\, V_{dc} \sin \omega t d(\omega t) - \left(\frac{\pi}{2} - \alpha_1^1 \right) \right) \right\} \tag{28}$$

Similarly, by equating $A_2{}^1$ to A_2, the angle α_2 can be expressed as,

$$\alpha_2 = \frac{\pi}{2} - \left\{ \int_0^{\frac{\pi}{2}} 3\, V_{dc} \sin \omega t d(\omega t) - \left(\int_{\alpha_2^1}^{\frac{\pi}{2}} 3\, V_{dc} \sin \omega t d(\omega t) - \left(\frac{\pi}{2} - \alpha_2^1 \right) \right) \right\} \tag{29}$$

α_3 can be obtained by equating $A_3{}^1$ to A_3 or by using Equation (30):

$$\int_0^{\frac{\pi}{2}} 3\, V_{dc} \sin \omega t d(\omega t) - \left(A_1^1 + A_2^1 \right) = \left(\frac{\pi}{2} - \alpha_3 \right) V_{dc} \tag{30}$$

For a seven-level DCLCHB inverter, the conducting angles are calculated by using Equations (31)–(33).

$$\alpha_1 = \frac{12 M_i}{\pi} \left\{ \cos\left(\sin^{-1}\left(\frac{\pi}{12 M_i} \right) \right) - 1 \right\} + \sin^{-1}\left(\frac{\pi}{12 M_i} \right) \tag{31}$$

$$\alpha_2 = \frac{12 M_i}{\pi} \left\{ \cos\left(\sin^{-1}\left(\frac{2\pi}{12 M_i} \right) \right) - \cos\left(\sin^{-1}\left(\frac{\pi}{12 M_i} \right) \right) \right\} + 2 \sin^{-1}\left(\frac{2\pi}{12 M_i} \right) - \sin^{-1}\left(\frac{\pi}{12 M_i} \right) \tag{32}$$

$$\alpha_3 = \frac{3\pi}{2} - \frac{12 M_i}{\pi} \cos\left(\sin^{-1}\left(\frac{2\pi}{12 M_i} \right) \right) - 2 \sin^{-1}\left(\frac{2\pi}{12 M_i} \right) \tag{33}$$

Therefore, for a seven-level stepped waveform using the SPWCAD technique, three main conducting angles need to determined using Equations (31)–(33), i.e., α_1, α_2 and α_3 and their values are 9.43°, 29.59°, and 55.88°, respectively, for the $M_i = 0.8$.

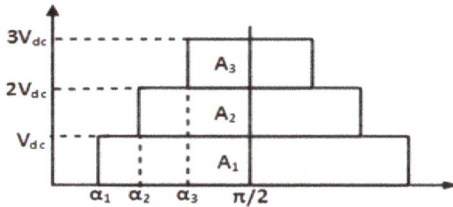

Figure 10. Output phase-voltage of a seven-level DCLCHB inverter in the positive half-cycle voltage.

5.3. Comparison of the SPWCAD and EPCAD Technique

The conducting angles acquired by the SPWCAD technique are different from those acquired by the EPCAD technique. In the SPWCAD technique, conducting angles are acquired based on the modulation index (M_i) whereas the EPCAD technique gives the same conducting angles irrespective of the M_i. The proposed SPWCAD technique method can be applied to different modulation indices. The range of M_i and the number of conducting angles are listed in Table 4. Also, the values of conducting angles for the proposed SPWCAD technique using various modulation indices are listed in Table 5 for the DCLCHB inverter.

Table 4. Number of conducting angles and steps in output waveform SPWCAD technique for various M_i.

Range of M_i	Number of Conducting Angles	Number of Steps in Output Waveform
$0 < M_i < 0.33$	1	3
$0.33 \leq M_i < 0.66$	2	5
$0.66 \leq M_i < 1$	3	7

Table 5. Conducting angles of SPWCAD technique for various M_i.

Conducting Angles (Degrees)	Modulation Indices (M_i)		
	0.3	0.6	0.8
α_1	27.17	12.7	9.439
α_2	–	41.65	29.59
α_3	–	–	55.88

5.4. THD Calculation of the Seven-Level Stepped Output Phase Voltage using CAD Techniques

The general THD expression for a periodic output phase voltage for a proposed DCLCHB inverter can be expressed as,

$$THD = \sqrt{\left(\frac{V_{rms}}{V_1}\right)^2 - 1} \tag{34}$$

where V_1 is the RMS (root mean square) value of the fundamental component and V_{rms} is the RMS value of the output phase voltage. For the proposed seven-level RV MLI, V_{rms} and V_1 can be obtained by using Equations (35) and (36).

$$V_{rms} = V_{dc}\sqrt{\left[\frac{2}{\pi}\cdot\left((\alpha_2 - \alpha_1) + 4(\alpha_3 - \alpha_2) + 9(\alpha_3 - \alpha_2)\right)\right]} \tag{35}$$

$$V_1 = \frac{4V_{dc}}{\pi\sqrt{2}}[(\text{Cos }\alpha_1 + \text{Cos }\alpha_2 + \text{Cos }\alpha_3)] \tag{36}$$

The output phase voltage THD expression for the proposed seven-level DCLCHB inverter can be obtained by substituting Equations (35) and (36) into Equation (34) and is given by:

$$V_1 = \frac{4V_{dc}}{\pi\sqrt{2}}[(\text{Cos }\alpha_1 + \text{Cos }\alpha_2 + \text{Cos }\alpha_3)] \tag{37}$$

Theoretical values of the output phase voltage THD for the seven-level DCLCHB inverter using EPCAD and the proposed SPWCAD techniques with the corresponding main conducting angles are given in Table 6.

Table 6. Conducting angles, theoretical output phase voltage THD, and V_{rms} values for a seven-level DCLCHB inverter (m = 7).

CAD Technique	Conducting Angles (in Degrees)			% THD (Theoretical)	V_{rms} (V)
	α_1	α_2	α_3		
EPCAD	25.71	51.43	77.14	31.05	165.8
SPWCAD	9.43	29.59	55.88	11.95	219.1

6. Simulation and Experimental Validation of the Proposed TLBDCLCHB Inverter System

The simulation of the proposed single-phase seven-level TLBDCLCHB inverter system is analyzed using MATLAB Simulink and validated experimentally through FPGA-based pulse generation.

6.1. TLBC Simulation Results in Open-Loop Mode

The simulation of the TLBC is carried out and analyzed in open-loop mode by considering the DC input voltage V_{dc} of 50 V, which should be boosted to a total DC-link voltage of 250 V, 300 V, and 375 V for the duty cycles of 0.4, 0.5, and 0.6, respectively, and the voltage across each of the capacitors at the output should be boosted to 83.33 V, 100 V, and 125 V for the duty cycles of 0.4, 0.5, and 0.6, respectively, as shown in Figures 11–13.

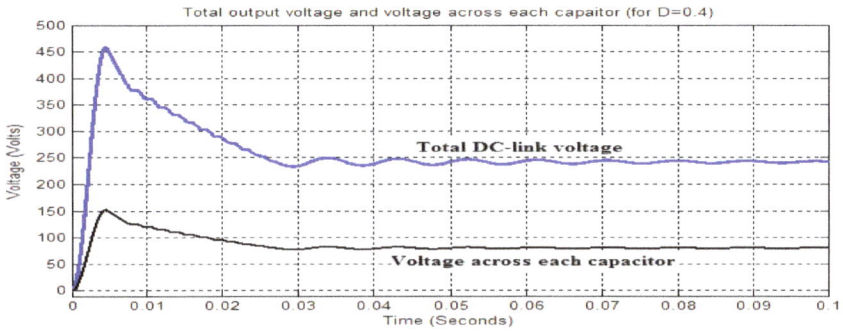

Figure 11. Output of TLBC in open-loop for $D = 0.4$.

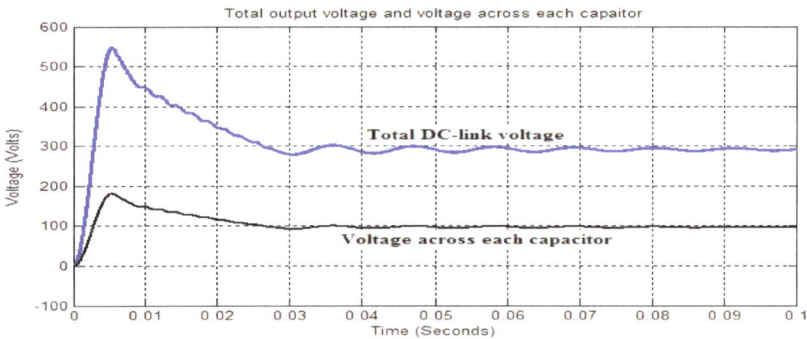

Figure 12. Output of TLBC in open-loop for $D = 0.5$.

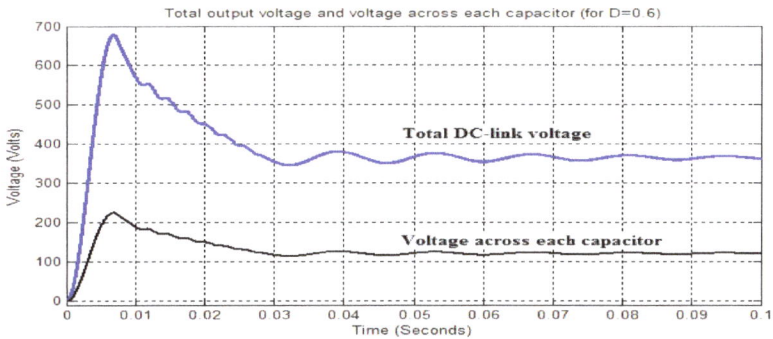

Figure 13. Output of TLBC in open-loop for $D = 0.6$.

From Figures 11–13, it is observed that the total DC-link output voltage for the duty cycles of 0.4, 0.5, and 0.6 has been boosted to 242.6, 291.4 V, and 361.4 V as opposed to 250 V, 300 V, and 375 V, respectively, due to the voltage drop across the switch and diodes. Similarly, the voltage drops across each of the capacitors are boosted to 82.17 V, 98.5 V, and 121.8 V as opposed to 83.33 V, 100 V, and 125 V. Also, it is noticed that the ripple in the total DC-link output voltage and voltage across each capacitor are increased by increasing the duty cycle from 0.4 to 0.6 to achieve the maximum DC-link output voltage with minimum ripple. Further, the open-loop mode of TLBC is extended to operate in closed-loop mode to maintain the required amount of DC-link output voltage by compensating the voltage drop across the switch and diodes, and, also, to reduce the ripple.

6.2. TLBC Simulation Results in Closed-Loop Mode

The proposed TLBC is implemented in closed-loop to maintain the constant output voltage using the voltage control loop. Here, the output of TLBC is measured and fed to an integral controller by comparing with the required reference output voltage to vary the duty cycle. Figure 14 shows the total DC-link output voltage and the voltage across each capacitor of the TLBC in closed-loop mode for $D = 0.5$. The corresponding change in duty cycle due to the reference output voltage is shown in Figure 15.

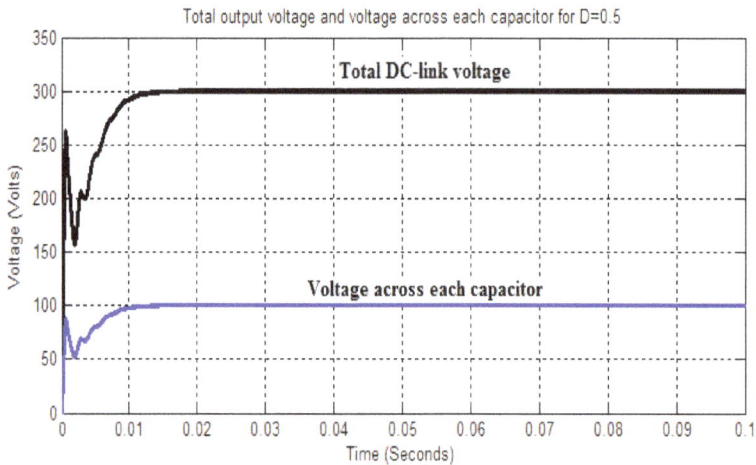

Figure 14. Output of TLBC in closed-loop for $D = 0.5$.

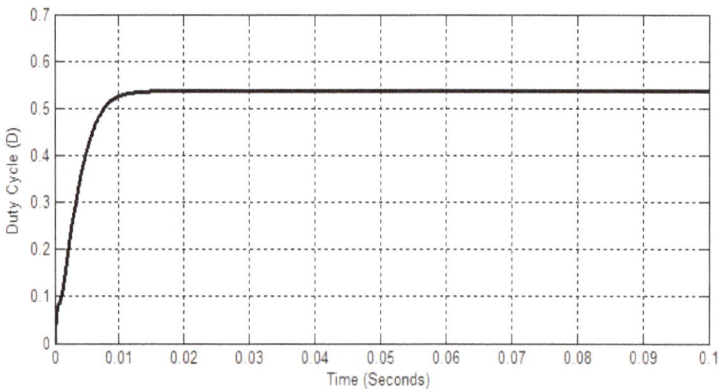

Figure 15. Change in duty cycle of TLBC in closed-loop.

From the Figures 14 and 15, it is observed that the total DC-link output voltage and the voltage across each capacitor of the TLBC is boosted to 300 V and 100 V, respectively, as per the reference output voltage by changing the duty cycle using an integral controller.

6.3. TLBDCLCHB Inverter Output using CAD Techniques

Simulation results of TLBC-fed DCLCHB inverter to generate a seven-level stepped output waveform using EPCAD and SPWCAD techniques and its THD analysis are shown in Figures 16–21 for $D = 0.5$.

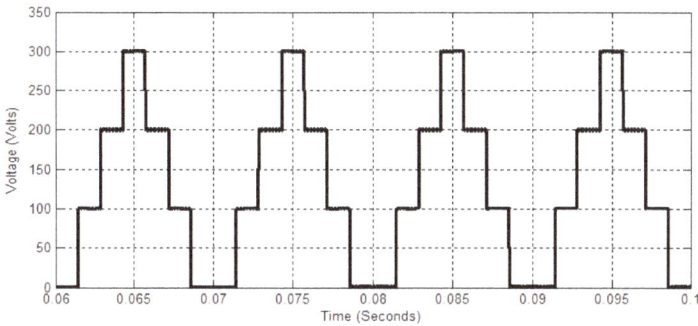

Figure 16. Inverter output across LGU using the EPCAD technique.

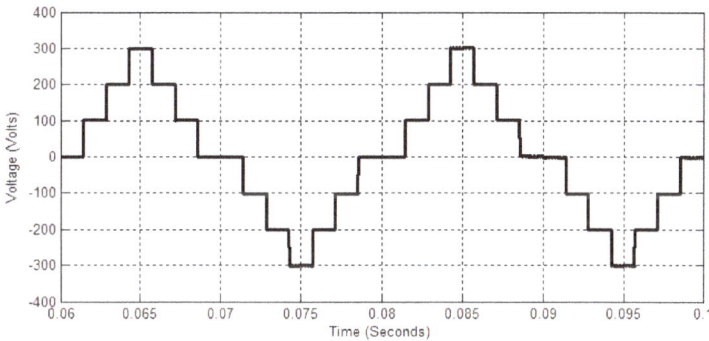

Figure 17. TLBDCLCHB inverter output using EPCAD technique.

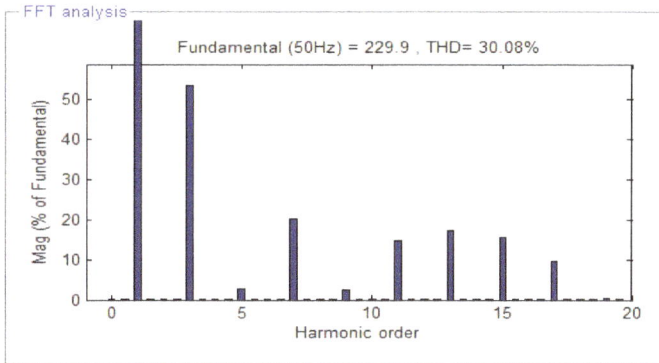

Figure 18. THD analysis of TLBDCLCHB inverter output using EPCAD technique.

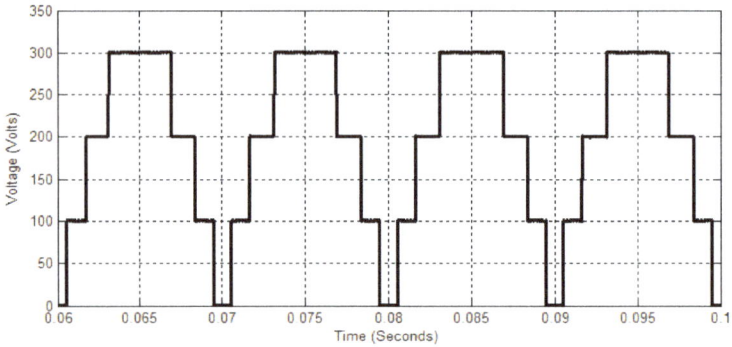

Figure 19. Inverter output across LGU using the SPWCAD technique.

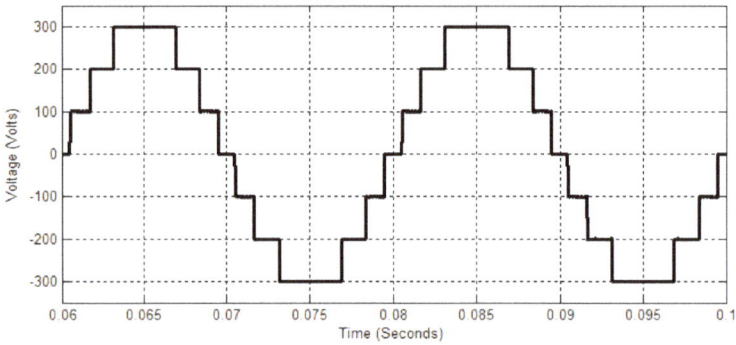

Figure 20. TLBDCLCHB inverter output using the SPWCAD technique.

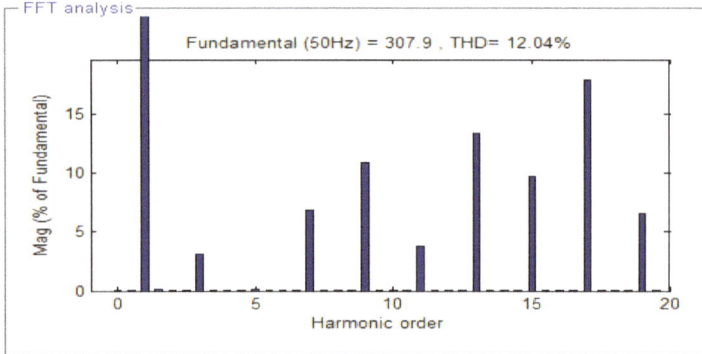

Figure 21. THD analysis of the TLBDCLCHB inverter output using the SWPCAD technique.

6.3.1. Using the EPCAD Technique

Figure 16 shows the output voltage across the LGU in the DCLCHB inverter. Referring to Figure 16, the LGU generates a unipolar stepped waveform, and it can be converted to a bipolar stepped wave using PSGU. Figures 17 and 18 show the seven-level stepped output phase voltage and its THD analysis of the TLBDCLCHB inverter using the EPCAD technique for the duty cycle of 0.5. It is observed that the magnitude of the fundamental output phase voltage and its RMS value is 229.9 V and 162.5 V, respectively. Also, the THD of the proposed TLBDCLCHB inverter output using the EPCAD technique is 30.08%.

6.3.2. Using the SPWCAD Technique

The unipolar output phase voltage across the LGU in the DCLCHB inverter using the SPWCAD technique is shown in Figure 19. Figures 20 and 21 show the seven-level stepped output phase voltage and its THD analysis of the TLBDCLCHB inverter using the SPWCAD technique for the duty cycle of 0.5 with M_i = 0.8. It is observed that the magnitude of the fundamental output phase voltage and its RMS value is 307.9 V and 217.7 V, respectively. Also, the THD of the proposed TLBDCLCHB inverter output using the SPWCAD technique is 12.04%.

6.4. Experimental Validation of the TLBDCLCHB Inverter System Using an FPGA-Based Pulse Generation

The model of the proposed TLBDCLCHB inverter system-fed R-load is implemented employing Xilinx Spartan FPGA-based pulse generation [19,20] to validate the Simulink results. The block diagram, Xilinx Spartan6 development board, and the prototype model of the TLBDCLCHB inverter system using an FPGA is shown in Figures 22–24, respectively. It consists of TLBC, the DCLCHB inverter, a personal computer (PC), an FPGA controller, R-load, buffer circuit, optocoupler, and driver circuit. The output of TLBC-fed DCLCHB inverter using an FPGA controller is shown in Figure 25.

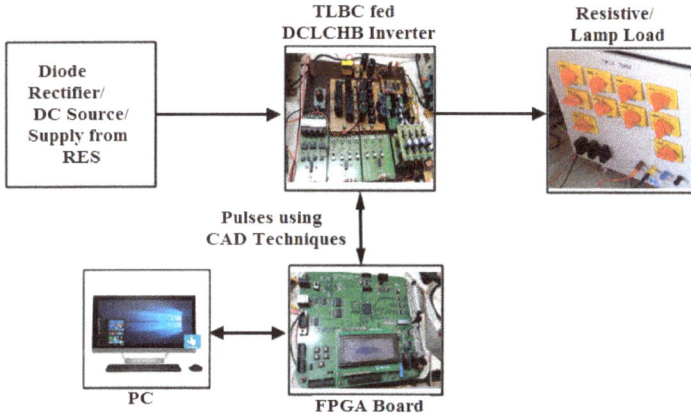

Figure 22. Block diagram of the TLBDCLCHB hardware implementation.

Figure 23. Hardware implementation of the TLBDCLCHB inverter system with an FPGA controller.

Figure 24. Xilinx Spartan6 development board.

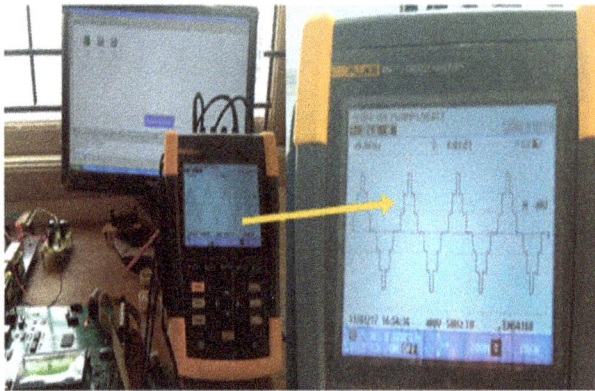

Figure 25. Seven-level stepped output voltage of TLBC-fed DCLCHB inverter system.

Referring to Figure 26, the TLBC output voltage across each capacitor and the total DC-link voltage are shown in Channel 1 (CH1) and Channel (CH2), respectively, for $D = 0.5$. It is observed that CH1 and CH2 voltages are boosted to 100 V and 300 V, respectively.

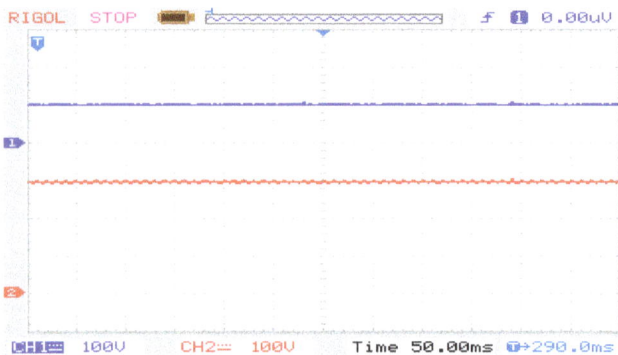

Figure 26. Voltage across the capacitor CH1 and total DC-link voltage of TLBC CH2 for $D = 0.5$.

Figures 27 and 28 show the generation of the pulse for the LGU switches (S_1 to S_6) in the DCLCHB inverter to generate the stepped output waveform using EPCAD and SPWCAD techniques, respectively, for the modulation index of 0.8. Figure 29 shows the generation of the pulse for the PSGU switches (P_1 to P_4) in the DCLCHB inverter to generate the positive and negative levels using an H-bridge inverter.

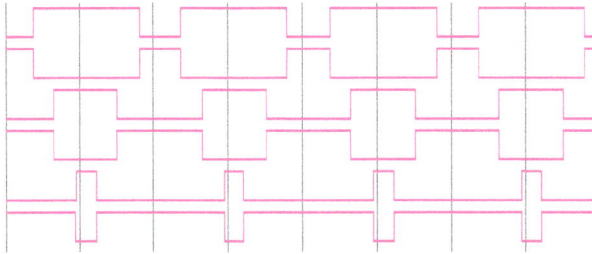

Figure 27. Generation of pulses for the LGU switches (S_1 to S_6) in the DCLCHB inverter using the EPCAD technique through Xilinx ISE.

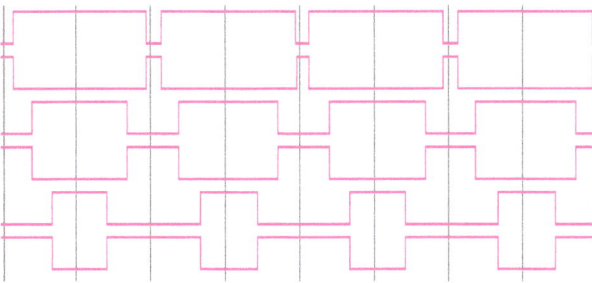

Figure 28. Generation of pulses for the LGU switches (S_1 to S_6) in the DCLCHB inverter using the SPWCAD technique through Xilinx ISE.

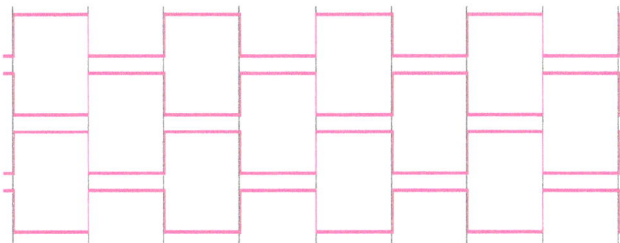

Figure 29. Generation of pulses for the PSGU switches (P_1 to P_4) in the DCLCHB inverter using a pulse generator through Xilinx ISE.

Figures 30 and 31 represent the TLBDCLCHB inverter system experimental output phase voltage and its harmonic spectrum using the EPCAD technique for the generation of a seven-level stepped output voltage. Referring to Figures 30 and 31, it is observed that the RMS value of the output phase voltage is 161.7 V and its THD is 31.5%.

Figures 32 and 33 represent the TLBDCLCHB inverter system experimental output phase voltage and its harmonic spectrum using the SPWCAD technique for the generation of a seven-level stepped output voltage. Referring to Figures 32 and 33, it is observed that the RMS value of the output phase voltage is 216 V and its THD is 11.5%.

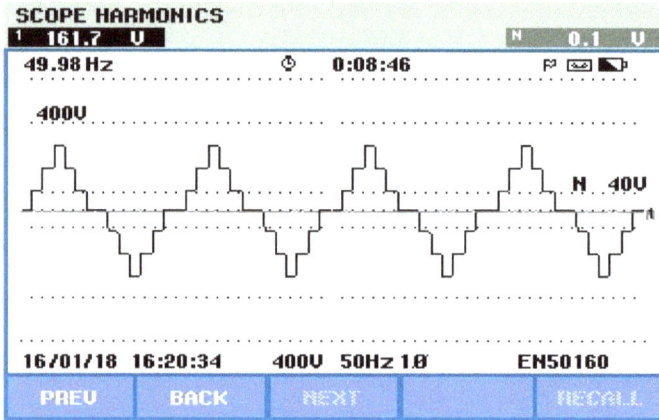

Figure 30. TLBDCLCHB inverter output using the EPCAD technique.

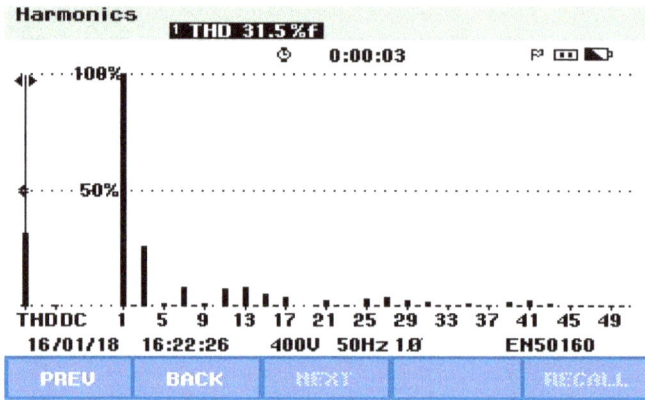

Figure 31. THD analysis of the TLBDCLCHB inverter output using the EPCAD technique.

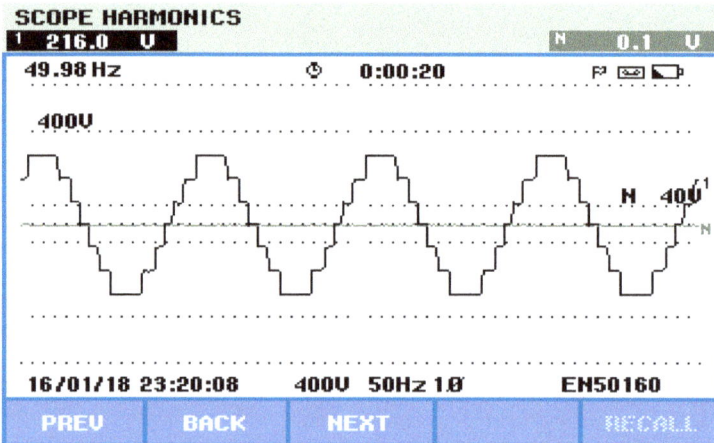

Figure 32. TLBDCLCHB inverter output using the SPWCAD technique.

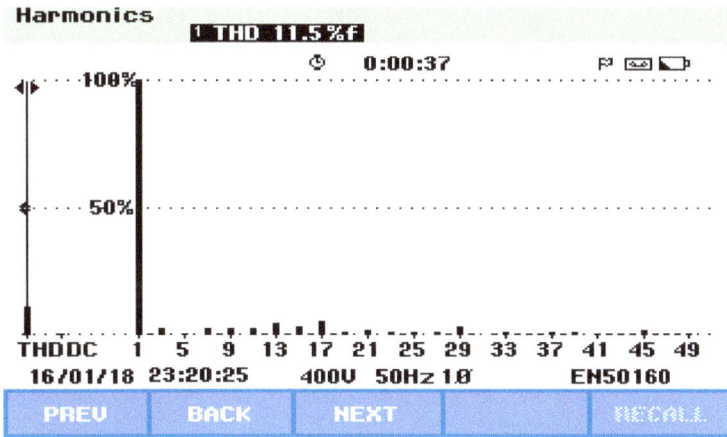

Figure 33. THD analysis of the TLBDCLCHB inverter output using the SPWCAD technique.

6.5. TLBDCLCHB Inverter System Analysis and Comparison Using CAD Techniques

In this study, theoretical and simulation results of the proposed TLBDCLCHB inverter system for the generation of a seven-level stepped output phase voltage using the EPCAD and SPWCAD techniques have been validated experimentally through an FPGA-based pulse generation. Tables 7 and 8 analyze the output phase voltage (V_{rms}) and THD of the TLBDCLCHB inverter system for different duty cycles, i.e., $D = 0.4, 0.5$, and 0.6. The THD of the prototype model for the proposed CAD techniques are conceded using a Fluke 435 power quality analyzer, and the results are presented in Figures 16–29 for the EPCAD and SPWCAD techniques.

Table 7. Simulation comparison of (V_{rms}) and % THD.

Duty Cycle (D)	EPCAD Technique		SPWCAD Technique	
	V_{rms} (V)	THD (%)	V_{rms} (V)	THD (%)
0.4	135.7	29.71	185.5	12.02
0.5	162.5	30.08	217.7	12.04
0.6	203.6	29.90	272	12.07

Table 8. Experimental comparison of (V_{rms}) and %THD.

Duty Cycle (D)	EPCAD Technique		SPWCAD Technique	
	V_{rms} (V)	THD (%)	V_{rms} (V)	THD (%)
0.4	134.6	31.8	184.7	11.8
0.5	161.7	31.5	216	11.5
0.6	202.1	31.5	270.9	11.7

Referring to Tables 6–8, it is inferred that the most extreme output phase voltage and lower THD are accomplished by utilizing the SPWCAD technique rather than the EPCAD technique. From the simulation, it is obtained that V_{rms} and %THD content in the EPCAD technique are 161.7 V and 31% for $D = 0.5$. Whereas, in the case of the proposed SPWCAD technique, V_{rms} is 216 V and %THD is only 11.5% for $D = 0.5$ by considering $M_i = 0.8$. From the experimental results, it is observed that the V_{rms} and %THD content in the EPCAD technique are 161.7 V and 31% for $D = 0.5$. For the proposed SPWCAD technique, V_{rms} is 216 V and %THD is only 11.5% for $D = 0.5$ by considering $M_i = 0.8$. Therefore, theoretical values V_{rms} and %THD shown in Table 6 are validated with the simulation and experimental results with a tolerable error of $\pm 2\%$. From the analysis, it is noticed that the magnitude of V_{rms} varies with respect to the duty cycle, but there is only a slight deviation in THD from 0.4 to 0.6.

7. Conclusions

In this paper, a TLBC-fed DCLCHB inverter system has been suggested to generate a seven-level stepped output phase voltage using a single DC source for better performance, efficiency, and reduced cost and size of the inverter. It also presented two control techniques for the DCLCHB inverter based on conducting angle determination, namely, EPCAD and SPWCAD techniques. Here, the SPWCAD technique gives the most extreme output phase voltage and lower THD compared to the EPCAD technique but the SPWCAD technique involves several trigonometric functions. However, same trigonometric functions are repeated; therefore, it is easy to acquire the conducting angles once the equations are derived based on the volt-second balance. In addition, TLBC has been suggested to achieve auto voltage balance and high voltage gain. Therefore, upon considering all the advantages, the proposed TLBDCLCHB inverter system is a good alternative for PV applications compared to the conventional boost-based MLI systems.

Author Contributions: A.K.D.V. and S.B.Ch. contributed to the main idea of this article. N.R.S. developed the Simulink model and performed the experiments, N.R.S., A.K.D.V. and S.B.Ch. all contributed to manuscript writing and revisions. All authors approved the final version to be published.

Funding: This research received no external funding.

Conflicts of Interest: The authors declare no conflict of interest.

Abbreviations

The following abbreviations are used in this manuscript:

CAD	conducting angle determination
DCLCHB	DC-link cascade H-bridge
EPCAD	equal-phase CAD
LGU	level generation unit
THD	total harmonic distortion
TLBC	three-level boost converter
TLBDCLCHB	three-level boost DC-link cascade H-bridge
MLI	multilevel inverter
PSGU	phase sequence generation unit
PWM	pulse width modulation
PV	photovoltaic
SPWCAD	step pulse wave CAD

References

1. Rodriguez, J.; Lai, J.S.; Peng, F.Z. Multilevel inverters: A survey of topologies, controls, and applications. *IEEE Trans. Ind. Electron.* **2002**, *49*, 724–738. [CrossRef]
2. Kjaer, S.B.; Pedersen, J.K.; Blaabjerg, F. A review of single-phase grid-connected inverters for photovoltaic modules. *IEEE Trans. Ind. Appl.* **2005**, *41*, 1292–1306. [CrossRef]
3. Rodríguez, J.; Bernet, S.; Wu, B.; Pontt, J.O.; Kouro, S. Multilevel voltage-source-converter topologies for industrial medium-voltage drives. *IEEE Trans. Ind. Electron.* **2007**, *54*, 2930–2945. [CrossRef]
4. Najafi, E.; Yatim, A.H. Design and implementation of a new multilevel inverter topology. *IEEE Trans. Ind. Electron.* **2012**, *59*, 4148–4154. [CrossRef]
5. Davari, P.; Yang, Y.; Zare, F.; Blaabjerg, F. A review of electronic inductor technique for power factor correction in three-phase adjustable speed drives. In Proceedings of the 2016 IEEE Energy Conversion Congress and Exposition (ECCE), Milwaukee, WI, USA, 18–22 September 2016; pp. 1–8.
6. Klumper, C.; Blaabjerg, F.; Thøgersen, P. Alternate ASDs: Evaluation of the converter topologies suited for integrated motor drives. *IEEE Ind. Appl. Mag.* **2006**, *12*, 71–83. [CrossRef]

7. Su, G.J. Multilevel DC-link inverter. *IEEE Trans. Ind. Appl.* **2005**, *41*, 848–854. [CrossRef]
8. Rao, S.N.; Kumar, D.A.; Babu, C.S. New multilevel inverter topology with reduced number of switches using advanced modulation strategies. In Proceedings of the 2013 International Conference on Power, Energy and Control (ICPEC), Sri Rangalatchum Dindigul, India, 6–8 February 2013; pp. 693–699.
9. Uthirasamy, R.; Chinnaiyan, V.K.; Ragupathy, U.S.; Karpagam, J. Investigation on three-phase seven-level cascaded DC-link converter using carrier level shifted modulation schemes for solar PV system applications. *IET Renew. Power Gener.* **2017**, *12*, 439–449. [CrossRef]
10. Prabaharan, N.; Palanisamy, K. A comprehensive review on reduced switch multilevel inverter topologies, modulation techniques and applications. *Renew. Sustain. Energy Rev.* **2017**, *76*, 1248–1482. [CrossRef]
11. Luo, F.L. Investigation on best switching angles to obtain lowest THD for multilevel DC/AC inverters. In Proceedings of the 2013 IEEE 8th Conference on Industrial Electronics and Applications (ICIEA), Melbourne, Australia, 19–21 June 2013; pp. 1814–1818.
12. Kang, D.W.; Kim, H.C.; Kim, T.J.; Hyun, D.S. A simple method for acquiring the conducting angle in a multilevel cascaded inverter using step pulse waves. *IEE Proc.-Electr. Power Appl.* **2005**, *152*, 103–111. [CrossRef]
13. Rao, S.N.; Kumar, D.A.; Babu, C.S. Implementation of Multilevel Boost DC-Link Cascade based Reversing Voltage Inverter for Low THD Operation. *J. Electr. Eng. Technol.* **2018**, *13*, 1527–1537.
14. Uthirasamy, R.; Ragupathy, U.S.; Chinnaiyan, V.K. Structure of boost DC-link cascaded multilevel inverter for uninterrupted power supply applications. *IET Power Electron.* **2015**, *8*, 2085–2096. [CrossRef]
15. Prasad, G.D.; Jegathesan, V.; Moorthy, V. Minimization of power loss in newfangled cascaded H-bridge multilevel inverter using in-phase disposition PWM and wavelet transform based fault diagnosis. *Ain Shams Eng. J.* **2016**. [CrossRef]
16. Rosas-Caro, J.C.; Ramírez, J.M.; García-Vite, P.M. Novel DC-DC multilevel boost converter. In Proceedings of the 2008 IEEE Power Electronics Specialists Conference, Rhodes, Greece, 15–19 June 2008; pp. 2146–2151.
17. Rosas-Caro, J.C.; Ramirez, J.M.; Valderrabano, A. Voltage balancing in DC/DC multilevel boost converters. In Proceedings of the 2008 40th North American Power Symposium, Calgary, AB, Canada, 28–30 September 2008; pp. 1–7.
18. Rosas-Caro, J.C.; Ramirez, J.M.; Peng, F.Z.; Valderrabano, A. A DC–DC multilevel boost converter. *IET Power Electron.* **2010**, *3*, 129–137. [CrossRef]
19. Maruthupandi, P.; Devarajan, N.; Sebasthirani, K.; Jose, J.K. Optimum control of total harmonic distortion in field programmable gate array-based cascaded multilevel inverter. *J. Vib. Control* **2015**, *21*, 1999–2005. [CrossRef]
20. Cong, J.; Liu, B.; Neuendorffer, S.; Noguera, J.; Vissers, K.; Zhang, Z. High-level synthesis for FPGAs: From prototyping to deployment. *IEEE Trans. Comput.-Aided Des. Integr. Circuits Syst.* **2011**, *30*, 473–491. [CrossRef]

electronics

MDPI

Article

All SiC Grid-Connected PV Supply with HF Link MPPT Converter: System Design Methodology and Development of a 20 kHz, 25 kVA Prototype

Serkan Öztürk [1], Mehmet Canver [2], Işık Çadırcı [1] and Muammer Ermiş [2,*]

[1] Department of Electrical and Electronics Engineering, Hacettepe University, Beytepe, Ankara 06800, Turkey; ozturk@ee.hacettepe.edu.tr (S.Ö.); cadirci@ee.hacettepe.edu.tr (I.Ç.)

[2] Department of Electrical and Electronics Engineering, Middle East Technical University, Ankara 06800, Turkey; mehmet.canver@artielektronik.com.tr

* Correspondence: ermis@metu.edu.tr; Tel.: +90-312-210-2364

Received: 7 May 2018; Accepted: 28 May 2018; Published: 31 May 2018

Abstract: Design methodology and implementation of an all SiC power semiconductor-based, grid-connected multi-string photovoltaic (PV) supply with an isolated high frequency (HF) link maximum power point tracker (MPPT) have been described. This system configuration makes possible the use of a simple and reliable two-level voltage source inverter (VSI) topology for grid connection, owing to the galvanic isolation provided by the HF transformer. This topology provides a viable alternative to the commonly used non-isolated PV supplies equipped with Si-based boost MPPT converters cascaded with relatively more complex inverter topologies, at competitive efficiency figures and a higher power density. A 20 kHz, 25 kVA prototype system was designed based on the dynamic model of the multi-string PV panels obtained from field tests. Design parameters such as input DC link capacitance, switching frequencies of MPPT converter and voltage source inverter, size and performance of HF transformer with nanocrystalline core, DC link voltage, and LCL filter of the VSI were optimized in view of the site dependent parameters such as the variation ranges of solar insolation, module surface temperature, and grid voltage. A modified synchronous reference frame control was implemented in the VSI by applying the grid voltage feedforward to the reference voltages in abc axes directly, so that zero-sequence components of grid voltages are taken into account in the case of an unbalanced grid. The system was implemented and the proposed design methodology verified satisfactorily in the field on a roof-mounted 23.7 kW multi-string PV system.

Keywords: dynamic PV model; grid-connected VSI; HF-link MPPT converter; nanocrystalline core; SiC PV Supply

1. Introduction

Various grid-connected photovoltaic system concepts and topologies have been summarized previously [1–4]: (i) micro inverters [5,6]; (ii) residential systems supplied from a PV string [7,8]; (iii) commercial/residential systems supplied from multiple PV strings having their own maximum power point tracking (MPPT) converters and a central inverter [9,10]; and (iv) commercial/utility-scale PV plants supplied from a common DC link with a central inverter [11]. The central inverter in [11] performs direct conversion of PV power to AC from multiple PV strings. MPPT and reliable and efficient conversion of PV power to AC at grid frequency are the major design issues for two stage grid-connected PV systems [12]. In the PV systems performing direct conversion of PV power to AC, the voltage of the common DC link is varied by the central inverter against the changes in solar insolation and panel temperature [13]. An MPPT algorithm integrated into a two-layer controller is recommended in [14] for direct conversion of PV power through a three-phase grid-connected VSI.

A neural network based MPPT algorithm is proposed in [15] to improve the dynamic performance of DC capacitor voltage at the input of VSI and to maintain faster tracking response against sudden changes in solar irradiance. Grid-connected PV supplies can be classified into two groups: (i) those having transformerless inverter topologies [16]; and (ii) those having magnetic transformers usually on the grid side [17]. Transformerless PV supplies require more complex and expensive inverter topologies as compared to PV supplies having transformers for galvanic isolation to prevent the flow of common-mode currents [18–22].

In the vast majority of commercially available PV supplies, new generation hybrid IGBTs are currently employed to synthesize grid frequency voltages and/or currents. In the next generation PV inverters, simpler circuit designs such as two-level instead of multi-level inverters are expected to be used for fewer components and hence higher reliability [23]. Wide bandgap (WBG) devices such as SiC power MOSFETs will be employed for higher switching frequency to reduce the size of passive components, heatsinks and hence the volume of the PV supply, as foreseen in [23]. WBG semiconductors show superior material properties, enabling potential power device operation at higher temperatures, voltages, and switching speeds than current Si technology [24].

In recent years, performances of all SiC PV supplies consisting of DC/DC converters and three-phase grid-connected inverters are assessed, as reported in [25–27]. These use either three-level T-type inverter topology [26] or cascaded multi-level inverter topology [27]. All SiC inverters such as single-phase H-bridge converter [28], neutral point clamped (NPC) T-type three-phase inverter [29], and three-phase five-level T-type inverter [30] are used to perform direct conversion of PV power to the grid. Design and performances of some non-application-specific two-level three-phase SiC inverters are presented in [31–34].

Since the power converters in PV supplies are presently switched at relatively higher frequencies, ranging from a few kHz to a few tens of kHz, sizing and design of these converters require the dynamic model of the PV cells, modules or arrays instead of the well-known static model. Some attempts have been made to obtain dynamic model parameters of various types of PV cells or modules by theoretical calculations and/or measurements [35–40]. The use of internal solar cell diffusion capacitance to replace the input capacitor in boost derived DC-DC converters has been proposed in [41] for energy harvesting applications. In almost all of the two-stage PV supplies, boost type DC-DC converters have been used for MPPT. Recently, in this area, small size SiC MOSFET [42–45] and SiC JFET based [46] MPPT converters are reported in the literature. A 1-MW solar power inverter with boost converter using all SiC power modules has been presented in [47]. A 3.5 kW, 100 kHz all SiC buck/boost interleaved MPPT converter is proposed in [48] for higher conversion efficiency. Design, construction, and testing of a general purpose 750-V, 100-kW, 20-kHz bidirectional dual active bridge DC–DC converter using all SiC dual modules with HF transformer is described in [49] to provide maximum conversion efficiency from the DC-input to the DC-output terminals.

Design of the AC filter on the grid side of a VSI is a compromise between performance and size. A step-by-step design procedure of an LCL filter for a grid-connected three-phase PWM voltage source converter is proposed in [50]. A new design method is recommended in [51] for LCL and LLCL filters with passive damping to be used in single-phase grid-connected VSIs. Peña-Alzola et al. [52] examined passive damping losses in LCL filters. Design methodology, analysis and performance of LCL filters for three-phase grid-connected inverters are also described in [53,54]. Stability regions of active damping control for LCL filters are discussed in [55]. A generalized LCL filter design algorithm for grid-connected VSIs is presented in [56].

The common control methods for three-phase grid-connected inverters are: (i) synchronous reference frame control (in dq-axes) [57–59]; (ii) stationary reference frame control (in $\alpha\beta$-axes) [60]; and (iii) natural reference frame control (in abc-axes) [61]. The main drawback of the synchronous reference frame control is the poor compensation capability of the low-order harmonics in the case of PI based controllers [60]. To cope with this drawback, the control system can be equipped with proportional resonant harmonic compensators, one pair for each prominent low order sidekick harmonics produced by the system itself [62]. In stationary reference frame control method, two proportional resonant controllers can be used in $\alpha\beta$-axes for the fundamental component to eliminate possible steady-state error which may arise in PI control. However, the number of PR controllers for low-order harmonic compensation are doubled. In *abc* control strategy, an individual controller is required for each line current on the grid side. In this control method, all types of controllers such as PI controller, PR controller, hysteresis controller, dead-beat controller and repetitive controller can be adopted [61].

This paper recommends a grid-connected, all SiC Power MOSFET based multi-string PV supply with a HF link MPPT converter in each string and a two-level central inverter. The design methodology described in the paper utilizes a custom dynamic model of the roof-mounted multi-string PV system, and site dependent parameters. These parameters are variation ranges of solar insolation, module surface temperature, and grid voltage. In the design procedure the following are optimized: (i) input DC link capacitance; (ii) switching frequencies of MPPT converter and VSI; (iii) the size and performance of HF transformer with nanocrystalline core; (iv) the DC link voltage; and (v) the LCL filter of VSI. Corresponding field test results of the implemented 20 kHz, 25 kVA prototype are presented. It has been shown that the HF link converter in cascade with a two-level VSI provides a viable alternative to multi-string PV supplies with a competitive efficiency, lower harmonic distortion and a much higher power density. This PV supply topology with a HF link MPPT converter makes possible the use of a simple and reliable, two-level, voltage source PV inverter for grid connection. In the case of PV supplies with non-isolated MPPT converters or those performing direct conversion of PV power to the grid, either complex inverter topologies or bulky grid-side common-mode filters or coupling transformers would be required to minimize the flow of undesirable common-mode currents.

2. System Description

2.1. General

Block diagram representation of all SiC grid-connected PV supply with HF link is as shown in Figure 1a. With the advents in SiC power MOSFETs, the kVA rating of three-phase, two-level voltage source all SiC central inverter can reach 175–400 kVA, by employing commercially available three half-bridge all SiC modules in Table 1 (by February 2018) for direct connection to 400 V l-to-l, 50 Hz utility grid. Each MPPT converter with DC/AC/DC link can be designed to process 15–25 kW PV power by employing SiC half-bridge modules, TO-package SiC Schottky diodes, and a nanocrystalline transformer core. The MPPT converter rating in the application described in this paper is limited to 25 kW by considering the partial shading risk of multi-string PV panels. In the experimental system, five mono-crystalline PV strings of 23.7 kW peak under standard test conditions, with adjustable tilt angles occupy nearly 375 m^2 in the roof area as shown in Figure 2. In this study, the experimental work was carried out on the prototype system given in Figure 1b.

Table 1. Commercially available all SiC power MOSFET half-bridge modules that can be employed in central inverter in Figure 1a/by February 2018.

Manufacturer	Half-Bridge Module	V_{DS}, V	I_D, A
ROHM [63]	BSM300D12P2E001	1200	300
CREE [64]	CAS325M12HM2	1200	325
SEMIKRON [65]	SKM500MB120SC	1200	541
Mitsubishi [66]	FMF800DX-24A	1200	800

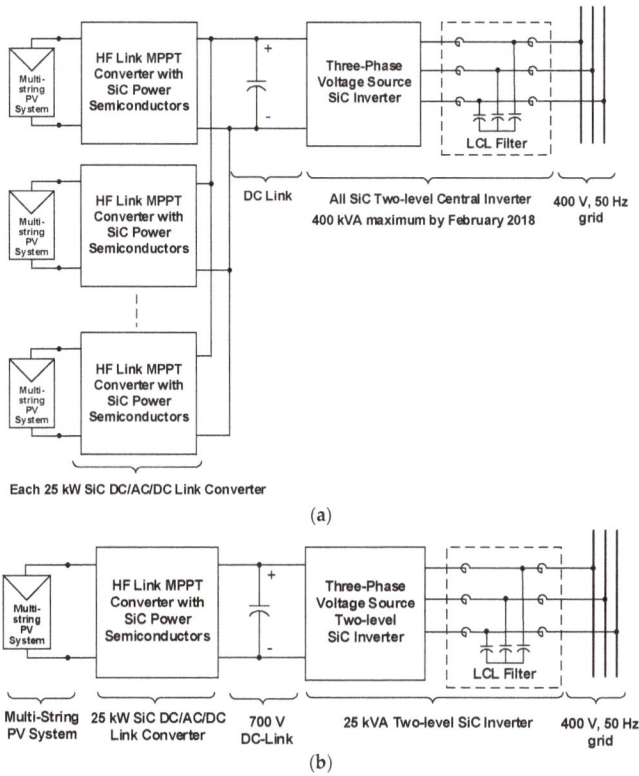

Figure 1. All SiC three-phase grid-connected PV supply with high-frequency link: (**a**) Multi-string PV inverter system; and (**b**) Experimental set-up.

Figure 2. Roof-mounted 23.7 kW at 1000 W/m^2 and 25 °C multi-string PV system containing 95 CSUN250-72M modules (five parallel connected strings of 19 series modules; with adjustable tilt angles of 15°, 30°, 45° and 60°); coordinates 39.890917, 32.782278; five modules in the photograph are idle; terminal box of the system inside the laboratory is shown at the right bottom corner of photograph while the pyranometer is at left bottom corner.

Advantages and disadvantages of the multi-string PV supply system in Figure 1a can be summarized as follows:

(i) In medium power applications, as a wide bandgap device, SiC power MOSFETs can be switched at higher frequencies (a few tens of kHz) in comparison with Si IGBTs. This ability brings us the following benefits:

- Higher efficiency at the same switching frequency;
- Much smaller HF transformer in DC/AC/DC link of the MPPT converter;
- Much smaller electrolytic or metal film DC link capacitor at the output of the MPPT converter;
- Much smaller distributed metal film DC link capacitor bank in the laminated bus/PCB of the inverter circuit;
- Smaller LCL filter bank at the output of the PV inverter;
- Silent operation;
- Permits the use of two-level three-phase bridge inverter topology to conform with the power quality regulations on the grid side; and
- Extends the range of high order current harmonics that can be eliminated.

(ii) They eliminate the need for a bulky, grid-side PV transformer to provide electrical isolation.

(iii) Separation of installed PV panels into multiple strings having MPPT converters minimizes the undesirable effects of partial shading.

(iv) The central inverter has higher efficiency and lower cost in comparison with the usage of several smaller scale inverters.

(v) In most countries, the PV system in Figure 1a can be directly connected to the low-voltage (LV) side of distribution transformers without the permission of distribution system operators. Although standard power ratings for distribution transformers are 100, 250, 400, and 630 kVA [67], higher ratings up to 2 MVA are also in service.

(vi) Since the HF link MPPT converter is cascaded with the grid-connected PV inverter, overall efficiency is 1–2% lower than that of PV converters performing direct conversion of PV power to the grid. However, the PV systems having non-isolated MPPT converters or performing direct conversion of PV power to the grid should have more complex converter topologies to minimize common-mode currents.

2.2. MPPT Converter with HF Link

A prototype of the MPPT converter with HF link shown in Figure 1b is designed and implemented. MPPT converter is composed of a SiC Power MOSFET based single-phase H-bridge converter, a HF transformer, and a SiC Schottky diode rectifier. The circuit diagram of the power stage, the control circuitry, and top and side views of the implemented MPPT converter prototype are as shown in Figure 3.

2.3. Three-Phase Two-Level Voltage Source Inverter

A prototype of the three-phase two-level voltage source inverter shown in Figure 1b was designed and implemented. This converter topology was chosen for the following reasons:

- The ability of SiC power MOSFETs at high switching frequencies for lower harmonic current content;
- Minimum power semiconductor count gives higher reliability in comparison with multi-level converters;
- Common-mode current is not a design concern because of the HF transformer in the MPPT converter; and
- PV inverter delivers power to one of the most common low voltage utility grid.

The circuit diagram of the power stage and its control circuitry, and top and side views of the implemented inverter prototype are shown in Figure 4.

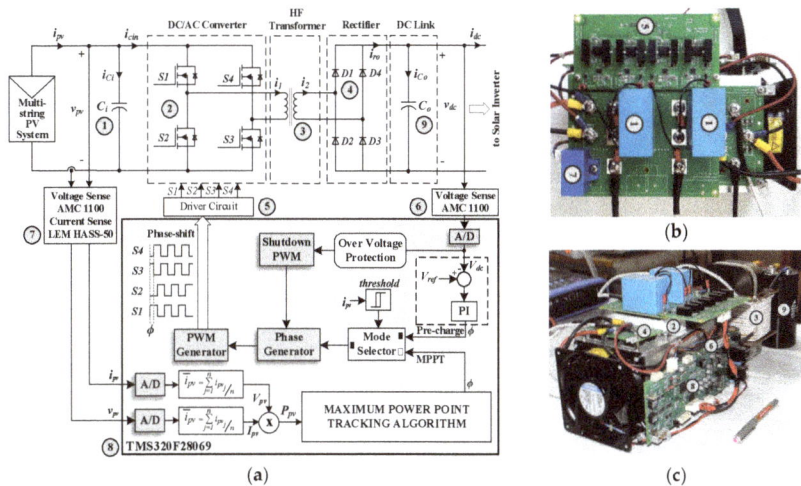

Figure 3. All SiC MPPT converter with high-frequency link: (**a**) power stage and its control circuitry; (**b**) yop view of the developed hardware; and (**c**) side view. (1) Metal Film 30//30 µF Input Capacitor Bank (EPCOS B32778G0306K); (2) All-SiC Half Bridge Module (CREE CAS120M12BM2); (3) Nanocrystalline Core (SU-102b) Based HF Transformer; (4) SiC Schottky Diode (CREE C4D40120D); (5) Dual Channel SiC MOSFET Driver (CREE CGD15HB62P); (6) Fully-Differential Isolation Amplifier (TI AMC-1100); (7) Hall-Effect Current Transducer (LEM HASS 50-S); (8) Microcontroller (TI TMS320F28069); (9) Electrolytic 3400 µF Output Capacitors (EPCOS B43456-A9688-M).

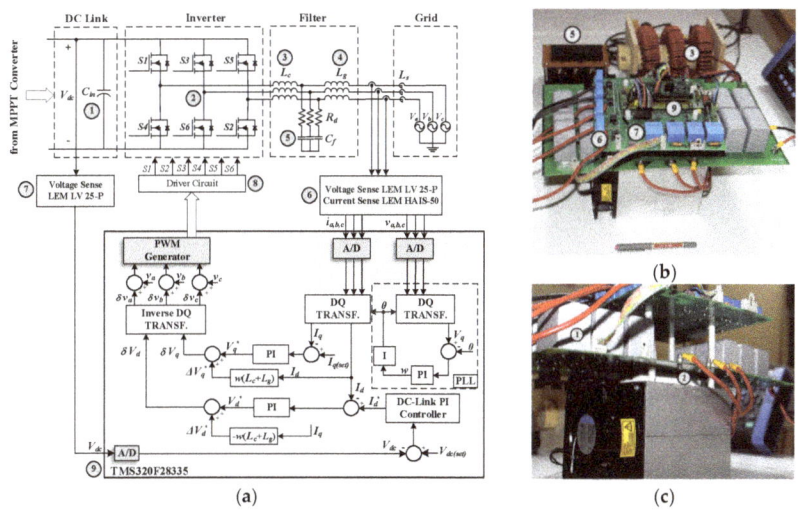

Figure 4. All SiC three-phase grid-connected two-level inverter: (**a**) power stage and its control circuitry; (**b**) top view of the developed hardware; and (**c**) side view. (1) Metal Film 12 × 40 µF Input Capacitor Bank (VISHAY MKP1848640094Y); (2) All-SiC Six-Pack Three-Phase Module (CREE CCS050M12CM2); (3,4) X-Flux L_c = 250 µH and L_g = 50 µH Filter Inductor (Magnetics 0078337A7); (5) Metal Film 15 µF Filter Capacitor (VISHAY MKP1848S61510JY5F); (6) Hall-Effect Current Transducer (LEM HAIS 50-P); (7) Voltage Transducer (LEM LV25-P); (8) Single Channel SiC MOSFET Driver (CREE CRD-001); (9) Experimenter Kit (TI TMDSDOCK28335).

3. Modeling and System Design

The recommended modeling and system design methodology is described in this section for the existing multi-string PV system shown in Figure 2. For another multi-string PV system configuration, the same design principles can be applied.

3.1. Dynamic Model of Multi-String PV System

Even in the steady-state operation of PV supplies, their power converters such as the MPPT converter and the inverter operate in the periodic transient state, instead of pure DC operation. Therefore, in the analysis and design of such systems, a proper dynamic model of the multi-string PV system is to be used. Several attempts have been made to obtain dynamic models of a PV module [35–41]. However, in this research work, the dynamic model parameters of the roof-mounted multi-string PV system shown in Figure 2 were obtained from a set of experimental data. It includes also all cabling and wirings up to the input terminals of the MPPT converter. These parameters were then combined with the static model of the CSUN250-72M modules available in MATLAB/Simscape/Power Systems R2016b for use in the design work.

Equivalent circuits of the multi-string PV system in Figure 2 consisting of 95 pieces of CSUN250-72M modules (5 × 19 modules) are as shown in Figure 5. Static model in Figure 5a is determined for the PV array in Figure 2 by using the PV array block developed by the National Renewable Energy Laboratory (NREL) System Advisor Model and available in MATLAB/Simscape/Power Systems R2016b. However, the design of the PV supply presented in this paper is based on the dynamic model of the PV system shown in Figure 5b.

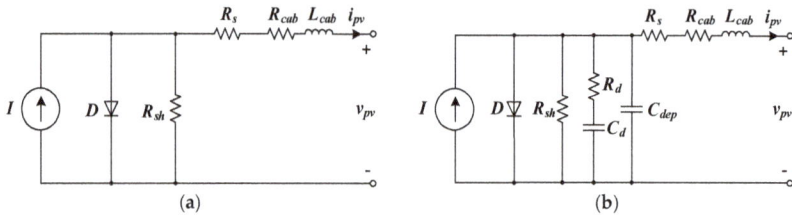

Figure 5. Equivalent circuit of the multi-string PV system in Figure 2 consisting of 95 pieces of CSUN250-72M modules (5 × 19 modules). (a) Static model: R_{sh} = 490 Ω and R_s = 1.33 Ω are, respectively, the equivalent shunt and series resistances calculated by MATLAB/Simscape/Power Systems; and R_{cab} =10 mΩ and L_{cab} = 45 µH are the equivalent parameters of all cabling and wiring and estimated from experimental results given in Figure 6. (b) Dynamic model: C_d = 4 µF is the equivalent diffusion capacitance and R_d = 3 Ω is its series resistance, and C_{dep} = 600 nF is the equivalent depletion layer capacitance, which were estimated from experimental results given in Figure 6.

The topology of this dynamic model is nearly the same as the one given in [41]. The dynamic model parameters are estimated from the results of two tests conducted on the multi-string PV system. In the first test, multi-string PV system is solidly short-circuited at the terminal box and the short-circuit current is recorded as shown in Figure 6a. In the second test, a slightly inductive resistive bank is suddenly connected to the open-circuited multi-string PV system terminals and the terminal voltage and the current are recorded as shown in Figure 6b. Parameters of the multi-string PV system estimated from the current and voltage records shown in Figure 6 are as given in the caption of Figure 5.

These experimental records are compared with the simulation results obtained by MATLAB/Simscape/Power Systems, for the static and dynamic models separately under the same test conditions, as given in Figure 6. It is clear from this figure that the dynamic model gives much better results than those of the static model, and it can therefore be satisfactorily used in the design of HF link MPPT converter.

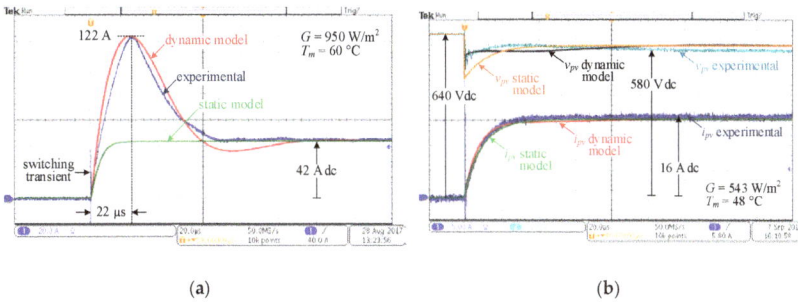

(a)   (b)

Figure 6. Transient response of the multi-string PV system at the input of the MPPT converter (Experimental: Tektronix MSO3034 oscilloscope, Tektronix TCPA300 current probe amplifier, Tektronix TCP404XL current probe; static model as given in Figure 5a; dynamic model as given in Figure 5b; Pyranometer EKO MS-410; Fluke Ti29 infrared camera): (**a**) from open-circuit to short-circuit; and (**b**) from open-circuit to partial resistive load (R_L = 42.5 Ω, L_L = 278 μH measured by GW INSTEK LCR-817 LCR meter at the resistor temperature of 72 °C).

3.2. Three-Phase Two-Level VSI

In this subsection, optimum DC link voltage and switching frequency are determined and the design of the inverter circuit, its control circuitry and LCL filter are described. The design of the HF link converter is directly affected by the optimum DC link voltage determined in this subsection.

3.2.1. Optimum DC Link Voltage

DC link voltage, V_{dc}, is kept constant by the inverter against the variations in PV power over the entire operating range of the PV supply. At the optimum value of V_{dc}, modulation index, M in Equation (1) of the inverter circuitry should vary in a range as close as possible to unity in order to minimize harmonic distortion of the output line currents.

$$M = (2\sqrt{2}/\sqrt{3})(V/V_{dc}) \tag{1}$$

where V is the l-to-l value of either grid voltage, V_s or consumers load voltage, V_t. Furthermore, in calculating the variation range of M, permissible changes in 400 V l-to-l, 50 Hz grid voltage are as specified in IEC 60038 2002-07 Standard Voltages [68]. This standard specifies the maximum changes in grid voltage, V_s as +6% to −0% while a further ±4% at the consumers load voltage, V_t. In view of these considerations, the variation ranges of M are calculated and plotted (Figure 7).

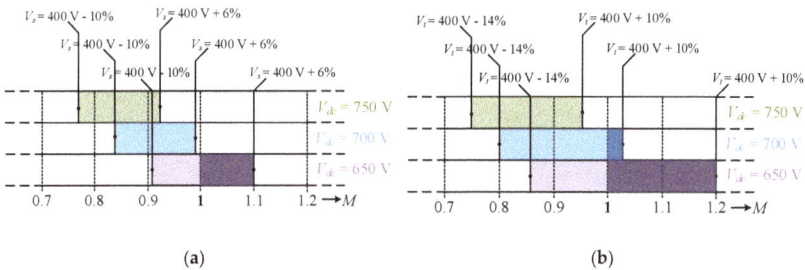

(a)   (b)

Figure 7. The variation range of modulation index, M as a function of chosen DC link voltage, V_{dc} and operating voltage (acceptable low voltage variation ranges for 230/400 V, 50 Hz systems are as specified in IEC 60038 2002-07 Standard Voltages): (**a**) for the supply side voltage range of V_s = 400 V +6% to −10%; and (**b**) for the utilization voltage range of V_t = 400 V +10% to −14%.

As can be understood from the results in Figure 7, the optimum value of V_{dc} is around 700 V. This choice may result in operation slightly in over-modulation region as given in Figure 7b. Since inverter rating is 25 kVA/22.3 kW for the available multi-string PV system, the inverter can deliver nearly 11 kVAr inductive or more to bring V_t back to V_s = 400 V + 6%. Application of third harmonic injection method [69,70] might be an alternative design approach in which the optimum value of V_{dc} is to be nearly 600 V for V_s = 400 V + 6%.

3.2.2. Optimum Switching Frequency

Higher switching frequencies for the two-level three-phase inverter with different modulation techniques such as SPWM, SVPWM, etc., excluding SHEM, result in low harmonic distortion for the line currents injected into the grid [71]. In this research work, sinusoidal PWM is chosen as the modulation technique because of its simplicity, and its ability to illustrate basic design guidelines. Since SiC Power MOSFETs can be switched at higher frequencies in comparison with Si IGBTs for the same power dissipation and solar inverter rating, power loss components of solar inverter with SPWM modulation are calculated for a reasonable operating frequency range, e.g., at f_{sw} = 10, 20, and 30 kHz, by using the expressions and manufacturers' design tools given in [72].

The associated pie-charts are shown in Figure 8. All wiring and cabling losses between discrete components, inverter, LCL filter, and grid are ignored in the preparation of Figure 8. In addition, extra power losses due to the switching ripple current through the power MOSFETs are not considered by the power loss calculation tools mentioned above in the calculation of conduction and switching losses. In summary, slightly higher loss content and a lower efficiency are expected for the solar inverter in the field tests.

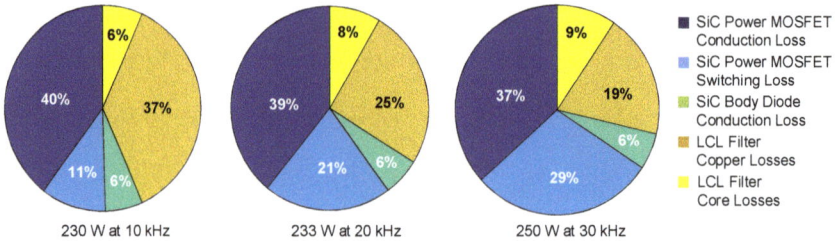

Figure 8. Power loss components of solar inverter for different switching frequencies at P_o = 22.3 kW. Numerical values are rounded.

In the design of LCL filter at different switching frequencies, only the converter side inductance, L_c, is optimized to keep its peak-to-peak current ripple constant at 25%. Power loss components in the associated LCL filters are then used in the preparation of pie charts in Figure 8. Although 10 kHz switching frequency reduces the power dissipation marginally in comparison with that of the 20 kHz, a considerably larger LCL filter size is to be used. Therefore, in the design and implementation of the solar inverter, f_{sw} = 20 kHz is chosen, which is a compromise between losses and LCL filter size.

3.2.3. LCL Filter Design

An LCL filter consisting of inverter side inductors L_c, shunt capacitors C_f, passive damping resistors R_d, and grid side inductors L_g are considered in design, as shown in Figure 4a. The LCL filter should be designed to have not more than 10 A peak-to-peak ripple superimposed on 36 A rms fundamental current in L_c at 25 kVA, and 400 V l-to-l. Peak-to-peak ripple remains nearly constant over the entire operating range of the all SiC three-phase grid-connected two-level inverter and is 25% at 25 kVA, 400 V l-to-l. The corner frequency of the LCL filter is chosen around 1/3rd of the 20 kHz switching frequency. These choices are consistent with the recommendations given in various papers [50–56]. The transfer function Bode plots of undamped LCL filter for three different L_c, C_f, and L_g parameter sets are given in Figure 9a. All of them provide nearly 100 dB attenuation at switching frequency. Among these, L_c = 250 µH, C_f =15 µF, and L_g = 50 µH parameter set is chosen for the implementation. Red colored parameter set (L_c =350 µH, C_f =20 µF, L_g = 50 µH) is not chosen because its L_c is nearly 40% greater than that of the optimum design and provides unnecessarily high attenuation. Although the green colored parameter set (L_c = 150 µH, C_f = 10 µF, and L_g = 50 µH) gives minimum LCL size, it is also not chosen in the implementation because it makes narrower the control range of the phase shift angle, and hence may cause undesirable oscillations in the output power and possible instability. On the other hand, field experience has shown that larger LCL filter size reduces harmonic content of the line currents and maintains stability of the inverter.

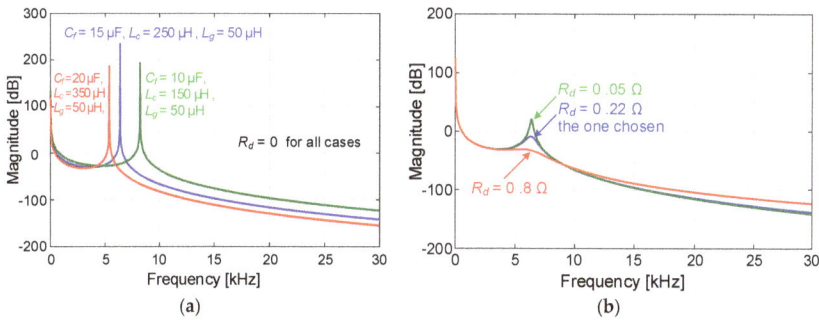

Figure 9. Transfer characteristics of various LCL filters shown in: (**a**) undamped filters, R_d = 0; and (**b**) effects of R_d on the transfer characteristic on the chosen LCL filter (L_c = 250 µH, C_f = 15 µF, and L_g = 50 µH).

Very high amplification of the current component at resonance frequency can be entirely eliminated by either passive or active damping technique. In this research work, passive damping is preferred and the damping resistance R_d is connected in series with C_f. In choosing the optimum value of R_d, a compromise is needed between copper losses and damping effect.

The effects of various damping resistors on the current transfer function bode plots of the chosen parameter set are as given in Figure 9b. Although lower R_d values are less dissipative, their damping effect is inadequate. On the other hand, higher R_d values provide strong damping at resonance frequency at the expense of higher losses and reduced attenuation at high frequencies. R_d = 0.22 Ω is therefore chosen for the implementation.

3.2.4. Controller Design

In this paper, active power delivered to the grid is controlled by using the rotating reference frame synchronized with the grid frequency by implementing a modified version of the control technique presented in [50]. The block diagram of the designed and implemented DSP (TMS320F28335) based controller is shown in Figure 4a. Line to neutral voltages $v_{a,b,c}$ and line currents $i_{a,b,c}$ on the grid side, and DC link voltage V_{dc} are the inputs to the controller. These quantities are sampled at 10 kHz/channel. Set values of the DC link voltage, $V_{dc(set)}$ and $I_{q(set)}$ are adjusted, respectively, to 700 V and 0 A in the control software. PWM signals applied to the driver circuits are the outputs of the control system. Control actions are achieved firstly in rotating DQ reference frame which is synchronized with the grid frequency by the PLL circuit, and then in ABC reference frame.

To be able to lock to the grid, reference value of V_q should be 0. PI controller in the PLL circuit calculates synchronous speed, ω which is equal to angular frequency of the grid voltage. ω is then integrated to give space angle θ, where θ defines relative position of synchronously rotating reference frame with respect to the stationary ABC reference frame, i.e., relative position of rotating d-axis with respect to stationary *a*-axis. I_q is compared with $I_{q(set)} = 0$ for zero reactive power and then processed in the PI controller to generate reference signal V_q^*. Cross-coupling term ΔV_q^* is then superimposed on V_q^* to compensate for potential drop on the total filter inductance and also for better transient response in the feed-forward form of cross-coupling terms. Actual value of DC link voltage V_{dc} is compared with its set value $V_{dc(set)}$ and resulting error signal is then processed in a PI controller to generate reference signal I_d^*. I_d^* is then compared with actual current I_d and processed in a PI controller to generate reference value, V_d^*. Cross-coupling term ΔV_d^* is superimposed on V_d^* to compensate for potential drop on the total filter inductance and also for better transient response.

The above operations yield main control signals δV_d and δV_q in synchronously rotating reference frame. Instead of superimposing δV_d and δV_q on *d*- and *q*-axis components, V_d and V_q of the grid voltages, δV_d and δV_q are transformed back to *abc*-axes, resulting in δ_{va}, δ_{vb}, and δ_{vc}, and these control signals are then superimposed on the actual grid voltage waveforms v_a, v_b, and v_c. This modification results in lower harmonic distortion in the line current waveforms because when the actual grid voltages are transformed to *dq* reference frame (instead of *dq0*) and then used in the control together with δV_d and δV_q, odd multiples of third harmonic (zero sequence component) would not be taken into account.

3.3. MPPT Converter with HF Link

In this subsection, switching frequency of the H-bridge converter and the transformer turns-ratio are optimized in view of the following design constraints:

(i) Multi-string PV system and its dynamic model should be known. For this purpose, the system in Figure 2 and its dynamic model in Figure 5b are prespecified for the experimental set-up in Figure 1b.

(ii) The variation range of global solar insolation, *G*, should be known for the geographical location at which the PV system is going to be installed, i.e., $G \leq 1000$ W/m^2 for the experimental set-up.

(iii) The variation range of module surface temperature, T_m, should be estimated, i.e., $10 \leq T_m \leq 70\,°C$ for the experimental set-up.

(iv) DC link voltage, V_{dc}, is kept constant at 700 V by the solar inverter in the experimental set-up.

(v) DC link capacitance, C_o in Figure 3a is taken to be 3400 µF.

All calculations are carried out on MATLAB/Simscape/Power Systems by using the equivalent circuit in Figure 3a in which leakage inductances of the HF transformer are assumed to be $L_p = 3.2$ µH and $L_s = 7.4$ µH, respectively, on the primary and secondary sides, and switching losses of all SiC power MOSFETs and Schottky diodes are neglected.

3.3.1. Optimum Transformer Turns-Ratio, *n*

The lowest and the highest maximum power point voltages for the multi-string PV system in Figure 2 are, respectively, $V_{mpp(min)} = 434$ V at $G = 50$ W/m^2, $T_m = 70\,°C$, and $V_{mpp(max)} = 600$ V at $G = 1000$ W/m^2, $T_m = 10\,°C$. Transformer turns-ratio is defined as $n = N_s/N_p$, where N_s and N_p are the number of series turns of the secondary and primary windings, respectively. The duty cycle, *D*, of the phase-shifted H-bridge converter is as defined in Equation (2).

$$D = t_{on}/(t_{on} + t_{off}) \qquad (2)$$

where t_{on} is the power transfer period of phase-shifted H-bridge, and t_{off} is the sum of freewheeling and no-conduction periods. It is desirable to maintain the operation of H-bridge converter at relatively high *D* values over the entire operating range to keep corresponding peak values of SiC power semiconductor and transformer currents relatively low. The variation ranges of *D* for two different *n* values and extreme operating conditions are shown in Figure 10.

On the other hand, for an ideal MPPT converter, the lowest MPPT voltage, $V_{mpp(min)}$, over the entire operating range can be related to the chosen DC link voltage, $V_{dc} = 700$ V in terms of n and D as given in Equation (3). As an example, for $D = 1.0$ and $V_{mpp(min)} = 434$ V at $G = 50$ W/m^2 and $T_m = 70$ °C, n is found to be 1.61 from Equation (3). However, an n value lower than 1.61 can be chosen, since the total leakage inductance of the transformer provides boosting action in a practical MPPT converter.

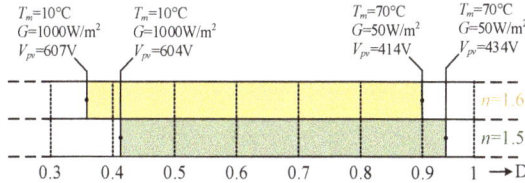

Figure 10. Variation range of duty cycle, D, for the practical MPPT converter as a function of n in between extreme operating conditions.

In the implemented MPPT converter, $n = 1.52$ is chosen for which $D = 0.94$ and $V_{mpp(min)} = 434$ V at $G = 50$ W/m^2 and $T_m = 70$ °C. This choice ensures power transfer even at the minimum G and maximum T_m conditions and provides a margin for better transient response.

$$V_{dc} = 700 \leq V_{mpp(min)} \, n \, D \tag{3}$$

3.3.2. Choice of DC Link Capacitor of H-Bridge Converter

Operation modes of the MPPT converter in Figure 3 are as defined in Figure 11. The controllable section of the MPPT converter is the phase-shifted H-bridge converter. The stray inductance of the implemented DC-bus on PCB is estimated as $L_{stray} = 15.2$ nH. Two discrete metallized film capacitors are connected across the DC link, one for each leg of the H-bridge converter. Total DC link capacitance is denoted by C_i in Figures 3 and 11. Suppose now that initially S_1 and S_3 are conducting in power transfer mode in the positive half-cycle as shown in Figure 11a.

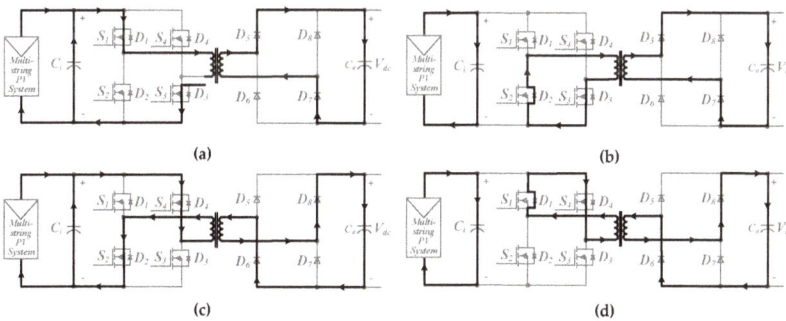

Figure 11. Operation modes of the MPPT converter for one complete cycle: (**a**) Mode 1: Power transfer in positive half-cycle; (**b**) Mode 2: Freewheeling in positive half-cycle; (**c**) Mode 3: Power transfer in negative half-cycle; and (**d**) Mode 4: Freewheeling in negative half-cycle.

When S_1 is turned off, current is commutated to D_2 which starts the freewheeling mode through S_3 and D_2 as shown in Figure 10b. The freewheeling mode is then followed by the OFF mode after the current decays to zero. For the negative half cycle, other diagonal switches S_2 and S_4 are turned on for power transfer mode which is followed by freewheeling mode through S_4 and D_1, respectively, shown in Figure 11c,d. Note that the converter is operated in the discontinuous conduction mode, owing to the absence of a lossy and bulky output filter inductor in the design.

Fall time of the SiC power MOSFET used in this research work is nearly 50 ns with increased gate resistance, R_g. During the commutation period, current closes its path mainly through C_i. The potential drop on L_{stray} is therefore $V_{stray} = L_{stray} \cdot (\Delta I / \Delta t) = 51$ V for maximum possible device current of $I_D = 200$ A at $f_{sw} = 20$ kHz when $G = 1000$ W/m^2, $T_m = 10$ °C on the predefined geographical site. V_{stray} will then be superimposed on drain-to-source voltage, v_{DS}, of outgoing SiC power MOSFET. Since open-circuit voltage of the multi-string PV system is 800 V, peak value of v_{DS} never exceeds 850 V in the worst case which is safely below the v_{DS} rating of the chosen SiC power MOSFETs.

A high C_i value is always desirable for better system performance at the expense of higher size and hence cost. Simulation studies have shown that a C_i value in the range from 20//20 µF to 40//40 µF can be chosen in the implemented H-bridge converter. Effects of C_i on peak-to-peak ripple content of i_{pv} and v_{pv}, $i_{ci(rms)}$, and form factor of i_{cin} are shown in Figure 12 for standard test conditions ($G = 1000$ W/m^2, $T_m = 25$ °C) and $f_{sw} = 20$ kHz. These curves show that:

- $i_{ci(rms)}$ and form factor of icin are not affected by C_i; and
- Peak-to-peak ripple content of i_{pv} and v_{pv} reduces as C_i is increased. Lower peak-to-peak content on the PV side is always desirable, not only for potential drop on all series inductances but also for MPPT efficiency.

In view of these characteristics, C_i greater than or equal to 30//30 µF seems to be suitable for the implemented H-bridge converter, provided that the commercially available metallized film capacitors can carry this rms current. In the implemented system $C_i = 30//30$ µF is chosen which is bigger than the DC link capacitor recommended by the SiC power MOSFET manufacturer [73]. To justify that 30//30 µF meets the entire operating range of the MPPT converter for the predefined geographical location and their commercial availability, characteristics in Figure 13 and manufacturer's data in Table 2 are given.

Table 2. Technical characteristics of some metallized film capacitors (1100 V DC at 70 °C) [74].

Capacity, µF	Code	I, A rms	ESR, mΩ
20	B32778T0206	13	11.9
30	B32778G0306	17.5	8.2
40	B32778G0406	21.5	6.2

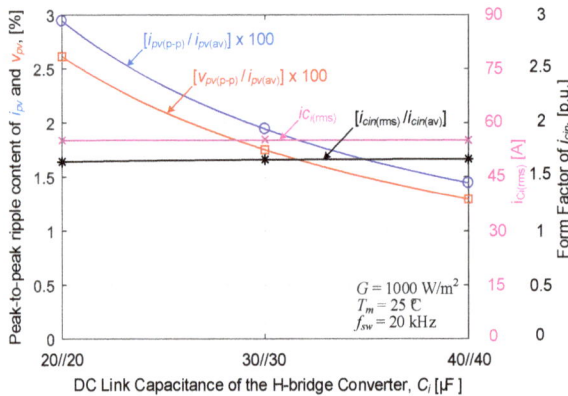

Figure 12. The variations in peak-to-peak ripple contents of i_{pv} and v_{pv}, $i_{ci(rms)}$, and form factor of i_{cin} against DC link capacitance of H-bridge converter for standard test conditions (these theoretical results were obtained using the dynamic model in Figure 5b; $f_{sw} = 20$ kHz assumed).

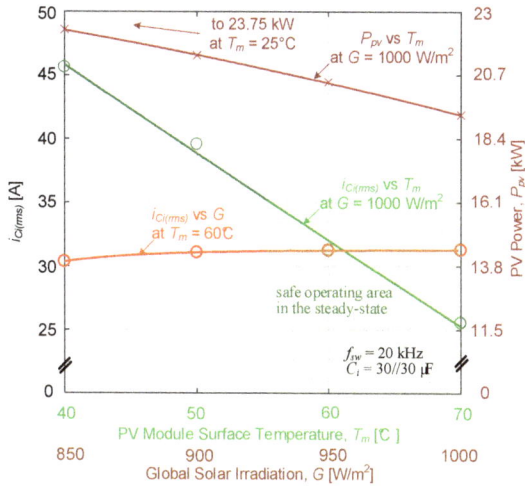

Figure 13. Effects of PV module surface temperature and global solar irradiation on $i_{ci(rms)}$ for 20 kHz switching frequency (these theoretical results were obtained from the dynamic model in Figure 5b; i_{ci} is defined in Figure 3a; $C_i = 30//30\ \mu\text{F}$ is assumed; variations in power delivered by the multi-string PV system in Figure 2 against PV module surface temperature are also given on this figure).

3.3.3. Optimum Switching Frequency of H-bridge Converter

Variations in ripple contents of v_{pv}, i_{pv}, i_{Ci}, and i_{cin} against switching frequency, f_{sw}, obtained by simulation studies at standard test conditions for the chosen $C_i = 30//30\ \mu\text{F}$ are as given in Figure 14.

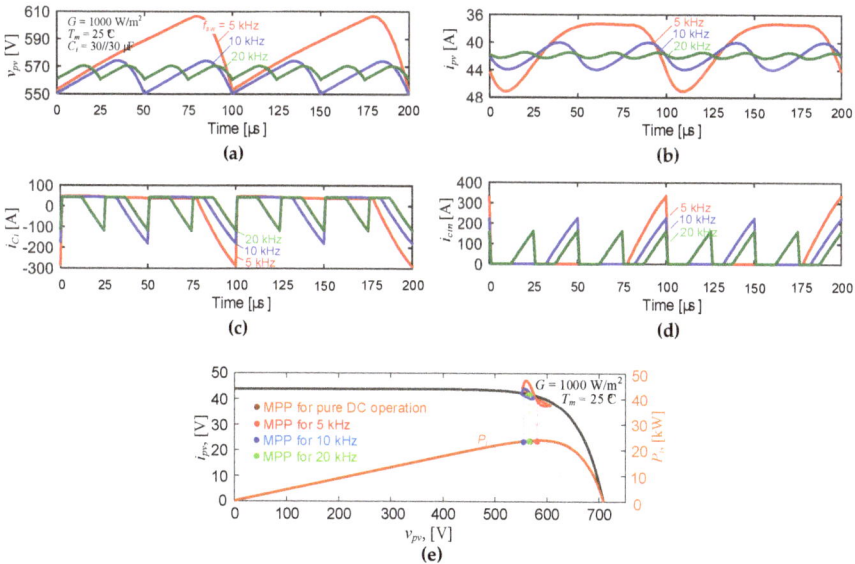

Figure 14. Variations in ripple content for different switching frequencies, f_{sw}: (**a**) PV panel voltage, v_{pv}; (**b**) PV panel current, i_{pv}; (**c**) input capacitor current, i_{Ci}; (**d**) total current transferred to MPPT converter; i_{cin}, and (**e**) effects of f_{sw} on maximum power point of multi-string PV system in steady-state.

As can be understood from these waveforms, operation at higher switching frequency reduces ripple contents and hence rms values of all currents. Lower rms values for i_{cin} and i_{Ci} are better in the selection not only of SiC power MOSFET but also of DC link capacitor, C_i. These variations are quantified and presented in graphical form as a function of f_{sw} in Figure 15. It can be concluded from these characteristics that the phase-shifted MPPT converter should be switched at a frequency greater than or equal to 15 kHz. Figure 14e justifies this statement. Dynamic variations in MPPs arising from ripple content of i_{pv} are marked on i_{pv}/v_{pv} characteristic given in Figure 14e. Calculated mean MPPs are also marked on the same figure. The ideal case is pure DC operation. As f_{sw} is increased dynamic MPP curve converges to that of pure DC operation, e.g., MPP power for $f_{sw} = 20$ kHz is nearly the same with that of pure DC operation.

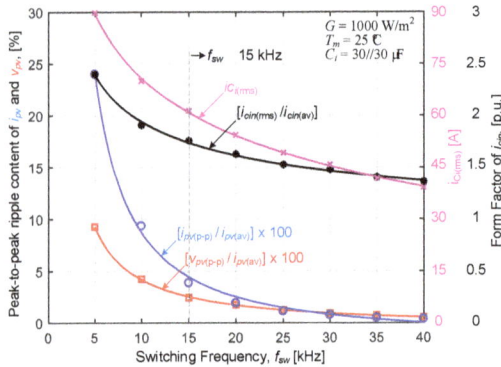

Figure 15. The variations in peak-to-peak ripple contents of i_{pv} and v_{pv}, $i_{Ci(rms)}$, and form factor of i_{cin} as a function of switching frequency (i_{pv}, v_{pv}, i_{Ci}, and i_{cin} are as defined in Figure 3a; these results were obtained by using the dynamic model in Figure 5b for the standard test conditions marked on the figure; DC link capacitance of the H-bridge converter in the MPPT converter is assumed to be $C_i = 30//30$ µF).

Thus far, in this subsection, only the factors for standard test conditions affecting the selection of f_{sw} have been considered. However, the entire operating range of the MPPT converter installed at the predefined geographical site gives more valuable information about the selection of f_{sw}. For this purpose, the variation range of duty ratio, D, for the H-bridge converter is calculated for different switching frequencies, and given in Figure 16. D should vary in a narrow range and at relatively high values to keep rms value of the semiconductor current, and hence its form factor at relatively low values. In view of these discussions, f_{sw} of the H-bridge converter should be at least 20 kHz.

In the selection of f_{sw}, the size of the HF transformer and the switching and conduction losses of H-bridge converter and conduction losses of Schottky diode rectifier should also be considered. Total semiconductor losses in the MPPT converter as a function of f_{sw} are given in Table 3. These losses exclude all wiring and cabling losses. It is seen in Table 3 that total power semiconductor losses in MPPT converter becomes minimum at $f_{sw} = 20$ kHz.

Table 3. Power semiconductor losses in MPPT converter against switching frequency (Operating Conditions: P_{pv} = 25.1 kW, G = 1000 W/m², T_m = 10 °C, T_j = 80 °C, R_g = 10 Ω).

H-Bridge Converter					Diode Bridge Rectifier		Total SiC MOSFET and SiC Diode Losses
f_{sw} (kHz)	I_{peak} (A)	I_{rms} (A)	P_{cond} (W)	P_{sw} (W)	I_{ave} (A)	P_{cond} (W)	P_{total} (W)
5	380	84	452	82	17.3	111	645
10	280	72	332	140	17.3	111	583
20	200	60	230	232	17.3	111	573
30	160	54	187	313	17.3	111	611
40	140	50	160	388	17.4	111	659

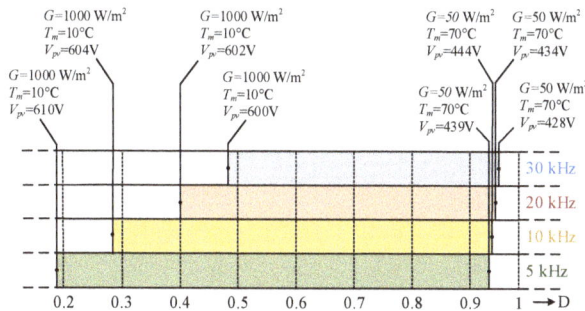

Figure 16. The variations in duty ratio, D of the H-bridge circuit in MPPT converter over the entire operating range for f_{sw} = 5, 10, 20, and 30 kHz, V_{dc} = 700 V, and $n = N_s/N_p$ = 1.52.

3.3.4. HF Transformer Design

In this application, advanced types of ferrite, amorphous metal cobalt-base, amorphous metal iron-base, and nanocrystalline core materials can be used. Their recommended peak flux densities for operating frequency around 20 kHz are 0.3, 0.5, 1.3, and 1.0 T, respectively. The nanocrystalline core material has lower core loss than ferrite and much lower core loss than high flux density, amorphous metal iron-base material operating at the same peak flux density and frequency. On the other hand, although core loss density of the amorphous metal cobalt-base material is comparable to that of nanocrystalline material, it requires higher core volume resulting in higher total core loss and higher cost because of lower operating flux density. Nanocrystalline core material is therefore chosen in the design of the HF transformer with natural air cooling.

Rated values of the target transformer are specified as 23 kW, 20 kHz rated frequency, 700 V peak secondary voltage, and $n = N_s/N_p$ = 1.52. SU102b nanocrystalline core is then used in the implementation of the HF transformer. Core loss, AC copper loss and total power loss of HF transformer against peak flux density are as given in Figure 17. Total power loss variation curve in Figure 17 shows that optimum design point can be chosen in between 0.25 T and 0.35 T. Design of the HF transformer is completed by choosing peak flux density as 0.3 T for minimum power dissipation at 20 kHz.

To test whether the design of MPPT converter is optimum, all power loss components excluding wiring and cabling losses are calculated and given in Figure 18 for three different operating frequencies and maximum PV power in the entire operating range of the MPPT converter installed at the predefined geographical site. Since f_{sw} = 20 kHz causes minimum power loss in the MPPT converter, the optimum design principles given in this subsection are justified.

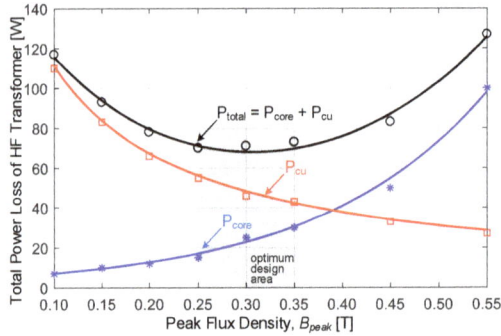

Figure 17. HF transformer power loss against peak flux density (23 kW, 700 V peak secondary voltage, f_{sw} = 20 kHz and n = 1.52 for SU102b nanocrystalline core).

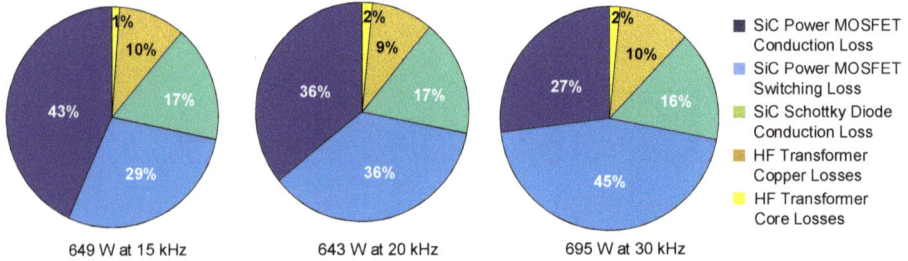

Figure 18. Power loss components of MPPT converter for different switching frequencies when P_{pv} = 25.1 kW. Numerical values are rounded.

3.3.5. Controller Design

The block diagram of the implemented DSP (TMS 320F28069) based controller is shown in Figure 3a. The controller is designed to fulfill the following tasks: (i) precharging the DC link capacitors, C_o, of the HF link MPPT converter; (ii) MPPT operation; and (iii) overvoltage protection and drain-to-source voltage monitoring for shoot-through protection. For this purpose, DC link voltage, V_{dc}, and PV panel voltage and current, V_{pv} and I_{pv}, are input to the controller, at a sampling rate of 40 kHz/channel. Voltage transducers used are of Fully-Differential Isolation Amplifier (TI AMC-1100) type for noise immunity, and the current transducer is Hall-Effect type (LEM HASS 50-S).

When the sun rises, the PI controller in Figure 3a starts to operate by applying narrow pulses to limit the charging current of C_o to a safe value. Precharging period is less than 30 s, during which D does not exceed 0.15. Set value of the DC link voltage, $V_{dc(set)}$, is specified to be 700 V in the control software. Whenever V_{dc} reaches 700 V, the inverter control system is activated to start the transfer of power to the grid. Just after the inverter operation, the controller enables MPPT algorithm based on an adaptive version of Perturb and Observe method [75], which runs only once every 300 ms, and stops whenever the error in active power between any two consecutive iterations is less than 0.1%. The algorithm then starts afresh after 1 s. In each iteration, the magnitude of duty ratio perturbation, Δd, is only $\pm 0.25\%$ of the previous duty ratio, D. Sampled I_{pv} and V_{pv} data are averaged over a period of 25 ms (256 samples), not only to filter out the measurement noise but also for the use in overvoltage protection software. The inverter tends to keep V_{dc} constant, and overvoltage protection facility in the controller of MPPT converter does not allow a rise in V_{dc} more than 10% in the case of loss of control. Furthermore, v_{DS} monitoring is carried out by an analog chip within the SiC driver [76].

4. Field Test Results

4.1. HF Link MPPT Converter

Variations in ac components of v_{pv} and i_{pv} while the MPPT converter is supplied from the multi-string PV system in Figure 2 are given in Figure 19, for operation at two different switching frequencies. The following observations can be made about these waveforms:

(i) Since the experimental results are the same with the theoretical ones based on the dynamic model of the PV array, the mathematical model and system design methodology can be successfully used in the design of the SiC power MOSFET-based HF link MPPT converter.

(ii) v_{pv} and i_{pv} are nearly pure DC at f_{sw} = 20 kHz, i.e., only 6 V p-p ripple is superimposed on $V_{pv(av)}$ = 533 V and 0.3 A p-p ripple is superimposed on $I_{pv(av)}$ = 15.9 A. This justifies the optimum switching frequency of around 20 kHz, for the SiC HF link converter.

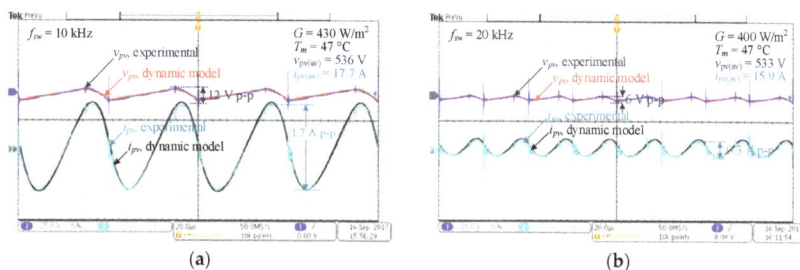

Figure 19. Variations in ac components of v_{pv} and i_{pv} while the MPPT converter supplied from multi-string PV system in Figure 2 is operating at two different switching frequencies in the steady-state: (a) f_{sw} = 10 kHz; and (b) f_{sw} = 20 kHz. (v_{pv} and i_{pv} are as defined in Figure 3a. Theoretical results have been obtained for the dynamic model in Figure 5b; v_{pv} and i_{pv} are recorded by using Tektronix MSO3034 oscilloscope, Tektronix P5205A high voltage differential probe, Tektronix TCP404XL current probe and Tektronix TCPA300 current probe amplifier.)

Drain-to-source voltage, v_{DS}, of SiC power MOSFET S_3 and the line current i_1 waveform of the H-bridge in Figure 3 are also recorded as shown in Figure 20. Positive half cycles of i_1 correspond to the drain current i_d waveform of S_3. Note that all SiC power MOSFETs turn-on at zero-current owing to the ramp current waveform of the discontinuous conduction mode. At the turn-off, however, only S_3 and S_4 are switched at zero current at the end of the OFF period as illustrated in Figure 11b,d, due to the phase shifted operation. S_1 and S_2, however, are switched off at the peak of the transformer primary current, i_1. The glitches superimposed on v_{DS} waveform of S_3 in Figure 20 are attributed to the noise coupled to the oscilloscope voltage probe during switching-off of the other SiC power MOSFETs.

HF transformer voltage and current waveforms on both the primary and the secondary sides are as shown in Figure 21, at nearly full-load. Operation modes of the HF link converter, as defined in Figure 11 are marked on various segments of the recorded voltage and current waveforms in Figure 21. The voltage spikes at the turn-off of S_1 and S_2 (just at the beginning of freewheeling modes of S_3-D_2 and S_4-D_1) in Figure 21a are caused by the ringing between switches' output capacitances (C_{oss} = 880 pF) and the primary stray inductance of the current path between C_i and S_1 or S_2 (calculated from layout as $L_{stray1} \approx 27$ nH). The corresponding oscillation frequency is measured as 33 MHz, as expected. This effect is more pronounced at the secondary side waveforms in Figure 21b, owing to the resonance between the two outgoing Schottky diode output capacitances ($C_j \approx 1000$ pF each) and the secondary leakage inductance ($L_s = 7.4$ μH), in series with the stray inductances ($L_{stray2} \approx 2.4$ μH) between C_o and the outgoing diodes, resulting in an oscillation frequency of 2.3 MHz. The slight voltage drops during freewheeling modes in Figure 21b are caused by L_{stray2}.

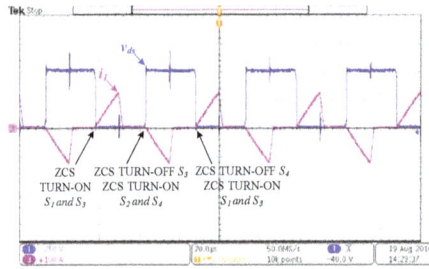

Figure 20. Drain-to-source voltage, v_{DS}, of SiC power MOSFET (S_3) and line current, i_1, waveforms in the H-bridge circuit recorded by Tektronix MSO3034 oscilloscope, Tektronix P5205A high voltage differential probe and Rogowski CWTUM/3/B current probe (f_{sw} = 20 kHz).

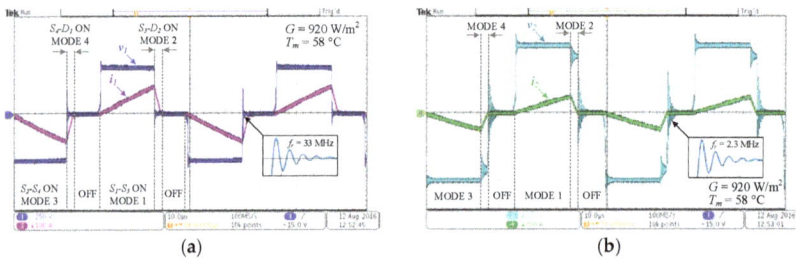

Figure 21. HF transformer voltage and current waveforms recorded by Tektronix MSO3034 oscilloscope, Tektronix P5205A high voltage differential probe and Rogowski CWTUM/3/B current probe (f_{sw} = 20 kHz, Test conditions are marked on the figure): (**a**) primary side; and (**b**) secondary side.

AC components of rectifier output current, i_{ro}, and the converter output current waveform, i_{dc}, in Figure 22 are recorded by a Rogowski current probe. Note that 72 A p-p rectifier output current ripple at full-load is filtered down to 12 A p-p by the low ESR DC link capacitor C_o.

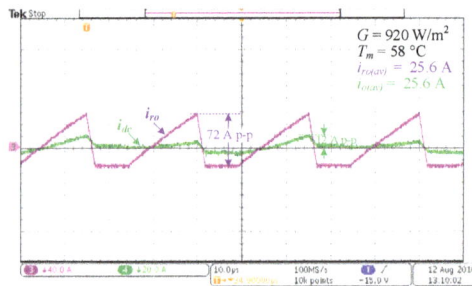

Figure 22. AC components of rectifier output current, i_{ro}, and the converter output current waveform, i_{dc}, recorded by Rogowski CWTUM/3/B current probe (average values of i_{ro} and i_{dc} and test conditions are marked on the figure; f_{sw} = 20 kHz).

4.2. Voltage Source Inverter

Since the three-phase two-level VSI is built by using a full SiC six-pack module, only the drain-to-source voltages v_{DS} and unfiltered line currents i_a, i_b, and i_c can be recorded. Figure 23 shows the circuit diagram of first leg of the inverter and the associated unfiltered line current, i_a.

To investigate effects of dead band on the turn-off performance of SiC power MOSFETs, v_{DS4} and i_a are recorded around the zero-crossing point of the unfiltered line-a current waveform for various dead times, as shown in Figure 24. These waveforms are as shown in Figure 24a when dead band is adjusted to 400 ns. Since current is very low, C_{oss4} and C_{oss1} are, respectively, charging and discharging slowly. When M_1 is turned on at the end of the dead band period, C_{oss4} is not charged yet to $V_{dc} = 700$ V and the C_{oss1} is not discharged entirely. The residual voltage on C_{oss1} will then be superimposed on DC link voltage which appears across the drain-source terminals of S_4 (v_{DS4}). A shorter dead band causes a larger overshoot on v_{DS4} waveform. This phenomenon does not occur around the peak value of unfiltered line-a current. This is because C_{oss4} and C_{oss1}, respectively, charges and discharges more rapidly since the current is high.

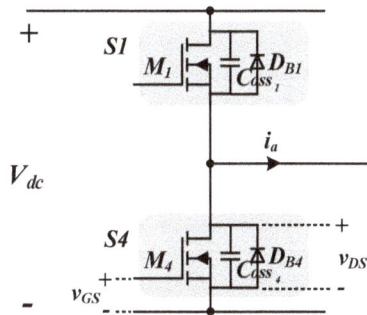

Figure 23. First leg of PV inverter and supply line-a (S:SiC power MOSFET, M:MOSFET part, C_{oss}: output capacitance, DB: body diode, and i_a: unfiltered line-a current).

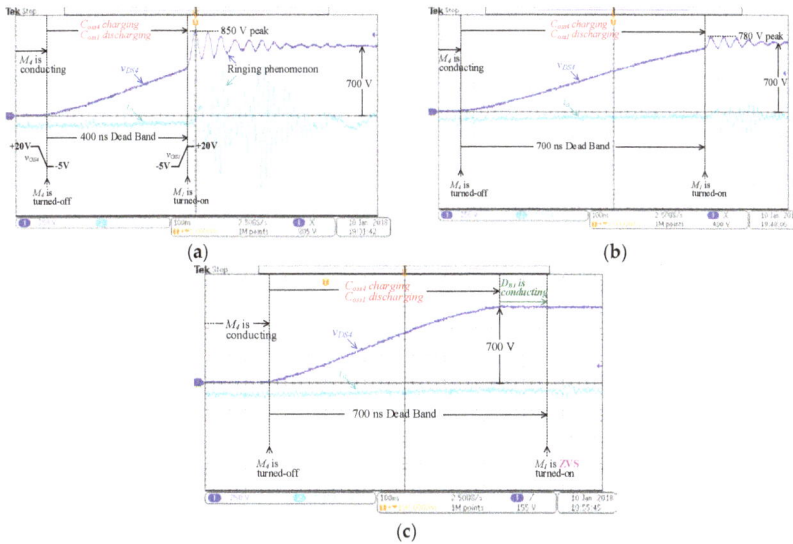

Figure 24. Effects of dead band on turn-off behavior of S_4 around zero crossing of the unfiltered line current (symbols are as defined in Figure 22 and v_{DS} and i_a are recorded by Tektronix MSO3034 oscilloscope, Tektronix P5205A high voltage differential probe, Tektronix TCP404XL current probe, Tektronix TCPA300 current probe amplifier): (**a**) 400 ns dead band; (**b**) 700 ns dead band; and (**c**) 700 ns dead band with ZVS turn-on of M_1.

On the other hand, ringing phenomenon is observed on v_{DS4} when S_1 is turned on due to the damped high frequency oscillation between the stray inductance of DC bus and C_{oss1}. In general, shorter dead time reduces low order harmonic distortion in line current waveforms at the expense of higher switching loss at turn-on and high frequency harmonic component. As can be seen in Figure 24b, a longer dead time (700 ns) reduces peak value of v_{DS4} and alleviates ringing phenomenon. At a current level slightly higher than that of Figure 24a,b, 700 ns dead time eliminates entirely ringing phenomenon and leads to ZVS turn-on of M_1 as shown in Figure 24c at the expense of higher low order harmonic distortion. In view of these considerations, 400 ns dead time is used in the implementation.

To investigate the transient performance of the VSI control system, multi-string PV system is suddenly disconnected from the input of the MPPT converter while the PV supply is delivering nearly 11 kW to the grid. The recorded filtered line current and the DC link voltage waveforms are as shown in Figure 25. Just after disconnection V_{dc} makes nearly 20% undershoot and then settles down to 98% of its rated value in nearly 480 ms. It is worth noting that, after reaching minimum V_{dc}, the voltage source converter starts to operate in rectification mode to allow power transfer from the grid to the DC link, thus maintaining V_{dc} at 700 V. Transient response is affected primarily by the size of the DC link capacitor and secondarily the LCL filter. V_{dc} in Figure 25 would decay more rapidly in the case of a smaller C_o, thus increasing undershoot in V_{dc}. To compensate for this phenomenon a larger size LCL filter could be used. A larger LCL filter would allow to increase the control range and hence the voltage source converter could settle down to the new operation state much more rapidly.

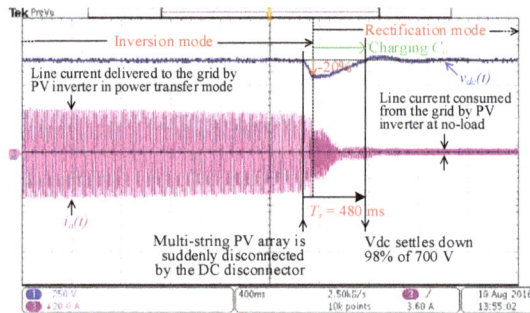

Figure 25. $v_{dc(t)}$ and $i_{a(t)}$ waveforms recorded by Tektronix MSO3034 oscilloscope, Tektronix P5205A high voltage differential probe Tektronix TCP404XL current probe and Tektronix TCPA300 current probe amplifier when the multi-string PV array is suddenly disconnected from MPPT converter.

4.3. Harmonic Distortion

Snapshots of the grid-side electrical quantities are given in Figure 26 for operation at full-load (Figure 26, left) and half-load (Figure 26, side). Figure 26a gives the three-phase voltages and line current waveforms, and Figure 26b the associated harmonic current contents according IEC61000-4-7:2002 harmonic measurement method [77], including both the line harmonics and the interharmonics. The rms quantities and output powers are recorded as shown in Figure 26c. The following observations can be made about these waveforms:

- The inverter operates connected to the 50-Hz AC grid with a total harmonic distortion, THDv ≈ 1.4% for the line-to-line voltages, and dominant 5th and 7th harmonics. The current THDi is recorded to be 3.8% at full-load, and 4.3% at half-load, with dominant 5th and 7th harmonics as in the AC grid. These current THD values correspond to current TDD values, respectively, of 3.4% and 1.9% by taking 25 kVA as the apparent power rating of the VSI.

- The resulting individual line current harmonics obtained experimentally are found to be within the recommended limits by the IEEE Std.-519-2014 [78] for all supply conditions, as can be seen from Figures 26b and 27 for a harmonic spectrum up to the 50th.
- The inverter operates successfully at unity pf, (pf $\approx$ 0.999 recorded), under both the full-load and the half-load conditions, according to the preset value, $I_{q(set)} = 0$.

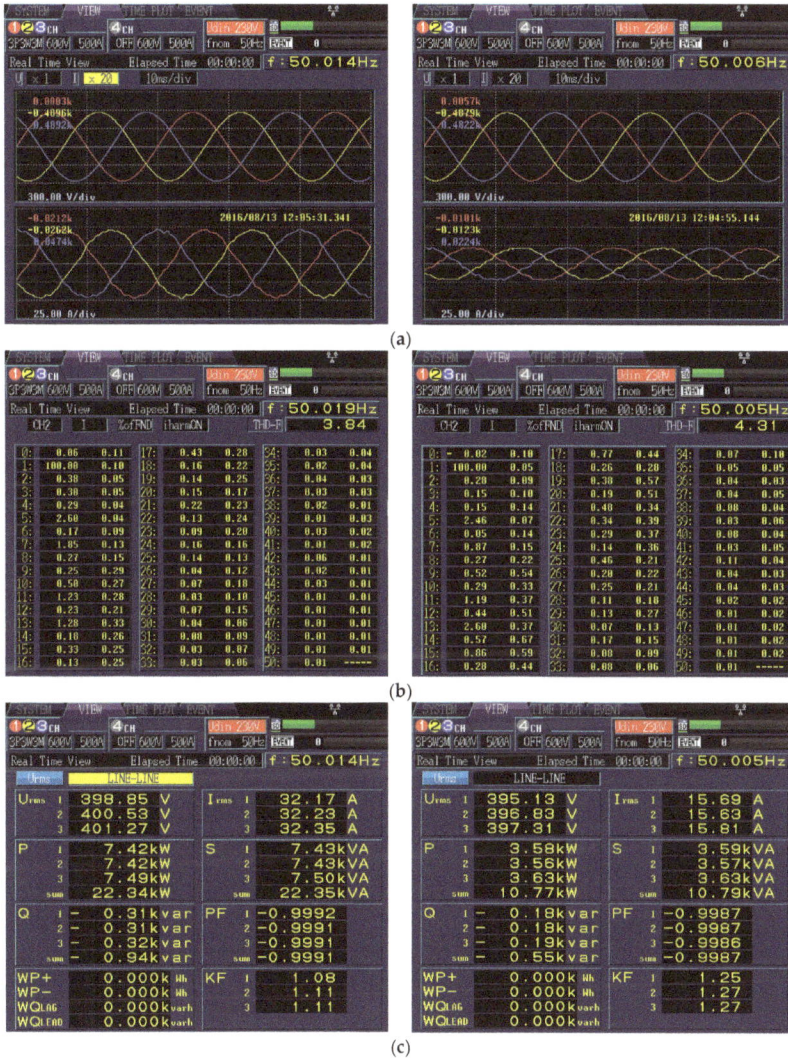

(a)

(b)

(c)

Figure 26. Snapshots of grid-side electrical quantities captured from the experimental set-up in Figure 1b by Hioki Power Analyzer PW3198 at P_o = 22 kW (left) and P_o = 11 kW (right): (**a**) line-to-line voltage and line current waveforms; (**b**) harmonic content of line current waveforms; and (**c**) output powers.

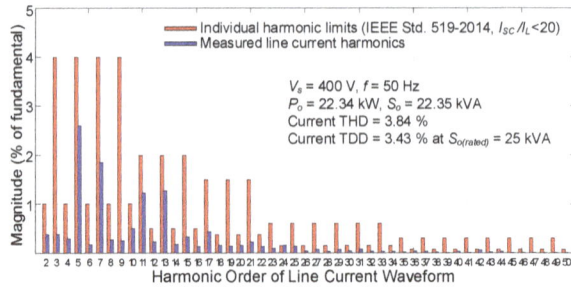

Figure 27. Sample harmonic spectra for the line current waveform injected by the inverter to the utility grid (deduced from the records of Hioki Power Analyzer PW3198 in Figure 26b).

In this research work, the LCL filter is optimized to yield minimum filter size and hence cost. To illustrate the effects of LCL filter size on the harmonic distortion of line current waveforms, L_g is increased from 50 µH to 1.5 mH, as preferred by several researchers in their implementations, and then the harmonic distortion record is repeated at 10 kVA. These records are given in Figure 28. THD and TDD values of the line current waveforms are 2.26% and 0.9%, respectively. It is seen that a larger size and hence more dissipative and costly LCL filter yields much lower harmonic distortion in line current waveforms.

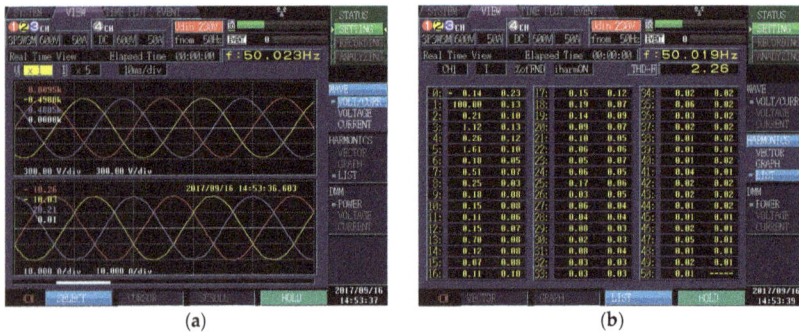

Figure 28. Snapshots of grid-side electrical quantities captured from the experimental set-up in Figure 1b by Hioki Power Analyzer PW3198 at P_o = 10 kW for L_g = 1.5 mH: (**a**) line-to-line voltage and line current waveforms; and (**b**) harmonic content of line current waveforms.

4.4. Efficiency

Efficiencies of HF link MPPT converter, VSI, and the overall grid-connected PV supply are obtained separately by field measurements for different operating conditions. Experimental results are given in comparison with theoretical values. For efficiency calculations, power components P_i, $P_{o(MPPT)}$, and P_o are as defined in Figure 29. For different operating conditions, $P_i = v_{pv(av)} \cdot i_{pv(av)}$ and $P_{o(MPPT)} = v_{dc(av)} \cdot i_{dc(av)}$ are calculated from measured $v_{pv(av)}$, $i_{pv(av)}$, $v_{dc(av)}$, and $i_{dc(av)}$ data as described in the caption of Figure 29. A sample set of $v_{pv(av)}$, $i_{pv(av)}$, $v_{dc(av)}$, and $i_{dc(av)}$ waveforms is given in Figure 30. Experimental efficiency values for the MPPT converter calculated from field data for different operating conditions are given in Table 4. For the corresponding insolation levels and module temperatures, the theoretical efficiency values are also calculated by using the MPPT converter model including the dynamic model of the multi-string PV system and running it on MATLAB Simulink at f_{sw} = 20 kHz. These theoretical values are also marked in Table 4. The following conclusions can be drawn from these results:

(i)   Maximum efficiency occurs at nearly half-load.

(ii) Full-load efficiency (97.3%) is only 0.5% lower than the maximum efficiency (97.7%).

(iii) Experimental values are slightly lower than corresponding theoretical values (discrepancies, $\delta\eta \leq 4\%$ at low power levels and $\delta\eta < 1\%$ at high powers).

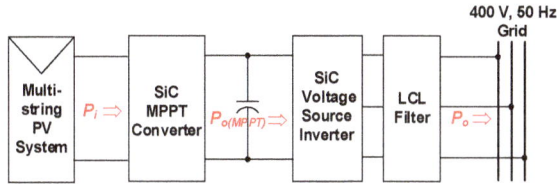

Figure 29. Definition of power components for efficiency calculations. P_i is calculated from the measured $v_{pv(av)}$ and $i_{pv(av)}$ data. $P_{o(MPPT)}$ is calculated from the measured $v_{dc(av)}$ and $i_{dc(av)}$ data. Measuring instruments are Tektronix MSO3034 oscilloscope together with its moving average filters, Tektronix P5205A high voltage differential probe, Tektronix TCP404XL current probe, Tektronix TCPA300 current probe amplifier, and Fluke 80i-110s current probe for i_{dc}. P_o is measured by Hioki Power Analyzer PW3198. Instantaneous values v_{pv}, i_{pv}, v_{dc} and i_{dc} are as defined in Figure 3 before averaging for steady-state operation.

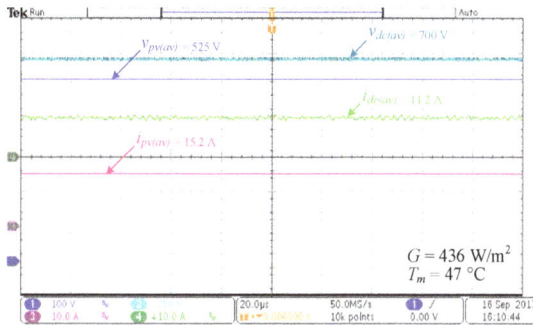

Figure 30. A sample set of $v_{pv(av)}$, $i_{pv(av)}$, $v_{dc(av)}$ and $i_{dc(av)}$ waveforms recorded at $G = 436$ W/m^2, and $T_m = 47\,^{\circ}$C.

Experimental efficiency values for the SiC VSI calculated from field data for different $P_{o(MPPT)}$ values are given in Table 5. Theoretical values of P_o are calculated by subtracting all inverter losses from experimental values of $P_{o(MPPT)}$. For the SiC VSI, computer simulations are carried out using the Wolfspeed SpeedFit design simulation software [72] to calculate SiC MOSFET losses, and Magnetics Inductor Design Tool [79] to determine the LCL filter losses. The following conclusions can be drawn from the results in Table 5:

(i)   Maximum efficiency (98.6%) occurs nearly at 40% of full 22.3 kW-load.

(ii) Efficiency is 98.1% at 88% of full kW-load.

(iii) Experimental values are slightly lower than corresponding theoretical values (discrepancies, $\delta\eta \leq 2\%$ at low power levels and $\delta\eta < 1\%$ at high powers).

The variations in efficiency of the all-SiC PV supply are calculated from field data and given in Figure 31 as a function of P_o. Maximum efficiency is observed to be 97%. Full-load efficiency is estimated to be slightly higher than 96%. At very low power levels such as 10% of the full-load, the overall efficiency is around 92%. These efficiency values are comparable with those of new generation PV supplies containing boost type MPPT converters and hybrid IGBT based inverters of the same power ratings and supplied from the existing multi-string PV system in Figure 2.

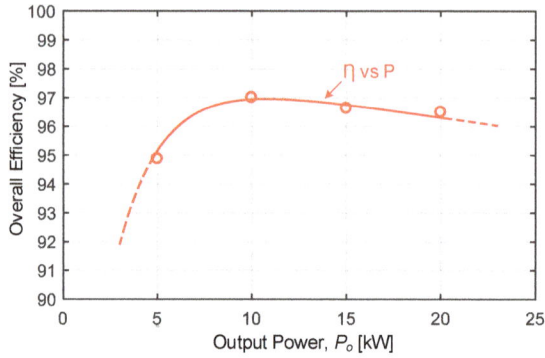

Figure 31. Efficiency of all SiC grid-connected PV supply against output power, $\eta = [P_o/P_i] \times 100$ (experimental values of P_o and P_i are obtained as described in the caption of Figure 29).

In the case where the multi-string PV system is initially not available and the types of the modules are not prespecified, the optimum multi-string PV system configuration and its technical characteristics will be a design issue for the overall grid-connected PV supply. The structure and technical characteristics of the multi-string PV system mainly affect the efficiency of the MPPT converter and hence efficiency of the overall system. To illustrate this fact, simulation studies for two different cases which employ 100 CSUN250-72M modules were carried out. In Case 1, 100 modules are connected to give 4×25 multi-string PV system to illustrate the effects of high operating voltage and low current on the efficiency of MPPT converter. In Case 2, the same modules are connected to give 5×20 multi-string PV system with lower operating voltage and higher current. The simulation results are as given in Figure 32. As can be understood from Figure 32, under standard test conditions, Case 1 gives 99% converter efficiency at nearly half-load and, when it is combined with inverter efficiency, the maximum efficiency of the overall system may reach 98.1%. This is because the operation of the MPPT converter with HF link in Figure 3 at a lower PV current, i_{pv}, reduces conduction loss components remarkably, thus improving the converter efficiency.

Table 4. Comparison of experimental and theoretical efficiency values of MPPT converter.

Test Condition *	P_i, kW		$P_{o(MPPT)}$, kW		Efficiency, $\eta = [P_{o(MPPT)}/P_i]$ 100, %	
G and T_m	Experimental *	Theoretical †	Experimental *	Theoretical °	Experimental	Theoretical
$G = 70$ W/m^2 $T_m = 32$ °C	1.66	1.60	1.49	1.55	89.76	93.47
$G = 150$ W/m^2 $T_m = 35$ °C	3.28	3.14	3.08	3.13	93.90	95.55
$G = 250$ W/m^2 $T_m = 38$ °C	4.51	5.19	4.32	5.01	95.78	96.69
$G = 370$ W/m^2 $T_m = 40$ °C	8.91	8.23	8.70	8.04	97.64	97.69
$G = 490$ W/m^2 $T_m = 49$ °C	10.61	10.21	10.37	10.02	97.73	98.20
$G = 730$ W/m^2 $T_m = 57$ °C	15.52	15.42	15.14	15.04	97.55	97.56
$G = 980$ W/m^2 $T_m = 60$ °C	20.14	20.61	19.59	20.05	97.26	97.28

* Experimental values of P_i and $P_{o(MPPT)}$ are determined as defined in Figure 29 for different test conditions.
† Theoretical values of P_i are calculated for the same test conditions by using the dynamic model in Figure 5b.
° Theoretical values of $P_{o(MPPT)}$ are calculated by adding all MPPT converter losses to theoretical values of P_i.

Table 5. Comparison of experimental and theoretical efficiency values of SiC inverter.

$P_{o(MPPT)}$, kW	P_o, kW		Efficiency, $\eta = [P_o/P_{o(MPPT)}]$ 100, %	
Experimental *	Experimental *	Theoretical †	Experimental	Theoretical
3.08	2.99	3.05	97.05	99.00
4.31	4.21	4.27	97.54	99.05
8.69	8.57	8.61	98.55	99.10
10.37	10.21	10.27	98.46	99.10
15.13	14.88	14.98	98.32	99.05
19.58	19.21	19.38	98.09	99.00

* Experimental values of $P_{o(MPPT)}$ and P_o are obtained as defined in Figure 29 for different operating conditions.
† Theoretical values of P_o are calculated by subtracting all inverter losses from experimental values of $P_{o(MPPT)}$.

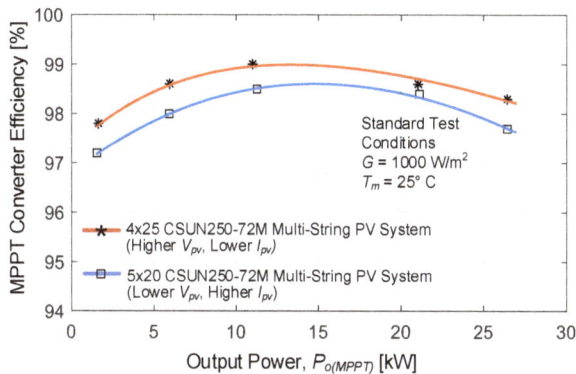

Figure 32. Effects of the configuration of multi-string PV system on the efficiency of MPPT converter with HF link in Figure 3 (Simulation results).

5. Conclusions

A system design methodology for an all SiC grid-connected PV supply with HF link MPPT converter has been proposed and a prototype of 25 kVA converter operating at 20 kHz has been implemented for verification. Owing to the very high dv/dt (>10 kV/μs) ratings of SiC power semiconductors, common-mode EMI is more pronounced in SiC based non-isolated converters. In this work, galvanic isolation of the proposed MPPT converter overcomes the common-mode EMI problem, thus enabling the grid connection using a simple and reliable three-phase, two-level inverter. In the design of the HF link MPPT converter operating at 20 kHz, a dynamic model of the multi-string PV system, parameters of which are obtained from the field test results, is used. More realistic MPPT converter parameters are shown to be obtained in the paper by using the dynamic PV model in the design procedure in comparison with the well-known static PV models.

The optimum switching frequency of the 25 kVA three-phase two-level inverter is determined as 20 kHz in the design procedure in view of inverter losses. The resulting 25 kVA, 20 kHz SiC VSI has 98.5% maximum efficiency which is slightly higher than or comparable with those of new-generation IGBT (Si IGBT + antiparallel SiC Schottky Diode) based counterparts for the existing multi-string PV system in Figure 2. This relatively high switching frequency not only reduces the size of the passive components, such as the LCL filter and the HF transformer but also the size of the cooling aggregates. LCL filter of the VSI, which is optimized by considering stability concerns of the controller in the design, provides nearly 100 dB attenuation at 20 kHz and its size is at least ten times smaller than those of LCL filter designs reported in the literature, even for lower size converters. TDD of the grid-connected VSI is measured to be 3.9% at nearly full-load and its individual current harmonics up to 50th conform with IEC Std. 61000-4-7:2002 even for the weakest grid. A higher grid-side inductance

(L_g = 1.5 mH) of the LCL filter lowers the current THDs considerably, which is measured to be 2.3% at nearly half-load.

The resulting SiC MPPT converter operating at 20 kHz and supplied from the existing 5 × 19 multi-string PV system in Figure 2 has 98% measured maximum efficiency, which is comparable with IGBT based and lower than SiC based boost type MPPT converters. Power densities are calculated as nearly 1.8 kW/lt and 1.6 kW/lt for forced air-cooled SiC MPPT converter and SiC grid-connected VSI, respectively. These figures are higher than those of forced air-cooled new-generation IGBT based converters and much higher than those of natural air-cooled new-generation IGBT based converters.

Author Contributions: M.E. and I.Ç. conceived the presented idea. M.C. and M.E. developed and implemented the SiC inverter. S.Ö. and I.Ç. developed and implemented the SiC MPPT converter. All authors discussed the results and contributed to the final manuscript.

Funding: This research was funded by Scientific Research Projects Office of Middle East Technical University for the development of SiC inverter with grant number BAP-03-01-2014-005, and ARTI Industrial Electronics Ltd. for the development of MPPT converter with HF link and multi-string PV system with grant number M.EE.P86.

Acknowledgments: The authors would like to acknowledge Energy Institute, TUBITAK Marmara Research Center, Ankara branch for the scholarship provided to M.C. for the research on SiC converters.

Conflicts of Interest: The authors declare no conflicts of interest.

References

1. Kjaer, S.B.; Pedersen, J.K.; Blaabjerg, F. A review of single-phase grid-connected inverters for photovoltaic modules. *IEEE Trans. Ind. Appl.* **2005**, *41*, 1292–1306. [CrossRef]
2. Mechouma, R.; Azoui, B.; Chaabane, M. Three-phase grid connected inverter for photovoltaic systems, a review. In Proceedings of the 2012 First International Conference on Renewable Energies and Vehicular Technology, Hammamet, Tunisia, 26–28 March 2012; pp. 37–42.
3. Orłowska-Kowalska, T.; Frede, B.; José, R. *Advanced and Intelligent Control in Power Electronics and Drives*; Springer: Berlin, Germany, 2014.
4. Barater, D.; Lorenzani, E.; Concari, C.; Franceschini, G.; Buticchi, G. Recent advances in single-phase transformerless photovoltaic inverters. *IET Renew. Power Gener.* **2016**, *10*, 260–273. [CrossRef]
5. Chen, S.M.; Liang, T.J.; Yang, L.S.; Chen, J.F. A Boost Converter with Capacitor Multiplier and Coupled Inductor for AC Module Applications. *IEEE Trans. Ind. Electron.* **2013**, *60*, 1503–1511. [CrossRef]
6. Liao, C.Y.; Lin, W.S.; Chen, Y.M.; Chou, C.Y. A PV Micro-inverter With PV Current Decoupling Strategy. *IEEE Trans. Power Electron.* **2017**, *32*, 6544–6557. [CrossRef]
7. Villanueva, E.; Correa, P.; Rodriguez, J.; Pacas, M. Control of a Single-Phase Cascaded H-Bridge Multilevel Inverter for Grid-Connected Photovoltaic Systems. *IEEE Trans. Ind. Electron.* **2009**, *56*, 4399–4406. [CrossRef]
8. Lorenzani, E.; Immovilli, F.; Migliazza, G.; Frigieri, M.; Bianchini, C.; Davoli, M. CSI7: A Modified Three-Phase Current-Source Inverter for Modular Photovoltaic Applications. *IEEE Trans. Ind. Electron.* **2017**, *64*, 5449–5459. [CrossRef]
9. Chen, Y.M.; Lo, K.Y.; Chang, Y.R. Multi-string single-stage grid-connected inverter for PV system. In Proceedings of the 2011 IEEE Energy Conversion Congress and Exposition, Phoenix, AZ, USA, 17–22 September 2011; pp. 2751–2756.
10. Edrington, C.S.; Balathandayuthapani, S.; Cao, J. Analysis and control of a multi-string photovoltaic (PV) system interfaced with a utility grid. In Proceedings of the IEEE PES General Meeting, Minneapolis, MN, USA, 25–29 July 2010; pp. 1–6.
11. SMA, SUNNY CENTRAL—High Tech Solution for Solar Power Stations. Products Category Brochure. Available online: http://www.sma-america.com/ (accessed on 21 February 2018).
12. Subudhi, B.; Pradhan, R. A Comparative Study on Maximum Power Point Tracking Techniques for Photovoltaic Power Systems. *IEEE Trans. Sustain. Energy* **2013**, *4*, 89–98. [CrossRef]
13. Müller, N.; Renaudineau, H.; Flores-Bahamonde, F.; Kouro, S.; Wheeler, P. Ultracapacitor storage enabled global MPPT for photovoltaic central inverters. In Proceedings of the 2017 IEEE 26th International Symposium on Industrial Electronics (ISIE), Edinburgh, UK, 19–21 June 2017; pp. 1046–1051.

14. Miguel, C.; Fernando, M.; Riganti, F.F.; Antonino, L.; Alessandro, S. A neural networks-based maximum power point tracker with improved dynamics for variable dc-link grid-connected photovoltaic power plants. *Int. J. Appl. Electromagn. Mech.* **2013**, *43*, 127–135.

15. Mancilla-David, F.; Arancibia, A.; Riganti-Fulginei, F.; Muljadi, E.; Cerroni, M. A maximum power point tracker variable-dc-link three-phase inverter for grid-connected PV panels. In Proceedings of the 2012 3rd IEEE PES Innovative Smart Grid Technologies Europe (ISGT Europe), Berlin, Germany, 14–17 October 2012; pp. 1–7.

16. Kerekes, T.; Teodorescu, R.; Liserre, M.; Klumpner, C.; Sumner, M. Evaluation of Three-Phase Transformerless Photovoltaic Inverter Topologies. *IEEE Trans. Power Electron.* **2009**, *24*, 2202–2211. [CrossRef]

17. Walker, G.R.; Sernia, P.C. Cascaded DC-DC converter connection of photovoltaic modules. *IEEE Trans. Power Electron.* **2004**, *19*, 1130–1139. [CrossRef]

18. Koutroulis, E.; Blaabjerg, F. Design Optimization of Transformerless Grid-Connected PV Inverters Including Reliability. *IEEE Trans. Power Electron.* **2013**, *28*, 325–335. [CrossRef]

19. Meneses, D.; Blaabjerg, F.; García, Ó.; Cobos, J.A. Review and Comparison of Step-Up Transformerless Topologies for Photovoltaic AC-Module Application. *IEEE Trans. Power Electron.* **2013**, *28*, 2649–2663. [CrossRef]

20. Araujo, S.V.; Zacharias, P.; Mallwitz, R. Highly Efficient Single-Phase Transformerless Inverters for Grid-Connected Photovoltaic Systems. *IEEE Trans. Ind. Electron.* **2010**, *57*, 3118–3128. [CrossRef]

21. Zhang, L.; Sun, K.; Xing, Y.; Xing, M. H6 Transformerless Full-Bridge PV Grid-Tied Inverters. *IEEE Trans. Power Electron.* **2014**, *29*, 1229–1238. [CrossRef]

22. Rodriguez, J.; Bernet, S.; Steimer, P.K.; Lizama, I.E. A Survey on Neutral-Point-Clamped Inverters. *IEEE Trans. Ind. Electron.* **2010**, *57*, 2219–2230. [CrossRef]

23. Bala, S. Next Gen PV Inverter Systems Using WBG Devices. In *High Pen PV Through Next-Gen PE Technologies Workshop*; NREL: Golden, CO, USA, 2016.

24. Millán, J.; Godignon, P.; Perpiñà, X.; Pérez-Tomás, A.; Rebollo, J. A Survey of Wide Bandgap Power Semiconductor Devices. *IEEE Trans. Power Electron.* **2014**, *29*, 2155–2163. [CrossRef]

25. Mookken, J.; Agrawal, B.; Liu, J. Efficient and Compact 50 kW Gen2 SiC Device Based PV String Inverter. In Proceedings of the International Exhibition and Conference for Power Electronics, Intelligent Motion, Renewable Energy and Energy Management, Nuremberg, Germany, 20–22 May 2014; pp. 1–7.

26. Wei, S.; He, F.; Yuan, L.; Zhao, Z.; Lu, T.; Ma, J. Design and implementation of high efficient two-stage three-phase/level isolated PV converter. In Proceedings of the 18th International Conference on Electrical Machines and Systems (ICEMS), Pattaya, Thailand, 25–28 October 2015; pp. 1649–1654.

27. Shi, Y.; Li, R.; Xue, Y.; Li, H. High-Frequency-Link-Based Grid-Tied PV System With Small DC-Link Capacitor and Low-Frequency Ripple-Free Maximum Power Point Tracking. *IEEE Trans. Power Electron.* **2016**, *31*, 328–339. [CrossRef]

28. Islam, M.; Mekhilef, S. Efficient Transformerless MOSFET Inverter for a Grid-Tied Photovoltaic System. *IEEE Trans. Power Electron.* **2016**, *31*, 6305–6316. [CrossRef]

29. Sintamarean, N.C.; Blaabjerg, F.; Wang, H.; Yang, Y. Real Field Mission Profile Oriented Design of a SiC-Based PV-Inverter Application. *IEEE Trans. Ind. Appl.* **2014**, *50*, 4082–4089. [CrossRef]

30. Shi, Y.; Wang, L.; Xie, R.; Shi, Y.; Li, H. A 60-kW 3-kW/kg Five-Level T-Type SiC PV Inverter With 99.2% Peak Efficiency. *IEEE Trans. Ind. Electron.* **2017**, *64*, 9144–9154. [CrossRef]

31. Colmenares, J.; Peftitsis, D.; Rabkowski, J.; Sadik, D.P.; Tolstoy, G.; Nee, H.P. High-Efficiency 312-kVA Three-Phase Inverter Using Parallel Connection of Silicon Carbide MOSFET Power Modules. *IEEE Trans. Ind. Appl.* **2015**, *51*, 4664–4676. [CrossRef]

32. Rabkowski, J.; Peftitsis, D.; Nee, H.P. Design steps towards a 40-kVA SiC inverter with an efficiency exceeding 99.5%. In Proceedings of the Twenty-Seventh Annual IEEE Applied Power Electronics Conference and Exposition (APEC), Orlando, FL, USA, 5–9 February 2012; pp. 1536–1543.

33. Yin, S.; Tseng, K.J.; Tong, C.F.; Simanjorang, R.; Gajanayake, C.J.; Gupta, A.K. A 99% efficiency SiC three-phase inverter using synchronous rectification. In Proceedings of the IEEE Applied Power Electronics Conference and Exposition (APEC), Long Beach, CA, USA, 20–24 March 2016; pp. 2942–2949.

34. Laird, Y.Y.I.; Yuan, X.; Scoltock, J.; Forsyth, A.J. A Design Optimization Tool for Maximizing the Power Density of 3-Phase DC–AC Converters Using Silicon Carbide (SiC) Devices. *IEEE Trans. Power Electron.* **2018**, *33*, 2913–2932. [CrossRef]

35. Payan, D.; Catani, J.P.; Schwander, D. Solar Array Dynamic Simulator Prevention of Solar Array Short-Circuits due to Electrostatic Discharge. In Proceedings of the Space Power, Sixth European Conference, Porto, Portugal, 6–10 May 2002; pp. 609–615.

36. Schwander, D. Dynamic Solar Cell Measurement Techniques: New Small Signal Measurement Techniques. In Proceedings of the Space Power, Sixth European Conference, Porto, Portugal, 6–10 May 2002; pp. 603–608.

37. Blok, R.; van den Berg, E.; Slootweg, D. Solar Cell Capacitance Measurement. In Proceedings of the Space Power, Sixth European Conference, Porto, Portugal, 6–10 May 2002; pp. 597–602.

38. Herman, M.; Jankovec, M.; Topic, M. Optimisation of the I-V measurement scan time through dynamic modelling of solar cells. *IET Renew. Power Gener.* **2013**, *7*, 63–70. [CrossRef]

39. Qin, L.; Xie, S.; Yang, C.; Cao, J. Dynamic model and dynamic characteristics of solar cell. In Proceedings of the IEEE ECCE Asia Downunder, Melbourne, Australia, 3–6 June 2013; pp. 659–663.

40. Bharadwaj, P.; Kulkarni, A.; John, V. Impedance estimation of photovoltaic modules for inverter start-up analysis. *Sadhana—Acad. Proc. Eng. Sci.* **2017**, *42*, 1377–1387. [CrossRef]

41. Huang, J.H.; Lehman, B.; Qian, T. Submodule integrated boost DC-DC converters with no external input capacitor or input inductor for low power photovoltaic applications. In Proceedings of the IEEE Energy Conversion Congress and Exposition (ECCE), Milwaukee, WI, USA, 18–22 September 2016; pp. 1–7.

42. Duan, S.; Yan, G.; Jin, L.; Ren, J.; Wu, W. Design of photovoltaic power generation mppt controller based on SIC MOSFET. In Proceedings of the TENCON 2015–IEEE Region 10 Conference, Macao, China, 1–4 November 2015; pp. 1–5.

43. Kim, T.; Jang, M.; Agelidis, V.G. Practical implementation of a silicon carbide-based 300 kHz, 1.2 kW hard-switching boost-converter and comparative thermal performance evaluation. *IET Power Electron.* **2015**, *8*, 333–341. [CrossRef]

44. Mouli, G.R.C.; Schijffelen, J.H.; Bauer, P.; Zeman, M. Design and Comparison of a 10-kW Interleaved Boost Converter for PV Application Using Si and SiC Devices. *IEEE J. Emerg. Sel. Top. Power Electron.* **2017**, *5*, 610–623. [CrossRef]

45. Anthon, Y.Y.A.; Zhang, Z.; Andersen, M.A.E. A high power boost converter for PV Systems operating up to 300 kHz using SiC devices. In Proceedings of the International Power Electronics and Application Conference and Exposition, Shanghai, China, 5–8 November 2014; pp. 302–307.

46. Mostaghimi, O.; Wright, N.; Horsfall, A. Design and performance evaluation of SiC based DC-DC converters for PV applications. In Proceedings of the IEEE Energy Conversion Congress and Exposition (ECCE), Raleigh, NC, USA, 15–20 September 2012; pp. 3956–3963.

47. Fujii, K.; Noto, Y.; Oshima, M.; Okuma, Y. 1-MW solar power inverter with boost converter using all SiC power module. In Proceedings of the 17th European Conference on Power Electronics and Applications (EPE'15 ECCE-Europe), Geneva, Switzerland, 8–10 September 2015; pp. 1–10.

48. Agamy, M.S.; Chi, S.; Elasser, A.; Harfman-Todorovic, M.; Jiang, Y.; Mueller, F.; Tao, F. A High-Power-Density DC–DC Converter for Distributed PV Architectures. *IEEE J. Photovolt.* **2013**, *3*, 791–798. [CrossRef]

49. Akagi, H.; Yamagishi, T.; Tan, N.M.L.; Kinouchi, S.I.; Miyazaki, Y.; Koyama, M. Power-Loss Breakdown of a 750-V 100-kW 20-kHz Bidirectional Isolated DC–DC Converter Using SiC-MOSFET/SBD Dual Modules. *IEEE Trans. Ind. Appl.* **2015**, *51*, 420–428. [CrossRef]

50. Liserre, M.; Blaabjerg, F.; Dell'Aquila, A. Step-by-step design procedure for a grid-connected three-phase PWM voltage source converter. *Int. J. Electron.* **2004**, *91*, 445–460. [CrossRef]

51. Wu, W.; He, Y.; Tang, T.; Blaabjerg, F. A New Design Method for the Passive Damped LCL and LLCL Filter-Based Single-Phase Grid-Tied Inverter. *IEEE Trans. Ind. Electron.* **2013**, *60*, 4339–4350. [CrossRef]

52. Peña-Alzola, R.; Liserre, M.; Blaabjerg, F.; Sebastián, R.; Dannehl, J.; Fuchs, F.W. Analysis of the Passive Damping Losses in LCL-Filter-Based Grid Converters. *IEEE Trans. Power Electron.* **2013**, *28*, 2642–2646. [CrossRef]

53. Reznik, Y.Y.A.; Simões, M.G.; Al-Durra, A.; Muyeen, S.M. LCL Filter Design and Performance Analysis for Grid-Interconnected Systems. *IEEE Trans. Ind. Appl.* **2014**, *50*, 1225–1232. [CrossRef]

54. Liu, J.; Zhou, L.; Yu, X.; Li, B.; Zheng, C. Design and analysis of an LCL circuit-based three-phase grid-connected inverter. *IET Power Electron.* **2017**, *10*, 232–239. [CrossRef]

55. Parker, S.G.; McGrath, B.P.; Holmes, D.G. Regions of Active Damping Control for LCL Filters. *IEEE Trans. Ind. Appl.* **2014**, *50*, 424–432. [CrossRef]

56. Jayalath, S.; Hanif, M. Generalized LCL-Filter Design Algorithm for Grid-Connected Voltage-Source Inverter. *IEEE Trans. Ind. Electron.* **2017**, *64*, 1905–1915. [CrossRef]

57. Figueres, E.; Garcera, G.; Sandia, J.; Gonzalez-Espin, F.; Rubio, J.C. Sensitivity Study of the Dynamics of Three-Phase Photovoltaic Inverters With an LCL Grid Filter. *IEEE Trans. Ind. Electron.* **2009**, *56*, 706–717. [CrossRef]

58. Timbus, Y.Y.A.; Liserre, M.; Teodorescu, R.; Rodriguez, P.; Blaabjerg, F. Evaluation of Current Controllers for Distributed Power Generation Systems. *IEEE Trans. Power Electron.* **2009**, *24*, 654–664. [CrossRef]

59. Blaabjerg, F.; Teodorescu, R.; Liserre, M.; Timbus, A.V. Overview of Control and Grid Synchronization for Distributed Power Generation Systems. *IEEE Trans. Ind. Electron.* **2006**, *53*, 1398–1409. [CrossRef]

60. Vasquez, J.C.; Guerrero, J.M.; Savaghebi, M.; Eloy-Garcia, J.; Teodorescu, R. Modeling, Analysis, and Design of Stationary-Reference-Frame Droop-Controlled Parallel Three-Phase Voltage Source Inverters. *IEEE Trans. Ind. Electron.* **2013**, *60*, 1271–1280. [CrossRef]

61. Bosch, S.; Steinhart, H. Active power filter with model based predictive current control in natural and dq frame. In Proceedings of the 18th European Conference on Power Electronics and Applications (EPE'16 ECCE Europe), Karlsruhe, Germany, 5–9 September 2016; pp. 1–10.

62. Liserre, M.; Teodorescu, R.; Blaabjerg, F. Multiple harmonics control for three-phase grid converter systems with the use of PI-RES current controller in a rotating frame. *IEEE Trans. Power Electron.* **2006**, *21*, 836–841. [CrossRef]

63. Silicon carbide Power Module-BSM300D12P2E001. Available online: http://www.rohm.com/web/global/products/-/product/BSM300D12P2E001 (accessed on 21 February 2018).

64. CAS325M12HM2 1200V, 325A, Silicon Carbide High-Performance-Wolfspeed. Available online: https://www.wolfspeed.com/cas325m12hm2 (accessed on 21 February 2018).

65. SKM500MB120SC–SEMIKRON. Available online: https://www.semikron.com/products/product-classes/sic/full-sic/detail/skm500mb120sc-21919770.html (accessed on 21 February 2018).

66. SiC POWER MODULES. Available online: http://www.mitsubishielectric.com/semiconductors/catalog/pdf/sicpowermodule_e_201505.pdf (accessed on 21 February 2018).

67. *IEC Standard for Power Transformers*; IEC Standard IEC 60076-1:2011; IEC Webstore; IEC: Geneva, Switzerland, 2011.

68. *IEC Standard for Standard Voltages*; IEC Standard IEC 60038 2002-07; IEC Webstore; IEC: Geneva, Switzerland, 2002.

69. Kimball, J.W.; Zawodniok, M. Reducing Common-Mode Voltage in Three-Phase Sine-Triangle PWM with Interleaved Carriers. *IEEE Trans. Power Electron.* **2011**, *26*, 2229–2236. [CrossRef]

70. Feng, J.; Wang, H.; Xu, J.; Su, M.; Gui, W.; Li, X. A Three-Phase Grid-Connected Micro-Inverter for AC Photovoltaic Module Applications. *IEEE Trans. Power Electron.* **2017**. [CrossRef]

71. Shi, Y.; Wang, L.; Li, H. Stability Analysis and Grid Disturbance Rejection for a 60 kW SiC based Filter-less Grid-connected PV Inverter. *IEEE Trans. Ind. Appl.* **2017**. [CrossRef]

72. SpeedFit Design Simulator™, USA: Wolfspeed. Available online: https://www.wolfspeed.com/speedfit (accessed on 24 February 2018).

73. Cree, Inc. *CREE Application Note, Design Considerations for Designing with Cree SiC Modules*; Cree: Durham, NC, USA, 2013.

74. Metallized Polypropylene Film Capacitors (MKP). Available online: http://www.mouser.com/ds/2/136/MKP_B32774_778-19326.pdf (accessed on 24 February 2018).

75. Piegari, L.; Rizzo, R. Adaptive perturb and observe algorithm for photovoltaic maximum power point tracking. *IET Renew. Power Gener.* **2010**, *4*, 317–328. [CrossRef]

76. 1ED020I12_F2-DS-v02. Available online: https://www.infineon.com/dgdl/Infineon-1ED020I12_F2-DS-v02_00-en.pdf?fileId=db3a304330f68606013122ce5f3649cb (accessed on 24 February 2018).

77. *IEC Standard for Electromagnetic Compatibility (EMC)—Part 4–7: Testing and Measurement Techniques—General Guide on Harmonics and Interharmonics Measurements and Instrumentation, for Power Supply Systems and Equipment Connected Thereto*; IEC Standard IEC 61000-4-7:2002; IEC Webstore; IEC: Geneva, Switzerland, 2002.

78. *IEEE Standard for Recommended Practice and Requirements for Harmonic Control in Electric Power Systems*; IEEE Std. 519-2014; IEEE Standards Association: Piscataway, NJ, USA, 2014.

79. Magnetics® Inductor Design Tool, USA: Magnetics. Available online: https://www.mag-inc.com/Design/Design-Tools/Inductor-Design (accessed on 24 February 2018).

electronics

MDPI

Article

Generalized Cascaded Symmetric and Level Doubling Multilevel Converter Topology with Reduced THD for Photovoltaic Applications

Karthikeyan D, Vijayakumar K * and Jagabar Sathik M

SRM Institute of Science and Technology, Kattankulathur 603203, India; karthipncl@gmail.com (K.D.);
mjsathik@ieee.org (J.S.K.)
* Correspondence: kvijay_srm@rediffmail.com

Received: 1 October 2018; Accepted: 11 January 2019; Published: 1 February 2019

Abstract: In this paper, two different converter topologies for a basic new switched capacitor diode converter with a reduced number of power electronics components, suitable for grid connected photovoltaic applications were proposed. The two different structures of switched diode multilevel converter proposed were: (i) cascaded switched diode and (ii) cascaded switched diode with doubling circuit. The switched-diode multilevel converter was compared with other recent converters. In addition, a new dc offset nearest level modulation technique was proposed. This proposed dc offset technique offers low voltage total harmonic distortion (THD) and high RMS output voltage. The proposed modulation technique was compared with conventional nearest level modulation (NLM) and modified NLM control techniques. The performance of the proposed dc offset modulation technique was implemented using a FPGA Spartan 3E controller and tested with a novel switched capacitor-diode multilevel converter. However, to prove the authenticity of the switched-diode multilevel converter and modulation technique, a laboratory-based prototype model for 7-level and 13-level converters was developed.

Keywords: multilevel inverter; cascaded topology; voltage doubling; switched capacitor; nearest level modulation (NLM); total harmonic distortion (THD)

---

1. Introduction

The multilevel converter is a promising power electronic converter for DC–AC high-power applications because it offers low electromagnetic interference (EMI), low dv/dt stress, and high efficiency. The conventional multilevel converters are (i) the neutral point clamped diode, (ii) flying capacitor, and (iii) cascaded H-bridge. The conventional multilevel converters have their own merits and demerits which are listed in References [1,2]. However, these conventional multilevel converters require a large number of power switches (IGBTs) for a higher number of levels, which is a major drawback. Many researchers are working towards the design of a novel multilevel converter with reduced power switches [3]. Several multilevel converters were recently proposed for are reduced number of dc sources and switches [4–8]. Another switched diode topology is presented in Reference [4], which requires single DC source and series connected DC-link capacitors to generate the maximum number of output voltage levels. These topologies have high blocking voltage stresses across the full bridge converter switches. To double the output voltage, a series of connected half-bridge circuits are presented in References [5,6]. These topologies produce a higher number of voltage levels with a greater number of switches and DC sources. The switched diode multilevel inverters for symmetric and asymmetric topologies with reduced DC sources and switches are presented in References [7,8]. Another recent topology is proposed with a lesser number of DC sources in Reference [9]. This topology generates a 11-level output. In order to increase the output voltage level, the cascaded connection is recommended. This topology produces a higher number of voltage levels with reduced switches, but it requires more isolated DC sources. The combined T-type and cross-connected topologies presented in Reference [10] with a self-capacitor voltage balancing. The proposed topology uses unidirectional switches and bidirectional switches without an anti-parallel diode. The presented topology requires a single DC source and multiple capacitors, therefore the switching complexity will increase as the number of levels increases. However, these topologies suffer from a greater number of power components like IGBTs, diodes, DC-link capacitors and complex pulse generation circuits. The full bridge inverter unit is used to change the polarity of the output voltage and current. As the number of levels increases, the voltage stress across the full bridge inverter with packed H-bridge inverter will also increase. This leads to many practical issues and hence increases the cost of the inverter [11].

The reduction of output voltage harmonics is still a challenge in power electronics DC/AC converters. In order to minimize the harmonics, different modulation techniques are developed by the researchers. The modulation techniques are classified into two categories: (i) low switching frequency and (ii) high switching frequency. The conventional modulation techniques are: multicarrier-based pulse width modulation (PD, POD, and APOD), hybrid modulation and hysteresis modulation. The multicarrier modulation technique will produce low THD by increasing the switching frequency, but it leads to high switching losses and a need for a complex cooling system which in turn degrades the reliability of the power switches. These modulation techniques are more suitable for medium power applications. For high power applications, low switching frequency modulation schemes are preferred. The low switching frequency modulation techniques are Nearest Space Vector Modulation (NSVM), Nearest Level Modulation (NLM) and space vector PWM (SVPWM) [12–19]. These modulation techniques produce low THD compared to conventional low switching frequency schemes, but the generation of gating pulses is more complicated as the number of level increases, except in the NLM technique. In addition, different offline PWM control strategies are employed for low switching frequency, which is used to find the optimum angles to minimize the harmonics in the output voltage [20,21]. However, these types of modulation techniques are used in open loop applications like uninterrupted power supplies (UPS), whereas it is not suitable for closed loop real-time applications like PV systems, since it takes more computational time.

The nearest level modulation technique [22] is another fundamental frequency method, which is operated at 50 Hz, but it is suitable for higher number of voltage levels. For a lower number of voltage levels, it produces high THD. The modified NLM technique has been presented [23], which has a DC offset value of 0.25 instead of 0.5 in the conventional NLM. The modified NLM increases the level and reduces the DC offset losses but is suitable for a modular multilevel converter. In this paper, a new multilevel converter with optimum nearest level modulation technique is presented. The DC offset value is chosen so that it minimizes the THD for a lower number of levels and increases the RMS voltage up to 17th level. This paper is organized as follows: In Section 2, the new multilevel converter basic unit is proposed with the various modes of operation. In order to increase the number of output voltage level, a cascaded connection of basic unit with and without half-bridge converter are presented and a comparison of the proposed multilevel converter with other recent topologies is made. In Section 3 Analysis of the proposed NLM technique with suitable illustrations is presented and the comparison of proposed modulation technique with other modulation methods is discussed. In Section 4 Experimental results are presented to confirm the objective of this paper. In Section 5 a suitable application and the advantages of the proposed methods are highlighted, in the Conclusions.

2. Proposed Multilevel Converter

2.1. Proposed Basic Unit

Figure 1 shows the proposed basic unit of the multilevel converter which consists of a single DC source with three DC-link capacitors (C_1, C_2, and C_3) connected parallel to the source. The upper and lower capacitor voltages were tapped through switches S_1 and S_2. Floating capacitor C_2 provides the current through diodes D_1 and D_2. The full bridge circuit (F_{11}–F_{14}) was used to alter the polarity of both the voltage and current paths through the load. Since the topology is symmetric, and it generates a maximum of seven levels of output voltage. The switch pairs (F_{11}, F_{13}) and (F_{12}, F_{14}) were not turned on simultaneously to avoid short circuit. The switching sequence for the basic unit is given in Table 1 and various modes of operation are illustrated in Figure 2.

Mode 0: The switch pairs, either (F_{11}, F_{12}) or (F_{13}, F_{14}), were turned on to produce the zero state. Mode 1: Diodes D_1 and D_2 were made to conduct to produce the level one output voltage level from the floating capacitor C_2. Mode 2: The floating capacitor C_2 was added with either a upper or lower capacitor by switching the S_1 or S_2 to generate the second level. Mode 3: Both S_1 and S_2 switches were turned on simultaneously to produce the level-3 voltage by connecting all the capacitors together. The full bridge converter switches (F_{11}, F_{14}) were turned on for the positive half cycle as shown in Figure 2a–c, and (F_{12}, F_{13}) were turned on for the negative half cycle in Figure 2d–f. Furthermore, the DC-link capacitor voltage balancing is another serious problem in series connected capacitors. This balancing was done either by switching techniques or by providing external circuits. In this paper, an external circuit, proposed in References [22,23], was used for balancing the capacitors.

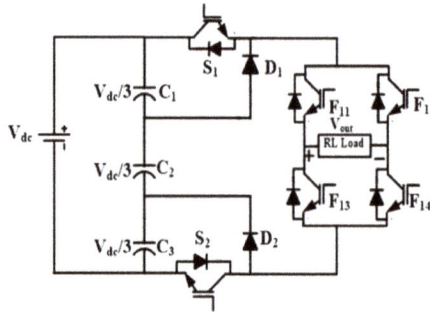

Figure 1. Proposed basic unit.

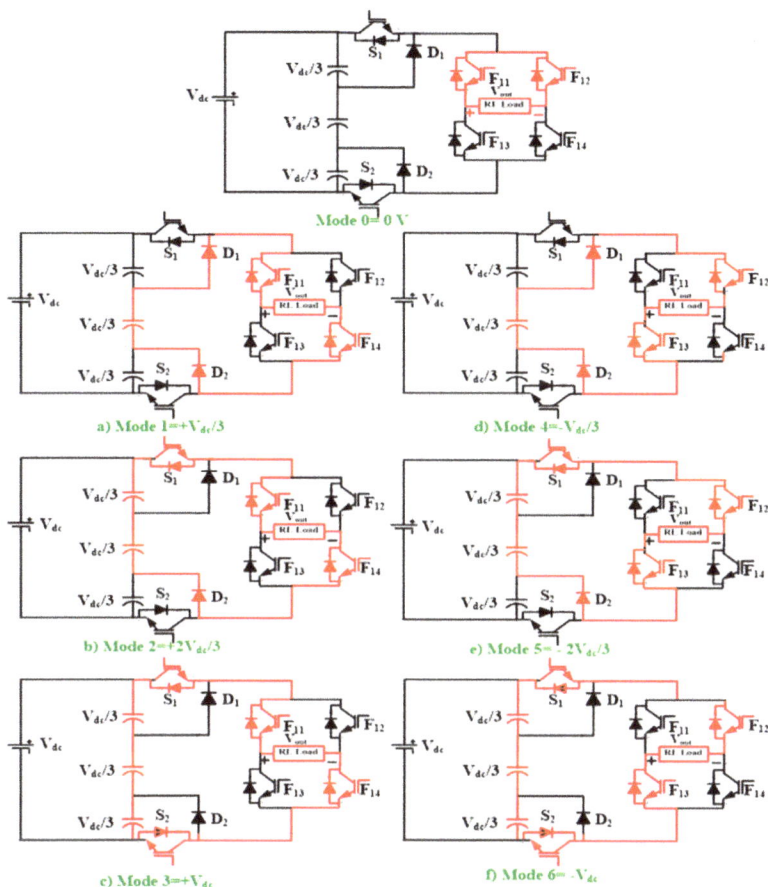

Figure 2. (a–f) Various modes of operation of the proposed basic unit.

Table 1. Switching sequence for the basic unit.

State	On State Switches	Full Bridge Switches	Output Voltage Level
0	—	F_{11},F_{12} F_{13},F_{14}	0 V
1	D_1, D_2	F_{11},F_{14} F_{12},F_{13}	$+V_{dc}/3$ $-V_{dc}/3$
2	(S_1, D_2) or (S_2, D_1)	F_{11},F_{14} F_{12},F_{13}	$+2V_{dc}/3$ $-2V_{dc}/3$
3	S_1, S_2	F_{11},F_{14} F_{12},F_{13}	$+V_{dc}$ $-V_{dc}$

2.2. Proposed Cascaded Topologies

The basic unit given was able to generate a maximum of 7-level output voltage with six switches and two diodes. To generate a higher number of voltagelevels, two different cascaded topologies were proposed and these topologies were named as switched capacitor-diode (SCD^{CAS}) cascaded multilevel converter, where *CAS* referred to cascaded topology.

The first topology given in Figure 3a consists of a series connection of "n" number of basic units and each unit requires a separate DC source with three dc-link capacitors. In unit-1, capacitor voltages were named as V_{11}, V_{12}, V_{13}. In unit-2 as V_{21}, V_{22}, V_{23} and in nth unit as V_{n1}, V_{n2}, V_{n3}. The maximum blocking voltage on the switches was the sum of all these dc source values. Another topology as a switched-diode half-bridge (SCDHB^{CAS}) multilevel converter and is shown in Figure 3b. The half-bridge inverter was cascaded with a series connection of basic units to double the output voltage level.

The expression to find the number of levels, number of switches, isolated dc sources, capacitors, maximum blocking voltage and total blocking voltage are given in Table 2. These topologies can also be configured in the asymmetric mode, but in this paper only symmetric configuration was considered. The output voltage of SCDHB^{CAS} topology was double that of the SCDC^{AS} topology with minimum switches. The SCDHB^{CAS} topology requires additional dc sources and reduce output voltage magnitude because the half-bridge circuit output voltage was always half that of the input voltage.

Table 2. Comparison of power components in the proposed topologies.

S.No	Various Parameters	(SCDC^{AS}) Topology	(SCDHB^{CAS}) Topology
1.	N^{Level}	$6n + 1$	$12n + 1$
2.	$N^{Switches}$	$6n$	$6n + 2$
3.	N^{diode}	$2n$	$2n$
4.	$N^{capacitors}$	$3n$	$3n + 1$
5.	N^{source}	n	$N + 1$
6.	Max^{block}	V_{dc}	V_{dc}
7.	T^{block}	$n6\,V_{dc}$	$(6n + 2)\,V_{dc}$

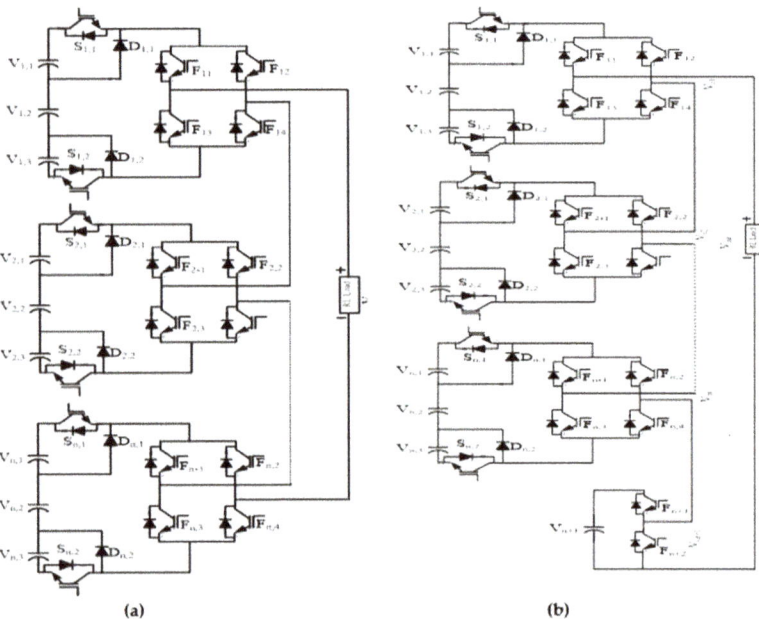

Figure 3. Proposed switched-diode cascaded topology; (a) without doubling circuit and (b) with doubling circuit.

The topologies presented in References [5–8] were considered for the comparison and are shown in Figure 4. Total blocking voltage was one of the important parameters that decides the cost of the converter. If the blocking voltage was less, the cost of the switch would be lower. As shown in Figure 4a, the proposed topology SCDHB^{CAS} has a lower total blocking voltage than the other topologies. In other words, the number of level was higher with reduced blocking voltage. The DC source is another parameter to be considered while designing a multilevel converter. From Figure 4b it is evident that the topologies referred in [7,8] requires more number of sources than the proposed converters. In Figure 4c, the number of switches is plotted against the number of levels. The proposed topology SCDHB^{CAS} requires a minimum number of switches when compared to other topologies.

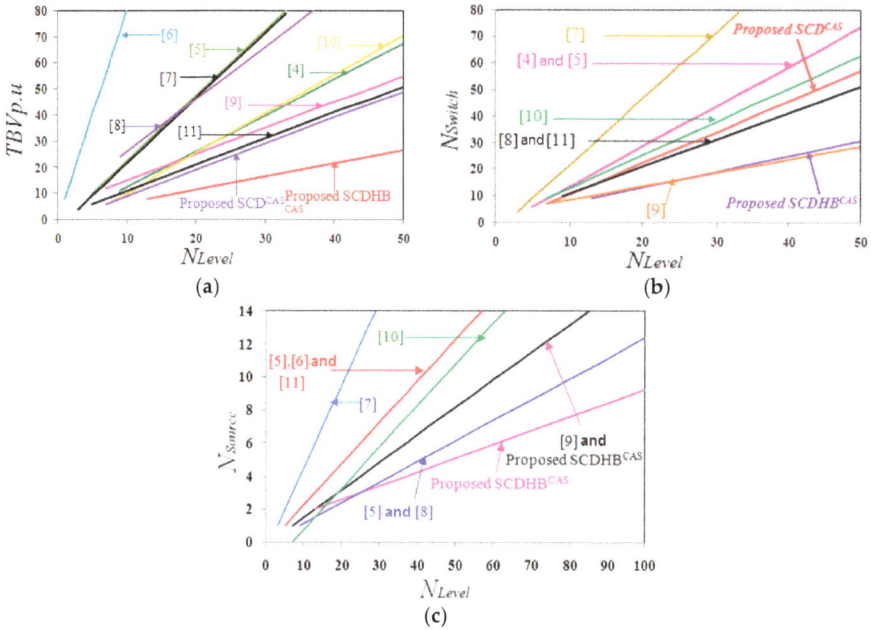

Figure 4. Comparison of proposed multilevel converter with other topologies (**a**) $N_{Switches}$ vs. N_{Level}; (**b**) TBVp.u vs. N_{Level} and (**c**) N_{Level} vs. N_{source}.

3. Modified DC-Offset Value in the NLM Method

The conventional NLM technique was more suitable for a higher number of levels because it produced a higher THD for a lower number of levels [2]. To get better THD and RMS output voltage, the optimum nearest modulation technique was proposed. In the NLM technique, the DC offset value was 0.5 (also called a half integer type), but in the case of optimal nearest level modulation (ONLM), the DC offset value was changed from 0.5 to 0.4.

$$V_{out} = m_a \times (N_{Level} - 1)/2 \times V_{dc} \times \cos(\omega t) \tag{1}$$

where N_{Level} was the number of output levels and 'm_a' was the modulation index. The output voltage for various modulation indices was calculated using Equation (1), where x_1, x_2 and x_3 were the variables and this was compared with the voltage reference signal, to generate the pulses as shown in Figure 4. The intersection of reference signal and variables x_1, x_2 and x_3 gavedifferent pulse widths with a duty cycle of D_1, D_2 and D_3. The voltage RMS of the converter depends on the duty cycle of the switches.

Here, the D_3 has maximum duty cycle when compared with D_2 and D_1. Each intersection point gave a different switching angle θ_1, θ_2 and θ_3 as shown in Figure 5. The variables x_1, x_2 and x_3 directly affected the switching angle and duty cycle. The mathematical expression to find the RMS value was given in Equation (2).

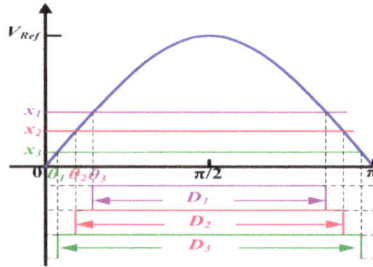

Figure 5. Variation of duty cycle with respect to DC offset value x_1, x_2 and x_3.

$$V_{rms} = \sqrt{\frac{1}{T} \int_0^T V(t).dt} \tag{2}$$

T was the time period of the pulse from zero and $V(t)$ was the magnitude of voltage. For simplicity $V(t)$ was replaced by 'a' and Equation (3) is rewritten as:

$$V_{rms}^2 = \frac{1}{T} \int_{t_1'}^{t_1} a^2 dt \tag{3}$$

Equation (3) was the voltage RMS of the pulses. RMS voltage for multilevel output was calculated by choosing the duty cycle of each level and multiplied with corresponding magnitude as discussed below:

$$V_{rms} = a_1 \sqrt{D_1} + a_2 \sqrt{D_2} + \ldots \ldots a_n \sqrt{D_n} \tag{4}$$

The magnitude between each level was symmetric and hence $a_1 = a_2 = a_3 = \ldots a_n = V_{dc}$.
The switching angle calculation for the proposed modulation technique was given below:

$$\theta = \sin^{-1}\left(\frac{i - 0.6}{x}\right) \quad \begin{array}{l} where\, i = 1, 2, \ldots \left(\frac{m-1}{2}\right) \\ x = \left(m - \frac{1}{2}\right) \end{array} \tag{5}$$

'm' is the number of levels. The different switching angle against the number of levels is shown in Figure 6. In this, if the number of level is increased, the switching angle (θ) value is reduced gradually. The key variables x_1, x_2 and x_3 decides the switching angles and duty cycles. Instead of x_1, x_2 and x_3 variables, constant DC offset was used in the modulation techniques. The DC offset value of 0.5 was used in the conventional NLM technique which is more suitable for a higher number of output voltage levels, but for lower voltage levels it produces high THD.

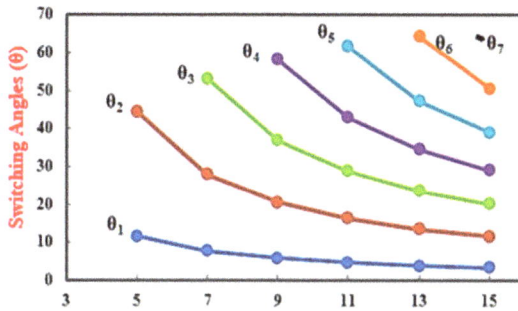

Figure 6. Number of levels vs switching angle.

The functional block diagram of the proposed system is shown in Figure 7a To minimize the THD in lower numbers of levels, a new DC offset value of 0.4 was proposed. The reason for choosing 0.4 as the DC offset was illustrated in Figure 7b with a reference waveform. The magnitude of individual harmonics orders is presented on the left side of the Y axis and on the right side, variations of voltage RMS were presented in per unit value. The X-axis was the variation of DC offset which is varied from 0.1 to 0.9. Up to the 15th odd harmonics orders were considered for selection of DC offset value.

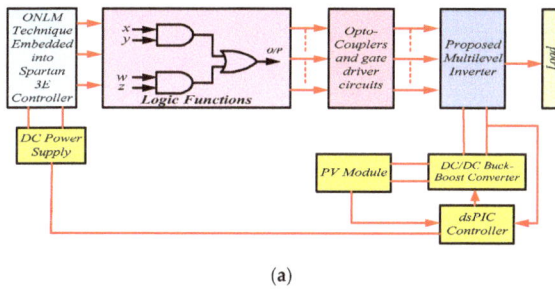

(a)

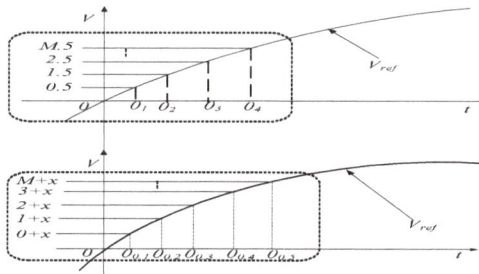

(b)

Figure 7. *Cont.*

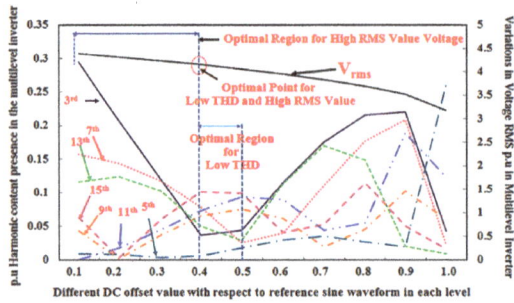

(c)

Figure 7. (a) A functional block diagram of the proposed system (b) DC offset 0.4 compared with reference waveform and pulse generation method; (c) the various DC offset values with corresponding THD.

The lower order harmonics like 3rd, 5th and 7th were lower in the region of DC offset from 0.4 to 0.5. The 3rd order harmonic voltage was low, whereas the RMS voltage was high for a DC offset of 0.4. Therefore, a DC offset of 0.4 was considered optimum in this paper. This DC offset value was valid for a lower number of output voltage levels, up to the 17th-level. In the half integer type method (conventional NLM), the error (DC losses) between the two levels was always maintained at $0.5V_{dc}$. In the proposed method, the error was minimized to $0.4V_{dc}$, as shown in Figure 7c. The RMS value of the output voltage and current was higher than the conventional methods.

4. Simulation and Experimental Results

In this, the FPGA Spartan 3E was used to generate the pulse for the proposed converter. In terms of hardware, the RC delay circuit was used to provide the dead time of 2 µs between the pulses to avoid short circuit. The subsystem of the PV Simulink module is shown in Figure 8. This PV model was designed based on a single diode model as used in Reference [19]. In this model, the temperature is kept at 298 K and due to the variations in irradiance; the output voltage and current were varied. To extract the maximum power from the PV panel, the basic perturb and observe MPPT method was implemented and the regulated output voltage was obtained. However, this is not in the scope of this paper. The PV model consists of one diode for cell polarization and series/shunt resistance for the losses. The I_{pv} (V_{pv}) of this model was calculated as follows:

$$I_{pv} = I_{ph} - I_d - I_{Rsh} \tag{6}$$

$$Ipv = Iph - Io\left[e^{\left(\frac{q(Vpv + Rs \,.Ipv)}{NsKT_j}\right)} - 1\right] - \frac{Vpv + Rs \,.Ipv}{Rsh} \tag{7}$$

Rs and Rsh were series and shunt resistance, I_{ph} was the photocurrent, I_o was the reverse saturation current of the diode, q was the electron charge, K was constant, T_j was the junction temperature of the panel, I_d was the intrinsic diode current, and Vpv and Ipv was the voltage and current in the panel. The DC/DC converter boosts the voltage from 40 V to 82 V. As per the single-phase grid, standard voltage is $Vdc > \sqrt{2} * V_{grid}$.

The multilevel inverter output peak voltage was given below:

$V_{pp} = 3 \times 110 = 330$ V, $V_{rms} = V_{pp}/\text{sqrt (2)}$

$V_{rms} = 330/\text{sqrt (2)} = 232.2$ V

To prove the dynamic performance of the proposed topology, the modulation scheme was implemented using MATLAB/Simulink and experimentally verified with the hardware developed. The details of the PV panel and other hardware ratings were listed in Table 3.

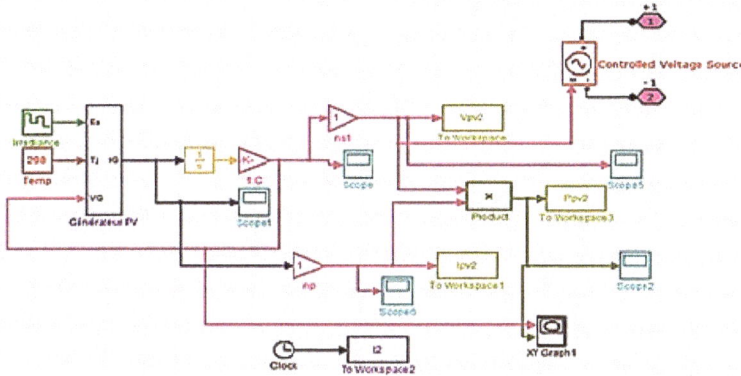

Figure 8. Simulation of PV module.

Table 3. Parameters used for hardware.

Description	Specifications
PV System	
PV Model 12100	04 Nos
Open Circuit Voltage	26.8 V
Short Circuit Current	6.2 A
Maximum Voltage (V_m)	21.8 V
Maximum Current (I_m)	5.62 A
Maximum system DC Voltage	1000 V
Power Tolerance	±5%
Load	
Resistance (R)	150 Ω & 80 Ω
Inductor (L)	70 mH & 80 mH
Multilevel Converter	
IRF 460 500 V/21 A	06 & 08 Nos
Gate Driver-HCPL316j	06 & 08 Nos
Capacitors	150 μF
FPGA Spartan3E	1
Snubber Circuits	RCD
Results	
Output Voltage	90 V & 120 V
Output Current	0.5 A & 1.51 A

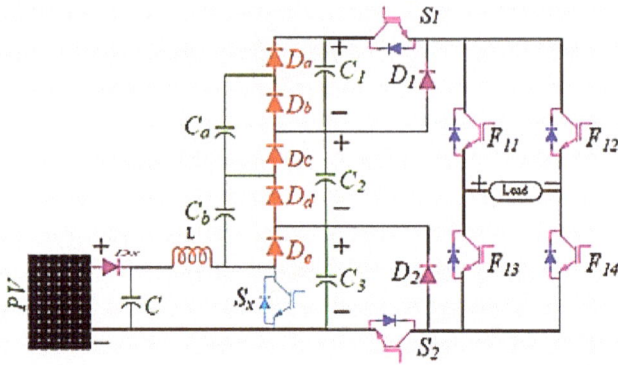

(a)

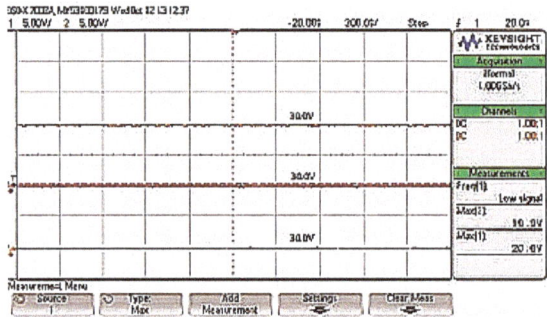

(b)

Figure 9. (a) Proposed seven-level inverter with voltage balancing circuit. (b) Proposed capacitor balancing circuit output voltage.

4.1. Seven Level Inverter

The proposed 7-level inverter with voltage balancing circuit is shown in Figure 9a along with its balance output voltage in Figure 9b. To determine the performance of the multilevel inverter and modulation technique, the dynamic (without DC/DC converter) and steady-state output voltage (with DC/DC converter) were shown in Figure 10a–i for different irradiance values. In the basic unit, the DC/DC boost converter was used to regulate the output voltage to 90 V and each capacitor was balanced to 30 V; when the full irradiance was present with a minimum duty ratio of 0.12. The corresponding output voltage and current waveforms are shown in Figure 10a. The current THD depended on the load inductance value, because this acted as a low pass filter to produce a sinusoidal waveform, but here the resistance value is higher than the inductance value. Here it is worth mentioning that the voltage THD of the proposed modulation technique was 11.81% as shown in Figure 10c, which was higher than in Reference [21]. It was confirmed that the proposed modulation technique was suitable for a lower number of voltage levels with increased output RMS voltage. To regulate the PV output voltage, the non-isolated DC/DC converter was used. The experimental output voltage and current waveforms were shown in Figure 11 with the THD spectrum. The load resistance was 150 Ω and inductance was 70 mH which gives the maximum output power of 20.938 W with a power factor of 0.89 as shown in Figure 11b. The voltage RMS of the 7-level inverter was 65.11 V and current RMS was 0.36 A, which was higher than the theoretical value of 63.63 for given input.

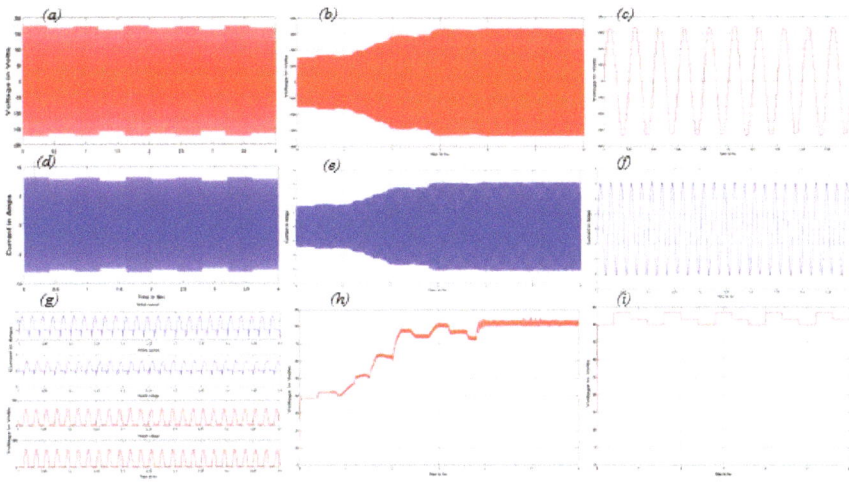

Figure 10. Multilevel inverter voltage waveform (**a**) without voltage balancing circuit; (**b**) with voltage balancing circuit; (**c**) steady-state voltage, multilevel inverter current waveform and (**d**) without voltage balancing circuit; (**e**) with voltage balancing circuit; (**f**) steady-state current; (**g**) current and voltage across the H-bridge switches; (**h**) output voltage and (**i**) variation in irradiance.

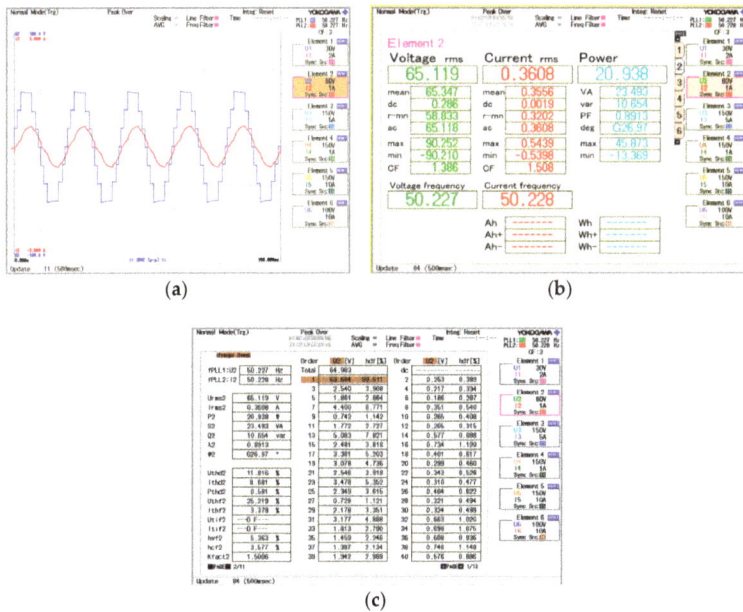

Figure 11. Experimental results for 7-level inverter (**a**) output voltage and current waveform (**b**) output from power quality analyzer and (**c**) voltage harmonics.

4.2. Thirteen-Level Inverter

In this configuration the SCDHB^{CAS} topology was used to generate the 13th-level. Each DC link capacitors voltage were regulated to 40 V and the half-bridge inverter voltage was 20 V.

For the 13-level inverter the maximum voltage RMS value was 126.28 V and current RMS was 1.05 A with a power factor of 0.94. The maximum output power was 86.35 W with the boost converter duty ratio of 0.35 to maintain the output voltage at 120 V with each DC-link capacitor voltage of 40 V, as shown in Figure 12a.The corresponding inverter output voltage and current waveforms were shown in Figure 12b. Voltage THD was 6.701% and current THD was 6.388%, the power factor was 0.94 which is close to unity, as illustrated in Figure 12c,d. In Figure 13, different loading conditions were used and the corresponding outputs were captured using DSO. In this way, the load values were changed ranging from purely resistive to inductive and a combination of both as illustrated in Figure 13a. A zoomed view of the load changes considered is shown in Figure 13b.

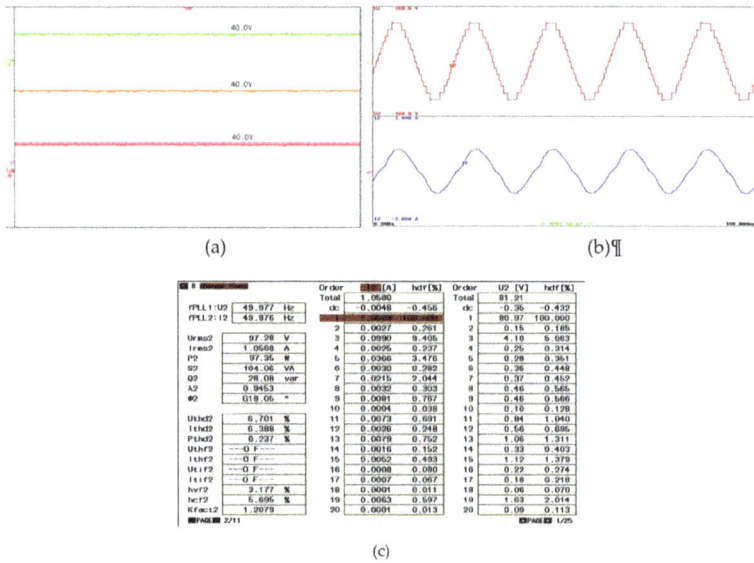

(a) (b)¶

(c)

Figure 12. Experimental results of thirteen-level converter (**a**) capacitor voltages after balancing (**b**) output voltage and current waveform and (**c**) voltage harmonics.

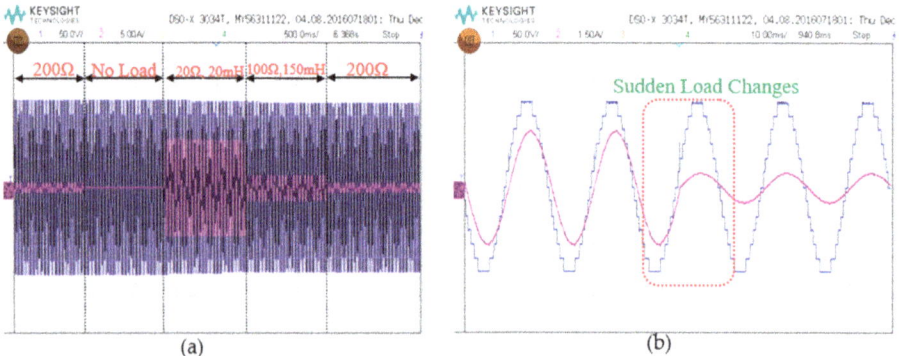

(a) (b)

Figure 13. Experimental results of thirteen-level converter; (**a**) different loading conditions and (**b**) zoomed view of load changes.

The proposed NLM technique produced high voltage RMS of 97.28 V, whereas 98.85 V was the theoretical value. From the proposed modulation it was confirmed that it generates low THD and high voltage RMS for a lower number of levels. The laboratory-based prototype model of the proposed multilevel converter for the 13th-level is shown in Figure 14 with PV as the input source.

Figure 14. Photograph of experimental setup.

5. Conclusions

The generalized cascaded multilevel converter was proposed with the optimum nearest level modulation technique in this paper. Basic units and the cascaded connection of basic units were discussed with half-bridge and without half-bridge circuits. With the half-bridge converter, it produced the maximum output voltage for higher voltage levels than it did without the half-bridge circuit. The proposed modified nearest level modulation technique was presented which generated the low THD and high voltage RMS. The experimental validation of the proposed converter and modified NLM technique was compared with the simulation results. The modified NLM technique was proved to be more suitable for a lower number of output voltage levels and was found suitable for photovoltaic applications.

Author Contributions: K.D: design, analysis, simulation and hardware implementation of proposed system; V.K: comparison and validation of results; J.S.M: formatting and English editing.

Funding: This research received no external funding.

Conflicts of Interest: The authors declare no conflict of interest.

References

1. Rodriguez, J.; Lai, J.-S.; Peng, F.Z. Multilevel inverters: A survey of topologies, controls, and applications. *IEEE Trans. Ind. Electron.* **2002**, *49*, 724–738. [CrossRef]
2. Franquelo, L.G.; Rodriguez, J.; Leon, J.I.; Kouro, S.; Portillo, R.; Prats, M.A.M. The age of multilevel converters arrives. *IEEE Trans. Ind. Electron. Mag.* **2008**, *2*, 28–39. [CrossRef]
3. Gupta, K.; Ranjan, A.; Bhatnagar, P.; Sahu, L.; Jain, S. Multilevel Inverter Topologies with Reduced Device Count: A Review. *IEEE Trans. Ind. Electron.* **2016**, *31*, 135–151. [CrossRef]
4. Lee, S.S. Single-Stage Switched-Capacitor Module (S3CM) Topology for Cascaded Multilevel Inverter. *IEEE Trans. Power Electron.* **2018**, *33*, 8204–8207. [CrossRef]
5. Liu, J.; Cheng, K.W.E.; Ye, Y. A Cascaded Multilevel Inverter Based on Switched-Capacitor for High-Frequency AC Power Distribution System. *IEEE Trans. Power Electron.* **2014**, *29*, 4219–4230. [CrossRef]
6. Ye, Y.; Cheng, K.W.E.; Liu, J.; Ding, K. A Step-Up Switched-Capacitor Multilevel Inverter with Self-Voltage Balancing. *IEEE Trans. Ind. Electron.* **2014**, *61*, 6672–6680. [CrossRef]
7. Taghvaie, J.A.; Rezanejad, M. A Self-Balanced Step-Up Multilevel Inverter Based on Switched-Capacitor Structure. *IEEE Trans. Power Electron.* **2018**, *33*, 199–209. [CrossRef]
8. Barzegarkhoo, R.; Moradzadeh, M.; Zamiri, E.; Kojabadi, H.M.; Blaabjerg, F. A New Boost Switched-Capacitor Multilevel Converter with Reduced Circuit Devices. *IEEE Trans. Power Electron.* **2018**, *33*, 6738–6754. [CrossRef]
9. Raman, S.R.; Cheng, K.W.E.; Ye, Y. Multi-Input Switched-Capacitor Multilevel Inverter for High-Frequency AC Power Distribution. *IEEE Trans. Power Electron.* **2018**, *33*, 5937–5948. [CrossRef]

10. Liu, J.; Wu, J.; Zeng, J. Symmetric/Asymmetric Hybrid Multilevel Inverters Integrating Switched-Capacitor Techniques. *IEEE J. Emerg. Sel. Top. Power Electron.* **2018**, *6*, 1616–1626. [CrossRef]
11. Prabaharan, N.; Palanisamy, K. Analysis of cascaded H-bridge multilevel inverter configuration with double level circuit. *IET Power Electron.* **2017**, *10*, 1023–1033. [CrossRef]
12. Carrara, G.; Gardella, S.; Marchesoni, M.; Salutari, R.; Sciutto, G. A new multilevel PWM method: A theoretical analysis. *IEEE Trans. Power Electron.* **1992**, *7*, 497–505. [CrossRef]
13. McGrath, B.P.; Holmes, D.G. Multicarrier PWM strategies for multilevel inverters. *IEEE Trans. Ind. Electron.* **2002**, *49*, 858–867. [CrossRef]
14. Franquelo, L.G.; Prats, M.A.M.; Portillo, R.C.; Galvan, J.I.L.; Perales, M.A.; Carrasco, J.M.; Diez, E.G.; Jimenez, J.L.M. Three dimensional space-vector modulation algorithm for four-leg multilevel converters using ABC coordinates. *IEEE Trans. Ind. Electron.* **2006**, *53*, 458–466. [CrossRef]
15. McGrath, B.P.; Holmes, D.G.; Lipo, T. Optimized space vector switching sequences for multilevel inverters. *IEEE Trans. Power Electron.* **2003**, *18*, 1293–1301. [CrossRef]
16. Govindaraju, C.; Baskaran, K. Efficient Sequential Switching Hybrid-Modulation Techniques for Cascaded Multilevel Inverters. *IEEE Trans. Power Electron.* **2011**, *26*, 1639–1648. [CrossRef]
17. Shukla, A.; Ghosh, A.; Joshi, A. Hysteresis Modulation of Multilevel Inverters. *IEEE Trans. Power Electron.* **2011**, *26*, 1396–1409. [CrossRef]
18. Rahim, N.A.; Selvaraj, J.; Chaniago, K. A novel PWM multilevel inverter for PV application. *IEICE Electron. Express* **2009**, *6*, 1105–1111. [CrossRef]
19. Rekioua, D.; Matagne, E. *Optimization of Photovoltaic Power Systems, Modelization, Simulation and Control*; Springer: London, UK, 2012; pp. 31–87.
20. Rathore, A.K.; Holtz, J.; Boller, T. Synchronous optimal pulsewidth modulation for low-switching-frequency control of medium-voltage multilevel inverters. *IEEE Trans. Ind. Electron.* **2010**, *57*, 2374–2381.
21. Rathore, A.; Holtz, J.; Boller, T. Generalized optimal pulsewidth modulation of multilevel inverters for low-switching-frequency control of medium-voltage high-power industrial ac drives. *IEEE Trans. Ind. Electron.* **2013**, *60*, 4215–4224. [CrossRef]
22. Karthikeyan, D.; Krishnasamy, V.; Sathik, M.A.J. Development of a switched diode asymmetric multilevel inverter topology. *J. Power Electron.* **2018**, *18*, 418–431.
23. JagabarSathik, M.A.; Abdel Aleem, S.H.E.; Kannan, R.; Zobaa, A.F. A New Switched DC-Link Capacitor-based Multi-level Converter (SDC2MLC). *Electr. Power Compon. Syst.* **2017**, *45*, 1001–1015. [CrossRef]

electronics

MDPI

Article

Optimized Design of Modular Multilevel DC De-Icer for High Voltage Transmission Lines

Jiazheng Lu, Qingjun Huang *, Xinguo Mao, Yanjun Tan, Siguo Zhu and Yuan Zhu

State Key Laboratory of Disaster Prevention and Reduction for Power Grid Transmission and Distribution Equipment, State Grid Hunan Electric Company Limited Disaster Prevention and Reduction Center, Changsha 410129, China; lujz1969@163.com (J.L.); huangqj@hust.edu.cn; maoxg_0@163.com (X.M.); zhengyuan2017307@126.com (Y.T.); zhusiguo2005@ 163.com (S.Z.); zhuyuan1278@163.com (Y.Z.)

* Correspondence: huangqj@hust.edu.cn; Tel.: +86-0731-8633-2088

Received: 17 August 2018; Accepted: 14 September 2018; Published: 17 September 2018

Abstract: Ice covering on overhead transmission lines would cause damage to transmission system and long-term power outage. Among various de-icing devices, a modular multilevel converter based direct-current (DC)de-icer (MMC-DDI) is recognized as a promising solution due to its excellent technical performance. Its principle feasibility has been well studied, but only a small amount of literature discusses its economy or hardware optimization. To fill this gap, this paper presents a quantitative analysis and calculation on the converter characteristics of MMC-DDI. It reveals that, for a given DC de-icing requirement, the converter rating varies greatly with its alternating-current (AC) -side voltage, and it sometimes far exceeds the melting power. To reduce converter rating and improve its economy, an optimized configuration is proposed in which a proper transformer should be configured on the input AC-side of converter under certain conditions. This configuration is verified in an MMC-DDI for a 500 kV transmission line as a case study. The result shows, in the case of outputting the same de-icing characteristics, the optimized converter is reduced from 151 MVA to 68 MVA, and the total cost of the MMC-DDI system is reduced by 48%. This conclusion is conducive to the design optimization of multilevel DC de-icer and then to its engineering application.

Keywords: converter; ice melting; modular multilevel converter (MMC); optimization design; transmission line; static var generator (SVG)

1. Introduction

Ice covering on overhead transmission lines is a serious threat to the safe operation of power grids. Overweight ice would break wires or collapse towers, and then cause disruption of power transmission and large-scale outage [1,2]. The ice storms in North America 1998 [3], Germany 2005 [4], and China 2008 [5] are good examples of such consequences. In order to protect the grid from ice disaster, dozens of anti-icing or de-icing methods have been proposed [1,3,5,6], such as thermal de-icing, mechanical de-icing, passive icephobic coatings, etc.

Among various de-icing methods, heating of ice-covered line conductors by electrical current is recognized as the most efficient engineering approach to minimize the catastrophic consequences of ice events [5–8] because it can eliminate the ice covered on hundreds of kilometers of line within an hour, without damaging the grid structure or polluting the environment. Both alternating-current (AC) and direct-current (DC) can be used to melt ice, but AC ice-melting is usually used for transmission lines up to 110 kV, while DC ice-melting is more recommended for high voltage lines up to 500 kV [3,4]. In a DC de-icing system, the most critical part is the DC de-icer (DDI), which generates the required DC voltage and current.

Nowadays, the most widely adopted de-icer is the thyristor-based line-commutated converter (LCC) [9–11], derived from the conventional high voltage direct current transmission (HVDC) technology. It can output a wide range of DC voltages by regulating the thyristor phase shift angle to meet the de-icing requirement of various lines; moreover, it can operate as a static var compensator (SVC) when there is no de-icing requirement. Thus, it has been widely used in Russia, Canada, China [5,12,13], etc. However, due to the inherent characteristics of thyristors, LCCabsorbs much reactive power and generate a lot of harmonics. Thus, it has to deploy an extra series of harmonic filters and many shunt capacitors to meet the grid requirements. Thus, it is bulky, inflexible, and costly. In order to overcome these shortcomings, some proposed constructing the DC de-icer using a voltage source converter (VSC). In [14], a multiple phase shift de-icer was proposed, but it needs a complex multi-winding transformer. In [15], a concept of DC de-icer constructing with a static synchronous compensator (STATCOM) was proposed, but it didn't give specific solutions. In [16], a 3-level STATCOM scheme was proposed for the de-icer application. It presents excellent harmonic and reactive power features, but it requires high-power 3-level converters up to 100 MVA, and such a high-power 3-level converter is difficult to manufacture. Moreover, its DC voltage has to exceed its AC voltage, thus it has a limited DC voltage range.

In the last few decades, modular multilevel converter (MMC) topology has been rapidly developed [17,18]. Since it was presented for the first time by Lesnicar and Marquardt in 2003 [19], it has been widely used in many high-voltage and medium-voltage applications [20–22]. It can output a smooth and nearly ideal sinusoidal voltage with little filters, and it has modularity and scalability, and is facile and flexible. The main application of MMC is VSC-based HVDC transmission [22,23]. In the last five years, dozens of large-capacity MMC-based HVDC systems have been built [22], their rated DC voltage is up to ±500 kV and their rated power is hundreds of MW or even 2000 MW. Another typical application of MMC is the STATCOM [24]. In recent years, most of the STATCOM above 10 Mvar have adopted the MMC structure.

For the de-icer application, an MMC-based DC de-icer (MMC-DDI) with full-bridge submodules (SM) was firstly presented in 2013 [25]. Its structure is similar to a pair of parallel star-configured static var generators (SVGs), and their neutral points are respectively led out as the DC positive and negative poles of DC de-icer. It inherits all the advantages of MMC topology. Moreover, since it employs the full-bridge SMs, it can provide both the buck and boost functions for the DC-link voltage [26]. Thus, it has a wide DC output voltage range to satisfy the de-icing requirements of different lines. In addition, it can be operated as SVG to provide reactive power compensation for the grid. Due to these advantages, the MMC-DDI is recognized as a promising de-icing solution [27]. Since MMC-DDI was first proposed in 2013 [25], its operation principle and control optimization have been further studied in [27–29]. In [28], the hardware selection of MMC-DDI was studied, and a quantitative comparison with an LCC-based de-icer was given. As is shown, both the electrical characteristics and the land occupation have more advantages. In [29], the control and modulation algorithms of MMC-DDI are described. In [27], the dynamic model of MMC-DDI and its harmonic features under phase-shifted carrier modulation are analyzed, and then a detailed control scheme is developed, and the MMC-DDI topology was experimentally verified by utilizing a downscaled prototype. The literature above mainly focus on the technical feasibility of MMC-DDI and pay little attention to its economy optimization.

Like most STATCOMs, the existing MMC-DDI is recommended to be directly connected to the substation distribution network without a transformer. This is considered as a major advantage of the MMC-DDI scheme because the absence of transformer is believed to make the whole device small, light, and compact. Under this configuration, the arm voltage and current of MMC are substantially determined by the grid-connected voltage in addition to the required DC melting voltage and current. For the common high-voltage transmission lines up to 500 kV, their DC melting current is 4000–5000 A or even higher, while their DC melting voltage is usually no more than 10 kV. When the distribution network voltage is unsuitable—for example, 35 kV for most 500 kV substations—the MMC in this configuration simultaneously withstands higher arm voltage and larger arm current.

Thus, the converter rating of MMC-DDI far exceeds its output ice-melting power, resulting in a poor economy to engineering apply.

To address this issue, this paper presents a quantitative analysis on the converter characteristics of MMC-DDI, and then calculates the required converter rating and its influencing factors. It reveals that, for a certain DC de-icing requirement, converter rating varies greatly with its AC-side voltage, and then an optimized design method is proposed to improve the economy of MMC-DDI. Finally, a design example and its corresponding simulation results are given. As this case shows, under the same de-icing outputting characteristics, the optimized converter rating is reduced from 151 MVA to 68 MVA, and the total cost of MMC-DDI system is reduced by 48%.

2. Circuit Configuration and Operation Principle

The circuit configuration of the MMC-DDI is shown in Figure 1. It contains two sets of star-configured arms and each arm has several full-bridge SMs along with a connection reactance. Structurally speaking, it can be viewed as a pair of three-phase star-configured SVGs. The AC terminals of these two SVGs are in parallel and connected to the grid, whereas their neutral points are respectively led out as the DC positive and negative poles of MMC-DDI, and then connected to the ice-covered overhead lines through a set of de-icing disconnectors.

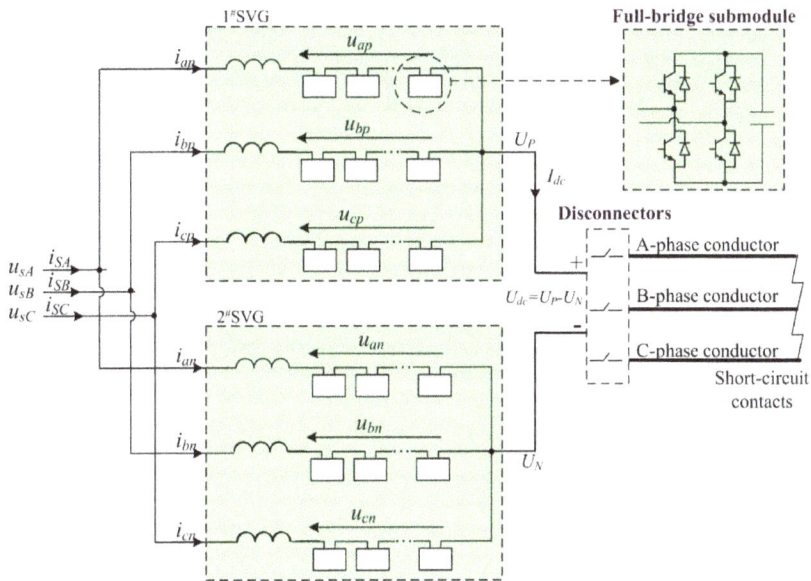

Figure 1. Circuit configuration of MMC based DC de-icer (MMC-DDI).

Since MMC-DDI can provide both buck and boost functions for DC-link voltage, it theoretically does not require a transformer to supply a wide and adjustable DC output voltage. In the existing literature, the AC terminal of MMC-DDI is directly connected to the distribution network with no transformer. This is considered as a major advantage of the MMC-DDI scheme because it can save the cost and floor area of a transformer, making the device small, light, and compact.

According to the grid requirements, MMC-DDI can have two different operation modes:

- Ice-melting Mode. When there is an ice-covered line to melt in the winter, the disconnectors are close to connect the MMC-DDI and the transmission line together, and the other terminal of the transmission line is artificially three-phase short-circuited to form a DC current loop. Then, the MMC-DDI provides a controlled DC voltage to generate the required current through the ice-covered line. At that time, the operation mode of MMC-DDI is similar to the MMC rectifier station in the VSC-HVDC transmission system, except that the DC-side output voltage almost remains unchanged in the VSC-HVDC system while it may vary with the line parameters in the MMC-DDI system. In addition, the typical control methods for the common MMC system are also applicable to MMC-DDI system, such as the capacitor voltage control, the active and reactive current control, the capacitor voltage balancing control, the circulating current control, etc.

- SVG Mode. When there is no icing line, the de-icing disconnectors can be open circuit. Then, the upper three arms and the lower three arms can operate as two parallel conventional SVGs, and provide reactive power compensation or alleviate other power quality problems.

3. Converter Characteristic of MMC-DDI

3.1. Arm Voltage and Current

Take the A-phase as an example, the dynamic equations of MMC-DDI can be expressed as:

$$u_{sA} = Ri_{ap} + L\frac{d}{dt}i_{ap} + u_{ap} + U_p \tag{1}$$

$$u_{sA} = Ri_{an} + L\frac{d}{dt}i_{an} + u_{an} + U_n \tag{2}$$

$$i_{sa} = i_{ap} + i_{an} \tag{3}$$

where u_{sA}, i_{sA} are the AC-side input phase voltage and current of converter. u_{ap}, u_{an} are respectively the output voltage of the upper arm and lower arm. i_{ap}, i_{an} are respectively the arm current of the upper and lower arms. U_p is the electric potential of the neutral point of 1#SVG, relative to the grid neutral point. U_n is the electric potential of the neutral point of 2#SVG. R and L represent the equivalent resistance and inductance of the connection reactance in each arm:

$$\begin{cases} I_{dc} = i_{ap} + i_{bp} + i_{cp} = -(i_{an} + i_{bn} + i_{cn}) \\ U_{dc} = U_p - U_n \end{cases} \tag{4}$$

where U_{dc} and I_{dc} are the DC-side output de-icing voltage and current of MMC-DDI.

Generally, the voltage and current of each arm are symmetrical, and the circulation current can be effectively suppressed with proper circulation current control method, and the voltage drop across the connection reactance is far less than other items in Equations (1) and (2). As a result, the A-phase arm voltages and currents can be expressed as:

$$u_{ap} = \sqrt{2}U_m \sin(\omega t) - 0.5U_{dc} \tag{5}$$

$$u_{an} = \sqrt{2}U_m \sin(\omega t) + 0.5U_{dc} \tag{6}$$

$$i_{ap} = \frac{\sqrt{2}}{2}I_m \sin(\omega t + \varphi) + \frac{I_{dc}}{3} \tag{7}$$

$$i_{an} = \frac{\sqrt{2}}{2}I_m \sin(\omega t + \varphi) - \frac{I_{dc}}{3} \tag{8}$$

where U_m, I_m are the root mean square (RMS) values of the AC-side input phase voltage and current of MMC converter. ω is the angular frequency of gird voltage while φ presents the AC-side power factor angle.

Similarly, the B-phase and C-phase arm voltage/current can also be expressed. As shown in Equations (5)–(8), the voltage/current of each arm contains both AC and DC components. Moreover, their peak values are the same for each arm, and can be expressed as

$$
\begin{cases}
I_{arm\_peak} = \frac{\sqrt{2}}{2} I_m + \frac{1}{3} I_{dc} \\
U_{arm\_peak} = \sqrt{2} U_m + 0.5 U_{dc}
\end{cases}
\tag{9}
$$

where I_{arm_peak}, U_{arm_peak} present the peak values of arm current and arm voltage.

According to Equations (5)–(8), the RMS values of arm voltage and current can be expressed as

$$
\begin{cases}
I_{arm\_RMS} = \sqrt{\frac{1}{2} I_m^2 + \frac{1}{9} I_{dc}^2} \\
U_{arm\_RMS} = \sqrt{2 U_m^2 + \frac{1}{4} U_{dc}^2}
\end{cases}
\tag{10}
$$

where I_{arm_RMS}, U_{arm_RMS} present the RMS values of the arm current and arm voltage.

Compared with that of common SVGs, the converter voltage/current of the MMC-DDI has different characteristics:

(1) The arm voltage/current of MMC-DDI contains both DC and AC components, while in the conventional SVG, there is only AC component.
(2) The arm voltage/current no longer equals the AC-side input voltage/current in MMC-DDI.
(3) The peak value of the arm voltage/current is no longer than $\sqrt{2}$ times of its RMS value.
(4) Due to these differences, although the MMC-DDI is structurally similar to a pair of common star-connected SVGs, their inner converter characteristics are quite different.

3.2. Influence of AC Side Input Voltage

Under normal operating conditions, the AC side input active power of the MMC converter is substantially equal to its DC side output power (neglecting tiny converter loss). According to the power balance between the AC and DC sides, the output DC ice-melting power can be obtained:

$$
P_{dc} = U_{dc} I_{dc} = 3 I_m U_m \cos \varphi
\tag{11}
$$

where P_{dc} is the output ice-melting power, $\cos\varphi$ is AC-side power factor and generally $\cos\varphi = 1.0$.

With (11), the AC-side input current of converter can be expressed as

$$
I_m = \frac{U_{dc}}{3 U_m \cos \varphi} I_{dc}.
\tag{12}
$$

Substituting (12) into (9), the peak values of arm voltage and arm current can be expressed as

$$
\begin{cases}
I_{arm\_peak} = \left(\frac{\sqrt{2}}{6\cos\varphi} \frac{U_{dc}}{U_m} + \frac{1}{3} \right) I_{dc} \\
U_{arm\_peak} = \left(\sqrt{2} \frac{U_m}{U_{dc}} + 0.5 \right) U_{dc}
\end{cases}
\tag{13}
$$

According to (13), the influence of AC side input voltage on the arm voltage and current peaks can be plotted and shown in Figure 2. As it shown, for a certain DC ice-melting requirement, with the increasing of AC-side voltage, arm voltage peak increases linearly (but not proportionally) while arm current peak decreases and tends to $1/3\, I_{dc}$. This is quite different from common SVG. In an SVG, in the case of a certain output reactive power, with the increasing of the AC-side voltage, the arm voltage peak increases proportionally while the arm current peak decreases and tends to 0.

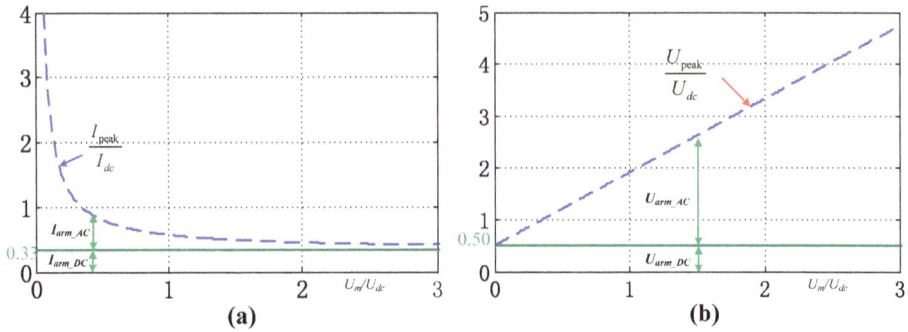

Figure 2. Influence of the AC side input voltage on the peaks of arm voltage and current (a) on the current (b) on arm voltage.

3.3. Converter Rating of MMC-DDI

In a power electronics system, the converter rating is an important technical indicator because device cost is closely related with the converter rating. For an MMC converter, its converter rating is mainly determined by the arm voltage peak and arm current peak because they largely determine the size and quantity of submodules, and then determines the main hardware of the converter. Therefore, the converter rating of the MMC-based devices can be collectively defined as

$$S_c = \sum_1^n \frac{U_{pi}I_{pi}}{2} \tag{14}$$

where S_c presents the converter rating. n presents the total number of arms. U_{pi}, I_{pi} are the output voltage and current peak of the i-th arm.

For a conventional star-connected SVG, there are three arms, and the current peak of each arm is approximately equal to the AC side phase current while arm voltage peak is approximately equal to the AC-side phase voltage (ignoring the voltage drop across the connection reactance). Then, its converter rating can be expressed as

$$S_c = 3\frac{U_p I_p}{2} = 3\frac{\sqrt{2}U_{sP} \times \sqrt{2}I_{sp}}{2} = 3U_{sp}I_{sp} = S_{out} \tag{15}$$

where U_{sp}, I_{sp} are respectively the RMS values of AC-side phase voltage and phase current, S_{out} presents the output apparent power of SVG.

Indeed, Equation (15) also applies to the delta-connected SVGs or an SVG group composed of several converters. In summary, for any SVG, the converter rating can be directly characterized by its rated output power.

For the MMC-DDI, the six arms share the same voltage and current peaks. Substituting Equation (9) into Equation (16), then the converter rating can be expressed as

$$S_c = 6\frac{U_{\text{arm\_peak}}I_{\text{arm\_peak}}}{2} = 3U_m I_m + \sqrt{2}U_m I_{dc} + \frac{3\sqrt{2}}{4}I_m U_{dc} + 0.5U_{dc}I_{dc} \tag{16}$$

Compared with equation (15), there are three other items in Equation (16), thus the converter rating characteristics of MMC-DDI are significantly different from that of common SVG.

Substituting Equation (13) into Equation (16) and considering cosϕ = 1.0, the converter rating can be simplified as

$$S_c = 3\left(\frac{\sqrt{2}}{6\cos\varphi}\frac{U_{dc}}{U_m} + \frac{1}{3}\right)I_{dc} \cdot \left(\sqrt{2}\frac{U_m}{U_{dc}} + 0.5\right)U_{dc} = \left(1.5 + \frac{\sqrt{2}}{4}\frac{U_{dc}}{U_m} + \sqrt{2}\frac{U_m}{U_{dc}}\right)P_{dc} \qquad (17)$$

With Equation (17), the relationship of the converter rating of MMC-DDI with its AC-side voltage can be calculated and shown as Figure 3. As it shown, under a certain DC-side output voltage and power requirement, the converter rating varies greatly with its AC input voltage. It can be analytically solved that when and only when $U_m = 0.5\,U_{dc}$, the converter rating gets its minimum value, and the minimum rating is 2.91 times the output ice-melting power. This conclusion can be expressed as

$$S_{c\_min} = \left(1.5 + \sqrt{2}\right)P_{dc} \quad \text{when} \quad U_m = 0.5U_{dc} \qquad (18)$$

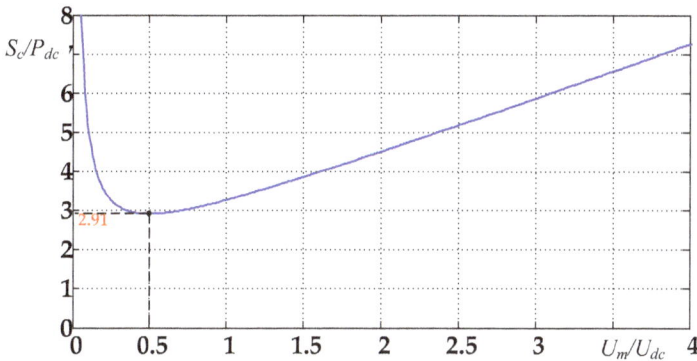

Figure 3. Relationship of the converter rating of MMC-DDI with its AC-side voltage.

4. The Proposed Optimization Design Method

4.1. General Design Process of IMD

For any type of DC ice melting device, its design process generally follows these steps:

Step 1: According to the line parameters and meteorological conditions of the transmission lines to be melted, calculate the required DC-side output de-icing current, voltage and power, and then determine the rated DC-side output parameters of IMD.

For a given transmission line, its required de-icing current depends on many parameters, such as conductor type, ambient temperature, wind velocity, ice thickness and de-icing duration, etc. The thermal behavior of overhead conductors has been well studied, and some formulas are given to calculate the de-icing current in many standards—for example, IEEE standard [30] and CIGRE standard [31]. Generally, the de-icing current should be greater than the minimum de-icing current and no more than the maximum endure current of the line conductor. For some typical conductor types used in China, the minimum de-icing current and the maximum endure current are shown as Table A1 (see Appendix A). In actual ice melting system, it generally tries to choose the intermediate value of the maximum and minimum values as the rated de-icing current.

After determining the de-icing current, the required de-icing DC voltage can be calculated as

$$U_{dc} = k_{icing} R_{line} I_{icing} \qquad (19)$$

where I_{icing} is the required de-icing current and R_{line} is the phase resistance of transmission line. k_{icing} corresponds to the ice-melting mode, $k_{icing} = 2$ when the de-icing current is passed down one phase conductor and back along another, and $k_{icing} = 1.5$ when down one and back along the other two [16].

When there are several lines to be melted, the de-icing DC current and voltage of each line can be calculated one by one, and then the rated DC-side output parameters of the IMD are determined by the output DC voltage range, the maximum de-icing current, and the maximum de-icing power.

Step 2: According to the optional voltage levels of the power substation as well as the rated IMD output power, select the proper access voltage of the IMD.

For typical transmission lines, their DC ice-melting power is generally among several MW and hundreds of MW. Within this range, the IMD is usually connected to the low-voltage distribution network of the substation, generally 10 kV or 35 kV in China.

Step 3: According to the DC-side output parameter requirements and the grid access voltage, design the internal structure and parameters of the IMD.

In the process of designing the internal IMD parameters, it is usually necessary to consider both the technical feasibility and the economy.

4.2. The Proposed Circuit Configuration and Its Economic Analysis

According to the above calculation, for a certain ice-melting requirement, the converter rating of MMC-DDI varies greatly with its AC-side voltage. Traditionally, MMC-DDI is directly connected to the grid, thus its AC-side input voltage always equals the grid voltage. This may correspond to a very high converter rating, resulting in poor economy. To solve such a problem, this paper proposes an optimization MMC-DDI configuration structure as shown in Figure 4, i.e., a transformer should be inserted between the grid and the converter under certain conditions. In order to realize this idea, there are two main questions:

(A) When should the transformer be desired and when is it undesired?
(B) If a transformer is inserted, what are the specifications and parameters of the transformer?

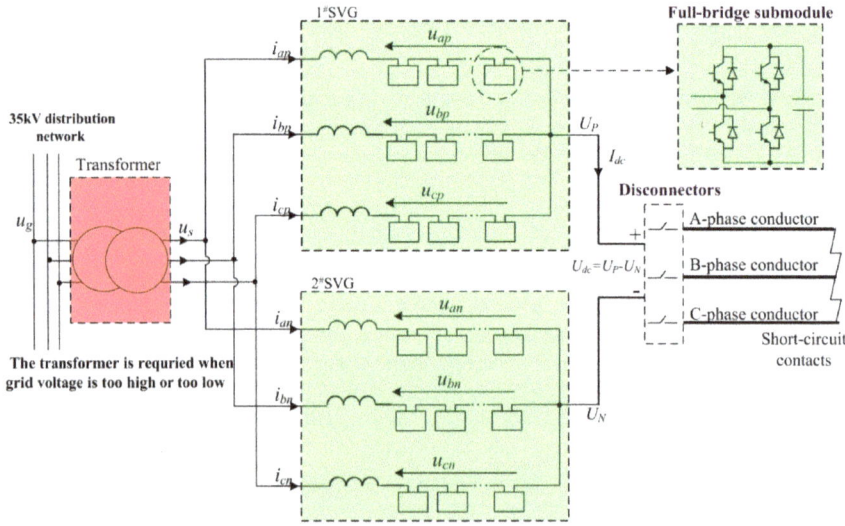

Figure 4. The proposed configuration structure of MMC-DDI.

According to (11), when the power factor is controlled as $\cos\phi = 1$, the AC-side input apparent power of MMC-DDI always equals its DC side output power regardless of the AC-side voltage. Therefore, if a transformer is inserted, its rating only needs to equal the output de-icing power rather than the converter rating. In order to get the minimum converter rating as shown in (18), the output phase voltage of the transformer can be set as $U_m = 0.5\,U_{dc}$, corresponding to a line voltage $\sqrt{3} \times 0.5U_{dc}$. In summary, the specification of the transformer can be determined as

$$\begin{cases} S_{Tran} = P_{dc} \\ T_r = U_g/(\sqrt{3} \times 0.5U_{dc}) \end{cases} \tag{20}$$

where S_{Tran} is the transformer rating, and T_r is the transformer rating voltage radio.

In order to get the timing of transformer insertion, the cost of converter and transformer should be compared. Since the MMC-DDI is rarely applied, it is difficult to obtain its market cost; here, its cost is estimated by referring to that of SVGs. This is due to three reasons: (1) MMC-DDI is structurally equivalent to a pair of star-connected SVGs, (2) SVG has been widely used and its cost is transparent, and (3) the rating range of common SVGs is wide enough to cover the potential MMC-DDI. Table A2 shows the deal prices of several high capacity SVGs built in China from 2013 to 2018.

As (15) shows, the converter rating of SVG is approximately equal to its rated output power, so the converter cost can be directly evaluated with the SVG deal price. As Table A2 shows, SVG cost is basically proportional to the rating, and its unit cost is around 15,000 \$/Mvar. For some SVGs over 60 Mvar, the unit cost is 40% higher. This is because there are only a few applications for such high-power SVGs, thus their R&D cost is higher. Moreover, such high-power SVG usually require higher reliability and larger configuration margin, and this also increases the device cost. For simplicity, here the MMC converter cost is estimated with the average unit price 15,000 \$/Mvar.

When a transformer is inserted as Figure 4, the transformer would bring a cost itself. Table A3 shows the deal prices of several 10 MVA-class rectifier transformers built in China. As is shown, the cost of 10 MVA rectifier transformer is about \$86,000, about half of the same rating SVG. With the rating growth of transformer, its unit cost decreases rapidly. For a 56 MVA transformer, its unit cost is 4400 \$/Mvar and about 1/3 of a similar rating SVG. For a 100 MVA transformer, its unit cost reduces to 3300 \$/Mvar and about 1/6 of the same rating SVG.

Based on these cost data, it can be obtained that the cost of a common transformer is much lower than that of the same-rating MMC converter.

In the proposed configuration of MMC-DDI, it can get a minimum MMC converter rating at the cost of an extra transformer. In order to quantitatively compare the economics of the proposed configuration, the costs of MMC-DDI with and without the transformer can be expressed as

$$\begin{cases} P_{no} = P_{con}(u_s = u_g) \\ P_{with} = P_{trans} + P_{con}(u_s = \sqrt{3} \times 0.5U_{dc}) \end{cases} \tag{21}$$

where P_{no} presents the cost of MMC-DDI with no transformer, and $P_{con}(u_s = u_g)$ presents the cost of the MMC converter when its AC-side voltage is equal to the grid voltage. P_{with} presents the cost of the MMC-DDI with a transformer; P_{trans} presents the transformer cost. $P_{con}(u_s = \sqrt{3} \times 0.5U_{dc})$ presents the cost of the MMC converter with an AC-side input voltage of $u_s = \sqrt{3} \times 0.5U_{dc}$.

As long as the cost of MMC-DDI with transformer is lower than that without a transformer, i.e., the reduced converter cost is greater than transformer cost, the proposed configuration structure is cost-effective. At this point, a transformer can be inserted on the AC side of converter to improve the system economy. Otherwise, this is no need to plug in the transformer.

4.3. Applicable Scope of the Proposed Configuration

Compared with the traditional MMC-DDI structure, the proposed MMC-DDI configuration structure requires an extra transformer. It seems that this would increase the cost of the total system, and partially offset the advantages of the MMC topology. However, in fact, the converter rating of traditional MMC-DDI varies greatly with its AC-side voltage, thus the insertion of transformer can sometimes reduce the converter rating and its cost. As long as the reduction of the converter cost is sufficient to offset the transformer cost, the proposed MMC-DDI structure is cost-effective.

According to the cost comparison data of the converter and transformer in the previous section, the unit cost of an MMC converter is generally much higher than that of a conventional transformer, especially for large-capacity converters above 50 MVA. Moreover, the reduced converter rating caused by an introduction of transformer is sometimes much higher than the transformer rating.

In order to obtain quantitative guidance, here an assumption is made of the cost of converter and SVG:

A   The converter cost is approximately considered to be proportional to the converter rating.
B   The transformer cost is a quarter of the same rating MMC converter cost.

Based on the above quantitative assumption, we can get the following conclusions:

a.   When the ratio of the grid line voltage to DC-side output voltage exceeds 2.0 or falls below 0.25, the overall cost of MMC-DDI with a transformer is less than that without transformer, i.e., a transformer can be inserted on the AC side of a converter to improve the system economy.
b.   When the ratio is between 0.25–2.0, the cost of the transformer exceeds its revenue. In that case, no transformer is required.

Indeed, for the common high-voltage transmission lines up to 500 kV, the required ice-melting voltage is generally less than 15 kV. Under such DC voltage range, if the MMC-DDI is connected to a 35 kV network, the grid voltage is more than two times the ice-melting DC voltage. In that case, the proposed MMC-DDI configuration is more applicable than the traditional one. However, if MMC-DDI is connected to a 10 kV distribution network, the grid voltage is usually among 0.25–2.0 times DC voltage, thus the traditional configuration is more applicable. In China, almost all of the distribution network voltage of 500 kV substations is 35 kV. Thus, at least for 500 kV transmission lines, the proposed MMC-DDI configuration is superior to the traditional configuration in most cases.

5. Design Example and Simulation Results

5.1. A Typical Design Example

In order to verify the above analysis and the proposed configuration, a design example of MMC-DDI is given here. For a 500 kV transmission line, the wire type is $4 \times$ LGJ-400, the line length is 40 km, and its single-phase resistance is 0.72 Ω. The minimum ambient temperature along the line is $-5\,^{\circ}$C, and the maximum wind speed in winter is about 5 m/s. In the 500 kV substation at one end of the transmission line, the distribution grid voltage is 35 kV, corresponding a 20.2 kV phase voltage.

With the data shown in Table A1, the required de-icing current of the above transmission line should be between 3475–4768 A. Within this range, the smaller the current, the longer the de-icing process lasts. Considering a balance between ice-melting rapidity and IMD economics, the rated DC de-icing current can be set as 4.0 kA. Then, with (19), the required de-icing voltage can be calculated as 5.76 kV (2×4.0 kA $\times 0.72\,\Omega$). Thus, the rated de-icing output power is 23.2 MW (= 5.76 kV $\times 4.0$ kA).

With the formulas in Chapter 3, the detailed electrical parameters of above MMC-DDI can be calculated and then listed in Table 1. The voltage and current peaks of the six arms are respectively 31.5 kV and 1.6 kV, thus the converter is equivalent to two conventional star-connected SVGs and each SVG has a 38.5 kV rated line voltage (31.5kV/$\sqrt{2} \times \sqrt{3}$), a 1.13 kA rated current (1.6kA/$\sqrt{2}$), and a 75.4 Mvar rating ($\sqrt{3} \times 38.5$ kV $\times 1.13$kA). Under the above total arm voltage and arm current, the specifications and numbers of MMC submodules can be freely selected within a certain range. As the 1700 V-level insulated gate bipolar transistor (IGBT) module is widely used in many medium-voltage engineering applications, here the submodule is construed with such IGBT, so the rated capacitor voltage of is set as 900 V and each arm contains 39 submodules. Referring to the SVG price list in Table A2, the converter cost can be estimated as 2.26 million dollar (15,000 $/Mvar × 75.4 × 2 = 2.26 million). With respect to its 23.2 MW output de-icing power, such high cost is too high to be acceptable.

Table 1. Electrical parameter comparison of the MMC-DDI under conventional configuration and optimized configuration.

Parameter	Symbol	Conventional Configuration (with No Transformer)	Optimized Configuration (with Transformer)
Rated DC voltage	U_{dc}	5.8 kV	5.8 kV
Rated DC current	I_{dc}	4.0 kA	4.0 kA
Rated output DC power	P_{dc}	23.2 MW	23.2 MW
AC-side phase voltage	U_m	20.2 kV	2.9 kV
AC-side phase current	I_m	0.38 A	4.6 kA
Arm voltage peak	U_{arm_peak}	31.5 kV	7.0 kV
Arm current peak	I_{arm_peak}	1.6 kA	3.2 kA
Converter rating	S_c	151 MVA	68 MVA
Transformer		None	23 MVA–35 kV/5 kV
Submodule number in each arm	N	39	9
Submodule capacitor voltage	U_{c0}	900 V	900 V
Submodule capacitance	C_c	10 mF	20 mF

If the proposed optimization method is adopted, a 23 MVA–35 kV/5 kV transformer should be inserted between the MMC converter and the 35 kV grid. At this time, the optimized MMC-DDI is mainly composed of an MMC converter and a transformer, and the detailed electrical parameters of MMC-DDI are also listed in Table 1. As Table 1shown, the voltage and current peaks of the six arms are 7.0 kV and 3.2 kV, thus the converter is equivalent to two common SVGs and each SVG has a rated line voltage 8.57 kV ($7.0 \text{ kV}/\sqrt{2} \times \sqrt{3}$), 2.26 kA rated current ($3.2 \text{ kA}/\sqrt{2}$) and 33.5 Mvar rating ($\sqrt{3} \times 8.57 \text{ kV} \times 2.26 \text{ kA}$). Considering the approximate SVG unit cost (15,000 $/Mvar), the converter cost can be estimated as 1.01 million dollar ($15,000 /Mvar × 33.5 Mvar × 2). In addition, in Table A3, the cost of a 24 MVA transformer is $166,000. Then, the total cost of the optimized MMC-DDI can be estimated as 1.18 million dollars.

The above cost comparison results are listed in Table 2. Compared with the cost of the original MMC-DDI with no transformer, the optimized cost of the IMD device has dropped by 48%.

Table 2. Cost comparison of the MMC-DDI under conventional configuration and optimized configuration.

Component	Original Cost (Million Dollar)	Optimized Cost (Million Dollar)
Converter	2.26	1.01
Transformer	-	0.17
Total	2.26	1.18

Besides the cost, the size and weight of the de-icer are also concerned in engineering applications. In practical projects, a complete MMC-DDI system contains not only the connection reactance, the converter valves and the disconnectors as shown in Figure 1, but also inlet cabinet, startup cabinet, control system, cooling subsystem, power distribution cabinet, cable and other auxiliary equipment. The equipment footprint not only includes the size of these devices, but also the insulation distance and other factors. Considering the fact that the main difference of the two MMC-DDI configurations is the converter and transformer, here only the converter chain and transformer are carefully compared.

For simplicity, here refers to a 100 Mvar SVG project built in Hunan in 2016 as a benchmark to compare the size and weight of the two topologies. This project consists of two Y-connected SVGs based on IGBT, and each SVG is 50 Mvar with a 20 kV rated voltage. Its arm current peak is 2125A, and 1.2 times that of that the conventional MMC-DDI configuration. Each SVG contains 63 power submodules, packed in 11 power cabinets. The converter hall is arranged on the first floor, while the cooling system is arranged on the roof. The floorplan of the SVG room is shown in Figure A1 and the main installation parameters of the SVGs are shown in Table A3. As Figure A1 shows, the SVG room covers an area 280 m^2, wherein the converter chain occupies 163 m^2 (17.6 m × 9.25 m).

The submodule current peak of the above SVG is about 1.2 times that of the conventional MMC-DDI. Here, we adopt the same submodule to form MMC-DDI. Considering that the arm current peak of the optimized MMC-DDI configuration is just twice that of the conventional one, the submodules of the optimized MMC-DDI configuration can be constructed with two parallel SVG submodules. Based on the above ideas, the conventional MMC-DDI requires 234 power modules while the optimized one requires 108 modules, and then their size and weight parameter can be calculated and shown in Table 3. The size and weight of the transformer are based on a 24 MVA rectifier produced for another project, the body size of the transformer is 5.4 m × 4.7 m, but its actual land occupation is set as 8 m × 9 m while considering the insulation distance and ancillary facilities.

Table 3. Size and weight comparisons of the MMC-DDI under conventional and optimized configuration.

Items	100 Mvar SVG	Conventional MMC-DDI	Optimized MMC-DDI
Main components	Converter (2 × 50 Mvar)	Converter (151 MVA)	Converter + Transformer (68 MVA) (24 MVA)
Number of power units	2 × 63	2 × 117	2 × 54
Number of power cabinets	2 × 11	2 × 20	2 × 10
Submodule capacitor voltage	900 V	900 V	900 V
Size of each submodule	0.7 m × 0.7 m × 0.8 m	0.7 m × 0.7 m × 0.8 m	0.7 m × 0.7 m × 0.8 m
Weight of each submodule	250 kg	250 kg	250 kg
Total weight of submodules	31.5 t	59 t	26 t
Transformer weight	-	-	38 t
Converter area	163 m^2	296 m^2	133 m^2
Transformer area	None	None	72 m^2
Other floor area	117 m^2	117 m^2	117 m^2
Total floor area	280 m^2	413 m^2	322 m^2
Total weight	31.5 t	59t	64 t

Compared with the conventional MMC-DDI, the optimized scheme required additional 72 m^2 to place the transformer, but the converter area is reduced from 296 m^2 to 133 m^2, namely a reduction of 163 m^2. As a result, the overall footprint of MMC-DDI system is reduced by 91 m^2, corresponding to a ratio of 22%. It shows that the optimized scheme also has an advantage in the land occupation. On the other hand, the optimized scheme requires a transformer with weight of 38 Ton, but its converter weight is reduced by 35 Ton, thus the total weight was slightly increased by 5 Ton. It shows that the optimized scheme have no advantage in weight. However, the DC de-icer built for high voltage transmission lines up to 500 kV is generally installed in the substations, so this weight disadvantage is still acceptable.

5.2. Simulation Results

To verify the above analysis and calculation on the converter characteristic, a corresponding MMC-DDI system is built in Matlab/Simulink (MathWorks, Natick City, MA, USA), and the simulation parameters are listed in Table A4.

For comparison, a dual-SVGs system (2 × 11.6 Mvar), which is similar to Figure 1 but has a zero DC-side output current reference, is also simulated. Since this article focuses on the converter rating characteristics, the number of submodules in each arm was set as $n = 4$ to speed up the simulation. This is also sufficient to compare the converter characteristics of the two schemes. The circuit image of the MMC-DDI simulation model built in Matlab/Simulink is shown in Figure A2, and its control block is shown in Figure A3. The simulation results of the dual-SVGs, the conventional MMC-DDI and the optimized MMC-DDI are shown in Figures 5–7, respectively.

As Figure 5a–c shows, in the dual-SVGs system, both the arm current and voltage are positive-negative symmetrical. The arm current equals half of the AC-side input current, while arm voltage is slightly higher than AC-side phase voltage due to the voltage drop across arm reactance. Their peaks are respectively 0.28 kA and 32.0 kV. With (15), the corresponding converter rating can be calculated as 26.9 MVA, just slightly higher than its output reactive power.

As Figure 5d,f shows, in the dual-SVGs system, the center point voltages of two SVGs only have tiny low-frequency component although they have obvious high frequency ripples. These ripples are mainly caused by the separate phase control method adopted in this simulation.

As Figure 5e shows, in the dual-SVGs system, the submodule capacitor voltages are around their set reference 8000 V and have tiny second harmonic fluctuations. The fluctuation amplitude is about 150 V, corresponding to 1% ripple factor.

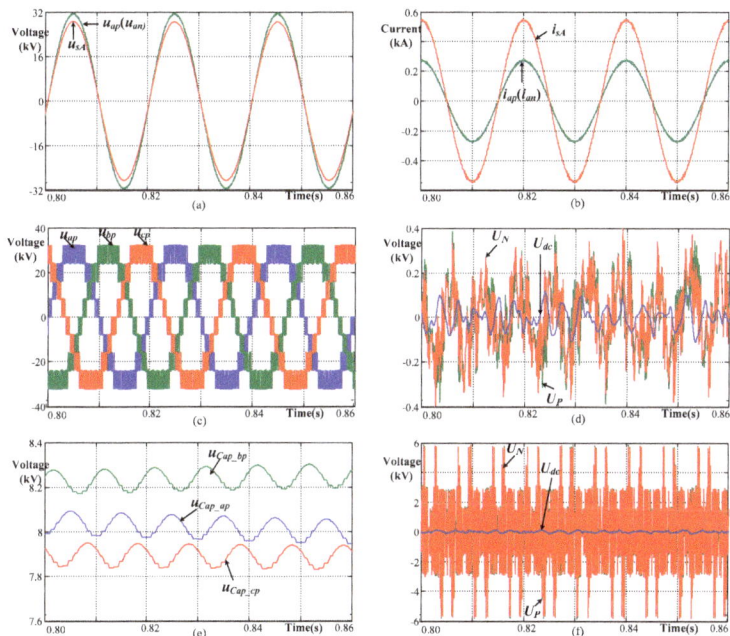

Figure 5. Simulation results of dual-SVGs system. (**a**) arm voltage (fileted high frequency ripple), (**b**) arm current, (**c**) arm voltage (unfiltered), (**d**) DC-side voltage (fileted high frequency ripple), (**e**) capacitor voltage of the first submodule in upper three arms, (**f**) original DC-side voltage (unfiltered).

As Figures 6g and 7g show, the DC-side output voltage and current in the conventional MMC-DDI and the optimized MMC-DDI are almost the same, and they rise gradually to their rated values during the melting-ice startup process. Correspondingly, both the DC and AC components in arm current rise slowly to the expected value. This indicates that the DC-side output voltage of MMC-DDI can be freely regulated within a range not exceeding its rated value, so that it can adapt to the different melting requirement of multiple transmission lines. As Figures 6d and 7d show, the DC-side output voltages are all around their expected value in both the conventional MMC-DDI and optimized MMC-DDI. The only difference is that the voltage ripple of the optimized MMC-DDI is smaller. Since the ice-melting process is mainly based on the Joule heat of the line current, this difference has little effect on the melting results.

As Figure 6a–c shows, in the conventional MMC-DDI that has no transformer, there is a visible DC component in the arm voltage and arm current. Especially in the arm current, the DC component far exceeds AC component. The arm voltage peak is 31.5 kV, which is slightly higher than AC phase voltage, while the arm current peak is 1.6 kA and far higher than the amplitude of its AC component. With (15), the corresponding converter rating can be calculated as 151 MVA, about 6.5 times the DC-side output power.

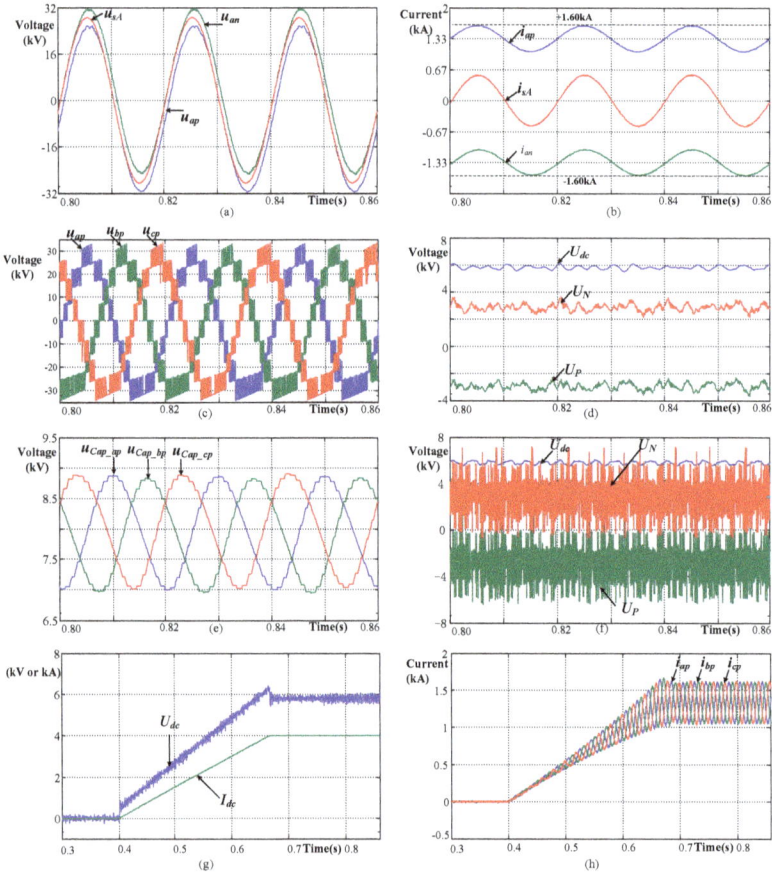

Figure 6. Simulation results of the conventional MMC-DDI. (**a**) arm voltage (fileted high-frequency ripple), (**b**) arm current, (**c**) arm voltage (unfiltered), (**d**) DC-side voltage (fileted high frequency ripple), (**e**) submodule capacitor voltage, (**f**) DC-side voltage (unfiltered), (**g**) DC-side output voltage and current during melting-ice startup process, (**h**) arm current during melting-ice startup process.

As Figure 7a–c shows, in the optimized MMC-DDI system that has a transformer, there is an obvious DC component in the arm voltage and current. The arm voltage and current peaks are respectively 7.0 kV and 3.2 kA, corresponding to a 67.2 MVA converter rating. Compared with the original MMC-DDI without a transformer, the arm current peak increases by 100% while the arm voltage peak reduces by 78%, thus the converter rating is only 44% of its original value.

The converter characteristics in such simulation results are consistent with the above analysis and calculation. In addition, the values of the converter voltage and current are also consistent with the theoretical results listed in Table 1. This proves the accuracy of the analysis and calculation on the MMC converter rating present in the paper.

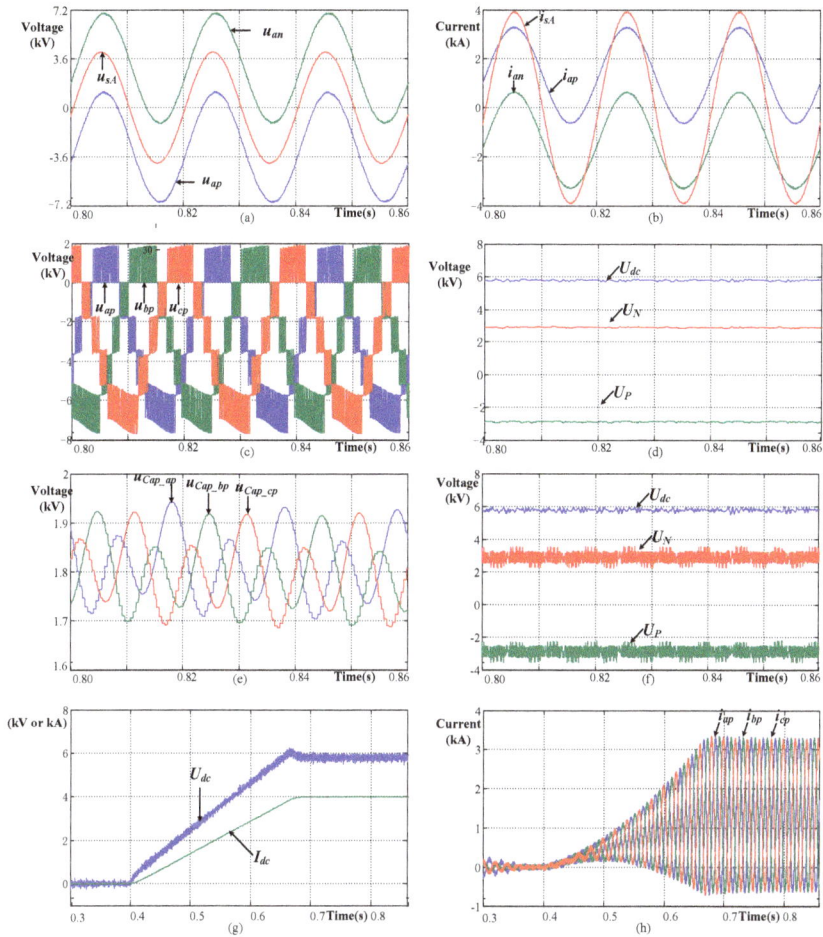

Figure 7. Simulation results of the optimized MMC-DDI. (**a**) arm voltage (fileted high-frequency ripple), (**b**) arm current, (**c**) arm voltage (unfiltered), (**d**) DC-side voltage (fileted high frequency ripple), (**e**) submodule capacitor voltage, (**f**) DC-side voltage (unfiltered), (**g**) DC-side output voltage and current during melting-ice startup process, (**h**) arm current during melting-ice startup process.

6. Discussion

Concerning the converter rating of MMC-DDI presented in this paper, the goal is to improve the economics of MMC-DDI while maintaining the same output de-icing characteristics. It turns out that, for a given DC ice-melting requirement, the converter rating of MMC-DDI varies greatly with its AC-side input voltage. Then, it is proposed to insert a transformer on the AC side of the MMC converter so that the converter rating as well as its cost can be significantly reduced, and then the economics of MMC-DDI can be improved.

It seems that this proposed configuration scheme is contradictory to traditional understanding of the MMC structure. Conventionally, in the common MMC system such as SVG, the AC side input transformers are expected to be avoided as much as possible.

This difference can be explained due to the converter characteristic of MMC-DDI having significant differences with that of the common MMC system:

(1) In an SVG, both the arm voltage and current contain only an AC component. As a result, in the case of a certain output power, the arm voltage is inversely proportional to arm current, thus the converter rating remains basically constant under any AC-side voltage. In that case, if a transformer was configured on the AC side of MMC converter, it has little influence on the converter rating while increasing a transformer. Therefore, in the common SVG, it tries to avoid a transformer.

(2) In the MMC-DDI, the arm voltage and arm current of converter contain both DC and AC components. As a result of the crossover between the DC and AC components, the converter rating of MMC-DDI varies greatly with its AC-side voltage. Due to such converter characteristics, a transformer can affect the converter rating. In this case, although the introduction of transformer will increase transformer cost, it can cause a cost increment or reduction of the converter. As long as the reduction of the converter cost is sufficient to offset the transformer cost, the introduction of the transformer is cost-effective. In addition, because the unit cost of MMC converter is generally much higher than that of the transformer, the above condition is easy to satisfy under the typical DC ice melting system parameters. Therefore, the optimized configuration scheme proposed in this paper is cost-effective in many cases.

It should be noted that the MMC-DDI can have two operation modes: ice-melting mode and SVG mode. This paper only considers the requirement of the ice melting mode, while not analyzing the operating characteristics of the SVG mode. In the optimization design process, the requirements of SVG mode have not been taken into account. This requirement can be further studied to get more comprehensive optimization results.

7. Conclusions

An MMC-based DC de-icer has been recognized as a promising de-icing solution. Conventionally, the MMC-DDI is recommended to be directly connected to the grid without a transformer.

In this paper, the converter rating of MMC-DDI was quantitatively analyzed. For a given DC ice-melting requirement, the converter rating varies greatly with its AC-side input voltage, and its minimum is 2.9 times the output ice-melting power. When the grid access point voltage is far more than DC de-icing voltage, the conventional MMC-DDI structure requires a far higher converter rating than its output de-icing power, thus the economy of MMC-DDI is very poor.

In order to improve the economy of MMC-DDI, this paper proposes an optimized MMC-DDI configuration structure in which a common two-winding transformer should be inserted at the AC-side of converter in some cases. Thus, the converter rating can be greatly reduced at the cost of an extra transformer. Since the cost of transformer is much lower than the same rating MMC converter, the introduction of transformer is cost-effective in many cases.Actually, for most 500 kV transmission lines, the optimized MMC-DDI configuration is superior to the transformerless MMC-DDI.

A design example and simulation results are given in this paper. In the case of outputting the same de-icing characteristics, the optimized converter rating is reduced from 151 MVA to 68 MVA, and the saved cost on the converter is much higher than the cost of the transformer, thus the total cost of MMC-DDI system is reduced by 48%. At the same time, the total floor space of MMC-DDI system is also greatly reduced by 22%, while, in total, the weight has a small increase.

This analysis and case show that, although the transformer is not technically necessary in an MMC-DDI, it can actually bring considerable benefits related to the total cost and space of MMC-DDI.

This conclusion is conducive to the optimized configuration of modular multilevel DC de-icer, and then to its engineering application for high voltage transmission lines.

Author Contributions: Conceptualization, J.L. and Q.H.; Formal Analysis, Q.H.; Data Curation, X.M., and Y.Z; Writing—Review and Editing, S.Z.; Supervision, Y.T.

Funding: This research was funded by the Science and Technology Project of State Grid Electric Corporation Grant No. 5216A016000P.

Conflicts of Interest: The authors declare no conflict of interest.

Appendix A

Table A1. The minimum de-icing current and maximum endure current for typical power lines [5].

Conductor Type	Min. De-Icing Current (A) $(-5\,^{\circ}\text{C}, 5\,\text{m/s}, 10\,\text{mm}, 1\,\text{h})$	Max. Endure Current(A) $(5\,^{\circ}\text{C}, 0.5\,\text{m/s}, \text{No Icing})$
LGJ-4 × 400/50	3475	4764
LGJ-2 × 500/45	1989	2698
LGJ-2 × 240/40	1218	1716
LGJ-1 × 240/40	609	858
LGJ-1 × 185/45	515	733
LGJ-1 × 150/35	441	633
LGJ-1 × 95/55	345	500

Table A2. Deal prices of several typical SVG projects in China from 2013 to 2018.

No.	Project Location	Rated Voltage (kV)	Rating (MVA)	Deal Price [1] ($1000)	Unit Cost (1000 $/MVA)
1	Kunming, Yunnan	35	10	154	15.4
2	Zhangjiakou, Hebei	35	12	175	14.6
3	Huimin, Shandong	35	15	215	14.4
4	Huangpi, Hubei	35	16	251	15.7
5	Tongyu, Gansu	35	20	269	13.5
6	Hua County, Henan	35	20	257	12.8
7	Chenzhou, Hunan	10	20	330	16.5
8	Qiaojia, Yunnan	35	30	385	12.8
9	Linwu, Ningxia	35	40	458	11.5
10	Dabancheng, Xinjiang	35	50	615	12.3
11	Yinan, Shandong	35	60	1023	17.1
12	Haixi, Xinjiang	35	60	1154	19.2
13	Hami, Xinjiang	35	80	1508	18.8
14	Huaping, Yunnan	35	100	2109	21.1
15	Xiangtan, Hunan	35	120	2615	21.8

[1] The deal price covers a complete set of SVG equipment (including the converter chain, connection reactance, startup circuit, cooling system, control system and other ancillary facilities) and its technical service.

Table A3. Deal prices of several 10 MVA-class rectifier transformers in China.

No.	Project Location	Rated Voltage (kV)	Rating (MVA)	Deal Price ($1000)	Unit Cost (1000 $/MVA)
1	Baoding, Hebei	10/5	10	86	8.6
2	Changsha, Hunan	10/7	14	110	7.8
3	Changsha, Hunan	35/6	24	166	6.9
4	Xinyu, JiangXi	35/12	56	246	4.4
5	Chongqing	35/15	86	284	3.3
6	Zhuzhou, Gansu	35/17	100	323	3.2
7	Hengyang, Hunan	35/19	120	361	3.0

Table A4. Simulation parameters of the two MMC-DDI and dual-SVG systems.

Parameter	Symbol	Dual-SVG	Conventional MMC-DDI	Optimized MMC-DDI
AC-side rated voltage	U_S	35 kV	35 kV	5 kV
AC-side rated current	I_M	(0.38 kA) [2]	(0.38 kA)	(2.68 kA)
AC-side rated power		+23.2 Mvar	(23.2 MW)	(23.2 MW)
Arm inductance	L	35 mH	35 mH	1 mH
Arm equivalent resistance	R	0.1 Ω	0.1 Ω	0.02 Ω
DC-side output de-icing voltage	U_{dc}	0	(5.8 kV)	(5.8 kV)
DC-side output de-icing current	I_{dc}	0	4.0 kA	4.0 kA
Resistance of de-icing line	R_{dc}	-	1.45 Ω	1.45 Ω
Inductance of de-icing line	L_{line}	-	32 mH	32 mH
Submodule number of each arm	N	4	4	4
Submodule capacitance	C_{cap}	4 mF	4 mF	10 mF
Submodule capacitor voltage	U_{cap}	9.0 kV	8.0 kV	1.8 kV
Switching frequency		500 Hz	500 Hz	500 Hz

[2] The parameters in parentheses indicate the calculated value, while the parameters in parentheses indicate the values directly set in the simulation.

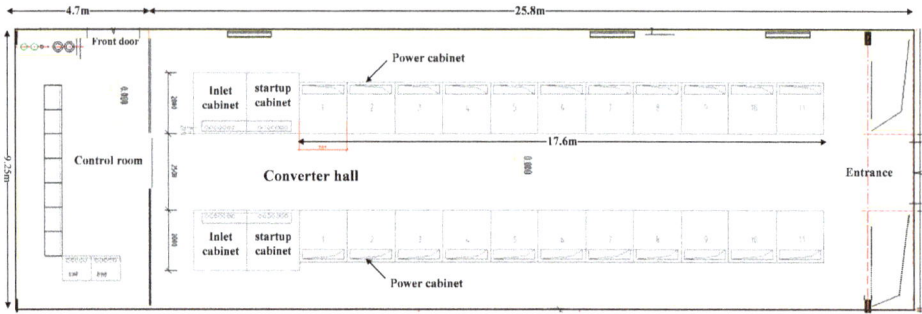

Figure A1. Floorplan of 100 Mvar SVG room in 500 kV Chuanshan substation.

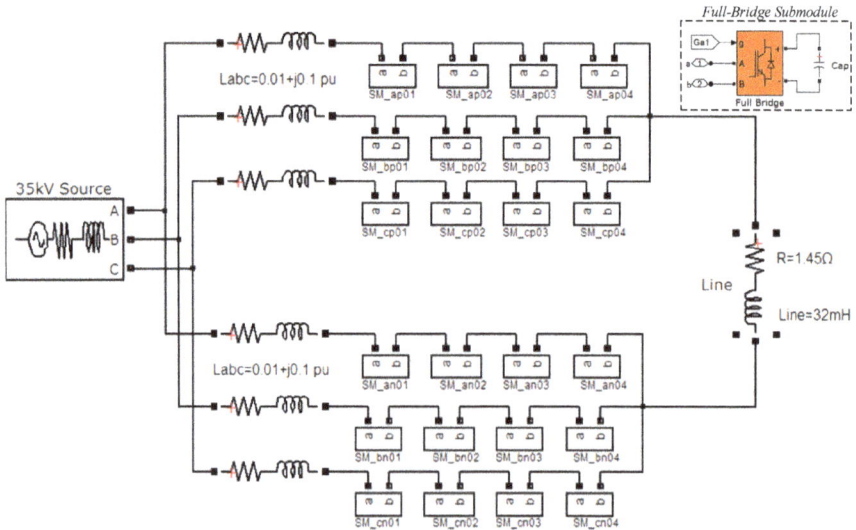

Figure A2. The circuit image of MMC-DDI simulation model built in Matlab/Simulink.

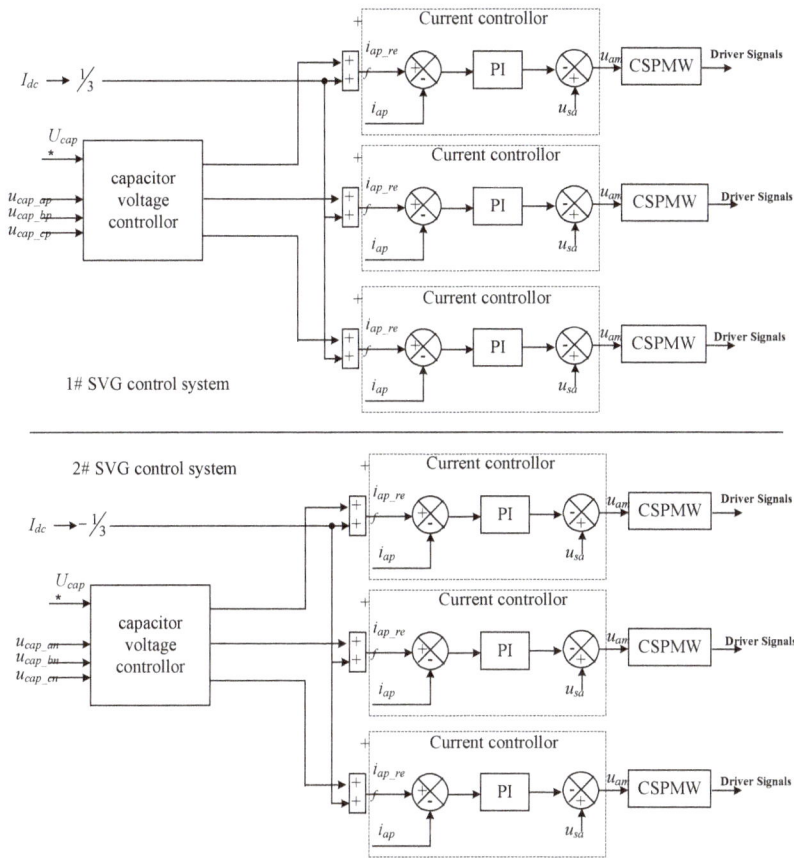

Figure A3. The control block of MMC-DDI simulation model built in Matlab/Simulink.

References

1. Joe, C.P.; Phillip, L. Present State-of-the-Art of Transmission Line Icing. *IEEE Trans. Power Appar. Syst.* **1982**, *8*, 2443–2450.
2. Farzaneh, M.; Savadjiev, K. Statistical analysis of field data for precipitation icing accretion on overhead power lines. *IEEE Trans. Power Deliv.* **2005**, *2*, 1080–1087. [CrossRef]
3. Volat, C.; Farzaneh, M.; Leblond, A. De-icing/Anti-icing Techniques for Power Lines: Current Methods and Future Direction. In Proceedings of the 11th International Workshop on Atmospheric Icing of Structures, Montreal, QC, Canada, 13–16 June 2005.
4. Brostrom, E.; Ahlberg, J.; Soder, L. Modelling of Ice Storms and their Impact Applied to a Part of the Swedish Transmission Network. In Proceedings of the 2007 IEEE Lausanne Power Tech 2007, Lausanne, Switzerland, 1–5 July 2007; pp. 1593–1598.
5. Wang, J.; Fu, C.; Chen, Y.; Rao, H.; Xu, S.; Yu, T.; Li, L. Research and application of DC de-icing technology in China southern power grid. *IEEE Trans. Power Deliv.* **2012**, *3*, 1234–1242. [CrossRef]
6. Laforte, J.L.; Allaire, M.A.; Laflamme, J. State-of-the-art on power line de-icing. *Atmos. Res.* **1998**, *46*, 143–158. [CrossRef]
7. Motlis, Y. Melting ice on overhead-line conductors by electrical current. In Proceedings of the CIGRE SC22/WG12, Paris, France, 26–30 August 2002.

8. Farzaneh, M.; Jakl, F.; Arabani, M.P.; Eliasson, A.J.; Fikke, S.M.; Gallego, A.; Haldar, A.; Isozaki, M.; Lake, R.; Leblond, L.; et al. *Systems for Prediction and Monitoring of Ice Shedding, Anti-Icing and De-Icing for Power Line Conductors And Ground Wires*; CIGRE: Paris, France, 2010.

9. Davidson, C.C.; Horwill, C.; Granger, M.; Dery, A. A power-electronics-based transmission line de-icing system. In Proceedings of the 8th IEE International Conference on AC and DC Power Transmission, London, UK, 28–31 March 2006; pp. 135–139.

10. Dery, A.; Granger, M.; Davidson, C.C.; Horwill, C.; Dery, A.; Granger, M.; Davidson, C.C.; Horwill, C. An Application of HVDC to the de-icing of Transmission Lines. In Proceedings of the 2005/2006 IEEE PES Transmission and Distribution Conference and Exhibition, Dallas, TX, USA, 21–24 May 2006; pp. 529–534.

11. Fu, C.; Rao, H.; Li, X.; Chao, J.; Tian, J.; Chen, S.; Zhao, L.; Xu, S.; Ma, X. Development and application of DC deicer. *Autom. Electr. Power Syst.* **2009**, *63*, 118–119.

12. Rao, H.; Li, L.; Li, X.; Fu, C. Study of DC based De-icing Technology in China Southern Power Grid. *South. Power Syst. Technol.* **2008**, *2*, 7–12.

13. Davidson, C.C.; Horwill, C.; Granger, M.; Dery, A. Thaw point. *Power Eng.* **2007**, *21*, 26–31. [CrossRef]

14. Jing, H.; Nian, X.; Fan, R.; Liu, D.; Deng, M. Control and switchover strategy of full-controlled ice-melting DC power for ice-covered power lines. *Autom. Electr. Power Syst.* **2012**, *36*, 86–91.

15. Zhao, G.S.; Li, X.-Y.; Fu, C.; Li, X.-L.; Wang, Y.H.; Xia, W. Overview of de-icing technology for transmission lines. *Power Syst. Prot. Control* **2011**, *39*, 148–154.

16. Bhattacharya, S.; Xi, Z.; Fardenesh, B.; Uzunovic, E. Control reconfiguration of VSC based STATCOM for de-icer application. In Proceedings of the 2008 IEEE Power and Energy Society General Meeting, Pittsburgh, PA, USA, 20–24 July 2008; pp. 1–7.

17. Hagiwara, M.; Akagi, H. Control and Experiment of Pulsewidth-Modulated Modular Multilevel Converters. *IEEE Trans. Power Electron.* **2009**, *24*, 1737–1746. [CrossRef]

18. Rohner, S.; Bernet, S.; Hiller, M.; Sommer, R. Modulation, Losses, and Semiconductor Requirements of Modular Multilevel Converters. *IEEE Trans. Ind. Electron.* **2010**, *57*, 2633–2642. [CrossRef]

19. Lesnicar, A.; Marquardt, R. An innovative modular multilevel converter topology suitable for a wide power range. In Proceedings of the 2003 IEEE Bologna Power Tech Conference Proceedings, Bologna, Italy, 23–26 June 2003.

20. Martinez-Rodrigo, F.; Ramirez, D.; Rey-Boue, A.B.; de Pablo, S.; Lucas, L.C.H. Modular Multilevel Converters: Control and Applications. *Energies* **2017**, *11*, 1709. [CrossRef]

21. Mehrasa, M.; Pouresmaeil, E.; Zabihi, S.; Mehrasa, M.; Pouresmaeil, E.; Zabihi, S.; Caballero, J.C.T.; Catalão, J.P.S. A Novel Modulation Function-Based Control of Modular Multilevel Converters for High Voltage Direct Current Transmission Systems. *Energies* **2016**, *9*, 867. [CrossRef]

22. Perez, M.A.; Bernet, S.; Rodriguez, J.; Kouro, S.; Lizana, R. Circuit topologies, modeling, control schemes, and applications of modular mul-tilevel converters. *IEEE Trans. Power Electron.* **2015**, *1*, 4–14. [CrossRef]

23. Flourentzou, N.; Agelidis, V.G.; Demetriades, G.D. VSC-Based HVDC Power Transmission Systems: An Overview. *IEEE Trans. Power Electron.* **2009**, *24*, 592–602. [CrossRef]

24. Mohammadi, H.P.; Bina, M.T. A Transformerless Medium-Voltage STATCOM Topology Based on Extended Modular Multilevel Converters. *IEEE Trans. Power Electron.* **2011**, *26*, 1534–1545.

25. Mei, H.; Liu, J. A Novel DC Ice-melting Equipment Based on Modular Multilevel Cascade Converter. *Autom. Power Syst.* **2013**, *37*, 96–102.

26. Thitichaiworakorn, N.; Hagiwara, M.; Akagi, H. Experimental verification of a modular multilevel cascade inverter based on double-star bridge cells. *IEEE Trans. Ind. Appl.* **2014**, *50*, 509–519. [CrossRef]

27. Li, B.; Shi, S.; Xu, D.; Wang, W. Control and Analysis of the Modular Multilevel DC De-Icer with STATCOM Functionality. *IEEE Trans. Ind. Electron.* **2016**, *9*, 5465–5476. [CrossRef]

28. Ning, Y.; Zheng, J.; Chen, Z. Application of star-connected cascaded STATCOM in DC ice-melting system. *Autom. Electr. Power Syst.* **2015**, *27*, 1920–1922.

29. Guo, Y.; Xu, J.; Guo, C.; Zhao, C.; Fu, C.; Zhou, Y.; Xu, S. Control of full-bridge modular multilevel converter for dc ice-melting application. In Proceedings of the 11th IET International Conference on AC and DC Power Transmission, Birmingham, UK, 10–12 February 2015; pp. 1–8.

Electronics **2018**, *7*, 204

30. *IEEE Standard for Calculating the Current-Temperature Relationship of Bare Overhead Conductors*; IEEE Std 738-2012; IEEE Standards Association: Piscataway, NJ, USA, 2013; pp. 1–72.
31. CIGRE. Thermal Behavior of Overhead Conductors. *Electra* **1992**, *144*, 107–125.

MDPI

St. Alban-Anlage 66

4052 Basel

Switzerland

Tel. +41 61 683 77 34

Fax +41 61 302 89 18

www.mdpi.com

Electronics Editorial Office

E-mail: electronics@mdpi.com

www.mdpi.com/journal/electronics